DERIVATIVES AND INTEGRALS

Basic Differentiation Rules

1. $\dfrac{d}{dx}[cu] = cu'$

2. $\dfrac{d}{dx}[u \pm v] = u' \pm v'$

3. $\dfrac{d}{dx}[uv] = uv' + vu'$

4. $\dfrac{d}{dx}\left[\dfrac{u}{v}\right] = \dfrac{vu' - uv'}{v^2}$

5. $\dfrac{d}{dx}[c] = 0$

6. $\dfrac{d}{dx}[u^n] = nu^{n-1}u'$

7. $\dfrac{d}{dx}[x] = 1$

8. $\dfrac{d}{dx}[|u|] = \dfrac{u}{|u|}(u'), \quad u \neq 0$

9. $\dfrac{d}{dx}[\ln u] = \dfrac{u'}{u}$

10. $\dfrac{d}{dx}[e^u] = e^u u'$

11. $\dfrac{d}{dx}[\log_a u] = \dfrac{u'}{(\ln a)u}$

12. $\dfrac{d}{dx}[a^u] = (\ln a)a^u u'$

13. $\dfrac{d}{dx}[\sin u] = (\cos u)u'$

14. $\dfrac{d}{dx}[\cos u] = -(\sin u)u'$

15. $\dfrac{d}{dx}[\tan u] = (\sec^2 u)u'$

16. $\dfrac{d}{dx}[\cot u] = -(\csc^2 u)u'$

17. $\dfrac{d}{dx}[\sec u] = (\sec u \tan u)u'$

18. $\dfrac{d}{dx}[\csc u] = -(\csc u \cot u)u'$

19. $\dfrac{d}{dx}[\arcsin u] = \dfrac{u'}{\sqrt{1 - u^2}}$

20. $\dfrac{d}{dx}[\arccos u] = \dfrac{-u'}{\sqrt{1 - u^2}}$

21. $\dfrac{d}{dx}[\arctan u] = \dfrac{u'}{1 + u^2}$

22. $\dfrac{d}{dx}[\operatorname{arccot} u] = \dfrac{-u'}{1 + u^2}$

23. $\dfrac{d}{dx}[\operatorname{arcsec} u] = \dfrac{u'}{|u|\sqrt{u^2 - 1}}$

24. $\dfrac{d}{dx}[\operatorname{arccsc} u] = \dfrac{-u'}{|u|\sqrt{u^2 - 1}}$

25. $\dfrac{d}{dx}[\sinh u] = (\cosh u)u'$

26. $\dfrac{d}{dx}[\cosh u] = (\sinh u)u'$

27. $\dfrac{d}{dx}[\tanh u] = (\operatorname{sech}^2 u)u'$

28. $\dfrac{d}{dx}[\coth u] = -(\operatorname{csch}^2 u)u'$

29. $\dfrac{d}{dx}[\operatorname{sech} u] = -(\operatorname{sech} u \tanh u)u'$

30. $\dfrac{d}{dx}[\operatorname{csch} u] = -(\operatorname{csch} u \coth u)u'$

31. $\dfrac{d}{dx}[\sinh^{-1} u] = \dfrac{u'}{\sqrt{u^2 + 1}}$

32. $\dfrac{d}{dx}[\cosh^{-1} u] = \dfrac{u'}{\sqrt{u^2 - 1}}$

33. $\dfrac{d}{dx}[\tanh^{-1} u] = \dfrac{u'}{1 - u^2}$

34. $\dfrac{d}{dx}[\coth^{-1} u] = \dfrac{u'}{1 - u^2}$

35. $\dfrac{d}{dx}[\operatorname{sech}^{-1} u] = \dfrac{-u'}{u\sqrt{1 - u^2}}$

36. $\dfrac{d}{dx}[\operatorname{csch}^{-1} u] = \dfrac{-u'}{|u|\sqrt{1 + u^2}}$

Basic Integration Formulas

1. $\int kf(u)\,du = k\int f(u)\,du$

2. $\int [f(u) \pm g(u)]\,du = \int f(u)\,du \pm \int g(u)\,du$

3. $\int du = u + C$

4. $\int a^u\,du = \left(\dfrac{1}{\ln a}\right)a^u + C$

5. $\int e^u\,du = e^u + C$

6. $\int \sin u\,du = -\cos u + C$

7. $\int \cos u\,du = \sin u + C$

8. $\int \tan u\,du = -\ln|\cos u| + C$

9. $\int \cot u\,du = \ln|\sin u| + C$

10. $\int \sec u\,du = \ln|\sec u + \tan u| + C$

11. $\int \csc u\,du = -\ln|\csc u + \cot u| + C$

12. $\int \sec^2 u\,du = \tan u + C$

13. $\int \csc^2 u\,du = -\cot u + C$

14. $\int \sec u \tan u\,du = \sec u + C$

15. $\int \csc u \cot u\,du = -\csc u + C$

16. $\int \dfrac{du}{\sqrt{a^2 - u^2}} = \arcsin \dfrac{u}{a} + C$

17. $\int \dfrac{du}{a^2 + u^2} = \dfrac{1}{a}\arctan \dfrac{u}{a} + C$

18. $\int \dfrac{du}{u\sqrt{u^2 - a^2}} = \dfrac{1}{a}\operatorname{arcsec} \dfrac{|u|}{a} + C$

© Houghton Mifflin Company, Inc.

TRIGONOMETRY

Definition of the Six Trigonometric Functions

Right triangle definitions, where $0 < \theta < \pi/2$.

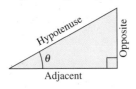

$\sin \theta = \dfrac{\text{opp}}{\text{hyp}} \quad \csc \theta = \dfrac{\text{hyp}}{\text{opp}}$

$\cos \theta = \dfrac{\text{adj}}{\text{hyp}} \quad \sec \theta = \dfrac{\text{hyp}}{\text{adj}}$

$\tan \theta = \dfrac{\text{opp}}{\text{adj}} \quad \cot \theta = \dfrac{\text{adj}}{\text{opp}}$

Circular function definitions, where θ is any angle.

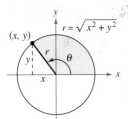

$\sin \theta = \dfrac{y}{r} \quad \csc \theta = \dfrac{r}{y}$

$\cos \theta = \dfrac{x}{r} \quad \sec \theta = \dfrac{r}{x}$

$\tan \theta = \dfrac{y}{x} \quad \cot \theta = \dfrac{x}{y}$

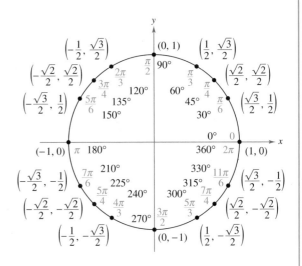

Reciprocal Identities

$\sin x = \dfrac{1}{\csc x} \quad \sec x = \dfrac{1}{\cos x} \quad \tan x = \dfrac{1}{\cot x}$

$\csc x = \dfrac{1}{\sin x} \quad \cos x = \dfrac{1}{\sec x} \quad \cot x = \dfrac{1}{\tan x}$

Tangent and Cotangent Identities

$\tan x = \dfrac{\sin x}{\cos x} \quad \cot x = \dfrac{\cos x}{\sin x}$

Pythagorean Identities

$\sin^2 x + \cos^2 x = 1$

$1 + \tan^2 x = \sec^2 x \qquad 1 + \cot^2 x = \csc^2 x$

Cofunction Identities

$\sin\left(\dfrac{\pi}{2} - x\right) = \cos x \quad \cos\left(\dfrac{\pi}{2} - x\right) = \sin x$

$\csc\left(\dfrac{\pi}{2} - x\right) = \sec x \quad \tan\left(\dfrac{\pi}{2} - x\right) = \cot x$

$\sec\left(\dfrac{\pi}{2} - x\right) = \csc x \quad \cot\left(\dfrac{\pi}{2} - x\right) = \tan x$

Reduction Formulas

$\sin(-x) = -\sin x \qquad \cos(-x) = \cos x$

$\csc(-x) = -\csc x \qquad \tan(-x) = -\tan x$

$\sec(-x) = \sec x \qquad \cot(-x) = -\cot x$

Sum and Difference Formulas

$\sin(u \pm v) = \sin u \cos v \pm \cos u \sin v$

$\cos(u \pm v) = \cos u \cos v \mp \sin u \sin v$

$\tan(u \pm v) = \dfrac{\tan u \pm \tan v}{1 \mp \tan u \tan v}$

Double-Angle Formulas

$\sin 2u = 2 \sin u \cos u$

$\cos 2u = \cos^2 u - \sin^2 u = 2 \cos^2 u - 1 = 1 - 2 \sin^2 u$

$\tan 2u = \dfrac{2 \tan u}{1 - \tan^2 u}$

Power-Reducing Formulas

$\sin^2 u = \dfrac{1 - \cos 2u}{2}$

$\cos^2 u = \dfrac{1 + \cos 2u}{2}$

$\tan^2 u = \dfrac{1 - \cos 2u}{1 + \cos 2u}$

Sum-to-Product Formulas

$\sin u + \sin v = 2 \sin\left(\dfrac{u+v}{2}\right) \cos\left(\dfrac{u-v}{2}\right)$

$\sin u - \sin v = 2 \cos\left(\dfrac{u+v}{2}\right) \sin\left(\dfrac{u-v}{2}\right)$

$\cos u + \cos v = 2 \cos\left(\dfrac{u+v}{2}\right) \cos\left(\dfrac{u-v}{2}\right)$

$\cos u - \cos v = -2 \sin\left(\dfrac{u+v}{2}\right) \sin\left(\dfrac{u-v}{2}\right)$

Product-to-Sum Formulas

$\sin u \sin v = \dfrac{1}{2}[\cos(u-v) - \cos(u+v)]$

$\cos u \cos v = \dfrac{1}{2}[\cos(u-v) + \cos(u+v)]$

$\sin u \cos v = \dfrac{1}{2}[\sin(u+v) + \sin(u-v)]$

$\cos u \sin v = \dfrac{1}{2}[\sin(u+v) - \sin(u-v)]$

© Houghton Mifflin Company, Inc.

Calculus II

Eighth Edition

Ron Larson
Robert P. Hostetler
The Pennsylvania State University
The Behrend College

Bruce H. Edwards
University of Florida

Houghton Mifflin Company   Boston   New York

Vice President and Publisher: Jack Shira
Associate Sponsoring Editor: Cathy Cantin
Development Manager: Maureen Ross
Senior Development Editor: Claire Boivin
Editorial Assistant: Elizabeth Kassab
Supervising Editor: Karen Carter
Senior Project Editor: Patty Bergin
Editorial Assistant: Allison Seymour/Julia Keller
Production Technology Supervisor: Gary Crespo
Executive Marketing Manager: Michael Busnach
Senior Marketing Manager: Danielle Potvin
Marketing Coordinator: Nicole Mollica
Senior Manufacturing Coordinator: Priscilla Bailey
Cover Design Manager: Tony Saizon

We have included examples and exercises that use real-life data as well as technology output from a variety of software. This would not have been possible without the help of many people and organizations. Our wholehearted thanks goes to all for their time and effort.

Cover photograph: "Music of the Spheres" by English sculptor John Robinson is a three-foot-tall sculpture in bronze that has one continuous edge. You can trace its edge three times around before returning to the starting point. To learn more about this and other works by John Robinson, see the Centre for the Popularisation of Mathematics, University of Wales, at
http://www.popmath.org.uk/sculpture/gallery2.html.

Trademark Acknowledgments: TI is a registered trademark of Texas Instruments, Inc. Mathcad is a registered trademark of MathSoft, Inc. Windows, Microsoft and MS-DOS are registered trademarks of Microsoft, Inc. Mathematica is a registered trademark of Wolfram Research, Inc. DERIVE is a registered trademark of Texas Instruments, Inc. IBM is a registered trademark of International Business Machines Corporation. Maple is a registered trademark of Waterloo Maple, Inc. HM ClassPrep is a trademark of Houghton Mifflin Company.

Copyright © 2006 by Houghton Mifflin Company. All rights reserved.

No part of this work may be reproduced or transmitted in any form or by any means, electronic or mechanical, including photocopying and recording, or by any information storage or retrieval system, without the prior written permission of Houghton Mifflin Company unless such copying is expressly permitted by federal copyright law. Address inquiries to College Permissions, Houghton Mifflin Company, 222 Berkeley Street, Boston, MA 02116-3764.

Printed in the U.S.A.

Library of Congress Control Number: 2004114152

ISBN: 0-618-51266-7

1 2 3 4 5 6 7 8 9-DOW-09 08 07 06 05

Contents

A Word from the Authors   vi

Integrated Learning System for Calculus   viii

Features   xiv

Chapter 7 — Applications of Integration   445

7.1 Area of a Region Between Two Curves   446
7.2 Volume: The Disk Method   456
7.3 Volume: The Shell Method   467
Section Project: Saturn   475
7.4 Arc Length and Surfaces of Revolution   476
7.5 Work   487
Section Project: Tidal Energy   495
7.6 Moments, Centers of Mass, and Centroids   496
7.7 Fluid Pressure and Fluid Force   507
  Review Exercises   513
P.S. Problem Solving   515

Chapter 8 — Integration Techniques, L'Hôpital's Rule, and Improper Integrals   517

8.1 Basic Integration Rules   518
8.2 Integration by Parts   525
8.3 Trigonometric Integrals   534
Section Project: Power Lines   542
8.4 Trigonometric Substitution   543
8.5 Partial Fractions   552
8.6 Integration by Tables and Other Integration Techniques   561
8.7 Indeterminate Forms and L'Hôpital's Rule   567
8.8 Improper Integrals   578
  Review Exercises   589
P.S. Problem Solving   591

Chapter 9 — Infinite Series 593

- 9.1 Sequences 594
- 9.2 Series and Convergence 606
- **Section Project:** Cantor's Disappearing Table 616
- 9.3 The Integral Test and p-Series 617
- **Section Project:** The Harmonic Series 623
- 9.4 Comparisons of Series 624
- **Section Project:** Solera Method 630
- 9.5 Alternating Series 631
- 9.6 The Ratio and Root Tests 639
- 9.7 Taylor Polynomials and Approximations 648
- 9.8 Power Series 659
- 9.9 Representation of Functions by Power Series 669
- 9.10 Taylor and Maclaurin Series 676
- Review Exercises 688
- *P.S. Problem Solving* 691

Chapter 10 — Conics, Parametric Equations, and Polar Coordinates 693

- 10.1 Conics and Calculus 694
- 10.2 Plane Curves and Parametric Equations 709
- **Section Project:** Cycloids 718
- 10.3 Parametric Equations and Calculus 719
- 10.4 Polar Coordinates and Polar Graphs 729
- **Section Project:** Anamorphic Art 738
- 10.5 Area and Arc Length in Polar Coordinates 739
- 10.6 Polar Equations of Conics and Kepler's Laws 748
- Review Exercises 756
- *P.S. Problem Solving* 759

Chapter 11 Vectors and the Geometry of Space   761

    11.1 Vectors in the Plane   762
    11.2 Space Coordinates and Vectors in Space   773
    11.3 The Dot Product of Two Vectors   781
    11.4 The Cross Product of Two Vectors in Space   790
    11.5 Lines and Planes in Space   798
    Section Project: Distances in Space   809
    11.6 Surfaces in Space   810
    11.7 Cylindrical and Spherical Coordinates   820
    Review Exercises   827
P.S. Problem Solving   829

Appendix A **Proofs of Selected Theorems**   **A1**

Appendix B **Integration Tables**   **A20**

    Answers to Odd-Numbered Exercises   A27
    Index of Applications   A67
    Index   A71

Additional Appendices The following appendices are available at the textbook website at *math.college.hmco.com*, on the HM mathSpace® Student CD-ROM, and the HM ClassPrep with HM Testing CD-ROM.

Appendix C **Additional Topics in Differential Equations**

    C.1 Exact First-Order Equations
    C.2 Second-Order Homogeneous Linear Equations
    C.3 Second-Order Nonhomogeneous Linear Equations
    C.4 Series Solutions of Differential Equations

Appendix D **Precalculus Review**

    D.1 Real Numbers and the Real Number Line
    D.2 The Cartesian Plane
    D.3 Review of Trigonometric Functions

Appendix E **Rotation and General Second-Degree Equation**

Appendix F **Complex Numbers**

Appendix G **Business and Economic Applications**

A Word from the Authors

Welcome to *Calculus II*, Eighth Edition. Much has changed since we wrote the first edition of *Calculus*, which we began writing in 1973—over 30 years ago. With each edition, we have listened to you, our users, and incorporated many of your suggestions for improvement.

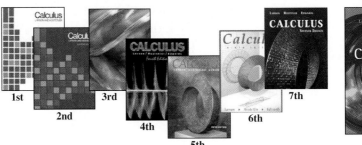
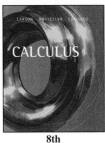

A Text Formed by Its Users

Through your support and suggestions, the text has evolved over eight editions to include these extensive enhancements:

- Comprehensive exercise sets containing a wide variety of problems such as skill-building exercises, applications, explorations, writing exercises, critical thinking exercises, and theoretical problems
- Abundant real-life applications that accurately represent the diverse uses of calculus
- Many open-ended activities and investigations
- Clear, uncluttered text presentation with full annotations and labels and a carefully planned page layout
- Comprehensive, four-color art program
- Comprehensive and mathematically rigorous text (The third semester of the Eighth Edition, in particular, is quite different when compared with the third semester of the First Edition.)
- Technology used throughout as both a problem-solving tool and an investigative tool
- A comprehensive program of additional resources available in print, on CD-ROM, and online
- With eight different volumes of the text available, you can choose the sequence, amount of content, and teaching approach that is best for you and your students (see pages viii–ix)
- References to the history of calculus and to the mathematicians who developed it, including over 50 biographical sketches available on the HM mathSpace® Student CD-ROM
- References to over 50 articles from mathematical journals are available at *www.MathArticles.com*

What's New and Different in the Eighth Edition

In the Eighth Edition, we continue to offer instructors and students a text that is pedagogically sound, mathematically precise, and still comprehensible. There are many changes in the mathematics, prose, art, and design; the more significant changes are noted here.

- *New Chapter Openers* Each Chapter Opener has two parts: a description of the concepts that are covered in the chapter and a thought-provoking question about a real-life application from the chapter.
- **Revised Exercise Sets** The exercise sets have been carefully and extensively examined to ensure they are rigorous and cover all topics suggested by our users. Many new skill-building and challenging exercises have been added.
- **Updated Data** All data in the examples and exercise sets have been updated.
- **EduSpace®** *Eduspace®* combines numerous dynamic resources with online homework and testing materials to create a comprehensive online learning system. Students benefit from having immediate access to algorithmic tutorial practice, videos, and resources such as a color graphing calculator. Instructors benefit from time-saving grading resources, as well as dynamic instructional tools such as animations, explorations, and Computer Algebra System Labs.
- *Student Study Guide* The worked-out solutions to the odd-numbered text exercises are now provided in a printed supplement, on a CD-ROM, in *Eduspace®*, and at *www.CalcChat.com*.

Although we carefully and thoroughly revised the text by enhancing the usefulness of some features and topics and by adding others, we did not change many of the things that our colleagues and the over two million students who have used this book have told us work for them. The *Calculus*, Eighth Edition, program offers comprehensive coverage of the material required by students in calculus courses, including carefully stated theories and proofs.

We hope you will enjoy the Eighth Edition. We welcome any comments, as well as suggestions for continued improvement.

Ron Larson   *Robert P. Hostetler*   *Bruce H. Edwards*

Integrated Learning System for Calculus

Over 25 Years of Success, Leadership, and Innovation

The best-selling program by Larson, Hostetler, and Edwards continues to offer instructors and students more flexible teaching and learning options for the calculus course.

Calculus Textbook Options

CALCULUS

The traditional calculus course is available in a variety of textbook configurations to address the different ways instructors teach—and students take—their classes.

Calculus I, II, and III

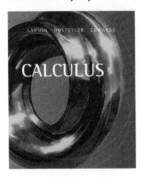

Calculus I and II

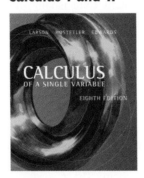

Calculus I

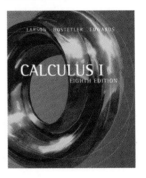

Calculus III

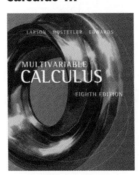

Calculus II

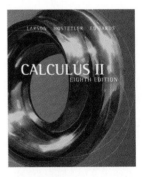

Also available for the *Calculus*, Eighth Edition, program by Larson, Hostetler, and Edwards

- Eduspace® online learning system
- HM mathSpace® Student CD-ROMs
- Instructional DVDs and videotapes

For more information on these—and more—electronic course materials, please turn to pages xi-xiii.

Calculus III

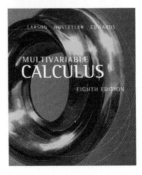

Also available from the Larson, Hostetler, and Edwards Calculus *Author Team*

CALCULUS WITH PRECALCULUS

To give more students access to calculus by easing the transition from precalculus, the following textbook sequence is available.

Precalculus and Calculus I

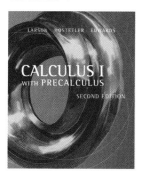

Calculus II

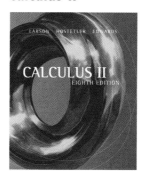

Calculus III

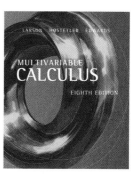

CALCULUS WITH EARLY TRANSCENDENTAL FUNCTIONS

For instructors who prefer to introduce the transcendental functions early in the calculus course, the following textbook sequence is available.

Calculus I, II, and III

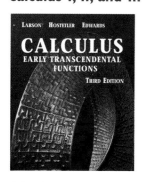

Calculus I and II

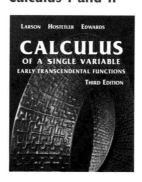

Calculus III

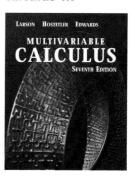

CALCULUS WITH LATE TRIGONOMETRY

For instructors who introduce the trigonometric functions in the second semester, the following textbook is available.

Calculus I, II, and III

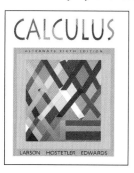

Integrated Learning System for Calculus

Comprehensive Calculus Resources

The Integrated Learning System for *Calculus*, Eighth Edition, addresses the changing needs of today's instructors and students. Recognizing that the calculus course is presented in a variety of teaching and learning environments, we offer extensive resources that support the textbook program in print, CD-ROM, and online formats.

- Online homework practice
- Testing
- Tutoring
- Graded homework
- Classroom management
- Online course
- Interactive resources

**ONE
INTEGRATED LEARNING SYSTEM**

The teaching and learning resources you need in the format you prefer

The Integrated Learning System for *Calculus*, Eighth Edition, offers dynamic teaching tools for instructors and interactive learning resources for students in the following flexible course delivery formats.

- Eduspace® online learning system
- HM mathSpace® Student CD-ROM
- Instructional DVDs and Videotapes
- HM ClassPrep with HM Testing CD-ROM
- Companion Textbook Websites
- Printed Resources

Enhanced! Eduspace® Online Calculus

Eduspace®, powered by Blackboard®, is ready to use and easy to integrate into the calculus course. It provides comprehensive homework exercises, tutorials, and testing keyed to the textbook by section.

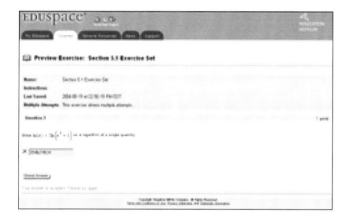

Features

- Algorithmically generated tutorial questions for unlimited practice
- Comprehensive exercise sets for graded homework
- Point-of-use links to additional tools, animations, simulations, videos, and exercises
- SMARTHINKING™ live, online tutoring for students
- Color graphing calculator
- Ample prerequisite skills review with customized student self-study plan
- Chapter tests
- Link to CalcChat
- Electronic version of all textbook exercises
- Links to detailed, stepped-out solutions to odd-numbered textbook exercises

Enhanced! HM mathSpace® CD-ROM

For the student, HM mathSpace® Student CD-ROM offers a wealth of learning resources keyed to the textbook by section.

Features

- Algorithmically generated tutorial questions for unlimited practice of prerequisite skills
- Point-of-use links to additional tools, animations, and simulations
- Link to CalcChat
- Color graphing calculator
- Chapter tests

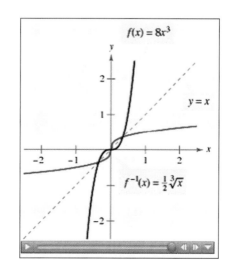

For additional information about the Larson, Hostetler, and Edwards Calculus program, go to *college.hmco.com/info/larsoncalculus*.

Integrated Learning System for Calculus

New! HM ClassPrep with HM Testing CD-ROM

This valuable CD-ROM contains an array of useful instructor resources keyed to the textbook.

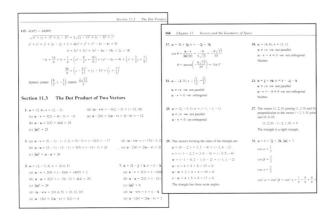

Features

- HM Testing test generator
- Solutions to all textbook exercises
- Digital textbook art
- Chapter summaries
- Section-level teaching strategies
- Additional presentations with exercises covering differential equations, precalculus review, rotation and the general second degree equation, complex numbers, and business and economic applications
- Graphing calculator programs

New! HM Testing

For the instructor, HM Testing is a robust test-generating system.

Features

- Comprehensive set of algorithmic test items
- Can produce chapter tests, cumulative tests, and final exams
- Online testing
- Gradebook function

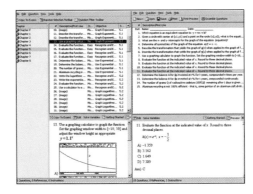

Enhanced! Instructional DVDs and Videos

These comprehensive video presentations complement the textbook topic coverage. This video presentation has a variety of uses, including supplementing an online or hybrid course, giving students the opportunity to catch up if they miss a class, and providing substantial course material for self-study and review.

Features

- Comprehensive topic coverage from Calculus I, II, and III
- Additional explanations of calculus concepts, sample problems, and applications
- Available in video and DVD format

Enhanced! Companion Textbook Website

The free Houghton Mifflin website at *math.college.hmco.com* contains an abundance of instructor and student resources.

Features

- Downloadable graphing calculator programs
- Textbook Appendices C–G, containing additional presentations with exercises covering differential equations, precalculus review, rotation and the general second-degree equation, complex numbers, and business and economic applications
- Algebra Review Summary
- Calculus Labs
- 3-D rotatable graphs

Printed Resources

For the convenience of both instructors and students, many of our teaching and learning resources are available as printed supplements, but are also available in electronic format.

For instructors
Complete Solutions Guide by Bruce Edwards
   This instructor's resource contains worked-out solutions to all textbook exercises. It is available in three volumes: Volume I covers Chapters P–6, Volume II covers Chapters 7–11, and Volume III covers Chapters 11–15.

Instructor's Resource Manual by Ann Rutledge Kraus
   This instructor's resource contains an abundance of resources keyed to the textbook by chapter and section, including chapter summaries, teaching strategies, multiple versions of chapter tests, final exams, and gateway tests, and suggested solutions to the Chapter Openers, Explorations, Section Projects, and Technology features in the text.

Test Item File
   The *Test Item File* contains a sample question for every algorithm in HM Testing.

For students
Study and Solutions Guide by Bruce Edwards
   This student resource contains detailed, worked-out solutions to all odd-numbered textbook exercises. It is available in two volumes: Volume I covers Chapters P–11 and Volume II covers Chapters 11–15.

For additional information about the Larson, Hostetler, and Edwards Calculus program, go to *college.hmco.com/info/larsoncalculus*.

Features

Chapter Openers

Each chapter opens with a real-life application of the concepts presented in the chapter, illustrated by a photograph. Open-ended and thought-provoking questions about the application encourage the student to consider how calculus concepts relate to real-life situations. A brief summary with a graphical component highlights the primary mathematical concepts presented in the chapter, and explains why they are important.

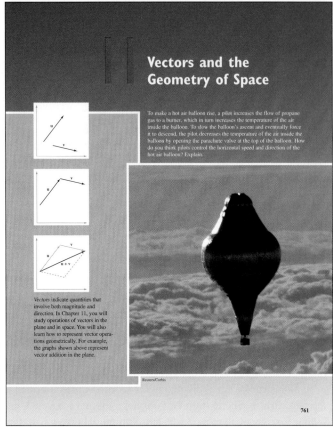

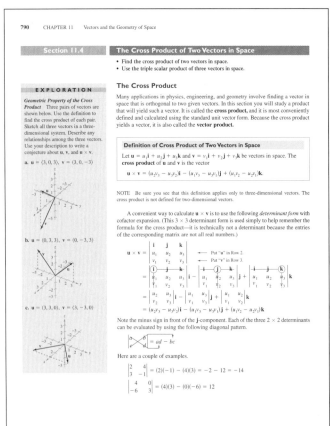

Section Openers

Every section begins with an outline of the key concepts covered in the section. This serves as a class planning resource for the instructor and a study and review guide for the student.

Explorations

For selected topics, Explorations offer the opportunity to discover calculus concepts before they are formally introduced in the text, thus enhancing student understanding. This optional feature can be omitted at the discretion of the instructor with no loss of continuity in the coverage of the material.

Historical Notes

Integrated throughout the text, Historical Notes help students grasp the basic mathematical foundations of calculus.

FEATURES

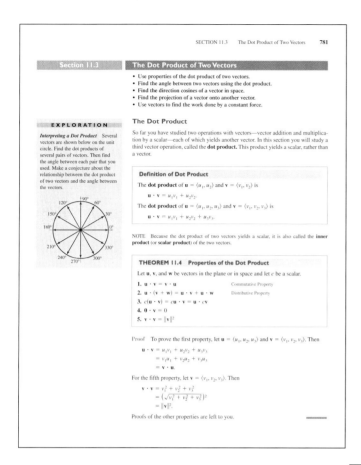

Theorems

All Theorems and Definitions are highlighted for emphasis and easy reference. Proofs are shown for selected theorems to enhance student understanding.

Study Tip

Located at point of use throughout the text, Study Tips advise students on how to avoid common errors, address special cases, and expand upon theoretical concepts.

Graphics

Numerous graphics throughout the text enhance student understanding of complex calculus concepts (especially in three-dimensional representations), as well as real-life applications.

Example

To enhance the usefulness of the text as a study and learning tool, the Eighth Edition contains numerous Examples. The detailed, worked-out Solutions (many with side comments to clarify the steps or the method) are presented graphically, analytically, and/or numerically to provide students with opportunities for practice and further insight into calculus concepts. Many Examples incorporate real-data analysis.

Open Exploration

Eduspace® contains Open Explorations, which investigate selected Examples using computer algebra systems (*Maple*, *Mathematica*, *Derive*, and *Mathcad*). The icon identifies these Examples.

Notes

Instructional Notes accompany many of the Theorems, Definitions, and Examples to offer additional insights or describe generalizations.

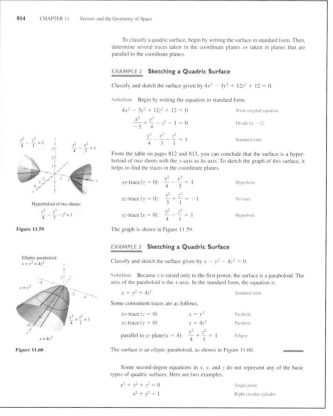

Exercises

The core of every calculus text, Exercises provide opportunities for exploration, practice, and comprehension. The Eighth Edition contains over 10,000 Section and Chapter Review Exercises, carefully graded in each set from skill-building to challenging. The extensive range of problem types includes true/false, writing, conceptual, real-data modeling, and graphical analysis.

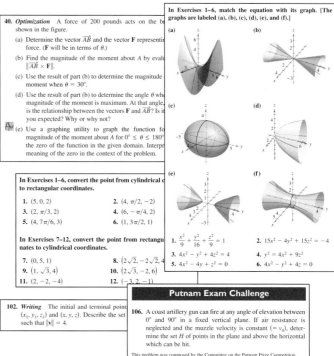

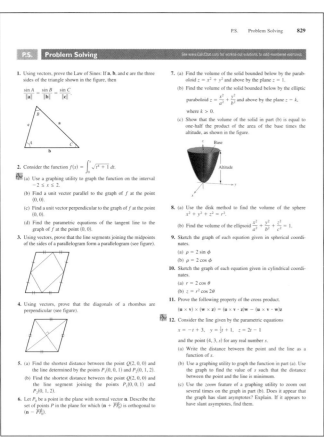

P.S. Problem Solving

Each chapter concludes with a set of thought-provoking and challenging exercises that provide opportunities for the student to explore the concepts in the chapter further.

Technology

Throughout the text, the use of a graphing utility or computer algebra system is suggested as appropriate for problem-solving as well as exploration and discovery. For example, students may choose to use a graphing utility to execute complicated computations, to visualize theoretical concepts, to discover alternative approaches, or to verify the results of other solution methods. However, students are not required to have access to a graphing utility to use this text effectively. In addition to describing the benefits of using technology to learn calculus, the text also addresses its possible misuse or misinterpretation.

Additional Features

Additional teaching and learning resources are integrated throughout the textbook, including Section Projects, journal references, and Writing About Concepts Exercises.

Acknowledgments

We would like to thank the many people who have helped us at various stages of this project over the last 30 years. Their encouragement, criticisms, and suggestions have been invaluable to us.

For the Eighth Edition

James Pommersheim
Reed College

Kevin J. Leith
Albuquerque Community College

Andrew J. Guzo
Chatham High School, NJ

Mary J. Quadrini
East Greenwich High School, RI

Jim Burton
Vernon Verona Sherrill High School, NY

Guillermo Barberena III
South Hills High School, TX

Susan A. Natale
The Ursuline School, NY

Patrick Ward
Illinois Central College

Donna J. Gorton
Butler County Community College

Diane Zych
Erie Community College

Guy Hogan
Norfolk State University

Michael Frantz
University of La Verne

Darren Narayan
Rochester Institute of Technology

Stanley J. Brzezicki
Iroquois High School, PA

Leland E. Rogers
Pepperdine University

Paul Seeburger
Monroe Community College

Ashok Kumar
Valdosta State University

Alexander Arhangelskii
Ohio University

James Braselton
Georgia Southern University

Harvey Braverman
Middlesex County College

Jianzhong Su
University of Texas at Arlington

P.S. Crooke
Vanderbilt University

Stan Adamski
Owens Community College

Edith A. Silver
Mercer County Community College

Seth G. Armstrong
Southern Utah University

Desmond Stephens
Florida A&M University

Terence H. Perciante
Wheaton College

Linda A. Bolte
Eastern Washington University

Sudhir Goel
Valdosta State University

Donna Flint
South Dakota State University

For the Eighth Edition Technology Program

John Gosselin
University of Georgia

Oiyin Pauline Chow
Harrisburg Area Community College

Murray Eisenberg
University of Massachusetts at Amherst

Douglas B. Meade
University of South Carolina

Tim Chappell
Penn Valley Community College

Howard Speier
Chandler-Gilbert Community College

Marcelle Bessman
Jacksonville University

Jim Ball
Indiana State University

Julie M. Clark
Hollins University

Teri Murphy
University of Oklahoma

Jim Dotzler
Nassau Community College

Arek Goetz
San Francisco State University

Shahryar Heydari
Piedmont College

Reviewers of Previous Editions

Dennis Alber, *Palm Beach Junior College*; James Angelos, *Central Michigan University*; Raymond Badalian, *Los Angeles City College*; Kerry D. Bailey, *Laramie County Community College*; Harry L. Baldwin, Jr., *San Diego State City College*; Homer F. Bechtell, *University of New Hampshire*; Keith Bergeron, *United States Air Force Academy*; Norman Birenes, *University of Regina*; Brian Blank, *Washington State University*; Andrew A. Bulleri, *Howard Community College*; Christopher Butler, *Case Western Reserve University*; Dane R. Camp, *New Trier High School, IL*; Paula Castagna, *Fresno City College*; Jack Ceder, *University of California-Santa Barbara*; Charles L. Cope, *Morehouse College*; Barbara Cortzen, *DePaul University*; Jorge Cossio, *Miami-Dade Community College*; Jack Courtney, *Michigan State University*; James Daniels, *Palomar College*; Kathy Davis, *University of Texas*; Paul W. Davis, *Worcester Polytechnic Institute*; Luz M. DeAlba, *Drake University*; Nicolae Dinculeanu, *University of Florida*; Rosario Diprizio, *Oakton Community College*; Garret J. Etgen, *University of Houston*; Russell Euler, *Northwest Missouri State University*; Phillip A. Ferguson, *Fresno City College*; Li Fong, *Johnson County Community College*; Michael Frantz, *University of La Verne*; William R. Fuller, *Purdue University*; Dewey Furness, *Ricks College*; Javier Garza, *Tarleton State University*; K. Elayn Gay, *University of New Orleans*; Thomas M. Green, *Contra Costa College*; Ali Hajjafar, *University of Akron*; Ruth A. Hartman, *Black Hawk College*; Irvin Roy Hentzel, *Iowa State University*; Kathy Hoke, *University of Richmond*; Howard E. Holcomb, *Monroe Community College*; Eric R. Immel, *Georgia Institute of Technology*; Arnold J. Insel, *Illinois State University*; Elgin Honston, *Iowa State University*; Hikeaki Kaneko, *Old Dominion University*; Toni Kasper, *Borough of Manhattan Community College*; William J. Keane, *Boston College*; Timothy J. Kearns, *Boston College*; Ronnie Khuri, *University of Florida*; Frank T. Kocher, Jr., *Pennsylvania State University*; Robert Kowalczyk, *University of Massachusetts-Dartmouth*; Joseph F. Krebs, *Boston College*; David C. Lantz, *Colgate University*; Norbert Lerner, *State University of New York at Cortland*; Maita Levine, *University of Cincinnati*; Murray Lieb, *New Jersey Institute of Technology*; Beth Long, *Pellissippi State Technical College*; Ransom Van B. Lynch, *Phillips Exeter Academy*; Bennet Manvel, *Colorado State University*; Mauricio Marroquin, *Los Angeles Valley College*; Robert L. Maynard, *Tidewater Community College*; Robert McMaster, *John Abbott College*; Gordon Melrose, *Old Dominion University*; Darrell Minor, *Columbus*

State Community College; Maurice Monahan, *South Dakota State University*; Michael Montaño, *Riverside Community College*; Philip Montgomery, *University of Kansas*; David C. Morency, *University of Vermont*; Gerald Mueller, *Columbus State Community College*; Duff A. Muir, *United States Air Force Academy*; Charlotte J. Newsom, *Tidewater Community College*; Terry J. Newton, *United States Air Force Academy*; Donna E. Nordstrom, *Pasadena City College*; Larry Norris, *North Carolina State University*; Robert A. Nowlan, *Southern Connecticut State University*; Luis Ortiz-Franco, *Chapman University*; Barbara L. Osofsky, *Rutgers University*; Judith A. Palagallo, *University of Akron*; Eleanor Palais, *Belmont High School*, MA; Wayne J. Peeples, *University of Texas*; Jorge A. Perez, *LaGuardia Community College*; Darrell J. Peterson, *Santa Monica College*; Donald Poulson, *Mesa Community College*; Lila Roberts, *Georgia Southern University*; Jean L. Rubin, *Purdue University*; John Santomas, *Villanova University*; Barry Sarnacki, *United States Air Force Academy*; N. James Schoonmaker, *University of Vermont*; George W. Schultz, *St. Petersburg Junior College*; Richard E. Shermoen, *Washburn University*; Thomas W. Shilgalis, *Illinois State University*; J. Philip Smith, *Southern Connecticut State University*; Lynn Smith, *Gloucester County College*; Frank Soler, *De Anza College*; Enid Steinbart, *University of New Orleans*; Michael Steuer, *Nassau Community College*; Mark Stevenson, *Oakland Community College*; Anthony Thomas, *University of Wisconsin-Platteville*; Lawrence A. Trivieri, *Mohawk Valley Community College*; John Tweed, *Old Dominion University*; Carol Urban, *College of DuPage*; Marjorie Valentine, *North Side ISD, San Antonio*; Robert J. Vojack, *Ridgewood High School, NJ*; Bert K. Waits, *The Ohio State University*; Florence A. Warfel, *University of Pittsburgh*; John R. Watret, *Embry-Riddle Aeronautical University*; Carroll G. Wells, *Western Kentucky University*; Charles Wheeler, *Montgomery College*; Jay Wiestling, *Palomar College*; Paul D. Zahn, *Borough of Manhattan Community College*; August J. Zarcone, *College of DuPage*

We would like to thank the staff at Larson Texts, Inc., who assisted in preparing the manuscript, rendering the art package, and typesetting and proofreading the pages and supplements.

A special note of thanks goes to the instructors who responded to our survey and to the over 2 million students who have used earlier editions of the text.

On a personal level, we are grateful to our wives, Deanna Gilbert Larson, Eloise Hostetler, and Consuelo Edwards, for their love, patience, and support. Also, a special note of thanks goes out to R. Scott O'Neil.

If you have suggestions for improving this text, please feel free to write to us. Over the years we have received many useful comments from both instructors and students, and we value these very much.

Ron Larson
Robert P. Hostetler
Bruce H. Edwards

7 Applications of Integration

The Atomium, located in Belgium, represents an iron crystal molecule magnified 165 billion times. The structure contains nine spheres connected with cylindrical tubes. The central sphere has one tube passing directly through its center. Explain how to find the volume of the portion of the central sphere that does not include the tube.

Andre Jenny/Alamy Images

The *disk method* is one method that is used to find the volume of a solid. This method requires finding the sum of the volumes of representative disks to approximate the volume of the solid. As you increase the number of disks, the approximation tends to become more accurate. In Section 7.2, you will use limits to write the exact volume of the solid as a definite integral.

446 CHAPTER 7 Applications of Integration

Section 7.1 Area of a Region Between Two Curves

- Find the area of a region between two curves using integration.
- Find the area of a region between intersecting curves using integration.
- Describe integration as an accumulation process.

Area of a Region Between Two Curves

With a few modifications you can extend the application of definite integrals from the area of a region *under* a curve to the area of a region *between* two curves. Consider two functions f and g that are continuous on the interval $[a, b]$. If, as in Figure 7.1, the graphs of both f and g lie above the x-axis, and the graph of g lies below the graph of f, you can geometrically interpret the area of the region between the graphs as the area of the region under the graph of g subtracted from the area of the region under the graph of f, as shown in Figure 7.2.

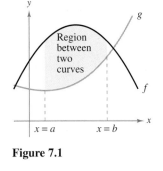

Figure 7.1

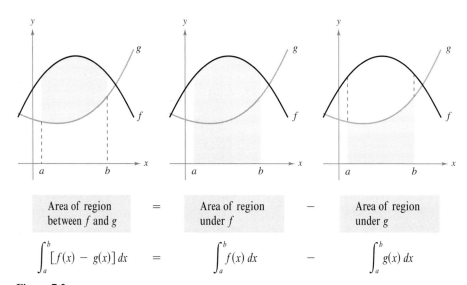

Figure 7.2

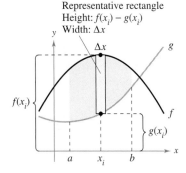

Figure 7.3

To verify the reasonableness of the result shown in Figure 7.2, you can partition the interval $[a, b]$ into n subintervals, each of width Δx. Then, as shown in Figure 7.3, sketch a **representative rectangle** of width Δx and height $f(x_i) - g(x_i)$, where x_i is in the ith interval. The area of this representative rectangle is

$$\Delta A_i = (\text{height})(\text{width}) = [f(x_i) - g(x_i)]\Delta x.$$

By adding the areas of the n rectangles and taking the limit as $\|\Delta\| \to 0$ $(n \to \infty)$, you obtain

$$\lim_{n \to \infty} \sum_{i=1}^{n} [f(x_i) - g(x_i)]\Delta x.$$

Because f and g are continuous on $[a, b]$, $f - g$ is also continuous on $[a, b]$ and the limit exists. So, the area of the given region is

$$\text{Area} = \lim_{n \to \infty} \sum_{i=1}^{n} [f(x_i) - g(x_i)]\Delta x$$

$$= \int_{a}^{b} [f(x) - g(x)]\, dx.$$

Area of a Region Between Two Curves

If f and g are continuous on $[a, b]$ and $g(x) \leq f(x)$ for all x in $[a, b]$, then the area of the region bounded by the graphs of f and g and the vertical lines $x = a$ and $x = b$ is

$$A = \int_a^b [f(x) - g(x)]\, dx.$$

In Figure 7.1, the graphs of f and g are shown above the x-axis. This, however, is not necessary. The same integrand $[f(x) - g(x)]$ can be used as long as f and g are continuous and $g(x) \leq f(x)$ for all x in the interval $[a, b]$. This result is summarized graphically in Figure 7.4.

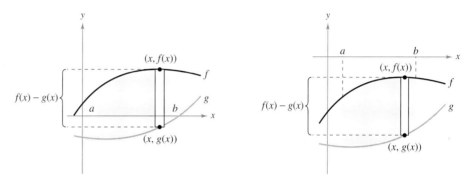

Figure 7.4

NOTE The height of a representative rectangle is $f(x) - g(x)$ regardless of the relative position of the x-axis, as shown in Figure 7.4.

Representative rectangles are used throughout this chapter in various applications of integration. A vertical rectangle (of width Δx) implies integration with respect to x, whereas a horizontal rectangle (of width Δy) implies integration with respect to y.

EXAMPLE 1 Finding the Area of a Region Between Two Curves

Find the area of the region bounded by the graphs of $y = x^2 + 2$, $y = -x$, $x = 0$, and $x = 1$.

Solution Let $g(x) = -x$ and $f(x) = x^2 + 2$. Then $g(x) \leq f(x)$ for all x in $[0, 1]$, as shown in Figure 7.5. So, the area of the representative rectangle is

$$\Delta A = [f(x) - g(x)]\, \Delta x$$
$$= [(x^2 + 2) - (-x)]\, \Delta x$$

and the area of the region is

$$A = \int_a^b [f(x) - g(x)]\, dx = \int_0^1 [(x^2 + 2) - (-x)]\, dx$$
$$= \left[\frac{x^3}{3} + \frac{x^2}{2} + 2x \right]_0^1$$
$$= \frac{1}{3} + \frac{1}{2} + 2$$
$$= \frac{17}{6}.$$

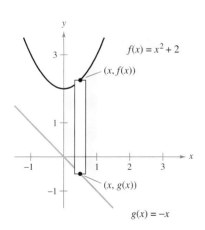

Region bounded by the graph of f, the graph of g, $x = 0$, and $x = 1$
Figure 7.5

Area of a Region Between Intersecting Curves

In Example 1, the graphs of $f(x) = x^2 + 2$ and $g(x) = -x$ do not intersect, and the values of a and b are given explicitly. A more common problem involves the area of a region bounded by two *intersecting* graphs, where the values of a and b must be calculated.

EXAMPLE 2 A Region Lying Between Two Intersecting Graphs

Find the area of the region bounded by the graphs of $f(x) = 2 - x^2$ and $g(x) = x$.

Solution In Figure 7.6, notice that the graphs of f and g have two points of intersection. To find the x-coordinates of these points, set $f(x)$ and $g(x)$ equal to each other and solve for x.

$$2 - x^2 = x \qquad \text{Set } f(x) \text{ equal to } g(x).$$
$$-x^2 - x + 2 = 0 \qquad \text{Write in general form.}$$
$$-(x + 2)(x - 1) = 0 \qquad \text{Factor.}$$
$$x = -2 \text{ or } 1 \qquad \text{Solve for } x.$$

So, $a = -2$ and $b = 1$. Because $g(x) \leq f(x)$ for all x in the interval $[-2, 1]$, the representative rectangle has an area of

$$\Delta A = [f(x) - g(x)] \Delta x$$
$$= [(2 - x^2) - x] \Delta x$$

and the area of the region is

$$A = \int_{-2}^{1} [(2 - x^2) - x] \, dx = \left[-\frac{x^3}{3} - \frac{x^2}{2} + 2x \right]_{-2}^{1}$$
$$= \frac{9}{2}.$$

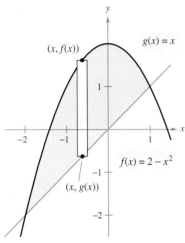

Region bounded by the graph of f and the graph of g
Figure 7.6

EXAMPLE 3 A Region Lying Between Two Intersecting Graphs

The sine and cosine curves intersect infinitely many times, bounding regions of equal areas, as shown in Figure 7.7. Find the area of one of these regions.

Solution

$$\sin x = \cos x \qquad \text{Set } f(x) \text{ equal to } g(x).$$
$$\frac{\sin x}{\cos x} = 1 \qquad \text{Divide each side by } \cos x.$$
$$\tan x = 1 \qquad \text{Trigonometric identity}$$
$$x = \frac{\pi}{4} \text{ or } \frac{5\pi}{4}, \qquad 0 \leq x \leq 2\pi \qquad \text{Solve for } x.$$

So, $a = \pi/4$ and $b = 5\pi/4$. Because $\sin x \geq \cos x$ for all x in the interval $[\pi/4, 5\pi/4]$, the area of the region is

$$A = \int_{\pi/4}^{5\pi/4} [\sin x - \cos x] \, dx = \left[-\cos x - \sin x \right]_{\pi/4}^{5\pi/4}$$
$$= 2\sqrt{2}.$$

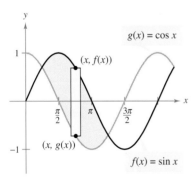

One of the regions bounded by the graphs of the sine and cosine functions
Figure 7.7

If two curves intersect at more than two points, then to find the area of the region between the curves, you must find all points of intersection and check to see which curve is above the other in each interval determined by these points.

EXAMPLE 4 Curves That Intersect at More Than Two Points

Find the area of the region between the graphs of $f(x) = 3x^3 - x^2 - 10x$ and $g(x) = -x^2 + 2x$.

Solution Begin by setting $f(x)$ and $g(x)$ equal to each other and solving for x. This yields the x-values at each point of intersection of the two graphs.

$$3x^3 - x^2 - 10x = -x^2 + 2x \qquad \text{Set } f(x) \text{ equal to } g(x).$$
$$3x^3 - 12x = 0 \qquad \text{Write in general form.}$$
$$3x(x - 2)(x + 2) = 0 \qquad \text{Factor.}$$
$$x = -2, 0, 2 \qquad \text{Solve for } x.$$

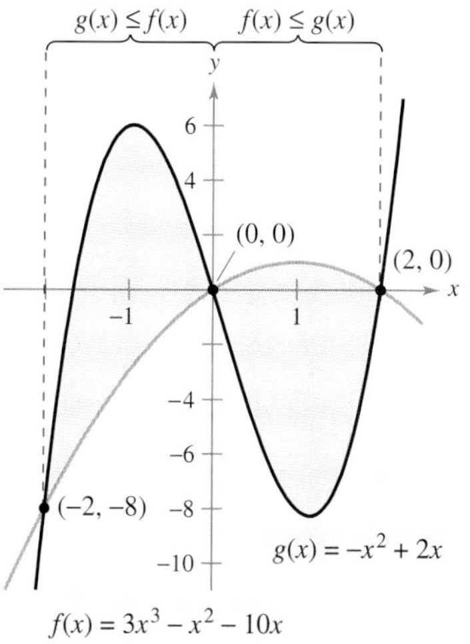

So, the two graphs intersect when $x = -2, 0,$ and 2. In Figure 7.8, notice that $g(x) \leq f(x)$ on the interval $[-2, 0]$. However, the two graphs switch at the origin, and $f(x) \leq g(x)$ on the interval $[0, 2]$. So, you need two integrals—one for the interval $[-2, 0]$ and one for the interval $[0, 2]$.

$$A = \int_{-2}^{0} [f(x) - g(x)] \, dx + \int_{0}^{2} [g(x) - f(x)] \, dx$$
$$= \int_{-2}^{0} (3x^3 - 12x) \, dx + \int_{0}^{2} (-3x^3 + 12x) \, dx$$
$$= \left[\frac{3x^4}{4} - 6x^2 \right]_{-2}^{0} + \left[\frac{-3x^4}{4} + 6x^2 \right]_{0}^{2}$$
$$= -(12 - 24) + (-12 + 24) = 24$$

On $[-2, 0]$, $g(x) \leq f(x)$, and on $[0, 2]$, $f(x) \leq g(x)$
Figure 7.8

NOTE In Example 4, notice that you obtain an incorrect result if you integrate from -2 to 2. Such integration produces

$$\int_{-2}^{2} [f(x) - g(x)] \, dx = \int_{-2}^{2} (3x^3 - 12x) \, dx = 0.$$

If the graph of a function of y is a boundary of a region, it is often convenient to use representative rectangles that are *horizontal* and find the area by integrating with respect to y. In general, to determine the area between two curves, you can use

$$A = \int_{x_1}^{x_2} \underbrace{[(\text{top curve}) - (\text{bottom curve})]}_{\text{in variable } x} dx \qquad \text{Vertical rectangles}$$

$$A = \int_{y_1}^{y_2} \underbrace{[(\text{right curve}) - (\text{left curve})]}_{\text{in variable } y} dy \qquad \text{Horizontal rectangles}$$

where (x_1, y_1) and (x_2, y_2) are either adjacent points of intersection of the two curves involved or points on the specified boundary lines.

---

indicates that in the **HM mathSpace® CD-ROM** *and the online* **Eduspace®** *system for this text, you will find an Open Exploration, which further explores this example using the computer algebra systems* **Maple, Mathcad, Mathematica,** *and* **Derive.**

EXAMPLE 5 Horizontal Representative Rectangles

Find the area of the region bounded by the graphs of $x = 3 - y^2$ and $x = y + 1$.

Solution Consider

$$g(y) = 3 - y^2 \quad \text{and} \quad f(y) = y + 1.$$

These two curves intersect when $y = -2$ and $y = 1$, as shown in Figure 7.9. Because $f(y) \le g(y)$ on this interval, you have

$$\Delta A = [g(y) - f(y)] \Delta y = [(3 - y^2) - (y + 1)] \Delta y.$$

So, the area is

$$A = \int_{-2}^{1} [(3 - y^2) - (y + 1)] \, dy$$

$$= \int_{-2}^{1} (-y^2 - y + 2) \, dy$$

$$= \left[\frac{-y^3}{3} - \frac{y^2}{2} + 2y \right]_{-2}^{1}$$

$$= \left(-\frac{1}{3} - \frac{1}{2} + 2 \right) - \left(\frac{8}{3} - 2 - 4 \right)$$

$$= \frac{9}{2}.$$

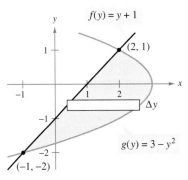

Horizontal rectangles (integration with respect to y)
Figure 7.9

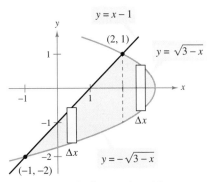

Vertical rectangles (integration with respect to x)
Figure 7.10

In Example 5, notice that by integrating with respect to y you need only one integral. If you had integrated with respect to x, you would have needed two integrals because the upper boundary would have changed at $x = 2$, as shown in Figure 7.10.

$$A = \int_{-1}^{2} \left[(x - 1) + \sqrt{3 - x} \right] dx + \int_{2}^{3} \left(\sqrt{3 - x} + \sqrt{3 - x} \right) dx$$

$$= \int_{-1}^{2} [x - 1 + (3 - x)^{1/2}] \, dx + 2 \int_{2}^{3} (3 - x)^{1/2} \, dx$$

$$= \left[\frac{x^2}{2} - x - \frac{(3 - x)^{3/2}}{3/2} \right]_{-1}^{2} - 2 \left[\frac{(3 - x)^{3/2}}{3/2} \right]_{2}^{3}$$

$$= \left(2 - 2 - \frac{2}{3} \right) - \left(\frac{1}{2} + 1 - \frac{16}{3} \right) - 2(0) + 2\left(\frac{2}{3} \right)$$

$$= \frac{9}{2}$$

Integration as an Accumulation Process

In this section, the integration formula for the area between two curves was developed by using a rectangle as the *representative element*. For each new application in the remaining sections of this chapter, an appropriate representative element will be constructed using precalculus formulas you already know. Each integration formula will then be obtained by summing or accumulating these representative elements.

Known precalculus formula ⇒ Representative element ⇒ New integration formula

For example, in this section the area formula was developed as follows.

$A = \text{(height)}\text{(width)}$ ⇒ $\Delta A = [f(x) - g(x)]\,\Delta x$ ⇒ $A = \int_a^b [f(x) - g(x)]\,dx$

EXAMPLE 6 Describing Integration as an Accumulation Process

Find the area of the region bounded by the graph of $y = 4 - x^2$ and the x-axis. Describe the integration as an accumulation process.

Solution The area of the region is given by

$$A = \int_{-2}^{2} (4 - x^2)\,dx.$$

You can think of the integration as an accumulation of the areas of the rectangles formed as the representative rectangle slides from $x = -2$ to $x = 2$, as shown in Figure 7.11.

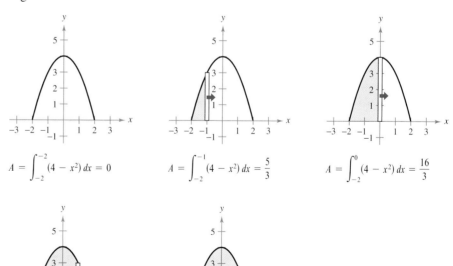

$A = \int_{-2}^{-2} (4 - x^2)\,dx = 0$ $A = \int_{-2}^{-1} (4 - x^2)\,dx = \frac{5}{3}$ $A = \int_{-2}^{0} (4 - x^2)\,dx = \frac{16}{3}$

$A = \int_{-2}^{1} (4 - x^2)\,dx = 9$ $A = \int_{-2}^{2} (4 - x^2)\,dx = \frac{32}{3}$

Figure 7.11

Exercises for Section 7.1

In Exercises 1–6, set up the definite integral that gives the area of the region.

1. $f(x) = x^2 - 6x$
$g(x) = 0$

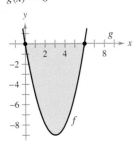

2. $f(x) = x^2 + 2x + 1$
$g(x) = 2x + 5$

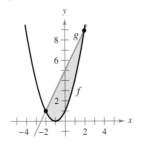

3. $f(x) = x^2 - 4x + 3$
$g(x) = -x^2 + 2x + 3$

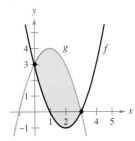

4. $f(x) = x^2$
$g(x) = x^3$

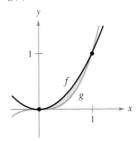

5. $f(x) = 3(x^3 - x)$
$g(x) = 0$

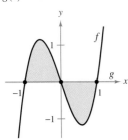

6. $f(x) = (x - 1)^3$
$g(x) = x - 1$

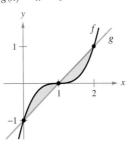

In Exercises 7–12, the integrand of the definite integral is a difference of two functions. Sketch the graph of each function and shade the region whose area is represented by the integral.

7. $\int_0^4 \left[(x+1) - \frac{x}{2} \right] dx$

8. $\int_{-1}^1 [(1 - x^2) - (x^2 - 1)] dx$

9. $\int_0^6 \left[4(2^{-x/3}) - \frac{x}{6} \right] dx$

10. $\int_2^3 \left[\left(\frac{x^3}{3} - x \right) - \frac{x}{3} \right] dx$

11. $\int_{-\pi/3}^{\pi/3} (2 - \sec x) \, dx$

12. $\int_{-\pi/4}^{\pi/4} (\sec^2 x - \cos x) \, dx$

In Exercises 13 and 14, find the area of the region by integrating (a) with respect to x and (b) with respect to y.

13. $x = 4 - y^2$
$x = y - 2$

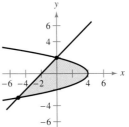

14. $y = x^2$
$y = 6 - x$

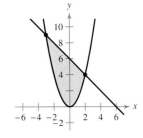

Think About It In Exercises 15 and 16, determine which value best approximates the area of the region bounded by the graphs of f and g. (Make your selection on the basis of a sketch of the region and not by performing any calculations.)

15. $f(x) = x + 1$, $g(x) = (x - 1)^2$
 (a) -2 (b) 2   (c) 10   (d) 4   (e) 8

16. $f(x) = 2 - \frac{1}{2}x$, $g(x) = 2 - \sqrt{x}$
 (a) 1   (b) 6   (c) -3 (d) 3   (e) 4

In Exercises 17–32, sketch the region bounded by the graphs of the algebraic functions and find the area of the region.

17. $y = \frac{1}{2}x^3 + 2$, $y = x + 1$, $x = 0$, $x = 2$
18. $y = -\frac{3}{8}x(x - 8)$, $y = 10 - \frac{1}{2}x$, $x = 2$, $x = 8$
19. $f(x) = x^2 - 4x$, $g(x) = 0$
20. $f(x) = -x^2 + 4x + 1$, $g(x) = x + 1$
21. $f(x) = x^2 + 2x + 1$, $g(x) = 3x + 3$
22. $f(x) = -x^2 + 4x + 2$, $g(x) = x + 2$
23. $y = x$, $y = 2 - x$, $y = 0$
24. $y = \frac{1}{x^2}$, $y = 0$, $x = 1$, $x = 5$
25. $f(x) = \sqrt{3x} + 1$, $g(x) = x + 1$
26. $f(x) = \sqrt[3]{x - 1}$, $g(x) = x - 1$
27. $f(y) = y^2$, $g(y) = y + 2$
28. $f(y) = y(2 - y)$, $g(y) = -y$
29. $f(y) = y^2 + 1$, $g(y) = 0$, $y = -1$, $y = 2$
30. $f(y) = \frac{y}{\sqrt{16 - y^2}}$, $g(y) = 0$, $y = 3$
31. $f(x) = \frac{10}{x}$, $x = 0$, $y = 2$, $y = 10$
32. $g(x) = \frac{4}{2 - x}$, $y = 4$, $x = 0$

In Exercises 33–42, (a) use a graphing utility to graph the region bounded by the graphs of the equations, (b) find the area of the region, and (c) use the integration capabilities of the graphing utility to verify your results.

33. $f(x) = x(x^2 - 3x + 3)$, $g(x) = x^2$
34. $f(x) = x^3 - 2x + 1$, $g(x) = -2x$, $x = 1$
35. $y = x^2 - 4x + 3$, $y = 3 + 4x - x^2$
36. $y = x^4 - 2x^2$, $y = 2x^2$
37. $f(x) = x^4 - 4x^2$, $g(x) = x^2 - 4$
38. $f(x) = x^4 - 4x^2$, $g(x) = x^3 - 4x$
39. $f(x) = 1/(1 + x^2)$, $g(x) = \frac{1}{2}x^2$
40. $f(x) = 6x/(x^2 + 1)$, $y = 0$, $0 \le x \le 3$
41. $y = \sqrt{1 + x^3}$, $y = \frac{1}{2}x + 2$, $x = 0$
42. $y = x\sqrt{\dfrac{4-x}{4+x}}$, $y = 0$, $x = 4$

In Exercises 43–48, sketch the region bounded by the graphs of the functions, and find the area of the region.

43. $f(x) = 2\sin x$, $g(x) = \tan x$, $-\dfrac{\pi}{3} \le x \le \dfrac{\pi}{3}$
44. $f(x) = \sin x$, $g(x) = \cos 2x$, $-\dfrac{\pi}{2} \le x \le \dfrac{\pi}{6}$
45. $f(x) = \cos x$, $g(x) = 2 - \cos x$, $0 \le x \le 2\pi$
46. $f(x) = \sec\dfrac{\pi x}{4} \tan\dfrac{\pi x}{4}$, $g(x) = (\sqrt{2} - 4)x + 4$, $x = 0$
47. $f(x) = xe^{-x^2}$, $y = 0$, $0 \le x \le 1$
48. $f(x) = 3^x$, $g(x) = 2x + 1$

In Exercises 49–52, (a) use a graphing utility to graph the region bounded by the graphs of the equations, (b) find the area of the region, and (c) use the integration capabilities of the graphing utility to verify your results.

49. $f(x) = 2\sin x + \sin 2x$, $y = 0$, $0 \le x \le \pi$
50. $f(x) = 2\sin x + \cos 2x$, $y = 0$, $0 < x \le \pi$
51. $f(x) = \dfrac{1}{x^2}e^{1/x}$, $y = 0$, $1 \le x \le 3$
52. $g(x) = \dfrac{4\ln x}{x}$, $y = 0$, $x = 5$

In Exercises 53–56, (a) use a graphing utility to graph the region bounded by the graphs of the equations, (b) explain why the area of the region is difficult to find by hand, and (c) use the integration capabilities of the graphing utility to approximate the area to four decimal places.

53. $y = \sqrt{\dfrac{x^3}{4-x}}$, $y = 0$, $x = 3$
54. $y = \sqrt{x}\, e^x$, $y = 0$, $x = 0$, $x = 1$
55. $y = x^2$, $y = 4\cos x$
56. $y = x^2$, $y = \sqrt{3+x}$

In Exercises 57–60, find the accumulation function F. Then evaluate F at each value of the independent variable and graphically show the area given by each value of F.

57. $F(x) = \displaystyle\int_0^x \left(\tfrac{1}{2}t + 1\right) dt$ (a) $F(0)$ (b) $F(2)$ (c) $F(6)$
58. $F(x) = \displaystyle\int_0^x \left(\tfrac{1}{2}t^2 + 2\right) dt$ (a) $F(0)$ (b) $F(4)$ (c) $F(6)$
59. $F(\alpha) = \displaystyle\int_{-1}^{\alpha} \cos\dfrac{\pi\theta}{2}\, d\theta$ (a) $F(-1)$ (b) $F(0)$ (c) $F(\tfrac{1}{2})$
60. $F(y) = \displaystyle\int_{-1}^{y} 4e^{x/2}\, dx$ (a) $F(-1)$ (b) $F(0)$ (c) $F(4)$

In Exercises 61–64, use integration to find the area of the figure having the given vertices.

61. $(2, -3), (4, 6), (6, 1)$
62. $(0, 0), (a, 0), (b, c)$
63. $(0, 2), (4, 2), (0, -2), (-4, -2)$
64. $(0, 0), (1, 2), (3, -2), (1, -3)$

65. *Numerical Integration*   Estimate the surface area of the golf green using (a) the Trapezoidal Rule and (b) Simpson's Rule.

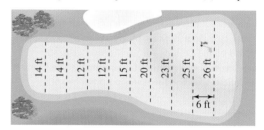

66. *Numerical Integration*   Estimate the surface area of the oil spill using (a) the Trapezoidal Rule and (b) Simpson's Rule.

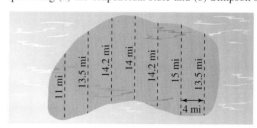

In Exercises 67–70, set up and evaluate the definite integral that gives the area of the region bounded by the graph of the function and the tangent line to the graph at the given point.

67. $f(x) = x^3$, $(1, 1)$
68. $y = x^3 - 2x$, $(-1, 1)$
69. $f(x) = \dfrac{1}{x^2 + 1}$, $\left(1, \tfrac{1}{2}\right)$
70. $y = \dfrac{2}{1 + 4x^2}$, $\left(\tfrac{1}{2}, 1\right)$

Writing About Concepts

71. The graphs of $y = x^4 - 2x^2 + 1$ and $y = 1 - x^2$ intersect at three points. However, the area between the curves *can* be found by a single integral. Explain why this is so, and write an integral for this area.

Writing About Concepts (continued)

72. The area of the region bounded by the graphs of $y = x^3$ and $y = x$ cannot be found by the single integral $\int_{-1}^{1}(x^3 - x)\,dx$. Explain why this is so. Use symmetry to write a single integral that does represent the area.

73. A college graduate has two job offers. The starting salary for each is $32,000, and after 8 years of service each will pay $54,000. The salary increase for each offer is shown in the figure. From a strictly monetary viewpoint, which is the better offer? Explain.

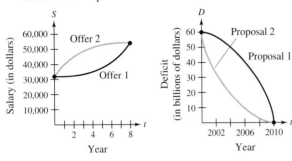

Figure for 73   Figure for 74

74. A state legislature is debating two proposals for eliminating the annual budget deficits by the year 2010. The rate of decrease of the deficits for each proposal is shown in the figure. From the viewpoint of minimizing the cumulative state deficit, which is the better proposal? Explain.

In Exercises 75 and 76, find b such that the line $y = b$ divides the region bounded by the graphs of the two equations into two regions of equal area.

75. $y = 9 - x^2$, $y = 0$ **76.** $y = 9 - |x|$, $y = 0$

In Exercises 77 and 78, find a such that the line $x = a$ divides the region bounded by the graphs of the equations into two regions of equal area.

77. $y = x$, $y = 4$, $x = 0$ **78.** $y^2 = 4 - x$, $x = 0$

In Exercises 79 and 80, evaluate the limit and sketch the graph of the region whose area is represented by the limit.

79. $\lim_{\|\Delta\| \to 0} \sum_{i=1}^{n}(x_i - x_i^2)\,\Delta x$, where $x_i = i/n$ and $\Delta x = 1/n$

80. $\lim_{\|\Delta\| \to 0} \sum_{i=1}^{n}(4 - x_i^2)\,\Delta x$, where $x_i = -2 + (4i/n)$ and $\Delta x = 4/n$

Revenue In Exercises 81 and 82, two models R_1 and R_2 are given for revenue (in billions of dollars per year) for a large corporation. The model R_1 gives projected annual revenues from 2000 to 2005, with $t = 0$ corresponding to 2000, and R_2 gives projected revenues if there is a decrease in the rate of growth of corporate sales over the period. Approximate the total reduction in revenue if corporate sales are actually closer to the model R_2.

81. $R_1 = 7.21 + 0.58t$
$R_2 = 7.21 + 0.45t$

82. $R_1 = 7.21 + 0.26t + 0.02t^2$
$R_2 = 7.21 + 0.1t + 0.01t^2$

 83. *Modeling Data* The table shows the total receipts R and total expenditures E for the Old-Age and Survivors Insurance Trust Fund (Social Security Trust Fund) in billions of dollars. The time t is given in years, with $t = 1$ corresponding to 1991. *(Source: Social Security Administration)*

t	1	2	3	4	5	6
R	299.3	311.2	323.3	328.3	342.8	363.7
E	245.6	259.9	273.1	284.1	297.8	308.2

t	7	8	9	10	11
R	397.2	424.8	457.0	490.5	518.1
E	322.1	332.3	339.9	358.3	377.5

(a) Use a graphing utility to fit an exponential model to the data for receipts. Plot the data and graph the model.

(b) Use a graphing utility to fit an exponential model to the data for expenditures. Plot the data and graph the model.

(c) If the models are assumed to be true for the years 2002 through 2007, use integration to approximate the surplus revenue generated during those years.

(d) Will the models found in parts (a) and (b) intersect? Explain. Based on your answer and news reports about the fund, will these models be accurate for long-term analysis?

84. *Lorenz Curve* Economists use *Lorenz* curves to illustrate the distribution of income in a country. A Lorenz curve, $y = f(x)$, represents the actual income distribution in the country. In this model, x represents percents of families in the country and y represents percents of total income. The model $y = x$ represents a country in which each family has the same income. The area between these two models, where $0 \le x \le 100$, indicates a country's "income inequality." The table lists percents of income y for selected percents of families x in a country.

x	10	20	30	40	50
y	3.35	6.07	9.17	13.39	19.45

x	60	70	80	90
y	28.03	39.77	55.28	75.12

(a) Use a graphing utility to find a quadratic model for the Lorenz curve.

(b) Plot the data and graph the model.

(c) Graph the model $y = x$. How does this model compare with the model in part (a)?

(d) Use the integration capabilities of a graphing utility to approximate the "income inequality."

85. Profit The chief financial officer of a company reports that profits for the past fiscal year were $893,000. The officer predicts that profits for the next 5 years will grow at a continuous annual rate somewhere between $3\frac{1}{2}\%$ and 5%. Estimate the cumulative difference in total profit over the 5 years based on the predicted range of growth rates.

86. Area The shaded region in the figure consists of all points whose distances from the center of the square are less than their distances from the edges of the square. Find the area of the region.

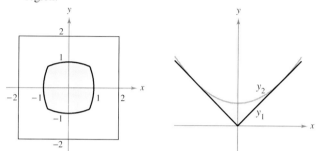

Figure for 86   **Figure for 87**

87. Mechanical Design The surface of a machine part is the region between the graphs of $y_1 = |x|$ and $y_2 = 0.08x^2 + k$ (see figure).

(a) Find k if the parabola is tangent to the graph of y_1.

(b) Find the area of the surface of the machine part.

88. Building Design Concrete sections for a new building have the dimensions (in meters) and shape shown in the figure.

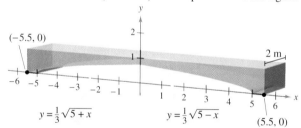

(a) Find the area of the face of the section superimposed on the rectangular coordinate system.

(b) Find the volume of concrete in one of the sections by multiplying the area in part (a) by 2 meters.

(c) One cubic meter of concrete weighs 5000 pounds. Find the weight of the section.

89. Building Design To decrease the weight and to aid in the hardening process, the concrete sections in Exercise 88 often are not solid. Rework Exercise 88 to allow for cylindrical openings such as those shown in the figure.

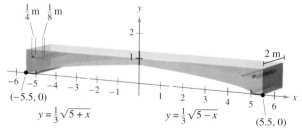

True or False? **In Exercises 90–92, determine whether the statement is true or false. If it is false, explain why or give an example that shows it is false.**

90. If the area of the region bounded by the graphs of f and g is 1, then the area of the region bounded by the graphs of $h(x) = f(x) + C$ and $k(x) = g(x) + C$ is also 1.

91. If $\int_a^b [f(x) - g(x)]\, dx = A$, then $\int_a^b [g(x) - f(x)]\, dx = -A$.

92. If the graphs of f and g intersect midway between $x = a$ and $x = b$, then

$$\int_a^b [f(x) - g(x)]\, dx = 0.$$

93. Area Find the area between the graph of $y = \sin x$ and the line segments joining the points $(0, 0)$ and $\left(\dfrac{7\pi}{6}, -\dfrac{1}{2}\right)$, as shown in the figure.

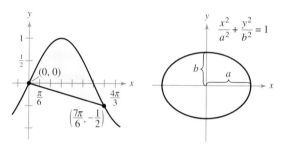

Figure for 93   **Figure for 94**

94. Area Let $a > 0$ and $b > 0$. Show that the area of the ellipse

$$\frac{x^2}{a^2} + \frac{y^2}{b^2} = 1$$

is πab (see figure).

Putnam Exam Challenge

95. The horizontal line $y = c$ intersects the curve $y = 2x - 3x^3$ in the first quadrant as shown in the figure. Find c so that the areas of the two shaded regions are equal.

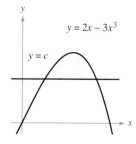

This problem was composed by the Committee on the Putnam Prize Competition. © The Mathematical Association of America. All rights reserved.

456 CHAPTER 7 Applications of Integration

Section 7.2 Volume: The Disk Method

- Find the volume of a solid of revolution using the disk method.
- Find the volume of a solid of revolution using the washer method.
- Find the volume of a solid with known cross sections.

The Disk Method

In Chapter 4 we mentioned that area is only one of the *many* applications of the definite integral. Another important application is its use in finding the volume of a three-dimensional solid. In this section you will study a particular type of three-dimensional solid—one whose cross sections are similar. Solids of revolution are used commonly in engineering and manufacturing. Some examples are axles, funnels, pills, bottles, and pistons, as shown in Figure 7.12.

Solids of revolution
Figure 7.12

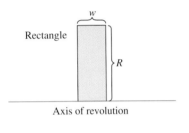

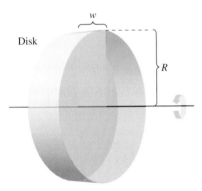

Volume of a disk: $\pi R^2 w$
Figure 7.13

If a region in the plane is revolved about a line, the resulting solid is a **solid of revolution**, and the line is called the **axis of revolution**. The simplest such solid is a right circular cylinder or **disk**, which is formed by revolving a rectangle about an axis adjacent to one side of the rectangle, as shown in Figure 7.13. The volume of such a disk is

Volume of disk = (area of disk)(width of disk)
$$= \pi R^2 w$$

where R is the radius of the disk and w is the width.

To see how to use the volume of a disk to find the volume of a general solid of revolution, consider a solid of revolution formed by revolving the plane region in Figure 7.14 about the indicated axis. To determine the volume of this solid, consider a representative rectangle in the plane region. When this rectangle is revolved about the axis of revolution, it generates a representative disk whose volume is

$$\Delta V = \pi R^2 \Delta x.$$

Approximating the volume of the solid by n such disks of width Δx and radius $R(x_i)$ produces

$$\text{Volume of solid} \approx \sum_{i=1}^{n} \pi [R(x_i)]^2 \Delta x$$
$$= \pi \sum_{i=1}^{n} [R(x_i)]^2 \Delta x.$$

SECTION 7.2   Volume: The Disk Method   457

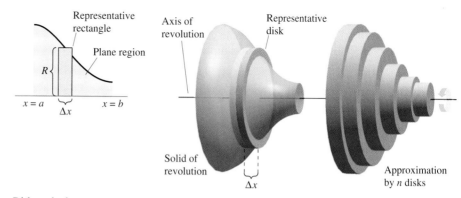

Disk method
Figure 7.14

This approximation appears to become better and better as $\|\Delta\| \to 0$ $(n \to \infty)$. So, you can define the volume of the solid as

$$\text{Volume of solid} = \lim_{\|\Delta\| \to 0} \pi \sum_{i=1}^{n} [R(x_i)]^2 \, \Delta x = \pi \int_a^b [R(x)]^2 \, dx.$$

Schematically, the disk method looks like this.

Known Precalculus Formula	*Representative Element*	*New Integration Formula*
Volume of disk $V = \pi R^2 w$	$\Delta V = \pi [R(x_i)]^2 \, \Delta x$	Solid of revolution $V = \pi \int_a^b [R(x)]^2 \, dx$

A similar formula can be derived if the axis of revolution is vertical.

The Disk Method

To find the volume of a solid of revolution with the **disk method,** use one of the following, as shown in Figure 7.15.

Horizontal Axis of Revolution

$$\text{Volume} = V = \pi \int_a^b [R(x)]^2 \, dx$$

Vertical Axis of Revolution

$$\text{Volume} = V = \pi \int_c^d [R(y)]^2 \, dy$$

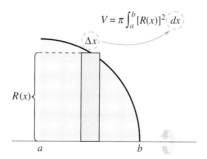

Horizontal axis of revolution
Figure 7.15

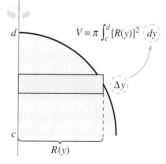

Vertical axis of revolution

NOTE In Figure 7.15, note that you can determine the variable of integration by placing a representative rectangle in the plane region "perpendicular" to the axis of revolution. If the width of the rectangle is Δx, integrate with respect to x, and if the width of the rectangle is Δy, integrate with respect to y.

458 CHAPTER 7 Applications of Integration

The simplest application of the disk method involves a plane region bounded by the graph of f and the x-axis. If the axis of revolution is the x-axis, the radius $R(x)$ is simply $f(x)$.

EXAMPLE 1 Using the Disk Method

Find the volume of the solid formed by revolving the region bounded by the graph of

$$f(x) = \sqrt{\sin x}$$

and the x-axis ($0 \leq x \leq \pi$) about the x-axis.

Solution From the representative rectangle in the upper graph in Figure 7.16, you can see that the radius of this solid is

$$R(x) = f(x)$$
$$= \sqrt{\sin x}.$$

So, the volume of the solid of revolution is

$$V = \pi \int_a^b [R(x)]^2\, dx = \pi \int_0^\pi \left(\sqrt{\sin x}\right)^2 dx \qquad \text{Apply disk method.}$$
$$= \pi \int_0^\pi \sin x\, dx \qquad \text{Simplify.}$$
$$= \pi \Big[-\cos x\Big]_0^\pi \qquad \text{Integrate.}$$
$$= \pi(1 + 1)$$
$$= 2\pi.$$

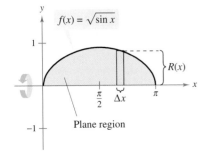
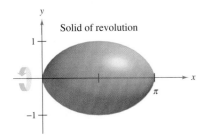

Figure 7.16

EXAMPLE 2 Revolving About a Line That Is Not a Coordinate Axis

Find the volume of the solid formed by revolving the region bounded by

$$f(x) = 2 - x^2$$

and $g(x) = 1$ about the line $y = 1$, as shown in Figure 7.17.

Solution By equating $f(x)$ and $g(x)$, you can determine that the two graphs intersect when $x = \pm 1$. To find the radius, subtract $g(x)$ from $f(x)$.

$$R(x) = f(x) - g(x)$$
$$= (2 - x^2) - 1$$
$$= 1 - x^2$$

Finally, integrate between -1 and 1 to find the volume.

$$V = \pi \int_a^b [R(x)]^2\, dx = \pi \int_{-1}^1 (1 - x^2)^2\, dx \qquad \text{Apply disk method.}$$
$$= \pi \int_{-1}^1 (1 - 2x^2 + x^4)\, dx \qquad \text{Simplify.}$$
$$= \pi \left[x - \frac{2x^3}{3} + \frac{x^5}{5} \right]_{-1}^1 \qquad \text{Integrate.}$$
$$= \frac{16\pi}{15}$$

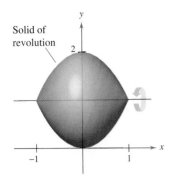

Figure 7.17

SECTION 7.2 Volume: The Disk Method 459

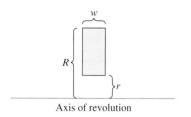

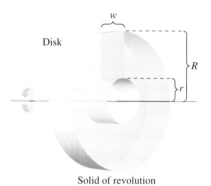

Figure 7.18

The Washer Method

The disk method can be extended to cover solids of revolution with holes by replacing the representative disk with a representative **washer.** The washer is formed by revolving a rectangle about an axis, as shown in Figure 7.18. If r and R are the inner and outer radii of the washer and w is the width of the washer, the volume is given by

Volume of washer $= \pi(R^2 - r^2)w.$

To see how this concept can be used to find the volume of a solid of revolution, consider a region bounded by an **outer radius** $R(x)$ and an **inner radius** $r(x)$, as shown in Figure 7.19. If the region is revolved about its axis of revolution, the volume of the resulting solid is given by

$$V = \pi \int_a^b ([R(x)]^2 - [r(x)]^2)\, dx. \qquad \text{Washer method}$$

Note that the integral involving the inner radius represents the volume of the hole and is *subtracted* from the integral involving the outer radius.

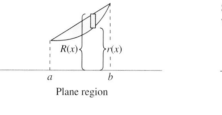

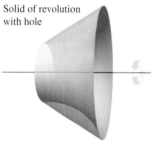

Figure 7.19

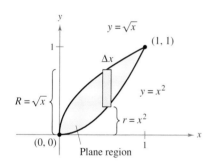

EXAMPLE 3 Using the Washer Method

Find the volume of the solid formed by revolving the region bounded by the graphs of $y = \sqrt{x}$ and $y = x^2$ about the x-axis, as shown in Figure 7.20.

Solution In Figure 7.20, you can see that the outer and inner radii are as follows.

$R(x) = \sqrt{x}$ Outer radius
$r(x) = x^2$ Inner radius

Integrating between 0 and 1 produces

$$V = \pi \int_a^b ([R(x)]^2 - [r(x)]^2)\, dx \qquad \text{Apply washer method.}$$
$$= \pi \int_0^1 \left[(\sqrt{x})^2 - (x^2)^2\right] dx$$
$$= \pi \int_0^1 (x - x^4)\, dx \qquad \text{Simplify.}$$
$$= \pi \left[\frac{x^2}{2} - \frac{x^5}{5}\right]_0^1 \qquad \text{Integrate.}$$
$$= \frac{3\pi}{10}.$$

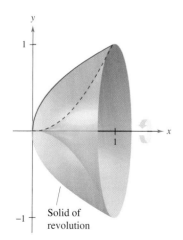

Solid of revolution
Figure 7.20

EXAMPLE 4 Integrating with Respect to y, Two-Integral Case

Find the volume of the solid formed by revolving the region bounded by the graphs of $y = x^2 + 1$, $y = 0$, $x = 0$, and $x = 1$ about the y-axis, as shown in Figure 7.21.

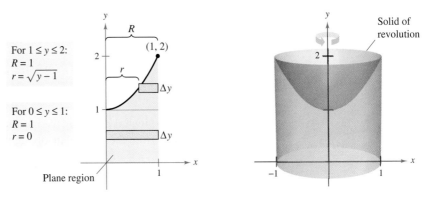

Figure 7.21

Solution For the region shown in Figure 7.21, the outer radius is simply $R = 1$. There is, however, no convenient formula that represents the inner radius. When $0 \le y \le 1$, $r = 0$, but when $1 \le y \le 2$, r is determined by the equation $y = x^2 + 1$, which implies that $r = \sqrt{y - 1}$.

$$r(y) = \begin{cases} 0, & 0 \le y \le 1 \\ \sqrt{y - 1}, & 1 \le y \le 2 \end{cases}$$

Using this definition of the inner radius, you can use two integrals to find the volume.

$$V = \pi \int_0^1 (1^2 - 0^2)\, dy + \pi \int_1^2 \left[1^2 - \left(\sqrt{y-1}\right)^2\right] dy \quad \text{Apply washer method.}$$

$$= \pi \int_0^1 1\, dy + \pi \int_1^2 (2 - y)\, dy \quad \text{Simplify.}$$

$$= \pi \Big[y\Big]_0^1 + \pi \left[2y - \frac{y^2}{2}\right]_1^2 \quad \text{Integrate.}$$

$$= \pi + \pi\left(4 - 2 - 2 + \frac{1}{2}\right) = \frac{3\pi}{2}$$

Note that the first integral $\pi \int_0^1 1\, dy$ represents the volume of a right circular cylinder of radius 1 and height 1. This portion of the volume could have been determined without using calculus.

TECHNOLOGY Some graphing utilities have the capability to generate (or have built-in software capable of generating) a solid of revolution. If you have access to such a utility, use it to graph some of the solids of revolution described in this section. For instance, the solid in Example 4 might appear like that shown in Figure 7.22.

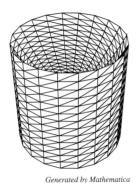

Generated by Mathematica

Figure 7.22

EXAMPLE 5 Manufacturing

A manufacturer drills a hole through the center of a metal sphere of radius 5 inches, as shown in Figure 7.23(a). The hole has a radius of 3 inches. What is the volume of the resulting metal ring?

Solution You can imagine the ring to be generated by a segment of the circle whose equation is $x^2 + y^2 = 25$, as shown in Figure 7.23(b). Because the radius of the hole is 3 inches, you can let $y = 3$ and solve the equation $x^2 + y^2 = 25$ to determine that the limits of integration are $x = \pm 4$. So, the inner and outer radii are $r(x) = 3$ and $R(x) = \sqrt{25 - x^2}$ and the volume is given by

$$V = \pi \int_a^b ([R(x)]^2 - [r(x)]^2)\, dx = \pi \int_{-4}^{4} \left[\left(\sqrt{25 - x^2}\right)^2 - (3)^2 \right] dx$$

$$= \pi \int_{-4}^{4} (16 - x^2)\, dx$$

$$= \pi \left[16x - \frac{x^3}{3} \right]_{-4}^{4}$$

$$= \frac{256\pi}{3} \text{ cubic inches.}$$

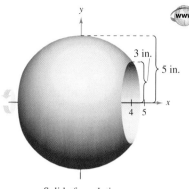

Solid of revolution

(a)

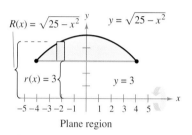

Plane region

(b)

Figure 7.23

Solids with Known Cross Sections

With the disk method, you can find the volume of a solid having a circular cross section whose area is $A = \pi R^2$. This method can be generalized to solids of any shape, as long as you know a formula for the area of an arbitrary cross section. Some common cross sections are squares, rectangles, triangles, semicircles, and trapezoids.

> ### Volumes of Solids with Known Cross Sections
>
> **1.** For cross sections of area $A(x)$ taken perpendicular to the x-axis,
>
> $$\text{Volume} = \int_a^b A(x)\, dx. \qquad \text{See Figure 7.24(a).}$$
>
> **2.** For cross sections of area $A(y)$ taken perpendicular to the y-axis,
>
> $$\text{Volume} = \int_c^d A(y)\, dy. \qquad \text{See Figure 7.24(b).}$$

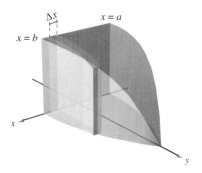

(a) Cross sections perpendicular to x-axis

Figure 7.24

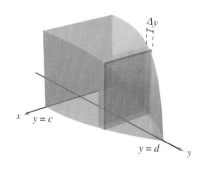

(b) Cross sections perpendicular to y-axis

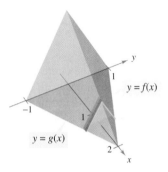

Cross sections are equilateral triangles.

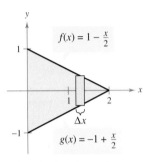

Triangular base in xy-plane
Figure 7.25

EXAMPLE 6 Triangular Cross Sections

Find the volume of the solid shown in Figure 7.25. The base of the solid is the region bounded by the lines

$$f(x) = 1 - \frac{x}{2}, \qquad g(x) = -1 + \frac{x}{2}, \qquad \text{and} \qquad x = 0.$$

The cross sections perpendicular to the x-axis are equilateral triangles.

Solution The base and area of each triangular cross section are as follows.

$$\text{Base} = \left(1 - \frac{x}{2}\right) - \left(-1 + \frac{x}{2}\right) = 2 - x \qquad \text{Length of base}$$

$$\text{Area} = \frac{\sqrt{3}}{4}(\text{base})^2 \qquad \text{Area of equilateral triangle}$$

$$A(x) = \frac{\sqrt{3}}{4}(2 - x)^2 \qquad \text{Area of cross section}$$

Because x ranges from 0 to 2, the volume of the solid is

$$V = \int_a^b A(x)\, dx = \int_0^2 \frac{\sqrt{3}}{4}(2 - x)^2\, dx$$

$$= -\frac{\sqrt{3}}{4}\left[\frac{(2 - x)^3}{3}\right]_0^2 = \frac{2\sqrt{3}}{3}.$$

EXAMPLE 7 An Application to Geometry

Prove that the volume of a pyramid with a square base is $V = \frac{1}{3}hB$, where h is the height of the pyramid and B is the area of the base.

Solution As shown in Figure 7.26, you can intersect the pyramid with a plane parallel to the base at height y to form a square cross section whose sides are of length b'. Using similar triangles, you can show that

$$\frac{b'}{b} = \frac{h - y}{h} \qquad \text{or} \qquad b' = \frac{b}{h}(h - y)$$

where b is the length of the sides of the base of the pyramid. So,

$$A(y) = (b')^2 = \frac{b^2}{h^2}(h - y)^2.$$

Integrating between 0 and h produces

$$V = \int_0^h A(y)\, dy = \int_0^h \frac{b^2}{h^2}(h - y)^2\, dy$$

$$= \frac{b^2}{h^2}\int_0^h (h - y)^2\, dy$$

$$= -\left(\frac{b^2}{h^2}\right)\left[\frac{(h - y)^3}{3}\right]_0^h$$

$$= \frac{b^2}{h^2}\left(\frac{h^3}{3}\right)$$

$$= \frac{1}{3}hB. \qquad B = b^2$$

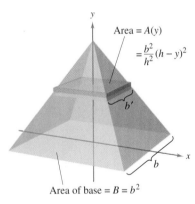

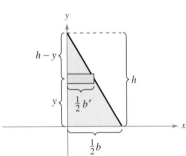

Figure 7.26

Exercises for Section 7.2

See www.CalcChat.com for worked-out solutions to odd-numbered exercises.

In Exercises 1–6, set up and evaluate the integral that gives the volume of the solid formed by revolving the region about the x-axis.

1. $y = -x + 1$

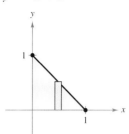

2. $y = 4 - x^2$

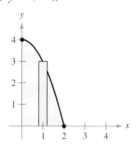

3. $y = \sqrt{x}$

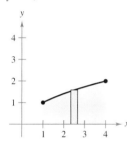

4. $y = \sqrt{9 - x^2}$

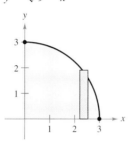

5. $y = x^2$, $y = x^3$

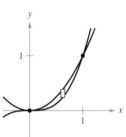

6. $y = 2$, $y = 4 - \dfrac{x^2}{4}$

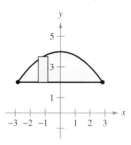

In Exercises 7–10, set up and evaluate the integral that gives the volume of the solid formed by revolving the region about the y-axis.

7. $y = x^2$

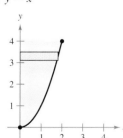

8. $y = \sqrt{16 - x^2}$

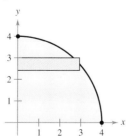

9. $y = x^{2/3}$

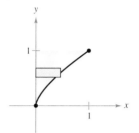

10. $x = -y^2 + 4y$

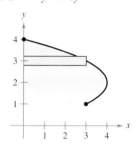

In Exercises 11–14, find the volume of the solid generated by revolving the region bounded by the graphs of the equations about the given lines.

11. $y = \sqrt{x}$, $y = 0$, $x = 4$
  (a) the x-axis (b) the y-axis
  (c) the line $x = 4$ (d) the line $x = 6$

12. $y = 2x^2$, $y = 0$, $x = 2$
  (a) the y-axis (b) the x-axis
  (c) the line $y = 8$ (d) the line $x = 2$

13. $y = x^2$, $y = 4x - x^2$
  (a) the x-axis (b) the line $y = 6$

14. $y = 6 - 2x - x^2$, $y = x + 6$
  (a) the x-axis (b) the line $y = 3$

In Exercises 15–18, find the volume of the solid generated by revolving the region bounded by the graphs of the equations about the line $y = 4$.

15. $y = x$, $y = 3$, $x = 0$ **16.** $y = \tfrac{1}{2}x^3$, $y = 4$, $x = 0$

17. $y = \dfrac{1}{1 + x}$, $y = 0$, $x = 0$, $x = 3$

18. $y = \sec x$, $y = 0$, $0 \le x \le \dfrac{\pi}{3}$

In Exercises 19–22, find the volume of the solid generated by revolving the region bounded by the graphs of the equations about the line $x = 6$.

19. $y = x$, $y = 0$, $y = 4$, $x = 6$
20. $y = 6 - x$, $y = 0$, $y = 4$, $x = 0$
21. $x = y^2$, $x = 4$
22. $xy = 6$, $y = 2$, $y = 6$, $x = 6$

In Exercises 23–30, find the volume of the solid generated by revolving the region bounded by the graphs of the equations about the x-axis.

23. $y = \dfrac{1}{\sqrt{x + 1}}$, $y = 0$, $x = 0$, $x = 3$

24. $y = x\sqrt{4 - x^2}$, $y = 0$

25. $y = \dfrac{1}{x}$, $y = 0$, $x = 1$, $x = 4$

26. $y = \dfrac{3}{x+1}$, $y = 0$, $x = 0$, $x = 8$

27. $y = e^{-x}$, $y = 0$, $x = 0$, $x = 1$

28. $y = e^{x/2}$, $y = 0$, $x = 0$, $x = 4$

29. $y = x^2 + 1$, $y = -x^2 + 2x + 5$, $x = 0$, $x = 3$

30. $y = \sqrt{x}$, $y = -\tfrac{1}{2}x + 4$, $x = 0$, $x = 8$

In Exercises 31 and 32, find the volume of the solid generated by revolving the region bounded by the graphs of the equations about the y-axis.

31. $y = 3(2 - x)$, $y = 0$, $x = 0$

32. $y = 9 - x^2$, $y = 0$, $x = 2$, $x = 3$

In Exercises 33–36, find the volume of the solid generated by revolving the region bounded by the graphs of the equations about the x-axis. Verify your results using the integration capabilities of a graphing utility.

33. $y = \sin x$, $y = 0$, $x = 0$, $x = \pi$

34. $y = \cos x$, $y = 0$, $x = 0$, $x = \dfrac{\pi}{2}$

35. $y = e^{x-1}$, $y = 0$, $x = 1$, $x = 2$

36. $y = e^{x/2} + e^{-x/2}$, $y = 0$, $x = -1$, $x = 2$

In Exercises 37–40, use the integration capabilities of a graphing utility to approximate the volume of the solid generated by revolving the region bounded by the graphs of the equations about the x-axis.

37. $y = e^{-x^2}$, $y = 0$, $x = 0$, $x = 2$

38. $y = \ln x$, $y = 0$, $x = 1$, $x = 3$

39. $y = 2\arctan(0.2x)$, $y = 0$, $x = 0$, $x = 5$

40. $y = \sqrt{2x}$, $y = x^2$

Writing About Concepts

In Exercises 41 and 42, the integral represents the volume of a solid. Describe the solid.

41. $\pi \displaystyle\int_0^{\pi/2} \sin^2 x \, dx$ 42. $\pi \displaystyle\int_2^4 y^4 \, dy$

Think About It **In Exercises 43 and 44, determine which value best approximates the volume of the solid generated by revolving the region bounded by the graphs of the equations about the x-axis. (Make your selection on the basis of a sketch of the solid and *not* by performing any calculations.)**

43. $y = e^{-x^2/2}$, $y = 0$, $x = 0$, $x = 2$

    (a) 3 (b) -5 (c) 10 (d) 7 (e) 20

44. $y = \arctan x$, $y = 0$, $x = 0$, $x = 1$

    (a) 10 (b) $\tfrac{3}{4}$ (c) 5 (d) -6 (e) 15

Writing About Concepts (continued)

45. A region bounded by the parabola $y = 4x - x^2$ and the x-axis is revolved about the x-axis. A second region bounded by the parabola $y = 4 - x^2$ and the x-axis is revolved about the x-axis. Without integrating, how do the volumes of the two solids compare? Explain.

46. The region in the figure is revolved about the indicated axes and line. Order the volumes of the resulting solids from least to greatest. Explain your reasoning.

    (a) x-axis (b) y-axis (c) $x = 8$

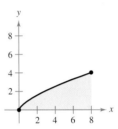

47. If the portion of the line $y = \tfrac{1}{2}x$ lying in the first quadrant is revolved about the x-axis, a cone is generated. Find the volume of the cone extending from $x = 0$ to $x = 6$.

48. Use the disk method to verify that the volume of a right circular cone is $\tfrac{1}{3}\pi r^2 h$, where r is the radius of the base and h is the height.

49. Use the disk method to verify that the volume of a sphere is $\tfrac{4}{3}\pi r^3$.

50. A sphere of radius r is cut by a plane h ($h < r$) units above the equator. Find the volume of the solid (spherical segment) above the plane.

51. A cone of height H with a base of radius r is cut by a plane parallel to and h units above the base. Find the volume of the solid (frustum of a cone) below the plane.

52. The region bounded by $y = \sqrt{x}$, $y = 0$, $x = 0$, and $x = 4$ is revolved about the x-axis.

    (a) Find the value of x in the interval $[0, 4]$ that divides the solid into two parts of equal volume.

    (b) Find the values of x in the interval $[0, 4]$ that divide the solid into three parts of equal volume.

53. ***Volume of a Fuel Tank*** A tank on the wing of a jet aircraft is formed by revolving the region bounded by the graph of $y = \tfrac{1}{8}x^2\sqrt{2 - x}$ and the x-axis about the x-axis (see figure), where x and y are measured in meters. Find the tank's volume.

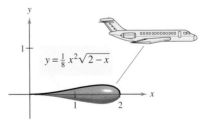

54. Volume of a Lab Glass A glass container can be modeled by revolving the graph of

$$y = \begin{cases} \sqrt{0.1x^3 - 2.2x^2 + 10.9x + 22.2}, & 0 \le x \le 11.5 \\ 2.95, & 11.5 < x \le 15 \end{cases}$$

about the x-axis, where x and y are measured in centimeters. Use a graphing utility to graph the function and find the volume of the container.

55. Find the volume of the solid generated if the upper half of the ellipse $9x^2 + 25y^2 = 225$ is revolved about (a) the x-axis to form a prolate spheroid (shaped like a football), and (b) the y-axis to form an oblate spheroid (shaped like half of a candy).

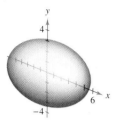

Figure for 55(a) Figure for 55(b)

56. Minimum Volume The arc of

$$y = 4 - \frac{x^2}{4}$$

on the interval $[0, 4]$ is revolved about the line $y = b$ (see figure).

(a) Find the volume of the resulting solid as a function of b.

(b) Use a graphing utility to graph the function in part (a), and use the graph to approximate the value of b that minimizes the volume of the solid.

(c) Use calculus to find the value of b that minimizes the volume of the solid, and compare the result with the answer to part (b).

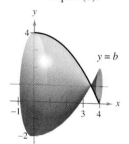 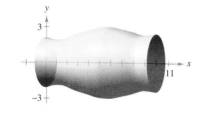

Figure for 56 Figure for 58

57. Water Depth in a Tank A tank on a water tower is a sphere of radius 50 feet. Determine the depths of the water when the tank is filled to one-fourth and three-fourths of its total capacity. (*Note:* Use the *zero* or *root* feature of a graphing utility after evaluating the definite integral.)

58. Modeling Data A draftsman is asked to determine the amount of material required to produce a machine part (see figure in first column). The diameters d of the part at equally spaced points x are listed in the table. The measurements are listed in centimeters.

x	0	1	2	3	4	5
d	4.2	3.8	4.2	4.7	5.2	5.7

x	6	7	8	9	10
d	5.8	5.4	4.9	4.4	4.6

(a) Use these data with Simpson's Rule to approximate the volume of the part.

(b) Use the regression capabilities of a graphing utility to find a fourth-degree polynomial through the points representing the radius of the solid. Plot the data and graph the model.

(c) Use a graphing utility to approximate the definite integral yielding the volume of the part. Compare the result with the answer to part (a).

59. Think About It Match each integral with the solid whose volume it represents, and give the dimensions of each solid.

(a) Right circular cylinder (b) Ellipsoid
(c) Sphere (d) Right circular cone (e) Torus

(i) $\pi \int_0^h \left(\frac{rx}{h}\right)^2 dx$ (ii) $\pi \int_0^h r^2 \, dx$

(iii) $\pi \int_{-r}^r \left(\sqrt{r^2 - x^2}\right)^2 dx$ (iv) $\pi \int_{-b}^b \left(a\sqrt{1 - \frac{x^2}{b^2}}\right)^2 dx$

(v) $\pi \int_{-r}^r \left[\left(R + \sqrt{r^2 - x^2}\right)^2 - \left(R - \sqrt{r^2 - x^2}\right)^2\right] dx$

60. Cavalieri's Theorem Prove that if two solids have equal altitudes and all plane sections parallel to their bases and at equal distances from their bases have equal areas, then the solids have the same volume (see figure).

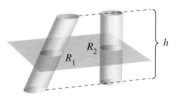

Area of R_1 = area of R_2

61. Find the volume of the solid whose base is bounded by the graphs of $y = x + 1$ and $y = x^2 - 1$, with the indicated cross sections taken perpendicular to the x-axis.

(a) Squares (b) Rectangles of height 1

62. Find the volume of the solid whose base is bounded by the circle

$$x^2 + y^2 = 4$$

with the indicated cross sections taken perpendicular to the x-axis.

(a) Squares (b) Equilateral triangles

(c) Semicircles (d) Isosceles right triangles

63. The base of a solid is bounded by $y = x^3$, $y = 0$, and $x = 1$. Find the volume of the solid for each of the following cross sections (taken perpendicular to the y-axis): (a) squares, (b) semicircles, (c) equilateral triangles, and (d) semiellipses whose heights are twice the lengths of their bases.

64. Find the volume of the solid of intersection (the solid common to both) of the two right circular cylinders of radius r whose axes meet at right angles (see figure).

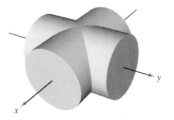

Two intersecting cylinders Solid of intersection

FOR FURTHER INFORMATION For more information on this problem, see the article "Estimating the Volumes of Solid Figures with Curved Surfaces" by Donald Cohen in *Mathematics Teacher*. To view this article, go to the website *www.matharticles.com*.

65. A manufacturer drills a hole through the center of a metal sphere of radius R. The hole has a radius r. Find the volume of the resulting ring.

66. For the metal sphere in Exercise 65, let $R = 5$. What value of r will produce a ring whose volume is exactly half the volume of the sphere?

In Exercises 67–74, find the volume generated by rotating the given region about the specified line.

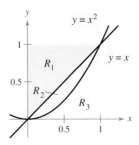

67. R_1 about $x = 0$ **68.** R_1 about $x = 1$
69. R_2 about $y = 0$ **70.** R_2 about $y = 1$
71. R_3 about $x = 0$ **72.** R_3 about $x = 1$
73. R_2 about $x = 0$ **74.** R_2 about $x = 1$

75. The solid shown in the figure has cross sections bounded by the graph of $|x|^a + |y|^a = 1$, where $1 \leq a \leq 2$.

(a) Describe the cross section when $a = 1$ and $a = 2$.

(b) Describe a procedure for approximating the volume of the solid.

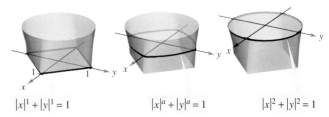

$|x|^1 + |y|^1 = 1$ $|x|^a + |y|^a = 1$ $|x|^2 + |y|^2 = 1$

76. Two planes cut a right circular cylinder to form a wedge. One plane is perpendicular to the axis of the cylinder and the second makes an angle of θ degrees with the first (see figure).

(a) Find the volume of the wedge if $\theta = 45°$.

(b) Find the volume of the wedge for an arbitrary angle θ. Assuming that the cylinder has sufficient length, how does the volume of the wedge change as θ increases from $0°$ to $90°$?

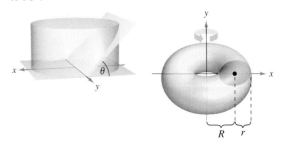

Figure for 76 Figure for 77

77. (a) Show that the volume of the torus shown is given by the integral $8\pi R \int_0^r \sqrt{r^2 - y^2}\, dy$, where $R > r > 0$.

(b) Find the volume of the torus.

Section 7.3 Volume: The Shell Method

- Find the volume of a solid of revolution using the shell method.
- Compare the uses of the disk method and the shell method.

The Shell Method

In this section, you will study an alternative method for finding the volume of a solid of revolution. This method is called the **shell method** because it uses cylindrical shells. A comparison of the advantages of the disk and shell methods is given later in this section.

To begin, consider a representative rectangle as shown in Figure 7.27, where w is the width of the rectangle, h is the height of the rectangle, and p is the distance between the axis of revolution and the *center* of the rectangle. When this rectangle is revolved about its axis of revolution, it forms a cylindrical shell (or tube) of thickness w. To find the volume of this shell, consider two cylinders. The radius of the larger cylinder corresponds to the outer radius of the shell, and the radius of the smaller cylinder corresponds to the inner radius of the shell. Because p is the average radius of the shell, you know the outer radius is $p + (w/2)$ and the inner radius is $p - (w/2)$.

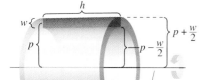

Figure 7.27

$$p + \frac{w}{2} \quad \text{Outer radius}$$

$$p - \frac{w}{2} \quad \text{Inner radius}$$

So, the volume of the shell is

$$\text{Volume of shell} = (\text{volume of cylinder}) - (\text{volume of hole})$$

$$= \pi\left(p + \frac{w}{2}\right)^2 h - \pi\left(p - \frac{w}{2}\right)^2 h$$

$$= 2\pi p h w$$

$$= 2\pi(\text{average radius})(\text{height})(\text{thickness}).$$

You can use this formula to find the volume of a solid of revolution. Assume that the plane region in Figure 7.28 is revolved about a line to form the indicated solid. If you consider a horizontal rectangle of width Δy, then, as the plane region is revolved about a line parallel to the x-axis, the rectangle generates a representative shell whose volume is

$$\Delta V = 2\pi[p(y)h(y)]\,\Delta y.$$

You can approximate the volume of the solid by n such shells of thickness Δy, height $h(y_i)$, and average radius $p(y_i)$.

$$\text{Volume of solid} \approx \sum_{i=1}^{n} 2\pi[p(y_i)h(y_i)]\,\Delta y = 2\pi \sum_{i=1}^{n}[p(y_i)h(y_i)]\,\Delta y$$

This approximation appears to become better and better as $\|\Delta\| \to 0$ ($n \to \infty$). So, the volume of the solid is

$$\text{Volume of solid} = \lim_{\|\Delta\| \to 0} 2\pi \sum_{i=1}^{n}[p(y_i)h(y_i)]\,\Delta y$$

$$= 2\pi \int_{c}^{d}[p(y)h(y)]\,dy.$$

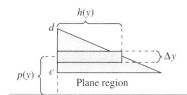
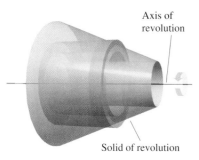

Figure 7.28

The Shell Method

To find the volume of a solid of revolution with the **shell method,** use one of the following, as shown in Figure 7.29.

Horizontal Axis of Revolution

$$\text{Volume} = V = 2\pi \int_c^d p(y)h(y)\, dy$$

Vertical Axis of Revolution

$$\text{Volume} = V = 2\pi \int_a^b p(x)h(x)\, dx$$

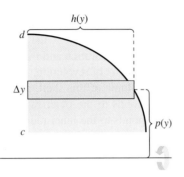
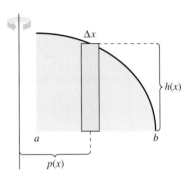

Horizontal axis of revolution

Vertical axis of revolution

Figure 7.29

EXAMPLE 1 Using the Shell Method to Find Volume

Find the volume of the solid of revolution formed by revolving the region bounded by

$$y = x - x^3$$

and the x-axis $(0 \leq x \leq 1)$ about the y-axis.

Solution Because the axis of revolution is vertical, use a vertical representative rectangle, as shown in Figure 7.30. The width Δx indicates that x is the variable of integration. The distance from the center of the rectangle to the axis of revolution is $p(x) = x$, and the height of the rectangle is

$$h(x) = x - x^3.$$

Because x ranges from 0 to 1, the volume of the solid is

$$\begin{aligned}
V &= 2\pi \int_a^b p(x)h(x)\, dx = 2\pi \int_0^1 x(x - x^3)\, dx &&\text{Apply shell method.}\\
&= 2\pi \int_0^1 (-x^4 + x^2)\, dx &&\text{Simplify.}\\
&= 2\pi \left[-\frac{x^5}{5} + \frac{x^3}{3} \right]_0^1 &&\text{Integrate.}\\
&= 2\pi \left(-\frac{1}{5} + \frac{1}{3} \right)\\
&= \frac{4\pi}{15}.
\end{aligned}$$

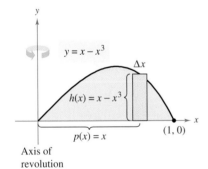

Axis of revolution

Figure 7.30

EXAMPLE 2 Using the Shell Method to Find Volume

Find the volume of the solid of revolution formed by revolving the region bounded by the graph of

$$x = e^{-y^2}$$

and the y-axis ($0 \leq y \leq 1$) about the x-axis.

Solution Because the axis of revolution is horizontal, use a horizontal representative rectangle, as shown in Figure 7.31. The width Δy indicates that y is the variable of integration. The distance from the center of the rectangle to the axis of revolution is $p(y) = y$, and the height of the rectangle is $h(y) = e^{-y^2}$. Because y ranges from 0 to 1, the volume of the solid is

$$\begin{aligned} V = 2\pi \int_c^d p(y)h(y)\,dy &= 2\pi \int_0^1 ye^{-y^2}\,dy & \text{Apply shell method.} \\ &= -\pi \left[e^{-y^2} \right]_0^1 & \text{Integrate.} \\ &= \pi \left(1 - \frac{1}{e} \right) \\ &\approx 1.986. \end{aligned}$$

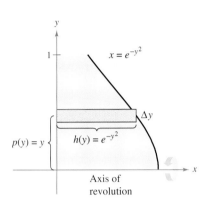

Figure 7.31

NOTE To see the advantage of using the shell method in Example 2, solve the equation $x = e^{-y^2}$ for y.

$$y = \begin{cases} 1, & 0 \leq x \leq 1/e \\ \sqrt{-\ln x}, & 1/e < x \leq 1 \end{cases}$$

Then use this equation to find the volume using the disk method.

Comparison of Disk and Shell Methods

The disk and shell methods can be distinguished as follows. For the disk method, the representative rectangle is always *perpendicular* to the axis of revolution, whereas for the shell method, the representative rectangle is always *parallel* to the axis of revolution, as shown in Figure 7.32.

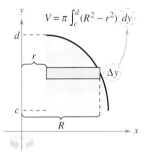 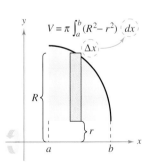 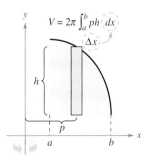 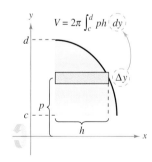

Vertical axis of revolution

Horizontal axis of revolution

Vertical axis of revolution

Horizontal axis of revolution

Disk method: Representative rectangle is perpendicular to the axis of revolution.

Shell method: Representative rectangle is parallel to the axis of revolution.

Figure 7.32

Often, one method is more convenient to use than the other. The following example illustrates a case in which the shell method is preferable.

EXAMPLE 3 Shell Method Preferable

Find the volume of the solid formed by revolving the region bounded by the graphs of

$$y = x^2 + 1, \quad y = 0, \quad x = 0, \quad \text{and} \quad x = 1$$

about the y-axis.

Solution In Example 4 in the preceding section, you saw that the washer method requires two integrals to determine the volume of this solid. See Figure 7.33(a).

$$V = \pi \int_0^1 (1^2 - 0^2) \, dy + \pi \int_1^2 \left[1^2 - \left(\sqrt{y-1}\right)^2\right] dy \quad \text{Apply washer method.}$$

$$= \pi \int_0^1 1 \, dy + \pi \int_1^2 (2 - y) \, dy \quad \text{Simplify.}$$

$$= \pi \Big[y\Big]_0^1 + \pi \left[2y - \frac{y^2}{2}\right]_1^2 \quad \text{Integrate.}$$

$$= \pi + \pi \left(4 - 2 - 2 + \frac{1}{2}\right)$$

$$= \frac{3\pi}{2}$$

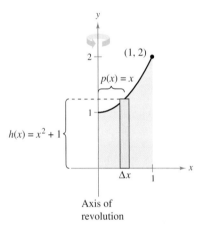

For $1 \le y \le 2$:
$R = 1$
$r = \sqrt{y-1}$

For $0 \le y \le 1$:
$R = 1$
$r = 0$

(a) Disk method

(b) Shell method

Figure 7.33

In Figure 7.33(b), you can see that the shell method requires only one integral to find the volume.

$$V = 2\pi \int_a^b p(x)h(x) \, dx \quad \text{Apply shell method.}$$

$$= 2\pi \int_0^1 x(x^2 + 1) \, dx$$

$$= 2\pi \left[\frac{x^4}{4} + \frac{x^2}{2}\right]_0^1 \quad \text{Integrate.}$$

$$= 2\pi \left(\frac{3}{4}\right)$$

$$= \frac{3\pi}{2}$$

Suppose the region in Example 3 were revolved about the vertical line $x = 1$. Would the resulting solid of revolution have a greater volume or a smaller volume than the solid in Example 3? Without integrating, you should be able to reason that the resulting solid would have a smaller volume because "more" of the revolved region would be closer to the axis of revolution. To confirm this, try solving the following integral, which gives the volume of the solid.

$$V = 2\pi \int_0^1 (1 - x)(x^2 + 1) \, dx \qquad p(x) = 1 - x$$

FOR FURTHER INFORMATION To learn more about the disk and shell methods, see the article "The Disk and Shell Method" by Charles A. Cable in *The American Mathematical Monthly*. To view this article, go to the website *www.matharticles.com*.

SECTION 7.3 Volume: The Shell Method 471

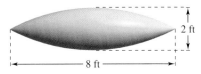

Figure 7.34

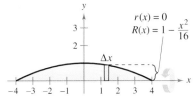

(a) Disk method

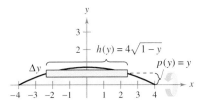

(b) Shell method

Figure 7.35

EXAMPLE 4 Volume of a Pontoon

A pontoon is to be made in the shape shown in Figure 7.34. The pontoon is designed by rotating the graph of

$$y = 1 - \frac{x^2}{16}, \qquad -4 \le x \le 4$$

about the x-axis, where x and y are measured in feet. Find the volume of the pontoon.

Solution Refer to Figure 7.35(a) and use the disk method as follows.

$$V = \pi \int_{-4}^{4} \left(1 - \frac{x^2}{16}\right)^2 dx \qquad \text{Apply disk method.}$$

$$= \pi \int_{-4}^{4} \left(1 - \frac{x^2}{8} + \frac{x^4}{256}\right) dx \qquad \text{Simplify.}$$

$$= \pi \left[x - \frac{x^3}{24} + \frac{x^5}{1280}\right]_{-4}^{4} \qquad \text{Integrate.}$$

$$= \frac{64\pi}{15} \approx 13.4 \text{ cubic feet}$$

Try using Figure 7.35(b) to set up the integral for the volume using the shell method. Does the integral seem more complicated?

For the shell method in Example 4, you would have to solve for x in terms of y in the equation

$$y = 1 - (x^2/16).$$

Sometimes, solving for x is very difficult (or even impossible). In such cases you must use a vertical rectangle (of width Δx), thus making x the variable of integration. The position (horizontal or vertical) of the axis of revolution then determines the method to be used. This is shown in Example 5.

EXAMPLE 5 Shell Method Necessary

Find the volume of the solid formed by revolving the region bounded by the graphs of $y = x^3 + x + 1$, $y = 1$, and $x = 1$ about the line $x = 2$, as shown in Figure 7.36.

Solution In the equation $y = x^3 + x + 1$, you cannot easily solve for x in terms of y. (See Section 3.8 on Newton's Method.) Therefore, the variable of integration must be x, and you should choose a vertical representative rectangle. Because the rectangle is parallel to the axis of revolution, use the shell method and obtain

$$V = 2\pi \int_a^b p(x)h(x)\, dx = 2\pi \int_0^1 (2-x)(x^3 + x + 1 - 1)\, dx \qquad \text{Apply shell method.}$$

$$= 2\pi \int_0^1 (-x^4 + 2x^3 - x^2 + 2x)\, dx \qquad \text{Simplify.}$$

$$= 2\pi \left[-\frac{x^5}{5} + \frac{x^4}{2} - \frac{x^3}{3} + x^2\right]_0^1 \qquad \text{Integrate.}$$

$$= 2\pi \left(-\frac{1}{5} + \frac{1}{2} - \frac{1}{3} + 1\right)$$

$$= \frac{29\pi}{15}.$$

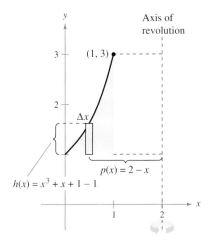

Figure 7.36

Exercises for Section 7.3

See www.CalcChat.com for worked-out solutions to odd-numbered exercises.

In Exercises 1–12, use the shell method to set up and evaluate the integral that gives the volume of the solid generated by revolving the plane region about the y-axis.

1. $y = x$

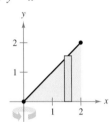

2. $y = 1 - x$

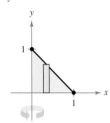

3. $y = \sqrt{x}$

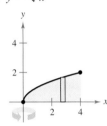

4. $y = x^2 + 4$

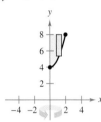

5. $y = x^2$, $y = 0$, $x = 2$

6. $y = \frac{1}{2}x^2$, $y = 0$, $x = 6$

7. $y = x^2$, $y = 4x - x^2$

8. $y = 4 - x^2$, $y = 0$

9. $y = 4x - x^2$, $x = 0$, $y = 4$

10. $y = 2x$, $y = 4$, $x = 0$

11. $y = \dfrac{1}{\sqrt{2\pi}} e^{-x^2/2}$, $y = 0$, $x = 0$, $x = 1$

12. $y = \begin{cases} \dfrac{\sin x}{x}, & x > 0 \\ 1, & x = 0 \end{cases}$, $y = 0$, $x = 0$, $x = \pi$

In Exercises 13–20, use the shell method to set up and evaluate the integral that gives the volume of the solid generated by revolving the plane region about the x-axis.

13. $y = x$

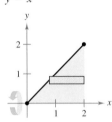

14. $y = 2 - x$

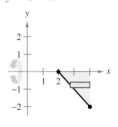

15. $y = \dfrac{1}{x}$

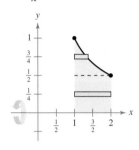

16. $x + y^2 = 16$

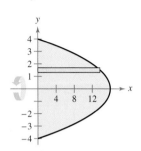

17. $y = x^3$, $x = 0$, $y = 8$

18. $y = x^2$, $x = 0$, $y = 9$

19. $x + y = 4$, $y = x$, $y = 0$

20. $y = \sqrt{x + 2}$, $y = x$, $y = 0$

In Exercises 21–24, use the shell method to find the volume of the solid generated by revolving the plane region about the given line.

21. $y = x^2$, $y = 4x - x^2$, about the line $x = 4$

22. $y = x^2$, $y = 4x - x^2$, about the line $x = 2$

23. $y = 4x - x^2$, $y = 0$, about the line $x = 5$

24. $y = \sqrt{x}$, $y = 0$, $x = 4$, about the line $x = 6$

In Exercises 25 and 26, decide whether it is more convenient to use the disk method or the shell method to find the volume of the solid of revolution. Explain your reasoning. (Do not find the volume.)

25. $(y - 2)^2 = 4 - x$

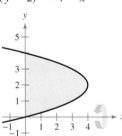

26. $y = 4 - e^x$

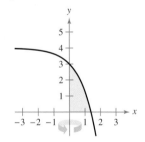

In Exercises 27–30, use the disk *or* the shell method to find the volume of the solid generated by revolving the region bounded by the graphs of the equations about each given line.

27. $y = x^3$, $y = 0$, $x = 2$
 (a) the x-axis (b) the y-axis (c) the line $x = 4$

28. $y = \dfrac{10}{x^2}$, $y = 0$, $x = 1$, $x = 5$
 (a) the x-axis (b) the y-axis (c) the line $y = 10$

29. $x^{1/2} + y^{1/2} = a^{1/2}$, $x = 0$, $y = 0$
 (a) the x-axis (b) the y-axis (c) the line $x = a$

30. $x^{2/3} + y^{2/3} = a^{2/3}$, $a > 0$ (hypocycloid)
    (a) the x-axis   (b) the y-axis

Writing About Concepts

31. Consider a solid that is generated by revolving a plane region about the y-axis. Describe the position of a representative rectangle when using (a) the shell method and (b) the disk method to find the volume of the solid.

32. The region in the figure is revolved about the indicated axes and line. Order the volumes of the resulting solids from least to greatest. Explain your reasoning.
    (a) x-axis   (b) y-axis   (c) $x = 5$

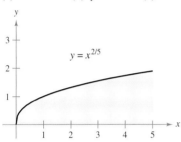

In Exercises 33 and 34, give a geometric argument that explains why the integrals have equal values.

33. $\pi \int_1^5 (x - 1)\, dx = 2\pi \int_0^2 y[5 - (y^2 + 1)]\, dy$

34. $\pi \int_0^2 [16 - (2y)^2]\, dy = 2\pi \int_0^4 x\left(\dfrac{x}{2}\right) dx$

In Exercises 35–38, (a) use a graphing utility to graph the plane region bounded by the graphs of the equations, and (b) use the integration capabilities of the graphing utility to approximate the volume of the solid generated by revolving the region about the y-axis.

35. $x^{4/3} + y^{4/3} = 1$, $x = 0$, $y = 0$, first quadrant
36. $y = \sqrt{1 - x^3}$, $y = 0$, $x = 0$
37. $y = \sqrt[3]{(x - 2)^2(x - 6)^2}$, $y = 0$, $x = 2$, $x = 6$
38. $y = \dfrac{2}{1 + e^{1/x}}$, $y = 0$, $x = 1$, $x = 3$

Think About It In Exercises 39 and 40, determine which value best approximates the volume of the solid generated by revolving the region bounded by the graphs of the equations about the y-axis. (Make your selection on the basis of a sketch of the solid and *not* by performing any calculations.)

39. $y = 2e^{-x}$, $y = 0$, $x = 0$, $x = 2$
    (a) $\dfrac{3}{2}$ (b) -2 (c) 4 (d) 7.5 (e) 15

40. $y = \tan x$, $y = 0$, $x = 0$, $x = \dfrac{\pi}{4}$
    (a) 3.5 (b) $-\dfrac{9}{4}$ (c) 8 (d) 10 (e) 1

41. *Machine Part* A solid is generated by revolving the region bounded by $y = \frac{1}{2}x^2$ and $y = 2$ about the y-axis. A hole, centered along the axis of revolution, is drilled through this solid so that one-fourth of the volume is removed. Find the diameter of the hole.

42. *Machine Part* A solid is generated by revolving the region bounded by $y = \sqrt{9 - x^2}$ and $y = 0$ about the y-axis. A hole, centered along the axis of revolution, is drilled through this solid so that one-third of the volume is removed. Find the diameter of the hole.

43. *Volume of a Torus* A torus is formed by revolving the region bounded by the circle $x^2 + y^2 = 1$ about the line $x = 2$ (see figure). Find the volume of this "doughnut-shaped" solid. (*Hint:* The integral $\int_{-1}^{1} \sqrt{1 - x^2}\, dx$ represents the area of a semicircle.)

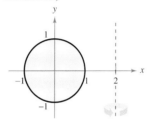

44. *Volume of a Torus* Repeat Exercise 43 for a torus formed by revolving the region bounded by the circle $x^2 + y^2 = r^2$ about the line $x = R$, where $r < R$.

45. (a) Use differentiation to verify that
    $$\int x \sin x\, dx = \sin x - x \cos x + C.$$
    (b) Use the result of part (a) to find the volume of the solid generated by revolving each plane region about the y-axis.

    (i) $y = \sin x$ (ii) $y = 2\sin x$, $y = -\sin x$

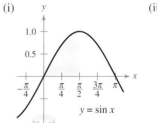
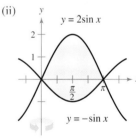

46. (a) Use differentiation to verify that
    $$\int x \cos x\, dx = \cos x + x \sin x + C.$$
    (b) Use the result of part (a) to find the volume of the solid generated by revolving each plane region about the y-axis. (*Hint:* Begin by approximating the points of intersection.)

    (i) $y = x^2$, $y = \cos x$ (ii) $y = 4\cos x$, $y = (x - 2)^2$

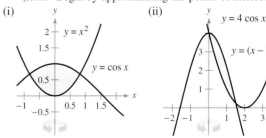

In Exercises 47–50, the integral represents the volume of a solid of revolution. Identify (a) the plane region that is revolved and (b) the axis of revolution.

47. $2\pi \int_0^2 x^3 \, dx$

48. $2\pi \int_0^1 y - y^{3/2} \, dy$

49. $2\pi \int_0^6 (y+2)\sqrt{6-y} \, dy$

50. $2\pi \int_0^1 (4-x)e^x \, dx$

51. **Volume of a Segment of a Sphere** Let a sphere of radius r be cut by a plane, thereby forming a segment of height h. Show that the volume of this segment is $\frac{1}{3}\pi h^2(3r - h)$.

52. **Volume of an Ellipsoid** Consider the plane region bounded by the graph of

$$\left(\frac{x}{a}\right)^2 + \left(\frac{y}{b}\right)^2 = 1$$

where $a > 0$ and $b > 0$. Show that the volume of the ellipsoid formed when this region revolves about the y-axis is $\dfrac{4\pi a^2 b}{3}$.

53. **Exploration** Consider the region bounded by the graphs of $y = ax^n$, $y = ab^n$, and $x = 0$ (see figure).

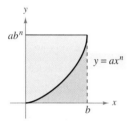

(a) Find the ratio $R_1(n)$ of the area of the region to the area of the circumscribed rectangle.

(b) Find $\lim_{n\to\infty} R_1(n)$ and compare the result with the area of the circumscribed rectangle.

(c) Find the volume of the solid of revolution formed by revolving the region about the y-axis. Find the ratio $R_2(n)$ of this volume to the volume of the circumscribed right circular cylinder.

(d) Find $\lim_{n\to\infty} R_2(n)$ and compare the result with the volume of the circumscribed cylinder.

(e) Use the results of parts (b) and (d) to make a conjecture about the shape of the graph of $y = ax^n$ ($0 \le x \le b$) as $n \to \infty$.

54. **Think About It** Match each integral with the solid whose volume it represents, and give the dimensions of each solid.

(a) Right circular cone (b) Torus (c) Sphere
(d) Right circular cylinder (e) Ellipsoid

(i) $2\pi \int_0^r hx \, dx$

(ii) $2\pi \int_0^r hx\left(1 - \dfrac{x}{r}\right) dx$

(iii) $2\pi \int_0^r 2x\sqrt{r^2 - x^2} \, dx$

(iv) $2\pi \int_0^b 2ax\sqrt{1 - \dfrac{x^2}{b^2}} \, dx$

(v) $2\pi \int_{-r}^r (R - x)\left(2\sqrt{r^2 - x^2}\right) dx$

55. **Volume of a Storage Shed** A storage shed has a circular base of diameter 80 feet (see figure). Starting at the center, the interior height is measured every 10 feet and recorded in the table.

x	0	10	20	30	40
Height	50	45	40	20	0

(a) Use Simpson's Rule to approximate the volume of the shed.

(b) Note that the roof line consists of two line segments. Find the equations of the line segments and use integration to find the volume of the shed.

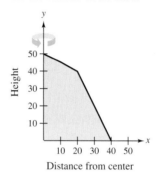

56. **Modeling Data** A pond is approximately circular, with a diameter of 400 feet (see figure). Starting at the center, the depth of the water is measured every 25 feet and recorded in the table.

x	0	25	50	75	100	125	150	175	200
Depth	20	19	19	17	15	14	10	6	0

(a) Use Simpson's Rule to approximate the volume of water in the pond.

(b) Use the regression capabilities of a graphing utility to find a quadratic model for the depths recorded in the table. Use the graphing utility to plot the depths and graph the model.

(c) Use the integration capabilities of a graphing utility and the model in part (b) to approximate the volume of water in the pond.

(d) Use the result of part (c) to approximate the number of gallons of water in the pond if 1 cubic foot of water is approximately 7.48 gallons.

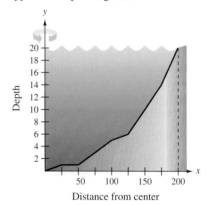

57. Consider the graph of $y^2 = x(4 - x)^2$ (see figure). Find the volumes of the solids that are generated when the loop of this graph is revolved around (a) the x-axis, (b) the y-axis, and (c) the line $x = 4$.

58. Consider the graph of $y^2 = x^2(x + 5)$ (see figure). Find the volume of the solid that is generated when the loop of this graph is revolved around (a) the x-axis, (b) the y-axis, and (c) the line $x = -5$.

59. Let V_1 and V_2 be the volumes of the solids that result when the plane region bounded by $y = 1/x$, $y = 0$, $x = \frac{1}{4}$, and $x = c \left(c > \frac{1}{4}\right)$ is revolved about the x-axis and y-axis, respectively. Find the value of c for which $V_1 = V_2$.

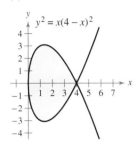

Figure for 57

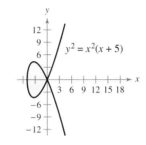

Figure for 58

Section Project: Saturn

The Oblateness of Saturn Saturn is the most oblate of the nine planets in our solar system. Its equatorial radius is 60,268 kilometers and its polar radius is 54,364 kilometers. The color enhanced photograph of Saturn was taken by Voyager 1. In the photograph, the oblateness of Saturn is clearly visible.

(a) Find the ratio of the volumes of the sphere and the oblate ellipsoid shown below.

(b) If a planet were spherical and had the same volume as Saturn, what would its radius be?

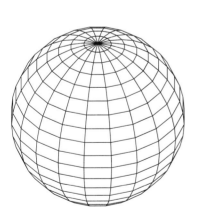

Computer model of "spherical Saturn," whose equatorial radius is equal to its polar radius. The equation of the cross section passing through the pole is

$x^2 + y^2 = 60{,}268^2.$

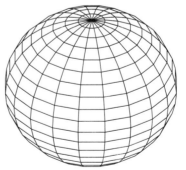

Computer model of "oblate Saturn," whose equatorial radius is greater than its polar radius. The equation of the cross section passing through the pole is

$$\frac{x^2}{60{,}268^2} + \frac{y^2}{54{,}364^2} = 1.$$

Section 7.4 Arc Length and Surfaces of Revolution

- Find the arc length of a smooth curve.
- Find the area of a surface of revolution.

Arc Length

In this section, definite integrals are used to find the arc lengths of curves and the areas of surfaces of revolution. In either case, an arc (a segment of a curve) is approximated by straight line segments whose lengths are given by the familiar Distance Formula

$$d = \sqrt{(x_2 - x_1)^2 + (y_2 - y_1)^2}.$$

A **rectifiable** curve is one that has a finite arc length. You will see that a sufficient condition for the graph of a function f to be rectifiable between $(a, f(a))$ and $(b, f(b))$ is that f' be continuous on $[a, b]$. Such a function is **continuously differentiable** on $[a, b]$, and its graph on the interval $[a, b]$ is a **smooth curve.**

Consider a function $y = f(x)$ that is continuously differentiable on the interval $[a, b]$. You can approximate the graph of f by n line segments whose endpoints are determined by the partition

$$a = x_0 < x_1 < x_2 < \cdots < x_n = b$$

as shown in Figure 7.37. By letting $\Delta x_i = x_i - x_{i-1}$ and $\Delta y_i = y_i - y_{i-1}$, you can approximate the length of the graph by

$$s \approx \sum_{i=1}^{n} \sqrt{(x_i - x_{i-1})^2 + (y_i - y_{i-1})^2}$$

$$= \sum_{i=1}^{n} \sqrt{(\Delta x_i)^2 + (\Delta y_i)^2}$$

$$= \sum_{i=1}^{n} \sqrt{(\Delta x_i)^2 + \left(\frac{\Delta y_i}{\Delta x_i}\right)^2 (\Delta x_i)^2}$$

$$= \sum_{i=1}^{n} \sqrt{1 + \left(\frac{\Delta y_i}{\Delta x_i}\right)^2} (\Delta x_i).$$

This approximation appears to become better and better as $\|\Delta\| \to 0$ $(n \to \infty)$. So, the length of the graph is

$$s = \lim_{\|\Delta\| \to 0} \sum_{i=1}^{n} \sqrt{1 + \left(\frac{\Delta y_i}{\Delta x_i}\right)^2} (\Delta x_i).$$

Because $f'(x)$ exists for each x in (x_{i-1}, x_i), the Mean Value Theorem guarantees the existence of c_i in (x_{i-1}, x_i) such that

$$f(x_i) - f(x_{i-1}) = f'(c_i)(x_i - x_{i-1})$$

$$\frac{\Delta y_i}{\Delta x_i} = f'(c_i).$$

Because f' is continuous on $[a, b]$, it follows that $\sqrt{1 + [f'(x)]^2}$ is also continuous (and therefore integrable) on $[a, b]$, which implies that

$$s = \lim_{\|\Delta\| \to 0} \sum_{i=1}^{n} \sqrt{1 + [f'(c_i)]^2} (\Delta x_i)$$

$$= \int_a^b \sqrt{1 + [f'(x)]^2}\, dx$$

where s is called the **arc length** of f between a and b.

CHRISTIAN HUYGENS (1629–1695)

The Dutch mathematician Christian Huygens, who invented the pendulum clock, and James Gregory (1638–1675), a Scottish mathematician, both made early contributions to the problem of finding the length of a rectifiable curve.

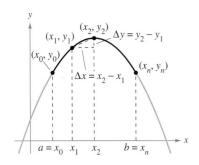

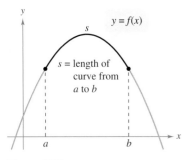

Figure 7.37

> **Definition of Arc Length**
>
> Let the function given by $y = f(x)$ represent a smooth curve on the interval $[a, b]$. The **arc length** of f between a and b is
>
> $$s = \int_a^b \sqrt{1 + [f'(x)]^2}\, dx.$$
>
> Similarly, for a smooth curve given by $x = g(y)$, the **arc length** of g between c and d is
>
> $$s = \int_c^d \sqrt{1 + [g'(y)]^2}\, dy.$$

Because the definition of arc length can be applied to a linear function, you can check to see that this new definition agrees with the standard Distance Formula for the length of a line segment. This is shown in Example 1.

EXAMPLE 1 The Length of a Line Segment

Find the arc length from (x_1, y_1) to (x_2, y_2) on the graph of $f(x) = mx + b$, as shown in Figure 7.38.

Solution Because

$$m = f'(x) = \frac{y_2 - y_1}{x_2 - x_1}$$

it follows that

$$\begin{aligned}
s &= \int_{x_1}^{x_2} \sqrt{1 + [f'(x)]^2}\, dx && \text{Formula for arc length} \\
&= \int_{x_1}^{x_2} \sqrt{1 + \left(\frac{y_2 - y_1}{x_2 - x_1}\right)^2}\, dx \\
&= \sqrt{\frac{(x_2 - x_1)^2 + (y_2 - y_1)^2}{(x_2 - x_1)^2}}\, (x) \Big]_{x_1}^{x_2} && \text{Integrate and simplify.} \\
&= \sqrt{\frac{(x_2 - x_1)^2 + (y_2 - y_1)^2}{(x_2 - x_1)^2}}\, (x_2 - x_1) \\
&= \sqrt{(x_2 - x_1)^2 + (y_2 - y_1)^2}
\end{aligned}$$

which is the formula for the distance between two points in the plane.

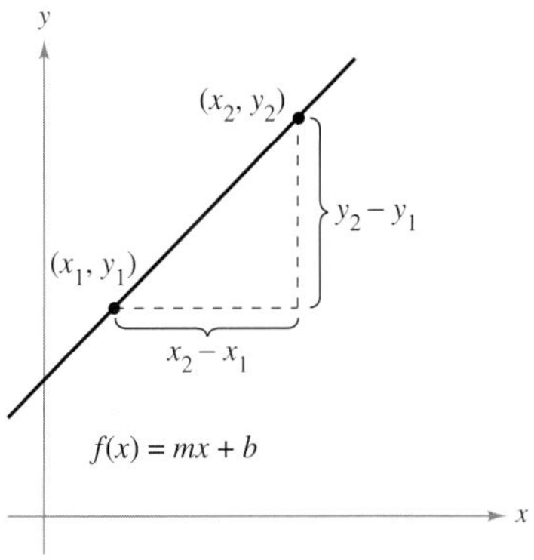

The arc length of the graph of f from (x_1, y_1) to (x_2, y_2) is the same as the standard Distance Formula.
Figure 7.38

> **TECHNOLOGY** Definite integrals representing arc length often are very difficult to evaluate. In this section, a few examples are presented. In the next chapter, with more advanced integration techniques, you will be able to tackle more difficult arc length problems. In the meantime, remember that you can always use a numerical integration program to approximate an arc length. For instance, use the *numerical integration* feature of a graphing utility to approximate the arc lengths in Examples 2 and 3.

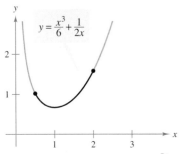

The arc length of the graph of y on $\left[\frac{1}{2}, 2\right]$
Figure 7.39

FOR FURTHER INFORMATION To see how arc length can be used to define trigonometric functions, see the article "Trigonometry Requires Calculus, Not Vice Versa" by Yves Nievergelt in *UMAP Modules*.

EXAMPLE 2 Finding Arc Length

Find the arc length of the graph of

$$y = \frac{x^3}{6} + \frac{1}{2x}$$

on the interval $\left[\frac{1}{2}, 2\right]$, as shown in Figure 7.39.

Solution Using

$$\frac{dy}{dx} = \frac{3x^2}{6} - \frac{1}{2x^2} = \frac{1}{2}\left(x^2 - \frac{1}{x^2}\right)$$

yields an arc length of

$$\begin{aligned}
s &= \int_a^b \sqrt{1 + \left(\frac{dy}{dx}\right)^2}\, dx = \int_{1/2}^2 \sqrt{1 + \left[\frac{1}{2}\left(x^2 - \frac{1}{x^2}\right)\right]^2}\, dx \quad \text{Formula for arc length} \\
&= \int_{1/2}^2 \sqrt{\frac{1}{4}\left(x^4 + 2 + \frac{1}{x^4}\right)}\, dx \\
&= \int_{1/2}^2 \frac{1}{2}\left(x^2 + \frac{1}{x^2}\right) dx \quad \text{Simplify.} \\
&= \frac{1}{2}\left[\frac{x^3}{3} - \frac{1}{x}\right]_{1/2}^2 \quad \text{Integrate.} \\
&= \frac{1}{2}\left(\frac{13}{6} + \frac{47}{24}\right) \\
&= \frac{33}{16}.
\end{aligned}$$

EXAMPLE 3 Finding Arc Length

Find the arc length of the graph of $(y - 1)^3 = x^2$ on the interval $[0, 8]$, as shown in Figure 7.40.

Solution Begin by solving for x in terms of y: $x = \pm(y - 1)^{3/2}$. Choosing the positive value of x produces

$$\frac{dx}{dy} = \frac{3}{2}(y - 1)^{1/2}.$$

The x-interval $[0, 8]$ corresponds to the y-interval $[1, 5]$, and the arc length is

$$\begin{aligned}
s &= \int_c^d \sqrt{1 + \left(\frac{dx}{dy}\right)^2}\, dy = \int_1^5 \sqrt{1 + \left[\frac{3}{2}(y-1)^{1/2}\right]^2}\, dy \quad \text{Formula for arc length} \\
&= \int_1^5 \sqrt{\frac{9}{4}y - \frac{5}{4}}\, dy \\
&= \frac{1}{2}\int_1^5 \sqrt{9y - 5}\, dy \quad \text{Simplify.} \\
&= \frac{1}{18}\left[\frac{(9y - 5)^{3/2}}{3/2}\right]_1^5 \quad \text{Integrate.} \\
&= \frac{1}{27}(40^{3/2} - 4^{3/2}) \\
&\approx 9.073.
\end{aligned}$$

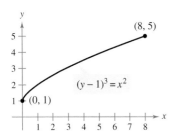

The arc length of the graph of y on $[0, 8]$
Figure 7.40

EXAMPLE 4 Finding Arc Length

Find the arc length of the graph of $y = \ln(\cos x)$ from $x = 0$ to $x = \pi/4$, as shown in Figure 7.41.

Solution Using

$$\frac{dy}{dx} = -\frac{\sin x}{\cos x} = -\tan x$$

yields an arc length of

$$s = \int_a^b \sqrt{1 + \left(\frac{dy}{dx}\right)^2}\, dx = \int_0^{\pi/4} \sqrt{1 + \tan^2 x}\, dx \qquad \text{Formula for arc length}$$

$$= \int_0^{\pi/4} \sqrt{\sec^2 x}\, dx \qquad \text{Trigonometric identity}$$

$$= \int_0^{\pi/4} \sec x\, dx \qquad \text{Simplify.}$$

$$= \Big[\ln|\sec x + \tan x|\Big]_0^{\pi/4} \qquad \text{Integrate.}$$

$$= \ln(\sqrt{2} + 1) - \ln 1$$

$$\approx 0.881.$$

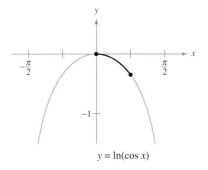

The arc length of the graph of y on $\left[0, \frac{\pi}{4}\right]$
Figure 7.41

EXAMPLE 5 Length of a Cable

An electric cable is hung between two towers that are 200 feet apart, as shown in Figure 7.42. The cable takes the shape of a catenary whose equation is

$$y = 75(e^{x/150} + e^{-x/150}) = 150 \cosh \frac{x}{150}.$$

Find the arc length of the cable between the two towers.

Solution Because $y' = \frac{1}{2}(e^{x/150} - e^{-x/150})$, you can write

$$(y')^2 = \frac{1}{4}(e^{x/75} - 2 + e^{-x/75})$$

and

$$1 + (y')^2 = \frac{1}{4}(e^{x/75} + 2 + e^{-x/75}) = \left[\frac{1}{2}(e^{x/150} + e^{-x/150})\right]^2.$$

Therefore, the arc length of the cable is

$$s = \int_a^b \sqrt{1 + (y')^2}\, dx = \frac{1}{2}\int_{-100}^{100} (e^{x/150} + e^{-x/150})\, dx \qquad \text{Formula for arc length}$$

$$= 75\Big[e^{x/150} - e^{-x/150}\Big]_{-100}^{100} \qquad \text{Integrate.}$$

$$= 150(e^{2/3} - e^{-2/3})$$

$$\approx 215 \text{ feet.}$$

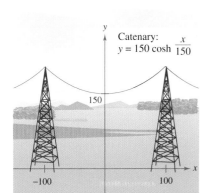

Catenary: $y = 150 \cosh \dfrac{x}{150}$

Figure 7.42

Area of a Surface of Revolution

In Sections 7.2 and 7.3, integration was used to calculate the volume of a solid of revolution. You will now look at a procedure for finding the area of a surface of revolution.

Definition of Surface of Revolution

If the graph of a continuous function is revolved about a line, the resulting surface is a **surface of revolution.**

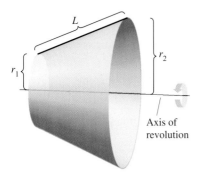

Figure 7.43

The area of a surface of revolution is derived from the formula for the lateral surface area of the frustum of a right circular cone. Consider the line segment in Figure 7.43, where L is the length of the line segment, r_1 is the radius at the left end of the line segment, and r_2 is the radius at the right end of the line segment. When the line segment is revolved about its axis of revolution, it forms a frustum of a right circular cone, with

$$S = 2\pi r L \qquad \text{Lateral surface area of frustum}$$

where

$$r = \frac{1}{2}(r_1 + r_2). \qquad \text{Average radius of frustum}$$

(In Exercise 60, you are asked to verify the formula for S.)

Suppose the graph of a function f, having a continuous derivative on the interval $[a, b]$, is revolved about the x-axis to form a surface of revolution, as shown in Figure 7.44. Let Δ be a partition of $[a, b]$, with subintervals of width Δx_i. Then the line segment of length

$$\Delta L_i = \sqrt{\Delta x_i^2 + \Delta y_i^2}$$

generates a frustum of a cone. Let r_i be the average radius of this frustum. By the Intermediate Value Theorem, a point d_i exists (in the ith subinterval) such that $r_i = f(d_i)$. The lateral surface area ΔS_i of the frustum is

$$\begin{aligned}\Delta S_i &= 2\pi r_i \Delta L_i \\ &= 2\pi f(d_i)\sqrt{\Delta x_i^2 + \Delta y_i^2} \\ &= 2\pi f(d_i)\sqrt{1 + \left(\frac{\Delta y_i}{\Delta x_i}\right)^2}\, \Delta x_i.\end{aligned}$$

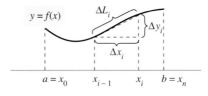
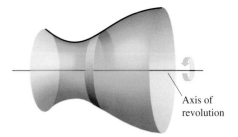

Figure 7.44

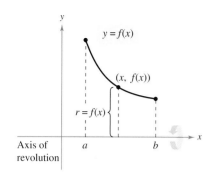
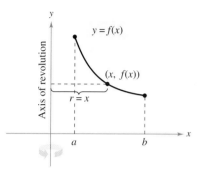

Figure 7.45

By the Mean Value Theorem, a point c_i exists in (x_{i-1}, x_i) such that

$$f'(c_i) = \frac{f(x_i) - f(x_{i-1})}{x_i - x_{i-1}}$$
$$= \frac{\Delta y_i}{\Delta x_i}.$$

So, $\Delta S_i = 2\pi f(d_i)\sqrt{1 + [f'(c_i)]^2}\,\Delta x_i$, and the total surface area can be approximated by

$$S \approx 2\pi \sum_{i=1}^{n} f(d_i)\sqrt{1 + [f'(c_i)]^2}\,\Delta x_i.$$

It can be shown that the limit of the right side as $\|\Delta\| \to 0$ $(n \to \infty)$ is

$$S = 2\pi \int_a^b f(x)\sqrt{1 + [f'(x)]^2}\,dx.$$

In a similar manner, if the graph of f is revolved about the y-axis, then S is

$$S = 2\pi \int_a^b x\sqrt{1 + [f'(x)]^2}\,dx.$$

In both formulas for S, you can regard the products $2\pi f(x)$ and $2\pi x$ as the circumference of the circle traced by a point (x, y) on the graph of f as it is revolved about the x- or y-axis (Figure 7.45). In one case the radius is $r = f(x)$, and in the other case the radius is $r = x$. Moreover, by appropriately adjusting r, you can generalize the formula for surface area to cover *any* horizontal or vertical axis of revolution, as indicated in the following definition.

Definition of the Area of a Surface of Revolution

Let $y = f(x)$ have a continuous derivative on the interval $[a, b]$. The area S of the surface of revolution formed by revolving the graph of f about a horizontal or vertical axis is

$$S = 2\pi \int_a^b r(x)\sqrt{1 + [f'(x)]^2}\,dx \qquad y \text{ is a function of } x.$$

where $r(x)$ is the distance between the graph of f and the axis of revolution. If $x = g(y)$ on the interval $[c, d]$, then the surface area is

$$S = 2\pi \int_c^d r(y)\sqrt{1 + [g'(y)]^2}\,dy \qquad x \text{ is a function of } y.$$

where $r(y)$ is the distance between the graph of g and the axis of revolution.

The formulas in this definition are sometimes written as

$$S = 2\pi \int_a^b r(x)\,ds \qquad y \text{ is a function of } x.$$

and

$$S = 2\pi \int_c^d r(y)\,ds \qquad x \text{ is a function of } y.$$

where $ds = \sqrt{1 + [f'(x)]^2}\,dx$ and $ds = \sqrt{1 + [g'(y)]^2}\,dy$, respectively.

EXAMPLE 6 The Area of a Surface of Revolution

Find the area of the surface formed by revolving the graph of

$$f(x) = x^3$$

on the interval $[0, 1]$ about the x-axis, as shown in Figure 7.46.

Solution The distance between the x-axis and the graph of f is $r(x) = f(x)$, and because $f'(x) = 3x^2$, the surface area is

$$\begin{aligned}
S &= 2\pi \int_a^b r(x)\sqrt{1 + [f'(x)]^2}\, dx && \text{Formula for surface area} \\
&= 2\pi \int_0^1 x^3\sqrt{1 + (3x^2)^2}\, dx \\
&= \frac{2\pi}{36}\int_0^1 (36x^3)(1 + 9x^4)^{1/2}\, dx && \text{Simplify.} \\
&= \frac{\pi}{18}\left[\frac{(1 + 9x^4)^{3/2}}{3/2}\right]_0^1 && \text{Integrate.} \\
&= \frac{\pi}{27}(10^{3/2} - 1) \\
&\approx 3.563.
\end{aligned}$$

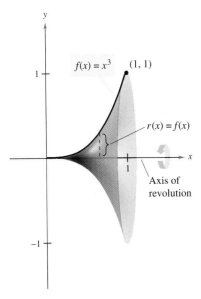

Figure 7.46

EXAMPLE 7 The Area of a Surface of Revolution

Find the area of the surface formed by revolving the graph of

$$f(x) = x^2$$

on the interval $\left[0, \sqrt{2}\right]$ about the y-axis, as shown in Figure 7.47.

Solution In this case, the distance between the graph of f and the y-axis is $r(x) = x$. Using $f'(x) = 2x$, you can determine that the surface area is

$$\begin{aligned}
S &= 2\pi \int_a^b r(x)\sqrt{1 + [f'(x)]^2}\, dx && \text{Formula for surface area} \\
&= 2\pi \int_0^{\sqrt{2}} x\sqrt{1 + (2x)^2}\, dx \\
&= \frac{2\pi}{8}\int_0^{\sqrt{2}} (1 + 4x^2)^{1/2}(8x)\, dx && \text{Simplify.} \\
&= \frac{\pi}{4}\left[\frac{(1 + 4x^2)^{3/2}}{3/2}\right]_0^{\sqrt{2}} && \text{Integrate.} \\
&= \frac{\pi}{6}\left[(1 + 8)^{3/2} - 1\right] \\
&= \frac{13\pi}{3} \\
&\approx 13.614.
\end{aligned}$$

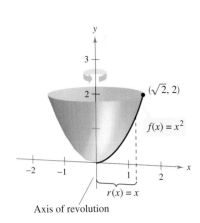

Figure 7.47

Exercises for Section 7.4

See www.CalcChat.com for worked-out solutions to odd-numbered exercises.

In Exercises 1 and 2, find the distance between the points using (a) the Distance Formula and (b) integration.

1. $(0, 0)$, $(5, 12)$
2. $(1, 2)$, $(7, 10)$

In Exercises 3–14, find the arc length of the graph of the function over the indicated interval.

3. $y = \dfrac{2}{3}x^{3/2} + 1$
4. $y = 2x^{3/2} + 3$

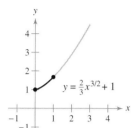
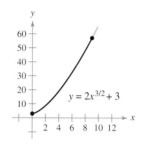

5. $y = \dfrac{3}{2}x^{2/3}$
6. $y = \dfrac{x^4}{8} + \dfrac{1}{4x^2}$

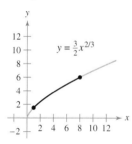
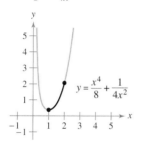

7. $y = \dfrac{x^5}{10} + \dfrac{1}{6x^3}$, $[1, 2]$
8. $y = \dfrac{3}{2}x^{2/3} + 4$, $[1, 27]$
9. $y = \ln(\sin x)$, $\left[\dfrac{\pi}{4}, \dfrac{3\pi}{4}\right]$
10. $y = \ln(\cos x)$, $\left[0, \dfrac{\pi}{3}\right]$
11. $y = \dfrac{1}{2}(e^x + e^{-x})$, $[0, 2]$
12. $y = \ln\left(\dfrac{e^x + 1}{e^x - 1}\right)$, $[\ln 2, \ln 3]$
13. $x = \dfrac{1}{3}(y^2 + 2)^{3/2}$, $0 \le y \le 4$
14. $x = \dfrac{1}{3}\sqrt{y}(y - 3)$, $1 \le y \le 4$

In Exercises 15–24, (a) graph the function, highlighting the part indicated by the given interval, (b) find a definite integral that represents the arc length of the curve over the indicated interval and observe that the integral cannot be evaluated with the techniques studied so far, and (c) use the integration capabilities of a graphing utility to approximate the arc length.

15. $y = 4 - x^2$, $0 \le x \le 2$
16. $y = x^2 + x - 2$, $-2 \le x \le 1$
17. $y = \dfrac{1}{x}$, $1 \le x \le 3$
18. $y = \dfrac{1}{x + 1}$, $0 \le x \le 1$
19. $y = \sin x$, $0 \le x \le \pi$
20. $y = \cos x$, $-\dfrac{\pi}{2} \le x \le \dfrac{\pi}{2}$
21. $x = e^{-y}$, $0 \le y \le 2$
22. $y = \ln x$, $1 \le x \le 5$
23. $y = 2 \arctan x$, $0 \le x \le 1$
24. $x = \sqrt{36 - y^2}$, $0 \le y \le 3$

Approximation In Exercises 25 and 26, determine which value best approximates the length of the arc represented by the integral. (Make your selection on the basis of a sketch of the arc and *not* by performing any calculations.)

25. $\displaystyle\int_0^2 \sqrt{1 + \left[\dfrac{d}{dx}\left(\dfrac{5}{x^2 + 1}\right)\right]^2}\, dx$

  (a) 25 (b) 5 (c) 2 (d) -4 (e) 3

26. $\displaystyle\int_0^{\pi/4} \sqrt{1 + \left[\dfrac{d}{dx}(\tan x)\right]^2}\, dx$

  (a) 3 (b) -2 (c) 4 (d) $\dfrac{4\pi}{3}$ (e) 1

Approximation In Exercises 27 and 28, approximate the arc length of the graph of the function over the interval [0, 4] in four ways. (a) Use the Distance Formula to find the distance between the endpoints of the arc. (b) Use the Distance Formula to find the lengths of the four line segments connecting the points on the arc when $x = 0$, $x = 1$, $x = 2$, $x = 3$, and $x = 4$. Find the sum of the four lengths. (c) Use Simpson's Rule with $n = 10$ to approximate the integral yielding the indicated arc length. (d) Use the integration capabilities of a graphing utility to approximate the integral yielding the indicated arc length.

27. $f(x) = x^3$
28. $f(x) = (x^2 - 4)^2$

29. (a) Use a graphing utility to graph the function $f(x) = x^{2/3}$.
    (b) Can you integrate with respect to x to find the arc length of the graph of f on the interval $[-1, 8]$? Explain.
    (c) Find the arc length of the graph of f on the interval $[-1, 8]$.

30. *Astroid* Find the total length of the graph of the astroid $x^{2/3} + y^{2/3} = 4$.

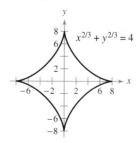

484 CHAPTER 7 Applications of Integration

31. *Think About It* The figure shows the graphs of the functions $y_1 = x$, $y_2 = \frac{1}{2}x^{3/2}$, $y_3 = \frac{1}{4}x^2$, and $y_4 = \frac{1}{8}x^{5/2}$ on the interval $[0, 4]$. To print an enlarged copy of the graph, go to the website www.mathgraphs.com.

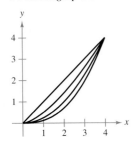

(a) Label the functions.

(b) List the functions in order of increasing arc length.

(c) Verify your answer in part (b) by approximating each arc length accurate to three decimal places.

32. *Think About It* Explain why the two integrals are equal.

$$\int_1^e \sqrt{1 + \frac{1}{x^2}}\, dx = \int_0^1 \sqrt{1 + e^{2x}}\, dx$$

Use the integration capabilities of a graphing utility to verify that the integrals are equal.

33. *Length of Pursuit* A fleeing object leaves the origin and moves up the y-axis (see figure). At the same time, a pursuer leaves the point $(1, 0)$ and always moves toward the fleeing object. The pursuer's speed is twice that of the fleeing object. The equation of the path is modeled by

$$y = \frac{1}{3}(x^{3/2} - 3x^{1/2} + 2).$$

How far has the fleeing object traveled when it is caught? Show that the pursuer has traveled twice as far.

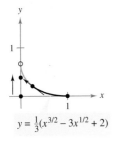

$y = \frac{1}{3}(x^{3/2} - 3x^{1/2} + 2)$ $y = 31 - 10(e^{x/20} + e^{-x/20})$

Figure for 33   **Figure for 34**

34. *Roof Area* A barn is 100 feet long and 40 feet wide (see figure). A cross section of the roof is the inverted catenary $y = 31 - 10(e^{x/20} + e^{-x/20})$. Find the number of square feet of roofing on the barn.

35. *Length of a Catenary* Electrical wires suspended between two towers form a catenary (see figure) modeled by the equation

$$y = 20 \cosh \frac{x}{20}, \quad -20 \le x \le 20$$

where x and y are measured in meters. The towers are 40 meters apart. Find the length of the suspended cable.

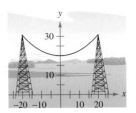
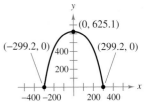

Figure for 35   **Figure for 36**

36. *Length of Gateway Arch* The Gateway Arch in St. Louis, Missouri, is modeled by

$$y = 693.8597 - 68.7672 \cosh 0.0100333x,$$
$$-299.2239 \le x \le 299.2239.$$

(See Section 5.8, Section Project: St. Louis Arch.) Find the length of this curve (see figure).

37. Find the arc length from $(0, 3)$ clockwise to $(2, \sqrt{5})$ along the circle $x^2 + y^2 = 9$.

38. Find the arc length from $(-3, 4)$ clockwise to $(4, 3)$ along the circle $x^2 + y^2 = 25$. Show that the result is one-fourth the circumference of the circle.

In Exercises 39–42, set up and evaluate the definite integral for the area of the surface generated by revolving the curve about the x-axis.

39. $y = \dfrac{1}{3}x^3$ **40.** $y = 2\sqrt{x}$

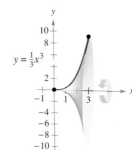
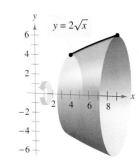

41. $y = \dfrac{x^3}{6} + \dfrac{1}{2x}$, $1 \le x \le 2$ **42.** $y = \dfrac{x}{2}$, $0 \le x \le 6$

In Exercises 43 and 44, set up and evaluate the definite integral for the area of the surface generated by revolving the curve about the y-axis.

43. $y = \sqrt[3]{x} + 2$ **44.** $y = 9 - x^2$

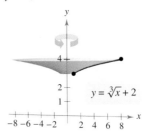
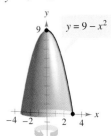

In Exercises 45 and 46, use the integration capabilities of a graphing utility to approximate the surface area of the solid of revolution.

Function	Interval
45. $y = \sin x$	$[0, \pi]$
revolved about the x-axis	
46. $y = \ln x$	$[1, e]$
revolved about the y-axis	

Writing About Concepts

47. Define a rectifiable curve.

48. What precalculus formula and representative element are used to develop the integration formula for arc length?

49. What precalculus formula and representative element are used to develop the integration formula for the area of a surface of revolution?

50. The graphs of the functions f_1 and f_2 on the interval $[a, b]$ are shown in the figure. The graph of each is revolved about the x-axis. Which surface of revolution has the greater surface area? Explain.

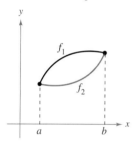

51. A right circular cone is generated by revolving the region bounded by $y = hx/r$, $y = h$, and $x = 0$ about the y-axis. Verify that the lateral surface area of the cone is
$$S = \pi r \sqrt{r^2 + h^2}.$$

52. A sphere of radius r is generated by revolving the graph of $y = \sqrt{r^2 - x^2}$ about the x-axis. Verify that the surface area of the sphere is $4\pi r^2$.

53. Find the area of the zone of a sphere formed by revolving the graph of $y = \sqrt{9 - x^2}$, $0 \le x \le 2$, about the y-axis.

54. Find the area of the zone of a sphere formed by revolving the graph of $y = \sqrt{r^2 - x^2}$, $0 \le x \le a$, about the y-axis. Assume that $a < r$.

55. *Bulb Design* An ornamental light bulb is designed by revolving the graph of
$$y = \tfrac{1}{3}x^{1/2} - x^{3/2}, \quad 0 \le x \le \tfrac{1}{3}$$
about the x-axis, where x and y are measured in feet (see figure). Find the surface area of the bulb and use the result to approximate the amount of glass needed to make the bulb. (Assume that the glass is 0.015 inch thick.)

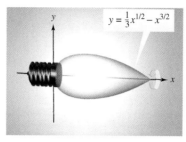

Figure for 55

56. *Think About It* Consider the equation $\dfrac{x^2}{9} + \dfrac{y^2}{4} = 1$.

(a) Use a graphing utility to graph the equation.

(b) Set up the definite integral for finding the first quadrant arc length of the graph in part (a).

(c) Compare the interval of integration in part (b) and the domain of the integrand. Is it possible to evaluate the definite integral? Is it possible to use Simpson's Rule to evaluate the definite integral? Explain. (You will learn how to evaluate this type of integral in Section 8.8.)

57. *Modeling Data* The circumference C (in inches) of a vase is measured at three-inch intervals starting at its base. The measurements are shown in the table, where y is the vertical distance in inches from the base.

y	0	3	6	9	12	15	18
C	50	65.5	70	66	58	51	48

(a) Use the data to approximate the volume of the vase by summing the volumes of approximating disks.

(b) Use the data to approximate the outside surface area (excluding the base) of the vase by summing the outside surface areas of approximating frustums of right circular cones.

(c) Use the regression capabilities of a graphing utility to find a cubic model for the points (y, r) where $r = C/(2\pi)$. Use the graphing utility to plot the points and graph the model.

(d) Use the model in part (c) and the integration capabilities of a graphing utility to approximate the volume and outside surface area of the vase. Compare the results with your answers in parts (a) and (b).

58. *Modeling Data* Property bounded by two perpendicular roads and a stream is shown in the figure on the next page. All distances are measured in feet.

(a) Use the regression capabilities of a graphing utility to fit a fourth-degree polynomial to the path of the stream.

(b) Use the model in part (a) to approximate the area of the property in acres.

(c) Use the integration capabilities of a graphing utility to find the length of the stream that bounds the property.

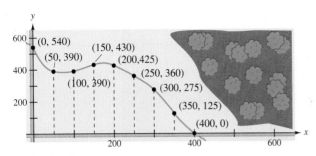

Figure for 58

59. Let R be the region bounded by $y = 1/x$, the x-axis, $x = 1$, and $x = b$, where $b > 1$. Let D be the solid formed when R is revolved about the x-axis.

(a) Find the volume V of D.

(b) Write the surface area S as an integral.

(c) Show that V approaches a finite limit as $b \to \infty$.

(d) Show that $S \to \infty$ as $b \to \infty$.

60. (a) Given a circular sector with radius L and central angle θ (see figure), show that the area of the sector is given by

$$S = \frac{1}{2}L^2\theta.$$

(b) By joining the straight line edges of the sector in part (a), a right circular cone is formed (see figure) and the lateral surface area of the cone is the same as the area of the sector. Show that the area is $S = \pi r L$, where r is the radius of the base of the cone. (*Hint:* The arc length of the sector equals the circumference of the base of the cone.)

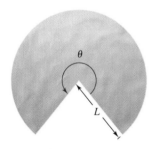

Figure for 60(a)   **Figure for 60(b)**

(c) Use the result of part (b) to verify that the formula for the lateral surface area of the frustum of a cone with slant height L and radii r_1 and r_2 (see figure) is $S = \pi(r_1 + r_2)L$. (*Note:* This formula was used to develop the integral for finding the surface area of a surface of revolution.)

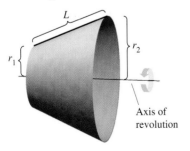

61. *Individual Project* Select a solid of revolution from everyday life. Measure the radius of the solid at a minimum of seven points along its axis. Use the data to approximate the volume of the solid and the surface area of the lateral sides of the solid.

62. *Writing* Read the article "Arc Length, Area and the Arcsine Function" by Andrew M. Rockett in *Mathematics Magazine*. Then write a paragraph explaining how the arcsine function can be defined in terms of an arc length. (To view this article, go to the website *www.matharticles.com*.)

63. *Astroid* Find the area of the surface formed by revolving the portion in the first quadrant of the graph of $x^{2/3} + y^{2/3} = 4$, $0 \le y \le 8$ about the y-axis.

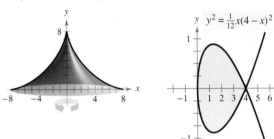

Figure for 63   **Figure for 64**

64. Consider the graph of $y^2 = \frac{1}{12}x(4-x)^2$ (see figure). Find the area of the surface formed when the loop of this graph is revolved around the x-axis.

65. *Suspension Bridge* A cable for a suspension bridge has the shape of a parabola with equation $y = kx^2$. Let h represent the height of the cable from its lowest point to its highest point and let $2w$ represent the total span of the bridge (see figure). Show that the length C of the cable is given by

$$C = 2\int_0^w \sqrt{1 + \frac{4h^2}{w^4}x^2}\, dx.$$

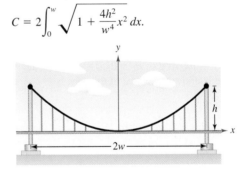

66. *Suspension Bridge* The Humber Bridge, located in the United Kingdom and opened in 1981, has a main span of about 1400 meters. Each of its towers has a height of about 155 meters. Use these dimensions, the integral in Exercise 65, and the integration capabilities of a graphing utility to approximate the length of a parabolic cable along the main span.

Putnam Exam Challenge

67. Find the length of the curve $y^2 = x^3$ from the origin to the point where the tangent makes an angle of $45°$ with the x-axis.

This problem was composed by the Committee on the Putnam Prize Competition. © The Mathematical Association of America. All rights reserved.

Section 7.5 Work

- Find the work done by a constant force.
- Find the work done by a variable force.

Work Done by a Constant Force

The concept of work is important to scientists and engineers for determining the energy needed to perform various jobs. For instance, it is useful to know the amount of work done when a crane lifts a steel girder, when a spring is compressed, when a rocket is propelled into the air, or when a truck pulls a load along a highway.

In general, **work** is done by a force when it moves an object. If the force applied to the object is *constant*, then the definition of work is as follows.

Definition of Work Done by a Constant Force

If an object is moved a distance D in the direction of an applied constant force F, then the **work** W done by the force is defined as $W = FD$.

There are many types of forces—centrifugal, electromotive, and gravitational, to name a few. A **force** can be thought of as a *push* or a *pull*; a force changes the state of rest or state of motion of a body. For gravitational forces on Earth, it is common to use units of measure corresponding to the weight of an object.

EXAMPLE 1 Lifting an Object

Determine the work done in lifting a 50-pound object 4 feet.

Solution The magnitude of the required force F is the weight of the object, as shown in Figure 7.48. So, the work done in lifting the object 4 feet is

$W = FD$ Work = (force)(distance)

$\quad = 50(4)$ Force = 50 pounds, distance = 4 feet

$\quad = 200$ foot-pounds.

In the U.S. measurement system, work is typically expressed in foot-pounds (ft-lb), inch-pounds, or foot-tons. In the centimeter-gram-second (C-G-S) system, the basic unit of force is the **dyne**—the force required to produce an acceleration of 1 centimeter per second per second on a mass of 1 gram. In this system, work is typically expressed in dyne-centimeters (ergs) or newton-meters (joules), where 1 joule = 10^7 ergs.

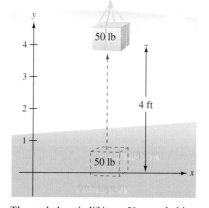

The work done in lifting a 50-pound object 4 feet is 200 foot-pounds.
Figure 7.48

EXPLORATION

How Much Work? In Example 1, 200 foot-pounds of work was needed to lift the 50-pound object 4 feet vertically off the ground. Suppose that once you lifted the object, you held it and walked a horizontal distance of 4 feet. Would this require an additional 200 foot-pounds of work? Explain your reasoning.

Work Done by a Variable Force

In Example 1, the force involved was *constant*. If a *variable* force is applied to an object, calculus is needed to determine the work done, because the amount of force changes as the object changes position. For instance, the force required to compress a spring increases as the spring is compressed.

Suppose that an object is moved along a straight line from $x = a$ to $x = b$ by a continuously varying force $F(x)$. Let Δ be a partition that divides the interval $[a, b]$ into n subintervals determined by

$$a = x_0 < x_1 < x_2 < \cdots < x_n = b$$

and let $\Delta x_i = x_i - x_{i-1}$. For each i, choose c_i such that $x_{i-1} \le c_i \le x_i$. Then at c_i the force is given by $F(c_i)$. Because F is continuous, you can approximate the work done in moving the object through the ith subinterval by the increment

$$\Delta W_i = F(c_i) \Delta x_i$$

as shown in Figure 7.49. So, the total work done as the object moves from a to b is approximated by

$$W \approx \sum_{i=1}^{n} \Delta W_i$$
$$= \sum_{i=1}^{n} F(c_i) \Delta x_i.$$

This approximation appears to become better and better as $\|\Delta\| \to 0$ ($n \to \infty$). So, the work done is

$$W = \lim_{\|\Delta\| \to 0} \sum_{i=1}^{n} F(c_i) \Delta x_i$$
$$= \int_{a}^{b} F(x) \, dx.$$

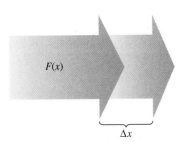

The amount of force changes as an object changes position (Δx).
Figure 7.49

Definition of Work Done by a Variable Force

If an object is moved along a straight line by a continuously varying force $F(x)$, then the **work** W done by the force as the object is moved from $x = a$ to $x = b$ is

$$W = \lim_{\|\Delta\| \to 0} \sum_{i=1}^{n} \Delta W_i$$
$$= \int_{a}^{b} F(x) \, dx.$$

The remaining examples in this section use some well-known physical laws. The discoveries of many of these laws occurred during the same period in which calculus was being developed. In fact, during the seventeenth and eighteenth centuries, there was little difference between physicists and mathematicians. One such physicist-mathematician was Emilie de Breteuil. Breteuil was instrumental in synthesizing the work of many other scientists, including Newton, Leibniz, Huygens, Kepler, and Descartes. Her physics text *Institutions* was widely used for many years.

EMILIE DE BRETEUIL (1706–1749)

Another major work by de Breteuil was the translation of Newton's "Philosophiae Naturalis Principia Mathematica" into French. Her translation and commentary greatly contributed to the acceptance of Newtonian science in Europe.

The following three laws of physics were developed by Robert Hooke (1635–1703), Isaac Newton (1642–1727), and Charles Coulomb (1736–1806).

1. **Hooke's Law:** The force F required to compress or stretch a spring (within its elastic limits) is proportional to the distance d that the spring is compressed or stretched from its original length. That is,

$$F = kd$$

where the constant of proportionality k (the spring constant) depends on the specific nature of the spring.

2. **Newton's Law of Universal Gravitation:** The force F of attraction between two particles of masses m_1 and m_2 is proportional to the product of the masses and inversely proportional to the square of the distance d between the two particles. That is,

$$F = k\frac{m_1 m_2}{d^2}.$$

If m_1 and m_2 are given in grams and d in centimeters, F will be in dynes for a value of $k = 6.670 \times 10^{-8}$ cubic centimeter per gram-second squared.

3. **Coulomb's Law:** The force between two charges q_1 and q_2 in a vacuum is proportional to the product of the charges and inversely proportional to the square of the distance d between the two charges. That is,

$$F = k\frac{q_1 q_2}{d^2}.$$

If q_1 and q_2 are given in electrostatic units and d in centimeters, F will be in dynes for a value of $k = 1$.

EXPLORATION

The work done in compressing the spring in Example 2 from $x = 3$ inches to $x = 6$ inches is 3375 inch-pounds. Should the work done in compressing the spring from $x = 0$ inches to $x = 3$ inches be more than, the same as, or less than this? Explain.

EXAMPLE 2 Compressing a Spring

A force of 750 pounds compresses a spring 3 inches from its natural length of 15 inches. Find the work done in compressing the spring an additional 3 inches.

Solution By Hooke's Law, the force $F(x)$ required to compress the spring x units (from its natural length) is $F(x) = kx$. Using the given data, it follows that $F(3) = 750 = (k)(3)$ and so $k = 250$ and $F(x) = 250x$, as shown in Figure 7.50. To find the increment of work, assume that the force required to compress the spring over a small increment Δx is nearly constant. So, the increment of work is

$$\Delta W = (\text{force})(\text{distance increment}) = (250x)\,\Delta x.$$

Because the spring is compressed from $x = 3$ to $x = 6$ inches less than its natural length, the work required is

$$W = \int_a^b F(x)\,dx = \int_3^6 250x\,dx \quad \text{Formula for work}$$

$$= 125x^2 \Big]_3^6 = 4500 - 1125 = 3375 \text{ inch-pounds}.$$

Note that you do *not* integrate from $x = 0$ to $x = 6$ because you were asked to determine the work done in compressing the spring an *additional* 3 inches (not including the first 3 inches).

Natural length ($F = 0$)

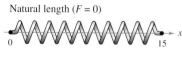

Compressed 3 inches ($F = 750$)

Compressed x inches ($F = 250x$)

Figure 7.50

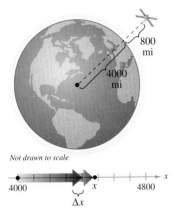

Figure 7.51

EXAMPLE 3 Moving a Space Module into Orbit

A space module weighs 15 metric tons on the surface of Earth. How much work is done in propelling the module to a height of 800 miles above Earth, as shown in Figure 7.51? (Use 4000 miles as the radius of Earth. Do not consider the effect of air resistance or the weight of the propellant.)

Solution Because the weight of a body varies inversely as the square of its distance from the center of Earth, the force $F(x)$ exerted by gravity is

$$F(x) = \frac{C}{x^2}. \qquad \text{\textit{C} is the constant of proportionality.}$$

Because the module weighs 15 metric tons on the surface of Earth and the radius of Earth is approximately 4000 miles, you have

$$15 = \frac{C}{(4000)^2}$$

$$240{,}000{,}000 = C.$$

So, the increment of work is

$$\Delta W = (\text{force})(\text{distance increment})$$

$$= \frac{240{,}000{,}000}{x^2} \Delta x.$$

Finally, because the module is propelled from $x = 4000$ to $x = 4800$ miles, the total work done is

$$W = \int_a^b F(x)\, dx = \int_{4000}^{4800} \frac{240{,}000{,}000}{x^2}\, dx \qquad \text{Formula for work}$$

$$= \frac{-240{,}000{,}000}{x} \Bigg]_{4000}^{4800} \qquad \text{Integrate.}$$

$$= -50{,}000 + 60{,}000$$

$$= 10{,}000 \text{ mile-tons}$$

$$\approx 1.164 \times 10^{11} \text{ foot-pounds.}$$

In the C-G-S system, using a conversion factor of 1 foot-pound $\approx$ 1.35582 joules, the work done is

$$W \approx 1.578 \times 10^{11} \text{ joules.}$$

The solutions to Examples 2 and 3 conform to our development of work as the summation of increments in the form

$$\Delta W = (\text{force})(\text{distance increment}) = (F)(\Delta x).$$

Another way to formulate the increment of work is

$$\Delta W = (\text{force increment})(\text{distance}) = (\Delta F)(x).$$

This second interpretation of ΔW is useful in problems involving the movement of nonrigid substances such as fluids and chains.

EXAMPLE 4 Emptying a Tank of Oil

A spherical tank of radius 8 feet is half full of oil that weighs 50 pounds per cubic foot. Find the work required to pump oil out through a hole in the top of the tank.

Solution Consider the oil to be subdivided into disks of thickness Δy and radius x, as shown in Figure 7.52. Because the increment of force for each disk is given by its weight, you have

$$\Delta F = \text{weight}$$
$$= \left(\frac{50 \text{ pounds}}{\text{cubic foot}}\right)(\text{volume})$$
$$= 50(\pi x^2 \Delta y) \text{ pounds.}$$

For a circle of radius 8 and center at $(0, 8)$, you have

$$x^2 + (y - 8)^2 = 8^2$$
$$x^2 = 16y - y^2$$

and you can write the force increment as

$$\Delta F = 50(\pi x^2 \Delta y)$$
$$= 50\pi(16y - y^2) \Delta y.$$

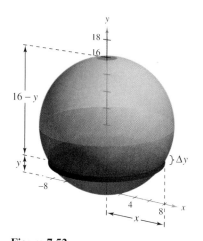

Figure 7.52

In Figure 7.52, note that a disk y feet from the bottom of the tank must be moved a distance of $(16 - y)$ feet. So, the increment of work is

$$\Delta W = \Delta F(16 - y)$$
$$= 50\pi(16y - y^2) \Delta y(16 - y)$$
$$= 50\pi(256y - 32y^2 + y^3) \Delta y.$$

Because the tank is half full, y ranges from 0 to 8, and the work required to empty the tank is

$$W = \int_0^8 50\pi(256y - 32y^2 + y^3) \, dy$$
$$= 50\pi \left[128y^2 - \frac{32}{3}y^3 + \frac{y^4}{4}\right]_0^8$$
$$= 50\pi \left(\frac{11,264}{3}\right)$$
$$\approx 589,782 \text{ foot-pounds.}$$

To estimate the reasonableness of the result in Example 4, consider that the weight of the oil in the tank is

$$\left(\frac{1}{2}\right)(\text{volume})(\text{density}) = \frac{1}{2}\left(\frac{4}{3}\pi 8^3\right)(50)$$
$$\approx 53,616.5 \text{ pounds.}$$

Lifting the entire half-tank of oil 8 feet would involve work of $8(53,616.5) \approx 428,932$ foot-pounds. Because the oil is actually lifted between 8 and 16 feet, it seems reasonable that the work done is 589,782 foot-pounds.

EXAMPLE 5 Lifting a Chain

A 20-foot chain weighing 5 pounds per foot is lying coiled on the ground. How much work is required to raise one end of the chain to a height of 20 feet so that it is fully extended, as shown in Figure 7.53?

Solution Imagine that the chain is divided into small sections, each of length Δy. Then the weight of each section is the increment of force

$$\Delta F = (\text{weight}) = \left(\frac{5 \text{ pounds}}{\text{foot}}\right)(\text{length}) = 5\Delta y.$$

Because a typical section (initially on the ground) is raised to a height of y, the increment of work is

$$\Delta W = (\text{force increment})(\text{distance}) = (5\,\Delta y)y = 5y\,\Delta y.$$

Because y ranges from 0 to 20, the total work is

$$W = \int_0^{20} 5y\,dy = \frac{5y^2}{2}\Big]_0^{20} = \frac{5(400)}{2} = 1000 \text{ foot-pounds.}$$

Work required to raise one end of the chain
Figure 7.53

In the next example you will consider a piston of radius r in a cylindrical casing, as shown in Figure 7.54. As the gas in the cylinder expands, the piston moves and work is done. If p represents the pressure of the gas (in pounds per square foot) against the piston head and V represents the volume of the gas (in cubic feet), the work increment involved in moving the piston Δx feet is

$$\Delta W = (\text{force})(\text{distance increment}) = F(\Delta x) = p(\pi r^2)\,\Delta x = p\,\Delta V.$$

So, as the volume of the gas expands from V_0 to V_1, the work done in moving the piston is

$$W = \int_{V_0}^{V_1} p\,dV.$$

Assuming the pressure of the gas to be inversely proportional to its volume, you have $p = k/V$ and the integral for work becomes

$$W = \int_{V_0}^{V_1} \frac{k}{V}\,dV.$$

Work done by expanding gas
Figure 7.54

EXAMPLE 6 Work Done by an Expanding Gas

A quantity of gas with an initial volume of 1 cubic foot and a pressure of 500 pounds per square foot expands to a volume of 2 cubic feet. Find the work done by the gas. (Assume that the pressure is inversely proportional to the volume.)

Solution Because $p = k/V$ and $p = 500$ when $V = 1$, you have $k = 500$. So, the work is

$$W = \int_{V_0}^{V_1} \frac{k}{V}\,dV$$

$$= \int_1^2 \frac{500}{V}\,dV$$

$$= 500 \ln|V|\Big]_1^2 \approx 346.6 \text{ foot-pounds.}$$

Exercises for Section 7.5

Constant Force In Exercises 1–4, determine the work done by the constant force.

1. A 100-pound bag of sugar is lifted 10 feet.
2. An electric hoist lifts a 2800-pound car 4 feet.
3. A force of 112 newtons is required to slide a cement block 4 meters in a construction project.
4. The locomotive of a freight train pulls its cars with a constant force of 9 tons a distance of one-half mile.

Writing About Concepts

5. State the definition of work done by a constant force.
6. State the definition of work done by a variable force.
7. The graphs show the force F_i (in pounds) required to move an object 9 feet along the x-axis. Order the force functions from the one that yields the least work to the one that yields the most work without doing any calculations. Explain your reasoning.

(a)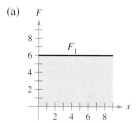

(b)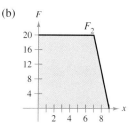

(c)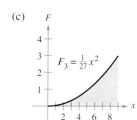

(d)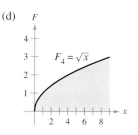

8. Verify your answer to Exercise 7 by calculating the work for each force function.

Hooke's Law In Exercises 9–16, use Hooke's Law to determine the variable force in the spring problem.

9. A force of 5 pounds compresses a 15-inch spring a total of 4 inches. How much work is done in compressing the spring 7 inches?
10. How much work is done in compressing the spring in Exercise 9 from a length of 10 inches to a length of 6 inches?
11. A force of 250 newtons stretches a spring 30 centimeters. How much work is done in stretching the spring from 20 centimeters to 50 centimeters?
12. A force of 800 newtons stretches a spring 70 centimeters on a mechanical device for driving fence posts. Find the work done in stretching the spring the required 70 centimeters.
13. A force of 20 pounds stretches a spring 9 inches in an exercise machine. Find the work done in stretching the spring 1 foot from its natural position.
14. An overhead garage door has two springs, one on each side of the door. A force of 15 pounds is required to stretch each spring 1 foot. Because of the pulley system, the springs stretch only one-half the distance the door travels. The door moves a total of 8 feet and the springs are at their natural length when the door is open. Find the work done by the pair of springs.
15. Eighteen foot-pounds of work is required to stretch a spring 4 inches from its natural length. Find the work required to stretch the spring an additional 3 inches.
16. Seven and one-half foot-pounds of work is required to compress a spring 2 inches from its natural length. Find the work required to compress the spring an additional one-half inch.
17. *Propulsion* Neglecting air resistance and the weight of the propellant, determine the work done in propelling a five-ton satellite to a height of
    (a) 100 miles above Earth.
    (b) 300 miles above Earth.
18. *Propulsion* Use the information in Exercise 17 to write the work W of the propulsion system as a function of the height h of the satellite above Earth. Find the limit (if it exists) of W as h approaches infinity.
19. *Propulsion* Neglecting air resistance and the weight of the propellant, determine the work done in propelling a 10-ton satellite to a height of
    (a) 11,000 miles above Earth.
    (b) 22,000 miles above Earth.
20. *Propulsion* A lunar module weighs 12 tons on the surface of Earth. How much work is done in propelling the module from the surface of the moon to a height of 50 miles? Consider the radius of the moon to be 1100 miles and its force of gravity to be one-sixth that of Earth.
21. *Pumping Water* A rectangular tank with a base 4 feet by 5 feet and a height of 4 feet is full of water (see figure). The water weighs 62.4 pounds per cubic foot. How much work is done in pumping water out over the top edge in order to empty (a) half of the tank? (b) all of the tank?

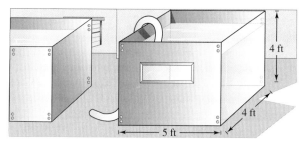

22. *Think About It* Explain why the answer in part (b) of Exercise 21 is not twice the answer in part (a).

23. *Pumping Water* A cylindrical water tank 4 meters high with a radius of 2 meters is buried so that the top of the tank is 1 meter below ground level (see figure). How much work is done in pumping a full tank of water up to ground level? (The water weighs 9800 newtons per cubic meter.)

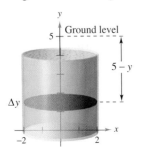
Figure for 23

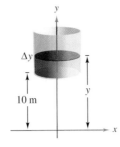
Figure for 24

24. *Pumping Water* Suppose the tank in Exercise 23 is located on a tower so that the bottom of the tank is 10 meters above the level of a stream (see figure). How much work is done in filling the tank half full of water through a hole in the bottom, using water from the stream?

25. *Pumping Water* An open tank has the shape of a right circular cone (see figure). The tank is 8 feet across the top and 6 feet high. How much work is done in emptying the tank by pumping the water over the top edge?

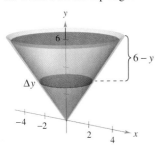
Figure for 25

Figure for 28

26. *Pumping Water* Water is pumped in through the bottom of the tank in Exercise 25. How much work is done to fill the tank

(a) to a depth of 2 feet?

(b) from a depth of 4 feet to a depth of 6 feet?

27. *Pumping Water* A hemispherical tank of radius 6 feet is positioned so that its base is circular. How much work is required to fill the tank with water through a hole in the base if the water source is at the base?

28. *Pumping Diesel Fuel* The fuel tank on a truck has trapezoidal cross sections with dimensions (in feet) shown in the figure. Assume that an engine is approximately 3 feet above the top of the fuel tank and that diesel fuel weighs approximately 53.1 pounds per cubic foot. Find the work done by the fuel pump in raising a full tank of fuel to the level of the engine.

Pumping Gasoline In Exercises 29 and 30, find the work done in pumping gasoline that weighs 42 pounds per cubic foot. (*Hint:* Evaluate one integral by a geometric formula and the other by observing that the integrand is an odd function.)

29. A cylindrical gasoline tank 3 feet in diameter and 4 feet long is carried on the back of a truck and is used to fuel tractors. The axis of the tank is horizontal. The opening on the tractor tank is 5 feet above the top of the tank in the truck. Find the work done in pumping the entire contents of the fuel tank into a tractor.

30. The top of a cylindrical storage tank for gasoline at a service station is 4 feet below ground level. The axis of the tank is horizontal and its diameter and length are 5 feet and 12 feet, respectively. Find the work done in pumping the entire contents of the full tank to a height of 3 feet above ground level.

Lifting a Chain In Exercises 31–34, consider a 15-foot chain that weighs 3 pounds per foot hanging from a winch 15 feet above ground level. Find the work done by the winch in winding up the specified amount of chain.

31. Wind up the entire chain.

32. Wind up one-third of the chain.

33. Run the winch until the bottom of the chain is at the 10-foot level.

34. Wind up the entire chain with a 500-pound load attached to it.

Lifting a Chain In Exercises 35 and 36, consider a 15-foot hanging chain that weighs 3 pounds per foot. Find the work done in lifting the chain vertically to the indicated position.

35. Take the bottom of the chain and raise it to the 15-foot level, leaving the chain doubled and still hanging vertically (see figure).

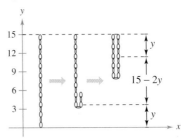

36. Repeat Exercise 35 raising the bottom of the chain to the 12-foot level.

Demolition Crane In Exercises 37 and 38, consider a demolition crane with a 500-pound ball suspended from a 40-foot cable that weighs 1 pound per foot.

37. Find the work required to wind up 15 feet of the apparatus.

38. Find the work required to wind up all 40 feet of the apparatus.

Boyle's Law In Exercises 39 and 40, find the work done by the gas for the given volume and pressure. Assume that the pressure is inversely proportional to the volume. (See Example 6.)

39. A quantity of gas with an initial volume of 2 cubic feet and a pressure of 1000 pounds per square foot expands to a volume of 3 cubic feet.

40. A quantity of gas with an initial volume of 1 cubic foot and a pressure of 2500 pounds per square foot expands to a volume of 3 cubic feet.

41. Electric Force Two electrons repel each other with a force that varies inversely as the square of the distance between them. One electron is fixed at the point $(2, 4)$. Find the work done in moving the second electron from $(-2, 4)$ to $(1, 4)$.

42. Modeling Data The hydraulic cylinder on a woodsplitter has a four-inch bore (diameter) and a stroke of 2 feet. The hydraulic pump creates a maximum pressure of 2000 pounds per square inch. Therefore, the maximum force created by the cylinder is $2000(\pi 2^2) = 8000\pi$ pounds.

(a) Find the work done through one extension of the cylinder given that the maximum force is required.

(b) The force exerted in splitting a piece of wood is variable. Measurements of the force obtained when a piece of wood was split are shown in the table. The variable x measures the extension of the cylinder in feet, and F is the force in pounds. Use Simpson's Rule to approximate the work done in splitting the piece of wood.

x	0	$\frac{1}{3}$	$\frac{2}{3}$	1	$\frac{4}{3}$	$\frac{5}{3}$	2
$F(x)$	0	20,000	22,000	15,000	10,000	5000	0

Table for 42(b)

(c) Use the regression capabilities of a graphing utility to find a fourth-degree polynomial model for the data. Plot the data and graph the model.

(d) Use the model in part (c) to approximate the extension of the cylinder when the force is maximum.

(e) Use the model in part (c) to approximate the work done in splitting the piece of wood.

Hydraulic Press In Exercises 43–46, use the integration capabilities of a graphing utility to approximate the work done by a press in a manufacturing process. A model for the variable force F (in pounds) and the distance x (in feet) the press moves is given.

Force	Interval
43. $F(x) = 1000[1.8 - \ln(x + 1)]$	$0 \le x \le 5$
44. $F(x) = \dfrac{e^{x^2} - 1}{100}$	$0 \le x \le 4$
45. $F(x) = 100x\sqrt{125 - x^3}$	$0 \le x \le 5$
46. $F(x) = 1000 \sinh x$	$0 \le x \le 2$

Section Project: Tidal Energy

Tidal power plants use "tidal energy" to produce electrical energy. To construct a tidal power plant, a dam is built to separate a basin from the sea. Electrical energy is produced as the water flows back and forth between the basin and the sea. The amount of "natural energy" produced depends on the volume of the basin and the tidal range—the vertical distance between high and low tides. (Several natural basin have tidal ranges in excess of 15 feet; the Bay of Fundy in Nova Scotia has a tidal range of 53 feet.)

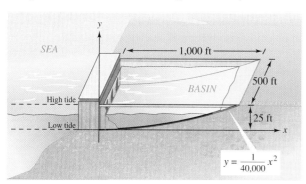

(a) Consider a basin with a rectangular base, as shown in the figure. The basin has a tidal range of 25 feet, with low tide corresponding to $y = 0$. How much water does the basin hold at high tide?

(b) The amount of energy produced during the filling (or the emptying) of the basin is proportional to the amount of work required to fill (or empty) the basin. How much work is required to fill the basin with seawater? (Use a seawater density of 64 pounds per cubic foot.)

The Bay of Fundy in Nova Scotia has an extreme tidal range, as displayed in the greatly contrasting photos above.

FOR FURTHER INFORMATION For more information on tidal power, see the article "LaRance: Six Years of Operating a Tidal Power Plant in France" by J. Cotillon in *Water Power Magazine*.

CHAPTER 7 Applications of Integration

Section 7.6 Moments, Centers of Mass, and Centroids

- Understand the definition of mass.
- Find the center of mass in a one-dimensional system.
- Find the center of mass in a two-dimensional system.
- Find the center of mass of a planar lamina.
- Use the Theorem of Pappus to find the volume of a solid of revolution.

Mass

In this section you will study several important applications of integration that are related to **mass.** Mass is a measure of a body's resistance to changes in motion, and is independent of the particular gravitational system in which the body is located. However, because so many applications involving mass occur on Earth's surface, an object's mass is sometimes equated with its weight. This is not technically correct. Weight is a type of force and as such is dependent on gravity. Force and mass are related by the equation

$$\text{Force} = (\text{mass})(\text{acceleration}).$$

The table below lists some commonly used measures of mass and force, together with their conversion factors.

System of Measurement	Measure of Mass	Measure of Force
U.S.	Slug	Pound = (slug)(ft/sec^2)
International	Kilogram	Newton = (kilogram)(m/sec^2)
C-G-S	Gram	Dyne = (gram)(cm/sec^2)
Conversions: 1 pound = 4.448 newtons 1 newton = 0.2248 pound 1 dyne = 0.000002248 pound 1 dyne = 0.00001 newton		1 slug = 14.59 kilograms 1 kilogram = 0.06852 slug 1 gram = 0.00006852 slug 1 foot = 0.3048 meter

EXAMPLE 1 Mass on the Surface of Earth

Find the mass (in slugs) of an object whose weight at sea level is 1 pound.

Solution Using 32 feet per second per second as the acceleration due to gravity produces

$$\text{Mass} = \frac{\text{force}}{\text{acceleration}} \qquad \text{Force} = (\text{mass})(\text{acceleration})$$

$$= \frac{1 \text{ pound}}{32 \text{ feet per second per second}}$$

$$= 0.03125 \frac{\text{pound}}{\text{foot per second per second}}$$

$$= 0.03125 \text{ slug}.$$

Because many applications involving mass occur on Earth's surface, this amount of mass is called a **pound mass.**

Center of Mass in a One-Dimensional System

You will now consider two types of moments of a mass—the **moment about a point** and the **moment about a line.** To define these two moments, consider an idealized situation in which a mass m is concentrated at a point. If x is the distance between this point mass and another point P, the **moment of m about the point P** is

$$\text{Moment} = mx$$

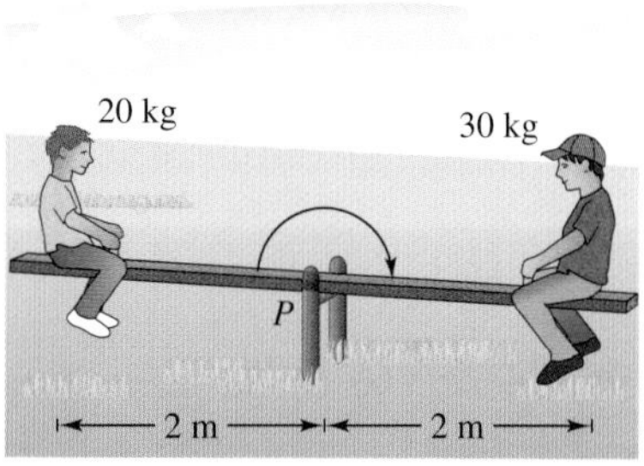

and x is the **length of the moment arm.**

The concept of moment can be demonstrated simply by a seesaw, as shown in Figure 7.55. A child of mass 20 kilograms sits 2 meters to the left of fulcrum P, and an older child of mass 30 kilograms sits 2 meters to the right of P. From experience, you know that the seesaw will begin to rotate clockwise, moving the larger child down. This rotation occurs because the moment produced by the child on the left is less than the moment produced by the child on the right.

Left moment $= (20)(2) = 40$ kilogram-meters
Right moment $= (30)(2) = 60$ kilogram-meters

To balance the seesaw, the two moments must be equal. For example, if the larger child moved to a position $\frac{4}{3}$ meters from the fulcrum, the seesaw would balance, because each child would produce a moment of 40 kilogram-meters.

To generalize this, you can introduce a coordinate line on which the origin corresponds to the fulcrum, as shown in Figure 7.56. Suppose several point masses are located on the x-axis. The measure of the tendency of this system to rotate about the origin is the **moment about the origin,** and it is defined as the sum of the n products $m_i x_i$.

$$M_0 = m_1 x_1 + m_2 x_2 + \cdots + m_n x_n$$

If $m_1 x_1 + m_2 x_2 + \cdots + m_n x_n = 0$, the system is in equilibrium.
Figure 7.56

The seesaw will balance when the left and the right moments are equal.
Figure 7.55

If M_0 is 0, the system is said to be in **equilibrium.**

For a system that is not in equilibrium, the **center of mass** is defined as the point $\bar{x}$ at which the fulcrum could be relocated to attain equilibrium. If the system were translated $\bar{x}$ units, each coordinate x_i would become $(x_i - \bar{x})$, and because the moment of the translated system is 0, you have

$$\sum_{i=1}^{n} m_i(x_i - \bar{x}) = \sum_{i=1}^{n} m_i x_i - \sum_{i=1}^{n} m_i \bar{x} = 0.$$

Solving for $\bar{x}$ produces

$$\bar{x} = \frac{\sum_{i=1}^{n} m_i x_i}{\sum_{i=1}^{n} m_i} = \frac{\text{moment of system about origin}}{\text{total mass of system}}.$$

If $m_1 x_1 + m_2 x_2 + \cdots + m_n x_n = 0$, the system is in equilibrium.

Moments and Center of Mass: One-Dimensional System

Let the point masses $m_1, m_2, \ldots, m_n$ be located at $x_1, x_2, \ldots, x_n$.

1. The **moment about the origin** is $M_0 = m_1 x_1 + m_2 x_2 + \cdots + m_n x_n$.
2. The **center of mass** is $\bar{x} = \dfrac{M_0}{m}$, where $m = m_1 + m_2 + \cdots + m_n$ is the **total mass** of the system.

EXAMPLE 2 The Center of Mass of a Linear System

Find the center of mass of the linear system shown in Figure 7.57.

Figure 7.57

Solution The moment about the origin is

$$M_0 = m_1 x_1 + m_2 x_2 + m_3 x_3 + m_4 x_4$$
$$= 10(-5) + 15(0) + 5(4) + 10(7)$$
$$= -50 + 0 + 20 + 70$$
$$= 40.$$

Because the total mass of the system is $m = 10 + 15 + 5 + 10 = 40$, the center of mass is

$$\bar{x} = \frac{M_0}{m} = \frac{40}{40} = 1.$$

NOTE In Example 2, where should you locate the fulcrum so that the point masses will be in equilibrium?

Rather than define the moment of a mass, you could define the moment of a *force*. In this context, the center of mass is called the **center of gravity**. Suppose that a system of point masses $m_1, m_2, \ldots, m_n$ is located at $x_1, x_2, \ldots, x_n$. Then, because force = (mass)(acceleration), the total force of the system is

$$F = m_1 a + m_2 a + \cdots + m_n a$$
$$= ma.$$

The **torque** (moment) about the origin is

$$T_0 = (m_1 a)x_1 + (m_2 a)x_2 + \cdots + (m_n a)x_n$$
$$= M_0 a$$

and the **center of gravity** is

$$\frac{T_0}{F} = \frac{M_0 a}{ma} = \frac{M_0}{m} = \bar{x}.$$

So, the center of gravity and the center of mass have the same location.

Center of Mass in a Two-Dimensional System

You can extend the concept of moment to two dimensions by considering a system of masses located in the xy-plane at the points $(x_1, y_1), (x_2, y_2), \ldots, (x_n, y_n)$, as shown in Figure 7.58. Rather than defining a single moment (with respect to the origin), two moments are defined—one with respect to the x-axis and one with respect to the y-axis.

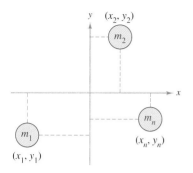

In a two-dimensional system, there is a moment about the y-axis, M_y, and a moment about the x-axis, M_x.
Figure 7.58

Moments and Center of Mass: Two-Dimensional System

Let the point masses $m_1, m_2, \ldots, m_n$ be located at $(x_1, y_1), (x_2, y_2), \ldots, (x_n, y_n)$.

1. The **moment about the y-axis** is $M_y = m_1 x_1 + m_2 x_2 + \cdots + m_n x_n$.
2. The **moment about the x-axis** is $M_x = m_1 y_1 + m_2 y_2 + \cdots + m_n y_n$.
3. The **center of mass** $(\bar{x}, \bar{y})$ (or **center of gravity**) is

$$\bar{x} = \frac{M_y}{m} \quad \text{and} \quad \bar{y} = \frac{M_x}{m}$$

where $m = m_1 + m_2 + \cdots + m_n$ is the **total mass** of the system.

The moment of a system of masses in the plane can be taken about any horizontal or vertical line. In general, the moment about a line is the sum of the product of the masses and the *directed distances* from the points to the line.

$$\text{Moment} = m_1(y_1 - b) + m_2(y_2 - b) + \cdots + m_n(y_n - b) \quad \text{Horizontal line } y = b$$

$$\text{Moment} = m_1(x_1 - a) + m_2(x_2 - a) + \cdots + m_n(x_n - a) \quad \text{Vertical line } x = a$$

EXAMPLE 3 The Center of Mass of a Two-Dimensional System

Find the center of mass of a system of point masses $m_1 = 6$, $m_2 = 3$, $m_3 = 2$, and $m_4 = 9$, located at

$$(3, -2), (0, 0), (-5, 3), \text{ and } (4, 2)$$

as shown in Figure 7.59.

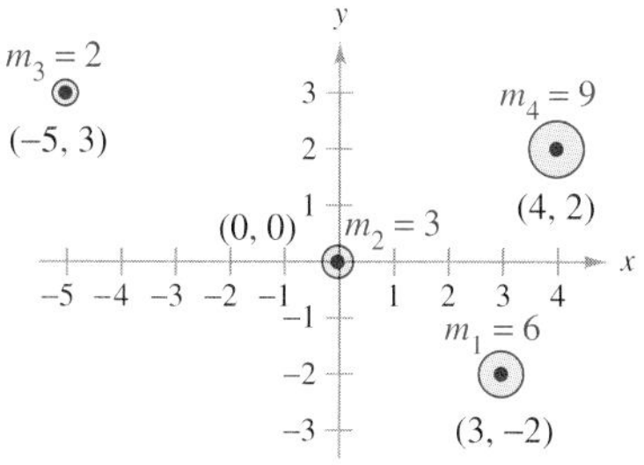

Figure 7.59

Solution

$$\begin{aligned}
m &= 6 \;\; + 3 \;\; + 2 \;\; + 9 \;\; = 20 & \text{Mass} \\
M_y &= 6(3) + 3(0) + 2(-5) + 9(4) = 44 & \text{Moment about } y\text{-axis} \\
M_x &= 6(-2) + 3(0) + 2(3) + 9(2) = 12 & \text{Moment about } x\text{-axis}
\end{aligned}$$

So,

$$\bar{x} = \frac{M_y}{m} = \frac{44}{20} = \frac{11}{5}$$

and

$$\bar{y} = \frac{M_x}{m} = \frac{12}{20} = \frac{3}{5}$$

and so the center of mass is $\left(\frac{11}{5}, \frac{3}{5}\right)$.

Center of Mass of a Planar Lamina

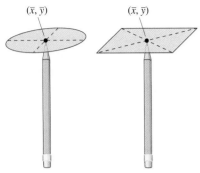

You can think of the center of mass $(\bar{x}, \bar{y})$ of a lamina as its balancing point. For a circular lamina, the center of mass is the center of the circle. For a rectangular lamina, the center of mass is the center of the rectangle.
Figure 7.60

So far in this section you have assumed the total mass of a system to be distributed at discrete points in a plane or on a line. Now consider a thin, flat plate of material of constant density called a **planar lamina** (see Figure 7.60). **Density** is a measure of mass per unit of volume, such as grams per cubic centimeter. For planar laminas, however, density is considered to be a measure of mass per unit of area. Density is denoted by ρ, the lowercase Greek letter rho.

Consider an irregularly shaped planar lamina of uniform density ρ, bounded by the graphs of $y = f(x)$, $y = g(x)$, and $a \leq x \leq b$, as shown in Figure 7.61. The mass of this region is given by

$$m = (\text{density})(\text{area})$$
$$= \rho \int_a^b [f(x) - g(x)] \, dx$$
$$= \rho A$$

where A is the area of the region. To find the center of mass of this lamina, partition the interval $[a, b]$ into n subintervals of equal width Δx. Let x_i be the center of the ith subinterval. You can approximate the portion of the lamina lying in the ith subinterval by a rectangle whose height is $h = f(x_i) - g(x_i)$. Because the density of the rectangle is ρ, its mass is

$$m_i = (\text{density})(\text{area})$$
$$= \underbrace{\rho}_{\text{Density}} \underbrace{[f(x_i) - g(x_i)]}_{\text{Height}} \underbrace{\Delta x}_{\text{Width}}.$$

Now, considering this mass to be located at the center (x_i, y_i) of the rectangle, the directed distance from the x-axis to (x_i, y_i) is $y_i = [f(x_i) + g(x_i)]/2$. So, the moment of m_i about the x-axis is

$$\text{Moment} = (\text{mass})(\text{distance})$$
$$= m_i y_i$$
$$= \rho [f(x_i) - g(x_i)] \Delta x \left[\frac{f(x_i) + g(x_i)}{2} \right].$$

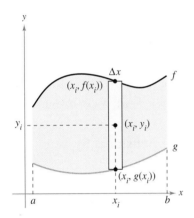

Planar lamina of uniform density ρ
Figure 7.61

Summing the moments and taking the limit as $n \to \infty$ suggest the definitions below.

Moments and Center of Mass of a Planar Lamina

Let f and g be continuous functions such that $f(x) \geq g(x)$ on $[a, b]$, and consider the planar lamina of uniform density ρ bounded by the graphs of $y = f(x)$, $y = g(x)$, and $a \leq x \leq b$.

1. The **moments about the x- and y-axes** are

$$M_x = \rho \int_a^b \left[\frac{f(x) + g(x)}{2} \right] [f(x) - g(x)] \, dx$$

$$M_y = \rho \int_a^b x[f(x) - g(x)] \, dx.$$

2. The **center of mass** $(\bar{x}, \bar{y})$ is given by $\bar{x} = \dfrac{M_y}{m}$ and $\bar{y} = \dfrac{M_x}{m}$, where $m = \rho \int_a^b [f(x) - g(x)] \, dx$ is the mass of the lamina.

EXAMPLE 4 The Center of Mass of a Planar Lamina

Find the center of mass of the lamina of uniform density ρ bounded by the graph of $f(x) = 4 - x^2$ and the x-axis.

Solution Because the center of mass lies on the axis of symmetry, you know that $\bar{x} = 0$. Moreover, the mass of the lamina is

$$m = \rho \int_{-2}^{2} (4 - x^2)\, dx$$

$$= \rho \left[4x - \frac{x^3}{3} \right]_{-2}^{2}$$

$$= \frac{32\rho}{3}.$$

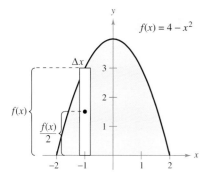

Figure 7.62

To find the moment about the x-axis, place a representative rectangle in the region, as shown in Figure 7.62. The distance from the x-axis to the center of this rectangle is

$$y_i = \frac{f(x)}{2} = \frac{4 - x^2}{2}.$$

Because the mass of the representative rectangle is

$$\rho f(x)\, \Delta x = \rho(4 - x^2)\, \Delta x$$

you have

$$M_x = \rho \int_{-2}^{2} \frac{4 - x^2}{2} (4 - x^2)\, dx$$

$$= \frac{\rho}{2} \int_{-2}^{2} (16 - 8x^2 + x^4)\, dx$$

$$= \frac{\rho}{2} \left[16x - \frac{8x^3}{3} + \frac{x^5}{5} \right]_{-2}^{2}$$

$$= \frac{256\rho}{15}$$

and $\bar{y}$ is given by

$$\bar{y} = \frac{M_x}{m} = \frac{256\rho/15}{32\rho/3} = \frac{8}{5}.$$

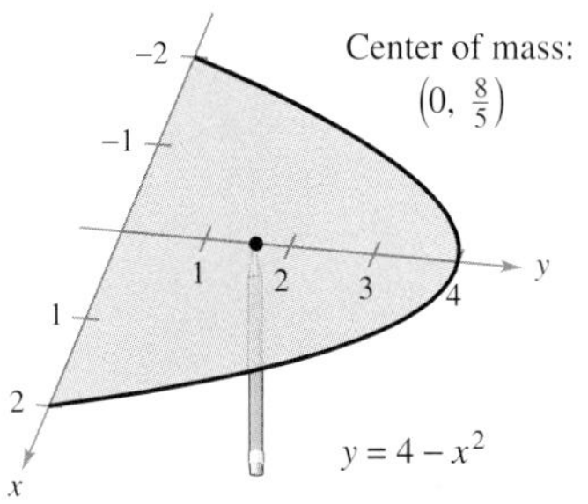

So, the center of mass (the balancing point) of the lamina is $\left(0, \frac{8}{5}\right)$, as shown in Figure 7.63.

The center of mass is the balancing point.
Figure 7.63

The density ρ in Example 4 is a common factor of both the moments and the mass, and as such divides out of the quotients representing the coordinates of the center of mass. So, the center of mass of a lamina of *uniform* density depends only on the shape of the lamina and not on its density. For this reason, the point

$(\bar{x}, \bar{y})$ Center of mass or centroid

is sometimes called the center of mass of a *region* in the plane, or the **centroid** of the region. In other words, to find the centroid of a region in the plane, you simply assume that the region has a constant density of $\rho = 1$ and compute the corresponding center of mass.

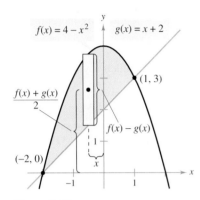

Figure 7.64

EXAMPLE 5 The Centroid of a Plane Region

Find the centroid of the region bounded by the graphs of $f(x) = 4 - x^2$ and $g(x) = x + 2$.

Solution The two graphs intersect at the points $(-2, 0)$ and $(1, 3)$, as shown in Figure 7.64. So, the area of the region is

$$A = \int_{-2}^{1} [f(x) - g(x)] \, dx = \int_{-2}^{1} (2 - x - x^2) \, dx = \frac{9}{2}.$$

The centroid $(\bar{x}, \bar{y})$ of the region has the following coordinates.

$$\bar{x} = \frac{1}{A} \int_{-2}^{1} x[(4 - x^2) - (x + 2)] \, dx = \frac{2}{9} \int_{-2}^{1} (-x^3 - x^2 + 2x) \, dx$$

$$= \frac{2}{9} \left[-\frac{x^4}{4} - \frac{x^3}{3} + x^2 \right]_{-2}^{1} = -\frac{1}{2}$$

$$\bar{y} = \frac{1}{A} \int_{-2}^{1} \left[\frac{(4 - x^2) + (x + 2)}{2} \right] [(4 - x^2) - (x + 2)] \, dx$$

$$= \frac{2}{9} \left(\frac{1}{2} \right) \int_{-2}^{1} (-x^2 + x + 6)(-x^2 - x + 2) \, dx$$

$$= \frac{1}{9} \int_{-2}^{1} (x^4 - 9x^2 - 4x + 12) \, dx$$

$$= \frac{1}{9} \left[\frac{x^5}{5} - 3x^3 - 2x^2 + 12x \right]_{-2}^{1} = \frac{12}{5}.$$

So, the centroid of the region is $(\bar{x}, \bar{y}) = \left(-\frac{1}{2}, \frac{12}{5} \right)$.

For simple plane regions, you may be able to find the centroids without resorting to integration.

EXPLORATION

Cut an irregular shape from a piece of cardboard.

a. Hold a pencil vertically and move the object on the pencil point until the centroid is located.

b. Divide the object into representative elements. Make the necessary measurements and numerically approximate the centroid. Compare your result with the result in part (a).

EXAMPLE 6 The Centroid of a Simple Plane Region

Find the centroid of the region shown in Figure 7.65(a).

Solution By superimposing a coordinate system on the region, as shown in Figure 7.65(b), you can locate the centroids of the three rectangles at

$$\left(\frac{1}{2}, \frac{3}{2} \right), \quad \left(\frac{5}{2}, \frac{1}{2} \right), \quad \text{and} \quad (5, 1).$$

Using these three points, you can find the centroid of the region.

$$A = \text{area of region} = 3 + 3 + 4 = 10$$

$$\bar{x} = \frac{(1/2)(3) + (5/2)(3) + (5)(4)}{10} = \frac{29}{10} = 2.9$$

$$\bar{y} = \frac{(3/2)(3) + (1/2)(3) + (1)(4)}{10} = \frac{10}{10} = 1$$

So, the centroid of the region is $(2.9, 1)$.

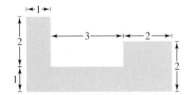

(a) Original region

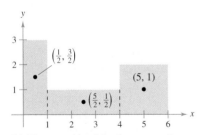

(b) The centroids of the three rectangles

Figure 7.65

NOTE In Example 6, notice that $(2.9, 1)$ is not the "average" of $\left(\frac{1}{2}, \frac{3}{2} \right)$, $\left(\frac{5}{2}, \frac{1}{2} \right)$, and $(5, 1)$.

SECTION 7.6   Moments, Centers of Mass, and Centroids   503

Theorem of Pappus

The final topic in this section is a useful theorem credited to Pappus of Alexandria (ca. 300 A.D.), a Greek mathematician whose eight-volume *Mathematical Collection* is a record of much of classical Greek mathematics. The proof of this theorem is given in Section 14.4.

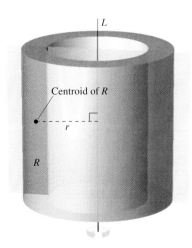

The volume V is $2\pi rA$, where A is the area of region R.
Figure 7.66

> **THEOREM 7.1   The Theorem of Pappus**
>
> Let R be a region in a plane and let L be a line in the same plane such that L does not intersect the interior of R, as shown in Figure 7.66. If r is the distance between the centroid of R and the line, then the volume V of the solid of revolution formed by revolving R about the line is
>
> $$V = 2\pi rA$$
>
> where A is the area of R. (Note that $2\pi r$ is the distance traveled by the centroid as the region is revolved about the line.)

The Theorem of Pappus can be used to find the volume of a torus, as shown in the following example. Recall that a torus is a doughnut-shaped solid formed by revolving a circular region about a line that lies in the same plane as the circle (but does not intersect the circle).

EXAMPLE 7   Finding Volume by the Theorem of Pappus

Find the volume of the torus shown in Figure 7.67(a), which was formed by revolving the circular region bounded by

$$(x - 2)^2 + y^2 = 1$$

about the y-axis, as shown in Figure 7.67(b).

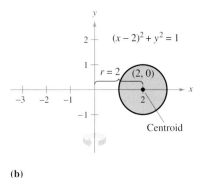

(a)   (b)
Figure 7.67

EXPLORATION

Use the shell method to show that the volume of the torus is given by

$$V = \int_1^3 4\pi x \sqrt{1 - (x - 2)^2}\, dx.$$

Evaluate this integral using a graphing utility. Does your answer agree with the one in Example 7?

Solution   In Figure 7.67(b), you can see that the centroid of the circular region is $(2, 0)$. So, the distance between the centroid and the axis of revolution is $r = 2$. Because the area of the circular region is $A = \pi$, the volume of the torus is

$$\begin{aligned} V &= 2\pi rA \\ &= 2\pi(2)(\pi) \\ &= 4\pi^2 \\ &\approx 39.5. \end{aligned}$$

Exercises for Section 7.6

In Exercises 1–4, find the center of mass of the point masses lying on the x-axis.

1. $m_1 = 6, m_2 = 3, m_3 = 5$
   $x_1 = -5, x_2 = 1, x_3 = 3$
2. $m_1 = 7, m_2 = 4, m_3 = 3, m_4 = 8$
   $x_1 = -3, x_2 = -2, x_3 = 5, x_4 = 6$
3. $m_1 = 1, m_2 = 1, m_3 = 1, m_4 = 1, m_5 = 1$
   $x_1 = 7, x_2 = 8, x_3 = 12, x_4 = 15, x_5 = 18$
4. $m_1 = 12, m_2 = 1, m_3 = 6, m_4 = 3, m_5 = 11$
   $x_1 = -6, x_2 = -4, x_3 = -2, x_4 = 0, x_5 = 8$

5. **Graphical Reasoning**
   (a) Translate each point mass in Exercise 3 to the right five units and determine the resulting center of mass.
   (b) Translate each point mass in Exercise 4 to the left three units and determine the resulting center of mass.

6. **Conjecture** Use the result of Exercise 5 to make a conjecture about the change in the center of mass that results when each point mass is translated k units horizontally.

Statics Problems In Exercises 7 and 8, consider a beam of length L with a fulcrum x feet from one end (see figure). There are objects with weights W_1 and W_2 placed on opposite ends of the beam. Find x such that the system is in equilibrium.

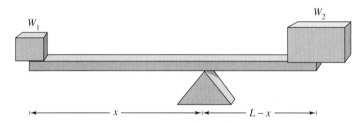

7. Two children weighing 50 pounds and 75 pounds are going to play on a seesaw that is 10 feet long.
8. In order to move a 550-pound rock, a person weighing 200 pounds wants to balance it on a beam that is 5 feet long.

In Exercise 9–12, find the center of mass of the given system of point masses.

9.
m_i	5	1	3
(x_1, y_1)	(2, 2)	(−3, 1)	(1, −4)

10.
m_i	10	2	5
(x_1, y_1)	(1, −1)	(5, 5)	(−4, 0)

11.
m_i	3	4
(x_1, y_1)	(−2, −3)	(5, 5)

m_i	2	1	6
(x_1, y_1)	(7, 1)	(0, 0)	(−3, 0)

12.
m_i	12	6	$\frac{15}{2}$	15
(x_1, y_1)	(2, 3)	(−1, 5)	(6, 8)	(2, −2)

In Exercises 13–24, find M_x, M_y, and $(\bar{x}, \bar{y})$ for the laminas of uniform density ρ bounded by the graphs of the equations.

13. $y = \sqrt{x}, y = 0, x = 4$
14. $y = \frac{1}{2}x^2, y = 0, x = 2$
15. $y = x^2, y = x^3$
16. $y = \sqrt{x}, y = x$
17. $y = -x^2 + 4x + 2, y = x + 2$
18. $y = \sqrt{x} + 1, y = \frac{1}{3}x + 1$
19. $y = x^{2/3}, y = 0, x = 8$
20. $y = x^{2/3}, y = 4$
21. $x = 4 - y^2, x = 0$
22. $x = 2y - y^2, x = 0$
23. $x = -y, x = 2y - y^2$
24. $x = y + 2, x = y^2$

In Exercises 25–28, set up and evaluate the integrals for finding the area and moments about the x- and y-axes for the region bounded by the graphs of the equations. (Assume $\rho = 1$.)

25. $y = x^2, y = x$
26. $y = \dfrac{1}{x}, y = 0, 1 \le x \le 4$
27. $y = 2x + 4, y = 0, 0 \le x \le 3$
28. $y = x^2 - 4, y = 0$

In Exercises 29–32, use a graphing utility to graph the region bounded by the graphs of the equations. Use the integration capabilities of the graphing utility to approximate the centroid of the region.

29. $y = 10x\sqrt{125 - x^3}, y = 0$
30. $y = xe^{-x/2}, y = 0, x = 0, x = 4$
31. *Prefabricated End Section of a Building*
    $y = 5\sqrt[3]{400 - x^2}, y = 0$
32. *Witch of Agnesi*
    $y = 8/(x^2 + 4), y = 0, x = -2, x = 2$

In Exercises 33–38, find and/or verify the centroid of the common region used in engineering.

33. Triangle Show that the centroid of the triangle with vertices $(-a, 0)$, $(a, 0)$, and (b, c) is the point of intersection of the medians (see figure).

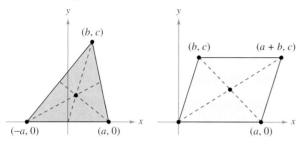

Figure for 33   Figure for 34

34. Parallelogram Show that the centroid of the parallelogram with vertices $(0, 0)$, $(a, 0)$, (b, c), and $(a + b, c)$ is the point of intersection of the diagonals (see figure).

35. Trapezoid Find the centroid of the trapezoid with vertices $(0, 0)$, $(0, a)$, (c, b), and $(c, 0)$. Show that it is the intersection of the line connecting the midpoints of the parallel sides and the line connecting the extended parallel sides, as shown in the figure.

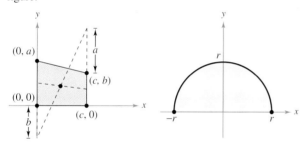

Figure for 35   Figure for 36

36. Semicircle Find the centroid of the region bounded by the graphs of $y = \sqrt{r^2 - x^2}$ and $y = 0$ (see figure).

37. Semiellipse Find the centroid of the region bounded by the graphs of $y = \frac{b}{a}\sqrt{a^2 - x^2}$ and $y = 0$ (see figure).

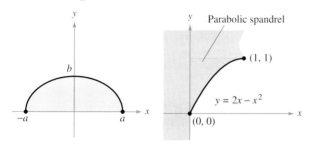

Figure for 37   Figure for 38

38. Parabolic Spandrel Find the centroid of the **parabolic spandrel** shown in the figure.

39. Graphical Reasoning Consider the region bounded by the graphs of $y = x^2$ and $y = b$, where $b > 0$.

(a) Sketch a graph of the region.

(b) Use the graph in part (a) to determine $\bar{x}$. Explain.

(c) Set up the integral for finding M_y. Because of the form of the integrand, the value of the integral can be obtained without integrating. What is the form of the integrand and what is the value of the integral? Compare with the result in part (b).

(d) Use the graph in part (a) to determine whether $\bar{y} > \frac{b}{2}$ or $\bar{y} < \frac{b}{2}$. Explain.

(e) Use integration to verify your answer in part (d).

40. Graphical and Numerical Reasoning Consider the region bounded by the graphs of $y = x^{2n}$ and $y = b$, where $b > 0$ and n is a positive integer.

(a) Set up the integral for finding M_y. Because of the form of the integrand, the value of the integral can be obtained without integrating. What is the form of the integrand and what is the value of the integral? Compare with the result in part (b).

(b) Is $\bar{y} > \frac{b}{2}$ or $\bar{y} < \frac{b}{2}$? Explain.

(c) Use integration to find $\bar{y}$ as a function of n.

(d) Use the result of part (c) to complete the table.

n	1	2	3	4
$\bar{y}$				

(e) Find $\lim\limits_{n \to \infty} \bar{y}$.

(f) Give a geometric explanation of the result in part (e).

41. Modeling Data The manufacturer of glass for a window in a conversion van needs to approximate its center of mass. A coordinate system is superimposed on a prototype of the glass (see figure). The measurements (in centimeters) for the right half of the symmetric piece of glass are shown in the table.

x	0	10	20	30	40
y	30	29	26	20	0

(a) Use Simpson's Rule to approximate the center of mass of the glass.

(b) Use the regression capabilities of a graphing utility to find a fourth-degree polynomial model for the data.

(c) Use the integration capabilities of a graphing utility and the model to approximate the center of mass of the glass. Compare with the result in part (a).

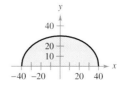

42. Modeling Data The manufacturer of a boat needs to approximate the center of mass of a section of the hull. A coordinate system is superimposed on a prototype (see figure). The measurements (in feet) for the right half of the symmetric prototype are listed in the table.

x	0	0.5	1.0	1.5	2
l	1.50	1.45	1.30	0.99	0
d	0.50	0.48	0.43	0.33	0

(a) Use Simpson's Rule to approximate the center of mass of the hull section.

(b) Use the regression capabilities of a graphing utility to find fourth-degree polynomial models for both curves shown in the figure. Plot the data and graph the models.

(c) Use the integration capabilities of a graphing utility and the model to approximate the center of mass of the hull section. Compare with the result in part (a).

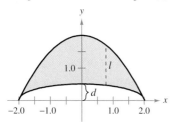

In Exercises 43–46, introduce an appropriate coordinate system and find the coordinates of the center of mass of the planar lamina. (The answer depends on the position of the coordinate system.)

43.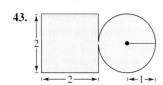

44.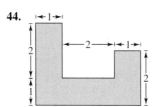

45.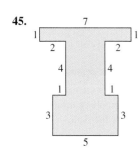

46.

47. Find the center of mass of the lamina in Exercise 43 if the circular portion of the lamina has twice the density of the square portion of the lamina.

48. Find the center of mass of the lamina in Exercise 43 if the square portion of the lamina has twice the density of the circular portion of the lamina.

In Exercises 49–52, use the Theorem of Pappus to find the volume of the solid of revolution.

49. The torus formed by revolving the circle $(x-5)^2 + y^2 = 16$ about the y-axis

50. The torus formed by revolving the circle $x^2 + (y-3)^2 = 4$ about the x-axis

51. The solid formed by revolving the region bounded by the graphs of $y = x$, $y = 4$, and $x = 0$ about the x-axis

52. The solid formed by revolving the region bounded by the graphs of $y = 2\sqrt{x-2}$, $y = 0$, and $x = 6$ about the y-axis

Writing About Concepts

53. Let the point masses $m_1, m_2, \ldots, m_n$ be located at (x_1, y_1), $(x_2, y_2), \ldots, (x_n, y_n)$. Define the center of mass $(\bar{x}, \bar{y})$.

54. What is a planar lamina? Describe what is meant by the center of mass $(\bar{x}, \bar{y})$ of a planar lamina.

55. The centroid of the plane region bounded by the graphs of $y = f(x)$, $y = 0$, $x = 0$, and $x = 1$ is $\left(\frac{5}{6}, \frac{5}{18}\right)$. Is it possible to find the centroid of each of the regions bounded by the graphs of the following sets of equations? If so, identify the centroid and explain your answer.

(a) $y = f(x) + 2$, $y = 2$, $x = 0$, and $x = 1$

(b) $y = f(x-2)$, $y = 0$, $x = 2$, and $x = 3$

(c) $y = -f(x)$, $y = 0$, $x = 0$, and $x = 1$

(d) $y = f(x)$, $y = 0$, $x = -1$, and $x = 1$

56. State the Theorem of Pappus.

In Exercises 57 and 58, use the *Second Theorem of Pappus*, which is stated as follows. If a segment of a plane curve C is revolved about an axis that does not intersect the curve (except possibly at its endpoints), the area S of the resulting surface of revolution is given by the product of the length of C times the distance d traveled by the centroid of C.

57. A sphere is formed by revolving the graph of $y = \sqrt{r^2 - x^2}$ about the x-axis. Use the formula for surface area, $S = 4\pi r^2$, to find the centroid of the semicircle $y = \sqrt{r^2 - x^2}$.

58. A torus is formed by revolving the graph of $(x-1)^2 + y^2 = 1$ about the y-axis. Find the surface area of the torus.

59. Let $n \geq 1$ be constant, and consider the region bounded by $f(x) = x^n$, the x-axis, and $x = 1$. Find the centroid of this region. As $n \to \infty$, what does the region look like, and where is its centroid?

Putnam Exam Challenge

60. Let V be the region in the cartesian plane consisting of all points (x, y) satisfying the simultaneous conditions

$|x| \leq y \leq |x| + 3 \quad \text{and} \quad y \leq 4.$

Find the centroid $(\bar{x}, \bar{y})$ of V.

This problem was composed by the Committee on the Putnam Prize Competition.
© The Mathematical Association of America. All rights reserved.

Section 7.7 Fluid Pressure and Fluid Force

- Find fluid pressure and fluid force.

Fluid Pressure and Fluid Force

Swimmers know that the deeper an object is submerged in a fluid, the greater the pressure on the object. **Pressure** is defined as the force per unit of area over the surface of a body. For example, because a column of water that is 10 feet in height and 1 inch square weighs 4.3 pounds, the *fluid pressure* at a depth of 10 feet of water is 4.3 pounds per square inch.* At 20 feet, this would increase to 8.6 pounds per square inch, and in general the pressure is proportional to the depth of the object in the fluid.

> **Definition of Fluid Pressure**
>
> The **pressure** on an object at depth h in a liquid is
>
> Pressure $= P = wh$
>
> where w is the weight-density of the liquid per unit of volume.

Below are some common weight-densities of fluids in pounds per cubic foot.

Fluid	Weight-density
Ethyl alcohol	49.4
Gasoline	41.0–43.0
Glycerin	78.6
Kerosene	51.2
Mercury	849.0
Seawater	64.0
Water	62.4

When calculating fluid pressure, you can use an important (and rather surprising) physical law called **Pascal's Principle,** named after the French mathematician Blaise Pascal. Pascal's Principle states that the pressure exerted by a fluid at a depth h is transmitted equally *in all directions*. For example, in Figure 7.68, the pressure at the indicated depth is the same for all three objects. Because fluid pressure is given in terms of force per unit area ($P = F/A$), the fluid force on a *submerged horizontal* surface of area A is

Fluid force $= F = PA =$ (pressure)(area).

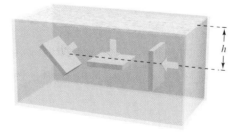

The pressure at h is the same for all three objects.
Figure 7.68

BLAISE PASCAL (1623–1662)

Pascal is well known for his work in many areas of mathematics and physics, and also for his influence on Leibniz. Although much of Pascal's work in calculus was intuitive and lacked the rigor of modern mathematics, he nevertheless anticipated many important results.

---

* *The total pressure on an object in 10 feet of water would also include the pressure due to Earth's atmosphere. At sea level, atmospheric pressure is approximately 14.7 pounds per square inch.*

EXAMPLE 1 Fluid Force on a Submerged Sheet

Find the fluid force on a rectangular metal sheet measuring 3 feet by 4 feet that is submerged in 6 feet of water, as shown in Figure 7.69.

Solution Because the weight-density of water is 62.4 pounds per cubic foot and the sheet is submerged in 6 feet of water, the fluid pressure is

$$P = (62.4)(6) \qquad P = wh$$
$$= 374.4 \text{ pounds per square foot.}$$

Because the total area of the sheet is $A = (3)(4) = 12$ square feet, the fluid force is

$$F = PA = \left(374.4 \ \frac{\text{pounds}}{\text{square foot}}\right)(12 \text{ square feet})$$
$$= 4492.8 \text{ pounds.}$$

This result is independent of the size of the body of water. The fluid force would be the same in a swimming pool or lake.

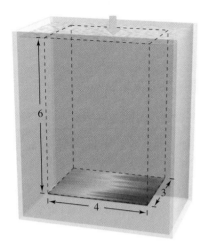

The fluid force on a horizontal metal sheet is equal to the fluid pressure times the area.
Figure 7.69

In Example 1, the fact that the sheet is rectangular and horizontal means that you do not need the methods of calculus to solve the problem. Consider a surface that is submerged vertically in a fluid. This problem is more difficult because the pressure is not constant over the surface.

Suppose a vertical plate is submerged in a fluid of weight-density w (per unit of volume), as shown in Figure 7.70. To determine the total force against *one side* of the region from depth c to depth d, you can subdivide the interval $[c, d]$ into n subintervals, each of width Δy. Next, consider the representative rectangle of width Δy and length $L(y_i)$, where y_i is in the ith subinterval. The force against this representative rectangle is

$$\Delta F_i = w(\text{depth})(\text{area})$$
$$= wh(y_i)L(y_i)\,\Delta y.$$

The force against n such rectangles is

$$\sum_{i=1}^{n} \Delta F_i = w \sum_{i=1}^{n} h(y_i)L(y_i)\,\Delta y.$$

Note that w is considered to be constant and is factored out of the summation. Therefore, taking the limit as $\|\Delta\| \to 0$ $(n \to \infty)$ suggests the following definition.

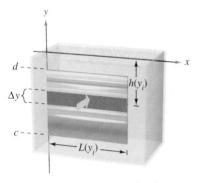

Calculus methods must be used to find the fluid force on a vertical metal plate.
Figure 7.70

Definition of Force Exerted by a Fluid

The **force F exerted by a fluid** of constant weight-density w (per unit of volume) against a submerged vertical plane region from $y = c$ to $y = d$ is

$$F = w \lim_{\|\Delta\| \to 0} \sum_{i=1}^{n} h(y_i)L(y_i)\,\Delta y$$
$$= w \int_{c}^{d} h(y)L(y)\,dy$$

where $h(y)$ is the depth of the fluid at y and $L(y)$ is the horizontal length of the region at y.

EXAMPLE 2 Fluid Force on a Vertical Surface

A vertical gate in a dam has the shape of an isosceles trapezoid 8 feet across the top and 6 feet across the bottom, with a height of 5 feet, as shown in Figure 7.71(a). What is the fluid force on the gate when the top of the gate is 4 feet below the surface of the water?

Solution In setting up a mathematical model for this problem, you are at liberty to locate the x- and y-axes in several different ways. A convenient approach is to let the y-axis bisect the gate and place the x-axis at the surface of the water, as shown in Figure 7.71(b). So, the depth of the water at y in feet is

$$\text{Depth} = h(y) = -y.$$

To find the length $L(y)$ of the region at y, find the equation of the line forming the right side of the gate. Because this line passes through the points $(3, -9)$ and $(4, -4)$, its equation is

$$y - (-9) = \frac{-4 - (-9)}{4 - 3}(x - 3)$$

$$y + 9 = 5(x - 3)$$

$$y = 5x - 24$$

$$x = \frac{y + 24}{5}.$$

In Figure 7.71(b) you can see that the length of the region at y is

$$\text{Length} = 2x$$

$$= \frac{2}{5}(y + 24)$$

$$= L(y).$$

Finally, by integrating from $y = -9$ to $y = -4$, you can calculate the fluid force to be

$$F = w \int_c^d h(y) L(y)\, dy$$

$$= 62.4 \int_{-9}^{-4} (-y)\left(\frac{2}{5}\right)(y + 24)\, dy$$

$$= -62.4 \left(\frac{2}{5}\right) \int_{-9}^{-4} (y^2 + 24y)\, dy$$

$$= -62.4 \left(\frac{2}{5}\right) \left[\frac{y^3}{3} + 12y^2\right]_{-9}^{-4}$$

$$= -62.4 \left(\frac{2}{5}\right)\left(\frac{-1675}{3}\right)$$

$$= 13{,}936 \text{ pounds.}$$

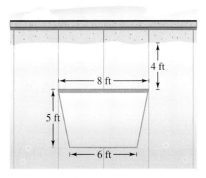

(a) Water gate in a dam

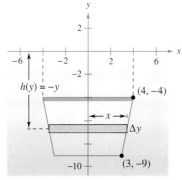

(b) The fluid force against the gate
Figure 7.71

NOTE In Example 2, the x-axis coincided with the surface of the water. This was convenient, but arbitrary. In choosing a coordinate system to represent a physical situation, you should consider various possibilities. Often you can simplify the calculations in a problem by locating the coordinate system to take advantage of special characteristics of the problem, such as symmetry.

510 CHAPTER 7 Applications of Integration

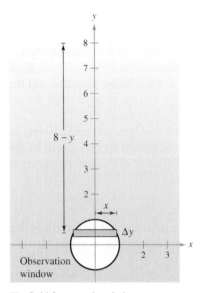

The fluid force on the window
Figure 7.72

EXAMPLE 3 Fluid Force on a Vertical Surface

A circular observation window on a marine science ship has a radius of 1 foot, and the center of the window is 8 feet below water level, as shown in Figure 7.72. What is the fluid force on the window?

Solution To take advantage of symmetry, locate a coordinate system such that the origin coincides with the center of the window, as shown in Figure 7.72. The depth at y is then

$$\text{Depth} = h(y) = 8 - y.$$

The horizontal length of the window is $2x$, and you can use the equation for the circle, $x^2 + y^2 = 1$, to solve for x as follows.

$$\text{Length} = 2x$$
$$= 2\sqrt{1 - y^2} = L(y)$$

Finally, because y ranges from -1 to 1, and using 64 pounds per cubic foot as the weight-density of seawater, you have

$$F = w \int_c^d h(y) L(y) \, dy$$
$$= 64 \int_{-1}^{1} (8 - y)(2)\sqrt{1 - y^2} \, dy.$$

Initially it looks as if this integral would be difficult to solve. However, if you break the integral into two parts and apply symmetry, the solution is simple.

$$F = 64(16) \int_{-1}^{1} \sqrt{1 - y^2} \, dy - 64(2) \int_{-1}^{1} y\sqrt{1 - y^2} \, dy$$

The second integral is 0 (because the integrand is odd and the limits of integration are symmetric to the origin). Moreover, by recognizing that the first integral represents the area of a semicircle of radius 1, you obtain

$$F = 64(16)\left(\frac{\pi}{2}\right) - 64(2)(0)$$
$$= 512\pi$$
$$\approx 1608.5 \text{ pounds}.$$

So, the fluid force on the window is 1608.5 pounds.

TECHNOLOGY To confirm the result obtained in Example 3, you might have considered using Simpson's Rule to approximate the value of

$$128 \int_{-1}^{1} (8 - x)\sqrt{1 - x^2} \, dx.$$

From the graph of

$$f(x) = (8 - x)\sqrt{1 - x^2}$$

however, you can see that f is not differentiable when $x = \pm 1$ (see Figure 7.73). This means that you cannot apply Theorem 4.19 from Section 4.6 to determine the potential error in Simpson's Rule. Without knowing the potential error, the approximation is of little value. Use a graphing utility to approximate the integral.

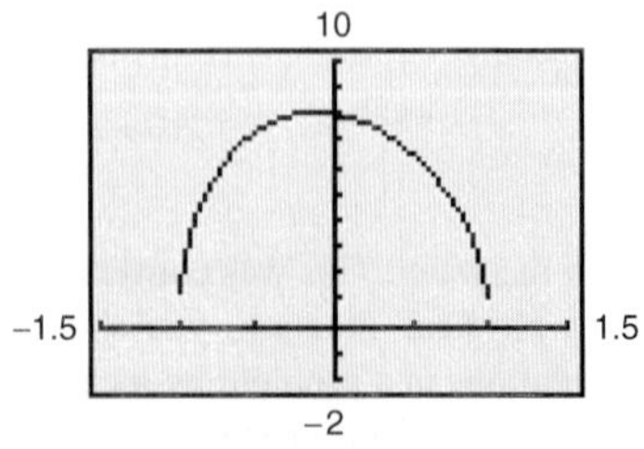

f is not differentiable at $x = \pm 1$.
Figure 7.73

Exercises for Section 7.7

Force on a Submerged Sheet In Exercises 1 and 2, the area of the top side of a piece of sheet metal is given. The sheet metal is submerged horizontally in 5 feet of water. Find the fluid force on the top side.

1. 3 square feet
2. 16 square feet

Buoyant Force In Exercises 3 and 4, find the buoyant force of a rectangular solid of the given dimensions submerged in water so that the top side is parallel to the surface of the water. The buoyant force is the difference between the fluid forces on the top and bottom sides of the solid.

3. 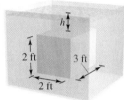   **4.**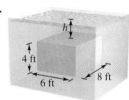

Fluid Force on a Tank Wall In Exercises 5–10, find the fluid force on the vertical side of the tank, where the dimensions are given in feet. Assume that the tank is full of water.

5. Rectangle   **6.** Triangle

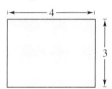

7. Trapezoid   **8.** Semicircle

   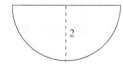

9. Parabola, $y = x^2$ **10.** Semiellipse, $y = -\frac{1}{2}\sqrt{36 - 9x^2}$

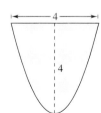

Fluid Force of Water In Exercises 11–14, find the fluid force on the vertical plate submerged in water, where the dimensions are given in meters and the weight-density of water is 9800 newtons per cubic meter.

11. Square   **12.** Square

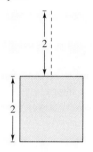   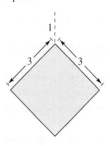

13. Triangle   **14.** Rectangle

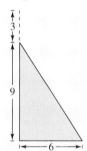   

Force on a Concrete Form In Exercises 15–18, the figure is the vertical side of a form for poured concrete that weighs 140.7 pounds per cubic foot. Determine the force on this part of the concrete form.

15. Rectangle   **16.** Semiellipse, $y = -\frac{3}{4}\sqrt{16 - x^2}$

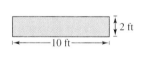   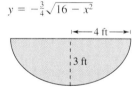

17. Rectangle   **18.** Triangle

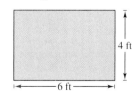   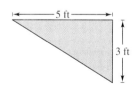

19. *Fluid Force of Gasoline* A cylindrical gasoline tank is placed so that the axis of the cylinder is horizontal. Find the fluid force on a circular end of the tank if the tank is half full, assuming that the diameter is 3 feet and the gasoline weighs 42 pounds per cubic foot.

20. Fluid Force of Gasoline Repeat Exercise 19 for a tank that is full. (Evaluate one integral by a geometric formula and the other by observing that the integrand is an odd function.)

21. Fluid Force on a Circular Plate A circular plate of radius r feet is submerged vertically in a tank of fluid that weighs w pounds per cubic foot. The center of the circle is k ($k > r$) feet below the surface of the fluid. Show that the fluid force on the surface of the plate is

$$F = wk(\pi r^2).$$

(Evaluate one integral by a geometric formula and the other by observing that the integrand is an odd function.)

22. Fluid Force on a Circular Plate Use the result of Exercise 21 to find the fluid force on the circular plate shown in each figure. Assume the plates are in the wall of a tank filled with water and the measurements are given in feet.

(a) 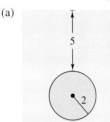 (b)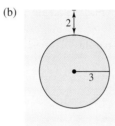

23. Fluid Force on a Rectangular Plate A rectangular plate of height h feet and base b feet is submerged vertically in a tank of fluid that weighs w pounds per cubic foot. The center is k feet below the surface of the fluid, where $h \leq k/2$. Show that the fluid force on the surface of the plate is

$$F = wkhb.$$

24. Fluid Force on a Rectangular Plate Use the result of Exercise 23 to find the fluid force on the rectangular plate shown in each figure. Assume the plates are in the wall of a tank filled with water and the measurements are given in feet.

(a) 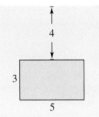 (b)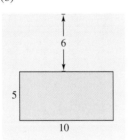

25. Submarine Porthole A porthole on a vertical side of a submarine (submerged in seawater) is 1 square foot. Find the fluid force on the porthole, assuming that the center of the square is 15 feet below the surface.

26. Submarine Porthole Repeat Exercise 25 for a circular porthole that has a diameter of 1 foot. The center is 15 feet below the surface.

27. Modeling Data The vertical stern of a boat with a superimposed coordinate system is shown in the figure. The table shows the width w of the stern at indicated values of y. Find the fluid force against the stern if the measurements are given in feet.

y	0	$\frac{1}{2}$	1	$\frac{3}{2}$	2	$\frac{5}{2}$	3	$\frac{7}{2}$	4
w	0	3	5	8	9	10	10.25	10.5	10.5

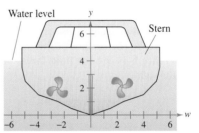

28. Irrigation Canal Gate The vertical cross section of an irrigation canal is modeled by

$$f(x) = \frac{5x^2}{x^2 + 4}$$

where x is measured in feet and $x = 0$ corresponds to the center of the canal. Use the integration capabilities of a graphing utility to approximate the fluid force against a vertical gate used to stop the flow of water if the water is 3 feet deep.

In Exercises 29 and 30, use the integration capabilities of a graphing utility to approximate the fluid force on the vertical plate bounded by the x-axis and the top half of the graph of the equation. Assume that the base of the plate is 12 feet beneath the surface of the water.

29. $x^{2/3} + y^{2/3} = 4^{2/3}$ **30.** $\dfrac{x^2}{28} + \dfrac{y^2}{16} = 1$

31. Think About It

(a) Approximate the depth of the water in the tank in Exercise 5 if the fluid force is one-half as great as when the tank is full.

(b) Explain why the answer in part (a) is not $\frac{3}{2}$.

Writing About Concepts

32. Define fluid pressure.

33. Define fluid force against a submerged vertical plane region.

34. Two identical semicircular windows are placed at the same depth in the vertical wall of an aquarium (see figure). Which has the greater fluid force? Explain.

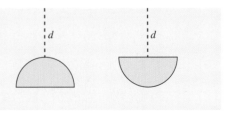

Review Exercises for Chapter 7

See www.CalcChat.com for worked-out solutions to odd-numbered exercises.

In Exercises 1–10, sketch the region bounded by the graphs of the equations, and determine the area of the region.

1. $y = \dfrac{1}{x^2}$, $y = 0$, $x = 1$, $x = 5$
2. $y = \dfrac{1}{x^2}$, $y = 4$, $x = 5$
3. $y = \dfrac{1}{x^2 + 1}$, $y = 0$, $x = -1$, $x = 1$
4. $x = y^2 - 2y$, $x = -1$, $y = 0$
5. $y = x$, $y = x^3$
6. $x = y^2 + 1$, $x = y + 3$
7. $y = e^x$, $y = e^2$, $x = 0$
8. $y = \csc x$, $y = 2$ (one region)
9. $y = \sin x$, $y = \cos x$, $\dfrac{\pi}{4} \le x \le \dfrac{5\pi}{4}$
10. $x = \cos y$, $x = \dfrac{1}{2}$, $\dfrac{\pi}{3} \le y \le \dfrac{7\pi}{3}$

In Exercises 11–14, use a graphing utility to graph the region bounded by the graphs of the functions, and use the integration capabilities of the graphing utility to find the area of the region.

11. $y = x^2 - 8x + 3$, $y = 3 + 8x - x^2$
12. $y = x^2 - 4x + 3$, $y = x^3$, $x = 0$
13. $\sqrt{x} + \sqrt{y} = 1$, $y = 0$, $x = 0$
14. $y = x^4 - 2x^2$, $y = 2x^2$

In Exercises 15–18, use vertical and horizontal representative rectangles to set up integrals for finding the area of the region bounded by the graphs of the equations. Find the area of the region by evaluating the easier of the two integrals.

15. $x = y^2 - 2y$, $x = 0$
16. $y = \sqrt{x - 1}$, $y = \dfrac{x - 1}{2}$
17. $y = 1 - \dfrac{x}{2}$, $y = x - 2$, $y = 1$
18. $y = \sqrt{x - 1}$, $y = 2$, $y = 0$, $x = 0$

19. *Think About It* A person has two job offers. The starting salary for each is $30,000, and after 10 years of service each will pay $56,000. The salary increases for each offer are shown in the figure. From a strictly monetary viewpoint, which is the better offer? Explain.

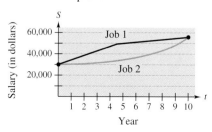

20. *Modeling Data* The table shows the annual service revenue R_1 in billions of dollars for the cellular telephone industry for the years 1995 through 2001. (*Source: Cellular Telecommunications & Internet Association*)

Year	1995	1996	1997	1998	1999	2000	2001
R_1	19.1	23.6	27.5	33.1	40.0	52.5	65.0

(a) Use the regression capabilities of a graphing utility to find an exponential model for the data. Let t represent the year, with $t = 5$ corresponding to 1995. Use the graphing utility to plot the data and graph the model in the same viewing window.

(b) A financial consultant believes that a model for service revenue for the years 2005 through 2010 is

$R_2 = 5 + 6.83e^{0.2t}$.

What is the difference in total service revenue between the two models for the years 2005 through 2010?

In Exercises 21–28, find the volume of the solid generated by revolving the plane region bounded by the equations about the indicated line(s).

21. $y = x$, $y = 0$, $x = 4$
    (a) the x-axis   (b) the y-axis
    (c) the line $x = 4$ (d) the line $x = 6$
22. $y = \sqrt{x}$, $y = 2$, $x = 0$
    (a) the x-axis   (b) the line $y = 2$
    (c) the y-axis   (d) the line $x = -1$
23. $\dfrac{x^2}{16} + \dfrac{y^2}{9} = 1$
    (a) the y-axis (oblate spheroid)
    (b) the x-axis (prolate spheroid)
24. $\dfrac{x^2}{a^2} + \dfrac{y^2}{b^2} = 1$
    (a) the y-axis (oblate spheroid)
    (b) the x-axis (prolate spheroid)
25. $y = \dfrac{1}{x^4 + 1}$, $y = 0$, $x = 0$, $x = 1$
    revolved about the y-axis
26. $y = \dfrac{1}{\sqrt{1 + x^2}}$, $y = 0$, $x = -1$, $x = 1$
    revolved about the x-axis
27. $y = 1/(1 + \sqrt{x - 2})$, $y = 0$, $x = 2$, $x = 6$
    revolved about the y-axis
28. $y = e^{-x}$, $y = 0$, $x = 0$, $x = 1$
    revolved about the x-axis

In Exercises 29 and 30, consider the region bounded by the graphs of the equations $y = x\sqrt{x + 1}$ and $y = 0$.

29. *Area* Find the area of the region.
30. *Volume* Find the volume of the solid generated by revolving the region about (a) the x-axis and (b) the y-axis.

31. Depth of Gasoline in a Tank A gasoline tank is an oblate spheroid generated by revolving the region bounded by the graph of $(x^2/16) + (y^2/9) = 1$ about the y-axis, where x and y are measured in feet. Find the depth of the gasoline in the tank when it is filled to one-fourth its capacity.

32. Magnitude of a Base The base of a solid is a circle of radius a, and its vertical cross sections are equilateral triangles. The volume of the solid is 10 cubic meters. Find the radius of the circle.

In Exercises 33 and 34, find the arc length of the graph of the function over the given interval.

33. $f(x) = \frac{4}{5}x^{5/4}$, $[0, 4]$ **34.** $y = \frac{1}{6}x^3 + \frac{1}{2x}$, $[1, 3]$

35. Length of a Catenary A cable of a suspension bridge forms a catenary modeled by the equation

$$y = 300 \cosh\left(\frac{x}{2000}\right) - 280, \quad -2000 \le x \le 2000$$

where x and y are measured in feet. Use a graphing utility to approximate the length of the cable.

36. Approximation Determine which value best approximates the length of the arc represented by the integral

$$\int_0^{\pi/4} \sqrt{1 + (\sec^2 x)^2}\, dx.$$

(Make your selection on the basis of a sketch of the arc and *not* by performing any calculations.)

(a) -2 (b) 1 (c) π (d) 4 (e) 3

37. Surface Area Use integration to find the lateral surface area of a right circular cone of height 4 and radius 3.

38. Surface Area The region bounded by the graphs of $y = 2\sqrt{x}$, $y = 0$, and $x = 3$ is revolved about the x-axis. Find the surface area of the solid generated.

39. Work A force of 4 pounds is needed to stretch a spring 1 inch from its natural position. Find the work done in stretching the spring from its natural length of 10 inches to a length of 15 inches.

40. Work The force required to stretch a spring is 50 pounds. Find the work done in stretching the spring from its natural length of 9 inches to double that length.

41. Work A water well has an eight-inch casing (diameter) and is 175 feet deep. The water is 25 feet from the top of the well. Determine the amount of work done in pumping the well dry, assuming that no water enters it while it is being pumped.

42. Work Repeat Exercise 41, assuming that water enters the well at a rate of 4 gallons per minute and the pump works at a rate of 12 gallons per minute. How many gallons are pumped in this case?

43. Work A chain 10 feet long weighs 5 pounds per foot and is hung from a platform 20 feet above the ground. How much work is required to raise the entire chain to the 20-foot level?

44. Work A windlass, 200 feet above ground level on the top of a building, uses a cable weighing 4 pounds per foot. Find the work done in winding up the cable if

(a) one end is at ground level.

(b) there is a 300-pound load attached to the end of the cable.

45. Work The work done by a variable force in a press is 80 foot-pounds. The press moves a distance of 4 feet and the force is a quadratic of the form $F = ax^2$. Find a.

46. Work Find the work done by the force F shown in the figure.

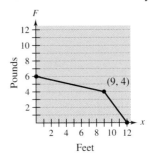

In Exercises 47–50, find the centroid of the region bounded by the graphs of the equations.

47. $\sqrt{x} + \sqrt{y} = \sqrt{a}$, $x = 0$, $y = 0$

48. $y = x^2$, $y = 2x + 3$

49. $y = a^2 - x^2$, $y = 0$

50. $y = x^{2/3}$, $y = \frac{1}{2}x$

51. Centroid A blade on an industrial fan has the configuration of a semicircle attached to a trapezoid (see figure). Find the centroid of the blade.

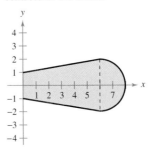

52. Fluid Force A swimming pool is 5 feet deep at one end and 10 feet deep at the other, and the bottom is an inclined plane. The length and width of the pool are 40 feet and 20 feet. If the pool is full of water, what is the fluid force on each of the vertical walls?

53. Fluid Force Show that the fluid force against any vertical region in a liquid is the product of the weight per cubic volume of the liquid, the area of the region, and the depth of the centroid of the region.

54. Fluid Force Using the result of Exercise 53, find the fluid force on one side of a vertical circular plate of radius 4 feet that is submerged in water so that its center is 5 feet below the surface.

P.S. Problem Solving

See www.CalcChat.com for worked-out solutions to odd-numbered exercises.

1. Let R be the area of the region in the first quadrant bounded by the parabola $y = x^2$ and the line $y = cx$, $c > 0$. Let T be the area of the triangle AOB. Calculate the limit

$$\lim_{c \to 0^+} \frac{T}{R}.$$

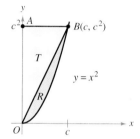

2. Let R be the region bounded by the parabola $y = x - x^2$ and the x-axis. Find the equation of the line $y = mx$ that divides this region into two regions of equal area.

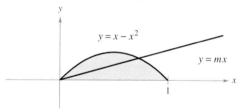

3. (a) A torus is formed by revolving the region bounded by the circle

   $(x - 2)^2 + y^2 = 1$

   about the y-axis (see figure). Use the disk method to calculate the volume of the torus.

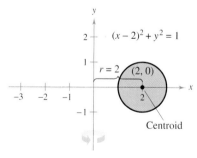

   (b) Use the disk method to find the volume of the general torus if the circle has radius r and its center is R units from the axis of rotation.

4. Graph the curve

   $8y^2 = x^2(1 - x^2)$.

   Use a computer algebra system to find the surface area of the solid of revolution obtained by revolving the curve about the x-axis.

5. A hole is cut through the center of a sphere of radius r (see figure). The height of the remaining spherical ring is h. Find the volume of the ring and show that it is independent of the radius of the sphere.

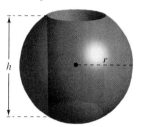

6. A rectangle R of length l and width w is revolved about the line L (see figure). Find the volume of the resulting solid of revolution.

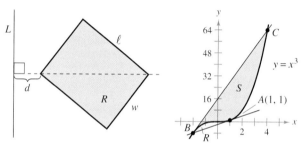

Figure for 6   **Figure for 7**

7. (a) The tangent line to the curve $y = x^3$ at the point $A(1, 1)$ intersects the curve at another point B. Let R be the area of the region bounded by the curve and the tangent line. The tangent line at B intersects the curve at another point C (see figure). Let S be the area of the region bounded by the curve and this second tangent line. How are the areas R and S related?

   (b) Repeat the construction in part (a) by selecting an arbitrary point A on the curve $y = x^3$. Show that the two areas R and S are always related in the same way.

8. The graph of $y = f(x)$ passes through the origin. The arc length of the curve from $(0, 0)$ to $(x, f(x))$ is given by

$$s(x) = \int_0^x \sqrt{1 + e^t}\, dt.$$

   Identify the function f.

9. Let f be rectifiable on the interval $[a, b]$, and let

$$s(x) = \int_a^x \sqrt{1 + [f'(t)]^2}\, dt.$$

   (a) Find $\dfrac{ds}{dx}$.

   (b) Find ds and $(ds)^2$.

   (c) If $f(t) = t^{3/2}$, find $s(x)$ on $[1, 3]$.

   (d) Calculate $s(2)$ and describe what it signifies.

10. The **Archimedes Principle** states that the upward or buoyant force on an object within a fluid is equal to the weight of the fluid that the object displaces. For a partially submerged object, you can obtain information about the relative densities of the floating object and the fluid by observing how much of the object is above and below the surface. You can also determine the size of a floating object if you know the amount that is above the surface and the relative densities. You can see the top of a floating iceberg (see figure). The density of ocean water is 1.03×10^3 kilograms per cubic meter, and that of ice is 0.92×10^3 kilograms per cubic meter. What percent of the total iceberg is below the surface?

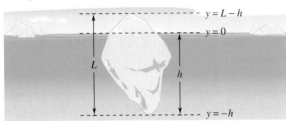

11. Sketch the region bounded on the left by $x = 1$, bounded above by $y = 1/x^3$, and bounded below by $y = -1/x^3$.

    (a) Find the centroid of the region for $1 \leq x \leq 6$.
    (b) Find the centroid of the region for $1 \leq x \leq b$.
    (c) Where is the centroid as $b \to \infty$?

12. Sketch the region to the right of the y-axis, bounded above by $y = 1/x^4$ and bounded below by $y = -1/x^4$.

    (a) Find the centroid of the region for $1 \leq x \leq 6$.
    (b) Find the centroid of the region for $1 \leq x \leq b$.
    (c) Where is the centroid as $b \to \infty$?

13. Find the work done by each force F.

    (a) 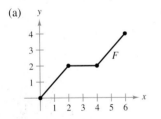 (b)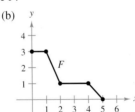

14. Estimate the surface area of the pond using (a) the Trapezoidal Rule and (b) Simpson's Rule.

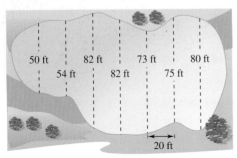

In Exercises 15 and 16, find the consumer surplus and producer surplus for the given demand $[p_1(x)]$ and supply $[p_2(x)]$ curves. The consumer surplus and producer surplus are represented by the areas shown in the figure.

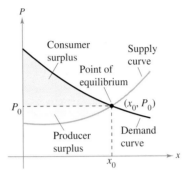

15. $p_1(x) = 50 - 0.5x$, $p_2(x) = 0.125x$
16. $p_1(x) = 1000 - 0.4x^2$, $p_2(x) = 42x$

17. A swimming pool is 20 feet wide, 40 feet long, 4 feet deep at one end, and 8 feet deep at the other end (see figure). The bottom is an inclined plane. Find the fluid force on each vertical wall.

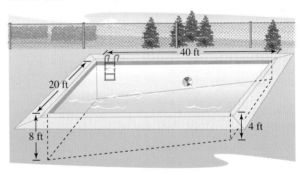

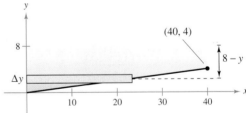

18. (a) Find at least two continuous functions f that satisfy each condition.

    (i) $f(x) \geq 0$ on $[0, 1]$ (ii) $f(0) = 0$ and $f(1) = 0$
    (iii) The area bounded by the graph of f and the x-axis for $0 \leq x \leq 1$ equals 1.

    (b) For each function found in part (a), approximate the arc length of the graph of the function on the interval $[0, 1]$. (Use a graphing utility if necessary.)

    (c) Can you find a function f that satisfies the conditions in part (a) and whose graph has an arc length of less than 3 on the interval $[0, 1]$?

Integration Techniques, L'Hôpital's Rule, and Improper Integrals

$\int_0^1 \frac{1}{\sqrt{x}}\,dx = 2$

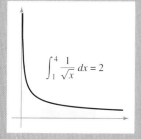

$\int_1^4 \frac{1}{\sqrt{x}}\,dx = 2$

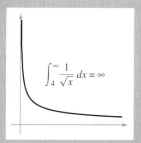

$\int_4^\infty \frac{1}{\sqrt{x}}\,dx = \infty$

The NASA Hubble Space Telescope image of a planetary nebula nicknamed the "Cat's Eye Nebula" gives just a glimpse of the kinds of things you might see if you could travel through space. Would it be possible to propel a spacecraft an unlimited distance away from Earth's surface? Why?

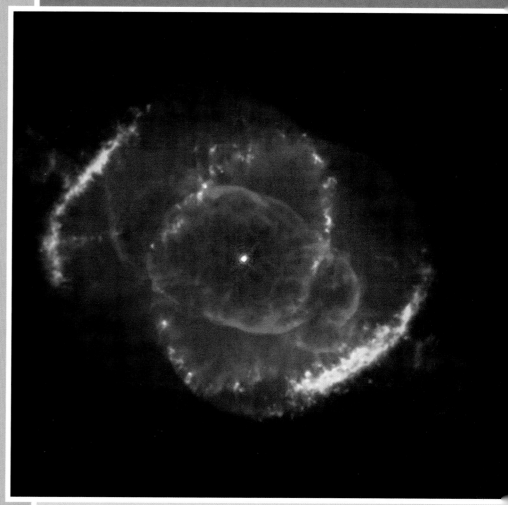

P. Harrington and K.J. Borkowski (University of Maryland), and NASA

From your studies of calculus thus far, you know that a definite integral has finite limits of integration and a continuous integrand. In Chapter 8, you will study *improper integrals*. Improper integrals have at least one infinite limit of integration or have an integrand with an infinite discontinuity. You will see that improper integrals either converge or diverge.

Section 8.1 Basic Integration Rules

- Review procedures for fitting an integrand to one of the basic integration rules.

Fitting Integrands to Basic Rules

In this chapter, you will study several integration techniques that greatly expand the set of integrals to which the basic integration rules can be applied. These rules are reviewed on page 520. A major step in solving any integration problem is recognizing which basic integration rule to use. As shown in Example 1, slight differences in the integrand can lead to very different solution techniques.

EXAMPLE 1 A Comparison of Three Similar Integrals

Find each integral.

a. $\displaystyle\int \frac{4}{x^2 + 9}\, dx$ **b.** $\displaystyle\int \frac{4x}{x^2 + 9}\, dx$ **c.** $\displaystyle\int \frac{4x^2}{x^2 + 9}\, dx$

Solution

a. Use the Arctangent Rule and let $u = x$ and $a = 3$.

$$\int \frac{4}{x^2 + 9}\, dx = 4 \int \frac{1}{x^2 + 3^2}\, dx \qquad \text{Constant Multiple Rule}$$

$$= 4\left(\frac{1}{3} \arctan \frac{x}{3}\right) + C \qquad \text{Arctangent Rule}$$

$$= \frac{4}{3} \arctan \frac{x}{3} + C \qquad \text{Simplify.}$$

b. Here the Arctangent Rule does not apply because the numerator contains a factor of x. Consider the Log Rule and let $u = x^2 + 9$. Then $du = 2x\, dx$, and you have

$$\int \frac{4x}{x^2 + 9}\, dx = 2 \int \frac{2x\, dx}{x^2 + 9} \qquad \text{Constant Multiple Rule}$$

$$= 2 \int \frac{du}{u} \qquad \text{Substitution: } u = x^2 + 9$$

$$= 2 \ln|u| + C = 2 \ln(x^2 + 9) + C. \qquad \text{Log Rule}$$

c. Because the degree of the numerator is equal to the degree of the denominator, you should first use division to rewrite the improper rational function as the sum of a polynomial and a proper rational function.

$$\int \frac{4x^2}{x^2 + 9}\, dx = \int \left(4 - \frac{36}{x^2 + 9}\right) dx \qquad \text{Rewrite using long division.}$$

$$= \int 4\, dx - 36 \int \frac{1}{x^2 + 9}\, dx \qquad \text{Write as two integrals.}$$

$$= 4x - 36\left(\frac{1}{3} \arctan \frac{x}{3}\right) + C \qquad \text{Integrate.}$$

$$= 4x - 12 \arctan \frac{x}{3} + C \qquad \text{Simplify.}$$

EXPLORATION

A Comparison of Three Similar Integrals Which, if any, of the following integrals can be evaluated using the 20 basic integration rules? For any that can be evaluated, do so. For any that can't, explain why.

a. $\displaystyle\int \frac{3}{\sqrt{1 - x^2}}\, dx$

b. $\displaystyle\int \frac{3x}{\sqrt{1 - x^2}}\, dx$

c. $\displaystyle\int \frac{3x^2}{\sqrt{1 - x^2}}\, dx$

NOTE Notice in Example 1(c) that some preliminary algebra is required before applying the rules for integration, and that subsequently more than one rule is needed to evaluate the resulting integral.

 indicates that in the **HM mathSpace®** CD-ROM and the online **Eduspace®** system for this text, you will find an Open Exploration, which further explores this example using the computer algebra systems Maple, Mathcad, Mathematica, and Derive.

SECTION 8.1 Basic Integration Rules 519

EXAMPLE 2 Using Two Basic Rules to Solve a Single Integral

Evaluate $\int_0^1 \frac{x+3}{\sqrt{4-x^2}} \, dx$.

Solution Begin by writing the integral as the sum of two integrals. Then apply the Power Rule and the Arcsine Rule as follows.

$$\int_0^1 \frac{x+3}{\sqrt{4-x^2}} \, dx = \int_0^1 \frac{x}{\sqrt{4-x^2}} \, dx + \int_0^1 \frac{3}{\sqrt{4-x^2}} \, dx$$

$$= -\frac{1}{2} \int_0^1 (4-x^2)^{-1/2}(-2x) \, dx + 3 \int_0^1 \frac{1}{\sqrt{2^2 - x^2}} \, dx$$

$$= \left[-(4-x^2)^{1/2} + 3 \arcsin \frac{x}{2} \right]_0^1$$

$$= \left(-\sqrt{3} + \frac{\pi}{2} \right) - (-2 + 0)$$

$$\approx 1.839$$

See Figure 8.1.

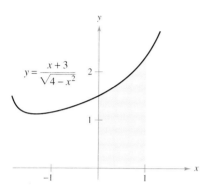

The area of the region is approximately 1.839.
Figure 8.1

TECHNOLOGY Simpson's Rule can be used to give a good approximation of the value of the integral in Example 2 (for $n = 10$, the approximation is 1.839). When using numerical integration, however, you should be aware that Simpson's Rule does not always give good approximations when one or both of the limits of integration are near a vertical asymptote. For instance, using the Fundamental Theorem of Calculus, you can obtain

$$\int_0^{1.99} \frac{x+3}{\sqrt{4-x^2}} \, dx \approx 6.213.$$

Applying Simpson's Rule (with $n = 10$) to this integral produces an approximation of 6.889.

EXAMPLE 3 A Substitution Involving $a^2 - u^2$

Find $\int \frac{x^2}{\sqrt{16 - x^6}} \, dx$.

Solution Because the radical in the denominator can be written in the form

$$\sqrt{a^2 - u^2} = \sqrt{4^2 - (x^3)^2}$$

you can try the substitution $u = x^3$. Then $du = 3x^2 \, dx$, and you have

$$\int \frac{x^2}{\sqrt{16 - x^6}} \, dx = \frac{1}{3} \int \frac{3x^2 \, dx}{\sqrt{16 - (x^3)^2}} \quad \text{Rewrite integral.}$$

$$= \frac{1}{3} \int \frac{du}{\sqrt{4^2 - u^2}} \quad \text{Substitution: } u = x^3$$

$$= \frac{1}{3} \arcsin \frac{u}{4} + C \quad \text{Arcsine Rule}$$

$$= \frac{1}{3} \arcsin \frac{x^3}{4} + C. \quad \text{Rewrite as a function of } x.$$

STUDY TIP Rules 18, 19, and 20 of the basic integration rules on the next page all have expressions involving the sum or difference of two squares:

$a^2 - u^2$
$a^2 + u^2$
$u^2 - a^2$

With such an expression, consider the substitution $u = f(x)$, as in Example 3.

Surprisingly, two of the most commonly overlooked integration rules are the Log Rule and the Power Rule. Notice in the next two examples how these two integration rules can be disguised.

EXAMPLE 4 A Disguised Form of the Log Rule

Find $\int \dfrac{1}{1+e^x}\,dx.$

Solution The integral does not appear to fit any of the basic rules. However, the quotient form suggests the Log Rule. If you let $u = 1 + e^x$, then $du = e^x\,dx$. You can obtain the required du by adding and subtracting e^x in the numerator, as follows.

$$\int \frac{1}{1+e^x}\,dx = \int \frac{1+e^x - e^x}{1+e^x}\,dx \qquad \text{Add and subtract } e^x \text{ in numerator.}$$

$$= \int \left(\frac{1+e^x}{1+e^x} - \frac{e^x}{1+e^x}\right)dx \qquad \text{Rewrite as two fractions.}$$

$$= \int dx - \int \frac{e^x\,dx}{1+e^x} \qquad \text{Rewrite as two integrals.}$$

$$= x - \ln(1+e^x) + C \qquad \text{Integrate.}$$

NOTE There is usually more than one way to solve an integration problem. For instance, in Example 4, try integrating by multiplying the numerator and denominator by e^{-x} to obtain an integral of the form $-\int du/u$. See if you can get the same answer by this procedure. (Be careful: the answer will appear in a different form.)

EXAMPLE 5 A Disguised Form of the Power Rule

Find $\int (\cot x)[\ln(\sin x)]\,dx.$

Solution Again, the integral does not appear to fit any of the basic rules. However, considering the two primary choices for u $[u = \cot x$ and $u = \ln(\sin x)]$, you can see that the second choice is the appropriate one because

$$u = \ln(\sin x) \quad \text{and} \quad du = \frac{\cos x}{\sin x}\,dx = \cot x\,dx.$$

So,

$$\int (\cot x)[\ln(\sin x)]\,dx = \int u\,du \qquad \text{Substitution: } u = \ln(\sin x)$$

$$= \frac{u^2}{2} + C \qquad \text{Integrate.}$$

$$= \frac{1}{2}[\ln(\sin x)]^2 + C. \qquad \text{Rewrite as a function of } x.$$

NOTE In Example 5, try *checking* that the derivative of

$$\frac{1}{2}[\ln(\sin x)]^2 + C$$

is the integrand of the original integral.

Review of Basic Integration Rules ($a > 0$)

1. $\int kf(u)\,du = k\int f(u)\,du$
2. $\int [f(u) \pm g(u)]\,du = \int f(u)\,du \pm \int g(u)\,du$
3. $\int du = u + C$
4. $\int u^n\,du = \dfrac{u^{n+1}}{n+1} + C,\ n \ne -1$
5. $\int \dfrac{du}{u} = \ln|u| + C$
6. $\int e^u\,du = e^u + C$
7. $\int a^u\,du = \left(\dfrac{1}{\ln a}\right)a^u + C$
8. $\int \sin u\,du = -\cos u + C$
9. $\int \cos u\,du = \sin u + C$
10. $\int \tan u\,du = -\ln|\cos u| + C$
11. $\int \cot u\,du = \ln|\sin u| + C$
12. $\int \sec u\,du = \ln|\sec u + \tan u| + C$
13. $\int \csc u\,du = -\ln|\csc u + \cot u| + C$
14. $\int \sec^2 u\,du = \tan u + C$
15. $\int \csc^2 u\,du = -\cot u + C$
16. $\int \sec u \tan u\,du = \sec u + C$
17. $\int \csc u \cot u\,du = -\csc u + C$
18. $\int \dfrac{du}{\sqrt{a^2 - u^2}} = \arcsin \dfrac{u}{a} + C$
19. $\int \dfrac{du}{a^2 + u^2} = \dfrac{1}{a}\arctan \dfrac{u}{a} + C$
20. $\int \dfrac{du}{u\sqrt{u^2 - a^2}} = \dfrac{1}{a}\operatorname{arcsec}\dfrac{|u|}{a} + C$

Trigonometric identities can often be used to fit integrals to one of the basic integration rules.

EXAMPLE 6 Using Trigonometric Identities

Find $\int \tan^2 2x\, dx$.

Solution Note that $\tan^2 u$ is not in the list of basic integration rules. However, $\sec^2 u$ is in the list. This suggests the trigonometric identity $\tan^2 u = \sec^2 u - 1$. If you let $u = 2x$, then $du = 2\, dx$ and

$$\int \tan^2 2x\, dx = \frac{1}{2} \int \tan^2 u\, du \qquad \text{Substitution: } u = 2x$$

$$= \frac{1}{2} \int (\sec^2 u - 1)\, du \qquad \text{Trigonometric identity}$$

$$= \frac{1}{2} \int \sec^2 u\, du - \frac{1}{2} \int du \qquad \text{Rewrite as two integrals.}$$

$$= \frac{1}{2} \tan u - \frac{u}{2} + C \qquad \text{Integrate.}$$

$$= \frac{1}{2} \tan 2x - x + C. \qquad \text{Rewrite as a function of } x.$$

TECHNOLOGY If you have access to a computer algebra system, try using it to evaluate the integrals in this section. Compare the *form* of the antiderivative given by the software with the form obtained by hand. Sometimes the forms will be the same, but often they will differ. For instance, why is the antiderivative $\ln 2x + C$ equivalent to the antiderivative $\ln x + C$?

This section concludes with a summary of the common procedures for fitting integrands to the basic integration rules.

Procedures for Fitting Integrands to Basic Rules

Technique	Example
Expand (numerator).	$(1 + e^x)^2 = 1 + 2e^x + e^{2x}$
Separate numerator.	$\dfrac{1 + x}{x^2 + 1} = \dfrac{1}{x^2 + 1} + \dfrac{x}{x^2 + 1}$
Complete the square.	$\dfrac{1}{\sqrt{2x - x^2}} = \dfrac{1}{\sqrt{1 - (x - 1)^2}}$
Divide improper rational function.	$\dfrac{x^2}{x^2 + 1} = 1 - \dfrac{1}{x^2 + 1}$
Add and subtract terms in numerator.	$\dfrac{2x}{x^2 + 2x + 1} = \dfrac{2x + 2 - 2}{x^2 + 2x + 1} = \dfrac{2x + 2}{x^2 + 2x + 1} - \dfrac{2}{(x + 1)^2}$
Use trigonometric identities.	$\cot^2 x = \csc^2 x - 1$
Multiply and divide by Pythagorean conjugate.	$\dfrac{1}{1 + \sin x} = \left(\dfrac{1}{1 + \sin x}\right)\left(\dfrac{1 - \sin x}{1 - \sin x}\right) = \dfrac{1 - \sin x}{1 - \sin^2 x}$
	$= \dfrac{1 - \sin x}{\cos^2 x} = \sec^2 x - \dfrac{\sin x}{\cos^2 x}$

NOTE Remember that you can separate numerators but not denominators. Watch out for this common error when fitting integrands to basic rules.

$$\dfrac{1}{x^2 + 1} \neq \dfrac{1}{x^2} + \dfrac{1}{1} \qquad \text{Do not separate denominators.}$$

Exercises for Section 8.1

See www.CalcChat.com for worked-out solutions to odd-numbered exercises.

In Exercises 1–4, select the correct antiderivative.

1. $\dfrac{dy}{dx} = \dfrac{x}{\sqrt{x^2+1}}$
   (a) $2\sqrt{x^2+1} + C$
   (b) $\sqrt{x^2+1} + C$
   (c) $\tfrac{1}{2}\sqrt{x^2+1} + C$
   (d) $\ln(x^2+1) + C$

2. $\dfrac{dy}{dx} = \dfrac{x}{x^2+1}$
   (a) $\ln\sqrt{x^2+1} + C$
   (b) $\dfrac{2x}{(x^2+1)^2} + C$
   (c) $\arctan x + C$
   (d) $\ln(x^2+1) + C$

3. $\dfrac{dy}{dx} = \dfrac{1}{x^2+1}$
   (a) $\ln\sqrt{x^2+1} + C$
   (b) $\dfrac{2x}{(x^2+1)^2} + C$
   (c) $\arctan x + C$
   (d) $\ln(x^2+1) + C$

4. $\dfrac{dy}{dx} = x\cos(x^2+1)$
   (a) $2x\sin(x^2+1) + C$
   (b) $-\tfrac{1}{2}\sin(x^2+1) + C$
   (c) $\tfrac{1}{2}\sin(x^2+1) + C$
   (d) $-2x\sin(x^2+1) + C$

In Exercises 5–14, select the basic integration formula you can use to find the integral, and identify u and a when appropriate.

5. $\displaystyle\int (3x-2)^4\, dx$

6. $\displaystyle\int \dfrac{2t-1}{t^2-t+2}\, dt$

7. $\displaystyle\int \dfrac{1}{\sqrt{x}(1-2\sqrt{x})}\, dx$

8. $\displaystyle\int \dfrac{2}{(2t-1)^2+4}\, dt$

9. $\displaystyle\int \dfrac{3}{\sqrt{1-t^2}}\, dt$

10. $\displaystyle\int \dfrac{-2x}{\sqrt{x^2-4}}\, dx$

11. $\displaystyle\int t\sin t^2\, dt$

12. $\displaystyle\int \sec 3x \tan 3x\, dx$

13. $\displaystyle\int (\cos x)e^{\sin x}\, dx$

14. $\displaystyle\int \dfrac{1}{x\sqrt{x^2-4}}\, dx$

In Exercises 15–50, find the indefinite integral.

15. $\displaystyle\int 6(x-4)^5\, dx$

16. $\displaystyle\int \dfrac{2}{(t-9)^2}\, dt$

17. $\displaystyle\int \dfrac{5}{(z-4)^5}\, dz$

18. $\displaystyle\int t^2\sqrt[3]{t^3-1}\, dt$

19. $\displaystyle\int \left[v + \dfrac{1}{(3v-1)^3}\right]\, dv$

20. $\displaystyle\int \left[x - \dfrac{3}{(2x+3)^2}\right]\, dx$

21. $\displaystyle\int \dfrac{t^2-3}{-t^3+9t+1}\, dt$

22. $\displaystyle\int \dfrac{x+1}{\sqrt{x^2+2x-4}}\, dx$

23. $\displaystyle\int \dfrac{x^2}{x-1}\, dx$

24. $\displaystyle\int \dfrac{2x}{x-4}\, dx$

25. $\displaystyle\int \dfrac{e^x}{1+e^x}\, dx$

26. $\displaystyle\int \left(\dfrac{1}{3x-1} - \dfrac{1}{3x+1}\right)\, dx$

27. $\displaystyle\int (1+2x^2)^2\, dx$

28. $\displaystyle\int x\left(1+\dfrac{1}{x}\right)^3\, dx$

29. $\displaystyle\int x\cos 2\pi x^2\, dx$

30. $\displaystyle\int \sec 4x\, dx$

31. $\displaystyle\int \csc \pi x \cot \pi x\, dx$

32. $\displaystyle\int \dfrac{\sin x}{\sqrt{\cos x}}\, dx$

33. $\displaystyle\int e^{5x}\, dx$

34. $\displaystyle\int \csc^2 x e^{\cot x}\, dx$

35. $\displaystyle\int \dfrac{2}{e^{-x}+1}\, dx$

36. $\displaystyle\int \dfrac{5}{3e^x-2}\, dx$

37. $\displaystyle\int \dfrac{\ln x^2}{x}\, dx$

38. $\displaystyle\int (\tan x)[\ln(\cos x)]\, dx$

39. $\displaystyle\int \dfrac{1+\sin x}{\cos x}\, dx$

40. $\displaystyle\int \dfrac{1+\cos\alpha}{\sin\alpha}\, d\alpha$

41. $\displaystyle\int \dfrac{1}{\cos\theta - 1}\, d\theta$

42. $\displaystyle\int \dfrac{2}{3(\sec x - 1)}\, dx$

43. $\displaystyle\int \dfrac{-1}{\sqrt{1-(2t-1)^2}}\, dt$

44. $\displaystyle\int \dfrac{1}{4+3x^2}\, dx$

45. $\displaystyle\int \dfrac{\tan(2/t)}{t^2}\, dt$

46. $\displaystyle\int \dfrac{e^{1/t}}{t^2}\, dt$

47. $\displaystyle\int \dfrac{3}{\sqrt{6x-x^2}}\, dx$

48. $\displaystyle\int \dfrac{1}{(x-1)\sqrt{4x^2-8x+3}}\, dx$

49. $\displaystyle\int \dfrac{4}{4x^2+4x+65}\, dx$

50. $\displaystyle\int \dfrac{1}{\sqrt{1-4x-x^2}}\, dx$

 Slope Fields In Exercises 51–54, a differential equation, a point, and a slope field are given. (a) Sketch two approximate solutions of the differential equation on the slope field, one of which passes through the given point. (b) Use integration to find the particular solution of the differential equation and use a graphing utility to graph the solution. Compare the result with the sketches in part (a). To print an enlarged copy of the graph, go to the website *www.mathgraphs.com*.

51. $\dfrac{ds}{dt} = \dfrac{t}{\sqrt{1-t^4}}$
    $\left(0, -\dfrac{1}{2}\right)$

52. $\dfrac{dy}{dx} = \tan^2(2x)$
    $(0, 0)$

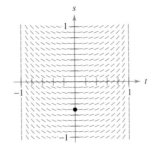
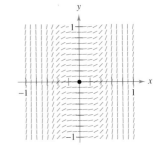

53. $\dfrac{dy}{dx} = (\sec x + \tan x)^2$

(0, 1)

54. $\dfrac{dy}{dx} = \dfrac{1}{\sqrt{4x - x^2}}$

$\left(2, \dfrac{1}{2}\right)$

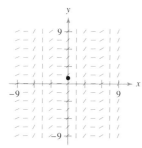
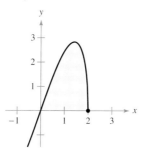

71. $y = \dfrac{3x + 2}{x^2 + 9}$

72. $y = \dfrac{3}{x^2 + 1}$

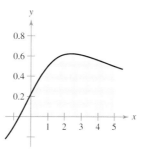
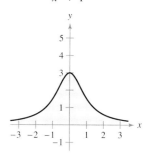

73. $y^2 = x^2(1 - x^2)$

74. $y = \sin 2x$

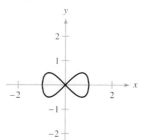
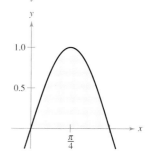

 Slope Fields In Exercises 55 and 56, use a computer algebra system to graph the slope field for the differential equation and graph the solution through the specified initial condition.

55. $\dfrac{dy}{dx} = 0.2y, \; y(0) = 3$

56. $\dfrac{dy}{dx} = 5 - y, \; y(0) = 1$

In Exercises 57–60, solve the differential equation.

57. $\dfrac{dy}{dx} = (1 + e^x)^2$

58. $\dfrac{dr}{dt} = \dfrac{(1 + e^t)^2}{e^t}$

59. $(4 + \tan^2 x)y' = \sec^2 x$

60. $y' = \dfrac{1}{x\sqrt{4x^2 - 1}}$

In Exercises 61–68, evaluate the definite integral. Use the integration capabilities of a graphing utility to verify your result.

61. $\displaystyle\int_0^{\pi/4} \cos 2x \, dx$

62. $\displaystyle\int_0^{\pi} \sin^2 t \cos t \, dt$

63. $\displaystyle\int_0^1 xe^{-x^2} \, dx$

64. $\displaystyle\int_1^e \dfrac{1 - \ln x}{x} \, dx$

65. $\displaystyle\int_0^4 \dfrac{2x}{\sqrt{x^2 + 9}} \, dx$

66. $\displaystyle\int_1^2 \dfrac{x - 2}{x} \, dx$

67. $\displaystyle\int_0^{2/\sqrt{3}} \dfrac{1}{4 + 9x^2} \, dx$

68. $\displaystyle\int_0^4 \dfrac{1}{\sqrt{25 - x^2}} \, dx$

Area In Exercises 69–74, find the area of the region.

69. $y = (-2x + 5)^{3/2}$

70. $y = x\sqrt{8 - 2x^2}$

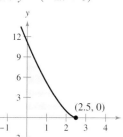

 In Exercises 75–78, use a computer algebra system to find the integral. Use the computer algebra system to graph two antiderivatives. Describe the relationship between the two graphs of the antiderivatives.

75. $\displaystyle\int \dfrac{1}{x^2 + 4x + 13} \, dx$

76. $\displaystyle\int \dfrac{x - 2}{x^2 + 4x + 13} \, dx$

77. $\displaystyle\int \dfrac{1}{1 + \sin \theta} \, d\theta$

78. $\displaystyle\int \left(\dfrac{e^x + e^{-x}}{2}\right)^3 dx$

Writing About Concepts

In Exercises 79–82, state the integration formula you would use to perform the integration. Explain why you chose that formula. Do not integrate.

79. $\displaystyle\int x(x^2 + 1)^3 \, dx$

80. $\displaystyle\int x \sec(x^2 + 1) \tan(x^2 + 1) \, dx$

81. $\displaystyle\int \dfrac{x}{x^2 + 1} \, dx$

82. $\displaystyle\int \dfrac{1}{x^2 + 1} \, dx$

83. Explain why the antiderivative $y_1 = e^{x + C_1}$ is equivalent to the antiderivative $y_2 = Ce^x$.

84. Explain why the antiderivative $y_1 = \sec^2 x + C_1$ is equivalent to the antiderivative $y_2 = \tan^2 x + C$.

85. Determine the constants a and b such that

$$\sin x + \cos x = a \sin(x + b).$$

Use this result to integrate $\displaystyle\int \frac{dx}{\sin x + \cos x}$.

86. Area The graphs of $f(x) = x$ and $g(x) = ax^2$ intersect at the points $(0, 0)$ and $(1/a, 1/a)$. Find a $(a > 0)$ such that the area of the region bounded by the graphs of these two functions is $\frac{2}{3}$.

87. Think About It Use a graphing utility to graph the function $f(x) = \frac{1}{5}(x^3 - 7x^2 + 10x)$. Use the graph to determine whether

$$\int_0^5 f(x)\, dx$$

is positive or negative. Explain.

88. Think About It When evaluating

$$\int_{-1}^{1} x^2\, dx$$

is it appropriate to substitute $u = x^2$, $x = \sqrt{u}$, and $dx = \dfrac{du}{2\sqrt{u}}$

to obtain

$$\frac{1}{2}\int_1^1 \sqrt{u}\, du = 0?$$

Explain.

Approximation In Exercises 89 and 90, determine which value best approximates the area of the region between the x-axis and the function over the given interval. (Make your selection on the basis of a sketch of the region and *not* by integrating.)

89. $f(x) = \dfrac{4x}{x^2 + 1},\ [0, 2]$

(a) 3 (b) 1 (c) -8 (d) 8 (e) 10

90. $f(x) = \dfrac{4}{x^2 + 1},\ [0, 2]$

(a) 3 (b) 1 (c) -4 (d) 4 (e) 10

Interpreting Integrals In Exercises 91 and 92, (a) sketch the region whose area is given by the integral, (b) sketch the solid whose volume is given by the integral if the disk method is used, and (c) sketch the solid whose volume is given by the integral if the shell method is used. (There is more than one correct answer for each part.)

91. $\displaystyle\int_0^2 2\pi x^2\, dx$

92. $\displaystyle\int_0^4 \pi y\, dy$

93. Volume The region bounded by $y = e^{-x^2}$, $y = 0$, $x = 0$, and $x = b\ (b > 0)$ is revolved about the y-axis.

(a) Find the volume of the solid generated if $b = 1$.

(b) Find b such that the volume of the generated solid is $\frac{4}{3}$ cubic units.

94. Arc Length Find the arc length of the graph of $y = \ln(\sin x)$ from $x = \pi/4$ to $x = \pi/2$.

95. Surface Area Find the area of the surface formed by revolving the graph of $y = 2\sqrt{x}$ on the interval $[0, 9]$ about the x-axis.

96. Centroid Find the x-coordinate of the centroid of the region bounded by the graphs of

$$y = \frac{5}{\sqrt{25 - x^2}},\quad y = 0,\quad x = 0,\quad \text{and}\quad x = 4.$$

In Exercises 97 and 98, find the average value of the function over the given interval.

97. $f(x) = \dfrac{1}{1 + x^2},\quad -3 \le x \le 3$

98. $f(x) = \sin nx,\quad 0 \le x \le \pi/n,\ n$ is a positive integer.

Arc Length In Exercises 99 and 100, use the integration capabilities of a graphing utility to approximate the arc length of the curve over the given interval.

99. $y = \tan \pi x,\ \left[0, \frac{1}{4}\right]$ **100.** $y = x^{2/3},\ [1, 8]$

101. Finding a Pattern

(a) Find $\displaystyle\int \cos^3 x\, dx$.

(b) Find $\displaystyle\int \cos^5 x\, dx$.

(c) Find $\displaystyle\int \cos^7 x\, dx$.

(d) Explain how to find $\displaystyle\int \cos^{15} x\, dx$ without actually integrating.

102. Finding a Pattern

(a) Write $\int \tan^3 x\, dx$ in terms of $\int \tan x\, dx$. Then find $\int \tan^3 x\, dx$.

(b) Write $\int \tan^5 x\, dx$ in terms of $\int \tan^3 x\, dx$.

(c) Write $\int \tan^{2k+1} x\, dx$, where k is a positive integer, in terms of $\int \tan^{2k-1} x\, dx$.

(d) Explain how to find $\int \tan^{15} x\, dx$ without actually integrating.

103. Methods of Integration Show that the following results are equivalent.

Integration by tables:

$$\int \sqrt{x^2 + 1}\, dx = \frac{1}{2}\left(x\sqrt{x^2 + 1} + \ln\left|x + \sqrt{x^2 + 1}\right|\right) + C$$

Integration by computer algebra system:

$$\int \sqrt{x^2 + 1}\, dx = \frac{1}{2}\left(x\sqrt{x^2 + 1} + \operatorname{arcsinh}(x)\right) + C$$

Putnam Exam Challenge

104. Evaluate $\displaystyle\int_2^4 \frac{\sqrt{\ln(9 - x)}\, dx}{\sqrt{\ln(9 - x)} + \sqrt{\ln(x + 3)}}$.

This problem was composed by the Committee on the Putnam Prize Competition. © The Mathematical Association of America. All rights reserved.

Section 8.2

Integration by Parts

- Find an antiderivative using integration by parts.
- Use a tabular method to perform integration by parts.

Integration by Parts

In this section you will study an important integration technique called **integration by parts**. This technique can be applied to a wide variety of functions and is particularly useful for integrands involving *products* of algebraic and transcendental functions. For instance, integration by parts works well with integrals such as

$$\int x \ln x \, dx, \quad \int x^2 e^x \, dx, \quad \text{and} \quad \int e^x \sin x \, dx.$$

Integration by parts is based on the formula for the derivative of a product

$$\frac{d}{dx}[uv] = u\frac{dv}{dx} + v\frac{du}{dx}$$
$$= uv' + vu'$$

where both u and v are differentiable functions of x. If u' and v' are continuous, you can integrate both sides of this equation to obtain

$$uv = \int uv' \, dx + \int vu' \, dx$$
$$= \int u \, dv + \int v \, du.$$

By rewriting this equation, you obtain the following theorem.

THEOREM 8.1 Integration by Parts

If u and v are functions of x and have continuous derivatives, then

$$\int u \, dv = uv - \int v \, du.$$

This formula expresses the original integral in terms of another integral. Depending on the choices of u and dv, it may be easier to evaluate the second integral than the original one. Because the choices of u and dv are critical in the integration by parts process, the following guidelines are provided.

Guidelines for Integration by Parts

1. Try letting dv be the most complicated portion of the integrand that fits a basic integration rule. Then u will be the remaining factor(s) of the integrand.
2. Try letting u be the portion of the integrand whose derivative is a function simpler than u. Then dv will be the remaining factor(s) of the integrand.

EXPLORATION

Proof Without Words Here is a different approach to proving the formula for integration by parts. Exercise taken from "Proof Without Words: Integration by Parts" by Roger B. Nelsen, *Mathematics Magazine*, April 1991, by permission of the author.

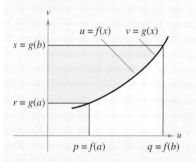

Area + Area = $qs - pr$

$$\int_r^s u \, dv + \int_q^p v \, du = \Big[uv\Big]_{(p,r)}^{(q,s)}$$

$$\int_r^s u \, dv = \Big[uv\Big]_{(p,r)}^{(q,s)} - \int_q^p v \, du$$

Explain how this graph proves the theorem. Which notation in this proof is unfamiliar? What do you think it means?

EXAMPLE 1 Integration by Parts

Find $\int xe^x \, dx$.

Solution To apply integration by parts, you need to write the integral in the form $\int u \, dv$. There are several ways to do this.

$$\int \underbrace{(x)}_{u}\underbrace{(e^x \, dx)}_{dv}, \quad \int \underbrace{(e^x)}_{u}\underbrace{(x \, dx)}_{dv}, \quad \int \underbrace{(1)}_{u}\underbrace{(xe^x \, dx)}_{dv}, \quad \int \underbrace{(xe^x)}_{u}\underbrace{(dx)}_{dv}$$

The guidelines on page 525 suggest choosing the first option because the derivative of $u = x$ is simpler than x, and $dv = e^x \, dx$ is the most complicated portion of the integrand that fits a basic integration formula.

$$dv = e^x \, dx \quad \Longrightarrow \quad v = \int dv = \int e^x \, dx = e^x$$

$$u = x \quad \Longrightarrow \quad du = dx$$

Now, integration by parts produces

$$\int u \, dv = uv - \int v \, du \qquad \text{Integration by parts formula}$$

$$\int xe^x \, dx = xe^x - \int e^x \, dx \qquad \text{Substitute.}$$

$$= xe^x - e^x + C. \qquad \text{Integrate.}$$

To check this, differentiate $xe^x - e^x + C$ to see that you obtain the original integrand.

NOTE In Example 1, note that it is not necessary to include a constant of integration when solving

$$v = \int e^x \, dx = e^x + C_1.$$

To illustrate this, replace $v = e^x$ by $v = e^x + C_1$ and apply integration by parts to see that you obtain the same result.

FOR FURTHER INFORMATION To see how integration by parts is used to prove Stirling's approximation

$$\ln(n!) = n \ln n - n$$

see the article "The Validity of Stirling's Approximation: A Physical Chemistry Project" by A. S. Wallner and K. A. Brandt in *Journal of Chemical Education*.

EXAMPLE 2 Integration by Parts

Find $\int x^2 \ln x \, dx$.

Solution In this case, x^2 is more easily integrated than $\ln x$. Furthermore, the derivative of $\ln x$ is simpler than $\ln x$. So, you should let $dv = x^2 \, dx$.

$$dv = x^2 \, dx \quad \Longrightarrow \quad v = \int x^2 \, dx = \frac{x^3}{3}$$

$$u = \ln x \quad \Longrightarrow \quad du = \frac{1}{x} \, dx$$

Integration by parts produces

$$\int u \, dv = uv - \int v \, du \qquad \text{Integration by parts formula}$$

$$\int x^2 \ln x \, dx = \frac{x^3}{3} \ln x - \int \left(\frac{x^3}{3}\right)\left(\frac{1}{x}\right) dx \qquad \text{Substitute.}$$

$$= \frac{x^3}{3} \ln x - \frac{1}{3} \int x^2 \, dx \qquad \text{Simplify.}$$

$$= \frac{x^3}{3} \ln x - \frac{x^3}{9} + C. \qquad \text{Integrate.}$$

You can check this result by differentiating.

$$\frac{d}{dx}\left[\frac{x^3}{3} \ln x - \frac{x^3}{9}\right] = \frac{x^3}{3}\left(\frac{1}{x}\right) + (\ln x)(x^2) - \frac{x^2}{3} = x^2 \ln x$$

TECHNOLOGY Try graphing

$$\int x^2 \ln x \, dx \quad \text{and} \quad \frac{x^3}{3} \ln x - \frac{x^3}{9}$$

on your graphing utility. Do you get the same graph? (This will take a while, so be patient.)

SECTION 8.2 Integration by Parts

One surprising application of integration by parts involves integrands consisting of a single term, such as $\int \ln x \, dx$ or $\int \arcsin x \, dx$. In these cases, try letting $dv = dx$, as shown in the next example.

EXAMPLE 3 An Integrand with a Single Term

Evaluate $\int_0^1 \arcsin x \, dx$.

Solution Let $dv = dx$.

$$dv = dx \implies v = \int dx = x$$

$$u = \arcsin x \implies du = \frac{1}{\sqrt{1-x^2}} \, dx$$

Integration by parts now produces

$$\int u \, dv = uv - \int v \, du \qquad \text{Integration by parts formula}$$

$$\int \arcsin x \, dx = x \arcsin x - \int \frac{x}{\sqrt{1-x^2}} \, dx \qquad \text{Substitute.}$$

$$= x \arcsin x + \frac{1}{2} \int (1-x^2)^{-1/2}(-2x) \, dx \qquad \text{Rewrite.}$$

$$= x \arcsin x + \sqrt{1-x^2} + C. \qquad \text{Integrate.}$$

Using this antiderivative, you can evaluate the definite integral as follows.

$$\int_0^1 \arcsin x \, dx = \left[x \arcsin x + \sqrt{1-x^2} \right]_0^1$$

$$= \frac{\pi}{2} - 1$$

$$\approx 0.571$$

The area represented by this definite integral is shown in Figure 8.2.

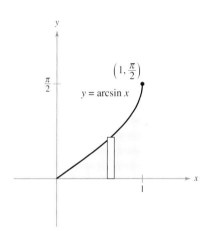

The area of the region is approximately 0.571.
Figure 8.2

TECHNOLOGY Remember that there are two ways to use technology to evaluate a definite integral: (1) you can use a numerical approximation such as the Trapezoidal Rule or Simpson's Rule, or (2) you can use a computer algebra system to find the antiderivative and then apply the Fundamental Theorem of Calculus. Both methods have shortcomings. To find the possible error when using a numerical method, the integrand must have a second derivative (Trapezoidal Rule) or a fourth derivative (Simpson's Rule) in the interval of integration: the integrand in Example 3 fails to meet either of these requirements. To apply the Fundamental Theorem of Calculus, the symbolic integration utility must be able to find the antiderivative.

Which method would you use to evaluate

$$\int_0^1 \arctan x \, dx?$$

Which method would you use to evaluate

$$\int_0^1 \arctan x^2 \, dx?$$

Some integrals require repeated use of the integration by parts formula.

EXAMPLE 4 Repeated Use of Integration by Parts

Find $\int x^2 \sin x \, dx$.

Solution The factors x^2 and $\sin x$ are equally easy to integrate. However, the derivative of x^2 becomes simpler, whereas the derivative of $\sin x$ does not. So, you should let $u = x^2$.

$$dv = \sin x \, dx \implies v = \int \sin x \, dx = -\cos x$$

$$u = x^2 \implies du = 2x \, dx$$

Now, integration by parts produces

$$\int x^2 \sin x \, dx = -x^2 \cos x + \int 2x \cos x \, dx. \qquad \text{First use of integration by parts}$$

This first use of integration by parts has succeeded in simplifying the original integral, but the integral on the right still doesn't fit a basic integration rule. To evaluate that integral, you can apply integration by parts again. This time, let $u = 2x$.

$$dv = \cos x \, dx \implies v = \int \cos x \, dx = \sin x$$

$$u = 2x \implies du = 2 \, dx$$

Now, integration by parts produces

$$\int 2x \cos x \, dx = 2x \sin x - \int 2 \sin x \, dx \qquad \text{Second use of integration by parts}$$

$$= 2x \sin x + 2 \cos x + C.$$

Combining these two results, you can write

$$\int x^2 \sin x \, dx = -x^2 \cos x + 2x \sin x + 2 \cos x + C.$$

When making repeated applications of integration by parts, you need to be careful not to interchange the substitutions in successive applications. For instance, in Example 4, the first substitution was $u = x^2$ and $dv = \sin x \, dx$. If, in the second application, you had switched the substitution to $u = \cos x$ and $dv = 2x$, you would have obtained

$$\int x^2 \sin x \, dx = -x^2 \cos x + \int 2x \cos x \, dx$$

$$= -x^2 \cos x + x^2 \cos x + \int x^2 \sin x \, dx$$

$$= \int x^2 \sin x \, dx$$

thereby undoing the previous integration and returning to the *original* integral. When making repeated applications of integration by parts, you should also watch for the appearance of a *constant multiple* of the original integral. For instance, this occurs when you use integration by parts to evaluate $\int e^x \cos 2x \, dx$, and also occurs in the next example.

EXPLORATION

Try to find

$$\int e^x \cos 2x \, dx$$

by letting $u = \cos 2x$ and $dv = e^x \, dx$ in the first substitution. For the second substitution, let $u = \sin 2x$ and $dv = e^x \, dx$.

SECTION 8.2 Integration by Parts

NOTE The integral in Example 5 is an important one. In Section 8.4 (Example 5), you will see that it is used to find the arc length of a parabolic segment.

EXAMPLE 5 Integration by Parts

Find $\int \sec^3 x \, dx$.

Solution The most complicated portion of the integrand that can be easily integrated is $\sec^2 x$, so you should let $dv = \sec^2 x \, dx$ and $u = \sec x$.

$$dv = \sec^2 x \, dx \implies v = \int \sec^2 x \, dx = \tan x$$

$$u = \sec x \implies du = \sec x \tan x \, dx$$

Integration by parts produces

$$\int u \, dv = uv - \int v \, du \qquad \text{Integration by parts formula}$$

$$\int \sec^3 x \, dx = \sec x \tan x - \int \sec x \tan^2 x \, dx \qquad \text{Substitute.}$$

$$\int \sec^3 x \, dx = \sec x \tan x - \int \sec x (\sec^2 x - 1) \, dx \qquad \text{Trigonometric identity}$$

$$\int \sec^3 x \, dx = \sec x \tan x - \int \sec^3 x \, dx + \int \sec x \, dx \qquad \text{Rewrite.}$$

$$2 \int \sec^3 x \, dx = \sec x \tan x + \int \sec x \, dx \qquad \text{Collect like integrals.}$$

$$\int \sec^3 x \, dx = \frac{1}{2} \sec x \tan x + \frac{1}{2} \ln|\sec x + \tan x| + C. \qquad \text{Integrate and divide by 2.}$$

STUDY TIP The trigonometric identities

$$\sin^2 x = \frac{1 - \cos 2x}{2}$$

$$\cos^2 x = \frac{1 + \cos 2x}{2}$$

play an important role in this chapter.

EXAMPLE 6 Finding a Centroid

A machine part is modeled by the region bounded by the graph of $y = \sin x$ and the x-axis, $0 \leq x \leq \pi/2$, as shown in Figure 8.3. Find the centroid of this region.

Solution Begin by finding the area of the region.

$$A = \int_0^{\pi/2} \sin x \, dx = \Big[-\cos x \Big]_0^{\pi/2} = 1$$

Now, you can find the coordinates of the centroid as follows.

$$\bar{y} = \frac{1}{A} \int_0^{\pi/2} \frac{\sin x}{2} (\sin x) \, dx = \frac{1}{4} \int_0^{\pi/2} (1 - \cos 2x) \, dx = \frac{1}{4} \Big[x - \frac{\sin 2x}{2} \Big]_0^{\pi/2} = \frac{\pi}{8}$$

You can evaluate the integral for $\bar{x}$, $(1/A) \int_0^{\pi/2} x \sin x \, dx$, with integration by parts. To do this, let $dv = \sin x \, dx$ and $u = x$. This produces $v = -\cos x$ and $du = dx$, and you can write

$$\int x \sin x \, dx = -x \cos x + \int \cos x \, dx$$

$$= -x \cos x + \sin x + C.$$

Finally, you can determine $\bar{x}$ to be

$$\bar{x} = \frac{1}{A} \int_0^{\pi/2} x \sin x \, dx = \Big[-x \cos x + \sin x \Big]_0^{\pi/2} = 1.$$

So, the centroid of the region is $(1, \pi/8)$.

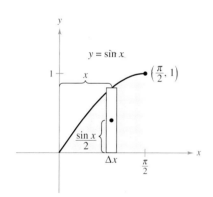

Figure 8.3

STUDY TIP You can use the acronym LIATE as a guideline for choosing u in integration by parts. In order, check the integrand for the following.

Is there a **L**ogarithmic part?
Is there an **I**nverse trigonometric part?
Is there an **A**lgebraic part?
Is there a **T**rigonometric part?
Is there an **E**xponential part?

As you gain experience in using integration by parts, your skill in determining u and dv will increase. The following summary lists several common integrals with suggestions for the choices of u and dv.

Summary of Common Integrals Using Integration by Parts

1. For integrals of the form

$$\int x^n e^{ax}\, dx, \quad \int x^n \sin ax\, dx, \quad \text{or} \quad \int x^n \cos ax\, dx$$

let $u = x^n$ and let $dv = e^{ax}\, dx$, $\sin ax\, dx$, or $\cos ax\, dx$.

2. For integrals of the form

$$\int x^n \ln x\, dx, \quad \int x^n \arcsin ax\, dx, \quad \text{or} \quad \int x^n \arctan ax\, dx$$

let $u = \ln x$, $\arcsin ax$, or $\arctan ax$ and let $dv = x^n\, dx$.

3. For integrals of the form

$$\int e^{ax} \sin bx\, dx \quad \text{or} \quad \int e^{ax} \cos bx\, dx$$

let $u = \sin bx$ or $\cos bx$ and let $dv = e^{ax}\, dx$.

Tabular Method

In problems involving repeated applications of integration by parts, a tabular method, illustrated in Example 7, can help to organize the work. This method works well for integrals of the form $\int x^n \sin ax\, dx$, $\int x^n \cos ax\, dx$, and $\int x^n e^{ax}\, dx$.

EXAMPLE 7 Using the Tabular Method

Find $\int x^2 \sin 4x\, dx$.

Solution Begin as usual by letting $u = x^2$ and $dv = v'\, dx = \sin 4x\, dx$. Next, create a table consisting of three columns, as shown.

Alternate Signs	u and Its Derivatives	v' and Its Antiderivatives
+	x^2	$\sin 4x$
−	$2x$	$-\frac{1}{4}\cos 4x$
+	2	$-\frac{1}{16}\sin 4x$
−	0	$\frac{1}{64}\cos 4x$

Differentiate until you obtain 0 as a derivative.

The solution is obtained by adding the signed products of the diagonal entries:

$$\int x^2 \sin 4x\, dx = -\frac{1}{4}x^2 \cos 4x + \frac{1}{8}x \sin 4x + \frac{1}{32}\cos 4x + C.$$

FOR FURTHER INFORMATION For more information on the tabular method, see the article "Tabular Integration by Parts" by David Horowitz in *The College Mathematics Journal*, and the article "More on Tabular Integration by Parts" by Leonard Gillman in *The College Mathematics Journal*. To view these articles, go to the website *www.matharticles.com*.

Exercises for Section 8.2

In Exercises 1–4, match the antiderivative with the correct integral. [Integrals are labeled (a), (b), (c), and (d).]

(a) $\int \ln x \, dx$ (b) $\int x \sin x \, dx$

(c) $\int x^2 e^x \, dx$ (d) $\int x^2 \cos x \, dx$

1. $y = \sin x - x \cos x$
2. $y = x^2 \sin x + 2x \cos x - 2 \sin x$
3. $y = x^2 e^x - 2xe^x + 2e^x$
4. $y = -x + x \ln x$

In Exercises 5–10, identify u and dv for finding the integral using integration by parts. (Do not evaluate the integral.)

5. $\int xe^{2x} \, dx$
6. $\int x^2 e^{2x} \, dx$
7. $\int (\ln x)^2 \, dx$
8. $\int \ln 3x \, dx$
9. $\int x \sec^2 x \, dx$
10. $\int x^2 \cos x \, dx$

In Exercises 11–36, find the integral. (*Note:* Solve by the simplest method—not all require integration by parts.)

11. $\int xe^{-2x} \, dx$
12. $\int \dfrac{2x}{e^x} \, dx$
13. $\int x^3 e^x \, dx$
14. $\int \dfrac{e^{1/t}}{t^2} \, dt$
15. $\int x^2 e^{x^3} \, dx$
16. $\int x^4 \ln x \, dx$
17. $\int t \ln(t + 1) \, dt$
18. $\int \dfrac{1}{x(\ln x)^3} \, dx$
19. $\int \dfrac{(\ln x)^2}{x} \, dx$
20. $\int \dfrac{\ln x}{x^2} \, dx$
21. $\int \dfrac{xe^{2x}}{(2x + 1)^2} \, dx$
22. $\int \dfrac{x^3 e^{x^2}}{(x^2 + 1)^2} \, dx$
23. $\int (x^2 - 1)e^x \, dx$
24. $\int \dfrac{\ln 2x}{x^2} \, dx$
25. $\int x\sqrt{x - 1} \, dx$
26. $\int \dfrac{x}{\sqrt{2 + 3x}} \, dx$
27. $\int x \cos x \, dx$
28. $\int x \sin x \, dx$
29. $\int x^3 \sin x \, dx$
30. $\int x^2 \cos x \, dx$
31. $\int t \csc t \cot t \, dt$
32. $\int \theta \sec \theta \tan \theta \, d\theta$
33. $\int \arctan x \, dx$
34. $\int 4 \arccos x \, dx$
35. $\int e^{2x} \sin x \, dx$
36. $\int e^x \cos 2x \, dx$

In Exercises 37–42, solve the differential equation.

37. $y' = xe^{x^2}$
38. $y' = \ln x$
39. $\dfrac{dy}{dt} = \dfrac{t^2}{\sqrt{2 + 3t}}$
40. $\dfrac{dy}{dx} = x^2\sqrt{x - 1}$
41. $(\cos y)y' = 2x$
42. $y' = \arctan \dfrac{x}{2}$

Slope Fields In Exercises 43 and 44, a differential equation, a point, and a slope field are given. (a) Sketch two approximate solutions of the differential equation on the slope field, one of which passes through the given point. (b) Use integration to find the particular solution of the differential equation and use a graphing utility to graph the solution. Compare the result with the sketches in part (a). To print an enlarged copy of the graph, go to the website *www.mathgraphs.com*.

43. $\dfrac{dy}{dx} = x\sqrt{y} \cos x, \ (0, 4)$ 44. $\dfrac{dy}{dx} = e^{-x/3} \sin 2x, \ \left(0, -\tfrac{18}{37}\right)$

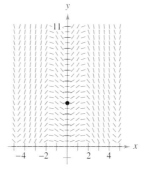 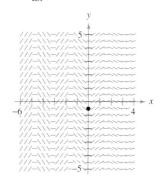

Slope Fields In Exercises 45 and 46, use a computer algebra system to graph the slope field for the differential equation and graph the solution through the specified initial condition.

45. $\dfrac{dy}{dx} = \dfrac{x}{y}e^{x/8}$
   46. $\dfrac{dy}{dx} = \dfrac{x}{y} \sin x$
$\ \ \ y(0) = 2$ $y(0) = 4$

In Exercises 47–58, evaluate the definite integral. Use a graphing utility to confirm your result.

47. $\displaystyle\int_0^4 xe^{-x/2} \, dx$
48. $\displaystyle\int_0^1 x^2 e^x \, dx$
49. $\displaystyle\int_0^{\pi/2} x \cos x \, dx$
50. $\displaystyle\int_0^{\pi} x \sin 2x \, dx$
51. $\displaystyle\int_0^{1/2} \arccos x \, dx$
52. $\displaystyle\int_0^1 x \arcsin x^2 \, dx$
53. $\displaystyle\int_0^1 e^x \sin x \, dx$
54. $\displaystyle\int_0^2 e^{-x} \cos x \, dx$
55. $\displaystyle\int_1^2 x^2 \ln x \, dx$
56. $\displaystyle\int_0^1 \ln(1 + x^2) \, dx$
57. $\displaystyle\int_2^4 x \, \text{arcsec} \, x \, dx$
58. $\displaystyle\int_0^{\pi/4} x \sec^2 x \, dx$

In Exercises 59–64, use the tabular method to find the integral.

59. $\int x^2 e^{2x}\, dx$

60. $\int x^3 e^{-2x}\, dx$

61. $\int x^3 \sin x\, dx$

62. $\int x^3 \cos 2x\, dx$

63. $\int x \sec^2 x\, dx$

64. $\int x^2(x-2)^{3/2}\, dx$

In Exercises 65–70, find or evaluate the integral using substitution first, then using integration by parts.

65. $\int \sin \sqrt{x}\, dx$

66. $\int 2x^3 \cos x^2\, dx$

67. $\int_0^4 x\sqrt{4-x}\, dx$

68. $\int_0^2 e^{\sqrt{2x}}\, dx$

69. $\int \cos(\ln x)\, dx$

70. $\int \ln(x^2+1)\, dx$

Writing About Concepts

71. Integration by parts is based on what differentiation rule? Explain.

72. In your own words, state guidelines for integration by parts.

In Exercises 73–78, state whether you would use integration by parts to evaluate the integral. If so, identify what you would use for u and dv. Explain your reasoning.

73. $\int \dfrac{\ln x}{x}\, dx$

74. $\int x \ln x\, dx$

75. $\int x^2 e^{2x}\, dx$

76. $\int 2x\, e^{x^2}\, dx$

77. $\int \dfrac{x}{\sqrt{x+1}}\, dx$

78. $\int \dfrac{x}{\sqrt{x^2+1}}\, dx$

In Exercises 79–82, use a computer algebra system to (a) find or evaluate the integral and (b) graph two antiderivatives. (c) Describe the relationship between the graphs of the antiderivatives.

79. $\int t^3 e^{-4t}\, dt$

80. $\int \alpha^4 \sin \pi\alpha\, d\alpha$

81. $\int_0^{\pi/2} e^{-2x} \sin 3x\, dx$

82. $\int_0^5 x^4(25-x^2)^{3/2}\, dx$

83. Integrate $\int 2x\sqrt{2x-3}\, dx$

(a) by parts, letting $dv = \sqrt{2x-3}\, dx$.

(b) by substitution, letting $u = 2x-3$.

84. Integrate $\int x\sqrt{4+x}\, dx$

(a) by parts, letting $dv = \sqrt{4+x}\, dx$.

(b) by substitution, letting $u = 4+x$.

85. Integrate $\int \dfrac{x^3}{\sqrt{4+x^2}}\, dx$

(a) by parts, letting $dv = \left(x/\sqrt{4+x^2}\right)dx$.

(b) by substitution, letting $u = 4+x^2$.

86. Integrate $\int x\sqrt{4-x}\, dx$

(a) by parts, letting $dv = \sqrt{4-x}\, dx$.

(b) by substitution, letting $u = 4-x$.

In Exercises 87 and 88, use a computer algebra system to find the integral for $n = 0, 1, 2,$ and 3. Use the result to obtain a general rule for the integral for any positive integer n and test your results for $n = 4$.

87. $\int x^n \ln x\, dx$

88. $\int x^n e^x\, dx$

In Exercises 89–94, use integration by parts to verify the formula. (For Exercises 89–92, assume that n is a positive integer.)

89. $\int x^n \sin x\, dx = -x^n \cos x + n \int x^{n-1} \cos x\, dx$

90. $\int x^n \cos x\, dx = x^n \sin x - n \int x^{n-1} \sin x\, dx$

91. $\int x^n \ln x\, dx = \dfrac{x^{n+1}}{(n+1)^2}[-1+(n+1)\ln x] + C$

92. $\int x^n e^{ax}\, dx = \dfrac{x^n e^{ax}}{a} - \dfrac{n}{a}\int x^{n-1} e^{ax}\, dx$

93. $\int e^{ax} \sin bx\, dx = \dfrac{e^{ax}(a \sin bx - b \cos bx)}{a^2+b^2} + C$

94. $\int e^{ax} \cos bx\, dx = \dfrac{e^{ax}(a \cos bx + b \sin bx)}{a^2+b^2} + C$

In Exercises 95–98, find the integral by using the appropriate formula from Exercises 89–94.

95. $\int x^3 \ln x\, dx$

96. $\int x^2 \cos x\, dx$

97. $\int e^{2x} \cos 3x\, dx$

98. $\int x^3 e^{2x}\, dx$

Area In Exercises 99–102, use a graphing utility to graph the region bounded by the graphs of the equations, and find the area of the region.

99. $y = xe^{-x},\ y = 0,\ x = 4$

100. $y = \tfrac{1}{9}xe^{-x/3},\ y = 0,\ x = 0,\ x = 3$

101. $y = e^{-x} \sin \pi x,\ y = 0,\ x = 0,\ x = 1$

102. $y = x \sin x,\ y = 0,\ x = 0,\ x = \pi$

103. *Area, Volume, and Centroid* Given the region bounded by the graphs of $y = \ln x$, $y = 0$, and $x = e$, find

(a) the area of the region.

(b) the volume of the solid generated by revolving the region about the x-axis.

(c) the volume of the solid generated by revolving the region about the y-axis.

(d) the centroid of the region.

104. *Volume and Centroid* Given the region bounded by the graphs of $y = x \sin x$, $y = 0$, $x = 0$, and $x = \pi$, find

(a) the volume of the solid generated by revolving the region about the x-axis.

(b) the volume of the solid generated by revolving the region about the y-axis.

(c) the centroid of the region.

105. *Centroid* Find the centroid of the region bounded by the graphs of $y = \arcsin x$, $x = 0$, and $y = \pi/2$. How is this problem related to Example 6 in this section?

106. *Centroid* Find the centroid of the region bounded by the graphs of $f(x) = x^2$, $g(x) = 2^x$, $x = 2$, and $x = 4$.

107. *Average Displacement* A damping force affects the vibration of a spring so that the displacement of the spring is given by $y = e^{-4t}(\cos 2t + 5 \sin 2t)$. Find the average value of y on the interval from $t = 0$ to $t = \pi$.

108. *Memory Model* A model for the ability M of a child to memorize, measured on a scale from 0 to 10, is given by $M = 1 + 1.6t \ln t$, $0 < t \leq 4$, where t is the child's age in years. Find the average value of this model

(a) between the child's first and second birthdays.

(b) between the child's third and fourth birthdays.

Present Value In Exercises 109 and 110, find the present value P of a continuous income flow of $c(t)$ dollars per year if

$$P = \int_0^{t_1} c(t)e^{-rt}\, dt$$

where t_1 is the time in years and r is the annual interest rate compounded continuously.

109. $c(t) = 100{,}000 + 4000t$, $r = 5\%$, $t_1 = 10$

110. $c(t) = 30{,}000 + 500t$, $r = 7\%$, $t_1 = 5$

Integrals Used to Find Fourier Coefficients In Exercises 111 and 112, verify the value of the definite integral, where n is a positive integer.

111. $\displaystyle\int_{-\pi}^{\pi} x \sin nx\, dx = \begin{cases} \dfrac{2\pi}{n}, & n \text{ is odd} \\ -\dfrac{2\pi}{n}, & n \text{ is even} \end{cases}$

112. $\displaystyle\int_{-\pi}^{\pi} x^2 \cos nx\, dx = \dfrac{(-1)^n 4\pi}{n^2}$

113. *Vibrating String* A string stretched between the two points $(0, 0)$ and $(2, 0)$ is plucked by displacing the string h units at its midpoint. The motion of the string is modeled by a **Fourier Sine Series** whose coefficients are given by

$$b_n = h\int_0^1 x \sin\frac{n\pi x}{2}\, dx + h\int_1^2 (-x + 2)\sin\frac{n\pi x}{2}\, dx.$$

Find b_n.

114. Find the fallacy in the following argument that $0 = 1$.

$$dv = dx \quad \Longrightarrow \quad v = \int dx = x$$

$$u = \frac{1}{x} \quad \Longrightarrow \quad du = -\frac{1}{x^2}\, dx$$

$$0 + \int \frac{dx}{x} = \left(\frac{1}{x}\right)(x) - \int \left(-\frac{1}{x^2}\right)(x)\, dx = 1 + \int \frac{dx}{x}$$

So, $0 = 1$.

115. Let $y = f(x)$ be positive and strictly increasing on the interval $0 < a \leq x \leq b$. Consider the region R bounded by the graphs of $y = f(x)$, $y = 0$, $x = a$, and $x = b$. If R is revolved about the y-axis, show that the disk method and shell method yield the same volume.

116. *Euler's Method* Consider the differential equation $f'(x) = xe^{-x}$ with the initial condition $f(0) = 0$.

(a) Use integration to solve the differential equation.

(b) Use a graphing utility to graph the solution of the differential equation.

(c) Use Euler's Method with $h = 0.05$, and the recursive capabilities of a graphing utility, to generate the first 80 points of the graph of the approximate solution. Use the graphing utility to plot the points. Compare the result with the graph in part (b).

(d) Repeat part (c) using $h = 0.1$ and generate the first 40 points.

(e) Why is the result in part (c) a better approximation of the solution than the result in part (d)?

Euler's Method In Exercises 117 and 118, consider the differential equation and repeat parts (a)–(d) of Exercise 116.

117. $f'(x) = 3x \sin(2x)$
$f(0) = 0$

118. $f'(x) = \cos\sqrt{x}$
$f(0) = 1$

119. *Think About It* Give a geometric explanation to explain why

$$\int_0^{\pi/2} x \sin x\, dx \leq \int_0^{\pi/2} x\, dx.$$

Verify the inequality by evaluating the integrals.

120. *Finding a Pattern* Find the area bounded by the graphs of $y = x \sin x$ and $y = 0$ over each interval.

(a) $[0, \pi]$ (b) $[\pi, 2\pi]$ (c) $[2\pi, 3\pi]$

Describe any patterns that you notice. What is the area between the graphs of $y = x \sin x$ and $y = 0$ over the interval $[n\pi, (n + 1)\pi]$, where n is any nonnegative integer? Explain.

Section 8.3 Trigonometric Integrals

- Solve trigonometric integrals involving powers of sine and cosine.
- Solve trigonometric integrals involving powers of secant and tangent.
- Solve trigonometric integrals involving sine-cosine products with different angles.

Integrals Involving Powers of Sine and Cosine

In this section you will study techniques for evaluating integrals of the form

$$\int \sin^m x \cos^n x \, dx \quad \text{and} \quad \int \sec^m x \tan^n x \, dx$$

where either m or n is a positive integer. To find antiderivatives for these forms, try to break them into combinations of trigonometric integrals to which you can apply the Power Rule.

For instance, you can evaluate $\int \sin^5 x \cos x \, dx$ with the Power Rule by letting $u = \sin x$. Then, $du = \cos x \, dx$ and you have

$$\int \sin^5 x \cos x \, dx = \int u^5 \, du = \frac{u^6}{6} + C = \frac{\sin^6 x}{6} + C.$$

To break up $\int \sin^m x \cos^n x \, dx$ into forms to which you can apply the Power Rule, use the following identities.

$\sin^2 x + \cos^2 x = 1$ Pythagorean identity

$\sin^2 x = \dfrac{1 - \cos 2x}{2}$ Half-angle identity for $\sin^2 x$

$\cos^2 x = \dfrac{1 + \cos 2x}{2}$ Half-angle identity for $\cos^2 x$

SHEILA SCOTT MACINTYRE (1910–1960)

Sheila Scott Macintyre published her first paper on the asymptotic periods of integral functions in 1935. She completed her doctorate work at Aberdeen University, where she taught. In 1958 she accepted a visiting research fellowship at the University of Cincinnati.

Guidelines for Evaluating Integrals Involving Sine and Cosine

1. If the power of the sine is odd and positive, save one sine factor and convert the remaining factors to cosines. Then, expand and integrate.

$$\int \sin^{2k+1} x \cos^n x \, dx = \int (\sin^2 x)^k \cos^n x \sin x \, dx = \int (1 - \cos^2 x)^k \cos^n x \sin x \, dx$$

(Odd; Convert to cosines; Save for du)

2. If the power of the cosine is odd and positive, save one cosine factor and convert the remaining factors to sines. Then, expand and integrate.

$$\int \sin^m x \cos^{2k+1} x \, dx = \int \sin^m x (\cos^2 x)^k \cos x \, dx = \int \sin^m x (1 - \sin^2 x)^k \cos x \, dx$$

(Odd; Convert to sines; Save for du)

3. If the powers of both the sine and cosine are even and nonnegative, make repeated use of the identities

$$\sin^2 x = \frac{1 - \cos 2x}{2} \quad \text{and} \quad \cos^2 x = \frac{1 + \cos 2x}{2}$$

to convert the integrand to odd powers of the cosine. Then proceed as in guideline 2.

TECHNOLOGY Use a computer algebra system to find the integral in Example 1. You should obtain

$$\int \sin^3 x \cos^4 x \, dx = -\cos^5 x \left(\frac{1}{7} \sin^2 x + \frac{2}{35} \right) + C.$$

Is this equivalent to the result obtained in Example 1?

EXAMPLE 1   Power of Sine Is Odd and Positive

Find $\int \sin^3 x \cos^4 x \, dx$.

Solution   Because you expect to use the Power Rule with $u = \cos x$, *save one sine factor* to form du and convert the remaining sine factors to cosines.

$$\int \sin^3 x \cos^4 x \, dx = \int \sin^2 x \cos^4 x (\sin x) \, dx \quad \text{Rewrite.}$$
$$= \int (1 - \cos^2 x) \cos^4 x \sin x \, dx \quad \text{Trigonometric identity}$$
$$= \int (\cos^4 x - \cos^6 x) \sin x \, dx \quad \text{Multiply.}$$
$$= \int \cos^4 x \sin x \, dx - \int \cos^6 x \sin x \, dx \quad \text{Rewrite.}$$
$$= -\int \cos^4 x (-\sin x) \, dx + \int \cos^6 x (-\sin x) \, dx$$
$$= -\frac{\cos^5 x}{5} + \frac{\cos^7 x}{7} + C \quad \text{Integrate.}$$

In Example 1, *both* of the powers m and n happened to be positive integers. However, the same strategy will work as long as either m or n is odd and positive. For instance, in the next example the power of the cosine is 3, but the power of the sine is $-\frac{1}{2}$.

EXAMPLE 2   Power of Cosine Is Odd and Positive

Evaluate $\int_{\pi/6}^{\pi/3} \frac{\cos^3 x}{\sqrt{\sin x}} \, dx$.

Solution   Because you expect to use the Power Rule with $u = \sin x$, *save one cosine factor* to form du and convert the remaining cosine factors to sines.

$$\int_{\pi/6}^{\pi/3} \frac{\cos^3 x}{\sqrt{\sin x}} \, dx = \int_{\pi/6}^{\pi/3} \frac{\cos^2 x \cos x}{\sqrt{\sin x}} \, dx$$
$$= \int_{\pi/6}^{\pi/3} \frac{(1 - \sin^2 x)(\cos x)}{\sqrt{\sin x}} \, dx$$
$$= \int_{\pi/6}^{\pi/3} [(\sin x)^{-1/2} \cos x - (\sin x)^{3/2} \cos x] \, dx$$
$$= \left[\frac{(\sin x)^{1/2}}{1/2} - \frac{(\sin x)^{5/2}}{5/2} \right]_{\pi/6}^{\pi/3}$$
$$= 2 \left(\frac{\sqrt{3}}{2} \right)^{1/2} - \frac{2}{5} \left(\frac{\sqrt{3}}{2} \right)^{5/2} - \sqrt{2} + \frac{\sqrt{32}}{80}$$
$$\approx 0.239$$

Figure 8.4 shows the region whose area is represented by this integral.

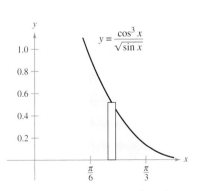

The area of the region is approximately 0.239.
Figure 8.4

EXAMPLE 3 Power of Cosine Is Even and Nonnegative

Find $\int \cos^4 x \, dx$.

Solution Because m and n are both even and nonnegative ($m = 0$), you can replace $\cos^4 x$ by $[(1 + \cos 2x)/2]^2$.

$$\int \cos^4 x \, dx = \int \left(\frac{1 + \cos 2x}{2}\right)^2 dx$$

$$= \int \left(\frac{1}{4} + \frac{\cos 2x}{2} + \frac{\cos^2 2x}{4}\right) dx$$

$$= \int \left[\frac{1}{4} + \frac{\cos 2x}{2} + \frac{1}{4}\left(\frac{1 + \cos 4x}{2}\right)\right] dx$$

$$= \frac{3}{8}\int dx + \frac{1}{4}\int 2 \cos 2x \, dx + \frac{1}{32}\int 4 \cos 4x \, dx$$

$$= \frac{3x}{8} + \frac{\sin 2x}{4} + \frac{\sin 4x}{32} + C$$

Use a symbolic differentiation utility to verify this. Can you simplify the derivative to obtain the original integrand?

In Example 3, if you were to evaluate the definite integral from 0 to $\pi/2$, you would obtain

$$\int_0^{\pi/2} \cos^4 x \, dx = \left[\frac{3x}{8} + \frac{\sin 2x}{4} + \frac{\sin 4x}{32}\right]_0^{\pi/2}$$

$$= \left(\frac{3\pi}{16} + 0 + 0\right) - (0 + 0 + 0)$$

$$= \frac{3\pi}{16}.$$

Note that the only term that contributes to the solution is $3x/8$. This observation is generalized in the following formulas developed by John Wallis.

JOHN WALLIS (1616–1703)

Wallis did much of his work in calculus prior to Newton and Leibniz, and he influenced the thinking of both of these men. Wallis is also credited with introducing the present symbol (∞) for infinity.

Wallis's Formulas

1. If n is odd ($n \geq 3$), then

$$\int_0^{\pi/2} \cos^n x \, dx = \left(\frac{2}{3}\right)\left(\frac{4}{5}\right)\left(\frac{6}{7}\right) \cdots \left(\frac{n-1}{n}\right).$$

2. If n is even ($n \geq 2$), then

$$\int_0^{\pi/2} \cos^n x \, dx = \left(\frac{1}{2}\right)\left(\frac{3}{4}\right)\left(\frac{5}{6}\right) \cdots \left(\frac{n-1}{n}\right)\left(\frac{\pi}{2}\right).$$

These formulas are also valid if $\cos^n x$ is replaced by $\sin^n x$. (You are asked to prove both formulas in Exercise 104.)

Integrals Involving Powers of Secant and Tangent

The following guidelines can help you evaluate integrals of the form

$$\int \sec^m x \tan^n x \, dx.$$

Guidelines for Evaluating Integrals Involving Secant and Tangent

1. If the power of the secant is even and positive, save a secant-squared factor and convert the remaining factors to tangents. Then expand and integrate.

$$\int \sec^{2k} x \tan^n x \, dx = \int \underbrace{(\sec^2 x)^{k-1}}_{\text{Convert to tangents}} \tan^n x \underbrace{\sec^2 x \, dx}_{\text{Save for } du} = \int (1 + \tan^2 x)^{k-1} \tan^n x \sec^2 x \, dx$$

2. If the power of the tangent is odd and positive, save a secant-tangent factor and convert the remaining factors to secants. Then expand and integrate.

$$\int \sec^m x \tan^{2k+1} x \, dx = \int \sec^{m-1} x \underbrace{(\tan^2 x)^k}_{\text{Convert to secants}} \underbrace{\sec x \tan x \, dx}_{\text{Save for } du} = \int \sec^{m-1} x (\sec^2 x - 1)^k \sec x \tan x \, dx$$

3. If there are no secant factors and the power of the tangent is even and positive, convert a tangent-squared factor to a secant-squared factor, then expand and repeat if necessary.

$$\int \tan^n x \, dx = \int \tan^{n-2} x \underbrace{(\tan^2 x)}_{\text{Convert to secants}} dx = \int \tan^{n-2} x (\sec^2 x - 1) \, dx$$

4. If the integral is of the form $\int \sec^m x \, dx$, where m is odd and positive, use integration by parts, as illustrated in Example 5 in the preceding section.

5. If none of the first four guidelines applies, try converting to sines and cosines.

EXAMPLE 4 Power of Tangent Is Odd and Positive

Find $\displaystyle\int \frac{\tan^3 x}{\sqrt{\sec x}} \, dx$.

Solution Because you expect to use the Power Rule with $u = \sec x$, *save a factor of* $(\sec x \tan x)$ to form du and convert the remaining tangent factors to secants.

$$\int \frac{\tan^3 x}{\sqrt{\sec x}} \, dx = \int (\sec x)^{-1/2} \tan^3 x \, dx$$

$$= \int (\sec x)^{-3/2} (\tan^2 x)(\sec x \tan x) \, dx$$

$$= \int (\sec x)^{-3/2} (\sec^2 x - 1)(\sec x \tan x) \, dx$$

$$= \int [(\sec x)^{1/2} - (\sec x)^{-3/2}](\sec x \tan x) \, dx$$

$$= \frac{2}{3}(\sec x)^{3/2} + 2(\sec x)^{-1/2} + C$$

NOTE In Example 5, the power of the tangent is odd and positive. So, you could also find the integral using the procedure described in guideline 2 on page 537. In Exercise 85, you are asked to show that the results obtained by these two procedures differ only by a constant.

EXAMPLE 5 **Power of Secant Is Even and Positive**

Find $\int \sec^4 3x \tan^3 3x \, dx$.

Solution Let $u = \tan 3x$, then $du = 3 \sec^2 3x \, dx$ and you can write

$$\int \sec^4 3x \tan^3 3x \, dx = \int \sec^2 3x \tan^3 3x (\sec^2 3x) \, dx$$

$$= \int (1 + \tan^2 3x) \tan^3 3x (\sec^2 3x) \, dx$$

$$= \frac{1}{3} \int (\tan^3 3x + \tan^5 3x)(3 \sec^2 3x) \, dx$$

$$= \frac{1}{3} \left(\frac{\tan^4 3x}{4} + \frac{\tan^6 3x}{6} \right) + C$$

$$= \frac{\tan^4 3x}{12} + \frac{\tan^6 3x}{18} + C.$$

EXAMPLE 6 **Power of Tangent Is Even**

Evaluate $\int_0^{\pi/4} \tan^4 x \, dx$.

Solution Because there are no secant factors, you can begin by converting a tangent-squared factor to a secant-squared factor.

$$\int \tan^4 x \, dx = \int \tan^2 x (\tan^2 x) \, dx$$

$$= \int \tan^2 x (\sec^2 x - 1) \, dx$$

$$= \int \tan^2 x \sec^2 x \, dx - \int \tan^2 x \, dx$$

$$= \int \tan^2 x \sec^2 x \, dx - \int (\sec^2 x - 1) \, dx$$

$$= \frac{\tan^3 x}{3} - \tan x + x + C$$

You can evaluate the definite integral as follows.

$$\int_0^{\pi/4} \tan^4 x \, dx = \left[\frac{\tan^3 x}{3} - \tan x + x \right]_0^{\pi/4}$$

$$= \frac{\pi}{4} - \frac{2}{3}$$

$$\approx 0.119$$

The area represented by the definite integral is shown in Figure 8.5. Try using Simpson's Rule to approximate this integral. With $n = 18$, you should obtain an approximation that is within 0.00001 of the actual value.

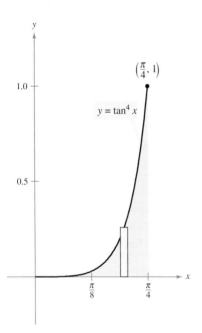

The area of the region is approximately 0.119.
Figure 8.5

For integrals involving powers of cotangents and cosecants, you can follow a strategy similar to that used for powers of tangents and secants. Also, when integrating trigonometric functions, remember that it sometimes helps to convert the entire integrand to powers of sines and cosines.

EXAMPLE 7 Converting to Sines and Cosines

Find $\int \dfrac{\sec x}{\tan^2 x}\, dx$.

Solution Because the first four guidelines on page 537 do not apply, try converting the integrand to sines and cosines. In this case, you are able to integrate the resulting powers of sine and cosine as follows.

$$\int \frac{\sec x}{\tan^2 x}\, dx = \int \left(\frac{1}{\cos x}\right)\left(\frac{\cos x}{\sin x}\right)^2 dx$$

$$= \int (\sin x)^{-2}(\cos x)\, dx$$

$$= -(\sin x)^{-1} + C$$

$$= -\csc x + C$$

Integrals Involving Sine-Cosine Products with Different Angles

Integrals involving the products of sines and cosines of two *different* angles occur in many applications. In such instances you can use the following product-to-sum identities.

$$\sin mx \sin nx = \frac{1}{2}(\cos[(m-n)x] - \cos[(m+n)x])$$

$$\sin mx \cos nx = \frac{1}{2}(\sin[(m-n)x] + \sin[(m+n)x])$$

$$\cos mx \cos nx = \frac{1}{2}(\cos[(m-n)x] + \cos[(m+n)x])$$

FOR FURTHER INFORMATION To learn more about integrals involving sine-cosine products with different angles, see the article "Integrals of Products of Sine and Cosine with Different Arguments" by Sherrie J. Nicol in *The College Mathematics Journal*. To view this article, go to the website *www.matharticles.com*.

EXAMPLE 8 Using Product-to-Sum Identities

Find $\int \sin 5x \cos 4x\, dx$.

Solution Considering the second product-to-sum identity above, you can write

$$\int \sin 5x \cos 4x\, dx = \frac{1}{2}\int (\sin x + \sin 9x)\, dx$$

$$= \frac{1}{2}\left(-\cos x - \frac{\cos 9x}{9}\right) + C$$

$$= -\frac{\cos x}{2} - \frac{\cos 9x}{18} + C.$$

Exercises for Section 8.3

In Exercises 1–4, use differentiation to match the antiderivative with the correct integral. [Integrals are labeled (a), (b), (c), and (d).]

(a) $\int \sin x \tan^2 x \, dx$ (b) $8 \int \cos^4 x \, dx$

(c) $\int \sin x \sec^2 x \, dx$ (d) $\int \tan^4 x \, dx$

1. $y = \sec x$
2. $y = \cos x + \sec x$
3. $y = x - \tan x + \frac{1}{3} \tan^3 x$
4. $y = 3x + 2 \sin x \cos^3 x + 3 \sin x \cos x$

In Exercises 5–18, find the integral.

5. $\int \cos^3 x \sin x \, dx$
6. $\int \cos^3 x \sin^4 x \, dx$
7. $\int \sin^5 2x \cos 2x \, dx$
8. $\int \sin^3 x \, dx$
9. $\int \sin^5 x \cos^2 x \, dx$
10. $\int \cos^3 \frac{x}{3} \, dx$
11. $\int \cos^3 \theta \sqrt{\sin \theta} \, d\theta$
12. $\int \frac{\sin^5 t}{\sqrt{\cos t}} \, dt$
13. $\int \cos^2 3x \, dx$
14. $\int \sin^2 2x \, dx$
15. $\int \sin^2 \alpha \cos^2 \alpha \, d\alpha$
16. $\int \sin^4 2\theta \, d\theta$
17. $\int x \sin^2 x \, dx$
18. $\int x^2 \sin^2 x \, dx$

In Exercises 19–24, use Wallis's Formulas to evaluate the integral.

19. $\int_0^{\pi/2} \cos^3 x \, dx$
20. $\int_0^{\pi/2} \cos^5 x \, dx$
21. $\int_0^{\pi/2} \cos^7 x \, dx$
22. $\int_0^{\pi/2} \sin^2 x \, dx$
23. $\int_0^{\pi/2} \sin^6 x \, dx$
24. $\int_0^{\pi/2} \sin^7 x \, dx$

In Exercises 25–42, find the integral involving secant and tangent.

25. $\int \sec 3x \, dx$
26. $\int \sec^2(2x - 1) \, dx$
27. $\int \sec^4 5x \, dx$
28. $\int \sec^6 3x \, dx$
29. $\int \sec^3 \pi x \, dx$
30. $\int \tan^2 x \, dx$
31. $\int \tan^5 \frac{x}{4} \, dx$
32. $\int \tan^3 \frac{\pi x}{2} \sec^2 \frac{\pi x}{2} \, dx$
33. $\int \sec^2 x \tan x \, dx$
34. $\int \tan^3 2t \sec^3 2t \, dt$
35. $\int \tan^2 x \sec^2 x \, dx$
36. $\int \tan^5 2x \sec^2 2x \, dx$
37. $\int \sec^6 4x \tan 4x \, dx$
38. $\int \sec^2 \frac{x}{2} \tan \frac{x}{2} \, dx$
39. $\int \sec^3 x \tan x \, dx$
40. $\int \tan^3 3x \, dx$
41. $\int \frac{\tan^2 x}{\sec x} \, dx$
42. $\int \frac{\tan^2 x}{\sec^5 x} \, dx$

In Exercises 43–46, solve the differential equation.

43. $\dfrac{dr}{d\theta} = \sin^4 \pi\theta$
44. $\dfrac{ds}{d\alpha} = \sin^2 \dfrac{\alpha}{2} \cos^2 \dfrac{\alpha}{2}$
45. $y' = \tan^3 3x \sec 3x$
46. $y' = \sqrt{\tan x} \sec^4 x$

Slope Fields In Exercises 47 and 48, a differential equation, a point, and a slope field are given. (a) Sketch two approximate solutions of the differential equation on the slope field, one of which passes through the given point. (b) Use integration to find the particular solution of the differential equation and use a graphing utility to graph the solution. Compare the result with the sketches in part (a). To print an enlarged copy of the graph, go to the website www.mathgraphs.com.

47. $\dfrac{dy}{dx} = \sin^2 x$, $(0, 0)$ 48. $\dfrac{dy}{dx} = \sec^2 x \tan^2 x$, $\left(0, -\dfrac{1}{4}\right)$

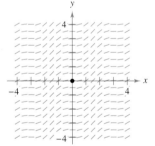
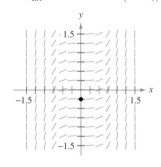

Slope Fields In Exercises 49 and 50, use a computer algebra system to graph the slope field for the differential equation, and graph the solution through the specified initial condition.

49. $\dfrac{dy}{dx} = \dfrac{3 \sin x}{y}$, $y(0) = 2$ 50. $\dfrac{dy}{dx} = 3\sqrt{y} \tan^2 x$, $y(0) = 3$

In Exercises 51–54, find the integral.

51. $\int \sin 3x \cos 2x \, dx$
52. $\int \cos 4\theta \cos(-3\theta) \, d\theta$
53. $\int \sin \theta \sin 3\theta \, d\theta$
54. $\int \sin(-4x) \cos 3x \, dx$

In Exercises 55–64, find the integral. Use a computer algebra system to confirm your result.

55. $\int \cot^3 2x \, dx$

56. $\int \tan^4 \frac{x}{2} \sec^4 \frac{x}{2} \, dx$

57. $\int \csc^4 \theta \, d\theta$

58. $\int \csc^2 3x \cot 3x \, dx$

59. $\int \frac{\cot^2 t}{\csc t} \, dt$

60. $\int \frac{\cot^3 t}{\csc t} \, dt$

61. $\int \frac{1}{\sec x \tan x} \, dx$

62. $\int \frac{\sin^2 x - \cos^2 x}{\cos x} \, dx$

63. $\int (\tan^4 t - \sec^4 t) \, dt$

64. $\int \frac{1 - \sec t}{\cos t - 1} \, dt$

In Exercises 65–72, evaluate the definite integral.

65. $\int_{-\pi}^{\pi} \sin^2 x \, dx$

66. $\int_0^{\pi/3} \tan^2 x \, dx$

67. $\int_0^{\pi/4} \tan^3 x \, dx$

68. $\int_0^{\pi/4} \sec^2 t \sqrt{\tan t} \, dt$

69. $\int_0^{\pi/2} \frac{\cos t}{1 + \sin t} \, dt$

70. $\int_{-\pi}^{\pi} \sin 3\theta \cos \theta \, d\theta$

71. $\int_{-\pi/2}^{\pi/2} \cos^3 x \, dx$

72. $\int_{-\pi/2}^{\pi/2} (\sin^2 x + 1) \, dx$

 In Exercises 73–78, use a computer algebra system to find the integral. Graph the antiderivatives for two different values of the constant of integration.

73. $\int \cos^4 \frac{x}{2} \, dx$

74. $\int \sin^2 x \cos^2 x \, dx$

75. $\int \sec^5 \pi x \, dx$

76. $\int \tan^3(1 - x) \, dx$

77. $\int \sec^5 \pi x \tan \pi x \, dx$

78. $\int \sec^4(1 - x) \tan(1 - x) \, dx$

 In Exercises 79–82, use a computer algebra system to evaluate the definite integral.

79. $\int_0^{\pi/4} \sin 2\theta \sin 3\theta \, d\theta$

80. $\int_0^{\pi/2} (1 - \cos \theta)^2 \, d\theta$

81. $\int_0^{\pi/2} \sin^4 x \, dx$

82. $\int_0^{\pi/2} \sin^6 x \, dx$

Writing About Concepts

83. In your own words, describe how you would integrate $\int \sin^m x \cos^n x \, dx$ for each condition.
    (a) m is positive and odd.
    (b) n is positive and odd.
    (c) m and n are both positive and even.

Writing About Concepts (continued)

84. In your own words, describe how you would integrate $\int \sec^m x \tan^n x \, dx$ for each condition.
    (a) m is positive and even.
    (b) n is positive and odd.
    (c) n is positive and even, and there are no secant factors.
    (d) m is positive and odd, and there are no tangent factors.

In Exercises 85 and 86, (a) find the indefinite integral in two different ways. (b) Use a graphing utility to graph the antiderivative (without the constant of integration) obtained by each method to show that the results differ only by a constant. (c) Verify analytically that the results differ only by a constant.

85. $\int \sec^4 3x \tan^3 3x \, dx$

86. $\int \sec^2 x \tan x \, dx$

Area In Exercises 87–90, find the area of the region bounded by the graphs of the equations.

87. $y = \sin x$, $y = \sin^3 x$, $x = 0$, $x = \pi/2$

88. $y = \sin^2 \pi x$, $y = 0$, $x = 0$, $x = 1$

89. $y = \cos^2 x$, $y = \sin^2 x$, $x = -\pi/4$, $x = \pi/4$

90. $y = \cos^2 x$, $y = \sin x \cos x$, $x = -\pi/2$, $x = \pi/4$

Volume In Exercises 91 and 92, find the volume of the solid generated by revolving the region bounded by the graphs of the equations about the x-axis.

91. $y = \tan x$, $y = 0$, $x = -\pi/4$, $x = \pi/4$

92. $y = \cos \frac{x}{2}$, $y = \sin \frac{x}{2}$, $x = 0$, $x = \pi/2$

Volume and Centroid In Exercises 93 and 94, for the region bounded by the graphs of the equations, find (a) the volume of the solid formed by revolving the region about the x-axis and (b) the centroid of the region.

93. $y = \sin x$, $y = 0$, $x = 0$, $x = \pi$

94. $y = \cos x$, $y = 0$, $x = 0$, $x = \pi/2$

In Exercises 95–98, use integration by parts to verify the reduction formula.

95. $\int \sin^n x \, dx = -\frac{\sin^{n-1} x \cos x}{n} + \frac{n-1}{n} \int \sin^{n-2} x \, dx$

96. $\int \cos^n x \, dx = \frac{\cos^{n-1} x \sin x}{n} + \frac{n-1}{n} \int \cos^{n-2} x \, dx$

97. $\int \cos^m x \sin^n x \, dx = -\frac{\cos^{m+1} x \sin^{n-1} x}{m+n} + \frac{n-1}{m+n} \int \cos^m x \sin^{n-2} x \, dx$

98. $\int \sec^n x \, dx = \frac{1}{n-1} \sec^{n-2} x \tan x + \frac{n-2}{n-1} \int \sec^{n-2} x \, dx$

In Exercises 99–102, use the results of Exercises 95–98 to find the integral.

99. $\int \sin^5 x \, dx$

100. $\int \cos^4 x \, dx$

101. $\int \sec^4 \dfrac{2\pi x}{5} \, dx$

102. $\int \sin^4 x \cos^2 x \, dx$

103. Modeling Data The table shows the normal maximum (high) and minimum (low) temperatures (in degrees Fahrenheit) for Erie, Pennsylvania for each month of the year. *(Source: NOAA)*

Month	Jan	Feb	Mar	Apr	May	Jun
Max	33.5	35.4	44.7	55.6	67.4	76.2
Min	20.3	20.9	28.2	37.9	48.7	58.5

Month	Jul	Aug	Sep	Oct	Nov	Dec
Max	80.4	79.0	72.0	61.0	49.3	38.6
Min	63.7	62.7	55.9	45.5	36.4	26.8

The maximum and minimum temperatures can be modeled by

$$f(t) = a_0 + a_1 \cos \dfrac{\pi t}{6} + b_1 \sin \dfrac{\pi t}{6}$$

where $t = 0$ corresponds to January and a_0, a_1, and b_1 are as follows.

$$a_0 = \dfrac{1}{12}\int_0^{12} f(t) \, dt$$

$$a_1 = \dfrac{1}{6}\int_0^{12} f(t) \cos \dfrac{\pi t}{6} \, dt$$

$$b_1 = \dfrac{1}{6}\int_0^{12} f(t) \sin \dfrac{\pi t}{6} \, dt$$

(a) Approximate the model $H(t)$ for the maximum temperatures. *(Hint:* Use Simpson's Rule to approximate the integrals and use the January data twice.*)*

(b) Repeat part (a) for a model $L(t)$ for the minimum temperature data.

(c) Use a graphing utility to compare each model with the actual data. During what part of the year is the difference between the maximum and minimum temperatures greatest?

104. Wallis's Formulas Use the result of Exercise 96 to prove the following versions of Wallis's Formulas.

(a) If n is odd $(n \geq 3)$, then

$$\int_0^{\pi/2} \cos^n x \, dx = \left(\dfrac{2}{3}\right)\left(\dfrac{4}{5}\right)\left(\dfrac{6}{7}\right) \cdots \left(\dfrac{n-1}{n}\right).$$

(b) If n is even $(n \geq 2)$, then

$$\int_0^{\pi/2} \cos^n x \, dx = \left(\dfrac{1}{2}\right)\left(\dfrac{3}{4}\right)\left(\dfrac{5}{6}\right) \cdots \left(\dfrac{n-1}{n}\right)\left(\dfrac{\pi}{2}\right).$$

105. The **inner product** of two functions f and g on $[a, b]$ is given by $\langle f, g \rangle = \int_a^b f(x)g(x) \, dx$. Two distinct functions f and g are said to be **orthogonal** if $\langle f, g \rangle = 0$. Show that the following set of functions is orthogonal on $[-\pi, \pi]$.

$\{\sin x, \sin 2x, \sin 3x, \ldots, \cos x, \cos 2x, \cos 3x, \ldots\}$

106. Fourier Series The following sum is a *finite Fourier series*.

$$f(x) = \sum_{i=1}^{N} a_i \sin ix$$
$$= a_1 \sin x + a_2 \sin 2x + a_3 \sin 3x + \cdots + a_N \sin Nx$$

(a) Use Exercise 105 to show that the nth coefficient a_n is given by $a_n = \dfrac{1}{\pi}\displaystyle\int_{-\pi}^{\pi} f(x) \sin nx \, dx$.

(b) Let $f(x) = x$. Find a_1, a_2, and a_3.

Section Project: Power Lines

Power lines are constructed by stringing wire between supports and adjusting the tension on each span. The wire hangs between supports in the shape of a catenary, as shown in the figure.

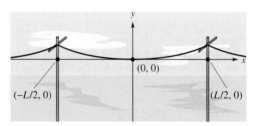

Let T be the tension (in pounds) on a span of wire, let u be the density (in pounds per foot), let $g \approx 32.2$ be the acceleration due to gravity (in feet per second per second), and let L be the distance (in feet) between the supports. Then the equation of the catenary is $y = \dfrac{T}{ug}\left(\cosh \dfrac{ugx}{T} - 1\right)$, where x and y are measured in feet.

(a) Find the length of the wire between two spans.

(b) To measure the tension in a span, power line workers use the *return wave method*. The wire is struck at one support, creating a wave in the line, and the time t (in seconds) it takes for the wave to make a round trip is measured. The velocity v (in feet per second) is given by $v = \sqrt{T/u}$. How long does it take the wave to make a round trip between supports?

(c) The sag s (in inches) can be obtained by evaluating y when $x = L/2$ in the equation for the catenary (and multiplying by 12). In practice, however, power line workers use the "lineman's equation" given by $s \approx 12.075t^2$. Use the fact that $[\cosh(ugL/2T) + 1] \approx 2$ to derive this equation.

FOR FURTHER INFORMATION To learn more about the mathematics of power lines, see the article "Constructing Power Lines" by Thomas O'Neil in *The UMAP Journal*.

Section 8.4

Trigonometric Substitution

- Use trigonometric substitution to solve an integral.
- Use integrals to model and solve real-life applications.

Trigonometric Substitution

Now that you can evaluate integrals involving powers of trigonometric functions, you can use **trigonometric substitution** to evaluate integrals involving the radicals

$$\sqrt{a^2 - u^2}, \qquad \sqrt{a^2 + u^2}, \qquad \text{and} \qquad \sqrt{u^2 - a^2}.$$

The objective with trigonometric substitution is to eliminate the radical in the integrand. You do this with the Pythagorean identities

$$\cos^2\theta = 1 - \sin^2\theta, \quad \sec^2\theta = 1 + \tan^2\theta, \quad \text{and} \quad \tan^2\theta = \sec^2\theta - 1.$$

For example, if $a > 0$, let $u = a\sin\theta$, where $-\pi/2 \leq \theta \leq \pi/2$. Then

$$\begin{aligned}
\sqrt{a^2 - u^2} &= \sqrt{a^2 - a^2\sin^2\theta} \\
&= \sqrt{a^2(1 - \sin^2\theta)} \\
&= \sqrt{a^2\cos^2\theta} \\
&= a\cos\theta.
\end{aligned}$$

Note that $\cos\theta \geq 0$, because $-\pi/2 \leq \theta \leq \pi/2$.

EXPLORATION

Integrating a Radical Function
Up to this point in the text, you have not evaluated the following integral.

$$\int_{-1}^{1} \sqrt{1 - x^2}\, dx$$

From geometry, you should be able to find the exact value of this integral—what is it? Using numerical integration, with Simpson's Rule or the Trapezoidal Rule, you can't be sure of the accuracy of the approximation. Why?

Try finding the exact value using the substitution

$$x = \sin\theta \quad \text{and} \quad dx = \cos\theta\, d\theta.$$

Does your answer agree with the value you obtained using geometry?

Trigonometric Substitution ($a > 0$)

1. For integrals involving $\sqrt{a^2 - u^2}$, let
   $$u = a\sin\theta.$$
   Then $\sqrt{a^2 - u^2} = a\cos\theta$, where $-\pi/2 \leq \theta \leq \pi/2$.

   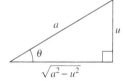

2. For integrals involving $\sqrt{a^2 + u^2}$, let
   $$u = a\tan\theta.$$
   Then $\sqrt{a^2 + u^2} = a\sec\theta$, where $-\pi/2 < \theta < \pi/2$.

   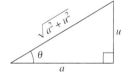

3. For integrals involving $\sqrt{u^2 - a^2}$, let
   $$u = a\sec\theta.$$
   Then $\sqrt{u^2 - a^2} = \pm a\tan\theta$, where $0 \leq \theta < \pi/2$ or $\pi/2 < \theta \leq \pi$.
   Use the positive value if $u > a$ and the negative value if $u < -a$.

   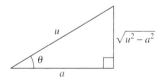

NOTE The restrictions on θ ensure that the function that defines the substitution is one-to-one. In fact, these are the same intervals over which the arcsine, arctangent, and arcsecant are defined.

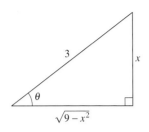

$\sin\theta = \dfrac{x}{3}$, $\cot\theta = \dfrac{\sqrt{9-x^2}}{x}$

Figure 8.6

EXAMPLE 1 Trigonometric Substitution: $u = a \sin\theta$

Find $\displaystyle\int \dfrac{dx}{x^2\sqrt{9-x^2}}$.

Solution First, note that none of the basic integration rules applies. To use trigonometric substitution, you should observe that $\sqrt{9-x^2}$ is of the form $\sqrt{a^2-u^2}$. So, you can use the substitution

$$x = a\sin\theta = 3\sin\theta.$$

Using differentiation and the triangle shown in Figure 8.6, you obtain

$$dx = 3\cos\theta\, d\theta, \qquad \sqrt{9-x^2} = 3\cos\theta, \qquad \text{and} \qquad x^2 = 9\sin^2\theta.$$

So, trigonometric substitution yields

$$\begin{aligned}
\int \dfrac{dx}{x^2\sqrt{9-x^2}} &= \int \dfrac{3\cos\theta\, d\theta}{(9\sin^2\theta)(3\cos\theta)} &&\text{Substitute.}\\
&= \dfrac{1}{9}\int \dfrac{d\theta}{\sin^2\theta} &&\text{Simplify.}\\
&= \dfrac{1}{9}\int \csc^2\theta\, d\theta &&\text{Trigonometric identity}\\
&= -\dfrac{1}{9}\cot\theta + C &&\text{Apply Cosecant Rule.}\\
&= -\dfrac{1}{9}\left(\dfrac{\sqrt{9-x^2}}{x}\right) + C &&\text{Substitute for } \cot\theta.\\
&= -\dfrac{\sqrt{9-x^2}}{9x} + C.
\end{aligned}$$

Note that the triangle in Figure 8.6 can be used to convert the θ's back to x's as follows.

$$\begin{aligned}
\cot\theta &= \dfrac{\text{adj.}}{\text{opp.}}\\
&= \dfrac{\sqrt{9-x^2}}{x}
\end{aligned}$$

TECHNOLOGY Use a computer algebra system to find each definite integral.

$$\int \dfrac{dx}{\sqrt{9-x^2}} \qquad \int \dfrac{dx}{x\sqrt{9-x^2}} \qquad \int \dfrac{dx}{x^2\sqrt{9-x^2}} \qquad \int \dfrac{dx}{x^3\sqrt{9-x^2}}$$

Then use trigonometric substitution to duplicate the results obtained with the computer algebra system.

In an earlier chapter, you saw how the inverse hyperbolic functions can be used to evaluate the integrals

$$\int \dfrac{du}{\sqrt{u^2 \pm a^2}}, \qquad \int \dfrac{du}{a^2 - u^2}, \qquad \text{and} \qquad \int \dfrac{du}{u\sqrt{a^2 \pm u^2}}.$$

You can also evaluate these integrals using trigonometric substitution. This is shown in the next example.

SECTION 8.4 Trigonometric Substitution   **545**

EXAMPLE 2 Trigonometric Substitution: $u = a \tan \theta$

Find $\displaystyle\int \frac{dx}{\sqrt{4x^2 + 1}}$.

Solution Let $u = 2x$, $a = 1$, and $2x = \tan \theta$, as shown in Figure 8.7. Then,

$$dx = \frac{1}{2} \sec^2 \theta \, d\theta \quad \text{and} \quad \sqrt{4x^2 + 1} = \sec \theta.$$

Trigonometric substitution produces

$$\int \frac{1}{\sqrt{4x^2 + 1}} dx = \frac{1}{2} \int \frac{\sec^2 \theta \, d\theta}{\sec \theta} \qquad \text{Substitute.}$$

$$= \frac{1}{2} \int \sec \theta \, d\theta \qquad \text{Simplify.}$$

$$= \frac{1}{2} \ln|\sec \theta + \tan \theta| + C \qquad \text{Apply Secant Rule.}$$

$$= \frac{1}{2} \ln\left|\sqrt{4x^2 + 1} + 2x\right| + C. \qquad \text{Back-substitute.}$$

Try checking this result with a computer algebra system. Is the result given in this form or in the form of an inverse hyperbolic function?

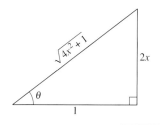

$\tan \theta = 2x$, $\sec \theta = \sqrt{4x^2 + 1}$
Figure 8.7

You can extend the use of trigonometric substitution to cover integrals involving expressions such as $(a^2 - u^2)^{n/2}$ by writing the expression as

$$(a^2 - u^2)^{n/2} = \left(\sqrt{a^2 - u^2}\right)^n.$$

EXAMPLE 3 Trigonometric Substitution: Rational Powers

Find $\displaystyle\int \frac{dx}{(x^2 + 1)^{3/2}}$.

Solution Begin by writing $(x^2 + 1)^{3/2}$ as $\left(\sqrt{x^2 + 1}\right)^3$. Then, let $a = 1$ and $u = x = \tan \theta$, as shown in Figure 8.8. Using

$$dx = \sec^2 \theta \, d\theta \quad \text{and} \quad \sqrt{x^2 + 1} = \sec \theta$$

you can apply trigonometric substitution as follows.

$$\int \frac{dx}{(x^2 + 1)^{3/2}} = \int \frac{dx}{\left(\sqrt{x^2 + 1}\right)^3} \qquad \text{Rewrite denominator.}$$

$$= \int \frac{\sec^2 \theta \, d\theta}{\sec^3 \theta} \qquad \text{Substitute.}$$

$$= \int \frac{d\theta}{\sec \theta} \qquad \text{Simplify.}$$

$$= \int \cos \theta \, d\theta \qquad \text{Trigonometric identity}$$

$$= \sin \theta + C \qquad \text{Apply Cosine Rule.}$$

$$= \frac{x}{\sqrt{x^2 + 1}} + C. \qquad \text{Back-substitute.}$$

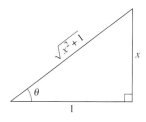

$\tan \theta = x$, $\sin \theta = \dfrac{x}{\sqrt{x^2 + 1}}$
Figure 8.8

For definite integrals, it is often convenient to determine the integration limits for θ that avoid converting back to x. You might want to review this procedure in Section 4.5, Examples 8 and 9.

EXAMPLE 4 Converting the Limits of Integration

Evaluate $\displaystyle\int_{\sqrt{3}}^{2} \frac{\sqrt{x^2-3}}{x}\,dx$.

Solution Because $\sqrt{x^2-3}$ has the form $\sqrt{u^2-a^2}$, you can consider

$$u = x, \quad a = \sqrt{3}, \quad \text{and} \quad x = \sqrt{3}\sec\theta$$

as shown in Figure 8.9. Then,

$$dx = \sqrt{3}\sec\theta\tan\theta\,d\theta \quad \text{and} \quad \sqrt{x^2-3} = \sqrt{3}\tan\theta.$$

To determine the upper and lower limits of integration, use the substitution $x = \sqrt{3}\sec\theta$, as follows.

Lower Limit	Upper Limit
When $x = \sqrt{3}$, $\sec\theta = 1$ and $\theta = 0$.	When $x = 2$, $\sec\theta = \dfrac{2}{\sqrt{3}}$ and $\theta = \dfrac{\pi}{6}$.

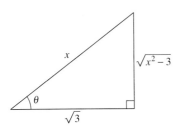

$\sec\theta = \dfrac{x}{\sqrt{3}}, \tan\theta = \dfrac{\sqrt{x^2-3}}{\sqrt{3}}$

Figure 8.9

So, you have

$$\int_{\sqrt{3}}^{2} \frac{\sqrt{x^2-3}}{x}\,dx = \int_{0}^{\pi/6} \frac{(\sqrt{3}\tan\theta)(\sqrt{3}\sec\theta\tan\theta)\,d\theta}{\sqrt{3}\sec\theta}$$

$$= \int_{0}^{\pi/6} \sqrt{3}\tan^2\theta\,d\theta$$

$$= \sqrt{3}\int_{0}^{\pi/6} (\sec^2\theta - 1)\,d\theta$$

$$= \sqrt{3}\Big[\tan\theta - \theta\Big]_{0}^{\pi/6}$$

$$= \sqrt{3}\left(\frac{1}{\sqrt{3}} - \frac{\pi}{6}\right)$$

$$= 1 - \frac{\sqrt{3}\pi}{6}$$

$$\approx 0.0931.$$

In Example 4, try converting back to the variable x and evaluating the antiderivative at the original limits of integration. You should obtain

$$\int_{\sqrt{3}}^{2} \frac{\sqrt{x^2-3}}{x}\,dx = \sqrt{3}\left[\frac{\sqrt{x^2-3}}{\sqrt{3}} - \operatorname{arcsec}\frac{x}{\sqrt{3}}\right]_{\sqrt{3}}^{2}.$$

When using trigonometric substitution to evaluate definite integrals, you must be careful to check that the values of θ lie in the intervals discussed at the beginning of this section. For instance, if in Example 4 you had been asked to evaluate the definite integral

$$\int_{-2}^{-\sqrt{3}} \frac{\sqrt{x^2-3}}{x}\, dx$$

then using $u = x$ and $a = \sqrt{3}$ in the interval $\left[-2, -\sqrt{3}\right]$ would imply that $u < -a$. So, when determining the upper and lower limits of integration, you would have to choose θ such that $\pi/2 < \theta \le \pi$. In this case the integral would be evaluated as follows.

$$\begin{aligned}
\int_{-2}^{-\sqrt{3}} \frac{\sqrt{x^2-3}}{x}\, dx &= \int_{5\pi/6}^{\pi} \frac{\left(-\sqrt{3}\tan\theta\right)\left(\sqrt{3}\sec\theta\tan\theta\right)d\theta}{\sqrt{3}\sec\theta} \\
&= \int_{5\pi/6}^{\pi} -\sqrt{3}\tan^2\theta\, d\theta \\
&= -\sqrt{3}\int_{5\pi/6}^{\pi} (\sec^2\theta - 1)\, d\theta \\
&= -\sqrt{3}\Big[\tan\theta - \theta\Big]_{5\pi/6}^{\pi} \\
&= -\sqrt{3}\left[(0-\pi) - \left(-\frac{1}{\sqrt{3}} - \frac{5\pi}{6}\right)\right] \\
&= -1 + \frac{\sqrt{3}\pi}{6} \\
&\approx -0.0931
\end{aligned}$$

Trigonometric substitution can be used with completing the square. For instance, try evaluating the following integral.

$$\int \sqrt{x^2 - 2x}\, dx$$

To begin, you could complete the square and write the integral as

$$\int \sqrt{(x-1)^2 - 1^2}\, dx.$$

Trigonometric substitution can be used to evaluate the three integrals listed in the following theorem. These integrals will be encountered several times in the remainder of the text. When this happens, we will simply refer to this theorem. (In Exercise 85, you are asked to verify the formulas given in the theorem.)

THEOREM 8.2 Special Integration Formulas ($a > 0$)

1. $\displaystyle\int \sqrt{a^2 - u^2}\, du = \frac{1}{2}\left(a^2 \arcsin\frac{u}{a} + u\sqrt{a^2 - u^2}\right) + C$

2. $\displaystyle\int \sqrt{u^2 - a^2}\, du = \frac{1}{2}\left(u\sqrt{u^2 - a^2} - a^2 \ln\left|u + \sqrt{u^2 - a^2}\right|\right) + C, \quad u > a$

3. $\displaystyle\int \sqrt{u^2 + a^2}\, du = \frac{1}{2}\left(u\sqrt{u^2 + a^2} + a^2 \ln\left|u + \sqrt{u^2 + a^2}\right|\right) + C$

Applications

EXAMPLE 5 Finding Arc Length

Find the arc length of the graph of $f(x) = \frac{1}{2}x^2$ from $x = 0$ to $x = 1$ (see Figure 8.10).

Solution Refer to the arc length formula in Section 7.4.

$$s = \int_0^1 \sqrt{1 + [f'(x)]^2}\, dx \qquad \text{Formula for arc length}$$

$$= \int_0^1 \sqrt{1 + x^2}\, dx \qquad f'(x) = x$$

$$= \int_0^{\pi/4} \sec^3 \theta\, d\theta \qquad \text{Let } a = 1 \text{ and } x = \tan \theta.$$

$$= \frac{1}{2}\Big[\sec \theta \tan \theta + \ln|\sec \theta + \tan \theta|\Big]_0^{\pi/4} \qquad \text{Example 5, Section 8.2}$$

$$= \frac{1}{2}\Big[\sqrt{2} + \ln(\sqrt{2} + 1)\Big] \approx 1.148$$

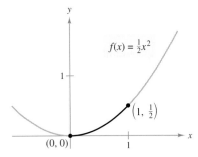

The arc length of the curve from $(0, 0)$ to $\left(1, \frac{1}{2}\right)$.
Figure 8.10

EXAMPLE 6 Comparing Two Fluid Forces

A sealed barrel of oil (weighing 48 pounds per cubic foot) is floating in seawater (weighing 64 pounds per cubic foot), as shown in Figures 8.11 and 8.12. (The barrel is not completely full of oil—on its side, the top 0.2 foot of the barrel is empty.) Compare the fluid forces against one end of the barrel from the inside and from the outside.

Solution In Figure 8.12, locate the coordinate system with the origin at the center of the circle given by $x^2 + y^2 = 1$. To find the fluid force against an end of the barrel *from the inside*, integrate between -1 and 0.8 (using a weight of $w = 48$).

$$F = w\int_c^d h(y)L(y)\, dy \qquad \text{General equation (see Section 7.7)}$$

$$F_{\text{inside}} = 48\int_{-1}^{0.8} (0.8 - y)(2)\sqrt{1 - y^2}\, dy$$

$$= 76.8\int_{-1}^{0.8} \sqrt{1 - y^2}\, dy - 96\int_{-1}^{0.8} y\sqrt{1 - y^2}\, dy$$

To find the fluid force *from the outside*, integrate between -1 and 0.4 (using a weight of $w = 64$).

$$F_{\text{outside}} = 64\int_{-1}^{0.4} (0.4 - y)(2)\sqrt{1 - y^2}\, dy$$

$$= 51.2\int_{-1}^{0.4} \sqrt{1 - y^2}\, dy - 128\int_{-1}^{0.4} y\sqrt{1 - y^2}\, dy$$

The details of integration are left for you to complete in Exercise 84. Intuitively, would you say that the force from the oil (the inside) or the force from the seawater (the outside) is greater? By evaluating these two integrals, you can determine that

$$F_{\text{inside}} \approx 121.3 \text{ pounds} \qquad \text{and} \qquad F_{\text{outside}} \approx 93.0 \text{ pounds.}$$

The barrel is not quite full of oil—the top 0.2 foot of the barrel is empty.
Figure 8.11

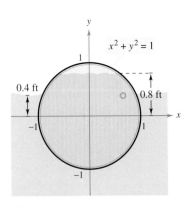

Figure 8.12

Exercises for Section 8.4

In Exercises 1–4, use differentiation to match the antiderivative with the correct integral. [Integrals are labeled (a), (b), (c), and (d).]

(a) $\int \dfrac{x^2}{\sqrt{16-x^2}}\,dx$
(b) $\int \dfrac{\sqrt{x^2+16}}{x}\,dx$
(c) $\int \sqrt{7+6x-x^2}\,dx$
(d) $\int \dfrac{x^2}{\sqrt{x^2-16}}\,dx$

1. $4\ln\left|\dfrac{\sqrt{x^2+16}-4}{x}\right| + \sqrt{x^2+16} + C$
2. $8\ln\left|\sqrt{x^2-16}+x\right| + \dfrac{x\sqrt{x^2-16}}{2} + C$
3. $8\arcsin\dfrac{x}{4} - \dfrac{x\sqrt{16-x^2}}{2} + C$
4. $8\arcsin\dfrac{x-3}{4} + \dfrac{(x-3)\sqrt{7+6x-x^2}}{2} + C$

In Exercises 5–8, find the indefinite integral using the substitution $x = 5\sin\theta$.

5. $\int \dfrac{1}{(25-x^2)^{3/2}}\,dx$
6. $\int \dfrac{10}{x^2\sqrt{25-x^2}}\,dx$
7. $\int \dfrac{\sqrt{25-x^2}}{x}\,dx$
8. $\int \dfrac{x^2}{\sqrt{25-x^2}}\,dx$

In Exercises 9–12, find the indefinite integral using the substitution $x = 2\sec\theta$.

9. $\int \dfrac{1}{\sqrt{x^2-4}}\,dx$
10. $\int \dfrac{\sqrt{x^2-4}}{x}\,dx$
11. $\int x^3\sqrt{x^2-4}\,dx$
12. $\int \dfrac{x^3}{\sqrt{x^2-4}}\,dx$

In Exercises 13–16, find the indefinite integral using the substitution $x = \tan\theta$.

13. $\int x\sqrt{1+x^2}\,dx$
14. $\int \dfrac{9x^3}{\sqrt{1+x^2}}\,dx$
15. $\int \dfrac{1}{(1+x^2)^2}\,dx$
16. $\int \dfrac{x^2}{(1+x^2)^2}\,dx$

In Exercises 17–20, use the Special Integration Formulas (Theorem 8.2) to find the integral.

17. $\int \sqrt{4+9x^2}\,dx$
18. $\int \sqrt{1+x^2}\,dx$
19. $\int \sqrt{25-4x^2}\,dx$
20. $\int \sqrt{2x^2-1}\,dx$

In Exercises 21–42, find the integral.

21. $\int \dfrac{x}{\sqrt{x^2+9}}\,dx$
22. $\int \dfrac{x}{\sqrt{9-x^2}}\,dx$
23. $\int \dfrac{1}{\sqrt{16-x^2}}\,dx$
24. $\int \dfrac{1}{\sqrt{25-x^2}}\,dx$
25. $\int \sqrt{16-4x^2}\,dx$
26. $\int x\sqrt{16-4x^2}\,dx$
27. $\int \dfrac{1}{\sqrt{x^2-9}}\,dx$
28. $\int \dfrac{t}{(1-t^2)^{3/2}}\,dt$
29. $\int \dfrac{\sqrt{1-x^2}}{x^4}\,dx$
30. $\int \dfrac{\sqrt{4x^2+9}}{x^4}\,dx$
31. $\int \dfrac{1}{x\sqrt{4x^2+9}}\,dx$
32. $\int \dfrac{1}{x\sqrt{4x^2+16}}\,dx$
33. $\int \dfrac{-5x}{(x^2+5)^{3/2}}\,dx$
34. $\int \dfrac{1}{(x^2+3)^{3/2}}\,dx$
35. $\int e^{2x}\sqrt{1+e^{2x}}\,dx$
36. $\int (x+1)\sqrt{x^2+2x+2}\,dx$
37. $\int e^x\sqrt{1-e^{2x}}\,dx$
38. $\int \dfrac{\sqrt{1-x}}{\sqrt{x}}\,dx$
39. $\int \dfrac{1}{4+4x^2+x^4}\,dx$
40. $\int \dfrac{x^3+x+1}{x^4+2x^2+1}\,dx$
41. $\int \operatorname{arcsec} 2x\,dx, \quad x > \dfrac{1}{2}$
42. $\int x\arcsin x\,dx$

In Exercises 43–46, complete the square and find the integral.

43. $\int \dfrac{1}{\sqrt{4x-x^2}}\,dx$
44. $\int \dfrac{x^2}{\sqrt{2x-x^2}}\,dx$
45. $\int \dfrac{x}{\sqrt{x^2+4x+8}}\,dx$
46. $\int \dfrac{x}{\sqrt{x^2-6x+5}}\,dx$

In Exercises 47–52, evaluate the integral using (a) the given integration limits and (b) the limits obtained by trigonometric substitution.

47. $\displaystyle\int_0^{\sqrt{3}/2} \dfrac{t^2}{(1-t^2)^{3/2}}\,dt$
48. $\displaystyle\int_0^{\sqrt{3}/2} \dfrac{1}{(1-t^2)^{5/2}}\,dt$
49. $\displaystyle\int_0^3 \dfrac{x^3}{\sqrt{x^2+9}}\,dx$
50. $\displaystyle\int_0^{3/5} \sqrt{9-25x^2}\,dx$
51. $\displaystyle\int_4^6 \dfrac{x^2}{\sqrt{x^2-9}}\,dx$
52. $\displaystyle\int_3^6 \dfrac{\sqrt{x^2-9}}{x^2}\,dx$

In Exercises 53 and 54, find the particular solution of the differential equation.

53. $x\dfrac{dy}{dx} = \sqrt{x^2-9}, \quad x \geq 3, \quad y(3) = 1$
54. $\sqrt{x^2+4}\,\dfrac{dy}{dx} = 1, \quad x \geq -2, \quad y(0) = 4$

In Exercises 55–58, use a computer algebra system to find the integral. Verify the result by differentiation.

55. $\int \dfrac{x^2}{\sqrt{x^2 + 10x + 9}}\,dx$

56. $\int (x^2 + 2x + 11)^{3/2}\,dx$

57. $\int \dfrac{x^2}{\sqrt{x^2 - 1}}\,dx$

58. $\int x^2\sqrt{x^2 - 4}\,dx$

Writing About Concepts

59. State the substitution you would make if you used trigonometric substitution and the integral involving the given radical, where $a > 0$. Explain your reasoning.
   (a) $\sqrt{a^2 - u^2}$ (b) $\sqrt{a^2 + u^2}$ (c) $\sqrt{u^2 - a^2}$

60. State the method of integration you would use to perform each integration. Explain why you chose that method. Do not integrate.
   (a) $\int x\sqrt{x^2 + 1}\,dx$ (b) $\int x^2\sqrt{x^2 - 1}\,dx$

61. Evaluate the integral $\int \dfrac{x}{x^2 + 9}\,dx$ using (a) u-substitution and (b) trigonometric substitution. Discuss the results.

62. Evaluate the integral $\int \dfrac{x^2}{x^2 + 9}\,dx$ (a) algebraically using $x^2 = (x^2 + 9) - 9$ and (b) using trigonometric substitution. Discuss the results.

True or False? In Exercises 63–66, determine whether the statement is true or false. If it is false, explain why or give an example that shows it is false.

63. If $x = \sin\theta$, then $\int \dfrac{dx}{\sqrt{1 - x^2}} = \int d\theta$.

64. If $x = \sec\theta$, then $\int \dfrac{\sqrt{x^2 - 1}}{x}\,dx = \int \sec\theta\tan\theta\,d\theta$.

65. If $x = \tan\theta$, then $\displaystyle\int_0^{\sqrt{3}} \dfrac{dx}{(1 + x^2)^{3/2}} = \int_0^{4\pi/3} \cos\theta\,d\theta$.

66. If $x = \sin\theta$, then $\displaystyle\int_{-1}^{1} x^2\sqrt{1 - x^2}\,dx = 2\int_0^{\pi/2} \sin^2\theta\cos^2\theta\,d\theta$.

67. *Area* Find the area enclosed by the ellipse shown in the figure.
$$\dfrac{x^2}{a^2} + \dfrac{y^2}{b^2} = 1$$

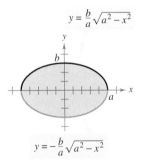

$y = \dfrac{b}{a}\sqrt{a^2 - x^2}$

$y = -\dfrac{b}{a}\sqrt{a^2 - x^2}$

68. *Area* Find the area of the shaded region of the circle of radius a, if the chord is h units $(0 < h < a)$ from the center of the circle (see figure).

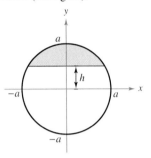

69. *Mechanical Design* The surface of a machine part is the region between the graphs of $y = |x|$ and $x^2 + (y - k)^2 = 25$ (see figure).

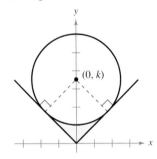

(a) Find k if the circle is tangent to the graph of $y = |x|$.

(b) Find the area of the surface of the machine part.

(c) Find the area of the surface of the machine part as a function of the radius r of the circle.

 70. *Volume* The axis of a storage tank in the form of a right circular cylinder is horizontal (see figure). The radius and length of the tank are 1 meter and 3 meters, respectively.

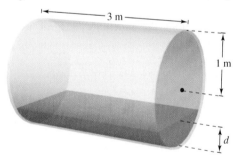

(a) Determine the volume of fluid in the tank as a function of its depth d.

(b) Use a graphing utility to graph the function in part (a).

(c) Design a dip stick for the tank with markings of $\tfrac{1}{4}$, $\tfrac{1}{2}$, and $\tfrac{3}{4}$.

(d) Fluid is entering the tank at a rate of $\tfrac{1}{4}$ cubic meter per minute. Determine the rate of change of the depth of the fluid as a function of its depth d.

(e) Use a graphing utility to graph the function in part (d). When will the rate of change of the depth be minimum? Does this agree with your intuition? Explain.

Volume of a Torus In Exercises 71 and 72, find the volume of the torus generated by revolving the region bounded by the graph of the circle about the *y*-axis.

71. $(x-3)^2 + y^2 = 1$ (see figure)

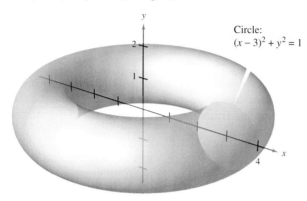

Circle: $(x-3)^2 + y^2 = 1$

72. $(x-h)^2 + y^2 = r^2$, $h > r$

Arc Length In Exercises 73 and 74, find the arc length of the curve over the given interval.

73. $y = \ln x$, $[1, 5]$
74. $y = \frac{1}{2}x^2$, $[0, 4]$

75. *Arc Length* Show that the length of one arch of the sine curve is equal to the length of one arch of the cosine curve.

76. *Conjecture*
  (a) Find formulas for the distance between $(0, 0)$ and (a, a^2) along the line between these points and along the parabola $y = x^2$.
  (b) Use the formulas from part (a) to find the distances for $a = 1$ and $a = 10$.
  (c) Make a conjecture about the difference between the two distances as a increases.

Projectile Motion In Exercises 77 and 78, (a) use a graphing utility to graph the path of a projectile that follows the path given by the graph of the equation, (b) determine the range of the projectile, and (c) use the integration capabilities of a graphing utility to determine the distance the projectile travels.

77. $y = x - 0.005x^2$
78. $y = x - \dfrac{x^2}{72}$

Centroid In Exercises 79 and 80, find the centroid of the region determined by the graphs of the inequalities.

79. $y \leq 3/\sqrt{x^2+9}$, $y \geq 0$, $x \geq -4$, $x \leq 4$
80. $y \leq \frac{1}{4}x^2$, $(x-4)^2 + y^2 \leq 16$, $y \geq 0$

81. *Surface Area* Find the surface area of the solid generated by revolving the region bounded by the graphs of $y = x^2$, $y = 0$, $x = 0$, and $x = \sqrt{2}$ about the *x*-axis.

82. *Field Strength* The field strength H of a magnet of length $2L$ on a particle r units from the center of the magnet is

$$H = \frac{2mL}{(r^2 + L^2)^{3/2}}$$

where $\pm m$ are the poles of the magnet (see figure). Find the average field strength as the particle moves from 0 to R units from the center by evaluating the integral

$$\frac{1}{R}\int_0^R \frac{2mL}{(r^2+L^2)^{3/2}}\, dr.$$

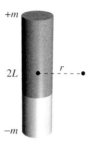

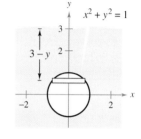

Figure for 82   **Figure for 83**

83. *Fluid Force* Find the fluid force on a circular observation window of radius 1 foot in a vertical wall of a large water-filled tank at a fish hatchery when the center of the window is (a) 3 feet and (b) d feet ($d > 1$) below the water's surface (see figure). Use trigonometric substitution to evaluate the one integral. (Recall that in Section 7.7 in a similar problem, you evaluated one integral by a geometric formula and the other by observing that the integrand was odd.)

84. *Fluid Force* Evaluate the following two integrals, which yield the fluid forces given in Example 6.

  (a) $F_{\text{inside}} = 48\displaystyle\int_{-1}^{0.8}(0.8 - y)(2)\sqrt{1-y^2}\, dy$

  (b) $F_{\text{outside}} = 64\displaystyle\int_{-1}^{0.4}(0.4 - y)(2)\sqrt{1-y^2}\, dy$

85. Use trigonometric substitution to verify the integration formulas given in Theorem 8.2.

86. *Arc Length* Show that the arc length of the graph of $y = \sin x$ on the interval $[0, 2\pi]$ is equal to the circumference of the ellipse $x^2 + 2y^2 = 2$ (see figure).

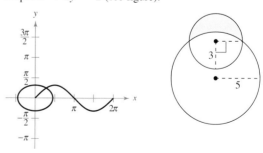

Figure for 86   **Figure for 87**

87. *Area of a Lune* The crescent-shaped region bounded by two circles forms a *lune* (see figure). Find the area of the lune given that the radius of the smaller circle is 3 and the radius of the larger circle is 5.

Section 8.5

Partial Fractions

- Understand the concept of a partial fraction decomposition.
- Use partial fraction decomposition with linear factors to integrate rational functions.
- Use partial fraction decomposition with quadratic factors to integrate rational functions.

Partial Fractions

This section examines a procedure for decomposing a rational function into simpler rational functions to which you can apply the basic integration formulas. This procedure is called the **method of partial fractions.** To see the benefit of the method of partial fractions, consider the integral

$$\int \frac{1}{x^2 - 5x + 6} \, dx.$$

To evaluate this integral *without* partial fractions, you can complete the square and use trigonometric substitution (see Figure 8.13) to obtain

$$\int \frac{1}{x^2 - 5x + 6} \, dx = \int \frac{dx}{(x - 5/2)^2 - (1/2)^2} \qquad a = \tfrac{1}{2}, x - \tfrac{5}{2} = \tfrac{1}{2} \sec \theta$$

$$= \int \frac{(1/2) \sec \theta \tan \theta \, d\theta}{(1/4) \tan^2 \theta} \qquad dx = \tfrac{1}{2} \sec \theta \tan \theta \, d\theta$$

$$= 2 \int \csc \theta \, d\theta$$

$$= 2 \ln|\csc \theta - \cot \theta| + C$$

$$= 2 \ln\left|\frac{2x - 5}{2\sqrt{x^2 - 5x + 6}} - \frac{1}{2\sqrt{x^2 - 5x + 6}}\right| + C$$

$$= 2 \ln\left|\frac{x - 3}{\sqrt{x^2 - 5x + 6}}\right| + C$$

$$= 2 \ln\left|\frac{\sqrt{x - 3}}{\sqrt{x - 2}}\right| + C$$

$$= \ln\left|\frac{x - 3}{x - 2}\right| + C$$

$$= \ln|x - 3| - \ln|x - 2| + C.$$

$\sec \theta = 2x - 5$
Figure 8.13

Now, suppose you had observed that

$$\frac{1}{x^2 - 5x + 6} = \frac{1}{x - 3} - \frac{1}{x - 2}. \qquad \text{Partial fraction decomposition}$$

Then you could evaluate the integral easily, as follows.

$$\int \frac{1}{x^2 - 5x + 6} \, dx = \int \left(\frac{1}{x - 3} - \frac{1}{x - 2}\right) dx$$

$$= \ln|x - 3| - \ln|x - 2| + C$$

This method is clearly preferable to trigonometric substitution. However, its use depends on the ability to factor the denominator, $x^2 - 5x + 6$, and to find the **partial fractions**

$$\frac{1}{x - 3} \quad \text{and} \quad -\frac{1}{x - 2}.$$

In this section, you will study techniques for finding partial fraction decompositions.

JOHN BERNOULLI (1667–1748)

The method of partial fractions was introduced by John Bernoulli, a Swiss mathematician who was instrumental in the early development of calculus. John Bernoulli was a professor at the University of Basel and taught many outstanding students, the most famous of whom was Leonhard Euler.

STUDY TIP In precalculus you learned how to combine functions such as

$$\frac{1}{x-2} + \frac{-1}{x+3} = \frac{5}{(x-2)(x+3)}.$$

The method of partial fractions shows you how to reverse this process.

$$\frac{5}{(x-2)(x+3)} = \frac{?}{x-2} + \frac{?}{x+3}$$

Recall from algebra that every polynomial with real coefficients can be factored into linear and irreducible quadratic factors.* For instance, the polynomial

$$x^5 + x^4 - x - 1$$

can be written as

$$\begin{aligned}
x^5 + x^4 - x - 1 &= x^4(x+1) - (x+1) \\
&= (x^4 - 1)(x+1) \\
&= (x^2 + 1)(x^2 - 1)(x+1) \\
&= (x^2 + 1)(x+1)(x-1)(x+1) \\
&= (x-1)(x+1)^2(x^2+1)
\end{aligned}$$

where $(x-1)$ is a linear factor, $(x+1)^2$ is a repeated linear factor, and (x^2+1) is an irreducible quadratic factor. Using this factorization, you can write the partial fraction decomposition of the rational expression

$$\frac{N(x)}{x^5 + x^4 - x - 1}$$

where $N(x)$ is a polynomial of degree less than 5, as follows.

$$\frac{N(x)}{(x-1)(x+1)^2(x^2+1)} = \frac{A}{x-1} + \frac{B}{x+1} + \frac{C}{(x+1)^2} + \frac{Dx + E}{x^2 + 1}$$

Decomposition of N(x)/D(x) into Partial Fractions

1. **Divide if improper:** If $N(x)/D(x)$ is an improper fraction (that is, if the degree of the numerator is greater than or equal to the degree of the denominator), divide the denominator into the numerator to obtain

$$\frac{N(x)}{D(x)} = (\text{a polynomial}) + \frac{N_1(x)}{D(x)}$$

where the degree of $N_1(x)$ is less than the degree of $D(x)$. Then apply Steps 2, 3, and 4 to the proper rational expression $N_1(x)/D(x)$.

2. **Factor denominator:** Completely factor the denominator into factors of the form

$$(px + q)^m \quad \text{and} \quad (ax^2 + bx + c)^n$$

where $ax^2 + bx + c$ is irreducible.

3. **Linear factors:** For each factor of the form $(px + q)^m$, the partial fraction decomposition must include the following sum of m fractions.

$$\frac{A_1}{(px+q)} + \frac{A_2}{(px+q)^2} + \cdots + \frac{A_m}{(px+q)^m}$$

4. **Quadratic factors:** For each factor of the form $(ax^2 + bx + c)^n$, the partial fraction decomposition must include the following sum of n fractions.

$$\frac{B_1 x + C_1}{ax^2 + bx + c} + \frac{B_2 x + C_2}{(ax^2 + bx + c)^2} + \cdots + \frac{B_n x + C_n}{(ax^2 + bx + c)^n}$$

---

*For a review of factorization techniques, see Precalculus, 6th edition, by Larson and Hostetler or Precalculus: A Graphing Approach, 4th edition, by Larson, Hostetler, and Edwards (Boston, Massachusetts: Houghton Mifflin, 2004 and 2005, respectively).

Linear Factors

Algebraic techniques for determining the constants in the numerators of a partial decomposition with linear or repeated linear factors are shown in Examples 1 and 2.

EXAMPLE 1 Distinct Linear Factors

Write the partial fraction decomposition for $\dfrac{1}{x^2 - 5x + 6}$.

Solution Because $x^2 - 5x + 6 = (x - 3)(x - 2)$, you should include one partial fraction for each factor and write

$$\frac{1}{x^2 - 5x + 6} = \frac{A}{x - 3} + \frac{B}{x - 2}$$

where A and B are to be determined. Multiplying this equation by the least common denominator $(x - 3)(x - 2)$ yields the **basic equation**

$$1 = A(x - 2) + B(x - 3). \quad \text{Basic equation}$$

Because this equation is to be true for all x, you can substitute any *convenient* values for x to obtain equations in A and B. The most convenient values are the ones that make particular factors equal to 0.

To solve for A, let $x = 3$ and obtain

$$1 = A(3 - 2) + B(3 - 3) \quad \text{Let } x = 3 \text{ in basic equation.}$$
$$1 = A(1) + B(0)$$
$$A = 1.$$

To solve for B, let $x = 2$ and obtain

$$1 = A(2 - 2) + B(2 - 3) \quad \text{Let } x = 2 \text{ in basic equation.}$$
$$1 = A(0) + B(-1)$$
$$B = -1.$$

So, the decomposition is

$$\frac{1}{x^2 - 5x + 6} = \frac{1}{x - 3} - \frac{1}{x - 2}$$

as shown at the beginning of this section.

NOTE Note that the substitutions for x in Example 1 are chosen for their convenience in determining values for A and B; $x = 2$ is chosen to eliminate the term $A(x - 2)$, and $x = 3$ is chosen to eliminate the term $B(x - 3)$. The goal is to make *convenient* substitutions whenever possible.

FOR FURTHER INFORMATION To learn a different method for finding the partial fraction decomposition, called the Heavyside Method, see the article "Calculus to Algebra Connections in Partial Fraction Decomposition" by Joseph Wiener and Will Watkins in *The AMATYC Review*.

Be sure you see that the method of partial fractions is practical only for integrals of rational functions whose denominators factor "nicely." For instance, if the denominator in Example 1 were changed to $x^2 - 5x + 5$, its factorization as

$$x^2 - 5x + 5 = \left[x + \frac{5 + \sqrt{5}}{2}\right]\left[x - \frac{5 - \sqrt{5}}{2}\right]$$

would be too cumbersome to use with partial fractions. In such cases, you should use completing the square or a computer algebra system to perform the integration. If you do this, you should obtain

$$\int \frac{1}{x^2 - 5x + 5}\, dx = \frac{\sqrt{5}}{5}\ln\left|2x - \sqrt{5} - 5\right| - \frac{\sqrt{5}}{5}\ln\left|2x + \sqrt{5} - 5\right| + C.$$

EXAMPLE 2 Repeated Linear Factors

Find $\int \dfrac{5x^2 + 20x + 6}{x^3 + 2x^2 + x}\, dx$.

Solution Because

$$x^3 + 2x^2 + x = x(x^2 + 2x + 1)$$
$$= x(x + 1)^2$$

you should include one fraction for *each power* of x and $(x + 1)$ and write

$$\dfrac{5x^2 + 20x + 6}{x(x + 1)^2} = \dfrac{A}{x} + \dfrac{B}{x + 1} + \dfrac{C}{(x + 1)^2}.$$

Multiplying by the least common denominator $x(x + 1)^2$ yields the *basic equation*

$$5x^2 + 20x + 6 = A(x + 1)^2 + Bx(x + 1) + Cx. \qquad \text{Basic equation}$$

To solve for A, let $x = 0$. This eliminates the B and C terms and yields

$$6 = A(1) + 0 + 0$$
$$A = 6.$$

To solve for C, let $x = -1$. This eliminates the A and B terms and yields

$$5 - 20 + 6 = 0 + 0 - C$$
$$C = 9.$$

The most convenient choices for x have been used, so to find the value of B, you can use *any other value* of x along with the calculated values of A and C. Using $x = 1$, $A = 6$, and $C = 9$ produces

$$5 + 20 + 6 = A(4) + B(2) + C$$
$$31 = 6(4) + 2B + 9$$
$$-2 = 2B$$
$$B = -1.$$

So, it follows that

$$\int \dfrac{5x^2 + 20x + 6}{x(x + 1)^2}\, dx = \int \left(\dfrac{6}{x} - \dfrac{1}{x + 1} + \dfrac{9}{(x + 1)^2} \right) dx$$
$$= 6 \ln|x| - \ln|x + 1| + 9 \dfrac{(x + 1)^{-1}}{-1} + C$$
$$= \ln \left| \dfrac{x^6}{x + 1} \right| - \dfrac{9}{x + 1} + C.$$

Try checking this result by differentiating. Include algebra in your check, simplifying the derivative until you have obtained the original integrand.

NOTE It is necessary to make as many substitutions for x as there are unknowns ($A, B, C, \ldots$) to be determined. For instance, in Example 2, three substitutions ($x = 0$, $x = -1$, and $x = 1$) were made to solve for A, B, and C.

FOR FURTHER INFORMATION For an alternative approach to using partial fractions, see the article "A Shortcut in Partial Fractions" by Xun-Cheng Huang in *The College Mathematics Journal*.

TECHNOLOGY Most computer algebra systems, such as *Derive*, *Maple*, *Mathcad*, *Mathematica*, and the *TI-89*, can be used to convert a rational function to its partial fraction decomposition. For instance, using *Maple*, you obtain the following.

> convert$\left(\dfrac{5x^2 + 20x + 6}{x^3 + 2x^2 + x}, \text{parfrac}, x \right)$

$\dfrac{6}{x} + \dfrac{9}{(x + 1)^2} - \dfrac{1}{x + 1}$

Quadratic Factors

When using the method of partial fractions with *linear* factors, a convenient choice of x immediately yields a value for one of the coefficients. With *quadratic* factors, a system of linear equations usually has to be solved, regardless of the choice of x.

EXAMPLE 3 Distinct Linear and Quadratic Factors

Find $\displaystyle\int \frac{2x^3 - 4x - 8}{(x^2 - x)(x^2 + 4)}\, dx.$

Solution Because

$$(x^2 - x)(x^2 + 4) = x(x - 1)(x^2 + 4)$$

you should include one partial fraction for each factor and write

$$\frac{2x^3 - 4x - 8}{x(x - 1)(x^2 + 4)} = \frac{A}{x} + \frac{B}{x - 1} + \frac{Cx + D}{x^2 + 4}.$$

Multiplying by the least common denominator $x(x - 1)(x^2 + 4)$ yields the *basic equation*

$$2x^3 - 4x - 8 = A(x - 1)(x^2 + 4) + Bx(x^2 + 4) + (Cx + D)(x)(x - 1).$$

To solve for A, let $x = 0$ and obtain

$$-8 = A(-1)(4) + 0 + 0 \quad \Longrightarrow \quad 2 = A.$$

To solve for B, let $x = 1$ and obtain

$$-10 = 0 + B(5) + 0 \quad \Longrightarrow \quad -2 = B.$$

At this point, C and D are yet to be determined. You can find these remaining constants by choosing two other values for x and solving the resulting system of linear equations. If $x = -1$, then, using $A = 2$ and $B = -2$, you can write

$$-6 = (2)(-2)(5) + (-2)(-1)(5) + (-C + D)(-1)(-2)$$
$$2 = -C + D.$$

If $x = 2$, you have

$$0 = (2)(1)(8) + (-2)(2)(8) + (2C + D)(2)(1)$$
$$8 = 2C + D.$$

Solving the linear system by subtracting the first equation from the second

$$-C + D = 2$$
$$2C + D = 8$$

yields $C = 2$. Consequently, $D = 4$, and it follows that

$$\int \frac{2x^3 - 4x - 8}{x(x - 1)(x^2 + 4)}\, dx = \int \left(\frac{2}{x} - \frac{2}{x - 1} + \frac{2x}{x^2 + 4} + \frac{4}{x^2 + 4}\right) dx$$

$$= 2\ln|x| - 2\ln|x - 1| + \ln(x^2 + 4) + 2\arctan\frac{x}{2} + C.$$

In Examples 1, 2, and 3, the solution of the basic equation began with substituting values of x that made the linear factors equal to 0. This method works well when the partial fraction decomposition involves linear factors. However, if the decomposition involves only quadratic factors, an alternative procedure is often more convenient.

EXAMPLE 4 Repeated Quadratic Factors

Find $\displaystyle\int \frac{8x^3 + 13x}{(x^2 + 2)^2}\, dx$.

Solution Include one partial fraction for each power of $(x^2 + 2)$ and write

$$\frac{8x^3 + 13x}{(x^2 + 2)^2} = \frac{Ax + B}{x^2 + 2} + \frac{Cx + D}{(x^2 + 2)^2}.$$

Multiplying by the least common denominator $(x^2 + 2)^2$ yields the *basic equation*

$$8x^3 + 13x = (Ax + B)(x^2 + 2) + Cx + D.$$

Expanding the basic equation and collecting like terms produces

$$8x^3 + 13x = Ax^3 + 2Ax + Bx^2 + 2B + Cx + D$$
$$8x^3 + 13x = Ax^3 + Bx^2 + (2A + C)x + (2B + D).$$

Now, you can equate the coefficients of like terms on opposite sides of the equation.

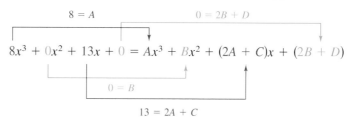

Using the known values $A = 8$ and $B = 0$, you can write

$$13 = 2A + C = 2(8) + C \implies C = -3$$
$$0 = 2B + D = 2(0) + D \implies D = 0.$$

Finally, you can conclude that

$$\int \frac{8x^3 + 13x}{(x^2 + 2)^2}\, dx = \int \left(\frac{8x}{x^2 + 2} + \frac{-3x}{(x^2 + 2)^2}\right) dx$$
$$= 4\ln(x^2 + 2) + \frac{3}{2(x^2 + 2)} + C.$$

TECHNOLOGY Use a computer algebra system to evaluate the integral in Example 4—you might find that the form of the antiderivative is different. For instance, when you use a computer algebra system to work Example 4, you obtain

$$\int \frac{8x^3 + 13x}{(x^2 + 2)^2}\, dx = \ln(x^8 + 8x^6 + 24x^4 + 32x^2 + 16) + \frac{3}{2(x^2 + 2)} + C.$$

Is this result equivalent to that obtained in Example 4?

When integrating rational expressions, keep in mind that for *improper* rational expressions such as

$$\frac{N(x)}{D(x)} = \frac{2x^3 + x^2 - 7x + 7}{x^2 + x - 2}$$

you must first divide to obtain

$$\frac{N(x)}{D(x)} = 2x - 1 + \frac{-2x + 5}{x^2 + x - 2}.$$

The proper rational expression is then decomposed into its partial fractions by the usual methods. Here are some guidelines for solving the basic equation that is obtained in a partial fraction decomposition.

Guidelines for Solving the Basic Equation

Linear Factors

1. Substitute the roots of the distinct linear factors into the basic equation.
2. For repeated linear factors, use the coefficients determined in guideline 1 to rewrite the basic equation. Then substitute other convenient values of x and solve for the remaining coefficients.

Quadratic Factors

1. Expand the basic equation.
2. Collect terms according to powers of x.
3. Equate the coefficients of like powers to obtain a system of linear equations involving A, B, C, and so on.
4. Solve the system of linear equations.

Before concluding this section, here are a few things you should remember. First, it is not necessary to use the partial fractions technique on all rational functions. For instance, the following integral is evaluated more easily by the Log Rule.

$$\int \frac{x^2 + 1}{x^3 + 3x - 4} dx = \frac{1}{3} \int \frac{3x^2 + 3}{x^3 + 3x - 4} dx$$
$$= \frac{1}{3} \ln|x^3 + 3x - 4| + C$$

Second, if the integrand is not in reduced form, reducing it may eliminate the need for partial fractions, as shown in the following integral.

$$\int \frac{x^2 - x - 2}{x^3 - 2x - 4} dx = \int \frac{(x + 1)(x - 2)}{(x - 2)(x^2 + 2x + 2)} dx$$
$$= \int \frac{x + 1}{x^2 + 2x + 2} dx$$
$$= \frac{1}{2} \ln|x^2 + 2x + 2| + C$$

Finally, partial fractions can be used with some quotients involving transcendental functions. For instance, the substitution $u = \sin x$ allows you to write

$$\int \frac{\cos x}{\sin x (\sin x - 1)} dx = \int \frac{du}{u(u - 1)}. \qquad u = \sin x, du = \cos x \, dx$$

Exercises for Section 8.5

In Exercises 1–6, write the form of the partial fraction decomposition of the rational expression. Do not solve for the constants.

1. $\dfrac{5}{x^2 - 10x}$
2. $\dfrac{4x^2 + 3}{(x - 5)^3}$
3. $\dfrac{2x - 3}{x^3 + 10x}$
4. $\dfrac{x - 2}{x^2 + 4x + 3}$
5. $\dfrac{16}{x^2 - 10x}$
6. $\dfrac{2x - 1}{x(x^2 + 1)^2}$

In Exercises 7–28, use partial fractions to find the integral.

7. $\displaystyle\int \dfrac{1}{x^2 - 1}\,dx$
8. $\displaystyle\int \dfrac{1}{4x^2 - 9}\,dx$
9. $\displaystyle\int \dfrac{3}{x^2 + x - 2}\,dx$
10. $\displaystyle\int \dfrac{x + 1}{x^2 + 4x + 3}\,dx$
11. $\displaystyle\int \dfrac{5 - x}{2x^2 + x - 1}\,dx$
12. $\displaystyle\int \dfrac{5x^2 - 12x - 12}{x^3 - 4x}\,dx$
13. $\displaystyle\int \dfrac{x^2 + 12x + 12}{x^3 - 4x}\,dx$
14. $\displaystyle\int \dfrac{x^3 - x + 3}{x^2 + x - 2}\,dx$
15. $\displaystyle\int \dfrac{2x^3 - 4x^2 - 15x + 5}{x^2 - 2x - 8}\,dx$
16. $\displaystyle\int \dfrac{x + 2}{x^2 - 4x}\,dx$
17. $\displaystyle\int \dfrac{4x^2 + 2x - 1}{x^3 + x^2}\,dx$
18. $\displaystyle\int \dfrac{2x - 3}{(x - 1)^2}\,dx$
19. $\displaystyle\int \dfrac{x^2 + 3x - 4}{x^3 - 4x^2 + 4x}\,dx$
20. $\displaystyle\int \dfrac{4x^2}{x^3 + x^2 - x - 1}\,dx$
21. $\displaystyle\int \dfrac{x^2 - 1}{x^3 + x}\,dx$
22. $\displaystyle\int \dfrac{6x}{x^3 - 8}\,dx$
23. $\displaystyle\int \dfrac{x^2}{x^4 - 2x^2 - 8}\,dx$
24. $\displaystyle\int \dfrac{x^2 - x + 9}{(x^2 + 9)^2}\,dx$
25. $\displaystyle\int \dfrac{x}{16x^4 - 1}\,dx$
26. $\displaystyle\int \dfrac{x^2 - 4x + 7}{x^3 - x^2 + x + 3}\,dx$
27. $\displaystyle\int \dfrac{x^2 + 5}{x^3 - x^2 + x + 3}\,dx$
28. $\displaystyle\int \dfrac{x^2 + x + 3}{x^4 + 6x^2 + 9}\,dx$

In Exercises 29–32, evaluate the definite integral. Use a graphing utility to verify your result.

29. $\displaystyle\int_0^1 \dfrac{3}{2x^2 + 5x + 2}\,dx$
30. $\displaystyle\int_1^5 \dfrac{x - 1}{x^2(x + 1)}\,dx$
31. $\displaystyle\int_1^2 \dfrac{x + 1}{x(x^2 + 1)}\,dx$
32. $\displaystyle\int_0^1 \dfrac{x^2 - x}{x^2 + x + 1}\,dx$

In Exercises 33–40, use a computer algebra system to determine the antiderivative that passes through the given point. Use the system to graph the resulting antiderivative.

33. $\displaystyle\int \dfrac{3x}{x^2 - 6x + 9}\,dx$, $(4, 0)$
34. $\displaystyle\int \dfrac{6x^2 + 1}{x^2(x - 1)^3}\,dx$, $(2, 1)$
35. $\displaystyle\int \dfrac{x^2 + x + 2}{(x^2 + 2)^2}\,dx$, $(0, 1)$
36. $\displaystyle\int \dfrac{x^3}{(x^2 - 4)^2}\,dx$, $(3, 4)$
37. $\displaystyle\int \dfrac{2x^2 - 2x + 3}{x^3 - x^2 - x - 2}\,dx$, $(3, 10)$
38. $\displaystyle\int \dfrac{x(2x - 9)}{x^3 - 6x^2 + 12x - 8}\,dx$, $(3, 2)$
39. $\displaystyle\int \dfrac{1}{x^2 - 4}\,dx$, $(6, 4)$
40. $\displaystyle\int \dfrac{x^2 - x + 2}{x^3 - x^2 + x - 1}\,dx$, $(2, 6)$

In Exercises 41–46, use substitution to find the integral.

41. $\displaystyle\int \dfrac{\sin x}{\cos x(\cos x - 1)}\,dx$
42. $\displaystyle\int \dfrac{\sin x}{\cos x + \cos^2 x}\,dx$
43. $\displaystyle\int \dfrac{3\cos x}{\sin^2 x + \sin x - 2}\,dx$
44. $\displaystyle\int \dfrac{\sec^2 x}{\tan x(\tan x + 1)}\,dx$
45. $\displaystyle\int \dfrac{e^x}{(e^x - 1)(e^x + 4)}\,dx$
46. $\displaystyle\int \dfrac{e^x}{(e^{2x} + 1)(e^x - 1)}\,dx$

In Exercises 47–50, use the method of partial fractions to verify the integration formula.

47. $\displaystyle\int \dfrac{1}{x(a + bx)}\,dx = \dfrac{1}{a}\ln\left|\dfrac{x}{a + bx}\right| + C$
48. $\displaystyle\int \dfrac{1}{a^2 - x^2}\,dx = \dfrac{1}{2a}\ln\left|\dfrac{a + x}{a - x}\right| + C$
49. $\displaystyle\int \dfrac{x}{(a + bx)^2}\,dx = \dfrac{1}{b^2}\left(\dfrac{a}{a + bx} + \ln|a + bx|\right) + C$
50. $\displaystyle\int \dfrac{1}{x^2(a + bx)}\,dx = -\dfrac{1}{ax} - \dfrac{b}{a^2}\ln\left|\dfrac{x}{a + bx}\right| + C$

Slope Fields In Exercises 51 and 52, use a computer algebra system to graph the slope field for the differential equation and graph the solution through the given initial condition.

51. $\dfrac{dy}{dx} = \dfrac{6}{4 - x^2}$
    $y(0) = 3$
52. $\dfrac{dy}{dx} = \dfrac{4}{x^2 - 2x - 3}$
    $y(0) = 5$

Writing About Concepts

53. What is the first step when integrating $\displaystyle\int \dfrac{x^3}{x - 5}\,dx$? Explain.
54. Describe the decomposition of the proper rational function $N(x)/D(x)$ (a) if $D(x) = (px + q)^m$, and (b) if $D(x) = (ax^2 + bx + c)^n$, where $ax^2 + bx + c$ is irreducible. Explain why you chose that method.
55. State the method you would use to evaluate each integral. Explain why you chose that method. Do not integrate.
    (a) $\displaystyle\int \dfrac{x + 1}{x^2 + 2x - 8}\,dx$
    (b) $\displaystyle\int \dfrac{7x + 4}{x^2 + 2x - 8}\,dx$
    (c) $\displaystyle\int \dfrac{4}{x^2 + 2x + 5}\,dx$

Writing About Concepts (continued)

56. Determine which value best approximates the area of the region between the x-axis and the graph of $f(x) = 10/[x(x^2 + 1)]$ over the interval $[1, 3]$. Make your selection on the basis of a sketch of the region and not by performing any calculations. Explain your reasoning.
(a) -6 (b) 6 (c) 3 (d) 5 (e) 8

57. *Area* Find the area of the region bounded by the graphs of $y = 12/(x^2 + 5x + 6)$, $y = 0$, $x = 0$, and $x = 1$.

58. *Area* Find the area of the region bounded by the graphs of $y = 7/(16 - x^2)$ and $y = 1$.

59. *Modeling Data* The predicted cost C (in hundreds of thousands of dollars) for a company to remove $p\%$ of a chemical from its waste water is shown in the table.

p	0	10	20	30	40	50	60	70	80	90
C	0	0.7	1.0	1.3	1.7	2.0	2.7	3.6	5.5	11.2

A model for the data is given by

$$C = \frac{124p}{(10 + p)(100 - p)}, \quad 0 \le p < 100.$$

Use the model to find the average cost for removing between 75% and 80% of the chemical.

60. *Logistic Growth* In Chapter 6, the exponential growth equation was derived from the assumption that the rate of growth was proportional to the existing quantity. In practice, there often exists some upper limit L past which growth cannot occur. In such cases, you assume the rate of growth to be proportional not only to the existing quantity, but also to the difference between the existing quantity y and the upper limit L. That is, $dy/dt = ky(L - y)$. In integral form, you can write this relationship as

$$\int \frac{dy}{y(L - y)} = \int k \, dt.$$

(a) A slope field for the differential equation $dy/dt = y(3 - y)$ is shown. Draw a possible solution to the differential equation if $y(0) = 5$, and another if $y(0) = \frac{1}{2}$. To print an enlarged copy of the graph, go to the website www.mathgraphs.com.

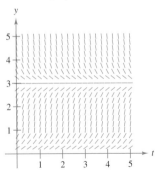

(b) Where $y(0)$ is greater than 3, what is the sign of the slope of the solution?

(c) For $y > 0$, find $\lim_{t \to \infty} y(t)$.

(d) Evaluate the two given integrals and solve for y as a function of t, where y_0 is the initial quantity.

 (e) Use the result of part (d) to find and graph the solutions in part (a). Use a graphing utility to graph the solutions and compare the results with the solutions in part (a).

(f) The graph of the function y is a **logistic curve**. Show that the rate of growth is maximum at the point of inflection, and that this occurs when $y = L/2$.

61. *Volume and Centroid* Consider the region bounded by the graphs of $y = 2x/(x^2 + 1)$, $y = 0$, $x = 0$, and $x = 3$. Find the volume of the solid generated by revolving the region about the x-axis. Find the centroid of the region.

62. *Volume* Consider the region bounded by the graph of $y^2 = (2 - x)^2/(1 + x)^2$ on the interval $[0, 1]$. Find the volume of the solid generated by revolving this region about the x-axis.

63. *Epidemic Model* A single infected individual enters a community of n susceptible individuals. Let x be the number of newly infected individuals at time t. The common epidemic model assumes that the disease spreads at a rate proportional to the product of the total number infected and the number not yet infected. So, $dx/dt = k(x + 1)(n - x)$ and you obtain

$$\int \frac{1}{(x + 1)(n - x)} \, dx = \int k \, dt.$$

Solve for x as a function of t.

64. *Chemical Reactions* In a chemical reaction, one unit of compound Y and one unit of compound Z are converted into a single unit of compound X. x is the amount of compound X formed, and the rate of formation of X is proportional to the product of the amounts of unconverted compounds Y and Z. So, $dx/dt = k(y_0 - x)(z_0 - x)$, where y_0 and z_0 are the initial amounts of compounds Y and Z. From this equation you obtain

$$\int \frac{1}{(y_0 - x)(z_0 - x)} \, dx = \int k \, dt.$$

(a) Perform the two integrations and solve for x in terms of t.

(b) Use the result of part (a) to find x as $t \to \infty$ if (1) $y_0 < z_0$, (2) $y_0 > z_0$, and (3) $y_0 = z_0$.

65. Evaluate

$$\int_0^1 \frac{x}{1 + x^4} \, dx$$

in two different ways, one of which is partial fractions.

Putnam Exam Challenge

66. Prove $\dfrac{22}{7} - \pi = \displaystyle\int_0^1 \dfrac{x^4(1 - x)^4}{1 + x^2} \, dx.$

This problem was composed by the Committee on the Putnam Prize Competition.
© The Mathematical Association of America. All rights reserved.

Section 8.6 Integration by Tables and Other Integration Techniques

- Evaluate an indefinite integral using a table of integrals.
- Evaluate an indefinite integral using reduction formulas.
- Evaluate an indefinite integral involving rational functions of sine and cosine.

Integration by Tables

So far in this chapter you have studied several integration techniques that can be used with the basic integration rules. But merely knowing *how* to use the various techniques is not enough. You also need to know *when* to use them. Integration is first and foremost a problem of recognition. That is, you must recognize which rule or technique to apply to obtain an antiderivative. Frequently, a slight alteration of an integrand will require a different integration technique (or produce a function whose antiderivative is not an elementary function), as shown below.

$$\int x \ln x \, dx = \frac{x^2}{2} \ln x - \frac{x^2}{4} + C \qquad \text{Integration by parts}$$

$$\int \frac{\ln x}{x} \, dx = \frac{(\ln x)^2}{2} + C \qquad \text{Power Rule}$$

$$\int \frac{1}{x \ln x} \, dx = \ln|\ln x| + C \qquad \text{Log Rule}$$

$$\int \frac{x}{\ln x} \, dx = ? \qquad \text{Not an elementary function}$$

TECHNOLOGY A computer algebra system consists, in part, of a database of integration formulas. The primary difference between using a computer algebra system and using tables of integrals is that with a computer algebra system the computer searches through the database to find a fit. With integration tables, *you* must do the searching.

Many people find tables of integrals to be a valuable supplement to the integration techniques discussed in this chapter. Tables of common integrals can be found in Appendix B. **Integration by tables** is not a "cure-all" for all of the difficulties that can accompany integration—using tables of integrals requires considerable thought and insight and often involves substitution.

Each integration formula in Appendix B can be developed using one or more of the techniques in this chapter. You should try to verify several of the formulas. For instance, Formula 4

$$\int \frac{u}{(a + bu)^2} \, du = \frac{1}{b^2} \left(\frac{a}{a + bu} + \ln|a + bu| \right) + C \qquad \text{Formula 4}$$

can be verified using the method of partial fractions, and Formula 19

$$\int \frac{\sqrt{a + bu}}{u} \, du = 2\sqrt{a + bu} + a \int \frac{du}{u\sqrt{a + bu}} \qquad \text{Formula 19}$$

can be verified using integration by parts. Note that the integrals in Appendix B are classified according to forms involving the following.

u^n	$(a + bu)$
$(a + bu + cu^2)$	$\sqrt{a + bu}$
$(a^2 \pm u^2)$	$\sqrt{u^2 \pm a^2}$
$\sqrt{a^2 - u^2}$	Trigonometric functions
Inverse trigonometric functions	Exponential functions
Logarithmic functions	

EXPLORATION

Use the tables of integrals in Appendix B and the substitution

$$u = \sqrt{x-1}$$

to evaluate the integral in Example 1. If you do this, you should obtain

$$\int \frac{dx}{x\sqrt{x-1}} = \int \frac{2\,du}{u^2+1}.$$

Does this produce the same result as that obtained in Example 1?

EXAMPLE 1 Integration by Tables

Find $\displaystyle\int \frac{dx}{x\sqrt{x-1}}$.

Solution Because the expression inside the radical is linear, you should consider forms involving $\sqrt{a+bu}$.

$$\int \frac{du}{u\sqrt{a+bu}} = \frac{2}{\sqrt{-a}}\arctan\sqrt{\frac{a+bu}{-a}} + C \qquad \text{Formula 17 } (a < 0)$$

Let $a = -1$, $b = 1$, and $u = x$. Then $du = dx$, and you can write

$$\int \frac{dx}{x\sqrt{x-1}} = 2\arctan\sqrt{x-1} + C.$$

EXAMPLE 2 Integration by Tables

Find $\displaystyle\int x\sqrt{x^4 - 9}\,dx$.

Solution Because the radical has the form $\sqrt{u^2 - a^2}$, you should consider Formula 26.

$$\int \sqrt{u^2 - a^2}\,du = \frac{1}{2}\left(u\sqrt{u^2-a^2} - a^2\ln\left|u + \sqrt{u^2-a^2}\right|\right) + C$$

Let $u = x^2$ and $a = 3$. Then $du = 2x\,dx$, and you have

$$\int x\sqrt{x^4 - 9}\,dx = \frac{1}{2}\int \sqrt{(x^2)^2 - 3^2}\,(2x)\,dx$$

$$= \frac{1}{4}\left(x^2\sqrt{x^4-9} - 9\ln\left|x^2 + \sqrt{x^4-9}\right|\right) + C.$$

EXAMPLE 3 Integration by Tables

Find $\displaystyle\int \frac{x}{1 + e^{-x^2}}\,dx$.

Solution Of the forms involving e^u, consider the following formula.

$$\int \frac{du}{1 + e^u} = u - \ln(1 + e^u) + C \qquad \text{Formula 84}$$

Let $u = -x^2$. Then $du = -2x\,dx$, and you have

$$\int \frac{x}{1 + e^{-x^2}}\,dx = -\frac{1}{2}\int \frac{-2x\,dx}{1 + e^{-x^2}}$$

$$= -\frac{1}{2}\left[-x^2 - \ln(1 + e^{-x^2})\right] + C$$

$$= \frac{1}{2}\left[x^2 + \ln(1 + e^{-x^2})\right] + C.$$

TECHNOLOGY Example 3 shows the importance of having several solution techniques at your disposal. This integral is not difficult to solve with a table, but when it was entered into a well-known computer algebra system, the utility was unable to find the antiderivative.

Reduction Formulas

Several of the integrals in the integration tables have the form $\int f(x)\,dx = g(x) + \int h(x)\,dx$. Such integration formulas are called **reduction formulas** because they reduce a given integral to the sum of a function and a simpler integral.

EXAMPLE 4 Using a Reduction Formula

Find $\int x^3 \sin x\,dx$.

Solution Consider the following three formulas.

$$\int u \sin u\,du = \sin u - u \cos u + C \qquad \text{Formula 52}$$

$$\int u^n \sin u\,du = -u^n \cos u + n \int u^{n-1} \cos u\,du \qquad \text{Formula 54}$$

$$\int u^n \cos u\,du = u^n \sin u - n \int u^{n-1} \sin u\,du \qquad \text{Formula 55}$$

Using Formula 54, Formula 55, and then Formula 52 produces

$$\int x^3 \sin x\,dx = -x^3 \cos x + 3 \int x^2 \cos x\,dx$$

$$= -x^3 \cos x + 3\left(x^2 \sin x - 2 \int x \sin x\,dx \right)$$

$$= -x^3 \cos x + 3x^2 \sin x + 6x \cos x - 6 \sin x + C.$$

EXAMPLE 5 Using a Reduction Formula

Find $\int \dfrac{\sqrt{3-5x}}{2x}\,dx$.

Solution Consider the following two formulas.

$$\int \frac{du}{u\sqrt{a+bu}} = \frac{1}{\sqrt{a}} \ln\left| \frac{\sqrt{a+bu} - \sqrt{a}}{\sqrt{a+bu} + \sqrt{a}} \right| + C \qquad \text{Formula 17 } (a > 0)$$

$$\int \frac{\sqrt{a+bu}}{u}\,du = 2\sqrt{a+bu} + a \int \frac{du}{u\sqrt{a+bu}} \qquad \text{Formula 19}$$

Using Formula 19, with $a = 3$, $b = -5$, and $u = x$, produces

$$\frac{1}{2} \int \frac{\sqrt{3-5x}}{x}\,dx = \frac{1}{2}\left(2\sqrt{3-5x} + 3 \int \frac{dx}{x\sqrt{3-5x}} \right)$$

$$= \sqrt{3-5x} + \frac{3}{2} \int \frac{dx}{x\sqrt{3-5x}}.$$

Using Formula 17, with $a = 3$, $b = -5$, and $u = x$, produces

$$\int \frac{\sqrt{3-5x}}{2x}\,dx = \sqrt{3-5x} + \frac{3}{2}\left(\frac{1}{\sqrt{3}} \ln\left| \frac{\sqrt{3-5x} - \sqrt{3}}{\sqrt{3-5x} + \sqrt{3}} \right| \right) + C$$

$$= \sqrt{3-5x} + \frac{\sqrt{3}}{2} \ln\left| \frac{\sqrt{3-5x} - \sqrt{3}}{\sqrt{3-5x} + \sqrt{3}} \right| + C.$$

TECHNOLOGY Sometimes when you use computer algebra systems you obtain results that look very different, but are actually equivalent. Here is how several different systems evaluated the integral in Example 5.

Maple

$\sqrt{3-5x} - \sqrt{3}\,\text{arctanh}\left(\tfrac{1}{3}\sqrt{3-5x}\sqrt{3}\right)$

Derive

$\sqrt{3} \ln\left[\dfrac{\sqrt{(3-5x)} - \sqrt{3}}{\sqrt{x}} \right] + \sqrt{(3-5x)}$

Mathematica

$\text{Sqrt}[3-5x] - \text{Sqrt}[3]\,\text{ArcTanh}\left[\dfrac{\text{Sqrt}[3-5x]}{\text{Sqrt}[3]} \right]$

Mathcad

$\sqrt{3-5x} + \tfrac{1}{2}\sqrt{3} \ln\left[-\tfrac{1}{5}\dfrac{(-6+5x+2\sqrt{3}\sqrt{3-5x})}{x} \right]$

Notice that computer algebra systems do not include a constant of integration.

Rational Functions of Sine and Cosine

EXAMPLE 6 Integration by Tables

Find $\displaystyle\int \frac{\sin 2x}{2 + \cos x}\, dx$.

Solution Substituting $2 \sin x \cos x$ for $\sin 2x$ produces

$$\int \frac{\sin 2x}{2 + \cos x}\, dx = 2 \int \frac{\sin x \cos x}{2 + \cos x}\, dx.$$

A check of the forms involving $\sin u$ or $\cos u$ in Appendix B shows that none of those listed applies. So, you can consider forms involving $a + bu$. For example,

$$\int \frac{u\, du}{a + bu} = \frac{1}{b^2}(bu - a \ln|a + bu|) + C. \qquad \text{Formula 3}$$

Let $a = 2$, $b = 1$, and $u = \cos x$. Then $du = -\sin x\, dx$, and you have

$$\begin{aligned}
2 \int \frac{\sin x \cos x}{2 + \cos x}\, dx &= -2 \int \frac{\cos x(-\sin x\, dx)}{2 + \cos x} \\
&= -2(\cos x - 2 \ln|2 + \cos x|) + C \\
&= -2 \cos x + 4 \ln|2 + \cos x| + C.
\end{aligned}$$

Example 6 involves a rational expression of $\sin x$ and $\cos x$. If you are unable to find an integral of this form in the integration tables, try using the following special substitution to convert the trigonometric expression to a standard rational expression.

Substitution for Rational Functions of Sine and Cosine

For integrals involving rational functions of sine and cosine, the substitution

$$u = \frac{\sin x}{1 + \cos x} = \tan \frac{x}{2}$$

yields

$$\cos x = \frac{1 - u^2}{1 + u^2}, \quad \sin x = \frac{2u}{1 + u^2}, \quad \text{and} \quad dx = \frac{2\, du}{1 + u^2}.$$

Proof From the substitution for u, it follows that

$$u^2 = \frac{\sin^2 x}{(1 + \cos x)^2} = \frac{1 - \cos^2 x}{(1 + \cos x)^2} = \frac{1 - \cos x}{1 + \cos x}.$$

Solving for $\cos x$ produces $\cos x = (1 - u^2)/(1 + u^2)$. To find $\sin x$, write $u = \sin x/(1 + \cos x)$ as

$$\sin x = u(1 + \cos x) = u\left(1 + \frac{1 - u^2}{1 + u^2}\right) = \frac{2u}{1 + u^2}.$$

Finally, to find dx, consider $u = \tan(x/2)$. Then you have $\arctan u = x/2$ and $dx = (2\, du)/(1 + u^2)$.

Exercises for Section 8.6

In Exercises 1 and 2, use a table of integrals with forms involving $a + bu$ to find the integral.

1. $\int \dfrac{x^2}{1 + x} \, dx$
2. $\int \dfrac{2}{3x^2(2x - 5)^2} \, dx$

In Exercises 3 and 4, use a table of integrals with forms involving $\sqrt{u^2 \pm a^2}$ to find the integral.

3. $\int e^x \sqrt{1 + e^{2x}} \, dx$
4. $\int \dfrac{\sqrt{x^2 - 9}}{3x} \, dx$

In Exercises 5 and 6, use a table of integrals with forms involving $\sqrt{a^2 - u^2}$ to find the integral.

5. $\int \dfrac{1}{x^2 \sqrt{1 - x^2}} \, dx$
6. $\int \dfrac{x}{\sqrt{9 - x^4}} \, dx$

In Exercises 7–10, use a table of integrals with forms involving the trigonometric functions to find the integral.

7. $\int \sin^4 2x \, dx$
8. $\int \dfrac{\cos^3 \sqrt{x}}{\sqrt{x}} \, dx$
9. $\int \dfrac{1}{\sqrt{x}\left(1 - \cos \sqrt{x}\right)} \, dx$
10. $\int \dfrac{1}{1 - \tan 5x} \, dx$

In Exercises 11 and 12, use a table of integrals with forms involving e^u to find the integral.

11. $\int \dfrac{1}{1 + e^{2x}} \, dx$
12. $\int e^{-x/2} \sin 2x \, dx$

In Exercises 13 and 14, use a table of integrals with forms involving $\ln u$ to find the integral.

13. $\int x^3 \ln x \, dx$
14. $\int (\ln x)^3 \, dx$

In Exercises 15–18, find the indefinite integral (a) using integration tables and (b) using the given method.

Integral	Method
15. $\int x^2 e^x \, dx$	Integration by parts
16. $\int x^4 \ln x \, dx$	Integration by parts
17. $\int \dfrac{1}{x^2(x + 1)} \, dx$	Partial fractions
18. $\int \dfrac{1}{x^2 - 75} \, dx$	Partial fractions

In Exercises 19–42, use integration tables to find the integral.

19. $\int x \operatorname{arcsec}(x^2 + 1) \, dx$
20. $\int \operatorname{arcsec} 2x \, dx$
21. $\int \dfrac{1}{x^2 \sqrt{x^2 - 4}} \, dx$
22. $\int \dfrac{1}{x^2 + 2x + 2} \, dx$
23. $\int \dfrac{2x}{(1 - 3x)^2} \, dx$
24. $\int \dfrac{\theta^2}{1 - \sin \theta^3} \, d\theta$
25. $\int e^x \arccos e^x \, dx$
26. $\int \dfrac{e^x}{1 - \tan e^x} \, dx$
27. $\int \dfrac{x}{1 - \sec x^2} \, dx$
28. $\int \dfrac{1}{t[1 + (\ln t)^2]} \, dt$
29. $\int \dfrac{\cos \theta}{3 + 2 \sin \theta + \sin^2 \theta} \, d\theta$
30. $\int x^2 \sqrt{2 + 9x^2} \, dx$
31. $\int \dfrac{1}{x^2 \sqrt{2 + 9x^2}} \, dx$
32. $\int \sqrt{x} \arctan x^{3/2} \, dx$
33. $\int \dfrac{\ln x}{x(3 + 2 \ln x)} \, dx$
34. $\int \dfrac{e^x}{(1 - e^{2x})^{3/2}} \, dx$
35. $\int \dfrac{x}{(x^2 - 6x + 10)^2} \, dx$
36. $\int (2x - 3)^2 \sqrt{(2x - 3)^2 + 4} \, dx$
37. $\int \dfrac{x}{\sqrt{x^4 - 6x^2 + 5}} \, dx$
38. $\int \dfrac{\cos x}{\sqrt{\sin^2 x + 1}} \, dx$
39. $\int \dfrac{x^3}{\sqrt{4 - x^2}} \, dx$
40. $\int \sqrt{\dfrac{3 - x}{3 + x}} \, dx$
41. $\int \dfrac{e^{3x}}{(1 + e^x)^3} \, dx$
42. $\int \tan^3 \theta \, d\theta$

In Exercises 43–50, use integration tables to evaluate the integral.

43. $\int_0^1 x e^{x^2} \, dx$
44. $\int_0^3 \dfrac{x}{\sqrt{1 + x}} \, dx$
45. $\int_1^3 x^2 \ln x \, dx$
46. $\int_0^\pi x \sin x \, dx$
47. $\int_{-\pi/2}^{\pi/2} \dfrac{\cos x}{1 + \sin^2 x} \, dx$
48. $\int_2^4 \dfrac{x^2}{(3x - 5)^2} \, dx$
49. $\int_0^{\pi/2} t^3 \cos t \, dt$
50. $\int_0^1 \sqrt{3 + x^2} \, dx$

In Exercises 51–56, verify the integration formula.

51. $\int \dfrac{u^2}{(a + bu)^2} \, du = \dfrac{1}{b^3}\left(bu - \dfrac{a^2}{a + bu} - 2a \ln|a + bu|\right) + C$
52. $\int \dfrac{u^n}{\sqrt{a + bu}} \, du = \dfrac{2}{(2n + 1)b}\left(u^n \sqrt{a + bu} - na \int \dfrac{u^{n-1}}{\sqrt{a + bu}} \, du\right)$
53. $\int \dfrac{1}{(u^2 \pm a^2)^{3/2}} \, du = \dfrac{\pm u}{a^2 \sqrt{u^2 \pm a^2}} + C$
54. $\int u^n \cos u \, du = u^n \sin u - n \int u^{n-1} \sin u \, du$
55. $\int \arctan u \, du = u \arctan u - \ln \sqrt{1 + u^2} + C$

56. $\int (\ln u)^n \, du = u(\ln u)^n - n \int (\ln u)^{n-1} \, du$

In Exercises 57–62, use a computer algebra system to determine the antiderivative that passes through the given point. Use the system to graph the resulting antiderivative.

57. $\int \dfrac{1}{x^{3/2}\sqrt{1-x}} \, dx, \ \left(\tfrac{1}{2}, 5\right)$ **58.** $\int x\sqrt{x^2 + 2x} \, dx, \ (0, 0)$

59. $\int \dfrac{1}{(x^2 - 6x + 10)^2} \, dx, \ (3, 0)$

60. $\int \dfrac{\sqrt{2 - 2x - x^2}}{x + 1} \, dx, \ (0, \sqrt{2})$

61. $\int \dfrac{1}{\sin \theta \tan \theta} \, d\theta, \ \left(\tfrac{\pi}{4}, 2\right)$

62. $\int \dfrac{\sin \theta}{(\cos \theta)(1 + \sin \theta)} \, d\theta, \ (0, 1)$

In Exercises 63–70, find or evaluate the integral.

63. $\int \dfrac{1}{2 - 3\sin \theta} \, d\theta$ **64.** $\int \dfrac{\sin \theta}{1 + \cos^2 \theta} \, d\theta$

65. $\int_0^{\pi/2} \dfrac{1}{1 + \sin \theta + \cos \theta} \, d\theta$ **66.** $\int_0^{\pi/2} \dfrac{1}{3 - 2\cos \theta} \, d\theta$

67. $\int \dfrac{\sin \theta}{3 - 2\cos \theta} \, d\theta$ **68.** $\int \dfrac{\cos \theta}{1 + \cos \theta} \, d\theta$

69. $\int \dfrac{\cos \sqrt{\theta}}{\sqrt{\theta}} \, d\theta$ **70.** $\int \dfrac{1}{\sec \theta - \tan \theta} \, d\theta$

Area In Exercises 71 and 72, find the area of the region bounded by the graphs of the equations.

71. $y = \dfrac{x}{\sqrt{x + 1}}, \ y = 0, \ x = 8$ **72.** $y = \dfrac{x}{1 + e^{x^2}}, \ y = 0, \ x = 2$

Writing About Concepts

In Exercises 73–78, state (if possible) the method or integration formula you would use to find the antiderivative. Explain why you chose that method or formula. Do not integrate.

73. $\int \dfrac{e^x}{e^{2x} + 1} \, dx$ **74.** $\int \dfrac{e^x}{e^x + 1} \, dx$

75. $\int x e^{x^2} \, dx$ **76.** $\int x e^x \, dx$

77. $\int e^{x^2} \, dx$ **78.** $\int e^{2x} \sqrt{e^{2x} + 1} \, dx$

79. (a) Evaluate $\int x^n \ln x \, dx$ for $n = 1, 2,$ and 3. Describe any patterns you notice.

(b) Write a general rule for evaluating the integral in part (a), for an integer $n \geq 1$.

80. Describe what is meant by a reduction formula. Give an example.

True or False? In Exercises 81 and 82, determine whether the statement is true or false. If it is false, explain why or give an example that shows it is false.

81. To use a table of integrals, the integral you are evaluating must appear in the table.

82. When using a table of integrals, you may have to make substitutions to rewrite your integral in the form in which it appears in the table.

83. Work A hydraulic cylinder on an industrial machine pushes a steel block a distance of x feet $(0 \leq x \leq 5)$, where the variable force required is $F(x) = 2000xe^{-x}$ pounds. Find the work done in pushing the block the full 5 feet through the machine.

84. Work Repeat Exercise 83, using $F(x) = \dfrac{500x}{\sqrt{26 - x^2}}$ pounds.

85. Building Design The cross section of a precast concrete beam for a building is bounded by the graphs of the equations

$$x = \dfrac{2}{\sqrt{1 + y^2}}, \ x = \dfrac{-2}{\sqrt{1 + y^2}}, \ y = 0, \text{ and } y = 3$$

where x and y are measured in feet. The length of the beam is 20 feet (see figure). (a) Find the volume V and the weight W of the beam. Assume the concrete weighs 148 pounds per cubic foot. (b) Then find the centroid of a cross section of the beam.

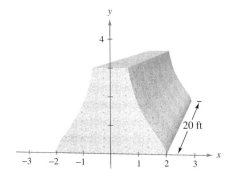

86. Population A population is growing according to the logistic model $N = \dfrac{5000}{1 + e^{4.8 - 1.9t}}$ where t is the time in days. Find the average population over the interval $[0, 2]$.

In Exercises 87 and 88, use a graphing utility to (a) solve the integral equation for the constant k and (b) graph the region whose area is given by the integral.

87. $\int_0^4 \dfrac{k}{2 + 3x} \, dx = 10$ **88.** $\int_0^k 6x^2 \, e^{-x/2} \, dx = 50$

Putnam Exam Challenge

89. Evaluate $\int_0^{\pi/2} \dfrac{dx}{1 + (\tan x)^{\sqrt{2}}}$.

This problem was composed by the Committee on the Putnam Prize Competition. © The Mathematical Association of America. All rights reserved.

Section 8.7 Indeterminate Forms and L'Hôpital's Rule

- Recognize limits that produce indeterminate forms.
- Apply L'Hôpital's Rule to evaluate a limit.

Indeterminate Forms

Recall from Chapters 1 and 3 that the forms $0/0$ and ∞/∞ are called *indeterminate* because they do not guarantee that a limit exists, nor do they indicate what the limit is, if one does exist. When you encountered one of these indeterminate forms earlier in the text, you attempted to rewrite the expression by using various algebraic techniques.

Indeterminate Form	Limit	Algebraic Technique
$\dfrac{0}{0}$	$\lim\limits_{x \to -1} \dfrac{2x^2 - 2}{x + 1} = \lim\limits_{x \to -1} 2(x - 1)$ $= -4$	Divide numerator and denominator by $(x + 1)$.
$\dfrac{\infty}{\infty}$	$\lim\limits_{x \to \infty} \dfrac{3x^2 - 1}{2x^2 + 1} = \lim\limits_{x \to \infty} \dfrac{3 - (1/x^2)}{2 + (1/x^2)}$ $= \dfrac{3}{2}$	Divide numerator and denominator by x^2.

Occasionally, you can extend these algebraic techniques to find limits of transcendental functions. For instance, the limit

$$\lim_{x \to 0} \frac{e^{2x} - 1}{e^x - 1}$$

produces the indeterminate form $0/0$. Factoring and then dividing produces

$$\lim_{x \to 0} \frac{e^{2x} - 1}{e^x - 1} = \lim_{x \to 0} \frac{(e^x + 1)(e^x - 1)}{e^x - 1} = \lim_{x \to 0} (e^x + 1) = 2.$$

However, not all indeterminate forms can be evaluated by algebraic manipulation. This is often true when *both* algebraic and transcendental functions are involved. For instance, the limit

$$\lim_{x \to 0} \frac{e^{2x} - 1}{x}$$

produces the indeterminate form $0/0$. Rewriting the expression to obtain

$$\lim_{x \to 0} \left(\frac{e^{2x}}{x} - \frac{1}{x} \right)$$

merely produces another indeterminate form, $\infty - \infty$. Of course, you could use technology to estimate the limit, as shown in the table and in Figure 8.14. From the table and the graph, the limit appears to be 2. (This limit will be verified in Example 1.)

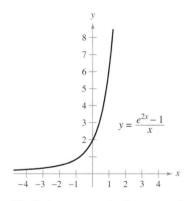

The limit as x approaches 0 appears to be 2.
Figure 8.14

x	-1	-0.1	-0.01	-0.001	0	0.001	0.01	0.1	1
$\dfrac{e^{2x} - 1}{x}$	0.865	1.813	1.980	1.998	?	2.002	2.020	2.214	6.389

L'Hôpital's Rule

To find the limit illustrated in Figure 8.14, you can use a theorem called **L'Hôpital's Rule.** This theorem states that under certain conditions the limit of the quotient $f(x)/g(x)$ is determined by the limit of the quotient of the derivatives

$$\frac{f'(x)}{g'(x)}.$$

To prove this theorem, you can use a more general result called the **Extended Mean Value Theorem.**

GUILLAUME L'HÔPITAL (1661–1704)

L'Hôpital's Rule is named after the French mathematician Guillaume François Antoine de L'Hôpital. L'Hôpital is credited with writing the first text on differential calculus (in 1696) in which the rule publicly appeared. It was recently discovered that the rule and its proof were written in a letter from John Bernoulli to L'Hôpital. "... I acknowledge that I owe very much to the bright minds of the Bernoulli brothers. ... I have made free use of their discoveries ...," said L'Hôpital.

THEOREM 8.3 The Extended Mean Value Theorem

If f and g are differentiable on an open interval (a, b) and continuous on $[a, b]$ such that $g'(x) \neq 0$ for any x in (a, b), then there exists a point c in (a, b) such that

$$\frac{f'(c)}{g'(c)} = \frac{f(b) - f(a)}{g(b) - g(a)}.$$

NOTE To see why this is called the Extended Mean Value Theorem, consider the special case in which $g(x) = x$. For this case, you obtain the "standard" Mean Value Theorem as presented in Section 3.2.

The Extended Mean Value Theorem and L'Hôpital's Rule are both proved in Appendix A.

THEOREM 8.4 L'Hôpital's Rule

Let f and g be functions that are differentiable on an open interval (a, b) containing c, except possibly at c itself. Assume that $g'(x) \neq 0$ for all x in (a, b), except possibly at c itself. If the limit of $f(x)/g(x)$ as x approaches c produces the indeterminate form $0/0$, then

$$\lim_{x \to c} \frac{f(x)}{g(x)} = \lim_{x \to c} \frac{f'(x)}{g'(x)}$$

provided the limit on the right exists (or is infinite). This result also applies if the limit of $f(x)/g(x)$ as x approaches c produces any one of the indeterminate forms ∞/∞, $(-\infty)/\infty$, $\infty/(-\infty)$, or $(-\infty)/(-\infty)$.

NOTE People occasionally use L'Hôpital's Rule incorrectly by applying the Quotient Rule to $f(x)/g(x)$. Be sure you see that the rule involves $f'(x)/g'(x)$, not the derivative of $f(x)/g(x)$.

L'Hôpital's Rule can also be applied to one-sided limits. For instance, if the limit of $f(x)/g(x)$ as x approaches c *from the right* produces the indeterminate form $0/0$, then

$$\lim_{x \to c^+} \frac{f(x)}{g(x)} = \lim_{x \to c^+} \frac{f'(x)}{g'(x)}$$

provided the limit exists (or is infinite).

FOR FURTHER INFORMATION To enhance your understanding of the necessity of the restriction that $g'(x)$ be nonzero for all x in (a, b), except possibly at c, see the article "Counterexamples to L'Hôpital's Rule" by R. P. Boas in *The American Mathematical Monthly*. To view this article, go to the website *www.matharticles.com*.

TECHNOLOGY *Numerical and Graphical Approaches* Use a numerical or a graphical approach to approximate each limit.

a. $\lim\limits_{x \to 0} \dfrac{2^{2x} - 1}{x}$

b. $\lim\limits_{x \to 0} \dfrac{3^{2x} - 1}{x}$

c. $\lim\limits_{x \to 0} \dfrac{4^{2x} - 1}{x}$

d. $\lim\limits_{x \to 0} \dfrac{5^{2x} - 1}{x}$

What pattern do you observe? Does an analytic approach have an advantage for these limits? If so, explain your reasoning.

EXAMPLE 1 Indeterminate Form 0/0

Evaluate $\lim\limits_{x \to 0} \dfrac{e^{2x} - 1}{x}$.

Solution Because direct substitution results in the indeterminate form 0/0

$$\lim_{x \to 0} \dfrac{e^{2x} - 1}{x} \quad \begin{matrix} \nearrow \lim\limits_{x \to 0} (e^{2x} - 1) = 0 \\ \searrow \lim\limits_{x \to 0} x = 0 \end{matrix}$$

you can apply L'Hôpital's Rule as shown below.

$$\lim_{x \to 0} \dfrac{e^{2x} - 1}{x} = \lim_{x \to 0} \dfrac{\dfrac{d}{dx}[e^{2x} - 1]}{\dfrac{d}{dx}[x]} \quad \text{Apply L'Hôpital's Rule.}$$

$$= \lim_{x \to 0} \dfrac{2e^{2x}}{1} \quad \text{Differentiate numerator and denominator.}$$

$$= 2 \quad \text{Evaluate the limit.}$$

NOTE In writing the string of equations in Example 1, you actually do not know that the first limit is equal to the second until you have shown that the second limit exists. In other words, if the second limit had not existed, it would not have been permissible to apply L'Hôpital's Rule.

Another form of L'Hôpital's Rule states that if the limit of $f(x)/g(x)$ as x approaches ∞ (or $-\infty$) produces the indeterminate form 0/0 or ∞/∞, then

$$\lim_{x \to \infty} \dfrac{f(x)}{g(x)} = \lim_{x \to \infty} \dfrac{f'(x)}{g'(x)}$$

provided the limit on the right exists.

EXAMPLE 2 Indeterminate Form ∞/∞

Evaluate $\lim\limits_{x \to \infty} \dfrac{\ln x}{x}$.

Solution Because direct substitution results in the indeterminate form ∞/∞, you can apply L'Hôpital's Rule to obtain

$$\lim_{x \to \infty} \dfrac{\ln x}{x} = \lim_{x \to \infty} \dfrac{\dfrac{d}{dx}[\ln x]}{\dfrac{d}{dx}[x]} \quad \text{Apply L'Hôpital's Rule.}$$

$$= \lim_{x \to \infty} \dfrac{1}{x} \quad \text{Differentiate numerator and denominator.}$$

$$= 0. \quad \text{Evaluate the limit.}$$

NOTE Try graphing $y_1 = \ln x$ and $y_2 = x$ in the same viewing window. Which function grows faster as x approaches ∞? How is this observation related to Example 2?

Occasionally it is necessary to apply L'Hôpital's Rule more than once to remove an indeterminate form, as shown in Example 3.

EXAMPLE 3 Applying L'Hôpital's Rule More Than Once

Evaluate $\lim\limits_{x \to -\infty} \dfrac{x^2}{e^{-x}}$.

Solution Because direct substitution results in the indeterminate form ∞/∞, you can apply L'Hôpital's Rule.

$$\lim_{x \to -\infty} \frac{x^2}{e^{-x}} = \lim_{x \to -\infty} \frac{\dfrac{d}{dx}[x^2]}{\dfrac{d}{dx}[e^{-x}]} = \lim_{x \to -\infty} \frac{2x}{-e^{-x}}$$

This limit yields the indeterminate form $(-\infty)/(-\infty)$, so you can apply L'Hôpital's Rule again to obtain

$$\lim_{x \to -\infty} \frac{2x}{-e^{-x}} = \lim_{x \to -\infty} \frac{\dfrac{d}{dx}[2x]}{\dfrac{d}{dx}[-e^{-x}]} = \lim_{x \to -\infty} \frac{2}{e^{-x}} = 0.$$

In addition to the forms $0/0$ and ∞/∞, there are other indeterminate forms such as $0 \cdot \infty$, 1^∞, ∞^0, 0^0, and $\infty - \infty$. For example, consider the following four limits that lead to the indeterminate form $0 \cdot \infty$.

$$\underbrace{\lim_{x \to 0} (x)\left(\frac{1}{x}\right)}_{\text{Limit is 1.}}, \quad \underbrace{\lim_{x \to 0} (x)\left(\frac{2}{x}\right)}_{\text{Limit is 2.}}, \quad \underbrace{\lim_{x \to \infty} (x)\left(\frac{1}{e^x}\right)}_{\text{Limit is 0.}}, \quad \underbrace{\lim_{x \to \infty} (e^x)\left(\frac{1}{x}\right)}_{\text{Limit is } \infty.}$$

Because each limit is different, it is clear that the form $0 \cdot \infty$ is indeterminate in the sense that it does not determine the value (or even the existence) of the limit. The following examples indicate methods for evaluating these forms. Basically, you attempt to convert each of these forms to $0/0$ or ∞/∞ so that L'Hôpital's Rule can be applied.

EXAMPLE 4 Indeterminate Form $0 \cdot \infty$

Evaluate $\lim\limits_{x \to \infty} e^{-x} \sqrt{x}$.

Solution Because direct substitution produces the indeterminate form $0 \cdot \infty$, you should try to rewrite the limit to fit the form $0/0$ or ∞/∞. In this case, you can rewrite the limit to fit the second form.

$$\lim_{x \to \infty} e^{-x} \sqrt{x} = \lim_{x \to \infty} \frac{\sqrt{x}}{e^x}$$

Now, by L'Hôpital's Rule, you have

$$\lim_{x \to \infty} \frac{\sqrt{x}}{e^x} = \lim_{x \to \infty} \frac{1/(2\sqrt{x})}{e^x} = \lim_{x \to \infty} \frac{1}{2\sqrt{x}\, e^x} = 0.$$

If rewriting a limit in one of the forms $0/0$ or ∞/∞ does not seem to work, try the other form. For instance, in Example 4 you can write the limit as

$$\lim_{x \to \infty} e^{-x} \sqrt{x} = \lim_{x \to \infty} \frac{e^{-x}}{x^{-1/2}}$$

which yields the indeterminate form $0/0$. As it happens, applying L'Hôpital's Rule to this limit produces

$$\lim_{x \to \infty} \frac{e^{-x}}{x^{-1/2}} = \lim_{x \to \infty} \frac{-e^{-x}}{-1/(2x^{3/2})}$$

which also yields the indeterminate form $0/0$.

The indeterminate forms 1^∞, ∞^0, and 0^0 arise from limits of functions that have variable bases and variable exponents. When you previously encountered this type of function, you used logarithmic differentiation to find the derivative. You can use a similar procedure when taking limits, as shown in the next example.

EXAMPLE 5 Indeterminate Form 1^∞

Evaluate $\lim\limits_{x \to \infty} \left(1 + \frac{1}{x}\right)^x$.

Solution Because direct substitution yields the indeterminate form 1^∞, you can proceed as follows. To begin, assume that the limit exists and is equal to y.

$$y = \lim_{x \to \infty} \left(1 + \frac{1}{x}\right)^x$$

Taking the natural logarithm of each side produces

$$\ln y = \ln \left[\lim_{x \to \infty} \left(1 + \frac{1}{x}\right)^x \right].$$

Because the natural logarithmic function is continuous, you can write

$$\ln y = \lim_{x \to \infty} \left[x \ln\left(1 + \frac{1}{x}\right) \right] \qquad \text{Indeterminate form } \infty \cdot 0$$

$$= \lim_{x \to \infty} \left(\frac{\ln[1 + (1/x)]}{1/x} \right) \qquad \text{Indeterminate form } 0/0$$

$$= \lim_{x \to \infty} \left(\frac{(-1/x^2)\{1/[1 + (1/x)]\}}{-1/x^2} \right) \qquad \text{L'Hôpital's Rule}$$

$$= \lim_{x \to \infty} \frac{1}{1 + (1/x)}$$

$$= 1.$$

Now, because you have shown that $\ln y = 1$, you can conclude that $y = e$ and obtain

$$\lim_{x \to \infty} \left(1 + \frac{1}{x}\right)^x = e.$$

You can use a graphing utility to confirm this result, as shown in Figure 8.15.

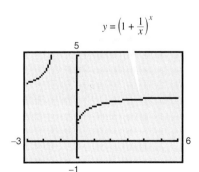

The limit of $[1 + (1/x)]^x$ as x approaches infinity is e.
Figure 8.15

L'Hôpital's Rule can also be applied to one-sided limits, as demonstrated in Examples 6 and 7.

EXAMPLE 6 Indeterminate Form 0^0

Find $\lim_{x \to 0^+} (\sin x)^x$.

Solution Because direct substitution produces the indeterminate form 0^0, you can proceed as shown below. To begin, assume that the limit exists and is equal to y.

$$y = \lim_{x \to 0^+} (\sin x)^x \qquad \text{Indeterminate form } 0^0$$

$$\ln y = \ln\left[\lim_{x \to 0^+} (\sin x)^x\right] \qquad \text{Take natural log of each side.}$$

$$= \lim_{x \to 0^+} \left[\ln(\sin x)^x\right] \qquad \text{Continuity}$$

$$= \lim_{x \to 0^+} \left[x \ln(\sin x)\right] \qquad \text{Indeterminate form } 0 \cdot (-\infty)$$

$$= \lim_{x \to 0^+} \frac{\ln(\sin x)}{1/x} \qquad \text{Indeterminate form } -\infty/\infty$$

$$= \lim_{x \to 0^+} \frac{\cot x}{-1/x^2} \qquad \text{L'Hôpital's Rule}$$

$$= \lim_{x \to 0^+} \frac{-x^2}{\tan x} \qquad \text{Indeterminate form } 0/0$$

$$= \lim_{x \to 0^+} \frac{-2x}{\sec^2 x} = 0 \qquad \text{L'Hôpital's Rule}$$

Now, because $\ln y = 0$, you can conclude that $y = e^0 = 1$, and it follows that

$$\lim_{x \to 0^+} (\sin x)^x = 1.$$

TECHNOLOGY When evaluating complicated limits such as the one in Example 6, it is helpful to check the reasonableness of the solution with a computer or with a graphing utility. For instance, the calculations in the following table and the graph in Figure 8.16 are consistent with the conclusion that $(\sin x)^x$ approaches 1 as x approaches 0 from the right.

x	1.0	0.1	0.01	0.001	0.0001	0.00001
$(\sin x)^x$	0.8415	0.7942	0.9550	0.9931	0.9991	0.9999

Use a computer algebra system or graphing utility to estimate the following limits:

$$\lim_{x \to 0} (1 - \cos x)^x$$

and

$$\lim_{x \to 0^+} (\tan x)^x.$$

Then see if you can verify your estimates analytically.

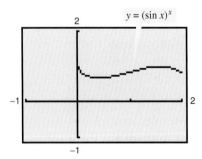

The limit of $(\sin x)^x$ is 1 as x approaches 0 from the right.
Figure 8.16

STUDY TIP In each of the examples presented in this section, L'Hôpital's Rule is used to find a limit that exists. It can also be used to conclude that a limit is infinite. For instance, try using L'Hôpital's Rule to show that

$$\lim_{x \to \infty} \frac{e^x}{x} = \infty.$$

EXAMPLE 7 Indeterminate Form $\infty - \infty$

Evaluate $\displaystyle\lim_{x \to 1^+} \left(\frac{1}{\ln x} - \frac{1}{x-1} \right)$.

Solution Because direct substitution yields the indeterminate form $\infty - \infty$, you should try to rewrite the expression to produce a form to which you can apply L'Hôpital's Rule. In this case, you can combine the two fractions to obtain

$$\lim_{x \to 1^+} \left(\frac{1}{\ln x} - \frac{1}{x-1} \right) = \lim_{x \to 1^+} \left[\frac{x - 1 - \ln x}{(x-1) \ln x} \right].$$

Now, because direct substitution produces the indeterminate form $0/0$, you can apply L'Hôpital's Rule to obtain

$$\lim_{x \to 1^+} \left(\frac{1}{\ln x} - \frac{1}{x-1} \right) = \lim_{x \to 1^+} \frac{\frac{d}{dx}[x - 1 - \ln x]}{\frac{d}{dx}[(x-1) \ln x]}$$

$$= \lim_{x \to 1^+} \left[\frac{1 - (1/x)}{(x-1)(1/x) + \ln x} \right]$$

$$= \lim_{x \to 1^+} \left(\frac{x - 1}{x - 1 + x \ln x} \right).$$

This limit also yields the indeterminate form $0/0$, so you can apply L'Hôpital's Rule again to obtain

$$\lim_{x \to 1^+} \left(\frac{1}{\ln x} - \frac{1}{x-1} \right) = \lim_{x \to 1^+} \left[\frac{1}{1 + x(1/x) + \ln x} \right]$$

$$= \frac{1}{2}.$$

The forms $0/0$, ∞/∞, $\infty - \infty$, $0 \cdot \infty$, 0^0, 1^∞, and ∞^0 have been identified as *indeterminate*. There are similar forms that you should recognize as "determinate."

$\infty + \infty \to \infty$ Limit is positive infinity.
$-\infty - \infty \to -\infty$ Limit is negative infinity.
$0^\infty \to 0$ Limit is zero.
$0^{-\infty} \to \infty$ Limit is positive infinity.

(You are asked to verify two of these in Exercises 106 and 107.)

As a final comment, remember that L'Hôpital's Rule can be applied only to quotients leading to the indeterminate forms $0/0$ and ∞/∞. For instance, the following application of L'Hôpital's Rule is *incorrect*.

$$\lim_{x \to 0} \frac{e^x}{x} \stackrel{?}{=} \lim_{x \to 0} \frac{e^x}{1} = 1 \qquad \text{Incorrect use of L'Hôpital's Rule}$$

The reason this application is incorrect is that, even though the limit of the denominator is 0, the limit of the numerator is 1, which means that the hypotheses of L'Hôpital's Rule have not been satisfied.

Exercises for Section 8.7

See www.CalcChat.com for worked-out solutions to odd-numbered exercises.

Numerical and Graphical Analysis In Exercises 1–4, complete the table and use the result to estimate the limit. Use a graphing utility to graph the function to support your result.

1. $\lim\limits_{x \to 0} \dfrac{\sin 5x}{\sin 2x}$

x	-0.1	-0.01	-0.001	0.001	0.01	0.1
$f(x)$						

2. $\lim\limits_{x \to 0} \dfrac{1 - e^x}{x}$

x	-0.1	-0.01	-0.001	0.001	0.01	0.1
$f(x)$						

3. $\lim\limits_{x \to \infty} x^5 e^{-x/100}$

x	1	10	10^2	10^3	10^4	10^5
$f(x)$						

4. $\lim\limits_{x \to \infty} \dfrac{6x}{\sqrt{3x^2 - 2x}}$

x	1	10	10^2	10^3	10^4	10^5
$f(x)$						

In Exercises 5–10, evaluate the limit (a) using techniques from Chapters 1 and 3 and (b) using L'Hôpital's Rule.

5. $\lim\limits_{x \to 3} \dfrac{2(x - 3)}{x^2 - 9}$

6. $\lim\limits_{x \to -1} \dfrac{2x^2 - x - 3}{x + 1}$

7. $\lim\limits_{x \to 3} \dfrac{\sqrt{x + 1} - 2}{x - 3}$

8. $\lim\limits_{x \to 0} \dfrac{\sin 4x}{2x}$

9. $\lim\limits_{x \to \infty} \dfrac{5x^2 - 3x + 1}{3x^2 - 5}$

10. $\lim\limits_{x \to \infty} \dfrac{2x + 1}{4x^2 + x}$

In Exercises 11–36, evaluate the limit, using L'Hôpital's Rule if necessary. (In Exercise 18, n is a positive integer.)

11. $\lim\limits_{x \to 2} \dfrac{x^2 - x - 2}{x - 2}$

12. $\lim\limits_{x \to -1} \dfrac{x^2 - x - 2}{x + 1}$

13. $\lim\limits_{x \to 0} \dfrac{\sqrt{4 - x^2} - 2}{x}$

14. $\lim\limits_{x \to 2^-} \dfrac{\sqrt{4 - x^2}}{x - 2}$

15. $\lim\limits_{x \to 0} \dfrac{e^x - (1 - x)}{x}$

16. $\lim\limits_{x \to 1} \dfrac{\ln x^2}{x^2 - 1}$

17. $\lim\limits_{x \to 0^+} \dfrac{e^x - (1 + x)}{x^3}$

18. $\lim\limits_{x \to 0^+} \dfrac{e^x - (1 + x)}{x^n}$

19. $\lim\limits_{x \to 0} \dfrac{\sin 2x}{\sin 3x}$

20. $\lim\limits_{x \to 0} \dfrac{\sin ax}{\sin bx}$

21. $\lim\limits_{x \to 0} \dfrac{\arcsin x}{x}$

22. $\lim\limits_{x \to 1} \dfrac{\arctan x - (\pi/4)}{x - 1}$

23. $\lim\limits_{x \to \infty} \dfrac{3x^2 - 2x + 1}{2x^2 + 3}$

24. $\lim\limits_{x \to \infty} \dfrac{x - 1}{x^2 + 2x + 3}$

25. $\lim\limits_{x \to \infty} \dfrac{x^2 + 2x + 3}{x - 1}$

26. $\lim\limits_{x \to \infty} \dfrac{x^3}{x + 2}$

27. $\lim\limits_{x \to \infty} \dfrac{x^3}{e^{x/2}}$

28. $\lim\limits_{x \to \infty} \dfrac{x^2}{e^x}$

29. $\lim\limits_{x \to \infty} \dfrac{x}{\sqrt{x^2 + 1}}$

30. $\lim\limits_{x \to \infty} \dfrac{x^2}{\sqrt{x^2 + 1}}$

31. $\lim\limits_{x \to \infty} \dfrac{\cos x}{x}$

32. $\lim\limits_{x \to \infty} \dfrac{\sin x}{x - \pi}$

33. $\lim\limits_{x \to \infty} \dfrac{\ln x}{x^2}$

34. $\lim\limits_{x \to \infty} \dfrac{\ln x^4}{x^3}$

35. $\lim\limits_{x \to \infty} \dfrac{e^x}{x^2}$

36. $\lim\limits_{x \to \infty} \dfrac{e^{x/2}}{x}$

 In Exercises 37–54, (a) describe the type of indeterminate form (if any) that is obtained by direct substitution. (b) Evaluate the limit, using L'Hôpital's Rule if necessary. (c) Use a graphing utility to graph the function and verify the result in part (b).

37. $\lim\limits_{x \to 0^+} x \ln x$

38. $\lim\limits_{x \to 0^+} x^3 \cot x$

39. $\lim\limits_{x \to \infty} \left(x \sin \dfrac{1}{x} \right)$

40. $\lim\limits_{x \to \infty} x \tan \dfrac{1}{x}$

41. $\lim\limits_{x \to 0^+} x^{1/x}$

42. $\lim\limits_{x \to 0^+} (e^x + x)^{2/x}$

43. $\lim\limits_{x \to \infty} x^{1/x}$

44. $\lim\limits_{x \to \infty} \left(1 + \dfrac{1}{x}\right)^x$

45. $\lim\limits_{x \to 0^+} (1 + x)^{1/x}$

46. $\lim\limits_{x \to \infty} (1 + x)^{1/x}$

47. $\lim\limits_{x \to 0^+} [3(x)^{x/2}]$

48. $\lim\limits_{x \to 4^+} [3(x - 4)]^{x-4}$

49. $\lim\limits_{x \to 1^+} (\ln x)^{x-1}$

50. $\lim\limits_{x \to 0^+} \left[\cos\left(\dfrac{\pi}{2} - x\right)\right]^x$

51. $\lim\limits_{x \to 2^+} \left(\dfrac{8}{x^2 - 4} - \dfrac{x}{x - 2} \right)$

52. $\lim\limits_{x \to 2^+} \left(\dfrac{1}{x^2 - 4} - \dfrac{\sqrt{x - 1}}{x^2 - 4} \right)$

53. $\lim\limits_{x \to 1^+} \left(\dfrac{3}{\ln x} - \dfrac{2}{x - 1} \right)$

54. $\lim\limits_{x \to 0^+} \left(\dfrac{10}{x} - \dfrac{3}{x^2} \right)$

 In Exercises 55–58, use a graphing utility to (a) graph the function and (b) find the required limit (if it exists).

55. $\lim\limits_{x \to 3} \dfrac{x - 3}{\ln(2x - 5)}$

56. $\lim\limits_{x \to 0^+} (\sin x)^x$

57. $\lim\limits_{x \to \infty} \left(\sqrt{x^2 + 5x + 2} - x \right)$

58. $\lim\limits_{x \to \infty} \dfrac{x^3}{e^{2x}}$

Writing About Concepts

59. List six different indeterminate forms.

60. State L'Hôpital's Rule.

61. Find the differentiable functions f and g that satisfy the specified condition such that
$$\lim_{x \to 5} f(x) = 0 \text{ and } \lim_{x \to 5} g(x) = 0.$$
Explain how you obtained your answers. (*Note:* There are many correct answers.)

(a) $\lim_{x \to 5} \dfrac{f(x)}{g(x)} = 10$ (b) $\lim_{x \to 5} \dfrac{f(x)}{g(x)} = 0$

(c) $\lim_{x \to 5} \dfrac{f(x)}{g(x)} = \infty$

62. Find differentiable functions f and g such that
$$\lim_{x \to \infty} f(x) = \lim_{x \to \infty} g(x) = \infty \text{ and}$$
$$\lim_{x \to \infty} [f(x) - g(x)] = 25.$$
Explain how you obtained your answers. (*Note:* There are many correct answers.)

63. *Numerical Approach* Complete the table to show that x eventually "overpowers" $(\ln x)^4$.

x	10	10^2	10^4	10^6	10^8	10^{10}
$\dfrac{(\ln x)^4}{x}$						

64. *Numerical Approach* Complete the table to show that e^x eventually "overpowers" x^5.

x	1	5	10	20	30	40	50	100
$\dfrac{e^x}{x^5}$								

Comparing Functions In Exercises 65–70, use L'Hôpital's Rule to determine the comparative rates of increase of the functions
$$f(x) = x^m, \quad g(x) = e^{nx}, \quad \text{and} \quad h(x) = (\ln x)^n$$
where $n > 0$, $m > 0$, and $x \to \infty$.

65. $\lim\limits_{x \to \infty} \dfrac{x^2}{e^{5x}}$ **66.** $\lim\limits_{x \to \infty} \dfrac{x^3}{e^{2x}}$

67. $\lim\limits_{x \to \infty} \dfrac{(\ln x)^3}{x}$ **68.** $\lim\limits_{x \to \infty} \dfrac{(\ln x)^2}{x^3}$

69. $\lim\limits_{x \to \infty} \dfrac{(\ln x)^n}{x^m}$ **70.** $\lim\limits_{x \to \infty} \dfrac{x^m}{e^{nx}}$

 In Exercises 71–74, find any asymptotes and relative extrema that may exist and use a graphing utility to graph the function. (*Hint:* Some of the limits required in finding asymptotes have been found in preceding exercises.)

71. $y = x^{1/x}, \quad x > 0$ **72.** $y = x^x, \quad x > 0$

73. $y = 2xe^{-x}$ **74.** $y = \dfrac{\ln x}{x}$

Think About It In Exercises 75–78, L'Hôpital's Rule is used incorrectly. Describe the error.

75. $\lim\limits_{x \to 0} \dfrac{e^{2x} - 1}{e^x} = \lim\limits_{x \to 0} \dfrac{2e^{2x}}{e^x} = \lim\limits_{x \to 0} 2e^x = 2$

76. $\lim\limits_{x \to 0} \dfrac{\sin \pi x - 1}{x} = \lim\limits_{x \to 0} \dfrac{\pi \cos \pi x}{1} = \pi$

77. $\lim\limits_{x \to \infty} x \cos \dfrac{1}{x} = \lim\limits_{x \to \infty} \dfrac{\cos(1/x)}{1/x}$
$= \lim\limits_{x \to \infty} \dfrac{[-\sin(1/x)](1/x^2)}{-1/x^2}$
$= 0$

78. $\lim\limits_{x \to \infty} \dfrac{e^{-x}}{1 + e^{-x}} = \lim\limits_{x \to \infty} \dfrac{-e^{-x}}{-e^{-x}}$
$= \lim\limits_{x \to \infty}$
$= 1$

 Analytical Approach In Exercises 79 and 80, (a) explain why L'Hôpital's Rule cannot be used to find the limit, (b) find the limit analytically, and (c) use a graphing utility to graph the function and approximate the limit from the graph. Compare the result with that in part (b).

79. $\lim\limits_{x \to \infty} \dfrac{x}{\sqrt{x^2 + 1}}$ **80.** $\lim\limits_{x \to \pi/2^-} \dfrac{\tan x}{\sec x}$

Graphical Analysis In Exercises 81 and 82, graph $f(x)/g(x)$ and $f'(x)/g'(x)$ near $x = 0$. What do you notice about these ratios as $x \to 0$? How does this illustrate L'Hôpital's Rule?

81. $f(x) = \sin 3x, \quad g(x) = \sin 4x$

82. $f(x) = e^{3x} - 1, \quad g(x) = x$

83. *Velocity in a Resisting Medium* The velocity v of an object falling through a resisting medium such as air or water is given by
$$v = \dfrac{32}{k}\left(1 - e^{-kt} + \dfrac{v_0 k e^{-kt}}{32}\right)$$
where v_0 is the initial velocity, t is the time in seconds, and k is the resistance constant of the medium. Use L'Hôpital's Rule to find the formula for the velocity of a falling body in a vacuum by fixing v_0 and t and letting k approach zero. (Assume that the downward direction is positive.)

84. Compound Interest The formula for the amount A in a savings account compounded n times per year for t years at an interest rate r and an initial deposit of P is given by

$$A = P\left(1 + \frac{r}{n}\right)^{nt}.$$

Use L'Hôpital's Rule to show that the limiting formula as the number of compoundings per year becomes infinite is given by $A = Pe^{rt}$.

85. The Gamma Function The Gamma Function $\Gamma(n)$ is defined in terms of the integral of the function given by $f(x) = x^{n-1}e^{-x}$, $n > 0$. Show that for any fixed value of n, the limit of $f(x)$ as x approaches infinity is zero.

86. Tractrix A person moves from the origin along the positive y-axis pulling a weight at the end of a 12-meter rope (see figure). Initially, the weight is located at the point $(12, 0)$.

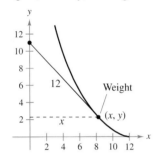

(a) Show that the slope of the tangent line of the path of the weight is
$$\frac{dy}{dx} = -\frac{\sqrt{144 - x^2}}{x}.$$

(b) Use the result of part (a) to find the equation of the path of the weight. Use a graphing utility to graph the path and compare it with the figure.

(c) Find any vertical asymptotes of the graph in part (b).

(d) When the person has reached the point $(0, 12)$, how far has the weight moved?

In Exercises 87–90, apply the Extended Mean Value Theorem to the functions f and g on the given interval. Find all values c in the interval (a, b) such that

$$\frac{f'(c)}{g'(c)} = \frac{f(b) - f(a)}{g(b) - g(a)}.$$

Functions	Interval
87. $f(x) = x^3$, $g(x) = x^2 + 1$	$[0, 1]$
88. $f(x) = \frac{1}{x}$, $g(x) = x^2 - 4$	$[1, 2]$
89. $f(x) = \sin x$, $g(x) = \cos x$	$\left[0, \frac{\pi}{2}\right]$
90. $f(x) = \ln x$, $g(x) = x^3$	$[1, 4]$

True or False? In Exercises 91–94, determine whether the statement is true or false. If it is false, explain why or give an example that shows it is false.

91. $\lim\limits_{x \to 0} \left[\dfrac{x^2 + x + 1}{x}\right] = \lim\limits_{x \to 0} \left[\dfrac{2x + 1}{1}\right] = 1$

92. If $y = e^x/x^2$, then $y' = e^x/2x$.

93. If $p(x)$ is a polynomial, then $\lim\limits_{x \to \infty} [p(x)/e^x] = 0$.

94. If $\lim\limits_{x \to \infty} \dfrac{f(x)}{g(x)} = 1$, then $\lim\limits_{x \to \infty} [f(x) - g(x)] = 0$.

95. Area Find the limit, as x approaches 0, of the ratio of the area of the triangle to the total shaded area in the figure.

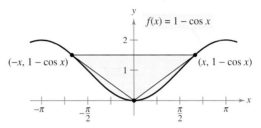

96. In Section 1.3, a geometric argument (see figure) was used to prove that
$$\lim_{\theta \to 0} \frac{\sin \theta}{\theta} = 1.$$

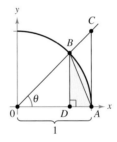

(a) Write the area of $\triangle ABD$ in terms of θ.

(b) Write the area of the shaded region in terms of θ.

(c) Write the ratio R of the area of $\triangle ABD$ to that of the shaded region.

(d) Find $\lim\limits_{\theta \to 0} R$.

Continuous Functions In Exercises 97 and 98, find the value of c that makes the function continuous at $x = 0$.

97. $f(x) = \begin{cases} \dfrac{4x - 2\sin 2x}{2x^3}, & x \neq 0 \\ c, & x = 0 \end{cases}$

98. $f(x) = \begin{cases} (e^x + x)^{1/x}, & x \neq 0 \\ c, & x = 0 \end{cases}$

99. Find the values of a and b such that $\lim\limits_{x \to 0} \dfrac{a - \cos bx}{x^2} = 2$.

100. Show that $\lim\limits_{x \to \infty} \dfrac{x^n}{e^x} = 0$ for any integer $n > 0$.

101. (a) Let $f'(x)$ be continuous. Show that
$$\lim_{h \to 0} \frac{f(x+h) - f(x-h)}{2h} = f'(x).$$
(b) Explain the result of part (a) graphically.

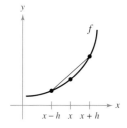

102. Let $f''(x)$ be continuous. Show that
$$\lim_{h \to 0} \frac{f(x+h) - 2f(x) + f(x-h)}{h^2} = f''(x).$$

103. Sketch the graph of
$$g(x) = \begin{cases} e^{-1/x^2}, & x \neq 0 \\ 0, & x = 0 \end{cases}$$
and determine $g'(0)$.

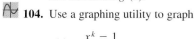 **104.** Use a graphing utility to graph
$$f(x) = \frac{x^k - 1}{k}$$
for $k = 1, 0.1$, and 0.01. Then evaluate the limit
$$\lim_{k \to 0^+} \frac{x^k - 1}{k}.$$

105. Consider the limit $\lim_{x \to 0^+} (-x \ln x)$.

(a) Describe the type of indeterminate form that is obtained by direct substitution.

(b) Evaluate the limit.

(c) Use a graphing utility to verify the result of part (b).

FOR FURTHER INFORMATION For a geometric approach to this exercise, see the article "A Geometric Proof of $\lim_{d \to 0^+} (-d \ln d) = 0$" by John H. Mathews in the *College Mathematics Journal*. To view this article, go to the website www.matharticles.com.

106. Prove that if $f(x) \geq 0$, $\lim_{x \to a} f(x) = 0$, and $\lim_{x \to a} g(x) = \infty$, then $\lim_{x \to a} f(x)^{g(x)} = 0$.

107. Prove that if $f(x) \geq 0$, $\lim_{x \to a} f(x) = 0$, and $\lim_{x \to a} g(x) = -\infty$, then $\lim_{x \to a} f(x)^{g(x)} = \infty$.

108. Prove the following generalization of the Mean Value Theorem. If f is twice differentiable on the closed interval $[a, b]$, then
$$f(b) - f(a) = f'(a)(b-a) - \int_a^b f''(t)(t-b)\, dt.$$

109. *Indeterminate Forms* Show that the indeterminate forms 0^0, ∞^0, and 1^∞ do not always have a value of 1 by evaluating each limit.

(a) $\lim_{x \to 0^+} x^{\ln 2/(1 + \ln x)}$

(b) $\lim_{x \to \infty} x^{\ln 2/(1 + \ln x)}$

(c) $\lim_{x \to 0} (x + 1)^{(\ln 2)/x}$

110. *Calculus History* In L'Hôpital's 1696 calculus textbook, he illustrated his rule using the limit of the function
$$f(x) = \frac{\sqrt{2a^3 x - x^4} - a\sqrt[3]{a^2 x}}{a - \sqrt[4]{ax^3}}$$
as x approaches a, $a > 0$. Find this limit.

111. Consider the function
$$h(x) = \frac{x + \sin x}{x}.$$

(a) Use a graphing utility to graph the function. Then use the *zoom* and *trace* features to investigate $\lim_{x \to \infty} h(x)$.

(b) Find $\lim_{x \to \infty} h(x)$ analytically by writing
$$h(x) = \frac{x}{x} + \frac{\sin x}{x}.$$

(c) Can you use L'Hôpital's Rule to find $\lim_{x \to \infty} h(x)$? Explain your reasoning.

Putnam Exam Challenge

112. Evaluate
$$\lim_{x \to \infty} \left[\frac{1}{x} \cdot \frac{a^x - 1}{a - 1} \right]^{1/x}$$
where $a > 0$, $a \neq 1$.

This problem was composed by the Committee on the Putnam Prize Competition.
© The Mathematical Association of America. All rights reserved.

Section 8.8 Improper Integrals

- Evaluate an improper integral that has an infinite limit of integration.
- Evaluate an improper integral that has an infinite discontinuity.

Improper Integrals with Infinite Limits of Integration

The definition of a definite integral

$$\int_a^b f(x)\, dx$$

requires that the interval $[a, b]$ be finite. Furthermore, the Fundamental Theorem of Calculus, by which you have been evaluating definite integrals, requires that f be continuous on $[a, b]$. In this section you will study a procedure for evaluating integrals that do not satisfy these requirements—usually because either one or both of the limits of integration are infinite, or f has a finite number of infinite discontinuities in the interval $[a, b]$. Integrals that possess either property are **improper integrals**. Note that a function f is said to have an **infinite discontinuity** at c if, *from the right or left*,

$$\lim_{x \to c} f(x) = \infty \quad \text{or} \quad \lim_{x \to c} f(x) = -\infty.$$

To get an idea of how to evaluate an improper integral, consider the integral

$$\int_1^b \frac{dx}{x^2} = -\frac{1}{x}\bigg]_1^b = -\frac{1}{b} + 1 = 1 - \frac{1}{b}$$

which can be interpreted as the area of the shaded region shown in Figure 8.17. Taking the limit as $b \to \infty$ produces

$$\int_1^\infty \frac{dx}{x^2} = \lim_{b \to \infty} \left(\int_1^b \frac{dx}{x^2} \right) = \lim_{b \to \infty} \left(1 - \frac{1}{b} \right) = 1.$$

This improper integral can be interpreted as the area of the *unbounded* region between the graph of $f(x) = 1/x^2$ and the x-axis (to the right of $x = 1$).

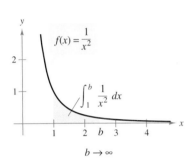

The unbounded region has an area of 1.
Figure 8.17

Definition of Improper Integrals with Infinite Integration Limits

1. If f is continuous on the interval $[a, \infty)$, then

$$\int_a^\infty f(x)\, dx = \lim_{b \to \infty} \int_a^b f(x)\, dx.$$

2. If f is continuous on the interval $(-\infty, b]$, then

$$\int_{-\infty}^b f(x)\, dx = \lim_{a \to -\infty} \int_a^b f(x)\, dx.$$

3. If f is continuous on the interval $(-\infty, \infty)$, then

$$\int_{-\infty}^\infty f(x)\, dx = \int_{-\infty}^c f(x)\, dx + \int_c^\infty f(x)\, dx$$

where c is any real number (see Exercise 110).

In the first two cases, the improper integral **converges** if the limit exists—otherwise, the improper integral **diverges**. In the third case, the improper integral on the left diverges if either of the improper integrals on the right diverges.

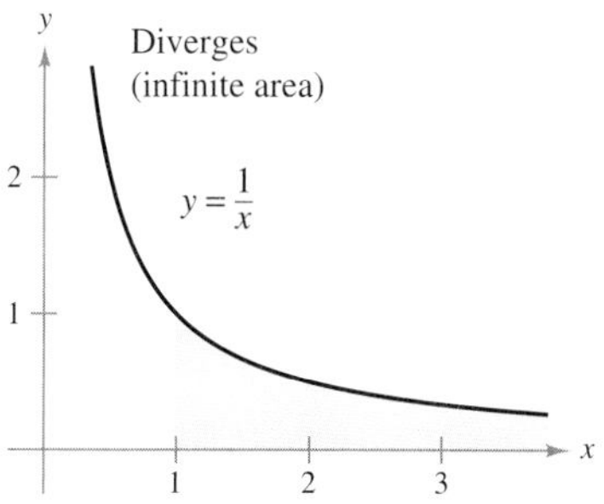

SECTION 8.8 Improper Integrals

EXAMPLE 1 An Improper Integral That Diverges

Evaluate $\int_1^\infty \dfrac{dx}{x}$.

Solution

$$\int_1^\infty \dfrac{dx}{x} = \lim_{b \to \infty} \int_1^b \dfrac{dx}{x} \qquad \text{Take limit as } b \to \infty.$$

$$= \lim_{b \to \infty} \Big[\ln x\Big]_1^b \qquad \text{Apply Log Rule.}$$

$$= \lim_{b \to \infty} (\ln b - 0) \qquad \text{Apply Fundamental Theorem of Calculus.}$$

$$= \infty \qquad \text{Evaluate limit.}$$

See Figure 8.18.

$\int_1^b \dfrac{1}{x}\, dx$

This unbounded region has an infinite area.
Figure 8.18

NOTE Try comparing the regions shown in Figures 8.17 and 8.18. They look similar, yet the region in Figure 8.17 has a finite area of 1 and the region in Figure 8.18 has an infinite area.

EXAMPLE 2 Improper Integrals That Converge

Evaluate each improper integral.

a. $\displaystyle\int_0^\infty e^{-x}\, dx$ **b.** $\displaystyle\int_0^\infty \dfrac{1}{x^2+1}\, dx$

Solution

a. $\displaystyle\int_0^\infty e^{-x}\, dx = \lim_{b \to \infty} \int_0^b e^{-x}\, dx$

$= \lim_{b \to \infty} \Big[-e^{-x}\Big]_0^b$

$= \lim_{b \to \infty} (-e^{-b} + 1)$

$= 1$

See Figure 8.19.

b. $\displaystyle\int_0^\infty \dfrac{1}{x^2+1}\, dx = \lim_{b \to \infty} \int_0^b \dfrac{1}{x^2+1}\, dx$

$= \lim_{b \to \infty} \Big[\arctan x\Big]_0^b$

$= \lim_{b \to \infty} \arctan b$

$= \dfrac{\pi}{2}$

See Figure 8.20.

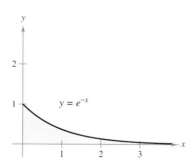

The area of the unbounded region is 1.
Figure 8.19

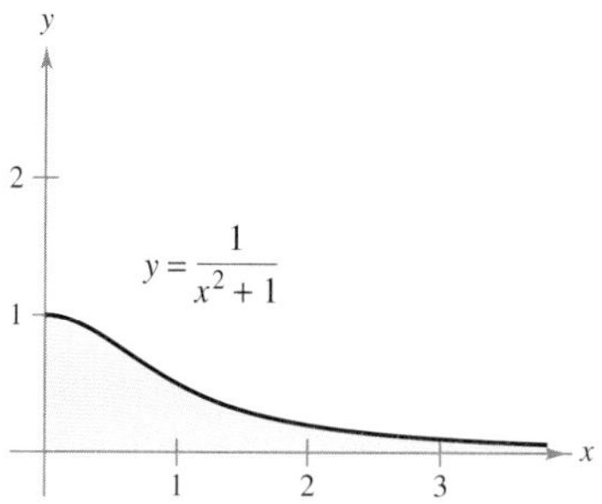

The area of the unbounded region is $\pi/2$.
Figure 8.20

In the following example, note how L'Hôpital's Rule can be used to evaluate an improper integral.

EXAMPLE 3 Using L'Hôpital's Rule with an Improper Integral

Evaluate $\int_1^\infty (1-x)e^{-x}\,dx$.

Solution Use integration by parts, with $dv = e^{-x}\,dx$ and $u = (1-x)$.

$$\int (1-x)e^{-x}\,dx = -e^{-x}(1-x) - \int e^{-x}\,dx$$
$$= -e^{-x} + xe^{-x} + e^{-x} + C$$
$$= xe^{-x} + C$$

Now, apply the definition of an improper integral.

$$\int_1^\infty (1-x)e^{-x}\,dx = \lim_{b\to\infty}\left[xe^{-x}\right]_1^b$$
$$= \left(\lim_{b\to\infty}\frac{b}{e^b}\right) - \frac{1}{e}$$

Finally, using L'Hôpital's Rule on the right-hand limit produces

$$\lim_{b\to\infty}\frac{b}{e^b} = \lim_{b\to\infty}\frac{1}{e^b} = 0$$

from which you can conclude that

$$\int_1^\infty (1-x)e^{-x}\,dx = -\frac{1}{e}.$$

See Figure 8.21.

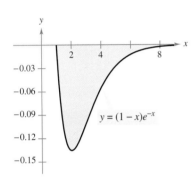

The area of the unbounded region is $|-1/e|$.
Figure 8.21

EXAMPLE 4 Infinite Upper and Lower Limits of Integration

Evaluate $\int_{-\infty}^\infty \dfrac{e^x}{1+e^{2x}}\,dx$.

Solution Note that the integrand is continuous on $(-\infty, \infty)$. To evaluate the integral, you can break it into two parts, choosing $c = 0$ as a convenient value.

$$\int_{-\infty}^\infty \frac{e^x}{1+e^{2x}}\,dx = \int_{-\infty}^0 \frac{e^x}{1+e^{2x}}\,dx + \int_0^\infty \frac{e^x}{1+e^{2x}}\,dx$$
$$= \lim_{b\to-\infty}\left[\arctan e^x\right]_b^0 + \lim_{b\to\infty}\left[\arctan e^x\right]_0^b$$
$$= \lim_{b\to-\infty}\left(\frac{\pi}{4} - \arctan e^b\right) + \lim_{b\to\infty}\left(\arctan e^b - \frac{\pi}{4}\right)$$
$$= \frac{\pi}{4} - 0 + \frac{\pi}{2} - \frac{\pi}{4}$$
$$= \frac{\pi}{2}$$

See Figure 8.22.

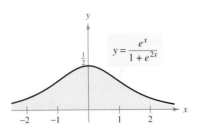

The area of the unbounded region is $\pi/2$.
Figure 8.22

EXAMPLE 5 Sending a Space Module into Orbit

In Example 3 of Section 7.5, you found that it would require 10,000 mile-tons of work to propel a 15-metric-ton space module to a height of 800 miles above Earth. How much work is required to propel the module an unlimited distance away from Earth's surface?

Solution At first you might think that an infinite amount of work would be required. But if this were the case, it would be impossible to send rockets into outer space. Because this has been done, the work required must be finite. You can determine the work in the following manner. Using the integral of Example 3, Section 7.5, replace the upper bound of 4800 miles by ∞ and write

$$W = \int_{4000}^{\infty} \frac{240{,}000{,}000}{x^2}\, dx$$

$$= \lim_{b \to \infty} \left[-\frac{240{,}000{,}000}{x} \right]_{4000}^{b}$$

$$= \lim_{b \to \infty} \left(-\frac{240{,}000{,}000}{b} + \frac{240{,}000{,}000}{4000} \right)$$

$$= 60{,}000 \text{ mile-tons}$$

$$\approx 6.984 \times 10^{11} \text{ foot-pounds.}$$

See Figure 8.23.

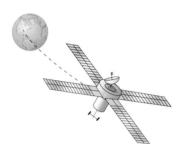

The work required to move a space module an unlimited distance away from Earth is approximately 6.984×10^{11} foot-pounds.
Figure 8.23

Improper Integrals with Infinite Discontinuities

The second basic type of improper integral is one that has an infinite discontinuity *at or between* the limits of integration.

Definition of Improper Integrals with Infinite Discontinuities

1. If f is continuous on the interval $[a, b)$ and has an infinite discontinuity at b, then
$$\int_a^b f(x)\, dx = \lim_{c \to b^-} \int_a^c f(x)\, dx.$$

2. If f is continuous on the interval $(a, b]$ and has an infinite discontinuity at a, then
$$\int_a^b f(x)\, dx = \lim_{c \to a^+} \int_c^b f(x)\, dx.$$

3. If f is continuous on the interval $[a, b]$, except for some c in (a, b) at which f has an infinite discontinuity, then
$$\int_a^b f(x)\, dx = \int_a^c f(x)\, dx + \int_c^b f(x)\, dx.$$

In the first two cases, the improper integral **converges** if the limit exists—otherwise, the improper integral **diverges**. In the third case, the improper integral on the left diverges if either of the improper integrals on the right diverges.

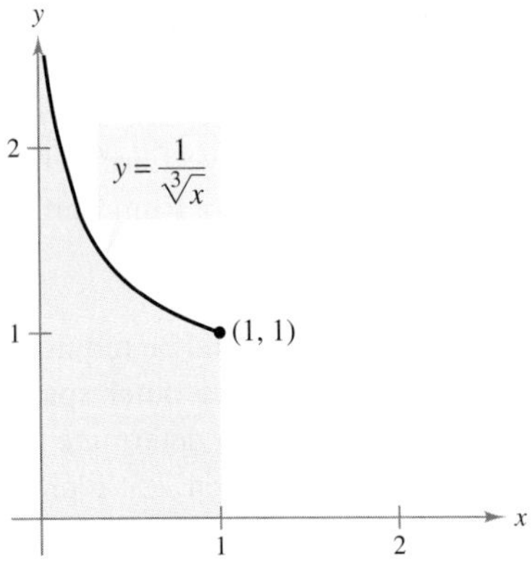

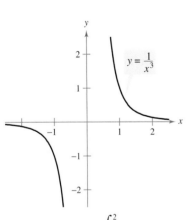

Infinite discontinuity at $x = 0$
Figure 8.24

EXAMPLE 6 An Improper Integral with an Infinite Discontinuity

Evaluate $\displaystyle\int_0^1 \frac{dx}{\sqrt[3]{x}}$.

Solution The integrand has an infinite discontinuity at $x = 0$, as shown in Figure 8.24. You can evaluate this integral as shown below.

$$\begin{aligned}\int_0^1 x^{-1/3}\,dx &= \lim_{b\to 0^+}\left[\frac{x^{2/3}}{2/3}\right]_b^1\\ &= \lim_{b\to 0^+}\frac{3}{2}(1 - b^{2/3})\\ &= \frac{3}{2}\end{aligned}$$

EXAMPLE 7 An Improper Integral That Diverges

Evaluate $\displaystyle\int_0^2 \frac{dx}{x^3}$.

Solution Because the integrand has an infinite discontinuity at $x = 0$, you can write

$$\begin{aligned}\int_0^2 \frac{dx}{x^3} &= \lim_{b\to 0^+}\left[-\frac{1}{2x^2}\right]_b^2\\ &= \lim_{b\to 0^+}\left(-\frac{1}{8} + \frac{1}{2b^2}\right)\\ &= \infty.\end{aligned}$$

So, you can conclude that the improper integral diverges.

EXAMPLE 8 An Improper Integral with an Interior Discontinuity

Evaluate $\displaystyle\int_{-1}^2 \frac{dx}{x^3}$.

Solution This integral is improper because the integrand has an infinite discontinuity at the interior point $x = 0$, as shown in Figure 8.25. So, you can write

$$\int_{-1}^2 \frac{dx}{x^3} = \int_{-1}^0 \frac{dx}{x^3} + \int_0^2 \frac{dx}{x^3}.$$

From Example 7 you know that the second integral diverges. So, the original improper integral also diverges.

The improper integral $\int_{-1}^{2} 1/x^3\,dx$ diverges.
Figure 8.25

NOTE Remember to check for infinite discontinuities at interior points as well as endpoints when determining whether an integral is improper. For instance, if you had not recognized that the integral in Example 8 was improper, you would have obtained the *incorrect* result

$$\int_{-1}^2 \frac{dx}{x^3} = \left[\frac{-1}{2x^2}\right]_{-1}^2 = -\frac{1}{8} + \frac{1}{2} = \frac{3}{8}.$$

Incorrect evaluation

The integral in the next example is improper for *two* reasons. One limit of integration is infinite, and the integrand has an infinite discontinuity at the outer limit of integration.

EXAMPLE 9   A Doubly Improper Integral

Evaluate $\int_0^\infty \dfrac{dx}{\sqrt{x}(x+1)}$.

Solution   To evaluate this integral, split it at a convenient point (say, $x = 1$) and write

$$\int_0^\infty \frac{dx}{\sqrt{x}(x+1)} = \int_0^1 \frac{dx}{\sqrt{x}(x+1)} + \int_1^\infty \frac{dx}{\sqrt{x}(x+1)}$$

$$= \lim_{b \to 0^+} \left[2 \arctan \sqrt{x} \right]_b^1 + \lim_{c \to \infty} \left[2 \arctan \sqrt{x} \right]_1^c$$

$$= 2\left(\frac{\pi}{4}\right) - 0 + 2\left(\frac{\pi}{2}\right) - 2\left(\frac{\pi}{4}\right)$$

$$= \pi.$$

See Figure 8.26.

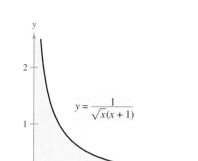

The area of the unbounded region is π.
Figure 8.26

EXAMPLE 10   An Application Involving Arc Length

Use the formula for arc length to show that the circumference of the circle $x^2 + y^2 = 1$ is 2π.

Solution   To simplify the work, consider the quarter circle given by $y = \sqrt{1-x^2}$, where $0 \le x \le 1$. The function y is differentiable for any x in this interval except $x = 1$. Therefore, the arc length of the quarter circle is given by the improper integral

$$s = \int_0^1 \sqrt{1 + (y')^2} \, dx$$

$$= \int_0^1 \sqrt{1 + \left(\frac{-x}{\sqrt{1-x^2}}\right)^2} \, dx$$

$$= \int_0^1 \frac{dx}{\sqrt{1-x^2}}.$$

This integral is improper because it has an infinite discontinuity at $x = 1$. So, you can write

$$s = \int_0^1 \frac{dx}{\sqrt{1-x^2}}$$

$$= \lim_{b \to 1^-} \left[\arcsin x \right]_0^b$$

$$= \frac{\pi}{2} - 0$$

$$= \frac{\pi}{2}.$$

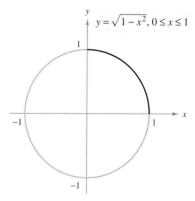

The circumference of the circle is 2π.
Figure 8.27

Finally, multiplying by 4, you can conclude that the circumference of the circle is $4s = 2\pi$, as shown in Figure 8.27.

This section concludes with a useful theorem describing the convergence or divergence of a common type of improper integral. The proof of this theorem is left as an exercise (see Exercise 49).

THEOREM 8.5   A Special Type of Improper Integral

$$\int_1^\infty \frac{dx}{x^p} = \begin{cases} \dfrac{1}{p-1}, & \text{if } p > 1 \\ \text{diverges}, & \text{if } p \leq 1 \end{cases}$$

EXAMPLE 11   An Application Involving A Solid of Revolution

The solid formed by revolving (about the x-axis) the *unbounded* region lying between the graph of $f(x) = 1/x$ and the x-axis ($x \geq 1$) is called **Gabriel's Horn.** (See Figure 8.28.) Show that this solid has a finite volume and an infinite surface area.

Solution   Using the disk method and Theorem 8.5, you can determine the volume to be

$$V = \pi \int_1^\infty \left(\frac{1}{x}\right)^2 dx \qquad \text{Theorem 8.5, } p = 2 > 1$$

$$= \pi \left(\frac{1}{2-1}\right) = \pi.$$

The surface area is given by

$$S = 2\pi \int_1^\infty f(x)\sqrt{1 + [f'(x)]^2}\, dx = 2\pi \int_1^\infty \frac{1}{x}\sqrt{1 + \frac{1}{x^4}}\, dx.$$

Because

$$\sqrt{1 + \frac{1}{x^4}} > 1$$

on the interval $[1, \infty)$, and the improper integral

$$\int_1^\infty \frac{1}{x}\, dx$$

diverges, you can conclude that the improper integral

$$\int_1^\infty \frac{1}{x}\sqrt{1 + \frac{1}{x^4}}\, dx$$

also diverges. (See Exercise 52.) So, the surface area is infinite.

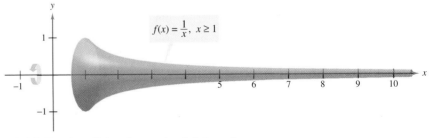

Gabriel's Horn has a finite volume and an infinite surface area.
Figure 8.28

FOR FURTHER INFORMATION For further investigation of solids that have finite volumes and infinite surface areas, see the article "Supersolids: Solids Having Finite Volume and Infinite Surfaces" by William P. Love in *Mathematics Teacher*. To view this article, go to the website *www.matharticles.com*.

FOR FURTHER INFORMATION To learn about another function that has a finite volume and an infinite surface area, see the article "Gabriel's Wedding Cake" by Julian F. Fleron in *The College Mathematics Journal*. To view this article, go to the website *www.matharticles.com*.

Exercises for Section 8.8

In Exercises 1–4, decide whether the integral is improper. Explain your reasoning.

1. $\int_0^1 \dfrac{dx}{3x-2}$
2. $\int_1^3 \dfrac{dx}{x^2}$
3. $\int_0^1 \dfrac{2x-5}{x^2-5x+6}\,dx$
4. $\int_1^\infty \ln(x^2)\,dx$

In Exercises 5–10, explain why the integral is improper and determine whether it diverges or converges. Evaluate the integral if it converges.

5. $\int_0^4 \dfrac{1}{\sqrt{x}}\,dx$
6. $\int_3^4 \dfrac{1}{(x-3)^{3/2}}\,dx$

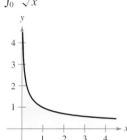
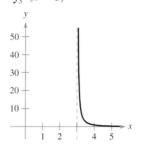

7. $\int_0^2 \dfrac{1}{(x-1)^2}\,dx$
8. $\int_0^2 \dfrac{1}{(x-1)^{2/3}}\,dx$

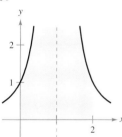
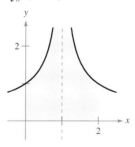

9. $\int_0^\infty e^{-x}\,dx$
10. $\int_{-\infty}^0 e^{2x}\,dx$

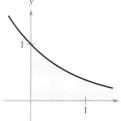
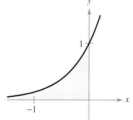

 Writing In Exercises 11–14, explain why the evaluation of the integral is *incorrect*. Use the integration capabilities of a graphing utility to attempt to evaluate the integral. Determine whether the utility gives the correct answer.

11. ~~$\int_{-1}^1 \dfrac{1}{x^2}\,dx = -2$~~
12. ~~$\int_{-2}^2 \dfrac{-2}{(x-1)^3}\,dx = \dfrac{8}{9}$~~
13. ~~$\int_0^\infty e^{-x}\,dx = 0$~~
14. ~~$\int_0^\pi \sec x\,dx = 0$~~

In Exercises 15–32, determine whether the improper integral diverges or converges. Evaluate the integral if it converges.

15. $\int_1^\infty \dfrac{1}{x^2}\,dx$
16. $\int_1^\infty \dfrac{5}{x^3}\,dx$
17. $\int_1^\infty \dfrac{3}{\sqrt[3]{x}}\,dx$
18. $\int_1^\infty \dfrac{4}{\sqrt[4]{x}}\,dx$
19. $\int_{-\infty}^0 xe^{-2x}\,dx$
20. $\int_0^\infty xe^{-x/2}\,dx$
21. $\int_0^\infty x^2 e^{-x}\,dx$
22. $\int_0^\infty (x-1)e^{-x}\,dx$
23. $\int_0^\infty e^{-x}\cos x\,dx$
24. $\int_0^\infty e^{-ax}\sin bx\,dx, \quad a>0$
25. $\int_4^\infty \dfrac{1}{x(\ln x)^3}\,dx$
26. $\int_1^\infty \dfrac{\ln x}{x}\,dx$
27. $\int_{-\infty}^\infty \dfrac{2}{4+x^2}\,dx$
28. $\int_0^\infty \dfrac{x^3}{(x^2+1)^2}\,dx$
29. $\int_0^\infty \dfrac{1}{e^x+e^{-x}}\,dx$
30. $\int_0^\infty \dfrac{e^x}{1+e^x}\,dx$
31. $\int_0^\infty \cos \pi x\,dx$
32. $\int_0^\infty \sin \dfrac{x}{2}\,dx$

In Exercises 33–48, determine whether the improper integral diverges or converges. Evaluate the integral if it converges, and check your results with the results obtained by using the integration capabilities of a graphing utility.

33. $\int_0^1 \dfrac{1}{x^2}\,dx$
34. $\int_0^4 \dfrac{8}{x}\,dx$
35. $\int_0^8 \dfrac{1}{\sqrt[3]{8-x}}\,dx$
36. $\int_0^6 \dfrac{4}{\sqrt{6-x}}\,dx$
37. $\int_0^1 x\ln x\,dx$
38. $\int_0^e \ln x^2\,dx$
39. $\int_0^{\pi/2} \tan\theta\,d\theta$
40. $\int_0^{\pi/2} \sec\theta\,d\theta$
41. $\int_2^4 \dfrac{2}{x\sqrt{x^2-4}}\,dx$
42. $\int_0^2 \dfrac{1}{\sqrt{4-x^2}}\,dx$
43. $\int_2^4 \dfrac{1}{\sqrt{x^2-4}}\,dx$
44. $\int_0^2 \dfrac{1}{4-x^2}\,dx$
45. $\int_0^2 \dfrac{1}{\sqrt[3]{x-1}}\,dx$
46. $\int_1^3 \dfrac{2}{(x-2)^{8/3}}\,dx$
47. $\int_0^\infty \dfrac{4}{\sqrt{x}(x+6)}\,dx$
48. $\int_1^\infty \dfrac{1}{x\ln x}\,dx$

In Exercises 49 and 50, determine all values of p for which the improper integral converges.

49. $\displaystyle\int_1^\infty \frac{1}{x^p}\,dx$

50. $\displaystyle\int_0^1 \frac{1}{x^p}\,dx$

51. Use mathematical induction to verify that the following integral converges for any positive integer n.

$$\int_0^\infty x^n e^{-x}\,dx$$

52. Given continuous functions f and g such that $0 \le f(x) \le g(x)$ on the interval $[a, \infty)$, prove the following.

(a) If $\int_a^\infty g(x)\,dx$ converges, then $\int_a^\infty f(x)\,dx$ converges.

(b) If $\int_a^\infty f(x)\,dx$ diverges, then $\int_a^\infty g(x)\,dx$ diverges.

In Exercises 53–62, use the results of Exercises 49–52 to determine whether the improper integral converges or diverges.

53. $\displaystyle\int_0^1 \frac{1}{x^3}\,dx$

54. $\displaystyle\int_0^1 \frac{1}{\sqrt[3]{x}}\,dx$

55. $\displaystyle\int_1^\infty \frac{1}{x^3}\,dx$

56. $\displaystyle\int_0^\infty x^4 e^{-x}\,dx$

57. $\displaystyle\int_1^\infty \frac{1}{x^2+5}\,dx$

58. $\displaystyle\int_2^\infty \frac{1}{\sqrt{x-1}}\,dx$

59. $\displaystyle\int_2^\infty \frac{1}{\sqrt[3]{x(x-1)}}\,dx$

60. $\displaystyle\int_1^\infty \frac{1}{\sqrt{x(x+1)}}\,dx$

61. $\displaystyle\int_0^\infty e^{-x^2}\,dx$

62. $\displaystyle\int_2^\infty \frac{1}{\sqrt{x}\ln x}\,dx$

Writing About Concepts

63. Describe the different types of improper integrals.

64. Define the terms *converges* and *diverges* when working with improper integrals.

65. Explain why $\displaystyle\int_{-1}^1 \frac{1}{x^3}\,dx \ne 0$.

66. Consider the integral

$$\int_0^3 \frac{10}{x^2 - 2x}\,dx.$$

To determine the convergence or divergence of the integral, how many improper integrals must be analyzed? What must be true of each of these integrals if the given integral converges?

Area In Exercises 67–70, find the area of the unbounded shaded region.

67. $y = e^x,\quad -\infty < x \le 1$

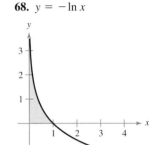

68. $y = -\ln x$

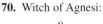

69. Witch of Agnesi:

$y = \dfrac{1}{x^2 + 1}$

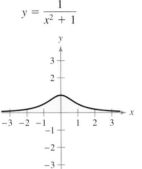

70. Witch of Agnesi:

$y = \dfrac{8}{x^2 + 4}$

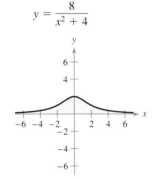

Area and Volume In Exercises 71 and 72, consider the region satisfying the inequalities. (a) Find the area of the region. (b) Find the volume of the solid generated by revolving the region about the x-axis. (c) Find the volume of the solid generated by revolving the region about the y-axis.

71. $y \le e^{-x},\ y \ge 0,\ x \ge 0$

72. $y \le \dfrac{1}{x^2},\ y \ge 0,\ x \ge 1$

73. *Arc Length* Sketch the graph of the hypocycloid of four cusps

$$x^{2/3} + y^{2/3} = 4$$

and find its perimeter.

74. *Arc Length* Find the arc length of the graph of

$$y = \sqrt{16 - x^2}$$

over the interval $[0, 4]$.

75. *Surface Area* The region bounded by

$$(x - 2)^2 + y^2 = 1$$

is revolved about the y-axis to form a torus. Find the surface area of the torus.

76. *Surface Area* Find the area of the surface formed by revolving the graph of $y = 2e^{-x}$ on the interval $[0, \infty)$ about the x-axis.

Propulsion In Exercises 77 and 78, use the weight of the rocket to answer each question. (Use 4000 miles as the radius of Earth and do not consider the effect of air resistance.)

(a) How much work is required to propel the rocket an unlimited distance away from Earth's surface?

(b) How far has the rocket traveled when half the total work has occurred?

77. 5-ton rocket **78.** 10-ton rocket

Probability A nonnegative function f is called a *probability density function* if

$$\int_{-\infty}^{\infty} f(t)\, dt = 1.$$

The probability that x lies between a and b is given by

$$P(a \le x \le b) = \int_a^b f(t)\, dt.$$

The expected value of x is given by

$$E(x) = \int_{-\infty}^{\infty} t f(t)\, dt.$$

In Exercises 79 and 80, (a) show that the nonnegative function is a probability density function, (b) find $P(0 \le x \le 4)$, and (c) find $E(x)$.

79. $f(t) = \begin{cases} \frac{1}{7} e^{-t/7}, & t \ge 0 \\ 0, & t < 0 \end{cases}$

80. $f(t) = \begin{cases} \frac{2}{5} e^{-2t/5}, & t \ge 0 \\ 0, & t < 0 \end{cases}$

Capitalized Cost In Exercises 81 and 82, find the capitalized cost C of an asset (a) for $n = 5$ years, (b) for $n = 10$ years, and (c) forever. The capitalized cost is given by

$$C = C_0 + \int_0^n c(t) e^{-rt}\, dt$$

where C_0 is the original investment, t is the time in years, r is the annual interest rate compounded continuously, and $c(t)$ is the annual cost of maintenance.

81. $C_0 = \$650{,}000$
$c(t) = \$25{,}000$
$r = 0.06$

82. $C_0 = \$650{,}000$
$c(t) = \$25{,}000(1 + 0.08t)$
$r = 0.06$

83. *Electromagnetic Theory* The magnetic potential P at a point on the axis of a circular coil is given by

$$P = \frac{2\pi NIr}{k} \int_c^{\infty} \frac{1}{(r^2 + x^2)^{3/2}}\, dx$$

where N, I, r, k, and c are constants. Find P.

84. *Gravitational Force* A "semi-infinite" uniform rod occupies the nonnegative x-axis. The rod has a linear density δ which means that a segment of length dx has a mass of $\delta\, dx$. A particle of mass m is located at the point $(-a, 0)$. The gravitational force F that the rod exerts on the mass is given by

$$F = \int_0^{\infty} \frac{GM\delta}{(a + x)^2}\, dx$$

where G is the gravitational constant. Find F.

True or False? In Exercises 85–88, determine whether the statement is true or false. If it is false, explain why or give an example that shows it is false.

85. If f is continuous on $[0, \infty)$ and $\lim_{x \to \infty} f(x) = 0$, then $\int_0^{\infty} f(x)\, dx$ converges.

86. If f is continuous on $[0, \infty)$ and $\int_0^{\infty} f(x)\, dx$ diverges, then $\lim_{x \to \infty} f(x) \ne 0$.

87. If f' is continuous on $[0, \infty)$ and $\lim_{x \to \infty} f(x) = 0$, then $\int_0^{\infty} f'(x)\, dx = -f(0)$.

88. If the graph of f is symmetric with respect to the origin or the y-axis, then $\int_0^{\infty} f(x)\, dx$ converges if and only if $\int_{-\infty}^{\infty} f(x)\, dx$ converges.

89. *Writing*

(a) The improper integrals

$$\int_1^{\infty} \frac{1}{x}\, dx \quad \text{and} \quad \int_1^{\infty} \frac{1}{x^2}\, dx$$

diverge and converge, respectively. Describe the essential differences between the integrands that cause one integral to converge and the other to diverge.

(b) Sketch a graph of the function $y = \sin x / x$ over the interval $(1, \infty)$. Use your knowledge of the definite integral to make an inference as to whether or not the integral

$$\int_1^{\infty} \frac{\sin x}{x}\, dx$$

converges. Give reasons for your answer.

(c) Use one iteration of integration by parts on the integral in part (b) to determine its divergence or convergence.

90. *Exploration* Consider the integral

$$\int_0^{\pi/2} \frac{4}{1 + (\tan x)^n}\, dx$$

where n is a positive integer.

(a) Is the integral improper? Explain.

(b) Use a graphing utility to graph the integrand for $n = 2, 4, 8,$ and 12.

(c) Use the graphs to approximate the integral as $n \to \infty$.

(d) Use a computer algebra system to evaluate the integral for the values of n in part (b). Make a conjecture about the value of the integral for any positive integer n. Compare your results with your answer in part (c).

91. The Gamma Function The Gamma Function $\Gamma(n)$ is defined by

$$\Gamma(n) = \int_0^\infty x^{n-1} e^{-x}\, dx, \quad n > 0.$$

(a) Find $\Gamma(1)$, $\Gamma(2)$, and $\Gamma(3)$.

(b) Use integration by parts to show that $\Gamma(n+1) = n\Gamma(n)$.

(c) Write $\Gamma(n)$ using factorial notation where n is a positive integer.

92. Prove that $I_n = \left(\dfrac{n-1}{n+2}\right) I_{n-1}$, where

$$I_n = \int_0^\infty \frac{x^{2n-1}}{(x^2+1)^{n+3}}\, dx, \quad n \geq 1.$$

Then evaluate each integral.

(a) $\displaystyle\int_0^\infty \frac{x}{(x^2+1)^4}\, dx$

(b) $\displaystyle\int_0^\infty \frac{x^3}{(x^2+1)^5}\, dx$

(c) $\displaystyle\int_0^\infty \frac{x^5}{(x^2+1)^6}\, dx$

Laplace Transforms Let $f(t)$ be a function defined for all positive values of t. The Laplace Transform of $f(t)$ is defined by

$$F(s) = \int_0^\infty e^{-st} f(t)\, dt$$

if the improper integral exists. Laplace Transforms are used to solve differential equations. In Exercises 93–100, find the Laplace Transform of the function.

93. $f(t) = 1$ \qquad **94.** $f(t) = t$

95. $f(t) = t^2$ \qquad **96.** $f(t) = e^{at}$

97. $f(t) = \cos at$ \qquad **98.** $f(t) = \sin at$

99. $f(t) = \cosh at$ \qquad **100.** $f(t) = \sinh at$

101. Normal Probability The mean height of American men between 18 and 24 years old is 70 inches, and the standard deviation is 3 inches. An 18- to 24-year-old man is chosen at random from the population. The probability that he is 6 feet tall or taller is

$$P(72 \leq x < \infty) = \int_{72}^\infty \frac{1}{3\sqrt{2\pi}} e^{-(x-70)^2/18}\, dx.$$

(Source: National Center for Health Statistics)

(a) Use a graphing utility to graph the integrand. Use the graphing utility to convince yourself that the area between the x-axis and the integrand is 1.

(b) Use a graphing utility to approximate $P(72 \leq x < \infty)$.

(c) Approximate $0.5 - P(70 \leq x \leq 72)$ using a graphing utility. Use the graph in part (a) to explain why this result is the same as the answer in part (b).

102. (a) Sketch the semicircle $y = \sqrt{4-x^2}$.

(b) Explain why

$$\int_{-2}^2 \frac{2\, dx}{\sqrt{4-x^2}} = \int_{-2}^2 \sqrt{4-x^2}\, dx$$

without evaluating either integral.

103. For what value of c does the integral

$$\int_0^\infty \left(\frac{1}{\sqrt{x^2+1}} - \frac{c}{x+1}\right) dx$$

converge? Evaluate the integral for this value of c.

104. For what value of c does the integral

$$\int_1^\infty \left(\frac{cx}{x^2+2} - \frac{1}{3x}\right) dx$$

converge? Evaluate the integral for this value of c.

105. Volume Find the volume of the solid generated by revolving the region bounded by the graph of f about the x-axis.

$$f(x) = \begin{cases} x \ln x, & 0 < x \leq 2 \\ 0, & x = 0 \end{cases}$$

106. Volume Find the volume of the solid generated by revolving the unbounded region lying between $y = -\ln x$ and the y-axis ($y \geq 0$) about the x-axis.

u-Substitution In Exercises 107 and 108, rewrite the improper integral as a proper integral using the given u-substitution. Then use the Trapezoidal Rule with $n = 5$ to approximate the integral.

107. $\displaystyle\int_0^1 \frac{\sin x}{\sqrt{x}}\, dx, \quad u = \sqrt{x}$

108. $\displaystyle\int_0^1 \frac{\cos x}{\sqrt{1-x}}\, dx, \quad u = \sqrt{1-x}$

109. (a) Use a graphing utility to graph the function $y = e^{-x^2}$.

(b) Show that $\displaystyle\int_0^\infty e^{-x^2}\, dx = \int_0^1 \sqrt{-\ln y}\, dy.$

110. Let $\displaystyle\int_{-\infty}^\infty f(x)\, dx$ be convergent and let a and b be real numbers where $a \neq b$. Show that

$$\int_{-\infty}^a f(x)\, dx + \int_a^\infty f(x)\, dx = \int_{-\infty}^b f(x)\, dx + \int_b^\infty f(x)\, dx.$$

Review Exercises for Chapter 8

See www.CalcChat.com for worked-out solutions to odd-numbered exercises.

In Exercises 1–8, use the basic integration rules to find or evaluate the integral.

1. $\int x\sqrt{x^2 - 1}\, dx$

2. $\int x e^{x^2 - 1}\, dx$

3. $\int \dfrac{x}{x^2 - 1}\, dx$

4. $\int \dfrac{x}{\sqrt{1 - x^2}}\, dx$

5. $\int_1^e \dfrac{\ln(2x)}{x}\, dx$

6. $\int_{3/2}^2 2x\sqrt{2x - 3}\, dx$

7. $\int \dfrac{16}{\sqrt{16 - x^2}}\, dx$

8. $\int \dfrac{x^4 + 2x^2 + x + 1}{(x^2 + 1)^2}\, dx$

In Exercises 9–16, use integration by parts to find the integral.

9. $\int e^{2x} \sin 3x\, dx$

10. $\int (x^2 - 1)e^x\, dx$

11. $\int x\sqrt{x - 5}\, dx$

12. $\int \arctan 2x\, dx$

13. $\int x^2 \sin 2x\, dx$

14. $\int \ln\sqrt{x^2 - 1}\, dx$

15. $\int x \arcsin 2x\, dx$

16. $\int e^x \arctan e^x\, dx$

In Exercises 17–22, find the trigonometric integral.

17. $\int \cos^3(\pi x - 1)\, dx$

18. $\int \sin^2 \dfrac{\pi x}{2}\, dx$

19. $\int \sec^4 \dfrac{x}{2}\, dx$

20. $\int \tan \theta \sec^4 \theta\, d\theta$

21. $\int \dfrac{1}{1 - \sin \theta}\, d\theta$

22. $\int \cos 2\theta(\sin \theta + \cos \theta)^2\, d\theta$

Area **In Exercises 23 and 24, find the area of the region.**

23. $y = \sin^4 x$

24. $y = \cos(3x)\cos x$

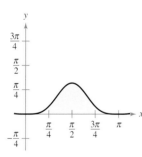

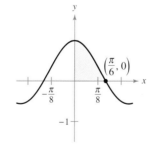

In Exercises 25–30, use trigonometric substitution to find or evaluate the integral.

25. $\int \dfrac{-12}{x^2\sqrt{4 - x^2}}\, dx$

26. $\int \dfrac{\sqrt{x^2 - 9}}{x}\, dx,\ x > 3$

27. $\int \dfrac{x^3}{\sqrt{4 + x^2}}\, dx$

28. $\int \sqrt{9 - 4x^2}\, dx$

29. $\int_{-2}^0 \sqrt{4 - x^2}\, dx$

30. $\int_0^{\pi/2} \dfrac{\sin \theta}{1 + 2\cos^2 \theta}\, d\theta$

In Exercises 31 and 32, find the integral using each method.

31. $\int \dfrac{x^3}{\sqrt{4 + x^2}}\, dx$

(a) Trigonometric substitution

(b) Substitution: $u^2 = 4 + x^2$

(c) Integration by parts: $dv = (x/\sqrt{4 + x^2})\, dx$

32. $\int x\sqrt{4 + x}\, dx$

(a) Trigonometric substitution

(b) Substitution: $u^2 = 4 + x$

(c) Substitution: $u = 4 + x$

(d) Integration by parts: $dv = \sqrt{4 + x}\, dx$

In Exercises 33–38, use partial fractions to find the integral.

33. $\int \dfrac{x - 28}{x^2 - x - 6}\, dx$

34. $\int \dfrac{2x^3 - 5x^2 + 4x - 4}{x^2 - x}\, dx$

35. $\int \dfrac{x^2 + 2x}{x^3 - x^2 + x - 1}\, dx$

36. $\int \dfrac{4x - 2}{3(x - 1)^2}\, dx$

37. $\int \dfrac{x^2}{x^2 + 2x - 15}\, dx$

38. $\int \dfrac{\sec^2 \theta}{\tan \theta(\tan \theta - 1)}\, d\theta$

In Exercises 39–46, use integration tables to find or evaluate the integral.

39. $\int \dfrac{x}{(2 + 3x)^2}\, dx$

40. $\int \dfrac{x}{\sqrt{2 + 3x}}\, dx$

41. $\int_0^{\sqrt{\pi/2}} \dfrac{x}{1 + \sin x^2}\, dx$

42. $\int_0^1 \dfrac{x}{1 + e^{x^2}}\, dx$

43. $\int \dfrac{x}{x^2 + 4x + 8}\, dx$

44. $\int \dfrac{3}{2x\sqrt{9x^2 - 1}}\, dx,\ x > \dfrac{1}{3}$

45. $\int \dfrac{1}{\sin \pi x \cos \pi x}\, dx$

46. $\int \dfrac{1}{1 + \tan \pi x}\, dx$

47. Verify the reduction formula

$$\int (\ln x)^n\, dx = x(\ln x)^n - n\int (\ln x)^{n-1}\, dx.$$

48. Verify the reduction formula

$$\int \tan^n x\, dx = \dfrac{1}{n - 1}\tan^{n-1} x - \int \tan^{n-2} x\, dx.$$

In Exercises 49–56, find the integral using any method.

49. $\int \theta \sin\theta \cos\theta \, d\theta$

50. $\int \dfrac{\csc\sqrt{2x}}{\sqrt{x}} \, dx$

51. $\int \dfrac{x^{1/4}}{1 + x^{1/2}} \, dx$

52. $\int \sqrt{1 + \sqrt{x}} \, dx$

53. $\int \sqrt{1 + \cos x} \, dx$

54. $\int \dfrac{3x^3 + 4x}{(x^2 + 1)^2} \, dx$

55. $\int \cos x \ln(\sin x) \, dx$

56. $\int (\sin\theta + \cos\theta)^2 \, d\theta$

In Exercises 57–60, solve the differential equation using any method.

57. $\dfrac{dy}{dx} = \dfrac{9}{x^2 - 9}$

58. $\dfrac{dy}{dx} = \dfrac{\sqrt{4 - x^2}}{2x}$

59. $y' = \ln(x^2 + x)$

60. $y' = \sqrt{1 - \cos\theta}$

In Exercises 61–66, evaluate the definite integral using any method. Use a graphing utility to verify your result.

61. $\int_{2}^{\sqrt{5}} x(x^2 - 4)^{3/2} \, dx$

62. $\int_{0}^{1} \dfrac{x}{(x - 2)(x - 4)} \, dx$

63. $\int_{1}^{4} \dfrac{\ln x}{x} \, dx$

64. $\int_{0}^{2} xe^{3x} \, dx$

65. $\int_{0}^{\pi} x \sin x \, dx$

66. $\int_{0}^{3} \dfrac{x}{\sqrt{1 + x}} \, dx$

Area In Exercises 67 and 68, find the area of the region.

67. $y = x\sqrt{4 - x}$

68. $y = \dfrac{1}{25 - x^2}$

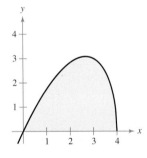
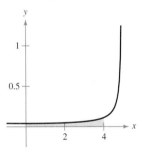

Centroid In Exercises 69 and 70, find the centroid of the region bounded by the graphs of the equations.

69. $y = \sqrt{1 - x^2}, \quad y = 0$

70. $(x - 1)^2 + y^2 = 1, \quad (x - 4)^2 + y^2 = 4$

Arc Length In Exercises 71 and 72, approximate to two decimal places the arc length of the curve over the given interval.

Function	Interval
71. $y = \sin x$	$[0, \pi]$
72. $y = \sin^2 x$	$[0, \pi]$

In Exercises 73–80, use L'Hôpital's Rule to evaluate the limit.

73. $\lim\limits_{x \to 1} \dfrac{(\ln x)^2}{x - 1}$

74. $\lim\limits_{x \to 0} \dfrac{\sin \pi x}{\sin 2\pi x}$

75. $\lim\limits_{x \to \infty} \dfrac{e^{2x}}{x^2}$

76. $\lim\limits_{x \to \infty} xe^{-x^2}$

77. $\lim\limits_{x \to \infty} (\ln x)^{2/x}$

78. $\lim\limits_{x \to 1^+} (x - 1)^{\ln x}$

79. $\lim\limits_{n \to \infty} 1000\left(1 + \dfrac{0.09}{n}\right)^n$

80. $\lim\limits_{x \to 1^+} \left(\dfrac{2}{\ln x} - \dfrac{2}{x - 1}\right)$

In Exercises 81–86, determine whether the improper integral diverges or converges. Evaluate the integral if it converges.

81. $\int_{0}^{16} \dfrac{1}{\sqrt[4]{x}} \, dx$

82. $\int_{0}^{1} \dfrac{6}{x - 1} \, dx$

83. $\int_{1}^{\infty} x^2 \ln x \, dx$

84. $\int_{0}^{\infty} \dfrac{e^{-1/x}}{x^2} \, dx$

85. $\int_{1}^{\infty} \dfrac{\ln x}{x^2} \, dx$

86. $\int_{1}^{\infty} \dfrac{1}{\sqrt[4]{x}} \, dx$

87. **Present Value** The board of directors of a corporation is calculating the price to pay for a business that is forecast to yield a continuous flow of profit of $500,000 per year. If money will earn a nominal rate of 5% per year compounded continuously, what is the present value of the business

  (a) for 20 years?

  (b) forever (in perpetuity)?

  (*Note:* The present value for t_0 years is $\int_{0}^{t_0} 500{,}000 e^{-0.05t} \, dt$.)

88. **Volume** Find the volume of the solid generated by revolving the region bounded by the graphs of $y = xe^{-x}$, $y = 0$, and $x = 0$ about the x-axis.

89. **Probability** The average lengths (from beak to tail) of different species of warblers in the eastern United States are approximately normally distributed with a mean of 12.9 centimeters and a standard deviation of 0.95 centimeter (see figure). The probability that a randomly selected warbler has a length between a and b centimeters is

$$P(a \leq x \leq b) = \dfrac{1}{0.95\sqrt{2\pi}} \int_{a}^{b} e^{-(x - 12.9)^2/2(0.95)^2} \, dx.$$

Use a graphing utility to approximate the probability that a randomly selected warbler has a length of (a) 13 centimeters or greater and (b) 15 centimeters or greater. (*Source: Peterson's Field Guide: Eastern Birds*)

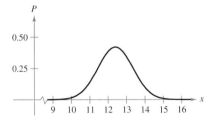

P.S. Problem Solving

1. (a) Evaluate the integrals
   $$\int_{-1}^{1} (1-x^2)\,dx \quad \text{and} \quad \int_{-1}^{1} (1-x^2)^2\,dx.$$
   (b) Use Wallis's Formulas to prove that
   $$\int_{-1}^{1} (1-x^2)^n\,dx = \frac{2^{2n+1}(n!)^2}{(2n+1)!}$$
   for all positive integers n.

2. (a) Evaluate the integrals $\int_0^1 \ln x\,dx$ and $\int_0^1 (\ln x)^2\,dx$.
   (b) Prove that
   $$\int_0^1 (\ln x)^n\,dx = (-1)^n n!$$
   for all positive integers n.

3. Find the value of the positive constant c such that
   $$\lim_{x\to\infty} \left(\frac{x+c}{x-c}\right)^x = 9.$$

4. Find the value of the positive constant c such that
   $$\lim_{x\to\infty} \left(\frac{x-c}{x+c}\right)^x = \frac{1}{4}.$$

5. In the figure, the line $x = 1$ is tangent to the unit circle at A. The length of segment QA equals the length of the circular arc $\overset{\frown}{PA}$. Show that the length of segment OR approaches 2 as P approaches A.

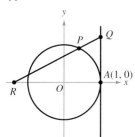

6. In the figure, the segment BD is the height of $\triangle OAB$. Let R be the ratio of the area of $\triangle DAB$ to that of the shaded region formed by deleting $\triangle OAB$ from the circular sector subtended by angle θ. Find $\lim_{\theta\to 0^+} R$.

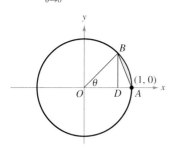

7. Consider the problem of finding the area of the region bounded by the x-axis, the line $x = 4$, and the curve
   $$y = \frac{x^2}{(x^2+9)^{3/2}}.$$

   (a) Use a graphing utility to graph the region and approximate its area.
   (b) Use an appropriate trigonometric substitution to find the exact area.
   (c) Use the substitution $x = 3\sinh u$ to find the exact area and verify that you obtain the same answer as in part (b).

8. Use the substitution $u = \tan\frac{x}{2}$ to find the area of the shaded region under the graph of $y = \dfrac{1}{2+\cos x}$, $0 \le x \le \pi/2$ (see figure).

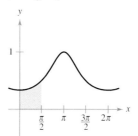

9. Find the arc length of the graph of the function $y = \ln(1-x^2)$ on the interval $0 \le x \le \frac{1}{2}$ (see figure).

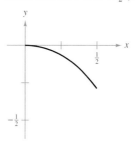

10. Find the centroid of the region above the x-axis and bounded above by the curve $y = e^{-c^2 x^2}$, where c is a positive constant (see figure).

    $\left(\text{Hint: Show that } \displaystyle\int_0^\infty e^{-c^2 x^2}\,dx = \frac{1}{c}\int_0^\infty e^{-x^2}\,dx.\right)$

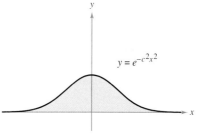

11. Some elementary functions, such as $f(x) = \sin(x^2)$, do not have antiderivatives that are elementary functions. Joseph Liouville proved that

$$\int \frac{e^x}{x}\, dx$$

does not have an elementary antiderivative. Use this fact to prove that

$$\int \frac{1}{\ln x}\, dx$$

is not elementary.

12. (a) Let $y = f^{-1}(x)$ be the inverse function of f. Use integration by parts to derive the formula

$$\int f^{-1}(x)\, dx = xf^{-1}(x) - \int f(y)\, dy.$$

(b) Use the formula in part (a) to find the integral

$$\int \arcsin x\, dx.$$

(c) Use the formula in part (a) to find the area under the graph of $y = \ln x$, $1 \le x \le e$ (see figure).

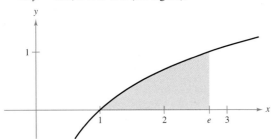

13. Factor the polynomial $p(x) = x^4 + 1$ and then find the area under the graph of $y = \dfrac{1}{x^4 + 1}$, $0 \le x \le 1$ (see figure).

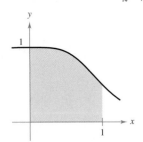

14. (a) Use the substitution $u = \dfrac{\pi}{2} - x$ to evaluate the integral

$$\int_0^{\pi/2} \frac{\sin x}{\cos x + \sin x}\, dx.$$

(b) Let n be a positive integer. Evaluate the integral

$$\int_0^{\pi/2} \frac{\sin^n x}{\cos^n x + \sin^n x}\, dx.$$

15. Use a graphing utility to estimate each limit. Then calculate each limit using L'Hôpital's Rule. What can you conclude about the indeterminate form $0 \cdot \infty$?

(a) $\displaystyle\lim_{x \to 0^+} \left(\cot x + \frac{1}{x}\right)$

(b) $\displaystyle\lim_{x \to 0^+} \left(\cot x - \frac{1}{x}\right)$

(c) $\displaystyle\lim_{x \to 0^+} \left[\left(\cot x + \frac{1}{x}\right)\left(\cot x - \frac{1}{x}\right)\right]$

16. Suppose the denominator of a rational function can be factored into distinct linear factors

$$D(x) = (x - c_1)(x - c_2) \cdots (x - c_n)$$

for a positive integer n and distinct real numbers $c_1, c_2, \ldots, c_n$. If N is a polynomial of degree less than n, show that

$$\frac{N(x)}{D(x)} = \frac{P_1}{x - c_1} + \frac{P_2}{x - c_2} + \cdots + \frac{P_n}{x - c_n}$$

where $P_k = N(c_k)/D'(c_k)$ for $k = 1, 2, \ldots, n$. Note that this is the partial fraction decomposition of $N(x)/D(x)$.

17. Use the results of Exercise 16 to find the partial fraction decomposition of

$$\frac{x^3 - 3x^2 + 1}{x^4 - 13x^2 + 12x}.$$

18. The velocity v (in feet per second) of a rocket whose initial mass (including fuel) is m is given by

$$v = gt + u\ln\frac{m}{m - rt},\quad t < \frac{m}{r}$$

where u is the expulsion speed of the fuel, r is the rate at which the fuel is consumed, and $g = -32$ feet per second per second is the acceleration due to gravity. Find the position equation for a rocket for which $m = 50{,}000$ pounds, $u = 12{,}000$ feet per second, and $r = 400$ pounds per second. What is the height of the rocket when $t = 100$ seconds? (Assume that the rocket was fired from ground level and is moving straight upward.)

19. Suppose that $f(a) = f(b) = g(a) = g(b) = 0$ and the second derivatives of f and g are continuous on the closed interval $[a, b]$. Prove that

$$\int_a^b f(x)g''(x)\, dx = \int_a^b f''(x)g(x)\, dx.$$

20. Suppose that $f(a) = f(b) = 0$ and the second derivatives of f exist on the closed interval $[a, b]$. Prove that

$$\int_a^b (x - a)(x - b)f''(x)\, dx = 2\int_a^b f(x)\, dx.$$

21. Using the inequality

$$\frac{1}{x^5} + \frac{1}{x^{10}} + \frac{1}{x^{15}} < \frac{1}{x^5 - 1} < \frac{1}{x^5} + \frac{1}{x^{10}} + \frac{2}{x^{15}}$$

for $x \ge 2$, approximate $\displaystyle\int_2^\infty \frac{1}{x^5 - 1}\, dx.$

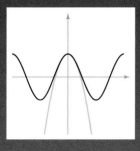

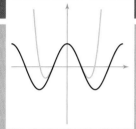

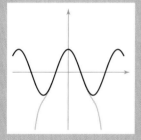

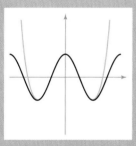

9 Infinite Series

A *ballista* was used as a portable rock-throwing machine. The propulsion mechanism was similar in appearance to a crossbow. Skilled artillerymen aimed and fired the ballista entirely by eye. What type of projectile path do you think these artillerymen preferred—a high, arching trajectory or a low, relatively level trajectory? Why?

Charles & Josette Lenars/Corbis

Maclaurin *polynomials* approximate a given function in an interval around $x = 0$. As you add terms to the Maclaurin polynomial, it becomes a better approximation of the given function near $x = 0$. In Section 9.10, you will see that a Maclaurin *series* is equivalent to the given function (under suitable conditions).

Section 9.1 Sequences

- List the terms of a sequence.
- Determine whether a sequence converges or diverges.
- Write a formula for the nth term of a sequence.
- Use properties of monotonic sequences and bounded sequences.

Sequences

In mathematics, the word "sequence" is used in much the same way as in ordinary English. To say that a collection of objects or events is *in sequence* usually means that the collection is ordered so that it has an identified first member, second member, third member, and so on.

Mathematically, a **sequence** is defined as a function whose domain is the set of positive integers. Although a sequence is a function, it is common to represent sequences by subscript notation rather than by the standard function notation. For instance, in the sequence

$$1, \quad 2, \quad 3, \quad 4, \quad \ldots, \quad n, \quad \ldots$$
$$\downarrow \quad \downarrow \quad \downarrow \quad \downarrow \quad \quad \downarrow \quad \quad \quad \text{Sequence}$$
$$a_1, \quad a_2, \quad a_3, \quad a_4, \quad \ldots, \quad a_n, \quad \ldots$$

1 is mapped onto a_1, 2 is mapped onto a_2, and so on. The numbers $a_1, a_2, a_3, \ldots, a_n, \ldots$ are the **terms** of the sequence. The number a_n is the **nth term** of the sequence, and the entire sequence is denoted by $\{a_n\}$.

EXPLORATION

Finding Patterns Describe a pattern for each of the following sequences. Then use your description to write a formula for the nth term of each sequence. As n increases, do the terms appear to be approaching a limit? Explain your reasoning.

a. $1, \frac{1}{2}, \frac{1}{4}, \frac{1}{8}, \frac{1}{16}, \ldots$

b. $1, \frac{1}{2}, \frac{1}{6}, \frac{1}{24}, \frac{1}{120}, \ldots$

c. $10, \frac{10}{3}, \frac{10}{6}, \frac{10}{10}, \frac{10}{15}, \ldots$

d. $\frac{1}{4}, \frac{4}{9}, \frac{9}{16}, \frac{16}{25}, \frac{25}{36}, \ldots$

e. $\frac{3}{7}, \frac{5}{10}, \frac{7}{13}, \frac{9}{16}, \frac{11}{19}, \ldots$

NOTE Occasionally, it is convenient to begin a sequence with a_0, so that the terms of the sequence become

$$a_0, a_1, a_2, a_3, \ldots, a_n, \ldots$$

EXAMPLE 1 Listing the Terms of a Sequence

a. The terms of the sequence $\{a_n\} = \{3 + (-1)^n\}$ are

$$3 + (-1)^1, \; 3 + (-1)^2, \; 3 + (-1)^3, \; 3 + (-1)^4, \ldots$$
$$2, \quad\quad 4, \quad\quad 2, \quad\quad 4, \quad\quad \ldots$$

b. The terms of the sequence $\{b_n\} = \left\{\dfrac{n}{1-2n}\right\}$ are

$$\frac{1}{1-2\cdot 1}, \; \frac{2}{1-2\cdot 2}, \; \frac{3}{1-2\cdot 3}, \; \frac{4}{1-2\cdot 4}, \ldots$$
$$-1, \quad -\frac{2}{3}, \quad -\frac{3}{5}, \quad -\frac{4}{7}, \quad \ldots$$

c. The terms of the sequence $\{c_n\} = \left\{\dfrac{n^2}{2^n - 1}\right\}$ are

$$\frac{1^2}{2^1 - 1}, \; \frac{2^2}{2^2 - 1}, \; \frac{3^2}{2^3 - 1}, \; \frac{4^2}{2^4 - 1}, \ldots$$
$$\frac{1}{1}, \quad \frac{4}{3}, \quad \frac{9}{7}, \quad \frac{16}{15}, \quad \ldots$$

d. The terms of the **recursively defined** sequence $\{d_n\}$, where $d_1 = 25$ and $d_{n+1} = d_n - 5$ are

$$25, \quad 25 - 5 = 20, \quad 20 - 5 = 15, \quad 15 - 5 = 10, \ldots$$

STUDY TIP Some sequences are defined recursively. To define a sequence recursively, you need to be given one or more of the first few terms. All other terms of the sequence are then defined using previous terms, as shown in Example 1(d).

Limit of a Sequence

The primary focus of this chapter concerns sequences whose terms approach limiting values. Such sequences are said to **converge.** For instance, the sequence $\{1/2^n\}$

$$\frac{1}{2}, \frac{1}{4}, \frac{1}{8}, \frac{1}{16}, \frac{1}{32}, \ldots$$

converges to 0, as indicated in the following definition.

Definition of the Limit of a Sequence

Let L be a real number. The **limit** of a sequence $\{a_n\}$ is L, written as

$$\lim_{n \to \infty} a_n = L$$

if for each $\varepsilon > 0$, there exists $M > 0$ such that $|a_n - L| < \varepsilon$ whenever $n > M$. If the limit L of a sequence exists, then the sequence **converges** to L. If the limit of a sequence does not exist, then the sequence **diverges.**

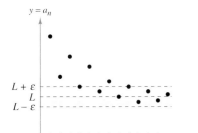

For $n > M$, the terms of the sequence all lie within ε units of L.
Figure 9.1

Graphically, this definition says that eventually (for $n > M$ and $\varepsilon > 0$) the terms of a sequence that converges to L will lie within the band between the lines $y = L + \varepsilon$ and $y = L - \varepsilon$, as shown in Figure 9.1.

If a sequence $\{a_n\}$ agrees with a function f at every positive integer, and if $f(x)$ approaches a limit L as $x \to \infty$, the sequence must converge to the same limit L.

THEOREM 9.1 Limit of a Sequence

Let L be a real number. Let f be a function of a real variable such that

$$\lim_{x \to \infty} f(x) = L.$$

If $\{a_n\}$ is a sequence such that $f(n) = a_n$ for every positive integer n, then

$$\lim_{n \to \infty} a_n = L.$$

EXAMPLE 2 Finding the Limit of a Sequence

Find the limit of the sequence whose nth term is

$$a_n = \left(1 + \frac{1}{n}\right)^n.$$

Solution In Theorem 5.15, you learned that

$$\lim_{x \to \infty} \left(1 + \frac{1}{x}\right)^x = e.$$

So, you can apply Theorem 9.1 to conclude that

$$\lim_{n \to \infty} a_n = \lim_{n \to \infty} \left(1 + \frac{1}{n}\right)^n$$
$$= e.$$

NOTE There are different ways in which a sequence can fail to have a limit. One way is that the terms of the sequence increase without bound or decrease without bound. These cases are written symbolically as follows.

Terms increase without bound:

$$\lim_{n \to \infty} a_n = \infty$$

Terms decrease without bound:

$$\lim_{n \to \infty} a_n = -\infty$$

The following properties of limits of sequences parallel those given for limits of functions of a real variable in Section 1.3.

THEOREM 9.2   Properties of Limits of Sequences

Let $\lim\limits_{n\to\infty} a_n = L$ and $\lim\limits_{n\to\infty} b_n = K$.

1. $\lim\limits_{n\to\infty} (a_n \pm b_n) = L \pm K$
2. $\lim\limits_{n\to\infty} ca_n = cL$, c is any real number
3. $\lim\limits_{n\to\infty} (a_n b_n) = LK$
4. $\lim\limits_{n\to\infty} \dfrac{a_n}{b_n} = \dfrac{L}{K}$, $b_n \neq 0$ and $K \neq 0$

EXAMPLE 3   Determining Convergence or Divergence

a. Because the sequence $\{a_n\} = \{3 + (-1)^n\}$ has terms

$$2, 4, 2, 4, \ldots$$

See Example 1(a), page 594.

that alternate between 2 and 4, the limit

$$\lim_{n\to\infty} a_n$$

does not exist. So, the sequence diverges.

b. For $\{b_n\} = \left\{\dfrac{n}{1 - 2n}\right\}$, divide the numerator and denominator by n to obtain

$$\lim_{n\to\infty} \frac{n}{1 - 2n} = \lim_{n\to\infty} \frac{1}{(1/n) - 2} = -\frac{1}{2}$$

See Example 1(b), page 594.

which implies that the sequence converges to $-\tfrac{1}{2}$.

EXAMPLE 4   Using L'Hôpital's Rule to Determine Convergence

Show that the sequence whose nth term is $a_n = \dfrac{n^2}{2^n - 1}$ converges.

Solution   Consider the function of a real variable

$$f(x) = \frac{x^2}{2^x - 1}.$$

Applying L'Hôpital's Rule twice produces

$$\lim_{x\to\infty} \frac{x^2}{2^x - 1} = \lim_{x\to\infty} \frac{2x}{(\ln 2)2^x} = \lim_{x\to\infty} \frac{2}{(\ln 2)^2 2^x} = 0.$$

Because $f(n) = a_n$ for every positive integer, you can apply Theorem 9.1 to conclude that

$$\lim_{n\to\infty} \frac{n^2}{2^n - 1} = 0.$$

See Example 1(c), page 594.

So, the sequence converges to 0.

TECHNOLOGY   Use a graphing utility to graph the function in Example 4. Notice that as x approaches infinity, the value of the function gets closer and closer to 0. If you have access to a graphing utility that can generate terms of a sequence, try using it to calculate the first 20 terms of the sequence in Example 4. Then view the terms to observe numerically that the sequence converges to 0.

---

www indicates that in the HM mathSpace® CD-ROM and the online Eduspace® system for this text, you will find an Open Exploration, which further explores this example using the computer algebra systems Maple, Mathcad, Mathematica, and Derive.

The symbol $n!$ (read "n factorial") is used to simplify some of the formulas developed in this chapter. Let n be a positive integer; then **n factorial** is defined as

$$n! = 1 \cdot 2 \cdot 3 \cdot 4 \cdots (n-1) \cdot n.$$

As a special case, **zero factorial** is defined as $0! = 1$. From this definition, you can see that $1! = 1$, $2! = 1 \cdot 2 = 2$, $3! = 1 \cdot 2 \cdot 3 = 6$, and so on. Factorials follow the same conventions for order of operations as exponents. That is, just as $2x^3$ and $(2x)^3$ imply different orders of operations, $2n!$ and $(2n)!$ imply the following orders.

$$2n! = 2(n!) = 2(1 \cdot 2 \cdot 3 \cdot 4 \cdots n)$$

and

$$(2n)! = 1 \cdot 2 \cdot 3 \cdot 4 \cdots n \cdot (n+1) \cdots 2n$$

Another useful limit theorem that can be rewritten for sequences is the Squeeze Theorem from Section 1.3.

THEOREM 9.3 Squeeze Theorem for Sequences

If

$$\lim_{n \to \infty} a_n = L = \lim_{n \to \infty} b_n$$

and there exists an integer N such that $a_n \leq c_n \leq b_n$ for all $n > N$, then

$$\lim_{n \to \infty} c_n = L.$$

EXAMPLE 5 Using the Squeeze Theorem

Show that the sequence $\{c_n\} = \left\{(-1)^n \dfrac{1}{n!}\right\}$ converges, and find its limit.

Solution To apply the Squeeze Theorem, you must find two convergent sequences that can be related to the given sequence. Two possibilities are $a_n = -1/2^n$ and $b_n = 1/2^n$, both of which converge to 0. By comparing the term $n!$ with 2^n, you can see that

$$n! = 1 \cdot 2 \cdot 3 \cdot 4 \cdot 5 \cdot 6 \cdots n = 24 \cdot \underbrace{5 \cdot 6 \cdots n}_{n-4 \text{ factors}} \qquad (n \geq 4)$$

and

$$2^n = 2 \cdot 2 \cdot 2 \cdot 2 \cdot 2 \cdot 2 \cdots 2 = 16 \cdot \underbrace{2 \cdot 2 \cdots 2}_{n-4 \text{ factors}}. \qquad (n \geq 4)$$

This implies that for $n \geq 4$, $2^n < n!$, and you have

$$\frac{-1}{2^n} \leq (-1)^n \frac{1}{n!} \leq \frac{1}{2^n}, \quad n \geq 4$$

as shown in Figure 9.2. So, by the Squeeze Theorem it follows that

$$\lim_{n \to \infty} (-1)^n \frac{1}{n!} = 0.$$

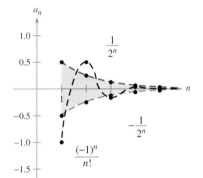

For $n \geq 4$, $(-1)^n/n!$ is squeezed between $-1/2^n$ and $1/2^n$.
Figure 9.2

NOTE Example 5 suggests something about the rate at which $n!$ increases as $n \to \infty$. As Figure 9.2 suggests, both $1/2^n$ and $1/n!$ approach 0 as $n \to \infty$. Yet $1/n!$ approaches 0 so much faster than $1/2^n$ does that

$$\lim_{n \to \infty} \frac{1/n!}{1/2^n} = \lim_{n \to \infty} \frac{2^n}{n!} = 0.$$

In fact, it can be shown that for any fixed number k,

$$\lim_{n \to \infty} \frac{k^n}{n!} = 0.$$

This means that *the factorial function grows faster than any exponential function.*

In Example 5, the sequence $\{c_n\}$ has both positive and negative terms. For this sequence, it happens that the sequence of absolute values, $\{|c_n|\}$, also converges to 0. You can show this by the Squeeze Theorem using the inequality

$$0 \leq \frac{1}{n!} \leq \frac{1}{2^n}, \quad n \geq 4.$$

In such cases, it is often convenient to consider the sequence of absolute values—and then apply Theorem 9.4, which states that if the absolute value sequence converges to 0, the original signed sequence also converges to 0.

THEOREM 9.4 Absolute Value Theorem

For the sequence $\{a_n\}$, if

$$\lim_{n \to \infty} |a_n| = 0 \quad \text{then} \quad \lim_{n \to \infty} a_n = 0.$$

Proof Consider the two sequences $\{|a_n|\}$ and $\{-|a_n|\}$. Because both of these sequences converge to 0 and

$$-|a_n| \leq a_n \leq |a_n|$$

you can use the Squeeze Theorem to conclude that $\{a_n\}$ converges to 0. ∎

Pattern Recognition for Sequences

Sometimes the terms of a sequence are generated by some rule that does not explicitly identify the nth term of the sequence. In such cases, you may be required to discover a *pattern* in the sequence and to describe the nth term. Once the nth term has been specified, you can investigate the convergence or divergence of the sequence.

EXAMPLE 6 Finding the nth Term of a Sequence

Find a sequence $\{a_n\}$ whose first five terms are

$$\frac{2}{1}, \frac{4}{3}, \frac{8}{5}, \frac{16}{7}, \frac{32}{9}, \ldots$$

and then determine whether the particular sequence you have chosen converges or diverges.

Solution First, note that the numerators are successive powers of 2, and the denominators form the sequence of positive odd integers. By comparing a_n with n, you have the following pattern.

$$\frac{2^1}{1}, \frac{2^2}{3}, \frac{2^3}{5}, \frac{2^4}{7}, \frac{2^5}{9}, \ldots, \frac{2^n}{2n-1}$$

Using L'Hôpital's Rule to evaluate the limit of $f(x) = 2^x/(2x-1)$, you obtain

$$\lim_{x \to \infty} \frac{2^x}{2x-1} = \lim_{x \to \infty} \frac{2^x(\ln 2)}{2} = \infty \implies \lim_{n \to \infty} \frac{2^n}{2n-1} = \infty.$$

So, the sequence diverges.

Without a specific rule for generating the terms of a sequence or some knowledge of the context in which the terms of the sequence are obtained, it is not possible to determine the convergence or divergence of the sequence merely from its first several terms. For instance, although the first three terms of the following four sequences are identical, the first two sequences converge to 0, the third sequence converges to $\frac{1}{9}$, and the fourth sequence diverges.

$$\{a_n\}: \frac{1}{2}, \frac{1}{4}, \frac{1}{8}, \frac{1}{16}, \ldots, \frac{1}{2^n}, \ldots$$

$$\{b_n\}: \frac{1}{2}, \frac{1}{4}, \frac{1}{8}, \frac{1}{15}, \ldots, \frac{6}{(n+1)(n^2-n+6)}, \ldots$$

$$\{c_n\}: \frac{1}{2}, \frac{1}{4}, \frac{1}{8}, \frac{7}{62}, \ldots, \frac{n^2-3n+3}{9n^2-25n+18}, \ldots$$

$$\{d_n\}: \frac{1}{2}, \frac{1}{4}, \frac{1}{8}, 0, \ldots, \frac{-n(n+1)(n-4)}{6(n^2+3n-2)}, \ldots$$

The process of determining an nth term from the pattern observed in the first several terms of a sequence is an example of *inductive reasoning*.

EXAMPLE 7 Finding the nth Term of a Sequence

Determine an nth term for a sequence whose first five terms are

$$-\frac{2}{1}, \frac{8}{2}, -\frac{26}{6}, \frac{80}{24}, -\frac{242}{120}, \ldots$$

and then decide whether the sequence converges or diverges.

Solution Note that the numerators are 1 less than 3^n. So, you can reason that the numerators are given by the rule $3^n - 1$. Factoring the denominators produces

$$1 = 1$$
$$2 = 1 \cdot 2$$
$$6 = 1 \cdot 2 \cdot 3$$
$$24 = 1 \cdot 2 \cdot 3 \cdot 4$$
$$120 = 1 \cdot 2 \cdot 3 \cdot 4 \cdot 5 \cdots.$$

This suggests that the denominators are represented by $n!$. Finally, because the signs alternate, you can write the nth term as

$$a_n = (-1)^n \left(\frac{3^n - 1}{n!} \right).$$

From the discussion about the growth of $n!$, it follows that

$$\lim_{n \to \infty} |a_n| = \lim_{n \to \infty} \frac{3^n - 1}{n!} = 0.$$

Applying Theorem 9.4, you can conclude that

$$\lim_{n \to \infty} a_n = 0.$$

So, the sequence $\{a_n\}$ converges to 0.

Monotonic Sequences and Bounded Sequences

So far you have determined the convergence of a sequence by finding its limit. Even if you cannot determine the limit of a particular sequence, it still may be useful to know whether the sequence converges. Theorem 9.5 provides a test for convergence of sequences without determining the limit. First, some preliminary definitions are given.

Definition of a Monotonic Sequence

A sequence $\{a_n\}$ is **monotonic** if its terms are nondecreasing

$$a_1 \leq a_2 \leq a_3 \leq \cdots \leq a_n \leq \cdots$$

or if its terms are nonincreasing

$$a_1 \geq a_2 \geq a_3 \geq \cdots \geq a_n \geq \cdots.$$

EXAMPLE 8 Determining Whether a Sequence Is Monotonic

Determine whether each sequence having the given nth term is monotonic.

a. $a_n = 3 + (-1)^n$ **b.** $b_n = \dfrac{2n}{1+n}$ **c.** $c_n = \dfrac{n^2}{2^n - 1}$

Solution

a. This sequence alternates between 2 and 4. So, it is not monotonic.

b. This sequence is monotonic because each successive term is larger than its predecessor. To see this, compare the terms b_n and b_{n+1}. [Note that, because n is positive, you can multiply each side of the inequality by $(1+n)$ and $(2+n)$ without reversing the inequality sign.]

$$b_n = \frac{2n}{1+n} \stackrel{?}{<} \frac{2(n+1)}{1+(n+1)} = b_{n+1}$$
$$2n(2+n) \stackrel{?}{<} (1+n)(2n+2)$$
$$4n + 2n^2 \stackrel{?}{<} 2 + 4n + 2n^2$$
$$0 < 2$$

Starting with the final inequality, which is valid, you can reverse the steps to conclude that the original inequality is also valid.

c. This sequence is not monotonic, because the second term is larger than the first term, and larger than the third. (Note that if you drop the first term, the remaining sequence $c_2, c_3, c_4, \ldots$ is monotonic.)

Figure 9.3 graphically illustrates these three sequences.

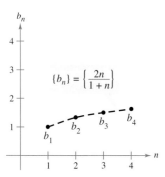

(a) Not monotonic

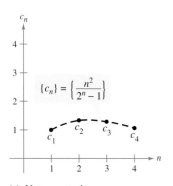

(b) Monotonic

(c) Not monotonic
Figure 9.3

NOTE In Example 8(b), another way to see that the sequence is monotonic is to argue that the derivative of the corresponding differentiable function $f(x) = 2x/(1+x)$ is positive for all x. This implies that f is increasing, which in turn implies that $\{a_n\}$ is increasing.

NOTE All three sequences shown in Figure 9.3 are bounded. To see this, consider the following.

$$2 \leq a_n \leq 4$$
$$1 \leq b_n \leq 2$$
$$0 \leq c_n \leq \frac{4}{3}$$

Definition of a Bounded Sequence

1. A sequence $\{a_n\}$ is **bounded above** if there is a real number M such that $a_n \leq M$ for all n. The number M is called an **upper bound** of the sequence.
2. A sequence $\{a_n\}$ is **bounded below** if there is a real number N such that $N \leq a_n$ for all n. The number N is called a **lower bound** of the sequence.
3. A sequence $\{a_n\}$ is **bounded** if it is bounded above and bounded below.

One important property of the real numbers is that they are **complete.** Informally, this means that there are no holes or gaps on the real number line. (The set of rational numbers does not have the completeness property.) The completeness axiom for real numbers can be used to conclude that if a sequence has an upper bound, it must have a **least upper bound** (an upper bound that is smaller than all other upper bounds for the sequence). For example, the least upper bound of the sequence $\{a_n\} = \{n/(n + 1)\}$,

$$\frac{1}{2}, \frac{2}{3}, \frac{3}{4}, \frac{4}{5}, \ldots, \frac{n}{n+1}, \ldots$$

is 1. The completeness axiom is used in the proof of Theorem 9.5.

THEOREM 9.5 Bounded Monotonic Sequences

If a sequence $\{a_n\}$ is bounded and monotonic, then it converges.

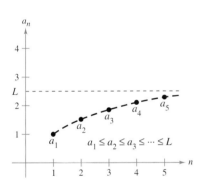

Every bounded nondecreasing sequence converges.
Figure 9.4

Proof Assume that the sequence is nondecreasing, as shown in Figure 9.4. For the sake of simplicity, also assume that each term in the sequence is positive. Because the sequence is bounded, there must exist an upper bound M such that

$$a_1 \leq a_2 \leq a_3 \leq \cdots \leq a_n \leq \cdots \leq M.$$

From the completeness axiom, it follows that there is a least upper bound L such that

$$a_1 \leq a_2 \leq a_3 \leq \cdots \leq a_n \leq \cdots \leq L.$$

For $\varepsilon > 0$, it follows that $L - \varepsilon < L$, and therefore $L - \varepsilon$ cannot be an upper bound for the sequence. Consequently, at least one term of $\{a_n\}$ is greater than $L - \varepsilon$. That is, $L - \varepsilon < a_N$ for some positive integer N. Because the terms of $\{a_n\}$ are nondecreasing, it follows that $a_N \leq a_n$ for $n > N$. You now know that $L - \varepsilon < a_N \leq a_n \leq L < L + \varepsilon$, for every $n > N$. It follows that $|a_n - L| < \varepsilon$ for $n > N$, which by definition means that $\{a_n\}$ converges to L. The proof for a nonincreasing sequence is similar.

EXAMPLE 9 Bounded and Monotonic Sequences

a. The sequence $\{a_n\} = \{1/n\}$ is both bounded and monotonic and so, by Theorem 9.5, must converge.

b. The divergent sequence $\{b_n\} = \{n^2/(n + 1)\}$ is monotonic, but not bounded. (It is bounded below.)

c. The divergent sequence $\{c_n\} = \{(-1)^n\}$ is bounded, but not monotonic.

Exercises for Section 9.1

See www.CalcChat.com for worked-out solutions to odd-numbered exercises.

In Exercises 1–10, write the first five terms of the sequence.

1. $a_n = 2^n$
2. $a_n = \dfrac{3^n}{n!}$
3. $a_n = \left(-\dfrac{1}{2}\right)^n$
4. $a_n = \left(-\dfrac{2}{3}\right)^n$
5. $a_n = \sin\dfrac{n\pi}{2}$
6. $a_n = \dfrac{2n}{n+3}$
7. $a_n = \dfrac{(-1)^{n(n+1)/2}}{n^2}$
8. $a_n = (-1)^{n+1}\left(\dfrac{2}{n}\right)$
9. $a_n = 5 - \dfrac{1}{n} + \dfrac{1}{n^2}$
10. $a_n = 10 + \dfrac{2}{n} + \dfrac{6}{n^2}$

In Exercises 11–14, write the first five terms of the recursively defined sequence.

11. $a_1 = 3,\ a_{k+1} = 2(a_k - 1)$
12. $a_1 = 4,\ a_{k+1} = \left(\dfrac{k+1}{2}\right)a_k$
13. $a_1 = 32,\ a_{k+1} = \tfrac{1}{2}a_k$
14. $a_1 = 6,\ a_{k+1} = \tfrac{1}{3}a_k^2$

In Exercises 15–20, match the sequence with its graph. [The graphs are labeled (a), (b), (c), (d), (e), and (f).]

(a)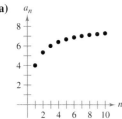

(b)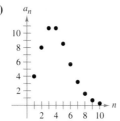

(c)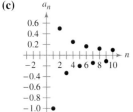

(d)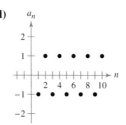

(e)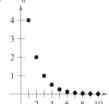

(f)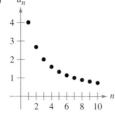

15. $a_n = \dfrac{8}{n+1}$
16. $a_n = \dfrac{8n}{n+1}$
17. $a_n = 4(0.5)^{n-1}$
18. $a_n = \dfrac{4^n}{n!}$
19. $a_n = (-1)^n$
20. $a_n = \dfrac{(-1)^n}{n}$

In Exercises 21–24, use a graphing utility to graph the first 10 terms of the sequence.

21. $a_n = \dfrac{2}{3}n$
22. $a_n = 2 - \dfrac{4}{n}$
23. $a_n = 16(-0.5)^{n-1}$
24. $a_n = \dfrac{2n}{n+1}$

In Exercises 25–30, write the next two apparent terms of the sequence. Describe the pattern you used to find these terms.

25. $2, 5, 8, 11, \ldots$
26. $\tfrac{7}{2}, 4, \tfrac{9}{2}, 5, \ldots$
27. $5, 10, 20, 40, \ldots$
28. $1, -\tfrac{1}{2}, \tfrac{1}{4}, -\tfrac{1}{8}, \ldots$
29. $3, -\tfrac{3}{2}, \tfrac{3}{4}, -\tfrac{3}{8}, \ldots$
30. $1, -\tfrac{3}{2}, \tfrac{9}{4}, -\tfrac{27}{8}, \ldots$

In Exercises 31–36, simplify the ratio of factorials.

31. $\dfrac{10!}{8!}$
32. $\dfrac{25!}{23!}$
33. $\dfrac{(n+1)!}{n!}$
34. $\dfrac{(n+2)!}{n!}$
35. $\dfrac{(2n-1)!}{(2n+1)!}$
36. $\dfrac{(2n+2)!}{(2n)!}$

In Exercises 37–42, find the limit (if possible) of the sequence.

37. $a_n = \dfrac{5n^2}{n^2+2}$
38. $a_n = 5 - \dfrac{1}{n^2}$
39. $a_n = \dfrac{2n}{\sqrt{n^2+1}}$
40. $a_n = \dfrac{5n}{\sqrt{n^2+4}}$
41. $a_n = \sin\dfrac{1}{n}$
42. $a_n = \cos\dfrac{2}{n}$

In Exercises 43–46, use a graphing utility to graph the first 10 terms of the sequence. Use the graph to make an inference about the convergence or divergence of the sequence. Verify your inference analytically and, if the sequence converges, find its limit.

43. $a_n = \dfrac{n+1}{n}$
44. $a_n = \dfrac{1}{n^{3/2}}$
45. $a_n = \cos\dfrac{n\pi}{2}$
46. $a_n = 3 - \dfrac{1}{2^n}$

In Exercises 47–68, determine the convergence or divergence of the sequence with the given nth term. If the sequence converges, find its limit.

47. $a_n = (-1)^n\left(\dfrac{n}{n+1}\right)$
48. $a_n = 1 + (-1)^n$
49. $a_n = \dfrac{3n^2 - n + 4}{2n^2 + 1}$
50. $a_n = \dfrac{\sqrt[3]{n}}{\sqrt[3]{n}+1}$
51. $a_n = \dfrac{1 \cdot 3 \cdot 5 \cdots (2n-1)}{(2n)^n}$
52. $a_n = \dfrac{1 \cdot 3 \cdot 5 \cdots (2n-1)}{n!}$

53. $a_n = \dfrac{1+(-1)^n}{n}$

54. $a_n = \dfrac{1+(-1)^n}{n^2}$

55. $a_n = \dfrac{\ln(n^3)}{2n}$

56. $a_n = \dfrac{\ln\sqrt{n}}{n}$

57. $a_n = \dfrac{3^n}{4^n}$

58. $a_n = (0.5)^n$

59. $a_n = \dfrac{(n+1)!}{n!}$

60. $a_n = \dfrac{(n-2)!}{n!}$

61. $a_n = \dfrac{n-1}{n} - \dfrac{n}{n-1},\ n \geq 2$

62. $a_n = \dfrac{n^2}{2n+1} - \dfrac{n^2}{2n-1}$

63. $a_n = \dfrac{n^p}{e^n},\ p > 0$

64. $a_n = n \sin \dfrac{1}{n}$

65. $a_n = \left(1 + \dfrac{k}{n}\right)^n$

66. $a_n = 2^{1/n}$

67. $a_n = \dfrac{\sin n}{n}$

68. $a_n = \dfrac{\cos \pi n}{n^2}$

In Exercises 69–82, write an expression for the nth term of the sequence. (There is more than one correct answer.)

69. 1, 4, 7, 10, . . .

70. 3, 7, 11, 15, . . .

71. −1, 2, 7, 14, 23, . . .

72. $1, -\dfrac{1}{4}, \dfrac{1}{9}, -\dfrac{1}{16}, \ldots$

73. $\dfrac{2}{3}, \dfrac{3}{4}, \dfrac{4}{5}, \dfrac{5}{6}, \ldots$

74. $2, -1, \dfrac{1}{2}, -\dfrac{1}{4}, \dfrac{1}{8}, \ldots$

75. $2, 1+\dfrac{1}{2}, 1+\dfrac{1}{3}, 1+\dfrac{1}{4}, 1+\dfrac{1}{5}, \ldots$

76. $1+\dfrac{1}{2}, 1+\dfrac{3}{4}, 1+\dfrac{7}{8}, 1+\dfrac{15}{16}, 1+\dfrac{31}{32}, \ldots$

77. $\dfrac{1}{2\cdot 3}, \dfrac{2}{3\cdot 4}, \dfrac{3}{4\cdot 5}, \dfrac{4}{5\cdot 6}, \ldots$

78. $1, \dfrac{1}{2}, \dfrac{1}{6}, \dfrac{1}{24}, \dfrac{1}{120}, \ldots$

79. $1, -\dfrac{1}{1\cdot 3}, \dfrac{1}{1\cdot 3\cdot 5}, -\dfrac{1}{1\cdot 3\cdot 5\cdot 7}, \ldots$

80. $1, x, \dfrac{x^2}{2}, \dfrac{x^3}{6}, \dfrac{x^4}{24}, \dfrac{x^5}{120}, \ldots$

81. 2, 24, 720, 40,320, 3,628,800, . . .

82. 1, 6, 120, 5040, 362,880, . . .

In Exercises 83–94, determine whether the sequence with the given nth term is monotonic. Discuss the boundedness of the sequence. Use a graphing utility to confirm your results.

83. $a_n = 4 - \dfrac{1}{n}$

84. $a_n = \dfrac{3n}{n+2}$

85. $a_n = \dfrac{n}{2^{n+2}}$

86. $a_n = ne^{-n/2}$

87. $a_n = (-1)^n \left(\dfrac{1}{n}\right)$

88. $a_n = \left(-\dfrac{2}{3}\right)^n$

89. $a_n = \left(\dfrac{2}{3}\right)^n$

90. $a_n = \left(\dfrac{3}{2}\right)^n$

91. $a_n = \sin \dfrac{n\pi}{6}$

92. $a_n = \cos\left(\dfrac{n\pi}{2}\right)$

93. $a_n = \dfrac{\cos n}{n}$

94. $a_n = \dfrac{\sin\sqrt{n}}{n}$

In Exercises 95–98, (a) use Theorem 9.5 to show that the sequence with the given nth term converges and (b) use a graphing utility to graph the first 10 terms of the sequence and find its limit.

95. $a_n = 5 + \dfrac{1}{n}$

96. $a_n = 4 - \dfrac{3}{n}$

97. $a_n = \dfrac{1}{3}\left(1 - \dfrac{1}{3^n}\right)$

98. $a_n = 4 + \dfrac{1}{2^n}$

99. Let $\{a_n\}$ be an increasing sequence such that $2 \leq a_n \leq 4$. Explain why $\{a_n\}$ has a limit. What can you conclude about the limit?

100. Let $\{a_n\}$ be a monotonic sequence such that $a_n \leq 1$. Discuss the convergence of $\{a_n\}$. If $\{a_n\}$ converges, what can you conclude about its limit?

101. *Compound Interest* Consider the sequence $\{A_n\}$ whose nth term is given by

$$A_n = P\left(1 + \dfrac{r}{12}\right)^n$$

where P is the principal, A_n is the account balance after n months, and r is the interest rate compounded annually.

(a) Is $\{A_n\}$ a convergent sequence? Explain.

(b) Find the first 10 terms of the sequence if $P = \$9000$ and $r = 0.055$.

102. *Compound Interest* A deposit of \$100 is made at the beginning of each month in an account at an annual interest rate of 3% compounded monthly. The balance in the account after n months is $A_n = 100(401)(1.0025^n - 1)$.

(a) Compute the first six terms of the sequence $\{A_n\}$.

(b) Find the balance in the account after 5 years by computing the 60th term of the sequence.

(c) Find the balance in the account after 20 years by computing the 240th term of the sequence.

Writing About Concepts

103. In your own words, define each of the following.

(a) Sequence (b) Convergence of a sequence

(c) Monotonic sequence (d) Bounded sequence

104. The graphs of two sequences are shown in the figures. Which graph represents the sequence with alternating signs? Explain.

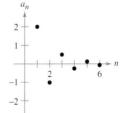 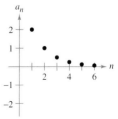

Writing About Concepts (continued)

In Exercises 105–108, give an example of a sequence satisfying the condition or explain why no such sequence exists. (Examples are not unique.)

105. A monotonically increasing sequence that converges to 10

106. A monotonically increasing bounded sequence that does not converge

107. A sequence that converges to $\frac{3}{4}$

108. An unbounded sequence that converges to 100

---

109. *Government Expenditures* A government program that currently costs taxpayers $2.5 billion per year is cut back by 20 percent per year.

(a) Write an expression for the amount budgeted for this program after n years.

(b) Compute the budgets for the first 4 years.

(c) Determine the convergence or divergence of the sequence of reduced budgets. If the sequence converges, find its limit.

110. *Inflation* If the rate of inflation is $4\frac{1}{2}\%$ per year and the average price of a car is currently $16,000, the average price after n years is

$$P_n = \$16{,}000(1.045)^n.$$

Compute the average prices for the next 5 years.

111. *Modeling Data* The number a_n of endangered and threatened species in the United States from 1996 through 2002 is shown in the table, where n represents the year, with $n = 6$ corresponding to 1996. *(Source: U.S. Fish and Wildlife Service)*

n	6	7	8	9	10	11	12
a_n	1053	1132	1194	1205	1244	1254	1263

(a) Use the regression capabilities of a graphing utility to find a model of the form

$$a_n = bn^2 + cn + d, \quad n = 6, 7, \ldots, 12$$

for the data. Use the graphing utility to plot the points and graph the model.

(b) Use the model to predict the number of endangered and threatened species in the year 2008.

112. *Modeling Data* The annual sales a_n (in millions of dollars) for Avon Products, Inc. from 1993 through 2002 are given below as ordered pairs of the form (n, a_n), where n represents the year, with $n = 3$ corresponding to 1993. *(Source: 2002 Avon Products, Inc. Annual Report)*

(3, 3844), (4, 4267), (5, 4492), (6, 4814), (7, 5079),

(8, 5213), (9, 5289), (10, 5682), (11, 5958), (12, 6171)

(a) Use the regression capabilities of a graphing utility to find a model of the form

$$a_n = bn + c, \quad n = 3, 4, \ldots, 12$$

for the data. Graphically compare the points and the model.

(b) Use the model to predict sales in the year 2008.

113. *Comparing Exponential and Factorial Growth* Consider the sequence $a_n = 10^n/n!$.

(a) Find two consecutive terms that are equal in magnitude.

(b) Are the terms following those found in part (a) increasing or decreasing?

(c) In Section 8.7, Exercises 65–70, it was shown that for "large" values of the independent variable an exponential function increases more rapidly than a polynomial function. From the result in part (b), what inference can you make about the rate of growth of an exponential function versus a factorial function for "large" integer values of n?

114. Compute the first six terms of the sequence

$$\{a_n\} = \left\{\left(1 + \frac{1}{n}\right)^n\right\}.$$

If the sequence converges, find its limit.

115. Compute the first six terms of the sequence $\{a_n\} = \{\sqrt[n]{n}\}$. If the sequence converges, find its limit.

116. Prove that if $\{s_n\}$ converges to L and $L > 0$, then there exists a number N such that $s_n > 0$ for $n > N$.

True or False? In Exercises 117–120, determine whether the statement is true or false. If it is false, explain why or give an example that shows it is false.

117. If $\{a_n\}$ converges to 3 and $\{b_n\}$ converges to 2, then $\{a_n + b_n\}$ converges to 5.

118. If $\{a_n\}$ converges, then $\lim_{n \to \infty} (a_n - a_{n+1}) = 0$.

119. If $n > 1$, then $n! = n(n-1)!$.

120. If $\{a_n\}$ converges, then $\{a_n/n\}$ converges to 0.

121. *Fibonacci Sequence* In a study of the progeny of rabbits, Fibonacci (ca. 1170–ca. 1240) encountered the sequence now bearing his name. It is defined recursively by

$$a_{n+2} = a_n + a_{n+1}, \quad \text{where} \quad a_1 = 1 \text{ and } a_2 = 1.$$

(a) Write the first 12 terms of the sequence.

(b) Write the first 10 terms of the sequence defined by

$$b_n = \frac{a_{n+1}}{a_n}, \quad n \geq 1.$$

(c) Using the definition in part (b), show that

$$b_n = 1 + \frac{1}{b_{n-1}}.$$

(d) The **golden ratio** ρ can be defined by $\lim_{n \to \infty} b_n = \rho$. Show that $\rho = 1 + 1/\rho$ and solve this equation for ρ.

122. Conjecture Let $x_0 = 1$ and consider the sequence x_n given by the formula

$$x_n = \frac{1}{2}x_{n-1} + \frac{1}{x_{n-1}}, \quad n = 1, 2, \ldots.$$

Use a graphing utility to compute the first 10 terms of the sequence and make a conjecture about the limit of the sequence.

123. Consider the sequence

$$\sqrt{2}, \sqrt{2 + \sqrt{2}}, \sqrt{2 + \sqrt{2 + \sqrt{2}}}, \ldots$$

(a) Compute the first five terms of this sequence.

(b) Write a recursion formula for a_n, for $n \geq 2$.

(c) Find $\lim_{n \to \infty} a_n$.

124. Consider the sequence

$$\sqrt{6}, \sqrt{6 + \sqrt{6}}, \sqrt{6 + \sqrt{6 + \sqrt{6}}}, \ldots$$

(a) Compute the first five terms of this sequence.

(b) Write a recursion formula for a_n, for $n \geq 2$.

(c) Find $\lim_{n \to \infty} a_n$.

125. Consider the sequence $\{a_n\}$ where $a_1 = \sqrt{k}$, $a_{n+1} = \sqrt{k + a_n}$, and $k > 0$.

(a) Show that $\{a_n\}$ is increasing and bounded.

(b) Prove that $\lim_{n \to \infty} a_n$ exists.

(c) Find $\lim_{n \to \infty} a_n$.

126. Arithmetic-Geometric Mean Let $a_0 > b_0 > 0$. Let a_1 be the arithmetic mean of a_0 and b_0 and let b_1 be the geometric mean of a_0 and b_0.

$$a_1 = \frac{a_0 + b_0}{2} \quad \text{Arithmetic mean}$$

$$b_1 = \sqrt{a_0 b_0} \quad \text{Geometric mean}$$

Now define the sequences $\{a_n\}$ and $\{b_n\}$ as follows.

$$a_n = \frac{a_{n-1} + b_{n-1}}{2} \quad b_n = \sqrt{a_{n-1} b_{n-1}}$$

(a) Let $a_0 = 10$ and $b_0 = 3$. Write out the first five terms of $\{a_n\}$ and $\{b_n\}$. Compare the terms of $\{b_n\}$. Compare a_n and b_n. What do you notice?

(b) Use induction to show that $a_n > a_{n+1} > b_{n+1} > b_n$, for $a_0 > b_0 > 0$.

(c) Explain why $\{a_n\}$ and $\{b_n\}$ are both convergent.

(d) Show that $\lim_{n \to \infty} a_n = \lim_{n \to \infty} b_n$.

127. (a) Let $f(x) = \sin x$ and $a_n = n \sin 1/n$. Show that

$$\lim_{n \to \infty} a_n = f'(0) = 1.$$

(b) Let $f(x)$ be differentiable on the interval $[0, 1]$ and $f(0) = 0$. Consider the sequence $\{a_n\}$, where $a_n = nf(1/n)$. Show that $\lim_{n \to \infty} a_n = f'(0)$.

128. Consider the sequence $\{a_n\} = \{nr^n\}$. Decide whether $\{a_n\}$ converges for each value of r.

(a) $r = \frac{1}{2}$ (b) $r = 1$ (c) $r = \frac{3}{2}$

(d) For what values or r does the sequence $\{nr^n\}$ converge?

129. (a) Show that $\int_1^n \ln x \, dx < \ln(n!)$ for $n \geq 2$.

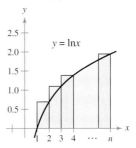

(b) Draw a graph similar to the one above that shows

$$\ln(n!) < \int_1^{n+1} \ln x \, dx.$$

(c) Use the results of parts (a) and (b) to show that

$$\frac{n^n}{e^{n-1}} < n! < \frac{(n+1)^{n+1}}{e^n}, \text{ for } n > 1.$$

(d) Use the Squeeze Theorem for Sequences and the result of part (c) to show that

$$\lim_{n \to \infty} \frac{\sqrt[n]{n!}}{n} = \frac{1}{e}.$$

(e) Test the result of part (d) for $n = 20, 50$, and 100.

130. Consider the sequence $\{a_n\} = \left\{\dfrac{1}{n} \sum_{k=1}^{n} \dfrac{1}{1 + (k/n)}\right\}$.

(a) Write the first five terms of $\{a_n\}$.

(b) Show that $\lim_{n \to \infty} a_n = \ln 2$ by interpreting a_n as a Riemann sum of a definite integral.

131. Prove, using the definition of the limit of a sequence, that

$$\lim_{n \to \infty} \frac{1}{n^3} = 0.$$

132. Prove, using the definition of the limit of a sequence, that

$$\lim_{n \to \infty} r^n = 0 \text{ for } -1 < r < 1.$$

133. Complete the proof of Theorem 9.5.

Putnam Exam Challenge

134. Let $\{x_n\}$, $n \geq 0$, be a sequence of nonzero real numbers such that $x_n^2 - x_{n-1}x_{n+1} = 1$ for $n = 1, 2, 3, \ldots$. Prove that there exists a real number a such that $x_{n+1} = ax_n - x_{n-1}$, for all $n \geq 1$.

135. Let $T_0 = 2$, $T_1 = 3$, $T_2 = 6$, and, for $n \geq 3$,

$$T_n = (n + 4)T_{n-1} - 4nT_{n-2} + (4n - 8)T_{n-3}.$$

The first 10 terms of the sequence are

2, 3, 6, 14, 40, 152, 784, 5168, 40,576, 363,392.

Find, with proof, a formula for T_n of the form $T_n = A_n + B_n$, where $\{A_n\}$ and $\{B_n\}$ are well-known sequences.

These problems were composed by the Committee on the Putnam Prize Competition.
© The Mathematical Association of America. All rights reserved.

Section 9.2 Series and Convergence

- Understand the definition of a convergent infinite series.
- Use properties of infinite geometric series.
- Use the *n*th-Term Test for Divergence of an infinite series.

Infinite Series

One important application of infinite sequences is in representing "infinite summations." Informally, if $\{a_n\}$ is an infinite sequence, then

$$\sum_{n=1}^{\infty} a_n = a_1 + a_2 + a_3 + \cdots + a_n + \cdots \qquad \text{Infinite series}$$

is an **infinite series** (or simply a **series**). The numbers $a_1, a_2, a_3,$ are the **terms** of the series. For some series it is convenient to begin the index at $n = 0$ (or some other integer). As a typesetting convention, it is common to represent an infinite series as simply $\Sigma\, a_n$. In such cases, the starting value for the index must be taken from the context of the statement.

To find the sum of an infinite series, consider the following **sequence of partial sums.**

$$S_1 = a_1$$
$$S_2 = a_1 + a_2$$
$$S_3 = a_1 + a_2 + a_3$$
$$\vdots$$
$$S_n = a_1 + a_2 + a_3 + \cdots + a_n$$

If this sequence of partial sums converges, the series is said to converge and has the sum indicated in the following definition.

Definitions of Convergent and Divergent Series

For the infinite series $\sum_{n=1}^{\infty} a_n$, the **nth partial sum** is given by

$$S_n = a_1 + a_2 + \cdots + a_n.$$

If the sequence of partial sums $\{S_n\}$ converges to S, then the series $\sum_{n=1}^{\infty} a_n$ **converges.** The limit S is called the **sum of the series.**

$$S = a_1 + a_2 + \cdots + a_n + \cdots$$

If $\{S_n\}$ diverges, then the series **diverges.**

INFINITE SERIES

The study of infinite series was considered a novelty in the fourteenth century. Logician Richard Suiseth, whose nickname was Calculator, solved this problem.

If throughout the first half of a given time interval a variation continues at a certain intensity, throughout the next quarter of the interval at double the intensity, throughout the following eighth at triple the intensity and so ad infinitum; then the average intensity for the whole interval will be the intensity of the variation during the second subinterval (or double the intensity).

This is the same as saying that the sum of the infinite series

$$\frac{1}{2} + \frac{2}{4} + \frac{3}{8} + \cdots + \frac{n}{2^n} + \cdots$$

is 2.

STUDY TIP As you study this chapter, you will see that there are two basic questions involving infinite series. Does a series converge or does it diverge? If a series converges, what is its sum? These questions are not always easy to answer, especially the second one.

EXPLORATION

Finding the Sum of an Infinite Series Find the sum of each infinite series. Explain your reasoning.

a. $0.1 + 0.01 + 0.001 + 0.0001 + \cdots$

b. $\frac{3}{10} + \frac{3}{100} + \frac{3}{1000} + \frac{3}{10,000} + \cdots$

c. $1 + \frac{1}{2} + \frac{1}{4} + \frac{1}{8} + \frac{1}{16} + \cdots$

d. $\frac{15}{100} + \frac{15}{10,000} + \frac{15}{1,000,000} + \cdots$

TECHNOLOGY Figure 9.5 shows the first 15 partial sums of the infinite series in Example 1(a). Notice how the values appear to approach the line $y = 1$.

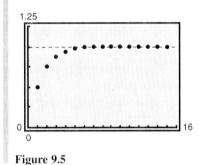

Figure 9.5

NOTE You can geometrically determine the partial sums of the series in Example 1(a) using Figure 9.6.

Figure 9.6

FOR FURTHER INFORMATION To learn more about the partial sums of infinite series, see the article "Six Ways to Sum a Series" by Dan Kalman in *The College Mathematics Journal*. To view this article, go to the website www.matharticles.com.

EXAMPLE 1 Convergent and Divergent Series

a. The series

$$\sum_{n=1}^{\infty} \frac{1}{2^n} = \frac{1}{2} + \frac{1}{4} + \frac{1}{8} + \frac{1}{16} + \cdots$$

has the following partial sums.

$$S_1 = \frac{1}{2}$$

$$S_2 = \frac{1}{2} + \frac{1}{4} = \frac{3}{4}$$

$$S_3 = \frac{1}{2} + \frac{1}{4} + \frac{1}{8} = \frac{7}{8}$$

$$\vdots$$

$$S_n = \frac{1}{2} + \frac{1}{4} + \frac{1}{8} + \cdots + \frac{1}{2^n} = \frac{2^n - 1}{2^n}$$

Because

$$\lim_{n \to \infty} \frac{2^n - 1}{2^n} = 1$$

it follows that the series converges and its sum is 1.

b. The nth partial sum of the series

$$\sum_{n=1}^{\infty} \left(\frac{1}{n} - \frac{1}{n+1} \right) = \left(1 - \frac{1}{2} \right) + \left(\frac{1}{2} - \frac{1}{3} \right) + \left(\frac{1}{3} - \frac{1}{4} \right) + \cdots$$

is given by

$$S_n = 1 - \frac{1}{n+1}.$$

Because the limit of S_n is 1, the series converges and its sum is 1.

c. The series

$$\sum_{n=1}^{\infty} 1 = 1 + 1 + 1 + 1 + \cdots$$

diverges because $S_n = n$ and the sequence of partial sums diverges.

The series in Example 1(b) is a **telescoping series** of the form

$$(b_1 - b_2) + (b_2 - b_3) + (b_3 - b_4) + (b_4 - b_5) + \cdots \quad \text{Telescoping series}$$

Note that b_2 is canceled by the second term, b_3 is canceled by the third term, and so on. Because the nth partial sum of this series is

$$S_n = b_1 - b_{n+1}$$

it follows that a telescoping series will converge if and only if b_n approaches a finite number as $n \to \infty$. Moreover, if the series converges, its sum is

$$S = b_1 - \lim_{n \to \infty} b_{n+1}.$$

EXAMPLE 2 Writing a Series in Telescoping Form

Find the sum of the series $\sum_{n=1}^{\infty} \dfrac{2}{4n^2 - 1}$.

Solution
Using partial fractions, you can write

$$a_n = \dfrac{2}{4n^2 - 1} = \dfrac{2}{(2n - 1)(2n + 1)} = \dfrac{1}{2n - 1} - \dfrac{1}{2n + 1}.$$

From this telescoping form, you can see that the nth partial sum is

$$S_n = \left(\dfrac{1}{1} - \dfrac{1}{3}\right) + \left(\dfrac{1}{3} - \dfrac{1}{5}\right) + \cdots + \left(\dfrac{1}{2n - 1} - \dfrac{1}{2n + 1}\right) = 1 - \dfrac{1}{2n + 1}.$$

So, the series converges and its sum is 1. That is,

$$\sum_{n=1}^{\infty} \dfrac{2}{4n^2 - 1} = \lim_{n \to \infty} S_n = \lim_{n \to \infty} \left(1 - \dfrac{1}{2n + 1}\right) = 1.$$

Geometric Series

The series given in Example 1(a) is a **geometric series.** In general, the series given by

$$\sum_{n=0}^{\infty} ar^n = a + ar + ar^2 + \cdots + ar^n + \cdots, \quad a \neq 0 \qquad \text{Geometric series}$$

is a **geometric series** with ratio r.

THEOREM 9.6 Convergence of a Geometric Series

A geometric series with ratio r diverges if $|r| \geq 1$. If $0 < |r| < 1$, then the series converges to the sum

$$\sum_{n=0}^{\infty} ar^n = \dfrac{a}{1 - r}, \quad 0 < |r| < 1.$$

Proof It is easy to see that the series diverges if $r = \pm 1$. If $r \neq \pm 1$, then $S_n = a + ar + ar^2 + \cdots + ar^{n-1}$. Multiplication by r yields

$$rS_n = ar + ar^2 + ar^3 + \cdots + ar^n.$$

Subtracting the second equation from the first produces $S_n - rS_n = a - ar^n$. Therefore, $S_n(1 - r) = a(1 - r^n)$, and the nth partial sum is

$$S_n = \dfrac{a}{1 - r}(1 - r^n).$$

If $0 < |r| < 1$, it follows that $r^n \to 0$ as $n \to \infty$, and you obtain

$$\lim_{n \to \infty} S_n = \lim_{n \to \infty} \left[\dfrac{a}{1 - r}(1 - r^n)\right] = \dfrac{a}{1 - r}\left[\lim_{n \to \infty}(1 - r^n)\right] = \dfrac{a}{1 - r}$$

which means that the series *converges* and its sum is $a/(1 - r)$. It is left to you to show that the series diverges if $|r| > 1$.

EXPLORATION

In "Proof Without Words," by Benjamin G. Klein and Irl C. Bivens, the authors present the following diagram. Explain why the final statement below the diagram is valid. How is this result related to Theorem 9.6?

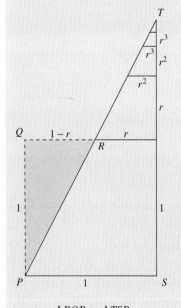

$\triangle PQR \approx \triangle TSP$

$1 + r + r^2 + r^3 + \cdots = \dfrac{1}{1 - r}$

Exercise taken from "Proof Without Words" by Benjamin G. Klein and Irl C. Bivens, *Mathematics Magazine*, October 1988, by permission of the authors.

TECHNOLOGY Try using a graphing utility or writing a computer program to compute the sum of the first 20 terms of the sequence in Example 3(a). You should obtain a sum of about 5.999994.

EXAMPLE 3 Convergent and Divergent Geometric Series

a. The geometric series

$$\sum_{n=0}^{\infty} \frac{3}{2^n} = \sum_{n=0}^{\infty} 3\left(\frac{1}{2}\right)^n$$

$$= 3(1) + 3\left(\frac{1}{2}\right) + 3\left(\frac{1}{2}\right)^2 + \cdots$$

has a ratio of $r = \frac{1}{2}$ with $a = 3$. Because $0 < |r| < 1$, the series converges and its sum is

$$S = \frac{a}{1-r} = \frac{3}{1-(1/2)} = 6.$$

b. The geometric series

$$\sum_{n=0}^{\infty} \left(\frac{3}{2}\right)^n = 1 + \frac{3}{2} + \frac{9}{4} + \frac{27}{8} + \cdots$$

has a ratio of $r = \frac{3}{2}$. Because $|r| \geq 1$, the series diverges.

The formula for the sum of a geometric series can be used to write a repeating decimal as the ratio of two integers, as demonstrated in the next example.

EXAMPLE 4 A Geometric Series for a Repeating Decimal

Use a geometric series to write $0.\overline{08}$ as the ratio of two integers.

Solution For the repeating decimal $0.\overline{08}$, you can write

$$0.080808\ldots = \frac{8}{10^2} + \frac{8}{10^4} + \frac{8}{10^6} + \frac{8}{10^8} + \cdots$$

$$= \sum_{n=0}^{\infty} \left(\frac{8}{10^2}\right)\left(\frac{1}{10^2}\right)^n.$$

For this series, you have $a = 8/10^2$ and $r = 1/10^2$. So,

$$0.080808\ldots = \frac{a}{1-r} = \frac{8/10^2}{1-(1/10^2)} = \frac{8}{99}.$$

Try dividing 8 by 99 on a calculator to see that it produces $0.\overline{08}$.

The convergence of a series is not affected by removal of a finite number of terms from the beginning of the series. For instance, the geometric series

$$\sum_{n=4}^{\infty} \left(\frac{1}{2}\right)^n \quad \text{and} \quad \sum_{n=0}^{\infty} \left(\frac{1}{2}\right)^n$$

both converge. Furthermore, because the sum of the second series is $a/(1-r) = 2$, you can conclude that the sum of the first series is

$$S = 2 - \left[\left(\frac{1}{2}\right)^0 + \left(\frac{1}{2}\right)^1 + \left(\frac{1}{2}\right)^2 + \left(\frac{1}{2}\right)^3\right]$$

$$= 2 - \frac{15}{8} = \frac{1}{8}.$$

STUDY TIP As you study this chapter, it is important to distinguish between an infinite series and a sequence. A sequence is an ordered collection of numbers

$$a_1, a_2, a_3, \ldots, a_n, \ldots$$

whereas a series is an infinite sum of terms from a sequence

$$a_1 + a_2 + \cdots + a_n + \cdots.$$

The following properties are direct consequences of the corresponding properties of limits of sequences.

THEOREM 9.7   Properties of Infinite Series

If $\sum a_n = A$, $\sum b_n = B$, and c is a real number, then the following series converge to the indicated sums.

1. $\displaystyle\sum_{n=1}^{\infty} ca_n = cA$

2. $\displaystyle\sum_{n=1}^{\infty} (a_n + b_n) = A + B$

3. $\displaystyle\sum_{n=1}^{\infty} (a_n - b_n) = A - B$

---

nth-Term Test for Divergence

The following theorem states that if a series converges, the limit of its nth term must be 0.

THEOREM 9.8   Limit of nth Term of a Convergent Series

If $\displaystyle\sum_{n=1}^{\infty} a_n$ converges, then $\displaystyle\lim_{n\to\infty} a_n = 0$.

NOTE Be sure you see that the converse of Theorem 9.8 is generally not true. That is, if the sequence $\{a_n\}$ converges to 0, then the series $\sum a_n$ may either converge or diverge.

Proof   Assume that

$$\sum_{n=1}^{\infty} a_n = \lim_{n\to\infty} S_n = L.$$

Then, because $S_n = S_{n-1} + a_n$ and

$$\lim_{n\to\infty} S_n = \lim_{n\to\infty} S_{n-1} = L$$

it follows that

$$L = \lim_{n\to\infty} S_n = \lim_{n\to\infty} (S_{n-1} + a_n)$$
$$= \lim_{n\to\infty} S_{n-1} + \lim_{n\to\infty} a_n$$
$$= L + \lim_{n\to\infty} a_n$$

which implies that $\{a_n\}$ converges to 0.

The contrapositive of Theorem 9.8 provides a useful test for *divergence*. This **nth-Term Test for Divergence** states that if the limit of the nth term of a series does *not* converge to 0, the series must diverge.

THEOREM 9.9   nth-Term Test for Divergence

If $\displaystyle\lim_{n\to\infty} a_n \neq 0$, then $\displaystyle\sum_{n=1}^{\infty} a_n$ diverges.

EXAMPLE 5 Using the *n*th-Term Test for Divergence

a. For the series $\sum_{n=0}^{\infty} 2^n$, you have

$$\lim_{n \to \infty} 2^n = \infty.$$

So, the limit of the *n*th term is not 0, and the series diverges.

b. For the series $\sum_{n=1}^{\infty} \frac{n!}{2n! + 1}$, you have

$$\lim_{n \to \infty} \frac{n!}{2n! + 1} = \frac{1}{2}.$$

So, the limit of the *n*th term is not 0, and the series diverges.

c. For the series $\sum_{n=1}^{\infty} \frac{1}{n}$, you have

$$\lim_{n \to \infty} \frac{1}{n} = 0.$$

Because the limit of the *n*th term is 0, the *n*th-Term Test for Divergence does *not* apply and you can draw no conclusions about convergence or divergence. (In the next section, you will see that this particular series diverges.)

STUDY TIP The series in Example 5(c) will play an important role in this chapter.

$$\sum_{n=1}^{\infty} \frac{1}{n} = 1 + \frac{1}{2} + \frac{1}{3} + \frac{1}{4} + \cdots$$

You will see that this series diverges even though the *n*th term approaches 0 as *n* approaches ∞.

EXAMPLE 6 Bouncing Ball Problem

A ball is dropped from a height of 6 feet and begins bouncing, as shown in Figure 9.7. The height of each bounce is three-fourths the height of the previous bounce. Find the total vertical distance traveled by the ball.

Solution When the ball hits the ground for the first time, it has traveled a distance of $D_1 = 6$ feet. For subsequent bounces, let D_i be the distance traveled up and down. For example, D_2 and D_3 are as follows.

$$D_2 = \underbrace{6\left(\tfrac{3}{4}\right)}_{\text{Up}} + \underbrace{6\left(\tfrac{3}{4}\right)}_{\text{Down}} = 12\left(\tfrac{3}{4}\right)$$

$$D_3 = \underbrace{6\left(\tfrac{3}{4}\right)\left(\tfrac{3}{4}\right)}_{\text{Up}} + \underbrace{6\left(\tfrac{3}{4}\right)\left(\tfrac{3}{4}\right)}_{\text{Down}} = 12\left(\tfrac{3}{4}\right)^2$$

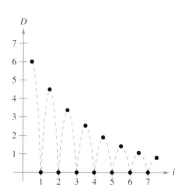

The height of each bounce is three-fourths the height of the preceding bounce.
Figure 9.7

By continuing this process, it can be determined that the total vertical distance is

$$D = 6 + 12\left(\tfrac{3}{4}\right) + 12\left(\tfrac{3}{4}\right)^2 + 12\left(\tfrac{3}{4}\right)^3 + \cdots$$

$$= 6 + 12 \sum_{n=0}^{\infty} \left(\tfrac{3}{4}\right)^{n+1}$$

$$= 6 + 12\left(\tfrac{3}{4}\right) \sum_{n=0}^{\infty} \left(\tfrac{3}{4}\right)^n$$

$$= 6 + 9\left(\frac{1}{1 - \tfrac{3}{4}}\right)$$

$$= 6 + 9(4)$$

$$= 42 \text{ feet.}$$

Exercises for Section 9.2

See www.CalcChat.com for worked-out solutions to odd-numbered exercises.

In Exercises 1–6, find the first five terms of the sequence of partial sums.

1. $1 + \frac{1}{4} + \frac{1}{9} + \frac{1}{16} + \frac{1}{25} + \cdots$
2. $\frac{1}{2 \cdot 3} + \frac{2}{3 \cdot 4} + \frac{3}{4 \cdot 5} + \frac{4}{5 \cdot 6} + \frac{5}{6 \cdot 7} + \cdots$
3. $3 - \frac{9}{2} + \frac{27}{4} - \frac{81}{8} + \frac{243}{16} - \cdots$
4. $\frac{1}{1} + \frac{1}{3} + \frac{1}{5} + \frac{1}{7} + \frac{1}{9} + \frac{1}{11} + \cdots$
5. $\sum_{n=1}^{\infty} \frac{3}{2^{n-1}}$
6. $\sum_{n=1}^{\infty} \frac{(-1)^{n+1}}{n!}$

In Exercises 7–16, verify that the infinite series diverges.

7. $\sum_{n=0}^{\infty} 3\left(\frac{3}{2}\right)^n$
8. $\sum_{n=0}^{\infty} \left(\frac{4}{3}\right)^n$
9. $\sum_{n=0}^{\infty} 1000(1.055)^n$
10. $\sum_{n=0}^{\infty} 2(-1.03)^n$
11. $\sum_{n=1}^{\infty} \frac{n}{n+1}$
12. $\sum_{n=1}^{\infty} \frac{n}{2n+3}$
13. $\sum_{n=1}^{\infty} \frac{n^2}{n^2+1}$
14. $\sum_{n=1}^{\infty} \frac{n}{\sqrt{n^2+1}}$
15. $\sum_{n=1}^{\infty} \frac{2^n + 1}{2^{n+1}}$
16. $\sum_{n=1}^{\infty} \frac{n!}{2^n}$

In Exercises 17–22, match the series with the graph of its sequence of partial sums. [The graphs are labeled (a), (b), (c), (d), (e), and (f).] Use the graph to estimate the sum of the series. Confirm your answer analytically.

(a)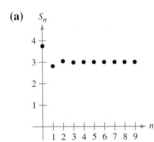
(b)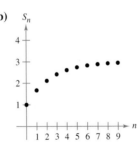
(c)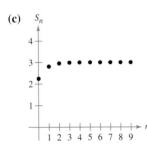
(d)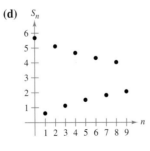
(e)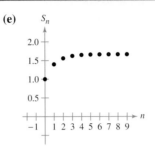
(f)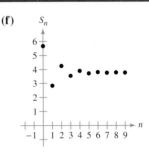

17. $\sum_{n=0}^{\infty} \frac{9}{4}\left(\frac{1}{4}\right)^n$
18. $\sum_{n=0}^{\infty} \left(\frac{2}{3}\right)^n$
19. $\sum_{n=0}^{\infty} \frac{15}{4}\left(-\frac{1}{4}\right)^n$
20. $\sum_{n=0}^{\infty} \frac{17}{3}\left(-\frac{8}{9}\right)^n$
21. $\sum_{n=0}^{\infty} \frac{17}{3}\left(-\frac{1}{2}\right)^n$
22. $\sum_{n=0}^{\infty} \left(\frac{2}{5}\right)^n$

In Exercises 23–28, verify that the infinite series converges.

23. $\sum_{n=1}^{\infty} \frac{1}{n(n+1)}$ (Use partial fractions.)
24. $\sum_{n=1}^{\infty} \frac{1}{n(n+2)}$ (Use partial fractions.)
25. $\sum_{n=0}^{\infty} 2\left(\frac{3}{4}\right)^n$
26. $\sum_{n=1}^{\infty} 2\left(-\frac{1}{2}\right)^n$
27. $\sum_{n=0}^{\infty} (0.9)^n = 1 + 0.9 + 0.81 + 0.729 + \cdots$
28. $\sum_{n=0}^{\infty} (-0.6)^n = 1 - 0.6 + 0.36 - 0.216 + \cdots$

Numerical, Graphical, and Analytic Analysis In Exercises 29–34, (a) find the sum of the series, (b) use a graphing utility to find the indicated partial sum S_n and complete the table, (c) use a graphing utility to graph the first 10 terms of the sequence of partial sums and a horizontal line representing the sum, and (d) explain the relationship between the magnitudes of the terms of the series and the rate at which the sequence of partial sums approaches the sum of the series.

n	5	10	20	50	100
S_n					

29. $\sum_{n=1}^{\infty} \frac{6}{n(n+3)}$
30. $\sum_{n=1}^{\infty} \frac{4}{n(n+4)}$
31. $\sum_{n=1}^{\infty} 2(0.9)^{n-1}$
32. $\sum_{n=1}^{\infty} 3(0.85)^{n-1}$
33. $\sum_{n=1}^{\infty} 10(0.25)^{n-1}$
34. $\sum_{n=1}^{\infty} 5\left(-\frac{1}{3}\right)^{n-1}$

In Exercises 35–50, find the sum of the convergent series.

35. $\sum_{n=2}^{\infty} \frac{1}{n^2 - 1}$
36. $\sum_{n=1}^{\infty} \frac{4}{n(n+2)}$

37. $\sum_{n=1}^{\infty} \dfrac{8}{(n+1)(n+2)}$
38. $\sum_{n=1}^{\infty} \dfrac{1}{(2n+1)(2n+3)}$

39. $\sum_{n=0}^{\infty} \left(\dfrac{1}{2}\right)^n$
40. $\sum_{n=0}^{\infty} 6\left(\dfrac{4}{5}\right)^n$

41. $\sum_{n=0}^{\infty} \left(-\dfrac{1}{2}\right)^n$
42. $\sum_{n=0}^{\infty} 2\left(-\dfrac{2}{3}\right)^n$

43. $1 + 0.1 + 0.01 + 0.001 + \cdots$

44. $8 + 6 + \dfrac{9}{2} + \dfrac{27}{8} + \cdots$

45. $3 - 1 + \dfrac{1}{3} - \dfrac{1}{9} + \cdots$

46. $4 - 2 + 1 - \dfrac{1}{2} + \cdots$

47. $\sum_{n=0}^{\infty} \left(\dfrac{1}{2^n} - \dfrac{1}{3^n}\right)$
48. $\sum_{n=1}^{\infty} [(0.7)^n + (0.9)^n]$

49. $\sum_{n=1}^{\infty} (\sin 1)^n$
50. $\sum_{n=1}^{\infty} \dfrac{1}{9n^2 + 3n - 2}$

In Exercises 51–56, (a) write the repeating decimal as a geometric series and (b) write its sum as the ratio of two integers.

51. $0.\overline{4}$
52. $0.\overline{9}$
53. $0.\overline{81}$
54. $0.\overline{01}$
55. $0.0\overline{75}$
56. $0.2\overline{15}$

In Exercises 57–72, determine the convergence or divergence of the series.

57. $\sum_{n=1}^{\infty} \dfrac{n+10}{10n+1}$
58. $\sum_{n=1}^{\infty} \dfrac{n+1}{2n-1}$

59. $\sum_{n=1}^{\infty} \left(\dfrac{1}{n} - \dfrac{1}{n+2}\right)$
60. $\sum_{n=1}^{\infty} \dfrac{1}{n(n+3)}$

61. $\sum_{n=1}^{\infty} \dfrac{3n-1}{2n+1}$
62. $\sum_{n=1}^{\infty} \dfrac{3^n}{n^3}$

63. $\sum_{n=0}^{\infty} \dfrac{4}{2^n}$
64. $\sum_{n=0}^{\infty} \dfrac{1}{4^n}$

65. $\sum_{n=0}^{\infty} (1.075)^n$
66. $\sum_{n=1}^{\infty} \dfrac{2^n}{100}$

67. $\sum_{n=2}^{\infty} \dfrac{n}{\ln n}$
68. $\sum_{n=1}^{\infty} \ln \dfrac{1}{n}$

69. $\sum_{n=1}^{\infty} \left(1 + \dfrac{k}{n}\right)^n$
70. $\sum_{n=1}^{\infty} e^{-n}$

71. $\sum_{n=1}^{\infty} \arctan n$
72. $\sum_{n=1}^{\infty} \ln\left(\dfrac{n+1}{n}\right)$

Writing About Concepts

73. State the definitions of convergent and divergent series.

74. Describe the difference between $\lim_{n \to \infty} a_n = 5$ and $\sum_{n=1}^{\infty} a_n = 5$.

75. Define a geometric series, state when it converges, and give the formula for the sum of a convergent geometric series.

Writing About Concepts (continued)

76. State the nth-Term Test for Divergence.

77. Let $a_n = \dfrac{n+1}{n}$. Discuss the convergence of $\{a_n\}$ and $\sum_{n=1}^{\infty} a_n$.

78. Explain any differences among the following series.

(a) $\sum_{n=1}^{\infty} a_n$ (b) $\sum_{k=1}^{\infty} a_k$ (c) $\sum_{n=1}^{\infty} a_k$

In Exercises 79–86, find all values of x for which the series converges. For these values of x, write the sum of the series as a function of x.

79. $\sum_{n=1}^{\infty} \dfrac{x^n}{2^n}$
80. $\sum_{n=1}^{\infty} (3x)^n$

81. $\sum_{n=1}^{\infty} (x-1)^n$
82. $\sum_{n=0}^{\infty} 4\left(\dfrac{x-3}{4}\right)^n$

83. $\sum_{n=0}^{\infty} (-1)^n x^n$
84. $\sum_{n=0}^{\infty} (-1)^n x^{2n}$

85. $\sum_{n=0}^{\infty} \left(\dfrac{1}{x}\right)^n$
86. $\sum_{n=1}^{\infty} \left(\dfrac{x^2}{x^2+4}\right)^n$

87. (a) You delete a finite number of terms from a divergent series. Will the new series still diverge? Explain your reasoning.

(b) You add a finite number of terms to a convergent series. Will the new series still converge? Explain your reasoning.

88. *Think About It* Consider the formula

$$\dfrac{1}{x-1} = 1 + x + x^2 + x^3 + \cdots.$$

Given $x = -1$ and $x = 2$, can you conclude that either of the following statements is true? Explain your reasoning.

(a) $\dfrac{1}{2} = 1 - 1 + 1 - 1 + \cdots$

(b) $-1 = 1 + 2 + 4 + 8 + \cdots$

In Exercises 89 and 90, (a) find the common ratio of the geometric series, (b) write the function that gives the sum of the series, and (c) use a graphing utility to graph the function and the partial sums S_3 and S_5. What do you notice?

89. $1 + x + x^2 + x^3 + \cdots$
90. $1 - \dfrac{x}{2} + \dfrac{x^2}{4} - \dfrac{x^3}{8} + \cdots$

In Exercises 91 and 92, use a graphing utility to graph the function. Identify the horizontal asymptote of the graph and determine its relationship to the sum of the series.

Function	Series
91. $f(x) = 3\left[\dfrac{1-(0.5)^x}{1-0.5}\right]$	$\sum_{n=0}^{\infty} 3\left(\dfrac{1}{2}\right)^n$
92. $f(x) = 2\left[\dfrac{1-(0.8)^x}{1-0.8}\right]$	$\sum_{n=0}^{\infty} 2\left(\dfrac{4}{5}\right)^n$

Writing In Exercises 93 and 94, use a graphing utility to determine the first term that is less than 0.0001 in each of the convergent series. Note that the answers are very different. Explain how this will affect the rate at which the series converges.

93. $\sum_{n=1}^{\infty} \frac{1}{n(n+1)}$, $\sum_{n=1}^{\infty} \left(\frac{1}{8}\right)^n$ 94. $\sum_{n=1}^{\infty} \frac{1}{2^n}$, $\sum_{n=1}^{\infty} (0.01)^n$

95. **Marketing** An electronic games manufacturer producing a new product estimates the annual sales to be 8000 units. Each year 10% of the units that have been sold will become inoperative. So, 8000 units will be in use after 1 year, [8000 + 0.9(8000)] units will be in use after 2 years, and so on. How many units will be in use after n years?

96. **Depreciation** A company buys a machine for $225,000 that depreciates at a rate of 30% per year. Find a formula for the value of the machine after n years. What is its value after 5 years?

97. **Multiplier Effect** The annual spending by tourists in a resort city is $100 million. Approximately 75% of that revenue is again spent in the resort city, and of that amount approximately 75% is again spent in the same city, and so on. Write the geometric series that gives the total amount of spending generated by the $100 million and find the sum of the series.

98. **Multiplier Effect** Repeat Exercise 97 if the percent of the revenue that is spent again in the city decreases to 60%.

99. **Distance** A ball is dropped from a height of 16 feet. Each time it drops h feet, it rebounds $0.81h$ feet. Find the total distance traveled by the ball.

100. **Time** The ball in Exercise 99 takes the following times for each fall.

$s_1 = -16t^2 + 16$, $s_1 = 0$ if $t = 1$

$s_2 = -16t^2 + 16(0.81)$, $s_2 = 0$ if $t = 0.9$

$s_3 = -16t^2 + 16(0.81)^2$, $s_3 = 0$ if $t = (0.9)^2$

$s_4 = -16t^2 + 16(0.81)^3$, $s_4 = 0$ if $t = (0.9)^3$

$\vdots$

$s_n = -16t^2 + 16(0.81)^{n-1}$, $s_n = 0$ if $t = (0.9)^{n-1}$

Beginning with s_2, the ball takes the same amount of time to bounce up as it does to fall, and so the total time elapsed before it comes to rest is given by

$$t = 1 + 2\sum_{n=1}^{\infty} (0.9)^n.$$

Find this total time.

Probability In Exercises 101 and 102, the random variable n represents the number of units of a product sold per day in a store. The probability distribution of n is given by $P(n)$. Find the probability that two units are sold in a given day $[P(2)]$ and show that $P(1) + P(2) + P(3) + \cdots = 1$.

101. $P(n) = \frac{1}{2}\left(\frac{1}{2}\right)^n$ 102. $P(n) = \frac{1}{3}\left(\frac{2}{3}\right)^n$

103. **Probability** A fair coin is tossed repeatedly. The probability that the first head occurs on the nth toss is given by $P(n) = \left(\frac{1}{2}\right)^n$, where $n \geq 1$.

(a) Show that $\sum_{n=1}^{\infty} \left(\frac{1}{2}\right)^n = 1$.

(b) The expected number of tosses required until the first head occurs in the experiment is given by

$$\sum_{n=1}^{\infty} n\left(\frac{1}{2}\right)^n.$$

Is this series geometric?

(c) Use a computer algebra system to find the sum in part (b).

104. **Probability** In an experiment, three people toss a fair coin one at a time until one of them tosses a head. Determine, for each person, the probability that he or she tosses the first head. Verify that the sum of the three probabilities is 1.

105. **Area** The sides of a square are 16 inches in length. A new square is formed by connecting the midpoints of the sides of the original square, and two of the triangles outside the second square are shaded (see figure). Determine the area of the shaded regions (a) if this process is continued five more times and (b) if this pattern of shading is continued infinitely.

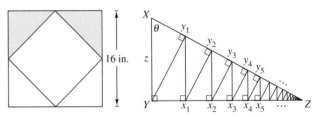

Figure for 105   Figure for 106

106. **Length** A right triangle XYZ is shown above where $|XY| = z$ and $\angle X = \theta$. Line segments are continually drawn to be perpendicular to the triangle, as shown in the figure.

(a) Find the total length of the perpendicular line segments $|Yy_1| + |x_1y_1| + |x_1y_2| + \cdots$ in terms of z and θ.

(b) If $z = 1$ and $\theta = \pi/6$, find the total length of the perpendicular line segments.

In Exercises 107–110, use the formula for the nth partial sum of a geometric series

$$\sum_{i=0}^{n-1} ar^i = \frac{a(1 - r^n)}{1 - r}.$$

107. **Present Value** The winner of a $1,000,000 sweepstakes will be paid $50,000 per year for 20 years. The money earns 6% interest per year. The present value of the winnings is

$$\sum_{n=1}^{20} 50,000\left(\frac{1}{1.06}\right)^n.$$

Compute the present value and interpret its meaning.

108. Sphereflake A sphereflake shown below is a computer-generated fractal that was created by Eric Haines. The radius of the large sphere is 1. To the large sphere, nine spheres of radius $\frac{1}{3}$ are attached. To each of these, nine spheres of radius $\frac{1}{9}$ are attached. This process is continued infinitely. Prove that the sphereflake has an infinite surface area.

Eric Haines

109. Salary You go to work at a company that pays $0.01 for the first day, $0.02 for the second day, $0.04 for the third day, and so on. If the daily wage keeps doubling, what would your total income be for working (a) 29 days, (b) 30 days, and (c) 31 days?

110. Annuities When an employee receives a paycheck at the end of each month, P dollars is invested in a retirement account. These deposits are made each month for t years and the account earns interest at the annual percentage rate r. If the interest is compounded monthly, the amount A in the account at the end of t years is

$$A = P + P\left(1 + \frac{r}{12}\right) + \cdots + P\left(1 + \frac{r}{12}\right)^{12t-1}$$

$$= P\left(\frac{12}{r}\right)\left[\left(1 + \frac{r}{12}\right)^{12t} - 1\right].$$

If the interest is compounded continuously, the amount A in the account after t years is

$$A = P + Pe^{r/12} + Pe^{2r/12} + Pe^{(12t-1)r/12}$$

$$= \frac{P(e^{rt} - 1)}{e^{r/12} - 1}.$$

Verify the formulas for the sums given above.

Annuities In Exercises 111–114, consider making monthly deposits of P dollars in a savings account at an annual interest rate r. Use the results of Exercise 110 to find the balance A after t years if the interest is compounded (a) monthly and (b) continuously.

111. $P = \$50$, $r = 3\%$, $t = 20$ years
112. $P = \$75$, $r = 5\%$, $t = 25$ years
113. $P = \$100$, $r = 4\%$, $t = 40$ years
114. $P = \$20$, $r = 6\%$, $t = 50$ years

115. Modeling Data The annual sales a_n (in millions of dollars) for Avon Products, Inc. from 1993 through 2002 are given below as ordered pairs of the form (n, a_n), where n represents the year, with $n = 3$ corresponding to 1993. (Source: 2002 Avon Products, Inc. Annual Report)

(3, 3844), (4, 4267), (5, 4492), (6, 4814), (7, 5079), (8, 5213), (9, 5289), (10, 5682), (11, 5958), (12, 6171)

(a) Use the regression capabilities of a graphing utility to find a model of the form

$$a_n = ce^{kn}, \quad n = 3, 4, 5, \ldots, 12$$

for the data. Graphically compare the points and the model.

(b) Use the data to find the total sales for the 10-year period.

(c) Approximate the total sales for the 10-year period using the formula for the sum of a geometric series. Compare the result with that in part (b).

116. Salary You accept a job that pays a salary of $40,000 for the first year. During the next 39 years you receive a 4% raise each year. What would be your total compensation over the 40-year period?

True or False? In Exercises 117–122, determine whether the statement is true or false. If it is false, explain why or give an example that shows it is false.

117. If $\lim_{n \to \infty} a_n = 0$, then $\sum_{n=1}^{\infty} a_n$ converges.

118. If $\sum_{n=1}^{\infty} a_n = L$, then $\sum_{n=0}^{\infty} a_n = L + a_0$.

119. If $|r| < 1$, then $\sum_{n=1}^{\infty} ar^n = \frac{a}{(1-r)}$.

120. The series $\sum_{n=1}^{\infty} \frac{n}{1000(n+1)}$ diverges.

121. $0.75 = 0.749999\ldots$.

122. Every decimal with a repeating pattern of digits is a rational number.

123. Show that the series $\sum_{n=1}^{\infty} a_n$ can be written in the telescoping form

$$\sum_{n=1}^{\infty} [(c - S_{n-1}) - (c - S_n)]$$

where $S_0 = 0$ and S_n is the nth partial sum.

124. Let Σa_n be a convergent series, and let

$$R_N = a_{N+1} + a_{N+2} + \cdots$$

be the remainder of the series after the first N terms. Prove that $\lim_{N \to \infty} R_N = 0$.

125. Find two divergent series Σa_n and Σb_n such that $\Sigma(a_n + b_n)$ converges.

126. Given two infinite series Σa_n and Σb_n such that Σa_n converges and Σb_n diverges, prove that $\Sigma(a_n + b_n)$ diverges.

127. Suppose that Σa_n diverges and c is a nonzero constant. Prove that Σca_n diverges.

128. If $\sum_{n=1}^{\infty} a_n$ converges where a_n is nonzero, show that $\sum_{n=1}^{\infty} \frac{1}{a_n}$ diverges.

129. The Fibonacci sequence is defined recursively by $a_{n+2} = a_n + a_{n+1}$, where $a_1 = 1$ and $a_2 = 1$.

(a) Show that $\dfrac{1}{a_{n+1} a_{n+3}} = \dfrac{1}{a_{n+1} a_{n+2}} - \dfrac{1}{a_{n+2} a_{n+3}}$.

(b) Show that $\sum_{n=0}^{\infty} \dfrac{1}{a_{n+1} a_{n+3}} = 1$.

130. Find the values of x for which the infinite series
$$1 + 2x + x^2 + 2x^3 + x^4 + 2x^5 + x^6 + \cdots$$
converges. What is the sum when the series converges?

131. Prove that $\dfrac{1}{r} + \dfrac{1}{r^2} + \dfrac{1}{r^3} + \cdots = \dfrac{1}{r-1}$, for $|r| > 1$.

132. *Writing* The figure below represents an informal way of showing that $\sum_{n=1}^{\infty} \dfrac{1}{n^2} < 2$. Explain how the figure implies this conclusion.

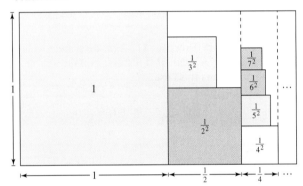

FOR FURTHER INFORMATION For more on this exercise, see the article "Convergence with Pictures" by P.J. Rippon in *American Mathematical Monthly*.

133. *Writing* Read the article "The Exponential-Decay Law Applied to Medical Dosages" by Gerald M. Armstrong and Calvin P. Midgley in *Mathematics Teacher*. (To view this article, go to the website *www.matharticles.com*.) Then write a paragraph on how a geometric sequence can be used to find the total amount of a drug that remains in a patient's system after n equal doses have been administered (at equal time intervals).

Putnam Exam Challenge

134. Write $\sum_{k=1}^{\infty} \dfrac{6^k}{(3^{k+1} - 2^{k+1})(3^k - 2^k)}$ as a rational number.

135. Let $f(n)$ be the sum of the first n terms of the sequence $0, 1, 1, 2, 2, 3, 3, 4, \ldots$, where the nth term is given by
$$a_n = \begin{cases} n/2, & \text{if } n \text{ is even} \\ (n-1)/2, & \text{if } n \text{ is odd} \end{cases}.$$
Show that if x and y are positive integers and $x > y$ then $xy = f(x + y) - f(x - y)$.

These problems were composed by the Committee on the Putnam Prize Competition.
© The Mathematical Association of America. All rights reserved.

Section Project: Cantor's Disappearing Table

The following procedure shows how to make a table disappear by removing only half of the table!

(a) Original table has a length of L.

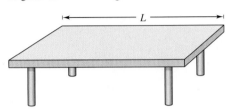

(b) Remove $\frac{1}{4}$ of the table centered at the midpoint. Each remaining piece has a length that is less than $\frac{1}{2}L$.

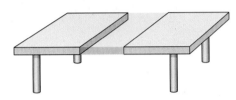

(c) Remove $\frac{1}{8}$ of the table by taking sections of length $\frac{1}{16}L$ from the centers of each of the two remaining pieces. Now, you have removed $\frac{1}{4} + \frac{1}{8}$ of the table. Each remaining piece has a length that is less than $\frac{1}{4}L$.

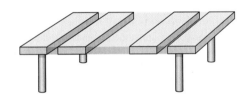

(d) Remove $\frac{1}{16}$ of the table by taking sections of length $\frac{1}{64}L$ from the centers of each of the four remaining pieces. Now, you have removed $\frac{1}{4} + \frac{1}{8} + \frac{1}{16}$ of the table. Each remaining piece has a length that is less than $\frac{1}{8}L$.

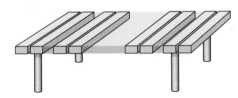

Will continuing this process cause the table to disappear, even though you have only removed half of the table? Why?

FOR FURTHER INFORMATION Read the article "Cantor's Disappearing Table" by Larry E. Knop in *The College Mathematics Journal*. To view this article, go to the website *www.matharticles.com*.

Section 9.3 The Integral Test and p-Series

- Use the Integral Test to determine whether an infinite series converges or diverges.
- Use properties of p-series and harmonic series.

The Integral Test

In this and the following section, you will study several convergence tests that apply to series with *positive* terms.

> **THEOREM 9.10 The Integral Test**
>
> If f is positive, continuous, and decreasing for $x \geq 1$ and $a_n = f(n)$, then
>
> $$\sum_{n=1}^{\infty} a_n \quad \text{and} \quad \int_1^{\infty} f(x)\, dx$$
>
> either both converge or both diverge.

Proof Begin by partitioning the interval $[1, n]$ into $n - 1$ unit intervals, as shown in Figure 9.8. The total areas of the inscribed rectangles and the circumscribed rectangles are as follows.

$$\sum_{i=2}^{n} f(i) = f(2) + f(3) + \cdots + f(n) \quad \text{Inscribed area}$$

$$\sum_{i=1}^{n-1} f(i) = f(1) + f(2) + \cdots + f(n-1) \quad \text{Circumscribed area}$$

The exact area under the graph of f from $x = 1$ to $x = n$ lies between the inscribed and circumscribed areas.

$$\sum_{i=2}^{n} f(i) \leq \int_1^n f(x)\, dx \leq \sum_{i=1}^{n-1} f(i)$$

Using the nth partial sum, $S_n = f(1) + f(2) + \cdots + f(n)$, you can write this inequality as

$$S_n - f(1) \leq \int_1^n f(x)\, dx \leq S_{n-1}.$$

Now, assuming that $\int_1^{\infty} f(x)\, dx$ converges to L, it follows that for $n \geq 1$

$$S_n - f(1) \leq L \quad \Longrightarrow \quad S_n \leq L + f(1).$$

Consequently, $\{S_n\}$ is bounded and monotonic, and by Theorem 9.5 it converges. So, Σa_n converges. For the other direction of the proof, assume that the improper integral diverges. Then $\int_1^n f(x)\, dx$ approaches infinity as $n \to \infty$, and the inequality $S_{n-1} \geq \int_1^n f(x)\, dx$ implies that $\{S_n\}$ diverges. So, Σa_n diverges.

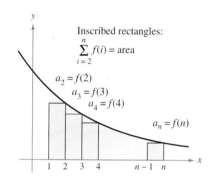
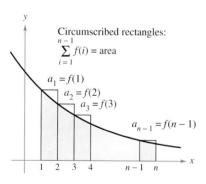

Figure 9.8

NOTE Remember that the convergence or divergence of Σa_n is not affected by deleting the first N terms. Similarly, if the conditions for the Integral Test are satisfied for all $x \geq N > 1$, you can simply use the integral $\int_N^{\infty} f(x)\, dx$ to test for convergence or divergence. (This is illustrated in Example 4.)

EXAMPLE 1 Using the Integral Test

Apply the Integral Test to the series $\sum_{n=1}^{\infty} \dfrac{n}{n^2 + 1}$.

Solution The function $f(x) = x/(x^2 + 1)$ is positive and continuous for $x \geq 1$. To determine whether f is decreasing, find the derivative.

$$f'(x) = \dfrac{(x^2 + 1)(1) - x(2x)}{(x^2 + 1)^2} = \dfrac{-x^2 + 1}{(x^2 + 1)^2}$$

So, $f'(x) < 0$ for $x > 1$ and it follows that f satisfies the conditions for the Integral Test. You can integrate to obtain

$$\int_1^{\infty} \dfrac{x}{x^2 + 1}\, dx = \dfrac{1}{2} \int_1^{\infty} \dfrac{2x}{x^2 + 1}\, dx$$

$$= \dfrac{1}{2} \lim_{b \to \infty} \int_1^b \dfrac{2x}{x^2 + 1}\, dx$$

$$= \dfrac{1}{2} \lim_{b \to \infty} \Big[\ln(x^2 + 1)\Big]_1^b$$

$$= \dfrac{1}{2} \lim_{b \to \infty} [\ln(b^2 + 1) - \ln 2]$$

$$= \infty.$$

So, the series *diverges*.

EXAMPLE 2 Using the Integral Test

Apply the Integral Test to the series $\sum_{n=1}^{\infty} \dfrac{1}{n^2 + 1}$.

Solution Because $f(x) = 1/(x^2 + 1)$ satisfies the conditions for the Integral Test (check this), you can integrate to obtain

$$\int_1^{\infty} \dfrac{1}{x^2 + 1}\, dx = \lim_{b \to \infty} \int_1^b \dfrac{1}{x^2 + 1}\, dx$$

$$= \lim_{b \to \infty} \Big[\arctan x\Big]_1^b$$

$$= \lim_{b \to \infty} (\arctan b - \arctan 1)$$

$$= \dfrac{\pi}{2} - \dfrac{\pi}{4} = \dfrac{\pi}{4}.$$

So, the series *converges* (see Figure 9.9).

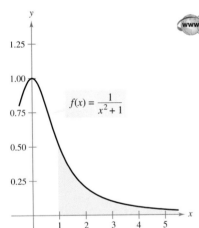

Because the improper integral converges, the infinite series also converges.
Figure 9.9

TECHNOLOGY
In Example 2, the fact that the improper integral converges to $\pi/4$ does not imply that the infinite series converges to $\pi/4$. To approximate the sum of the series, you can use the inequality

$$\sum_{n=1}^{N} \dfrac{1}{n^2 + 1} \leq \sum_{n=1}^{\infty} \dfrac{1}{n^2 + 1} \leq \sum_{n=1}^{N} \dfrac{1}{n^2 + 1} + \int_N^{\infty} \dfrac{1}{x^2 + 1}\, dx.$$

(See Exercise 60.) The larger the value of N, the better the approximation. For instance, using $N = 200$ produces $1.072 \leq \Sigma 1/(n^2 + 1) \leq 1.077$.

HARMONIC SERIES

Pythagoras and his students paid close attention to the development of music as an abstract science. This led to the discovery of the relationship between the tone and the length of the vibrating string. It was observed that the most beautiful musical harmonies corresponded to the simplest ratios of whole numbers. Later mathematicians developed this idea into the harmonic series, where the terms in the harmonic series correspond to the nodes on a vibrating string that produce multiples of the fundamental frequency. For example, $\frac{1}{2}$ is twice the fundamental frequency, $\frac{1}{3}$ is three times the fundamental frequency, and so on.

p-Series and Harmonic Series

In the remainder of this section, you will investigate a second type of series that has a simple arithmetic test for convergence or divergence. A series of the form

$$\sum_{n=1}^{\infty} \frac{1}{n^p} = \frac{1}{1^p} + \frac{1}{2^p} + \frac{1}{3^p} + \cdots \qquad \text{p-series}$$

is a **p-series,** where p is a positive constant. For $p = 1$, the series

$$\sum_{n=1}^{\infty} \frac{1}{n} = 1 + \frac{1}{2} + \frac{1}{3} + \cdots \qquad \text{Harmonic series}$$

is the **harmonic** series. A **general harmonic series** is of the form $\Sigma 1/(an + b)$. In music, strings of the same material, diameter, and tension, whose lengths form a harmonic series, produce harmonic tones.

The Integral Test is convenient for establishing the convergence or divergence of p-series. This is shown in the proof of Theorem 9.11.

THEOREM 9.11 Convergence of p-Series

The p-series

$$\sum_{n=1}^{\infty} \frac{1}{n^p} = \frac{1}{1^p} + \frac{1}{2^p} + \frac{1}{3^p} + \frac{1}{4^p} + \cdots$$

1. converges if $p > 1$, and
2. diverges if $0 < p \leq 1$.

Proof The proof follows from the Integral Test and from Theorem 8.5, which states that

$$\int_1^{\infty} \frac{1}{x^p} \, dx$$

converges if $p > 1$ and diverges if $0 < p \leq 1$.

EXAMPLE 3 Convergent and Divergent p-Series

Discuss the convergence or divergence of (a) the harmonic series and (b) the p-series with $p = 2$.

Solution

a. From Theorem 9.11, it follows that the harmonic series

$$\sum_{n=1}^{\infty} \frac{1}{n} = \frac{1}{1} + \frac{1}{2} + \frac{1}{3} + \cdots \qquad p = 1$$

diverges.

b. From Theorem 9.11, it follows that the p-series

$$\sum_{n=1}^{\infty} \frac{1}{n^2} = \frac{1}{1^2} + \frac{1}{2^2} + \frac{1}{3^2} + \cdots \qquad p = 2$$

converges.

NOTE The sum of the series in Example 3(b) can be shown to be $\pi^2/6$. (This was proved by Leonhard Euler, but the proof is too difficult to present here.) Be sure you see that the Integral Test does not tell you that the sum of the series is equal to the value of the integral. For instance, the sum of the series in Example 3(b) is

$$\sum_{n=1}^{\infty} \frac{1}{n^2} = \frac{\pi^2}{6} \approx 1.645$$

but the value of the corresponding improper integral is

$$\int_1^{\infty} \frac{1}{x^2} \, dx = 1.$$

EXAMPLE 4 Testing a Series for Convergence

Determine whether the following series converges or diverges.

$$\sum_{n=2}^{\infty} \frac{1}{n \ln n}$$

Solution This series is similar to the divergent harmonic series. If its terms were larger than those of the harmonic series, you would expect it to diverge. However, because its terms are smaller, you are not sure what to expect. The function $f(x) = 1/(x \ln x)$ is positive and continuous for $x \geq 2$. To determine whether f is decreasing, first rewrite f as $f(x) = (x \ln x)^{-1}$ and then find its derivative.

$$f'(x) = (-1)(x \ln x)^{-2}(1 + \ln x) = -\frac{1 + \ln x}{x^2 (\ln x)^2}$$

So, $f'(x) < 0$ for $x > 2$ and it follows that f satisfies the conditions for the Integral Test.

$$\int_2^{\infty} \frac{1}{x \ln x} dx = \int_2^{\infty} \frac{1/x}{\ln x} dx$$
$$= \lim_{b \to \infty} \left[\ln(\ln x) \right]_2^b$$
$$= \lim_{b \to \infty} \left[\ln(\ln b) - \ln(\ln 2) \right] = \infty$$

The series diverges.

NOTE The infinite series in Example 4 diverges very slowly. For instance, the sum of the first 10 terms is approximately 1.6878196, whereas the sum of the first 100 terms is just slightly larger: 2.3250871. In fact, the sum of the first 10,000 terms is approximately 3.015021704. You can see that although the infinite series "adds up to infinity," it does so very slowly.

Exercises for Section 9.3

See www.CalcChat.com for worked-out solutions to odd-numbered exercises.

In Exercises 1–18, use the Integral Test to determine the convergence or divergence of the series.

1. $\sum_{n=1}^{\infty} \frac{1}{n+1}$

2. $\sum_{n=1}^{\infty} \frac{2}{3n+5}$

3. $\sum_{n=1}^{\infty} e^{-n}$

4. $\sum_{n=1}^{\infty} ne^{-n/2}$

5. $\frac{1}{2} + \frac{1}{5} + \frac{1}{10} + \frac{1}{17} + \frac{1}{26} + \cdots$

6. $\frac{1}{3} + \frac{1}{5} + \frac{1}{7} + \frac{1}{9} + \frac{1}{11} + \cdots$

7. $\frac{\ln 2}{2} + \frac{\ln 3}{3} + \frac{\ln 4}{4} + \frac{\ln 5}{5} + \frac{\ln 6}{6} + \cdots$

8. $\frac{\ln 2}{\sqrt{2}} + \frac{\ln 3}{\sqrt{3}} + \frac{\ln 4}{\sqrt{4}} + \frac{\ln 5}{\sqrt{5}} + \frac{\ln 6}{\sqrt{6}} + \cdots$

9. $\frac{1}{\sqrt{1}(\sqrt{1}+1)} + \frac{1}{\sqrt{2}(\sqrt{2}+1)} + \frac{1}{\sqrt{3}(\sqrt{3}+1)} + \cdots + \frac{1}{\sqrt{n}(\sqrt{n}+1)} + \cdots$

10. $\frac{1}{4} + \frac{2}{7} + \frac{3}{12} + \cdots + \frac{n}{n^2 + 3} + \cdots$

11. $\sum_{n=1}^{\infty} \frac{1}{\sqrt{n+1}}$

12. $\sum_{n=2}^{\infty} \frac{\ln n}{n^3}$

13. $\sum_{n=1}^{\infty} \frac{\ln n}{n^2}$

14. $\sum_{n=2}^{\infty} \frac{1}{n\sqrt{\ln n}}$

15. $\sum_{n=1}^{\infty} \frac{\arctan n}{n^2 + 1}$

16. $\sum_{n=3}^{\infty} \frac{1}{n \ln n \ln(\ln n)}$

17. $\sum_{n=1}^{\infty} \frac{2n}{n^2 + 1}$

18. $\sum_{n=1}^{\infty} \frac{n}{n^4 + 1}$

In Exercises 19 and 20, use the Integral Test to determine the convergence or divergence of the series, where k is a positive integer.

19. $\sum_{n=1}^{\infty} \frac{n^{k-1}}{n^k + c}$

20. $\sum_{n=1}^{\infty} n^k e^{-n}$

In Exercises 21–24, explain why the Integral Test does not apply to the series.

21. $\sum_{n=1}^{\infty} \dfrac{(-1)^n}{n}$

22. $\sum_{n=1}^{\infty} e^{-n} \cos n$

23. $\sum_{n=1}^{\infty} \dfrac{2 + \sin n}{n}$

24. $\sum_{n=1}^{\infty} \left(\dfrac{\sin n}{n}\right)^2$

In Exercises 25–28, use the Integral Test to determine the convergence or divergence of the p-series.

25. $\sum_{n=1}^{\infty} \dfrac{1}{n^3}$

26. $\sum_{n=1}^{\infty} \dfrac{1}{n^{1/3}}$

27. $\sum_{n=1}^{\infty} \dfrac{1}{\sqrt{n}}$

28. $\sum_{n=1}^{\infty} \dfrac{1}{n^2}$

In Exercises 29–36, use Theorem 9.11 to determine the convergence or divergence of the p-series.

29. $\sum_{n=1}^{\infty} \dfrac{1}{\sqrt[5]{n}}$

30. $\sum_{n=1}^{\infty} \dfrac{3}{n^{5/3}}$

31. $1 + \dfrac{1}{\sqrt{2}} + \dfrac{1}{\sqrt{3}} + \dfrac{1}{\sqrt{4}} + \cdots$

32. $1 + \dfrac{1}{4} + \dfrac{1}{9} + \dfrac{1}{16} + \dfrac{1}{25} + \cdots$

33. $1 + \dfrac{1}{2\sqrt{2}} + \dfrac{1}{3\sqrt{3}} + \dfrac{1}{4\sqrt{4}} + \dfrac{1}{5\sqrt{5}} + \cdots$

34. $1 + \dfrac{1}{\sqrt[3]{4}} + \dfrac{1}{\sqrt[3]{9}} + \dfrac{1}{\sqrt[3]{16}} + \dfrac{1}{\sqrt[3]{25}} + \cdots$

35. $\sum_{n=1}^{\infty} \dfrac{1}{n^{1.04}}$

36. $\sum_{n=1}^{\infty} \dfrac{1}{n^{\pi}}$

In Exercises 37–42, match the series with the graph of its sequence of partial sums. [The graphs are labeled (a), (b), (c), (d), (e), and (f).] Determine the convergence or divergence of the series.

(a)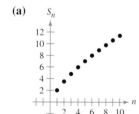

(b)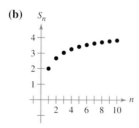

(c)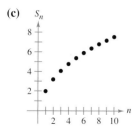

(d)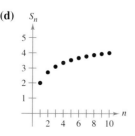

(e)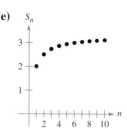

(f)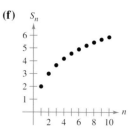

37. $\sum_{n=1}^{\infty} \dfrac{2}{\sqrt[4]{n^3}}$

38. $\sum_{n=1}^{\infty} \dfrac{2}{n}$

39. $\sum_{n=1}^{\infty} \dfrac{2}{\sqrt{n^{\pi}}}$

40. $\sum_{n=1}^{\infty} \dfrac{2}{\sqrt[5]{n^2}}$

41. $\sum_{n=1}^{\infty} \dfrac{2}{n\sqrt{n}}$

42. $\sum_{n=1}^{\infty} \dfrac{2}{n^2}$

43. **Numerical and Graphical Analysis** Use a graphing utility to find the indicated partial sum S_n and complete the table. Then use a graphing utility to graph the first 10 terms of the sequence of partial sums. Compare the rate at which the sequence of partial sums approaches the sum of the series for each series.

n	5	10	20	50	100
S_n					

(a) $\sum_{n=1}^{\infty} 3\left(\dfrac{1}{5}\right)^{n-1} = \dfrac{15}{4}$
(b) $\sum_{n=1}^{\infty} \dfrac{1}{n^2} = \dfrac{\pi^2}{6}$

44. **Numerical Reasoning** Because the harmonic series diverges, it follows that for any positive real number M there exists a positive integer N such that the partial sum

$$\sum_{n=1}^{N} \dfrac{1}{n} > M.$$

(a) Use a graphing utility to complete the table.

M	2	4	6	8
N				

(b) As the real number M increases in equal increments, does the number N increase in equal increments? Explain.

Writing About Concepts

45. State the Integral Test and give an example of its use.

46. Define a p-series and state the requirements for its convergence.

47. A friend in your calculus class tells you that the following series converges because the terms are very small and approach 0 rapidly. Is your friend correct? Explain.

$$\dfrac{1}{10{,}000} + \dfrac{1}{10{,}001} + \dfrac{1}{10{,}002} + \cdots$$

Writing About Concepts (continued)

48. In Exercises 37–42, $\lim_{n \to \infty} a_n = 0$ for each series but they do not all converge. Is this a contradiction of Theorem 9.9? Why do you think some converge and others diverge? Explain.

49. Use a graph to show that

$$\sum_{n=1}^{\infty} \frac{1}{\sqrt{n}} > \int_{1}^{\infty} \frac{1}{\sqrt{x}}\, dx.$$

What can you conclude about the convergence or divergence of the series? Explain.

50. Let f be a positive, continuous, and decreasing function for $x \geq 1$, such that $a_n = f(n)$. Use a graph to rank the following quantities in decreasing order. Explain your reasoning.

(a) $\sum_{n=2}^{7} a_n$ (b) $\int_{1}^{7} f(x)\, dx$ (c) $\sum_{n=1}^{6} a_n$

In Exercises 51–54, find the positive values of p for which the series converges.

51. $\sum_{n=2}^{\infty} \frac{1}{n(\ln n)^p}$

52. $\sum_{n=2}^{\infty} \frac{\ln n}{n^p}$

53. $\sum_{n=1}^{\infty} \frac{n}{(1+n^2)^p}$

54. $\sum_{n=1}^{\infty} n(1+n^2)^p$

In Exercises 55–58, use the result of Exercise 51 to determine the convergence or divergence of the series.

55. $\sum_{n=2}^{\infty} \frac{1}{n \ln n}$

56. $\sum_{n=2}^{\infty} \frac{1}{n \sqrt[3]{(\ln n)^2}}$

57. $\sum_{n=2}^{\infty} \frac{1}{n(\ln n)^2}$

58. $\sum_{n=2}^{\infty} \frac{1}{n \ln(n^2)}$

59. Let f be a positive, continuous, and decreasing function for $x \geq 1$, such that $a_n = f(n)$. Prove that if the series

$$\sum_{n=1}^{\infty} a_n$$

converges to S, then the remainder $R_N = S - S_N$ is bounded by

$$0 \leq R_N \leq \int_{N}^{\infty} f(x)\, dx.$$

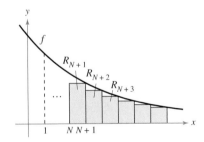

60. Show that the result of Exercise 59 can be written as

$$\sum_{n=1}^{N} a_n \leq \sum_{n=1}^{\infty} a_n \leq \sum_{n=1}^{N} a_n + \int_{N}^{\infty} f(x)\, dx.$$

In Exercises 61–66, use the result of Exercise 59 to approximate the sum of the convergent series using the indicated number of terms. Include an estimate of the maximum error for your approximation.

61. $\sum_{n=1}^{\infty} \frac{1}{n^4}$, six terms

62. $\sum_{n=1}^{\infty} \frac{1}{n^5}$, four terms

63. $\sum_{n=1}^{\infty} \frac{1}{n^2 + 1}$, ten terms

64. $\sum_{n=1}^{\infty} \frac{1}{(n+1)[\ln(n+1)]^3}$, ten terms

65. $\sum_{n=1}^{\infty} ne^{-n^2}$, four terms

66. $\sum_{n=1}^{\infty} e^{-n}$, four terms

In Exercises 67–72, use the result of Exercise 59 to find N such that $R_N \leq 0.001$ for the convergent series.

67. $\sum_{n=1}^{\infty} \frac{1}{n^4}$

68. $\sum_{n=1}^{\infty} \frac{1}{n^{3/2}}$

69. $\sum_{n=1}^{\infty} e^{-5n}$

70. $\sum_{n=1}^{\infty} e^{-n/2}$

71. $\sum_{n=1}^{\infty} \frac{1}{n^2 + 1}$

72. $\sum_{n=1}^{\infty} \frac{2}{n^2 + 5}$

73. (a) Show that $\sum_{n=2}^{\infty} \frac{1}{n^{1.1}}$ converges and $\sum_{n=2}^{\infty} \frac{1}{n \ln n}$ diverges.

(b) Compare the first five terms of each series in part (a).

(c) Find $n > 3$ such that

$$\frac{1}{n^{1.1}} < \frac{1}{n \ln n}.$$

74. Ten terms are used to approximate a convergent p-series. Therefore, the remainder is a function of p and is

$$0 \leq R_{10}(p) \leq \int_{10}^{\infty} \frac{1}{x^p}\, dx, \quad p > 1.$$

(a) Perform the integration in the inequality.

(b) Use a graphing utility to represent the inequality graphically.

(c) Identify any asymptotes of the error function and interpret their meaning.

75. Euler's Constant Let
$$S_n = \sum_{k=1}^{n} \frac{1}{k} = 1 + \frac{1}{2} + \cdots + \frac{1}{n}.$$

(a) Show that $\ln(n + 1) \le S_n \le 1 + \ln n$.

(b) Show that the sequence $\{a_n\} = \{S_n - \ln n\}$ is bounded.

(c) Show that the sequence $\{a_n\}$ is decreasing.

(d) Show that a_n converges to a limit γ (called Euler's constant).

(e) Approximate γ using a_{100}.

76. Find the sum of the series $\sum_{n=2}^{\infty} \ln\left(1 - \frac{1}{n^2}\right)$.

77. Consider the series
$$\sum_{n=2}^{\infty} x^{\ln n}.$$

(a) Determine the convergence or divergence of the series for $x = 1$.

(b) Determine the convergence or divergence of the series for $x = 1/e$.

(c) Find the positive values of x for which the series converges.

78. The **Riemann zeta function** for real numbers is defined for all x for which the series
$$\zeta(x) = \sum_{n=1}^{\infty} n^{-x}$$
converges. Find the domain of the function.

Review In Exercises 79–90, determine the convergence or divergence of the series.

79. $\sum_{n=1}^{\infty} \frac{1}{2n - 1}$

80. $\sum_{n=2}^{\infty} \frac{1}{n\sqrt{n^2 - 1}}$

81. $\sum_{n=1}^{\infty} \frac{1}{n\sqrt[4]{n}}$

82. $3 \sum_{n=1}^{\infty} \frac{1}{n^{0.95}}$

83. $\sum_{n=0}^{\infty} \left(\frac{2}{3}\right)^n$

84. $\sum_{n=0}^{\infty} (1.075)^n$

85. $\sum_{n=1}^{\infty} \frac{n}{\sqrt{n^2 + 1}}$

86. $\sum_{n=1}^{\infty} \left(\frac{1}{n^2} - \frac{1}{n^3}\right)$

87. $\sum_{n=1}^{\infty} \left(1 + \frac{1}{n}\right)^n$

88. $\sum_{n=2}^{\infty} \ln n$

89. $\sum_{n=2}^{\infty} \frac{1}{n(\ln n)^3}$

90. $\sum_{n=2}^{\infty} \frac{\ln n}{n^3}$

Section Project: The Harmonic Series

The harmonic series
$$\sum_{n=1}^{\infty} \frac{1}{n} = 1 + \frac{1}{2} + \frac{1}{3} + \frac{1}{4} + \cdots + \frac{1}{n} + \cdots$$
is one of the most important series in this chapter. Even though its terms tend to zero as n increases,
$$\lim_{n \to \infty} \frac{1}{n} = 0$$
the harmonic series diverges. In other words, even though the terms are getting smaller and smaller, the sum "adds up to infinity."

(a) One way to show that the harmonic series diverges is attributed to Jakob Bernoulli. He grouped the terms of the harmonic series as follows:
$$1 + \frac{1}{2} + \underbrace{\frac{1}{3} + \frac{1}{4}}_{> \frac{1}{2}} + \underbrace{\frac{1}{5} + \cdots + \frac{1}{8}}_{> \frac{1}{2}} + \underbrace{\frac{1}{9} + \cdots + \frac{1}{16}}_{> \frac{1}{2}} +$$
$$\underbrace{\frac{1}{17} + \cdots + \frac{1}{32}}_{> \frac{1}{2}} + \cdots$$

Write a short paragraph explaining how you can use this grouping to show that the harmonic series diverges.

(b) Use the proof of the Integral Test, Theorem 9.10, to show that
$$\ln(n + 1) \le 1 + \frac{1}{2} + \frac{1}{3} + \frac{1}{4} + \cdots + \frac{1}{n} \le 1 + \ln n.$$

(c) Use part (b) to determine how many terms M you would need so that
$$\sum_{n=1}^{M} \frac{1}{n} > 50.$$

(d) Show that the sum of the first million terms of the harmonic series is less than 15.

(e) Show that the following inequalities are valid.
$$\ln \frac{21}{10} \le \frac{1}{10} + \frac{1}{11} + \cdots + \frac{1}{20} \le \ln \frac{20}{9}$$
$$\ln \frac{201}{100} \le \frac{1}{100} + \frac{1}{101} + \cdots + \frac{1}{200} \le \ln \frac{200}{99}$$

(f) Use the ideas in part (e) to find the limit
$$\lim_{m \to \infty} \sum_{n=m}^{2m} \frac{1}{n}.$$

Section 9.4 Comparisons of Series

- Use the Direct Comparison Test to determine whether a series converges or diverges.
- Use the Limit Comparison Test to determine whether a series converges or diverges.

Direct Comparison Test

For the convergence tests developed so far, the terms of the series have to be fairly simple and the series must have special characteristics in order for the convergence tests to be applied. A slight deviation from these special characteristics can make a test nonapplicable. For example, in the following pairs, the second series cannot be tested by the same convergence test as the first series even though it is similar to the first.

1. $\sum_{n=0}^{\infty} \frac{1}{2^n}$ is geometric, but $\sum_{n=0}^{\infty} \frac{n}{2^n}$ is not.

2. $\sum_{n=1}^{\infty} \frac{1}{n^3}$ is a p-series, but $\sum_{n=1}^{\infty} \frac{1}{n^3 + 1}$ is not.

3. $a_n = \frac{n}{(n^2 + 3)^2}$ is easily integrated, but $b_n = \frac{n^2}{(n^2 + 3)^2}$ is not.

In this section you will study two additional tests for positive-term series. These two tests greatly expand the variety of series you are able to test for convergence or divergence. They allow you to *compare* a series having complicated terms with a simpler series whose convergence or divergence is known.

THEOREM 9.12 Direct Comparison Test

Let $0 < a_n \leq b_n$ for all n.

1. If $\sum_{n=1}^{\infty} b_n$ converges, then $\sum_{n=1}^{\infty} a_n$ converges.

2. If $\sum_{n=1}^{\infty} a_n$ diverges, then $\sum_{n=1}^{\infty} b_n$ diverges.

Proof To prove the first property, let $L = \sum_{n=1}^{\infty} b_n$ and let

$$S_n = a_1 + a_2 + \cdots + a_n.$$

Because $0 < a_n \leq b_n$, the sequence $S_1, S_2, S_3, \ldots$ is nondecreasing and bounded above by L; so, it must converge. Because

$$\lim_{n \to \infty} S_n = \sum_{n=1}^{\infty} a_n$$

it follows that Σa_n converges. The second property is logically equivalent to the first.

NOTE As stated, the Direct Comparison Test requires that $0 < a_n \leq b_n$ for all n. Because the convergence of a series is not dependent on its first several terms, you could modify the test to require only that $0 < a_n \leq b_n$ for all n greater than some integer N.

EXAMPLE 1 Using the Direct Comparison Test

Determine the convergence or divergence of

$$\sum_{n=1}^{\infty} \frac{1}{2 + 3^n}.$$

Solution This series resembles

$$\sum_{n=1}^{\infty} \frac{1}{3^n}. \quad \text{Convergent geometric series}$$

Term-by-term comparison yields

$$a_n = \frac{1}{2 + 3^n} < \frac{1}{3^n} = b_n, \quad n \geq 1.$$

So, by the Direct Comparison Test, the series converges.

 ## EXAMPLE 2 Using the Direct Comparison Test

Determine the convergence or divergence of

$$\sum_{n=1}^{\infty} \frac{1}{2 + \sqrt{n}}.$$

Solution This series resembles

$$\sum_{n=1}^{\infty} \frac{1}{n^{1/2}}. \quad \text{Divergent } p\text{-series}$$

Term-by-term comparison yields

$$\frac{1}{2 + \sqrt{n}} \leq \frac{1}{\sqrt{n}}, \quad n \geq 1$$

which *does not* meet the requirements for divergence. (Remember that if term-by-term comparison reveals a series that is *smaller* than a divergent series, the Direct Comparison Test tells you nothing.) Still expecting the series to diverge, you can compare the given series with

$$\sum_{n=1}^{\infty} \frac{1}{n}. \quad \text{Divergent harmonic series}$$

In this case, term-by-term comparison yields

$$a_n = \frac{1}{n} \leq \frac{1}{2 + \sqrt{n}} = b_n, \quad n \geq 4$$

and, by the Direct Comparison Test, the given series diverges.

NOTE To verify the last inequality in Example 2, try showing that $2 + \sqrt{n} \leq n$ whenever $n \geq 4$.

Remember that both parts of the Direct Comparison Test require that $0 < a_n \leq b_n$. Informally, the test says the following about the two series with nonnegative terms.

1. If the "larger" series converges, the "smaller" series must also converge.
2. If the "smaller" series diverges, the "larger" series must also diverge.

Limit Comparison Test

Often a given series closely resembles a *p*-series or a geometric series, yet you cannot establish the term-by-term comparison necessary to apply the Direct Comparison Test. Under these circumstances you may be able to apply a second comparison test, called the **Limit Comparison Test.**

> **THEOREM 9.13 Limit Comparison Test**
>
> Suppose that $a_n > 0$, $b_n > 0$, and
> $$\lim_{n \to \infty} \left(\frac{a_n}{b_n}\right) = L$$
> where L is *finite and positive*. Then the two series $\Sigma \, a_n$ and $\Sigma \, b_n$ either both converge or both diverge.

NOTE As with the Direct Comparison Test, the Limit Comparison Test could be modified to require only that a_n and b_n be positive for all n greater than some integer N.

Proof Because $a_n > 0$, $b_n > 0$, and
$$\lim_{n \to \infty} \left(\frac{a_n}{b_n}\right) = L$$
there exists $N > 0$ such that
$$0 < \frac{a_n}{b_n} < L + 1, \quad \text{for } n \geq N.$$
This implies that
$$0 < a_n < (L + 1)b_n.$$
So, by the Direct Comparison Test, the convergence of $\Sigma \, b_n$ implies the convergence of $\Sigma \, a_n$. Similarly, the fact that
$$\lim_{n \to \infty} \left(\frac{b_n}{a_n}\right) = \frac{1}{L}$$
can be used to show that the convergence of $\Sigma \, a_n$ implies the convergence of $\Sigma \, b_n$.

EXAMPLE 3 Using the Limit Comparison Test

Show that the following general harmonic series diverges.

$$\sum_{n=1}^{\infty} \frac{1}{an + b}, \quad a > 0, \quad b > 0$$

Solution By comparison with

$$\sum_{n=1}^{\infty} \frac{1}{n} \qquad \text{Divergent harmonic series}$$

you have

$$\lim_{n \to \infty} \frac{1/(an + b)}{1/n} = \lim_{n \to \infty} \frac{n}{an + b} = \frac{1}{a}.$$

Because this limit is greater than 0, you can conclude from the Limit Comparison Test that the given series diverges.

The Limit Comparison Test works well for comparing a "messy" algebraic series with a p-series. In choosing an appropriate p-series, you must choose one with an nth term of the same magnitude as the nth term of the given series.

Given Series	Comparison Series	Conclusion
$\sum_{n=1}^{\infty} \dfrac{1}{3n^2 - 4n + 5}$	$\sum_{n=1}^{\infty} \dfrac{1}{n^2}$	Both series converge.
$\sum_{n=1}^{\infty} \dfrac{1}{\sqrt{3n-2}}$	$\sum_{n=1}^{\infty} \dfrac{1}{\sqrt{n}}$	Both series diverge.
$\sum_{n=1}^{\infty} \dfrac{n^2 - 10}{4n^5 + n^3}$	$\sum_{n=1}^{\infty} \dfrac{n^2}{n^5} = \sum_{n=1}^{\infty} \dfrac{1}{n^3}$	Both series converge.

In other words, when choosing a series for comparison, you can disregard all but the *highest powers of n* in both the numerator and the denominator.

EXAMPLE 4 Using the Limit Comparison Test

Determine the convergence or divergence of

$$\sum_{n=1}^{\infty} \frac{\sqrt{n}}{n^2 + 1}.$$

Solution Disregarding all but the highest powers of n in the numerator and the denominator, you can compare the series with

$$\sum_{n=1}^{\infty} \frac{\sqrt{n}}{n^2} = \sum_{n=1}^{\infty} \frac{1}{n^{3/2}}. \qquad \text{Convergent } p\text{-series}$$

Because

$$\lim_{n \to \infty} \frac{a_n}{b_n} = \lim_{n \to \infty} \left(\frac{\sqrt{n}}{n^2 + 1}\right)\left(\frac{n^{3/2}}{1}\right)$$

$$= \lim_{n \to \infty} \frac{n^2}{n^2 + 1} = 1$$

you can conclude by the Limit Comparison Test that the given series converges.

EXAMPLE 5 Using the Limit Comparison Test

Determine the convergence or divergence of

$$\sum_{n=1}^{\infty} \frac{n 2^n}{4n^3 + 1}.$$

Solution A reasonable comparison would be with the series

$$\sum_{n=1}^{\infty} \frac{2^n}{n^2}. \qquad \text{Divergent series}$$

Note that this series diverges by the nth-Term Test. From the limit

$$\lim_{n \to \infty} \frac{a_n}{b_n} = \lim_{n \to \infty} \left(\frac{n 2^n}{4n^3 + 1}\right)\left(\frac{n^2}{2^n}\right)$$

$$= \lim_{n \to \infty} \frac{1}{4 + (1/n^3)} = \frac{1}{4}$$

you can conclude that the given series diverges.

Exercises for Section 9.4

See www.CalcChat.com for worked-out solutions to odd-numbered exercises.

1. Graphical Analysis The figures show the graphs of the first 10 terms, and the graphs of the first 10 terms of the sequence of partial sums, of each series.

$$\sum_{n=1}^{\infty} \frac{6}{n^{3/2}}, \quad \sum_{n=1}^{\infty} \frac{6}{n^{3/2}+3}, \quad \text{and} \quad \sum_{n=1}^{\infty} \frac{6}{n\sqrt{n^2+0.5}}$$

(a) Identify the series in each figure.

(b) Which series is a *p*-series? Does it converge or diverge?

(c) For the series that are not *p*-series, how do the magnitudes of the terms compare with the magnitudes of the terms of the *p*-series? What conclusion can you draw about the convergence or divergence of the series?

(d) Explain the relationship between the magnitudes of the terms of the series and the magnitudes of the terms of the partial sums.

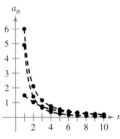
Graphs of terms

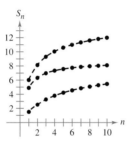
Graphs of partial sums

2. Graphical Analysis The figures show the graphs of the first 10 terms, and the graphs of the first 10 terms of the sequence of partial sums, of each series.

$$\sum_{n=1}^{\infty} \frac{2}{\sqrt{n}}, \quad \sum_{n=1}^{\infty} \frac{2}{\sqrt{n}-0.5}, \quad \text{and} \quad \sum_{n=1}^{\infty} \frac{4}{\sqrt{n}+0.5}$$

(a) Identify the series in each figure.

(b) Which series is a *p*-series? Does it converge or diverge?

(c) For the series that are not *p*-series, how do the magnitudes of the terms compare with the magnitudes of the terms of the *p*-series? What conclusion can you draw about the convergence or divergence of the series?

(d) Explain the relationship between the magnitudes of the terms of the series and the magnitudes of the terms of the partial sums.

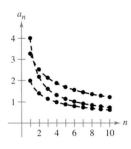
Graphs of terms

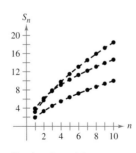
Graphs of partial sums

In Exercises 3–14, use the Direct Comparison Test to determine the convergence or divergence of the series.

3. $\sum_{n=1}^{\infty} \frac{1}{n^2+1}$

4. $\sum_{n=1}^{\infty} \frac{1}{3n^2+2}$

5. $\sum_{n=2}^{\infty} \frac{1}{n-1}$

6. $\sum_{n=2}^{\infty} \frac{1}{\sqrt{n}-1}$

7. $\sum_{n=0}^{\infty} \frac{1}{3^n+1}$

8. $\sum_{n=0}^{\infty} \frac{3^n}{4^n+5}$

9. $\sum_{n=2}^{\infty} \frac{\ln n}{n+1}$

10. $\sum_{n=1}^{\infty} \frac{1}{\sqrt{n^3+1}}$

11. $\sum_{n=0}^{\infty} \frac{1}{n!}$

12. $\sum_{n=1}^{\infty} \frac{1}{4\sqrt[3]{n}-1}$

13. $\sum_{n=0}^{\infty} e^{-n^2}$

14. $\sum_{n=1}^{\infty} \frac{4^n}{3^n-1}$

In Exercises 15–28, use the Limit Comparison Test to determine the convergence or divergence of the series.

15. $\sum_{n=1}^{\infty} \frac{n}{n^2+1}$

16. $\sum_{n=1}^{\infty} \frac{2}{3^n-5}$

17. $\sum_{n=0}^{\infty} \frac{1}{\sqrt{n^2+1}}$

18. $\sum_{n=3}^{\infty} \frac{3}{\sqrt{n^2-4}}$

19. $\sum_{n=1}^{\infty} \frac{2n^2-1}{3n^5+2n+1}$

20. $\sum_{n=1}^{\infty} \frac{5n-3}{n^2-2n+5}$

21. $\sum_{n=1}^{\infty} \frac{n+3}{n(n+2)}$

22. $\sum_{n=1}^{\infty} \frac{1}{n(n^2+1)}$

23. $\sum_{n=1}^{\infty} \frac{1}{n\sqrt{n^2+1}}$

24. $\sum_{n=1}^{\infty} \frac{n}{(n+1)2^{n-1}}$

25. $\sum_{n=1}^{\infty} \frac{n^{k-1}}{n^k+1}, \quad k>2$

26. $\sum_{n=1}^{\infty} \frac{5}{n+\sqrt{n^2+4}}$

27. $\sum_{n=1}^{\infty} \sin \frac{1}{n}$

28. $\sum_{n=1}^{\infty} \tan \frac{1}{n}$

In Exercises 29–36, test for convergence or divergence, using each test at least once. Identify which test was used.

(a) *n*th-Term Test (b) Geometric Series Test

(c) *p*-Series Test (d) Telescoping Series Test

(e) Integral Test (f) Direct Comparison Test

(g) Limit Comparison Test

29. $\sum_{n=1}^{\infty} \frac{\sqrt{n}}{n}$

30. $\sum_{n=0}^{\infty} 5\left(-\frac{1}{5}\right)^n$

31. $\sum_{n=1}^{\infty} \frac{1}{3^n+2}$

32. $\sum_{n=4}^{\infty} \frac{1}{3n^2-2n-15}$

33. $\sum_{n=1}^{\infty} \frac{n}{2n+3}$

34. $\sum_{n=1}^{\infty} \left(\frac{1}{n+1}-\frac{1}{n+2}\right)$

35. $\sum_{n=1}^{\infty} \frac{n}{(n^2+1)^2}$

36. $\sum_{n=1}^{\infty} \frac{3}{n(n+3)}$

37. Use the Limit Comparison Test with the harmonic series to show that the series Σa_n (where $0 < a_n < a_{n-1}$) diverges if $\lim\limits_{n \to \infty} na_n$ is finite and nonzero.

38. Prove that, if $P(n)$ and $Q(n)$ are polynomials of degree j and k, respectively, then the series

$$\sum_{n=1}^{\infty} \frac{P(n)}{Q(n)}$$

converges if $j < k - 1$ and diverges if $j \geq k - 1$.

In Exercises 39–42, use the polynomial test given in Exercise 38 to determine whether the series converges or diverges.

39. $\frac{1}{2} + \frac{2}{5} + \frac{3}{10} + \frac{4}{17} + \frac{5}{26} + \cdots$

40. $\frac{1}{3} + \frac{1}{8} + \frac{1}{15} + \frac{1}{24} + \frac{1}{35} + \cdots$

41. $\sum_{n=1}^{\infty} \frac{1}{n^3 + 1}$ **42.** $\sum_{n=1}^{\infty} \frac{n^2}{n^3 + 1}$

In Exercises 43 and 44, use the divergence test given in Exercise 37 to show that the series diverges.

43. $\sum_{n=1}^{\infty} \frac{n^3}{5n^4 + 3}$ **44.** $\sum_{n=2}^{\infty} \frac{1}{\ln n}$

In Exercises 45–48, determine the convergence or divergence of the series.

45. $\frac{1}{200} + \frac{1}{400} + \frac{1}{600} + \frac{1}{800} + \cdots$

46. $\frac{1}{200} + \frac{1}{210} + \frac{1}{220} + \frac{1}{230} + \cdots$

47. $\frac{1}{201} + \frac{1}{204} + \frac{1}{209} + \frac{1}{216} + \cdots$

48. $\frac{1}{201} + \frac{1}{208} + \frac{1}{227} + \frac{1}{264} + \cdots$

Writing About Concepts

49. Review the results of Exercises 45–48. Explain why careful analysis is required to determine the convergence or divergence of a series and why only considering the magnitudes of the terms of a series could be misleading.

50. State the Direct Comparison Test and give an example of its use.

51. State the Limit Comparison Test and give an example of its use.

52. It appears that the terms of the series

$$\frac{1}{1000} + \frac{1}{1001} + \frac{1}{1002} + \frac{1}{1003} + \cdots$$

are less than the corresponding terms of the convergent series

$$1 + \frac{1}{4} + \frac{1}{9} + \frac{1}{16} + \cdots.$$

If the statement above is correct, the first series converges. Is this correct? Why or why not? Make a statement about how the divergence or convergence of a series is affected by inclusion or exclusion of the first finite number of terms.

Writing About Concepts (continued)

53. The figure shows the first 20 terms of the convergent series $\sum_{n=1}^{\infty} a_n$ and the first 20 terms of the series $\sum_{n=1}^{\infty} a_n^2$. Identify the two series and explain your reasoning in making the selection.

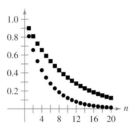

54. Consider the series $\sum_{n=1}^{\infty} \frac{1}{(2n-1)^2}$.

(a) Verify that the series converges.

(b) Use a graphing utility to complete the table.

n	5	10	20	50	100
S_n					

(c) The sum of the series is $\pi^2/8$. Find the sum of the series

$$\sum_{n=3}^{\infty} \frac{1}{(2n-1)^2}.$$

(d) Use a graphing utility to find the sum of the series

$$\sum_{n=10}^{\infty} \frac{1}{(2n-1)^2}.$$

True or False? In Exercises 55–60, determine whether the statement is true or false. If it is false, explain why or give an example that shows it is false.

55. If $0 < a_n \leq b_n$ and $\sum_{n=1}^{\infty} a_n$ converges, then $\sum_{n=1}^{\infty} b_n$ diverges.

56. If $0 < a_{n+10} \leq b_n$ and $\sum_{n=1}^{\infty} b_n$ converges, then $\sum_{n=1}^{\infty} a_n$ converges.

57. If $a_n + b_n \leq c_n$ and $\sum_{n=1}^{\infty} c_n$ converges, then the series $\sum_{n=1}^{\infty} a_n$ and $\sum_{n=1}^{\infty} b_n$ both converge. (Assume that the terms of all three series are positive.)

58. If $a_n \leq b_n + c_n$ and $\sum_{n=1}^{\infty} a_n$ diverges, then the series $\sum_{n=1}^{\infty} b_n$ and $\sum_{n=1}^{\infty} c_n$ both diverge. (Assume that the terms of all three series are positive.)

59. If $0 < a_n \leq b_n$ and $\sum_{n=1}^{\infty} a_n$ diverges, then $\sum_{n=1}^{\infty} b_n$ diverges.

60. If $0 < a_n \le b_n$ and $\sum_{n=1}^{\infty} b_n$ diverges, then $\sum_{n=1}^{\infty} a_n$ diverges.

61. Prove that if the nonnegative series

$$\sum_{n=1}^{\infty} a_n \quad \text{and} \quad \sum_{n=1}^{\infty} b_n$$

converge, then so does the series

$$\sum_{n=1}^{\infty} a_n b_n.$$

62. Use the result of Exercise 61 to prove that if the nonnegative series

$$\sum_{n=1}^{\infty} a_n$$

converges, then so does the series

$$\sum_{n=1}^{\infty} a_n^2.$$

63. Find two series that demonstrate the result of Exercise 61.

64. Find two series that demonstrate the result of Exercise 62.

65. Suppose that Σa_n and Σb_n are series with positive terms. Prove that if $\lim_{n \to \infty} \dfrac{a_n}{b_n} = 0$ and Σb_n converges, Σa_n also converges.

66. Suppose that Σa_n and Σb_n are series with positive terms. Prove that if $\lim_{n \to \infty} \dfrac{a_n}{b_n} = \infty$ and Σb_n diverges, Σa_n also diverges.

67. Use the result of Exercise 65 to show that each series converges.

(a) $\displaystyle\sum_{n=1}^{\infty} \dfrac{1}{(n+1)^3}$ (b) $\displaystyle\sum_{n=1}^{\infty} \dfrac{1}{\sqrt{n}\pi^n}$

68. Use the result of Exercise 66 to show that each series diverges.

(a) $\displaystyle\sum_{n=1}^{\infty} \dfrac{\ln n}{n}$ (b) $\displaystyle\sum_{n=2}^{\infty} \dfrac{1}{\ln n}$

69. Suppose that Σa_n is a series with positive terms. Prove that if Σa_n converges, then $\Sigma \sin a_n$ also converges.

70. Prove that the series $\displaystyle\sum_{n=1}^{\infty} \dfrac{1}{1+2+3+\cdots+n}$ converges.

Putnam Exam Challenge

71. Is the infinite series

$$\sum_{n=1}^{\infty} \dfrac{1}{n^{(n+1)/n}}$$

convergent? Prove your statement.

72. Prove that if $\displaystyle\sum_{n=1}^{\infty} a_n$ is a convergent series of positive real numbers, then so is

$$\sum_{n=1}^{\infty} (a_n)^{n/(n+1)}.$$

These problems were composed by the Committee on the Putnam Prize Competition. © The Mathematical Association of America. All rights reserved.

Section Project: Solera Method

Most wines are produced entirely from grapes grown in a single year. Sherry, however, is a complex mixture of older wines with new wines. This is done with a sequence of barrels (called a solera) stacked on top of each other, as shown in the photo.

Everton/The Image Works

The oldest wine is in the bottom tier of barrels, and the newest is in the top tier. Each year, half of each barrel in the bottom tier is bottled as sherry. The bottom barrels are then refilled with the wine from the barrels above. This process is repeated throughout the solera, with new wine being added to the top barrels. A mathematical model for the amount of n-year-old wine that is removed from a solera (with k tiers) each year is

$$f(n, k) = \binom{n-1}{k-1}\left(\dfrac{1}{2}\right)^{n+1}, \quad k \le n.$$

(a) Consider a solera that has five tiers, numbered $k = 1, 2, 3, 4,$ and 5. In 1990 ($n = 0$), half of each barrel in the top tier (tier 1) was refilled with new wine. How much of this wine was removed from the solera in 1991? In 1992? In 1993? . . . In 2005? During which year(s) was the greatest amount of the 1990 wine removed from the solera?

(b) In part (a), let a_n be the amount of 1990 wine that is removed from the solera in year n. Evaluate

$$\sum_{n=0}^{\infty} a_n.$$

FOR FURTHER INFORMATION See the article "Finding Vintage Concentrations in a Sherry Solera" by Rhodes Peele and John T. MacQueen in the *UMAP Modules*.

Section 9.5

Alternating Series

- Use the Alternating Series Test to determine whether an infinite series converges.
- Use the Alternating Series Remainder to approximate the sum of an alternating series.
- Classify a convergent series as absolutely or conditionally convergent.
- Rearrange an infinite series to obtain a different sum.

Alternating Series

So far, most series you have dealt with have had positive terms. In this section and the following section, you will study series that contain both positive and negative terms. The simplest such series is an **alternating series,** whose terms alternate in sign. For example, the geometric series

$$\sum_{n=0}^{\infty} \left(-\frac{1}{2}\right)^n = \sum_{n=0}^{\infty} (-1)^n \frac{1}{2^n}$$
$$= 1 - \frac{1}{2} + \frac{1}{4} - \frac{1}{8} + \frac{1}{16} - \cdots$$

is an *alternating geometric series* with $r = -\frac{1}{2}$. Alternating series occur in two ways: either the odd terms are negative or the even terms are negative.

THEOREM 9.14   Alternating Series Test

Let $a_n > 0$. The alternating series

$$\sum_{n=1}^{\infty} (-1)^n a_n \quad \text{and} \quad \sum_{n=1}^{\infty} (-1)^{n+1} a_n$$

converge if the following two conditions are met.

1. $\lim\limits_{n \to \infty} a_n = 0$ 2. $a_{n+1} \leq a_n$, for all n

---

Proof   Consider the alternating series $\sum (-1)^{n+1} a_n$. For this series, the partial sum (where $2n$ is even)

$$S_{2n} = (a_1 - a_2) + (a_3 - a_4) + (a_5 - a_6) + \cdots + (a_{2n-1} - a_{2n})$$

has all nonnegative terms, and therefore $\{S_{2n}\}$ is a nondecreasing sequence. But you can also write

$$S_{2n} = a_1 - (a_2 - a_3) - (a_4 - a_5) - \cdots - (a_{2n-2} - a_{2n-1}) - a_{2n}$$

which implies that $S_{2n} \leq a_1$ for every integer n. So, $\{S_{2n}\}$ is a bounded, nondecreasing sequence that converges to some value L. Because $S_{2n-1} - a_{2n} = S_{2n}$ and $a_{2n} \to 0$, you have

$$\lim_{n \to \infty} S_{2n-1} = \lim_{n \to \infty} S_{2n} + \lim_{n \to \infty} a_{2n}$$
$$= L + \lim_{n \to \infty} a_{2n} = L.$$

Because both S_{2n} and S_{2n-1} converge to the same limit L, it follows that $\{S_n\}$ also converges to L. Consequently, the given alternating series converges. ■

NOTE   The second condition in the Alternating Series Test can be modified to require only that $0 < a_{n+1} \leq a_n$ for all n greater than some integer N.

632 CHAPTER 9 Infinite Series

NOTE The series in Example 1 is called the *alternating harmonic series*—more is said about this series in Example 7.

EXAMPLE 1 Using the Alternating Series Test

Determine the convergence or divergence of $\sum_{n=1}^{\infty} (-1)^{n+1} \frac{1}{n}$.

Solution Note that $\lim_{n \to \infty} a_n = \lim_{n \to \infty} \frac{1}{n} = 0$. So, the first condition of Theorem 9.14 is satisfied. Also note that the second condition of Theorem 9.14 is satisfied because

$$a_{n+1} = \frac{1}{n+1} \leq \frac{1}{n} = a_n$$

for all n. So, applying the Alternating Series Test, you can conclude that the series converges.

EXAMPLE 2 Using the Alternating Series Test

Determine the convergence or divergence of $\sum_{n=1}^{\infty} \frac{n}{(-2)^{n-1}}$.

Solution To apply the Alternating Series Test, note that, for $n \geq 1$,

$$\frac{1}{2} \leq \frac{n}{n+1}$$

$$\frac{2^{n-1}}{2^n} \leq \frac{n}{n+1}$$

$$(n+1)2^{n-1} \leq n2^n$$

$$\frac{n+1}{2^n} \leq \frac{n}{2^{n-1}}.$$

So, $a_{n+1} = (n+1)/2^n \leq n/2^{n-1} = a_n$ for all n. Furthermore, by L'Hôpital's Rule,

$$\lim_{x \to \infty} \frac{x}{2^{x-1}} = \lim_{x \to \infty} \frac{1}{2^{x-1}(\ln 2)} = 0 \implies \lim_{n \to \infty} \frac{n}{2^{n-1}} = 0.$$

Therefore, by the Alternating Series Test, the series converges.

EXAMPLE 3 Cases for Which the Alternating Series Test Fails

NOTE In Example 3(a), remember that whenever a series does not pass the first condition of the Alternating Series Test, you can use the nth-Term Test for Divergence to conclude that the series diverges.

a. The alternating series

$$\sum_{n=1}^{\infty} \frac{(-1)^{n+1}(n+1)}{n} = \frac{2}{1} - \frac{3}{2} + \frac{4}{3} - \frac{5}{4} + \frac{6}{5} - \cdots$$

passes the second condition of the Alternating Series Test because $a_{n+1} \leq a_n$ for all n. You cannot apply the Alternating Series Test, however, because the series does not pass the first condition. In fact, the series diverges.

b. The alternating series

$$\frac{2}{1} - \frac{1}{1} + \frac{2}{2} - \frac{1}{2} + \frac{2}{3} - \frac{1}{3} + \frac{2}{4} - \frac{1}{4} + \cdots$$

passes the first condition because a_n approaches 0 as $n \to \infty$. You cannot apply the Alternating Series Test, however, because the series does not pass the second condition. To conclude that the series diverges, you can argue that S_{2N} equals the Nth partial sum of the divergent harmonic series. This implies that the sequence of partial sums diverges. So, the series diverges.

Alternating Series Remainder

For a convergent alternating series, the partial sum S_N can be a useful approximation for the sum S of the series. The error involved in using $S \approx S_N$ is the remainder $R_N = S - S_N$.

THEOREM 9.15 Alternating Series Remainder

If a convergent alternating series satisfies the condition $a_{n+1} \leq a_n$, then the absolute value of the remainder R_N involved in approximating the sum S by S_N is less than (or equal to) the first neglected term. That is,
$$|S - S_N| = |R_N| \leq a_{N+1}.$$

Proof The series obtained by deleting the first N terms of the given series satisfies the conditions of the Alternating Series Test and has a sum of R_N.

$$R_N = S - S_N = \sum_{n=1}^{\infty} (-1)^{n+1} a_n - \sum_{n=1}^{N} (-1)^{n+1} a_n$$
$$= (-1)^N a_{N+1} + (-1)^{N+1} a_{N+2} + (-1)^{N+2} a_{N+3} + \cdots$$
$$= (-1)^N (a_{N+1} - a_{N+2} + a_{N+3} - \cdots)$$
$$|R_N| = a_{N+1} - a_{N+2} + a_{N+3} - a_{N+4} + a_{N+5} - \cdots$$
$$= a_{N+1} - (a_{N+2} - a_{N+3}) - (a_{N+4} - a_{N+5}) - \cdots \leq a_{N+1}$$

Consequently, $|S - S_N| = |R_N| \leq a_{N+1}$, which establishes the theorem.

EXAMPLE 4 Approximating the Sum of an Alternating Series

Approximate the sum of the following series by its first six terms.
$$\sum_{n=1}^{\infty} (-1)^{n+1} \left(\frac{1}{n!}\right) = \frac{1}{1!} - \frac{1}{2!} + \frac{1}{3!} - \frac{1}{4!} + \frac{1}{5!} - \frac{1}{6!} + \cdots$$

Solution The series converges by the Alternating Series Test because
$$\frac{1}{(n+1)!} \leq \frac{1}{n!} \quad \text{and} \quad \lim_{n \to \infty} \frac{1}{n!} = 0.$$

The sum of the first six terms is
$$S_6 = 1 - \frac{1}{2} + \frac{1}{6} - \frac{1}{24} + \frac{1}{120} - \frac{1}{720} = \frac{91}{144} \approx 0.63194$$

and, by the Alternating Series Remainder, you have
$$|S - S_6| = |R_6| \leq a_7 = \frac{1}{5040} \approx 0.0002.$$

So, the sum S lies between $0.63194 - 0.0002$ and $0.63194 + 0.0002$, and you have
$$0.63174 \leq S \leq 0.63214.$$

TECHNOLOGY Later, in Section 9.10, you will be able to show that the series in Example 4 converges to
$$\frac{e-1}{e} \approx 0.63212.$$

For now, try using a computer to obtain an approximation of the sum of the series. How many terms do you need to obtain an approximation that is within 0.00001 unit of the actual sum?

Absolute and Conditional Convergence

Occasionally, a series may have both positive and negative terms and not be an alternating series. For instance, the series

$$\sum_{n=1}^{\infty} \frac{\sin n}{n^2} = \frac{\sin 1}{1} + \frac{\sin 2}{4} + \frac{\sin 3}{9} + \cdots$$

has both positive and negative terms, yet it is not an alternating series. One way to obtain some information about the convergence of this series is to investigate the convergence of the series

$$\sum_{n=1}^{\infty} \left| \frac{\sin n}{n^2} \right|.$$

By direct comparison, you have $|\sin n| \leq 1$ for all n, so

$$\left| \frac{\sin n}{n^2} \right| \leq \frac{1}{n^2}, \quad n \geq 1.$$

Therefore, by the Direct Comparison Test, the series $\sum \left| \frac{\sin n}{n^2} \right|$ converges. The next theorem tells you that the original series also converges.

THEOREM 9.16 Absolute Convergence

If the series $\sum |a_n|$ converges, then the series $\sum a_n$ also converges.

Proof Because $0 \leq a_n + |a_n| \leq 2|a_n|$ for all n, the series

$$\sum_{n=1}^{\infty} (a_n + |a_n|)$$

converges by comparison with the convergent series

$$\sum_{n=1}^{\infty} 2|a_n|.$$

Furthermore, because $a_n = (a_n + |a_n|) - |a_n|$, you can write

$$\sum_{n=1}^{\infty} a_n = \sum_{n=1}^{\infty} (a_n + |a_n|) - \sum_{n=1}^{\infty} |a_n|$$

where both series on the right converge. So, it follows that $\sum a_n$ converges.

The converse of Theorem 9.16 is not true. For instance, the **alternating harmonic series**

$$\sum_{n=1}^{\infty} \frac{(-1)^{n+1}}{n} = \frac{1}{1} - \frac{1}{2} + \frac{1}{3} - \frac{1}{4} + \cdots$$

converges by the Alternating Series Test. Yet the harmonic series diverges. This type of convergence is called **conditional.**

Definitions of Absolute and Conditional Convergence

1. $\sum a_n$ is **absolutely convergent** if $\sum |a_n|$ converges.
2. $\sum a_n$ is **conditionally convergent** if $\sum a_n$ converges but $\sum |a_n|$ diverges.

EXAMPLE 5 Absolute and Conditional Convergence

Determine whether each of the series is convergent or divergent. Classify any convergent series as absolutely or conditionally convergent.

a. $\displaystyle\sum_{n=0}^{\infty} \frac{(-1)^n n!}{2^n} = \frac{0!}{2^0} - \frac{1!}{2^1} + \frac{2!}{2^2} - \frac{3!}{2^3} + \cdots$

b. $\displaystyle\sum_{n=1}^{\infty} \frac{(-1)^n}{\sqrt{n}} = -\frac{1}{\sqrt{1}} + \frac{1}{\sqrt{2}} - \frac{1}{\sqrt{3}} + \frac{1}{\sqrt{4}} - \cdots$

Solution

a. By the nth-Term Test for Divergence, you can conclude that this series diverges.

b. The given series can be shown to be convergent by the Alternating Series Test. Moreover, because the p-series

$$\sum_{n=1}^{\infty} \left|\frac{(-1)^n}{\sqrt{n}}\right| = \frac{1}{\sqrt{1}} + \frac{1}{\sqrt{2}} + \frac{1}{\sqrt{3}} + \frac{1}{\sqrt{4}} + \cdots$$

diverges, the given series is *conditionally* convergent.

EXAMPLE 6 Absolute and Conditional Convergence

Determine whether each of the series is convergent or divergent. Classify any convergent series as absolutely or conditionally convergent.

a. $\displaystyle\sum_{n=1}^{\infty} \frac{(-1)^{n(n+1)/2}}{3^n} = -\frac{1}{3} - \frac{1}{9} + \frac{1}{27} + \frac{1}{81} - \cdots$

b. $\displaystyle\sum_{n=1}^{\infty} \frac{(-1)^n}{\ln(n+1)} = -\frac{1}{\ln 2} + \frac{1}{\ln 3} - \frac{1}{\ln 4} + \frac{1}{\ln 5} - \cdots$

Solution

a. This is *not* an alternating series. However, because

$$\sum_{n=1}^{\infty} \left|\frac{(-1)^{n(n+1)/2}}{3^n}\right| = \sum_{n=1}^{\infty} \frac{1}{3^n}$$

is a convergent geometric series, you can apply Theorem 9.16 to conclude that the given series is *absolutely* convergent (and therefore convergent).

b. In this case, the Alternating Series Test indicates that the given series converges. However, the series

$$\sum_{n=1}^{\infty} \left|\frac{(-1)^n}{\ln(n+1)}\right| = \frac{1}{\ln 2} + \frac{1}{\ln 3} + \frac{1}{\ln 4} + \cdots$$

diverges by direct comparison with the terms of the harmonic series. Therefore, the given series is *conditionally* convergent.

Rearrangement of Series

A finite sum such as $(1 + 3 - 2 + 5 - 4)$ can be rearranged without changing the value of the sum. This is not necessarily true of an infinite series—it depends on whether the series is absolutely convergent (every rearrangement has the same sum) or conditionally convergent.

EXAMPLE 7 Rearrangement of a Series

The alternating harmonic series converges to ln 2. That is,

$$\sum_{n=1}^{\infty} (-1)^{n+1} \frac{1}{n} = 1 - \frac{1}{2} + \frac{1}{3} - \frac{1}{4} + \cdots = \ln 2.$$

(See Exercise 49, Section 9.10.)

Rearrange the series to produce a different sum.

Solution Consider the following rearrangement.

$$1 - \frac{1}{2} - \frac{1}{4} + \frac{1}{3} - \frac{1}{6} - \frac{1}{8} + \frac{1}{5} - \frac{1}{10} - \frac{1}{12} + \frac{1}{7} - \frac{1}{14} - \cdots$$

$$= \left(1 - \frac{1}{2}\right) - \frac{1}{4} + \left(\frac{1}{3} - \frac{1}{6}\right) - \frac{1}{8} + \left(\frac{1}{5} - \frac{1}{10}\right) - \frac{1}{12} + \left(\frac{1}{7} - \frac{1}{14}\right) - \cdots$$

$$= \frac{1}{2} - \frac{1}{4} + \frac{1}{6} - \frac{1}{8} + \frac{1}{10} - \frac{1}{12} + \frac{1}{14} - \cdots$$

$$= \frac{1}{2}\left(1 - \frac{1}{2} + \frac{1}{3} - \frac{1}{4} + \frac{1}{5} - \frac{1}{6} + \frac{1}{7} - \cdots\right) = \frac{1}{2}(\ln 2)$$

By rearranging the terms, you obtain a sum that is half the original sum.

FOR FURTHER INFORMATION Georg Friedrich Riemann (1826–1866) proved that if $\Sigma\, a_n$ is conditionally convergent and S is any real number, the terms of the series can be rearranged to converge to S. For more on this topic, see the article "Riemann's Rearrangement Theorem" by Stewart Galanor in *Mathematics Teacher*. To view this article, go to the website *www.matharticles.com*.

Exercises for Section 9.5

See www.CalcChat.com for worked-out solutions to odd-numbered exercises.

In Exercises 1–6, match the series with the graph of its sequence of partial sums. [The graphs are labeled (a), (b), (c), (d), (e), and (f).]

(a)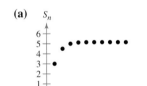

(b)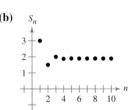

(c)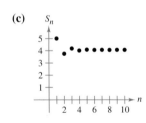

(d)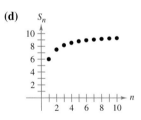

(e)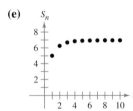

(f)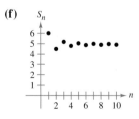

1. $\sum_{n=1}^{\infty} \frac{6}{n^2}$

2. $\sum_{n=1}^{\infty} \frac{(-1)^{n-1} 6}{n^2}$

3. $\sum_{n=1}^{\infty} \frac{3}{n!}$

4. $\sum_{n=1}^{\infty} \frac{(-1)^{n-1} 3}{n!}$

5. $\sum_{n=1}^{\infty} \frac{10}{n 2^n}$

6. $\sum_{n=1}^{\infty} \frac{(-1)^{n-1} 10}{n 2^n}$

 Numerical and Graphical Analysis In Exercises 7–10, explore the Alternating Series Remainder.

(a) Use a graphing utility to find the indicated partial sum S_n and complete the table.

n	1	2	3	4	5	6	7	8	9	10
S_n										

(b) Use a graphing utility to graph the first 10 terms of the sequence of partial sums and a horizontal line representing the sum.

(c) What pattern exists between the plot of the successive points in part (b) relative to the horizontal line representing the sum of the series? Do the distances between the successive points and the horizontal line increase or decrease?

(d) Discuss the relationship between the answers in part (c) and the Alternating Series Remainder as given in Theorem 9.15.

7. $\sum_{n=1}^{\infty} \frac{(-1)^{n-1}}{2n-1} = \frac{\pi}{4}$

8. $\sum_{n=1}^{\infty} \frac{(-1)^{n-1}}{(n-1)!} = \frac{1}{e}$

9. $\sum_{n=1}^{\infty} \frac{(-1)^{n-1}}{n^2} = \frac{\pi^2}{12}$

10. $\sum_{n=1}^{\infty} \frac{(-1)^{n-1}}{(2n-1)!} = \sin 1$

In Exercises 11–32, determine the convergence or divergence of the series.

11. $\sum_{n=1}^{\infty} \frac{(-1)^{n+1}}{n}$

12. $\sum_{n=1}^{\infty} \frac{(-1)^{n+1} n}{2n-1}$

13. $\sum_{n=1}^{\infty} \frac{(-1)^{n+1}}{2n-1}$

14. $\sum_{n=1}^{\infty} \frac{(-1)^n}{\ln(n+1)}$

15. $\sum_{n=1}^{\infty} \frac{(-1)^n n^2}{n^2+1}$

16. $\sum_{n=1}^{\infty} \frac{(-1)^{n+1} n}{n^2+1}$

17. $\sum_{n=1}^{\infty} \frac{(-1)^n}{\sqrt{n}}$

18. $\sum_{n=1}^{\infty} \frac{(-1)^{n+1} n^2}{n^2+5}$

19. $\sum_{n=1}^{\infty} \frac{(-1)^{n+1}(n+1)}{\ln(n+1)}$

20. $\sum_{n=1}^{\infty} \frac{(-1)^{n+1} \ln(n+1)}{n+1}$

21. $\sum_{n=1}^{\infty} \sin \frac{(2n-1)\pi}{2}$

22. $\sum_{n=1}^{\infty} \frac{1}{n} \sin \frac{(2n-1)\pi}{2}$

23. $\sum_{n=1}^{\infty} \cos n\pi$

24. $\sum_{n=1}^{\infty} \frac{1}{n} \cos n\pi$

25. $\sum_{n=0}^{\infty} \frac{(-1)^n}{n!}$

26. $\sum_{n=0}^{\infty} \frac{(-1)^n}{(2n+1)!}$

27. $\sum_{n=1}^{\infty} \frac{(-1)^{n+1} \sqrt{n}}{n+2}$

28. $\sum_{n=1}^{\infty} \frac{(-1)^{n+1} \sqrt{n}}{\sqrt[3]{n}}$

29. $\sum_{n=1}^{\infty} \frac{(-1)^{n+1} n!}{1 \cdot 3 \cdot 5 \cdots (2n-1)}$

30. $\sum_{n=1}^{\infty} (-1)^{n+1} \frac{1 \cdot 3 \cdot 5 \cdots (2n-1)}{1 \cdot 4 \cdot 7 \cdots (3n-2)}$

31. $\sum_{n=1}^{\infty} \frac{2(-1)^{n+1}}{e^n - e^{-n}} = \sum_{n=1}^{\infty} (-1)^{n+1} \operatorname{csch} n$

32. $\sum_{n=1}^{\infty} \frac{2(-1)^{n+1}}{e^n + e^{-n}} = \sum_{n=1}^{\infty} (-1)^{n+1} \operatorname{sech} n$

In Exercises 33–36, approximate the sum of the series by using the first six terms. (See Example 4.)

33. $\sum_{n=1}^{\infty} \frac{(-1)^{n+1} 3}{n^2}$

34. $\sum_{n=1}^{\infty} \frac{(-1)^{n+1} 4}{\ln(n+1)}$

35. $\sum_{n=0}^{\infty} \frac{(-1)^n 2}{n!}$

36. $\sum_{n=1}^{\infty} \frac{(-1)^{n+1} n}{2^n}$

In Exercises 37–42, (a) use Theorem 9.15 to determine the number of terms required to approximate the sum of the convergent series with an error of less than 0.001, and (b) use a graphing utility to approximate the sum of the series with an error of less than 0.001.

37. $\sum_{n=0}^{\infty} \frac{(-1)^n}{n!} = \frac{1}{e}$

38. $\sum_{n=0}^{\infty} \frac{(-1)^n}{2^n n!} = \frac{1}{\sqrt{e}}$

39. $\sum_{n=0}^{\infty} \frac{(-1)^n}{(2n+1)!} = \sin 1$

40. $\sum_{n=0}^{\infty} \frac{(-1)^n}{(2n)!} = \cos 1$

41. $\sum_{n=1}^{\infty} \frac{(-1)^{n+1}}{n} = \ln 2$

42. $\sum_{n=1}^{\infty} \frac{(-1)^{n+1}}{n4^n} = \ln \frac{5}{4}$

In Exercises 43–46, use Theorem 9.15 to determine the number of terms required to approximate the sum of the series with an error of less than 0.001.

43. $\sum_{n=1}^{\infty} \frac{(-1)^{n+1}}{n^3}$

44. $\sum_{n=1}^{\infty} \frac{(-1)^{n+1}}{n^2}$

45. $\sum_{n=1}^{\infty} \frac{(-1)^{n+1}}{2n^3 - 1}$

46. $\sum_{n=1}^{\infty} \frac{(-1)^{n+1}}{n^4}$

In Exercises 47–62, determine whether the series converges conditionally or absolutely, or diverges.

47. $\sum_{n=1}^{\infty} \frac{(-1)^{n+1}}{(n+1)^2}$

48. $\sum_{n=1}^{\infty} \frac{(-1)^{n+1}}{n+1}$

49. $\sum_{n=1}^{\infty} \frac{(-1)^{n+1}}{\sqrt{n}}$

50. $\sum_{n=1}^{\infty} \frac{(-1)^{n+1}}{n\sqrt{n}}$

51. $\sum_{n=1}^{\infty} \frac{(-1)^{n+1} n^2}{(n+1)^2}$

52. $\sum_{n=1}^{\infty} \frac{(-1)^{n+1}(2n+3)}{n+10}$

53. $\sum_{n=2}^{\infty} \frac{(-1)^n}{\ln n}$

54. $\sum_{n=0}^{\infty} (-1)^n e^{-n^2}$

55. $\sum_{n=2}^{\infty} \frac{(-1)^n n}{n^3 - 1}$

56. $\sum_{n=1}^{\infty} \frac{(-1)^{n+1}}{n^{1.5}}$

57. $\sum_{n=0}^{\infty} \frac{(-1)^n}{(2n+1)!}$

58. $\sum_{n=0}^{\infty} \frac{(-1)^n}{\sqrt{n+4}}$

59. $\sum_{n=0}^{\infty} \frac{\cos n\pi}{n+1}$

60. $\sum_{n=1}^{\infty} (-1)^{n+1} \arctan n$

61. $\sum_{n=1}^{\infty} \frac{\cos n\pi}{n^2}$

62. $\sum_{n=1}^{\infty} \frac{\sin[(2n-1)\pi/2]}{n}$

Writing About Concepts

63. Define an alternating series and state the Alternating Series Test.

64. Give the remainder after N terms of a convergent alternating series.

65. In your own words, state the difference between absolute and conditional convergence of an alternating series.

66. The graphs of the sequences of partial sums of two series are shown in the figures. Which graph represents the partial sums of an alternating series? Explain.

(a)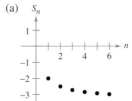

(b)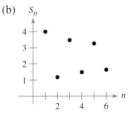

True or False? In Exercises 67–70, determine whether the statement is true or false. If it is false, explain why or give an example that shows it is false.

67. If both $\Sigma \, a_n$ and $\Sigma \, (-a_n)$ converge, then $\Sigma \, |a_n|$ converges.
68. If $\Sigma \, a_n$ diverges, then $\Sigma \, |a_n|$ diverges.
69. For the alternating series $\sum_{n=1}^{\infty} \frac{(-1)^n}{n}$, the partial sum S_{100} is an overestimate of the sum of the series.
70. If $\Sigma \, a_n$ and $\Sigma \, b_n$ both converge, then $\Sigma \, a_n b_n$ converges.

In Exercises 71 and 72, find the values of p for which the series converges.

71. $\sum_{n=1}^{\infty} (-1)^n \left(\frac{1}{n^p} \right)$
72. $\sum_{n=1}^{\infty} (-1)^n \left(\frac{1}{n+p} \right)$

73. Prove that if $\Sigma \, |a_n|$ converges, then $\Sigma \, a_n^2$ converges. Is the converse true? If not, give an example that shows it is false.
74. Use the result of Exercise 71 to give an example of an alternating p-series that converges, but whose corresponding p-series diverges.
75. Give an example of a series that demonstrates the statement you proved in Exercise 73.
76. Find all values of x for which the series $\Sigma \, (x^n/n)$ (a) converges absolutely and (b) converges conditionally.
77. Consider the following series.

$$\frac{1}{2} - \frac{1}{3} + \frac{1}{4} - \frac{1}{9} + \frac{1}{8} - \frac{1}{27} + \cdots + \frac{1}{2^n} - \frac{1}{3^n} + \cdots$$

   (a) Does the series meet the conditions of Theorem 9.14? Explain why or why not.
   (b) Does the series converge? If so, what is the sum?

78. Consider the following series.

$$\sum_{n=1}^{\infty} (-1)^{n+1} a_n, \quad a_n = \begin{cases} \frac{1}{\sqrt{n}}, & \text{if } n \text{ is odd} \\ \frac{1}{n^3}, & \text{if } n \text{ is even} \end{cases}$$

   (a) Does the series meet the conditions of Theorem 9.14? Explain why or why not.
   (b) Does the series converge? If so, what is the sum?

Review In Exercises 79–88, test for convergence or divergence and identify the test used.

79. $\sum_{n=1}^{\infty} \frac{10}{n^{3/2}}$
80. $\sum_{n=1}^{\infty} \frac{3}{n^2 + 5}$
81. $\sum_{n=1}^{\infty} \frac{3^n}{n^2}$
82. $\sum_{n=1}^{\infty} \frac{1}{2^n + 1}$
83. $\sum_{n=0}^{\infty} 5 \left(\frac{7}{8} \right)^n$
84. $\sum_{n=1}^{\infty} \frac{3n^2}{2n^2 + 1}$
85. $\sum_{n=1}^{\infty} 100 e^{-n/2}$
86. $\sum_{n=0}^{\infty} \frac{(-1)^n}{n + 4}$
87. $\sum_{n=1}^{\infty} \frac{(-1)^{n+1} 4}{3n^2 - 1}$
88. $\sum_{n=2}^{\infty} \frac{\ln n}{n}$

89. The following argument, that $0 = 1$, is *incorrect*. Describe the error.

$$0 = 0 + 0 + 0 + \cdots$$
$$= (1 - 1) + (1 - 1) + (1 - 1) + \cdots$$
$$= 1 + (-1 + 1) + (-1 + 1) + \cdots$$
$$= 1 + 0 + 0 + \cdots$$
$$= 1$$

90. The following argument, $2 = 1$, is *incorrect*. Describe the error. Multiply each side of the alternating harmonic series

$$S = 1 - \frac{1}{2} + \frac{1}{3} - \frac{1}{4} + \frac{1}{5} - \frac{1}{6} + \frac{1}{7} - \frac{1}{8} + \frac{1}{9} - \frac{1}{10} + \cdots$$

by 2 to get

$$2S = 2 - 1 + \frac{2}{3} - \frac{1}{2} + \frac{2}{5} - \frac{1}{3} + \frac{2}{7} - \frac{1}{4} + \frac{2}{9} - \frac{1}{5} + \cdots$$

Now collect terms with like denominators (as indicated by the arrows) to get

$$2S = 1 - \frac{1}{2} + \frac{1}{3} - \frac{1}{4} + \frac{1}{5} + \cdots$$

The resulting series is the same one that you started with. So, $2S = S$ and divide each side by S to get $2 = 1$.

FOR FURTHER INFORMATION For more on this exercise, see the article "Riemann's Rearrangement Theorem" by Stewart Galanor in *Mathematics Teacher*. To view this article, go to the website *www.matharticles.com*.

Putnam Exam Challenge

91. Assume as known the (true) fact that the alternating harmonic series

    (1) $1 - \frac{1}{2} + \frac{1}{3} - \frac{1}{4} + \frac{1}{5} - \frac{1}{6} + \frac{1}{7} - \frac{1}{8} + \cdots$

    is convergent, and denote its sum by s. Rearrange the series (1) as follows:

    (2) $1 + \frac{1}{3} - \frac{1}{2} + \frac{1}{5} + \frac{1}{7} - \frac{1}{4} + \frac{1}{9} + \frac{1}{11} - \frac{1}{6} + \cdots$

    Assume as known the (true) fact that the series (2) is also convergent, and denote its sum by S. Denote by s_k, S_k the kth partial sum of the series (1) and (2), respectively. Prove each statement.

    (i) $S_{3n} = s_{4n} + \frac{1}{2} s_{2n}$, (ii) $S \neq s$

This problem was composed by the Committee on the Putnam Prize Competition.
© The Mathematical Association of America. All rights reserved.

Section 9.6 The Ratio and Root Tests

- Use the Ratio Test to determine whether a series converges or diverges.
- Use the Root Test to determine whether a series converges or diverges.
- Review the tests for convergence and divergence of an infinite series.

The Ratio Test

This section begins with a test for absolute convergence—the **Ratio Test**.

THEOREM 9.17 Ratio Test

Let $\Sigma\, a_n$ be a series with nonzero terms.

1. $\Sigma\, a_n$ converges absolutely if $\lim\limits_{n \to \infty} \left| \dfrac{a_{n+1}}{a_n} \right| < 1$.
2. $\Sigma\, a_n$ diverges if $\lim\limits_{n \to \infty} \left| \dfrac{a_{n+1}}{a_n} \right| > 1$ or $\lim\limits_{n \to \infty} \left| \dfrac{a_{n+1}}{a_n} \right| = \infty$.
3. The Ratio Test is inconclusive if $\lim\limits_{n \to \infty} \left| \dfrac{a_{n+1}}{a_n} \right| = 1$.

Proof   To prove Property 1, assume that

$$\lim_{n \to \infty} \left| \frac{a_{n+1}}{a_n} \right| = r < 1$$

and choose R such that $0 \le r < R < 1$. By the definition of the limit of a sequence, there exists some $N > 0$ such that $|a_{n+1}/a_n| < R$ for all $n > N$. Therefore, you can write the following inequalities.

$$|a_{N+1}| < |a_N|R$$
$$|a_{N+2}| < |a_{N+1}|R < |a_N|R^2$$
$$|a_{N+3}| < |a_{N+2}|R < |a_{N+1}|R^2 < |a_N|R^3$$
$$\vdots$$

The geometric series $\Sigma\, |a_N|R^n = |a_N|R + |a_N|R^2 + \cdots + |a_N|R^n + \cdots$ converges, and so, by the Direct Comparison Test, the series

$$\sum_{n=1}^{\infty} |a_{N+n}| = |a_{N+1}| + |a_{N+2}| + \cdots + |a_{N+n}| + \cdots$$

also converges. This in turn implies that the series $\Sigma\, |a_n|$ converges, because discarding a finite number of terms ($n = N - 1$) does not affect convergence. Consequently, by Theorem 9.16, the series $\Sigma\, a_n$ converges absolutely. The proof of Property 2 is similar and is left as an exercise (see Exercise 98).

NOTE   The fact that the Ratio Test is inconclusive when $|a_{n+1}/a_n| \to 1$ can be seen by comparing the two series $\Sigma\, (1/n)$ and $\Sigma\, (1/n^2)$. The first series diverges and the second one converges, but in both cases

$$\lim_{n \to \infty} \left| \frac{a_{n+1}}{a_n} \right| = 1.$$

EXPLORATION

Writing a Series   One of the following conditions guarantees that a series will diverge, two conditions guarantee that a series will converge, and one has no guarantee—the series can either converge or diverge. Which is which? Explain your reasoning.

a. $\lim\limits_{n \to \infty} \left| \dfrac{a_{n+1}}{a_n} \right| = 0$

b. $\lim\limits_{n \to \infty} \left| \dfrac{a_{n+1}}{a_n} \right| = \dfrac{1}{2}$

c. $\lim\limits_{n \to \infty} \left| \dfrac{a_{n+1}}{a_n} \right| = 1$

d. $\lim\limits_{n \to \infty} \left| \dfrac{a_{n+1}}{a_n} \right| = 2$

Although the Ratio Test is not a cure for all ills related to tests for convergence, it is particularly useful for series that *converge rapidly*. Series involving factorials or exponentials are frequently of this type.

EXAMPLE 1 Using the Ratio Test

Determine the convergence or divergence of

$$\sum_{n=0}^{\infty} \frac{2^n}{n!}.$$

Solution Because $a_n = 2^n/n!$, you can write the following.

$$\lim_{n\to\infty} \left| \frac{a_{n+1}}{a_n} \right| = \lim_{n\to\infty} \left[\frac{2^{n+1}}{(n+1)!} \div \frac{2^n}{n!} \right]$$

$$= \lim_{n\to\infty} \left[\frac{2^{n+1}}{(n+1)!} \cdot \frac{n!}{2^n} \right]$$

$$= \lim_{n\to\infty} \frac{2}{n+1}$$

$$= 0$$

Therefore, the series converges.

STUDY TIP A step frequently used in applications of the Ratio Test involves simplifying quotients of factorials. In Example 1, for instance, notice that

$$\frac{n!}{(n+1)!} = \frac{n!}{(n+1)n!} = \frac{1}{n+1}.$$

EXAMPLE 2 Using the Ratio Test

Determine whether each series converges or diverges.

a. $\sum_{n=0}^{\infty} \frac{n^2 2^{n+1}}{3^n}$ **b.** $\sum_{n=1}^{\infty} \frac{n^n}{n!}$

Solution

a. This series converges because the limit of $|a_{n+1}/a_n|$ is less than 1.

$$\lim_{n\to\infty} \left| \frac{a_{n+1}}{a_n} \right| = \lim_{n\to\infty} \left[(n+1)^2 \left(\frac{2^{n+2}}{3^{n+1}} \right) \left(\frac{3^n}{n^2 2^{n+1}} \right) \right]$$

$$= \lim_{n\to\infty} \frac{2(n+1)^2}{3n^2}$$

$$= \frac{2}{3} < 1$$

b. This series diverges because the limit of $|a_{n+1}/a_n|$ is greater than 1.

$$\lim_{n\to\infty} \left| \frac{a_{n+1}}{a_n} \right| = \lim_{n\to\infty} \left[\frac{(n+1)^{n+1}}{(n+1)!} \left(\frac{n!}{n^n} \right) \right]$$

$$= \lim_{n\to\infty} \left[\frac{(n+1)^{n+1}}{(n+1)} \left(\frac{1}{n^n} \right) \right]$$

$$= \lim_{n\to\infty} \frac{(n+1)^n}{n^n}$$

$$= \lim_{n\to\infty} \left(1 + \frac{1}{n} \right)^n$$

$$= e > 1$$

EXAMPLE 3 A Failure of the Ratio Test

Determine the convergence or divergence of $\sum_{n=1}^{\infty} (-1)^n \dfrac{\sqrt{n}}{n+1}$.

Solution The limit of $|a_{n+1}/a_n|$ is equal to 1.

$$\lim_{n\to\infty} \left|\frac{a_{n+1}}{a_n}\right| = \lim_{n\to\infty} \left[\left(\frac{\sqrt{n+1}}{n+2}\right)\left(\frac{n+1}{\sqrt{n}}\right)\right]$$

$$= \lim_{n\to\infty} \left[\sqrt{\frac{n+1}{n}}\left(\frac{n+1}{n+2}\right)\right]$$

$$= \sqrt{1}(1)$$

$$= 1$$

NOTE The Ratio Test is also inconclusive for any p-series.

So, the Ratio Test is inconclusive. To determine whether the series converges, you need to try a different test. In this case, you can apply the Alternating Series Test. To show that $a_{n+1} \leq a_n$, let

$$f(x) = \frac{\sqrt{x}}{x+1}.$$

Then the derivative is

$$f'(x) = \frac{-x+1}{2\sqrt{x}(x+1)^2}.$$

Because the derivative is negative for $x > 1$, you know that f is a decreasing function. Also, by L'Hôpital's Rule,

$$\lim_{x\to\infty} \frac{\sqrt{x}}{x+1} = \lim_{x\to\infty} \frac{1/(2\sqrt{x})}{1}$$

$$= \lim_{x\to\infty} \frac{1}{2\sqrt{x}}$$

$$= 0.$$

Therefore, by the Alternating Series Test, the series converges.

The series in Example 3 is *conditionally convergent*. This follows from the fact that the series

$$\sum_{n=1}^{\infty} |a_n|$$

diverges (by the Limit Comparison Test with $\sum 1/\sqrt{n}$), but the series

$$\sum_{n=1}^{\infty} a_n$$

converges.

TECHNOLOGY A computer or programmable calculator can reinforce the conclusion that the series in Example 3 converges *conditionally*. By adding the first 100 terms of the series, you obtain a sum of about -0.2. (The sum of the first 100 terms of the series $\sum |a_n|$ is about 17.)

The Root Test

The next test for convergence or divergence of series works especially well for series involving nth powers. The proof of this theorem is similar to that given for the Ratio Test, and is left as an exercise (see Exercise 99).

THEOREM 9.18 Root Test

Let $\Sigma\, a_n$ be a series.

1. $\Sigma\, a_n$ converges absolutely if $\lim_{n\to\infty} \sqrt[n]{|a_n|} < 1$.
2. $\Sigma\, a_n$ diverges if $\lim_{n\to\infty} \sqrt[n]{|a_n|} > 1$ or $\lim_{n\to\infty} \sqrt[n]{|a_n|} = \infty$.
3. The Root Test is inconclusive if $\lim_{n\to\infty} \sqrt[n]{|a_n|} = 1$.

EXAMPLE 4 Using the Root Test

Determine the convergence or divergence of

$$\sum_{n=1}^{\infty} \frac{e^{2n}}{n^n}.$$

Solution   You can apply the Root Test as follows.

$$\lim_{n\to\infty} \sqrt[n]{|a_n|} = \lim_{n\to\infty} \sqrt[n]{\frac{e^{2n}}{n^n}}$$

$$= \lim_{n\to\infty} \frac{e^{2n/n}}{n^{n/n}}$$

$$= \lim_{n\to\infty} \frac{e^2}{n}$$

$$= 0 < 1$$

Because this limit is less than 1, you can conclude that the series converges absolutely (and therefore converges).

NOTE   The Root Test is always inconclusive for any p-series.

To see the usefulness of the Root Test for the series in Example 4, try applying the Ratio Test to that series. When you do this, you obtain the following.

$$\lim_{n\to\infty} \left|\frac{a_{n+1}}{a_n}\right| = \lim_{n\to\infty} \left[\frac{e^{2(n+1)}}{(n+1)^{n+1}} \div \frac{e^{2n}}{n^n}\right]$$

$$= \lim_{n\to\infty} e^2 \frac{n^n}{(n+1)^{n+1}}$$

$$= \lim_{n\to\infty} e^2 \left(\frac{n}{n+1}\right)^n \left(\frac{1}{n+1}\right)$$

$$= 0$$

Note that this limit is not as easily evaluated as the limit obtained by the Root Test in Example 4.

FOR FURTHER INFORMATION   For more information on the usefulness of the Root Test, see the article "N! and the Root Test" by Charles C. Mumma II in *The American Mathematical Monthly*. To view this article, go to the website *www.matharticles.com*.

Strategies for Testing Series

You have now studied 10 tests for determining the convergence or divergence of an infinite series. (See the summary in the table on page 644.) Skill in choosing and applying the various tests will come only with practice. Below is a set of guidelines for choosing an appropriate test.

> **Guidelines for Testing a Series for Convergence or Divergence**
>
> 1. Does the nth term approach 0? If not, the series diverges.
> 2. Is the series one of the special types—geometric, p-series, telescoping, or alternating?
> 3. Can the Integral Test, the Root Test, or the Ratio Test be applied?
> 4. Can the series be compared favorably to one of the special types?

In some instances, more than one test is applicable. However, your objective should be to learn to choose the most efficient test.

EXAMPLE 5 Applying the Strategies for Testing Series

Determine the convergence or divergence of each series.

a. $\sum_{n=1}^{\infty} \dfrac{n+1}{3n+1}$
b. $\sum_{n=1}^{\infty} \left(\dfrac{\pi}{6}\right)^n$
c. $\sum_{n=1}^{\infty} n e^{-n^2}$
d. $\sum_{n=1}^{\infty} \dfrac{1}{3n+1}$
e. $\sum_{n=1}^{\infty} (-1)^n \dfrac{3}{4n+1}$
f. $\sum_{n=1}^{\infty} \dfrac{n!}{10^n}$
g. $\sum_{n=1}^{\infty} \left(\dfrac{n+1}{2n+1}\right)^n$

Solution

a. For this series, the limit of the nth term is not 0 $\left(a_n \to \tfrac{1}{3} \text{ as } n \to \infty\right)$. So, by the nth-Term Test, the series diverges.

b. This series is geometric. Moreover, because the ratio $r = \pi/6$ of the terms is less than 1 in absolute value, you can conclude that the series converges.

c. Because the function $f(x) = xe^{-x^2}$ is easily integrated, you can use the Integral Test to conclude that the series converges.

d. The nth term of this series can be compared to the nth term of the harmonic series. After using the Limit Comparison Test, you can conclude that the series diverges.

e. This is an alternating series whose nth term approaches 0. Because $a_{n+1} \leq a_n$, you can use the Alternating Series Test to conclude that the series converges.

f. The nth term of this series involves a factorial, which indicates that the Ratio Test may work well. After applying the Ratio Test, you can conclude that the series diverges.

g. The nth term of this series involves a variable that is raised to the nth power, which indicates that the Root Test may work well. After applying the Root Test, you can conclude that the series converges.

Summary of Tests for Series

Test	Series	Condition(s) of Convergence	Condition(s) of Divergence	Comment
nth-Term	$\sum_{n=1}^{\infty} a_n$		$\lim_{n \to \infty} a_n \neq 0$	This test cannot be used to show convergence.
Geometric Series	$\sum_{n=0}^{\infty} ar^n$	$\|r\| < 1$	$\|r\| \geq 1$	Sum: $S = \dfrac{a}{1-r}$
Telescoping Series	$\sum_{n=1}^{\infty} (b_n - b_{n+1})$	$\lim_{n \to \infty} b_n = L$		Sum: $S = b_1 - L$
p-Series	$\sum_{n=1}^{\infty} \dfrac{1}{n^p}$	$p > 1$	$p \leq 1$	
Alternating Series	$\sum_{n=1}^{\infty} (-1)^{n-1} a_n$	$0 < a_{n+1} \leq a_n$ and $\lim_{n \to \infty} a_n = 0$		Remainder: $\|R_N\| \leq a_{N+1}$
Integral (f is continuous, positive, and decreasing)	$\sum_{n=1}^{\infty} a_n$, $a_n = f(n) \geq 0$	$\int_1^{\infty} f(x)\,dx$ converges	$\int_1^{\infty} f(x)\,dx$ diverges	Remainder: $0 < R_N < \int_N^{\infty} f(x)\,dx$
Root	$\sum_{n=1}^{\infty} a_n$	$\lim_{n \to \infty} \sqrt[n]{\|a_n\|} < 1$	$\lim_{n \to \infty} \sqrt[n]{\|a_n\|} > 1$	Test is inconclusive if $\lim_{n \to \infty} \sqrt[n]{\|a_n\|} = 1$.
Ratio	$\sum_{n=1}^{\infty} a_n$	$\lim_{n \to \infty} \left\|\dfrac{a_{n+1}}{a_n}\right\| < 1$	$\lim_{n \to \infty} \left\|\dfrac{a_{n+1}}{a_n}\right\| > 1$	Test is inconclusive if $\lim_{n \to \infty} \left\|\dfrac{a_{n+1}}{a_n}\right\| = 1$.
Direct Comparison ($a_n, b_n > 0$)	$\sum_{n=1}^{\infty} a_n$	$0 < a_n \leq b_n$ and $\sum_{n=1}^{\infty} b_n$ converges	$0 < b_n \leq a_n$ and $\sum_{n=1}^{\infty} b_n$ diverges	
Limit Comparison ($a_n, b_n > 0$)	$\sum_{n=1}^{\infty} a_n$	$\lim_{n \to \infty} \dfrac{a_n}{b_n} = L > 0$ and $\sum_{n=1}^{\infty} b_n$ converges	$\lim_{n \to \infty} \dfrac{a_n}{b_n} = L > 0$ and $\sum_{n=1}^{\infty} b_n$ diverges	

Exercises for Section 9.6

See www.CalcChat.com for worked-out solutions to odd-numbered exercises.

In Exercises 1–4, verify the formula.

1. $\dfrac{(n+1)!}{(n-2)!} = (n+1)(n)(n-1)$

2. $\dfrac{(2k-2)!}{(2k)!} = \dfrac{1}{(2k)(2k-1)}$

3. $1 \cdot 3 \cdot 5 \cdots (2k-1) = \dfrac{(2k)!}{2^k k!}$

4. $\dfrac{1}{1 \cdot 3 \cdot 5 \cdots (2k-5)} = \dfrac{2^k k! (2k-3)(2k-1)}{(2k)!}, \quad k \geq 3$

In Exercises 5–10, match the series with the graph of its sequence of partial sums. [The graphs are labeled (a), (b), (c), (d), (e), and (f).]

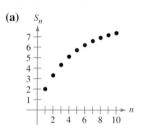

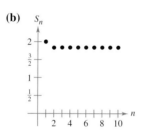

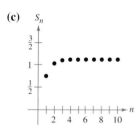

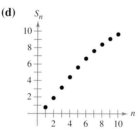

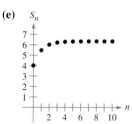

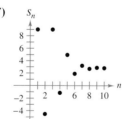

5. $\sum_{n=1}^{\infty} n\left(\dfrac{3}{4}\right)^n$

6. $\sum_{n=1}^{\infty} \left(\dfrac{3}{4}\right)^n \left(\dfrac{1}{n!}\right)$

7. $\sum_{n=1}^{\infty} \dfrac{(-3)^{n+1}}{n!}$

8. $\sum_{n=1}^{\infty} \dfrac{(-1)^{n-1} 4}{(2n)!}$

9. $\sum_{n=1}^{\infty} \left(\dfrac{4n}{5n-3}\right)^n$

10. $\sum_{n=0}^{\infty} 4e^{-n}$

Numerical, Graphical, and Analytic Analysis In Exercises 11 and 12, (a) verify that the series converges. (b) Use a graphing utility to find the indicated partial sum S_n and complete the table. (c) Use a graphing utility to graph the first 10 terms of the sequence of partial sums. (d) Use the table to estimate the sum of the series. (e) Explain the relationship between the magnitudes of the terms of the series and the rate at which the sequence of partial sums approaches the sum of the series.

n	5	10	15	20	25
S_n					

11. $\sum_{n=1}^{\infty} n^2 \left(\dfrac{5}{8}\right)^n$

12. $\sum_{n=1}^{\infty} \dfrac{n^2+1}{n!}$

In Exercises 13–32, use the Ratio Test to determine the convergence or divergence of the series.

13. $\sum_{n=0}^{\infty} \dfrac{n!}{3^n}$

14. $\sum_{n=0}^{\infty} \dfrac{3^n}{n!}$

15. $\sum_{n=1}^{\infty} n\left(\dfrac{3}{4}\right)^n$

16. $\sum_{n=1}^{\infty} n\left(\dfrac{3}{2}\right)^n$

17. $\sum_{n=1}^{\infty} \dfrac{n}{2^n}$

18. $\sum_{n=1}^{\infty} \dfrac{n^3}{2^n}$

19. $\sum_{n=1}^{\infty} \dfrac{2^n}{n^2}$

20. $\sum_{n=1}^{\infty} \dfrac{(-1)^{n+1}(n+2)}{n(n+1)}$

21. $\sum_{n=0}^{\infty} \dfrac{(-1)^n 2^n}{n!}$

22. $\sum_{n=1}^{\infty} \dfrac{(-1)^{n-1}(3/2)^n}{n^2}$

23. $\sum_{n=1}^{\infty} \dfrac{n!}{n3^n}$

24. $\sum_{n=1}^{\infty} \dfrac{(2n)!}{n^5}$

25. $\sum_{n=0}^{\infty} \dfrac{4^n}{n!}$

26. $\sum_{n=1}^{\infty} \dfrac{n^n}{n!}$

27. $\sum_{n=0}^{\infty} \dfrac{3^n}{(n+1)^n}$

28. $\sum_{n=0}^{\infty} \dfrac{(n!)^2}{(3n)!}$

29. $\sum_{n=1}^{\infty} \dfrac{4^n}{3^n+1}$

30. $\sum_{n=0}^{\infty} \dfrac{(-1)^n 2^{4n}}{(2n+1)!}$

31. $\sum_{n=0}^{\infty} \dfrac{(-1)^{n+1} n!}{1 \cdot 3 \cdot 5 \cdots (2n+1)}$

32. $\sum_{n=1}^{\infty} \dfrac{(-1)^n [2 \cdot 4 \cdot 6 \cdots (2n)]}{2 \cdot 5 \cdot 8 \cdots (3n-1)}$

In Exercises 33–36, verify that the Ratio Test is inconclusive for the p-series.

33. $\sum_{n=1}^{\infty} \dfrac{1}{n^{3/2}}$

34. $\sum_{n=1}^{\infty} \dfrac{1}{n^{1/2}}$

35. $\sum_{n=1}^{\infty} \dfrac{1}{n^4}$

36. $\sum_{n=1}^{\infty} \dfrac{1}{n^p}$

In Exercises 37–50, use the Root Test to determine the convergence or divergence of the series.

37. $\sum_{n=1}^{\infty} \left(\dfrac{n}{2n+1}\right)^n$

38. $\sum_{n=1}^{\infty} \left(\dfrac{2n}{n+1}\right)^n$

39. $\sum_{n=2}^{\infty} \left(\dfrac{2n+1}{n-1}\right)^n$

40. $\sum_{n=1}^{\infty} \left(\dfrac{4n+3}{2n-1}\right)^n$

41. $\sum_{n=2}^{\infty} \dfrac{(-1)^n}{(\ln n)^n}$

42. $\sum_{n=1}^{\infty} \left(\dfrac{-3n}{2n+1}\right)^{3n}$

43. $\sum_{n=1}^{\infty} \left(2\sqrt[n]{n}+1\right)^n$

44. $\sum_{n=0}^{\infty} e^{-n}$

45. $\sum_{n=1}^{\infty} \dfrac{n}{4^n}$

46. $\sum_{n=1}^{\infty} \left(\dfrac{n}{500}\right)^n$

47. $\sum_{n=1}^{\infty} \left(\dfrac{1}{n}-\dfrac{1}{n^2}\right)^n$

48. $\sum_{n=1}^{\infty} \left(\dfrac{\ln n}{n}\right)^n$

49. $\sum_{n=2}^{\infty} \dfrac{n}{(\ln n)^n}$

50. $\sum_{n=1}^{\infty} \dfrac{(n!)^n}{(n^n)^2}$

In Exercises 51–68, determine the convergence or divergence of the series using any appropriate test from this chapter. Identify the test used.

51. $\sum_{n=1}^{\infty} \dfrac{(-1)^{n+1}5}{n}$

52. $\sum_{n=1}^{\infty} \dfrac{5}{n}$

53. $\sum_{n=1}^{\infty} \dfrac{3}{n\sqrt{n}}$

54. $\sum_{n=1}^{\infty} \left(\dfrac{\pi}{4}\right)^n$

55. $\sum_{n=1}^{\infty} \dfrac{2n}{n+1}$

56. $\sum_{n=1}^{\infty} \dfrac{n}{2n^2+1}$

57. $\sum_{n=1}^{\infty} \dfrac{(-1)^n 3^{n-2}}{2^n}$

58. $\sum_{n=1}^{\infty} \dfrac{10}{3\sqrt{n^3}}$

59. $\sum_{n=1}^{\infty} \dfrac{10n+3}{n2^n}$

60. $\sum_{n=1}^{\infty} \dfrac{2^n}{4n^2-1}$

61. $\sum_{n=1}^{\infty} \dfrac{\cos n}{2^n}$

62. $\sum_{n=2}^{\infty} \dfrac{(-1)^n}{n \ln n}$

63. $\sum_{n=1}^{\infty} \dfrac{n7^n}{n!}$

64. $\sum_{n=1}^{\infty} \dfrac{\ln n}{n^2}$

65. $\sum_{n=1}^{\infty} \dfrac{(-1)^n 3^{n-1}}{n!}$

66. $\sum_{n=1}^{\infty} \dfrac{(-1)^n 3^n}{n2^n}$

67. $\sum_{n=1}^{\infty} \dfrac{(-3)^n}{3 \cdot 5 \cdot 7 \cdots (2n+1)}$

68. $\sum_{n=1}^{\infty} \dfrac{3 \cdot 5 \cdot 7 \cdots (2n+1)}{18^n (2n-1)n!}$

In Exercises 69–72, identify the two series that are the same.

69. (a) $\sum_{n=1}^{\infty} \dfrac{n5^n}{n!}$

(b) $\sum_{n=0}^{\infty} \dfrac{n5^n}{(n+1)!}$

(c) $\sum_{n=0}^{\infty} \dfrac{(n+1)5^{n+1}}{(n+1)!}$

70. (a) $\sum_{n=4}^{\infty} n\left(\dfrac{3}{4}\right)^n$

(b) $\sum_{n=0}^{\infty} (n+1)\left(\dfrac{3}{4}\right)^n$

(c) $\sum_{n=1}^{\infty} n\left(\dfrac{3}{4}\right)^{n-1}$

71. (a) $\sum_{n=0}^{\infty} \dfrac{(-1)^n}{(2n+1)!}$

(b) $\sum_{n=1}^{\infty} \dfrac{(-1)^{n-1}}{(2n-1)!}$

(c) $\sum_{n=1}^{\infty} \dfrac{(-1)^{n-1}}{(2n+1)!}$

72. (a) $\sum_{n=2}^{\infty} \dfrac{(-1)^n}{(n-1)2^{n-1}}$

(b) $\sum_{n=1}^{\infty} \dfrac{(-1)^{n+1}}{n2^n}$

(c) $\sum_{n=0}^{\infty} \dfrac{(-1)^{n+1}}{(n+1)2^n}$

In Exercises 73 and 74, write an equivalent series with the index of summation beginning at $n = 0$.

73. $\sum_{n=1}^{\infty} \dfrac{n}{4^n}$

74. $\sum_{n=2}^{\infty} \dfrac{2^n}{(n-2)!}$

In Exercises 75 and 76, (a) determine the number of terms required to approximate the sum of the series with an error less than 0.0001, and (b) use a graphing utility to approximate the sum of the series with an error less than 0.0001.

75. $\sum_{k=1}^{\infty} \dfrac{(-3)^k}{2^k k!}$

76. $\sum_{k=0}^{\infty} \dfrac{(-3)^k}{1 \cdot 3 \cdot 5 \cdots (2k+1)}$

In Exercises 77–82, the terms of a series $\sum_{n=1}^{\infty} a_n$ are defined recursively. Determine the convergence or divergence of the series. Explain your reasoning.

77. $a_1 = \dfrac{1}{2},\ a_{n+1} = \dfrac{4n-1}{3n+2}a_n$

78. $a_1 = 2,\ a_{n+1} = \dfrac{2n+1}{5n-4}a_n$

79. $a_1 = 1,\ a_{n+1} = \dfrac{\sin n + 1}{\sqrt{n}}a_n$

80. $a_1 = \dfrac{1}{5},\ a_{n+1} = \dfrac{\cos n + 1}{n}a_n$

81. $a_1 = \dfrac{1}{3},\ a_{n+1} = \left(1+\dfrac{1}{n}\right)a_n$

82. $a_1 = \dfrac{1}{4},\ a_{n+1} = \sqrt[n]{a_n}$

In Exercises 83–86, use the Ratio Test or the Root Test to determine the convergence or divergence of the series.

83. $1 + \dfrac{1 \cdot 2}{1 \cdot 3} + \dfrac{1 \cdot 2 \cdot 3}{1 \cdot 3 \cdot 5} + \dfrac{1 \cdot 2 \cdot 3 \cdot 4}{1 \cdot 3 \cdot 5 \cdot 7} + \cdots$

84. $1 + \dfrac{2}{3} + \dfrac{3}{3^2} + \dfrac{4}{3^3} + \dfrac{5}{3^4} + \dfrac{6}{3^5} + \cdots$

85. $\dfrac{1}{(\ln 3)^3} + \dfrac{1}{(\ln 4)^4} + \dfrac{1}{(\ln 5)^5} + \dfrac{1}{(\ln 6)^6} + \cdots$

86. $1 + \dfrac{1 \cdot 3}{1 \cdot 2 \cdot 3} + \dfrac{1 \cdot 3 \cdot 5}{1 \cdot 2 \cdot 3 \cdot 4 \cdot 5}$
$+ \dfrac{1 \cdot 3 \cdot 5 \cdot 7}{1 \cdot 2 \cdot 3 \cdot 4 \cdot 5 \cdot 6 \cdot 7} + \cdots$

In Exercises 87–92, find the values of x for which the series converges.

87. $\sum_{n=0}^{\infty} 2\left(\dfrac{x}{3}\right)^n$

88. $\sum_{n=0}^{\infty} \left(\dfrac{x+1}{4}\right)^n$

89. $\sum_{n=1}^{\infty} \dfrac{(-1)^n (x+1)^n}{n}$

90. $\sum_{n=0}^{\infty} 2(x-1)^n$

91. $\sum_{n=0}^{\infty} n!\left(\dfrac{x}{2}\right)^n$

92. $\sum_{n=0}^{\infty} \dfrac{(x+1)^n}{n!}$

Writing About Concepts

93. State the Ratio Test.

94. State the Root Test.

95. You are told that the terms of a positive series appear to approach zero rapidly as n approaches infinity. In fact, $a_7 \leq 0.0001$. Given no other information, does this imply that the series converges? Support your conclusion with examples.

96. The graph shows the first 10 terms of the sequence of partial sums of the convergent series

$$\sum_{n=1}^{\infty} \left(\dfrac{2n}{3n+2}\right)^n.$$

Find a series such that the terms of its sequence of partial sums are less than the corresponding terms of the sequence in the figure, but such that the series diverges. Explain your reasoning.

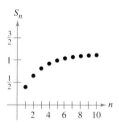

97. Using the Ratio Test, it is determined that an alternating series converges. Does the series converge conditionally or absolutely? Explain.

98. Prove Property 2 of Theorem 9.17.

99. Prove Theorem 9.18. (*Hint for Property 1:* If the limit equals $r < 1$, choose a real number R such that $r < R < 1$. By the definitions of the limit, there exists some $N > 0$ such that $\sqrt[n]{|a_n|} < R$ for $n > N$.)

100. Show that the Root Test is inconclusive for the p-series

$$\sum_{n=1}^{\infty} \dfrac{1}{n^p}.$$

101. Show that the Ratio Test and the Root Test are both inconclusive for the logarithmic p-series

$$\sum_{n=2}^{\infty} \dfrac{1}{n(\ln n)^p}.$$

102. Determine the convergence or divergence of the series

$$\sum_{n=1}^{\infty} \dfrac{(n!)^2}{(xn)!}$$

when (a) $x = 1$, (b) $x = 2$, (c) $x = 3$, and (d) x is a positive integer.

103. Show that if $\sum_{n=1}^{\infty} a_n$ is absolutely convergent, then

$$\left| \sum_{n=1}^{\infty} a_n \right| \leq \sum_{n=1}^{\infty} |a_n|.$$

104. *Writing* Read the article "A Differentiation Test for Absolute Convergence" by Yaser S. Abu-Mostafa in *Mathematics Magazine*. Then write a paragraph that describes the test. Include examples of series that converge and examples of series that diverge.

Putnam Exam Challenge

105. Is the following series convergent or divergent?

$$1 + \dfrac{1}{2}\cdot\dfrac{19}{7} + \dfrac{2!}{3^2}\left(\dfrac{19}{7}\right)^2 + \dfrac{3!}{4^3}\left(\dfrac{19}{7}\right)^3 + \dfrac{4!}{5^4}\left(\dfrac{19}{7}\right)^4 + \cdots$$

106. Show that if the series

$$a_1 + a_2 + a_3 + \cdots + a_n + \cdots$$

converges, then the series

$$a_1 + \dfrac{a_2}{2} + \dfrac{a_3}{3} + \cdots + \dfrac{a_n}{n} + \cdots$$

converges also.

These problems were composed by the Committee on the Putnam Prize Competition.
© The Mathematical Association of America. All rights reserved.

Section 9.7 Taylor Polynomials and Approximations

- Find polynomial approximations of elementary functions and compare them with the elementary functions.
- Find Taylor and Maclaurin polynomial approximations of elementary functions.
- Use the remainder of a Taylor polynomial.

Polynomial Approximations of Elementary Functions

The goal of this section is to show how polynomial functions can be used as approximations for other elementary functions. To find a polynomial function P that approximates another function f, begin by choosing a number c in the domain of f at which f and P have the same value. That is,

$$P(c) = f(c). \qquad \text{Graphs of } f \text{ and } P \text{ pass through } (c, f(c)).$$

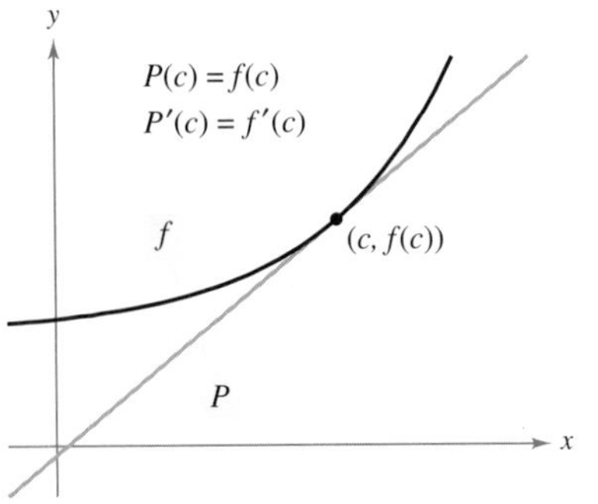

The approximating polynomial is said to be **expanded about** c or **centered at** c. Geometrically, the requirement that $P(c) = f(c)$ means that the graph of P passes through the point $(c, f(c))$. Of course, there are many polynomials whose graphs pass through the point $(c, f(c))$. Your task is to find a polynomial whose graph resembles the graph of f near this point. One way to do this is to impose the additional requirement that the slope of the polynomial function be the same as the slope of the graph of f at the point $(c, f(c))$.

$$P'(c) = f'(c) \qquad \text{Graphs of } f \text{ and } P \text{ have the same slope at } (c, f(c)).$$

With these two requirements, you can obtain a simple linear approximation of f, as shown in Figure 9.10.

Near $(c, f(c))$, the graph of P can be used to approximate the graph of f.
Figure 9.10

EXAMPLE 1 First-Degree Polynomial Approximation of $f(x) = e^x$

For the function $f(x) = e^x$, find a first-degree polynomial function

$$P_1(x) = a_0 + a_1 x$$

whose value and slope agree with the value and slope of f at $x = 0$.

Solution Because $f(x) = e^x$ and $f'(x) = e^x$, the value and the slope of f, at $x = 0$, are given by

$$f(0) = e^0 = 1$$

and

$$f'(0) = e^0 = 1.$$

Because $P_1(x) = a_0 + a_1 x$, you can use the condition that $P_1(0) = f(0)$ to conclude that $a_0 = 1$. Moreover, because $P_1{}'(x) = a_1$, you can use the condition that $P_1{}'(0) = f'(0)$ to conclude that $a_1 = 1$. Therefore,

$$P_1(x) = 1 + x.$$

Figure 9.11 shows the graphs of $P_1(x) = 1 + x$ and $f(x) = e^x$.

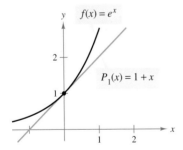

P_1 is the first-degree polynomial approximation of $f(x) = e^x$.
Figure 9.11

NOTE Example 1 isn't the first time you have used a linear function to approximate another function. The same procedure was used as the basis for Newton's Method.

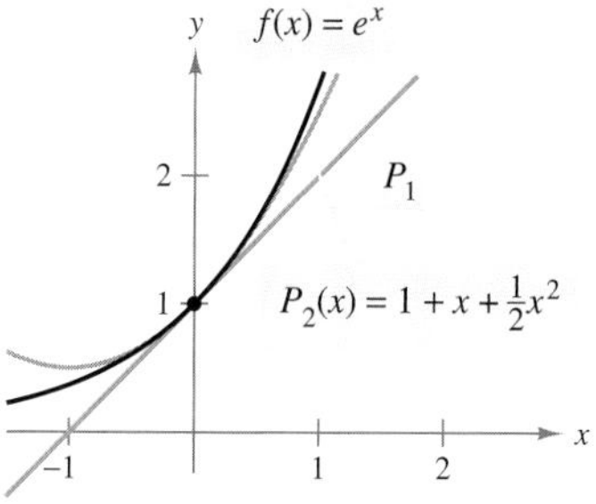

In Figure 9.12 you can see that, at points near (0, 1), the graph of

$$P_1(x) = 1 + x \qquad \text{1st-degree approximation}$$

is reasonably close to the graph of $f(x) = e^x$. However, as you move away from (0, 1), the graphs move farther from each other and the accuracy of the approximation decreases. To improve the approximation, you can impose yet another requirement—that the values of the second derivatives of P and f agree when $x = 0$. The polynomial, P_2, of least degree that satisfies all three requirements $P_2(0) = f(0)$, $P_2'(0) = f'(0)$, and $P_2''(0) = f''(0)$ can be shown to be

$$P_2(x) = 1 + x + \frac{1}{2}x^2. \qquad \text{2nd-degree approximation}$$

Moreover, in Figure 9.12, you can see that P_2 is a better approximation of f than P_1. If you continue this pattern, requiring that the values of $P_n(x)$ and its first n derivatives match those of $f(x) = e^x$ at $x = 0$, you obtain the following.

$$P_n(x) = 1 + x + \frac{1}{2}x^2 + \frac{1}{3!}x^3 + \cdots + \frac{1}{n!}x^n \qquad \text{nth-degree approximation}$$
$$\approx e^x$$

$f(x) = e^x$

$P_2(x) = 1 + x + \frac{1}{2}x^2$

P_2 is the second-degree polynomial approximation of $f(x) = e^x$.
Figure 9.12

EXAMPLE 2 Third-Degree Polynomial Approximation of $f(x) = e^x$

Construct a table comparing the values of the polynomial

$$P_3(x) = 1 + x + \frac{1}{2}x^2 + \frac{1}{3!}x^3 \qquad \text{3rd-degree approximation}$$

with $f(x) = e^x$ for several values of x near 0.

Solution Using a calculator or a computer, you can obtain the results shown in the table. Note that for $x = 0$, the two functions have the same value, but that as x moves farther away from 0, the accuracy of the approximating polynomial $P_3(x)$ decreases.

x	-1.0	-0.2	-0.1	0	0.1	0.2	1.0
e^x	0.3679	0.81873	0.904837	1	1.105171	1.22140	2.7183
$P_3(x)$	0.3333	0.81867	0.904833	1	1.105167	1.22133	2.6667

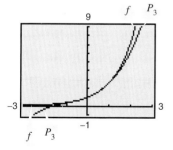

P_3 is the third-degree polynomial approximation of $f(x) = e^x$.
Figure 9.13

TECHNOLOGY A graphing utility can be used to compare the graph of the approximating polynomial with the graph of the function f. For instance, in Figure 9.13, the graph of

$$P_3(x) = 1 + x + \tfrac{1}{2}x^2 + \tfrac{1}{6}x^3 \qquad \text{3rd-degree approximation}$$

is compared with the graph of $f(x) = e^x$. If you have access to a graphing utility, try comparing the graphs of

$$P_4(x) = 1 + x + \tfrac{1}{2}x^2 + \tfrac{1}{6}x^3 + \tfrac{1}{24}x^4 \qquad \text{4th-degree approximation}$$
$$P_5(x) = 1 + x + \tfrac{1}{2}x^2 + \tfrac{1}{6}x^3 + \tfrac{1}{24}x^4 + \tfrac{1}{120}x^5 \qquad \text{5th-degree approximation}$$
$$P_6(x) = 1 + x + \tfrac{1}{2}x^2 + \tfrac{1}{6}x^3 + \tfrac{1}{24}x^4 + \tfrac{1}{120}x^5 + \tfrac{1}{720}x^6 \qquad \text{6th-degree approximation}$$

with the graph of f. What do you notice?

Taylor and Maclaurin Polynomials

The polynomial approximation of $f(x) = e^x$ given in Example 2 is expanded about $c = 0$. For expansions about an arbitrary value of c, it is convenient to write the polynomial in the form

$$P_n(x) = a_0 + a_1(x - c) + a_2(x - c)^2 + a_3(x - c)^3 + \cdots + a_n(x - c)^n.$$

In this form, repeated differentiation produces

$$P_n'(x) = a_1 + 2a_2(x - c) + 3a_3(x - c)^2 + \cdots + na_n(x - c)^{n-1}$$
$$P_n''(x) = 2a_2 + 2(3a_3)(x - c) + \cdots + n(n - 1)a_n(x - c)^{n-2}$$
$$P_n'''(x) = 2(3a_3) + \cdots + n(n - 1)(n - 2)a_n(x - c)^{n-3}$$
$$\vdots$$
$$P_n^{(n)}(x) = n(n - 1)(n - 2) \cdots (2)(1)a_n.$$

Letting $x = c$, you then obtain

$$P_n(c) = a_0, \quad P_n'(c) = a_1, \quad P_n''(c) = 2a_2, \quad \ldots, \quad P_n^{(n)}(c) = n!a_n$$

and because the value of f and its first n derivatives must agree with the value of P_n and its first n derivatives at $x = c$, it follows that

$$f(c) = a_0, \quad f'(c) = a_1, \quad \frac{f''(c)}{2!} = a_2, \quad \ldots, \quad \frac{f^{(n)}(c)}{n!} = a_n.$$

With these coefficients, you can obtain the following definition of **Taylor polynomials**, named after the English mathematician Brook Taylor, and **Maclaurin polynomials**, named after the English mathematician Colin Maclaurin (1698–1746).

Definitions of *n*th Taylor Polynomial and *n*th Maclaurin Polynomial

If f has n derivatives at c, then the polynomial

$$P_n(x) = f(c) + f'(c)(x - c) + \frac{f''(c)}{2!}(x - c)^2 + \cdots + \frac{f^{(n)}(c)}{n!}(x - c)^n$$

is called the **nth Taylor polynomial for f at c**. If $c = 0$, then

$$P_n(x) = f(0) + f'(0)x + \frac{f''(0)}{2!}x^2 + \frac{f'''(0)}{3!}x^3 + \cdots + \frac{f^{(n)}(0)}{n!}x^n$$

is also called the **nth Maclaurin polynomial for f**.

Brook Taylor (1685–1731)

Although Taylor was not the first to seek polynomial approximations of transcendental functions, his account published in 1715 was one of the first comprehensive works on the subject.

NOTE Maclaurin polynomials are special types of Taylor polynomials for which $c = 0$.

EXAMPLE 3 A Maclaurin Polynomial for $f(x) = e^x$

Find the nth Maclaurin polynomial for $f(x) = e^x$.

Solution From the discussion on page 649, the nth Maclaurin polynomial for

$$f(x) = e^x$$

is given by

$$P_n(x) = 1 + x + \frac{1}{2!}x^2 + \frac{1}{3!}x^3 + \cdots + \frac{1}{n!}x^n.$$

FOR FURTHER INFORMATION To see how to use series to obtain other approximations to e, see the article "Novel Series-based Approximations to e" by John Knox and Harlan J. Brothers in *The College Mathematics Journal*. To view this article, go to the website *www.matharticles.com*.

EXAMPLE 4 Finding Taylor Polynomials for ln x

Find the Taylor polynomials P_0, P_1, P_2, P_3, and P_4 for $f(x) = \ln x$ centered at $c = 1$.

Solution Expanding about $c = 1$ yields the following.

$$f(x) = \ln x \qquad f(1) = \ln 1 = 0$$
$$f'(x) = \frac{1}{x} \qquad f'(1) = \frac{1}{1} = 1$$
$$f''(x) = -\frac{1}{x^2} \qquad f''(1) = -\frac{1}{1^2} = -1$$
$$f'''(x) = \frac{2!}{x^3} \qquad f'''(1) = \frac{2!}{1^3} = 2$$
$$f^{(4)}(x) = -\frac{3!}{x^4} \qquad f^{(4)}(1) = -\frac{3!}{1^4} = -6$$

Therefore, the Taylor polynomials are as follows.

$$P_0(x) = f(1) = 0$$
$$P_1(x) = f(1) + f'(1)(x - 1) = (x - 1)$$
$$P_2(x) = f(1) + f'(1)(x - 1) + \frac{f''(1)}{2!}(x - 1)^2$$
$$= (x - 1) - \frac{1}{2}(x - 1)^2$$
$$P_3(x) = f(1) + f'(1)(x - 1) + \frac{f''(1)}{2!}(x - 1)^2 + \frac{f'''(1)}{3!}(x - 1)^3$$
$$= (x - 1) - \frac{1}{2}(x - 1)^2 + \frac{1}{3}(x - 1)^3$$
$$P_4(x) = f(1) + f'(1)(x - 1) + \frac{f''(1)}{2!}(x - 1)^2 + \frac{f'''(1)}{3!}(x - 1)^3$$
$$+ \frac{f^{(4)}(1)}{4!}(x - 1)^4$$
$$= (x - 1) - \frac{1}{2}(x - 1)^2 + \frac{1}{3}(x - 1)^3 - \frac{1}{4}(x - 1)^4$$

Figure 9.14 compares the graphs of P_1, P_2, P_3, and P_4 with the graph of $f(x) = \ln x$. Note that near $x = 1$ the graphs are nearly indistinguishable. For instance, $P_4(0.9) \approx -0.105358$ and $\ln(0.9) \approx -0.105361$.

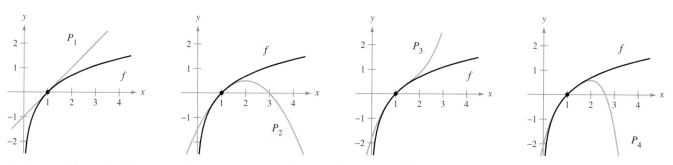

As *n* increases, the graph of P_n becomes a better and better approximation of the graph of $f(x) = \ln x$ near $x = 1$.
Figure 9.14

EXAMPLE 5 Finding Maclaurin Polynomials for cos x

Find the Maclaurin polynomials P_0, P_2, P_4, and P_6 for $f(x) = \cos x$. Use $P_6(x)$ to approximate the value of $\cos(0.1)$.

Solution Expanding about $c = 0$ yields the following.

$$f(x) = \cos x \qquad f(0) = \cos 0 = 1$$
$$f'(x) = -\sin x \qquad f'(0) = -\sin 0 = 0$$
$$f''(x) = -\cos x \qquad f''(0) = -\cos 0 = -1$$
$$f'''(x) = \sin x \qquad f'''(0) = \sin 0 = 0$$

Through repeated differentiation, you can see that the pattern 1, 0, -1, 0 continues, and you obtain the following Maclaurin polynomials.

$$P_0(x) = 1, \quad P_2(x) = 1 - \frac{1}{2!}x^2,$$

$$P_4(x) = 1 - \frac{1}{2!}x^2 + \frac{1}{4!}x^4, \quad P_6(x) = 1 - \frac{1}{2!}x^2 + \frac{1}{4!}x^4 - \frac{1}{6!}x^6$$

Using $P_6(x)$, you obtain the approximation $\cos(0.1) \approx 0.995004165$, which coincides with the calculator value to nine decimal places. Figure 9.15 compares the graphs of $f(x) = \cos x$ and P_6.

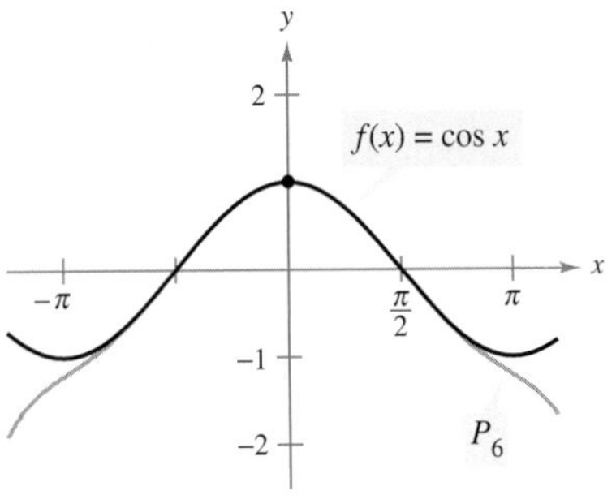

Near (0, 1), the graph of P_6 can be used to approximate the graph of $f(x) = \cos x$.
Figure 9.15

Note in Example 5 that the Maclaurin polynomials for $\cos x$ have only even powers of x. Similarly, the Maclaurin polynomials for $\sin x$ have only odd powers of x (see Exercise 17). This is not generally true of the Taylor polynomials for $\sin x$ and $\cos x$ expanded about $c \neq 0$, as you can see in the next example.

EXAMPLE 6 Finding a Taylor Polynomial for sin x

Find the third Taylor polynomial for $f(x) = \sin x$, expanded about $c = \pi/6$.

Solution Expanding about $c = \pi/6$ yields the following.

$$f(x) = \sin x \qquad f\left(\frac{\pi}{6}\right) = \sin\frac{\pi}{6} = \frac{1}{2}$$
$$f'(x) = \cos x \qquad f'\left(\frac{\pi}{6}\right) = \cos\frac{\pi}{6} = \frac{\sqrt{3}}{2}$$
$$f''(x) = -\sin x \qquad f''\left(\frac{\pi}{6}\right) = -\sin\frac{\pi}{6} = -\frac{1}{2}$$
$$f'''(x) = -\cos x \qquad f'''\left(\frac{\pi}{6}\right) = -\cos\frac{\pi}{6} = -\frac{\sqrt{3}}{2}$$

So, the third Taylor polynomial for $f(x) = \sin x$, expanded about $c = \pi/6$, is

$$P_3(x) = f\left(\frac{\pi}{6}\right) + f'\left(\frac{\pi}{6}\right)\left(x - \frac{\pi}{6}\right) + \frac{f''\left(\frac{\pi}{6}\right)}{2!}\left(x - \frac{\pi}{6}\right)^2 + \frac{f'''\left(\frac{\pi}{6}\right)}{3!}\left(x - \frac{\pi}{6}\right)^3$$

$$= \frac{1}{2} + \frac{\sqrt{3}}{2}\left(x - \frac{\pi}{6}\right) - \frac{1}{2(2!)}\left(x - \frac{\pi}{6}\right)^2 - \frac{\sqrt{3}}{2(3!)}\left(x - \frac{\pi}{6}\right)^3.$$

Figure 9.16 compares the graphs of $f(x) = \sin x$ and P_3.

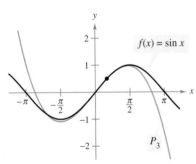

Near $(\pi/6, 1/2)$, the graph of P_3 can be used to approximate the graph of $f(x) = \sin x$.
Figure 9.16

Taylor polynomials and Maclaurin polynomials can be used to approximate the value of a function at a specific point. For instance, to approximate the value of $\ln(1.1)$, you can use Taylor polynomials for $f(x) = \ln x$ expanded about $c = 1$, as shown in Example 4, or you can use Maclaurin polynomials, as shown in Example 7.

EXAMPLE 7 Approximation Using Maclaurin Polynomials

Use a fourth Maclaurin polynomial to approximate the value of $\ln(1.1)$.

Solution Because 1.1 is closer to 1 than to 0, you should consider Maclaurin polynomials for the function $g(x) = \ln(1 + x)$.

$$g(x) = \ln(1 + x) \qquad g(0) = \ln(1 + 0) = 0$$
$$g'(x) = (1 + x)^{-1} \qquad g'(0) = (1 + 0)^{-1} = 1$$
$$g''(x) = -(1 + x)^{-2} \qquad g''(0) = -(1 + 0)^{-2} = -1$$
$$g'''(x) = 2(1 + x)^{-3} \qquad g'''(0) = 2(1 + 0)^{-3} = 2$$
$$g^{(4)}(x) = -6(1 + x)^{-4} \qquad g^{(4)}(0) = -6(1 + 0)^{-4} = -6$$

Note that you obtain the same coefficients as in Example 4. Therefore, the fourth Maclaurin polynomial for $g(x) = \ln(1 + x)$ is

$$P_4(x) = g(0) + g'(0)x + \frac{g''(0)}{2!}x^2 + \frac{g'''(0)}{3!}x^3 + \frac{g^{(4)}(0)}{4!}x^4$$
$$= x - \frac{1}{2}x^2 + \frac{1}{3}x^3 - \frac{1}{4}x^4.$$

Consequently,

$$\ln(1.1) = \ln(1 + 0.1) \approx P_4(0.1) \approx 0.0953083.$$

Check to see that the fourth Taylor polynomial (from Example 4), evaluated at $x = 1.1$, yields the same result.

n	$P_n(0.1)$
1	0.1000000
2	0.0950000
3	0.0953333
4	0.0953083

The table at the left illustrates the accuracy of the Taylor polynomial approximation of the calculator value of $\ln(1.1)$. You can see that as n becomes larger, $P_n(0.1)$ approaches the calculator value of 0.0953102.

On the other hand, the table below illustrates that as you move away from the expansion point $c = 1$, the accuracy of the approximation decreases.

Fourth Taylor Polynomial Approximation of $\ln(1 + x)$

x	0	0.1	0.5	0.75	1.0
$\ln(1 + x)$	0	0.0953102	0.4054651	0.5596158	0.6931472
$P_4(x)$	0	0.0953083	0.4010417	0.5302734	0.5833333

These two tables illustrate two very important points about the accuracy of Taylor (or Maclaurin) polynomials for use in approximations.

1. The approximation is usually better at x-values close to c than at x-values far from c.

2. The approximation is usually better for higher-degree Taylor (or Maclaurin) polynomials than for those of lower degree.

Remainder of a Taylor Polynomial

An approximation technique is of little value without some idea of its accuracy. To measure the accuracy of approximating a function value $f(x)$ by the Taylor polynomial $P_n(x)$, you can use the concept of a **remainder** $R_n(x)$, defined as follows.

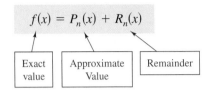

So, $R_n(x) = f(x) - P_n(x)$. The absolute value of $R_n(x)$ is called the **error** associated with the approximation. That is,

$$\text{Error} = |R_n(x)| = |f(x) - P_n(x)|.$$

The next theorem gives a general procedure for estimating the remainder associated with a Taylor polynomial. This important theorem is called **Taylor's Theorem,** and the remainder given in the theorem is called the **Lagrange form of the remainder.** (The proof of the theorem is lengthy, and is given in Appendix A.)

THEOREM 9.19 Taylor's Theorem

If a function f is differentiable through order $n + 1$ in an interval I containing c, then, for each x in I, there exists z between x and c such that

$$f(x) = f(c) + f'(c)(x - c) + \frac{f''(c)}{2!}(x - c)^2 + \cdots + \frac{f^{(n)}(c)}{n!}(x - c)^n + R_n(x)$$

where

$$R_n(x) = \frac{f^{(n+1)}(z)}{(n+1)!}(x - c)^{n+1}.$$

NOTE One useful consequence of Taylor's Theorem is that

$$|R_n(x)| \le \frac{|x - c|^{n+1}}{(n+1)!} \max |f^{(n+1)}(z)|$$

where $\max |f^{(n+1)}(z)|$ is the maximum value of $f^{(n+1)}(z)$ between x and c.

For $n = 0$, Taylor's Theorem states that if f is differentiable in an interval I containing c, then, for each x in I, there exists z between x and c such that

$$f(x) = f(c) + f'(z)(x - c) \quad \text{or} \quad f'(z) = \frac{f(x) - f(c)}{x - c}.$$

Do you recognize this special case of Taylor's Theorem? (It is the Mean Value Theorem.)

When applying Taylor's Theorem, you should not expect to be able to find the exact value of z. (If you could do this, an approximation would not be necessary.) Rather, you try to find bounds for $f^{(n+1)}(z)$ from which you are able to tell how large the remainder $R_n(x)$ is.

EXAMPLE 8 Determining the Accuracy of an Approximation

The third Maclaurin polynomial for sin x is given by

$$P_3(x) = x - \frac{x^3}{3!}.$$

Use Taylor's Theorem to approximate $\sin(0.1)$ by $P_3(0.1)$ and determine the accuracy of the approximation.

Solution Using Taylor's Theorem, you have

$$\sin x = x - \frac{x^3}{3!} + R_3(x) = x - \frac{x^3}{3!} + \frac{f^{(4)}(z)}{4!} x^4$$

where $0 < z < 0.1$. Therefore,

$$\sin(0.1) \approx 0.1 - \frac{(0.1)^3}{3!} \approx 0.1 - 0.000167 = 0.099833.$$

Because $f^{(4)}(z) = \sin z$, it follows that the error $|R_3(0.1)|$ can be bounded as follows.

$$0 < R_3(0.1) = \frac{\sin z}{4!}(0.1)^4 < \frac{0.0001}{4!} \approx 0.000004$$

This implies that

$$0.099833 < \sin(0.1) = 0.099833 + R_3(x) < 0.099833 + 0.000004$$
$$0.099833 < \sin(0.1) < 0.099837.$$

NOTE Try using a calculator to verify the results obtained in Examples 8 and 9. For Example 8, you obtain

$$\sin(0.1) \approx 0.0998334.$$

For Example 9, you obtain

$$P_3(1.2) \approx 0.1827$$

and

$$\ln(1.2) \approx 0.1823.$$

EXAMPLE 9 Approximating a Value to a Desired Accuracy

Determine the degree of the Taylor polynomial $P_n(x)$ expanded about $c = 1$ that should be used to approximate $\ln(1.2)$ so that the error is less than 0.001.

Solution Following the pattern of Example 4, you can see that the $(n + 1)$st derivative of $f(x) = \ln x$ is given by

$$f^{(n+1)}(x) = (-1)^n \frac{n!}{x^{n+1}}.$$

Using Taylor's Theorem, you know that the error $|R_n(1.2)|$ is given by

$$|R_n(1.2)| = \left| \frac{f^{(n+1)}(z)}{(n+1)!}(1.2 - 1)^{n+1} \right| = \frac{n!}{z^{n+1}} \left[\frac{1}{(n+1)!} \right] (0.2)^{n+1}$$
$$= \frac{(0.2)^{n+1}}{z^{n+1}(n+1)}$$

where $1 < z < 1.2$. In this interval, $(0.2)^{n+1}/[z^{n+1}(n+1)]$ is less than $(0.2)^{n+1}/(n+1)$. So, you are seeking a value of n such that

$$\frac{(0.2)^{n+1}}{(n+1)} < 0.001 \quad \Longrightarrow \quad 1000 < (n+1)5^{n+1}.$$

By trial and error, you can determine that the smallest value of n that satisfies this inequality is $n = 3$. So, you would need the third Taylor polynomial to achieve the desired accuracy in approximating $\ln(1.2)$.

Exercises for Section 9.7

In Exercises 1–4, match the Taylor polynomial approximation of the function $f(x) = e^{-x^2/2}$ with the correct graph. [The graphs are labeled (a), (b), (c), and (d).]

(a)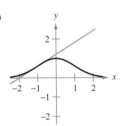

(b)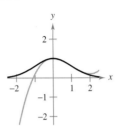

(c)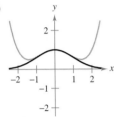

(d)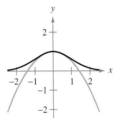

1. $g(x) = -\frac{1}{2}x^2 + 1$
2. $g(x) = \frac{1}{8}x^4 - \frac{1}{2}x^2 + 1$
3. $g(x) = e^{-1/2}[(x + 1) + 1]$
4. $g(x) = e^{-1/2}\left[\frac{1}{3}(x - 1)^3 - (x - 1) + 1\right]$

In Exercises 5–8, find a first-degree polynomial function P_1 whose value and slope agree with the value and slope of f at $x = c$. Use a graphing utility to graph f and P_1. What is P_1 called?

5. $f(x) = \dfrac{4}{\sqrt{x}}$, $c = 1$
6. $f(x) = \dfrac{4}{\sqrt[3]{x}}$, $c = 8$
7. $f(x) = \sec x$, $c = \dfrac{\pi}{4}$
8. $f(x) = \tan x$, $c = \dfrac{\pi}{4}$

Graphical and Numerical Analysis In Exercises 9 and 10, use a graphing utility to graph f and its second-degree polynomial approximation P_2 at $x = c$. Complete the table comparing the values of f and P_2.

9. $f(x) = \dfrac{4}{\sqrt{x}}$, $c = 1$

$P_2(x) = 4 - 2(x - 1) + \frac{3}{2}(x - 1)^2$

x	0	0.8	0.9	1	1.1	1.2	2
$f(x)$							
$P_2(x)$							

10. $f(x) = \sec x$, $c = \dfrac{\pi}{4}$

$P_2(x) = \sqrt{2} + \sqrt{2}\left(x - \dfrac{\pi}{4}\right) + \dfrac{3}{2}\sqrt{2}\left(x - \dfrac{\pi}{4}\right)^2$

x	-2.15	0.585	0.685	$\dfrac{\pi}{4}$	0.885	0.985	1.785
$f(x)$							
$P_2(x)$							

11. **Conjecture** Consider the function $f(x) = \cos x$ and its Maclaurin polynomials P_2, P_4, and P_6 (see Example 5).
    (a) Use a graphing utility to graph f and the indicated polynomial approximations.
    (b) Evaluate and compare the values of $f^{(n)}(0)$ and $P_n^{(n)}(0)$ for $n = 2, 4,$ and 6.
    (c) Use the results in part (b) to make a conjecture about $f^{(n)}(0)$ and $P_n^{(n)}(0)$.

12. **Conjecture** Consider the function $f(x) = x^2 e^x$.
    (a) Find the Maclaurin polynomials P_2, P_3, and P_4 for f.
    (b) Use a graphing utility to graph f, P_2, P_3, and P_4.
    (c) Evaluate and compare the values of $f^{(n)}(0)$ and $P_n^{(n)}(0)$ for $n = 2, 3,$ and 4.
    (d) Use the results in part (c) to make a conjecture about $f^{(n)}(0)$ and $P_n^{(n)}(0)$.

In Exercises 13–24, find the Maclaurin polynomial of degree n for the function.

13. $f(x) = e^{-x}$, $n = 3$
14. $f(x) = e^{-x}$, $n = 5$
15. $f(x) = e^{2x}$, $n = 4$
16. $f(x) = e^{3x}$, $n = 4$
17. $f(x) = \sin x$, $n = 5$
18. $f(x) = \sin \pi x$, $n = 3$
19. $f(x) = xe^x$, $n = 4$
20. $f(x) = x^2 e^{-x}$, $n = 4$
21. $f(x) = \dfrac{1}{x + 1}$, $n = 4$
22. $f(x) = \dfrac{x}{x + 1}$, $n = 4$
23. $f(x) = \sec x$, $n = 2$
24. $f(x) = \tan x$, $n = 3$

In Exercises 25–30, find the nth Taylor polynomial centered at c.

25. $f(x) = \dfrac{1}{x}$, $n = 4$, $c = 1$
26. $f(x) = \dfrac{2}{x^2}$, $n = 4$, $c = 2$
27. $f(x) = \sqrt{x}$, $n = 4$, $c = 1$
28. $f(x) = \sqrt[3]{x}$, $n = 3$, $c = 8$
29. $f(x) = \ln x$, $n = 4$, $c = 1$
30. $f(x) = x^2 \cos x$, $n = 2$, $c = \pi$

In Exercises 31 and 32, use a computer algebra system to find the indicated Taylor polynomials for the function f. Graph the function and the Taylor polynomials.

31. $f(x) = \tan x$

(a) $n = 3$, $c = 0$

(b) $n = 3$, $c = \pi/4$

32. $f(x) = \dfrac{1}{x^2 + 1}$

(a) $n = 4$, $c = 0$

(b) $n = 4$, $c = 1$

 33. Numerical and Graphical Approximations

(a) Use the Maclaurin polynomials $P_1(x)$, $P_3(x)$, and $P_5(x)$ for $f(x) = \sin x$ to complete the table.

x	0	0.25	0.50	0.75	1.00
$\sin x$	0	0.2474	0.4794	0.6816	0.8415
$P_1(x)$					
$P_3(x)$					
$P_5(x)$					

(b) Use a graphing utility to graph $f(x) = \sin x$ and the Maclaurin polynomials in part (a).

(c) Describe the change in accuracy of a polynomial approximation as the distance from the point where the polynomial is centered increases.

34. Numerical and Graphical Approximations

(a) Use the Taylor polynomials $P_1(x)$, $P_2(x)$, and $P_4(x)$ for $f(x) = \ln x$ centered at $c = 1$ to complete the table.

x	1.00	1.25	1.50	1.75	2.00
$\ln x$	0	0.2231	0.4055	0.5596	0.6931
$P_1(x)$					
$P_2(x)$					
$P_4(x)$					

(b) Use a graphing utility to graph $f(x) = \ln x$ and the Taylor polynomials in part (a).

(c) Describe the change in accuracy of polynomial approximations as the degree increases.

Numerical and Graphical Approximations In Exercises 35 and 36, (a) find the Maclaurin polynomial $P_3(x)$ for $f(x)$, (b) complete the table for $f(x)$ and $P_3(x)$, and (c) sketch the graphs of $f(x)$ and $P_3(x)$ on the same set of coordinate axes.

x	-0.75	-0.50	-0.25	0	0.25	0.50	0.75
$f(x)$							
$P_3(x)$							

35. $f(x) = \arcsin x$

36. $f(x) = \arctan x$

In Exercises 37–40, the graph of $y = f(x)$ is shown with four of its Maclaurin polynomials. Identify the Maclaurin polynomials and use a graphing utility to confirm your results.

37.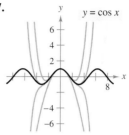

38.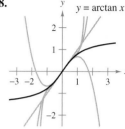

39.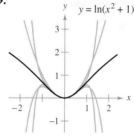

40.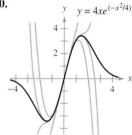

In Exercises 41–44, approximate the function at the given value of x, using the polynomial found in the indicated exercise.

41. $f(x) = e^{-x}$, $f\left(\tfrac{1}{2}\right)$, Exercise 13

42. $f(x) = x^2 e^{-x}$, $f\left(\tfrac{1}{5}\right)$, Exercise 20

43. $f(x) = \ln x$, $f(1.2)$, Exercise 29

44. $f(x) = x^2 \cos x$, $f\left(\dfrac{7\pi}{8}\right)$, Exercise 30

In Exercises 45–48, use Taylor's Theorem to obtain an upper bound for the error of the approximation. Then calculate the exact value of the error.

45. $\cos(0.3) \approx 1 - \dfrac{(0.3)^2}{2!} + \dfrac{(0.3)^4}{4!}$

46. $e \approx 1 + 1 + \dfrac{1^2}{2!} + \dfrac{1^3}{3!} + \dfrac{1^4}{4!} + \dfrac{1^5}{5!}$

47. $\arcsin(0.4) \approx 0.4 + \dfrac{(0.4)^3}{2 \cdot 3}$

48. $\arctan(0.4) \approx 0.4 - \dfrac{(0.4)^3}{3}$

In Exercises 49–52, determine the degree of the Maclaurin polynomial required for the error in the approximation of the function at the indicated value of x to be less than 0.001.

49. $\sin(0.3)$

50. $\cos(0.1)$

51. $e^{0.6}$

52. $e^{0.3}$

 In Exercises 53–56, determine the degree of the Maclaurin polynomial required for the error in the approximation of the function at the indicated value of x to be less than 0.0001. Use a computer algebra system to obtain and evaluate the required derivatives.

53. $f(x) = \ln(x + 1)$, approximate $f(0.5)$.

54. $f(x) = \cos(\pi x^2)$, approximate $f(0.6)$.
55. $f(x) = e^{-\pi x}$, approximate $f(1.3)$.
56. $f(x) = e^{-x}$, approximate $f(1)$.

In Exercises 57–60, determine the values of x for which the function can be replaced by the Taylor polynomial if the error cannot exceed 0.001.

57. $f(x) = e^x \approx 1 + x + \dfrac{x^2}{2!} + \dfrac{x^3}{3!}, \quad x < 0$

58. $f(x) = \sin x \approx x - \dfrac{x^3}{3!}$

59. $f(x) = \cos x \approx 1 - \dfrac{x^2}{2!} + \dfrac{x^4}{4!}$

60. $f(x) = e^{-2x} \approx 1 - 2x + 2x^2 - \dfrac{4}{3}x^3$

Writing About Concepts

61. An elementary function is approximated by a polynomial. In your own words, describe what is meant by saying that the polynomial is *expanded about c* or *centered at c*.

62. When an elementary function f is approximated by a second-degree polynomial P_2 centered at c, what is known about f and P_2 at c? Explain your reasoning.

63. State the definition of an nth-degree Taylor polynomial of f centered at c.

64. Describe the accuracy of the nth-degree Taylor polynomial of f centered at c as the distance between c and x increases.

65. In general, how does the accuracy of a Taylor polynomial change as the degree of the polynomial is increased? Explain your reasoning.

66. The graphs show first-, second-, and third-degree polynomial approximations P_1, P_2, and P_3 of a function f. Label the graphs of P_1, P_2, and P_3. To print an enlarged copy of the graph, go to the website *www.mathgraphs.com*.

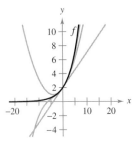

67. *Comparing Maclaurin Polynomials*

(a) Compare the Maclaurin polynomials of degree 4 and degree 5, respectively, for the functions $f(x) = e^x$ and $g(x) = xe^x$. What is the relationship between them?

(b) Use the result in part (a) and the Maclaurin polynomial of degree 5 for $f(x) = \sin x$ to find a Maclaurin polynomial of degree 6 for the function $g(x) = x \sin x$.

(c) Use the result in part (a) and the Maclaurin polynomial of degree 5 for $f(x) = \sin x$ to find a Maclaurin polynomial of degree 4 for the function $g(x) = (\sin x)/x$.

68. *Differentiating Maclaurin Polynomials*

(a) Differentiate the Maclaurin polynomial of degree 5 for $f(x) = \sin x$ and compare the result with the Maclaurin polynomial of degree 4 for $g(x) = \cos x$.

(b) Differentiate the Maclaurin polynomial of degree 6 for $f(x) = \cos x$ and compare the result with the Maclaurin polynomial of degree 5 for $g(x) = \sin x$.

(c) Differentiate the Maclaurin polynomial of degree 4 for $f(x) = e^x$. Describe the relationship between the two series.

69. *Graphical Reasoning* The figure shows the graph of the function

$$f(x) = \sin\left(\dfrac{\pi x}{4}\right)$$

and the second-degree Taylor polynomial

$$P_2(x) = 1 - \dfrac{\pi^2}{32}(x - 2)^2$$

centered at $x = 2$.

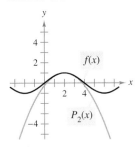

(a) Use the symmetry of the graph of f to write the second-degree Taylor polynomial for f centered at $x = -2$.

(b) Use a horizontal translation of the result in part (a) to find the second-degree Taylor polynomial for f centered at $x = 6$.

(c) Is it possible to use a horizontal translation of the result in part (a) to write a second-degree Taylor polynomial for f centered at $x = 4$? Explain.

70. Prove that if f is an odd function, then its nth Maclaurin polynomial contains only terms with odd powers of x.

71. Prove that if f is an even function, then its nth Maclaurin polynomial contains only terms with even powers of x.

72. Let $P_n(x)$ be the nth Taylor polynomial for f at c. Prove that $P_n(c) = f(c)$ and $P^{(k)}(c) = f^{(k)}(c)$ for $1 \leq k \leq n$. (See Exercises 9 and 10.)

73. *Writing* The proof in Exercise 72 guarantees that the Taylor polynomial and its derivatives agree with the function and its derivatives at $x = c$. Use the graphs and tables in Exercises 33–36 to discuss what happens to the accuracy of the Taylor polynomial as you move away from $x = c$.

Section 9.8 Power Series

- Understand the definition of a power series.
- Find the radius and interval of convergence of a power series.
- Determine the endpoint convergence of a power series.
- Differentiate and integrate a power series.

Power Series

In Section 9.7, you were introduced to the concept of approximating functions by Taylor polynomials. For instance, the function $f(x) = e^x$ can be *approximated* by its Maclaurin polynomials as follows.

$e^x \approx 1 + x$ 1st-degree polynomial

$e^x \approx 1 + x + \dfrac{x^2}{2!}$ 2nd-degree polynomial

$e^x \approx 1 + x + \dfrac{x^2}{2!} + \dfrac{x^3}{3!}$ 3rd-degree polynomial

$e^x \approx 1 + x + \dfrac{x^2}{2!} + \dfrac{x^3}{3!} + \dfrac{x^4}{4!}$ 4th-degree polynomial

$e^x \approx 1 + x + \dfrac{x^2}{2!} + \dfrac{x^3}{3!} + \dfrac{x^4}{4!} + \dfrac{x^5}{5!}$ 5th-degree polynomial

In that section, you saw that the higher the degree of the approximating polynomial, the better the approximation becomes.

In this and the next two sections, you will see that several important types of functions, including

$$f(x) = e^x$$

can be represented *exactly* by an infinite series called a **power series.** For example, the power series representation for e^x is

$$e^x = 1 + x + \dfrac{x^2}{2!} + \dfrac{x^3}{3!} + \cdots + \dfrac{x^n}{n!} + \cdots.$$

For each real number x, it can be shown that the infinite series on the right converges to the number e^x. Before doing this, however, some preliminary results dealing with power series will be discussed—beginning with the following definition.

EXPLORATION

Graphical Reasoning Use a graphing utility to approximate the graph of each power series near $x = 0$. (Use the first several terms of each series.) Each series represents a well-known function. What is the function?

a. $\displaystyle\sum_{n=0}^{\infty} \dfrac{(-1)^n x^n}{n!}$

b. $\displaystyle\sum_{n=0}^{\infty} \dfrac{(-1)^n x^{2n}}{(2n)!}$

c. $\displaystyle\sum_{n=0}^{\infty} \dfrac{(-1)^n x^{2n+1}}{(2n+1)!}$

d. $\displaystyle\sum_{n=0}^{\infty} \dfrac{(-1)^n x^{2n+1}}{2n+1}$

e. $\displaystyle\sum_{n=0}^{\infty} \dfrac{2^n x^n}{n!}$

Definition of Power Series

If x is a variable, then an infinite series of the form

$$\sum_{n=0}^{\infty} a_n x^n = a_0 + a_1 x + a_2 x^2 + a_3 x^3 + \cdots + a_n x^n + \cdots$$

is called a **power series.** More generally, an infinite series of the form

$$\sum_{n=0}^{\infty} a_n (x-c)^n = a_0 + a_1(x-c) + a_2(x-c)^2 + \cdots + a_n(x-c)^n + \cdots$$

is called a **power series centered at c,** where c is a constant.

NOTE To simplify the notation for power series, we agree that $(x - c)^0 = 1$, even if $x = c$.

EXAMPLE 1 Power Series

a. The following power series is centered at 0.
$$\sum_{n=0}^{\infty} \frac{x^n}{n!} = 1 + x + \frac{x^2}{2} + \frac{x^3}{3!} + \cdots$$

b. The following power series is centered at -1.
$$\sum_{n=0}^{\infty} (-1)^n (x+1)^n = 1 - (x+1) + (x+1)^2 - (x+1)^3 + \cdots$$

c. The following power series is centered at 1.
$$\sum_{n=1}^{\infty} \frac{1}{n}(x-1)^n = (x-1) + \frac{1}{2}(x-1)^2 + \frac{1}{3}(x-1)^3 + \cdots$$

Radius and Interval of Convergence

A power series in x can be viewed as a function of x

$$f(x) = \sum_{n=0}^{\infty} a_n (x-c)^n$$

where the *domain of f* is the set of all x for which the power series converges. Determination of the domain of a power series is the primary concern in this section. Of course, every power series converges at its center c because

$$f(c) = \sum_{n=0}^{\infty} a_n (c-c)^n$$
$$= a_0(1) + 0 + 0 + \cdots + 0 + \cdots$$
$$= a_0.$$

So, c always lies in the domain of f. The following important theorem states that the domain of a power series can take three basic forms: a single point, an interval centered at c, or the entire real line, as shown in Figure 9.17. A proof is given in Appendix A.

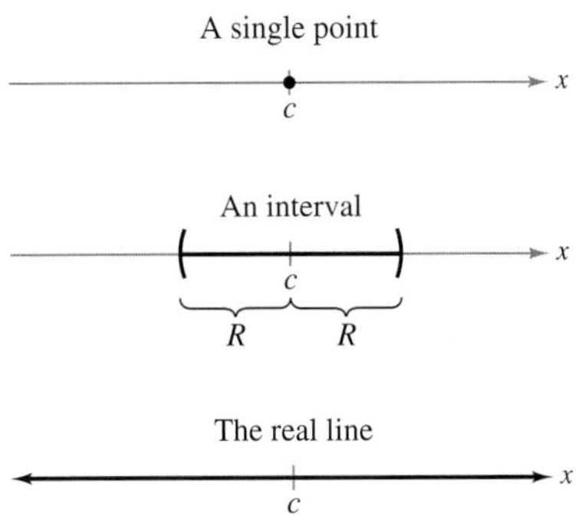

The domain of a power series has only three basic forms: a single point, an interval centered at c, or the entire real line.
Figure 9.17

THEOREM 9.20 Convergence of a Power Series

For a power series centered at c, precisely one of the following is true.

1. The series converges only at c.
2. There exists a real number $R > 0$ such that the series converges absolutely for $|x - c| < R$, and diverges for $|x - c| > R$.
3. The series converges absolutely for all x.

The number R is the **radius of convergence** of the power series. If the series converges only at c, the radius of convergence is $R = 0$, and if the series converges for all x, the radius of convergence is $R = \infty$. The set of all values of x for which the power series converges is the **interval of convergence** of the power series.

STUDY TIP To determine the radius of convergence of a power series, use the Ratio Test, as demonstrated in Examples 2, 3, and 4.

EXAMPLE 2 Finding the Radius of Convergence

Find the radius of convergence of $\sum_{n=0}^{\infty} n!x^n$.

Solution For $x = 0$, you obtain

$$f(0) = \sum_{n=0}^{\infty} n!0^n = 1 + 0 + 0 + \cdots = 1.$$

For any fixed value of x such that $|x| > 0$, let $u_n = n!x^n$. Then

$$\lim_{n \to \infty} \left| \frac{u_{n+1}}{u_n} \right| = \lim_{n \to \infty} \left| \frac{(n+1)!x^{n+1}}{n!x^n} \right|$$
$$= |x| \lim_{n \to \infty} (n+1)$$
$$= \infty.$$

Therefore, by the Ratio Test, the series diverges for $|x| > 0$ and converges only at its center, 0. So, the radius of convergence is $R = 0$.

EXAMPLE 3 Finding the Radius of Convergence

Find the radius of convergence of

$$\sum_{n=0}^{\infty} 3(x-2)^n.$$

Solution For $x \neq 2$, let $u_n = 3(x-2)^n$. Then

$$\lim_{n \to \infty} \left| \frac{u_{n+1}}{u_n} \right| = \lim_{n \to \infty} \left| \frac{3(x-2)^{n+1}}{3(x-2)^n} \right|$$
$$= \lim_{n \to \infty} |x - 2|$$
$$= |x - 2|.$$

By the Ratio Test, the series converges if $|x - 2| < 1$ and diverges if $|x - 2| > 1$. Therefore, the radius of convergence of the series is $R = 1$.

EXAMPLE 4 Finding the Radius of Convergence

Find the radius of convergence of

$$\sum_{n=0}^{\infty} \frac{(-1)^n x^{2n+1}}{(2n+1)!}.$$

Solution Let $u_n = (-1)^n x^{2n+1}/(2n+1)!$. Then

$$\lim_{n \to \infty} \left| \frac{u_{n+1}}{u_n} \right| = \lim_{n \to \infty} \left| \frac{\dfrac{(-1)^{n+1} x^{2n+3}}{(2n+3)!}}{\dfrac{(-1)^n x^{2n+1}}{(2n-1)!}} \right|$$
$$= \lim_{n \to \infty} \frac{x^2}{(2n+3)(2n+2)}.$$

For any *fixed* value of x, this limit is 0. So, by the Ratio Test, the series converges for all x. Therefore, the radius of convergence is $R = \infty$.

Endpoint Convergence

Note that for a power series whose radius of convergence is a finite number R, Theorem 9.20 says nothing about the convergence at the *endpoints* of the interval of convergence. Each endpoint must be tested separately for convergence or divergence. As a result, the interval of convergence of a power series can take any one of the six forms shown in Figure 9.18.

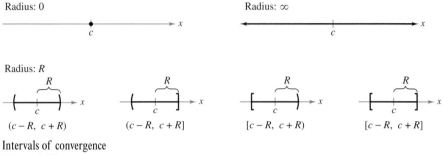

Intervals of convergence
Figure 9.18

EXAMPLE 5 Finding the Interval of Convergence

Find the interval of convergence of $\sum_{n=1}^{\infty} \dfrac{x^n}{n}$.

Solution Letting $u_n = x^n/n$ produces

$$\lim_{n \to \infty} \left| \frac{u_{n+1}}{u_n} \right| = \lim_{n \to \infty} \left| \frac{\dfrac{x^{n+1}}{(n+1)}}{\dfrac{x^n}{n}} \right|$$

$$= \lim_{n \to \infty} \left| \frac{nx}{n+1} \right|$$

$$= |x|.$$

So, by the Ratio Test, the radius of convergence is $R = 1$. Moreover, because the series is centered at 0, it converges in the interval $(-1, 1)$. This interval, however, is not necessarily the *interval of convergence*. To determine this, you must test for convergence at each endpoint. When $x = 1$, you obtain the *divergent* harmonic series

$$\sum_{n=1}^{\infty} \frac{1}{n} = \frac{1}{1} + \frac{1}{2} + \frac{1}{3} + \cdots . \qquad \text{Diverges when } x = 1$$

When $x = -1$, you obtain the *convergent* alternating harmonic series

$$\sum_{n=1}^{\infty} \frac{(-1)^n}{n} = -1 + \frac{1}{2} - \frac{1}{3} + \frac{1}{4} - \cdots . \qquad \text{Converges when } x = -1$$

So, the interval of convergence for the series is $[-1, 1)$, as shown in Figure 9.19.

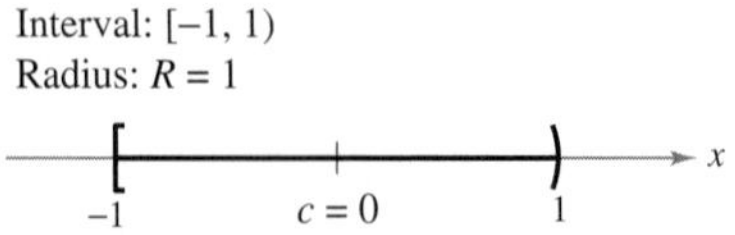

Interval: $[-1, 1)$
Radius: $R = 1$

Figure 9.19

EXAMPLE 6 Finding the Interval of Convergence

Find the interval of convergence of $\sum_{n=0}^{\infty} \frac{(-1)^n (x+1)^n}{2^n}$.

Solution Letting $u_n = (-1)^n(x+1)^n/2^n$ produces

$$\lim_{n\to\infty} \left|\frac{u_{n+1}}{u_n}\right| = \lim_{n\to\infty} \left|\frac{\frac{(-1)^{n+1}(x+1)^{n+1}}{2^{n+1}}}{\frac{(-1)^n(x+1)^n}{2^n}}\right|$$

$$= \lim_{n\to\infty} \left|\frac{2^n(x+1)}{2^{n+1}}\right|$$

$$= \left|\frac{x+1}{2}\right|.$$

By the Ratio Test, the series converges if $|(x+1)/2| < 1$ or $|x+1| < 2$. So, the radius of convergence is $R = 2$. Because the series is centered at $x = -1$, it will converge in the interval $(-3, 1)$. Furthermore, at the endpoints you have

$$\sum_{n=0}^{\infty} \frac{(-1)^n(-2)^n}{2^n} = \sum_{n=0}^{\infty} \frac{2^n}{2^n} = \sum_{n=0}^{\infty} 1 \qquad \text{Diverges when } x = -3$$

and

$$\sum_{n=0}^{\infty} \frac{(-1)^n(2)^n}{2^n} = \sum_{n=0}^{\infty} (-1)^n \qquad \text{Diverges when } x = 1$$

both of which diverge. So, the interval of convergence is $(-3, 1)$, as shown in Figure 9.20.

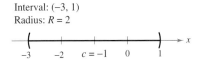

Interval: $(-3, 1)$
Radius: $R = 2$

Figure 9.20

EXAMPLE 7 Finding the Interval of Convergence

Find the interval of convergence of

$$\sum_{n=1}^{\infty} \frac{x^n}{n^2}.$$

Solution Letting $u_n = x^n/n^2$ produces

$$\lim_{n\to\infty} \left|\frac{u_{n+1}}{u_n}\right| = \lim_{n\to\infty} \left|\frac{x^{n+1}/(n+1)^2}{x^n/n^2}\right|$$

$$= \lim_{n\to\infty} \left|\frac{n^2 x}{(n+1)^2}\right| = |x|.$$

So, the radius of convergence is $R = 1$. Because the series is centered at $x = 0$, it converges in the interval $(-1, 1)$. When $x = 1$, you obtain the convergent p-series

$$\sum_{n=1}^{\infty} \frac{1}{n^2} = \frac{1}{1^2} + \frac{1}{2^2} + \frac{1}{3^2} + \frac{1}{4^2} + \cdots. \qquad \text{Converges when } x = 1$$

When $x = -1$, you obtain the convergent alternating series

$$\sum_{n=1}^{\infty} \frac{(-1)^n}{n^2} = -\frac{1}{1^2} + \frac{1}{2^2} - \frac{1}{3^2} + \frac{1}{4^2} - \cdots. \qquad \text{Converges when } x = -1$$

Therefore, the interval of convergence for the given series is $[-1, 1]$.

Differentiation and Integration of Power Series

Power series representation of functions has played an important role in the development of calculus. In fact, much of Newton's work with differentiation and integration was done in the context of power series—especially his work with complicated algebraic functions and transcendental functions. Euler, Lagrange, Leibniz, and the Bernoullis all used power series extensively in calculus.

Once you have defined a function with a power series, it is natural to wonder how you can determine the characteristics of the function. Is it continuous? Differentiable? Theorem 9.21, which is stated without proof, answers these questions.

JAMES GREGORY (1638–1675)

One of the earliest mathematicians to work with power series was a Scotsman, James Gregory. He developed a power series method for interpolating table values—a method that was later used by Brook Taylor in the development of Taylor polynomials and Taylor series.

THEOREM 9.21 Properties of Functions Defined by Power Series

If the function given by

$$f(x) = \sum_{n=0}^{\infty} a_n(x-c)^n$$
$$= a_0 + a_1(x-c) + a_2(x-c)^2 + a_3(x-c)^3 + \cdots$$

has a radius of convergence of $R > 0$, then, on the interval $(c - R, c + R)$, f is differentiable (and therefore continuous). Moreover, the derivative and antiderivative of f are as follows.

1. $f'(x) = \sum_{n=1}^{\infty} na_n(x-c)^{n-1}$
$= a_1 + 2a_2(x-c) + 3a_3(x-c)^2 + \cdots$

2. $\int f(x)\, dx = C + \sum_{n=0}^{\infty} a_n \dfrac{(x-c)^{n+1}}{n+1}$
$= C + a_0(x-c) + a_1 \dfrac{(x-c)^2}{2} + a_2 \dfrac{(x-c)^3}{3} + \cdots$

The *radius of convergence* of the series obtained by differentiating or integrating a power series is the same as that of the original power series. The *interval of convergence*, however, may differ as a result of the behavior at the endpoints.

Theorem 9.21 states that, in many ways, a function defined by a power series behaves like a polynomial. It is continuous in its interval of convergence, and both its derivative and its antiderivative can be determined by differentiating and integrating each term of the given power series. For instance, the derivative of the power series

$$f(x) = \sum_{n=0}^{\infty} \frac{x^n}{n!}$$
$$= 1 + x + \frac{x^2}{2} + \frac{x^3}{3!} + \frac{x^4}{4!} + \cdots$$

is

$$f'(x) = 1 + (2)\frac{x}{2} + (3)\frac{x^2}{3!} + (4)\frac{x^3}{4!} + \cdots$$
$$= 1 + x + \frac{x^2}{2} + \frac{x^3}{3!} + \frac{x^4}{4!} + \cdots$$
$$= f(x).$$

Notice that $f'(x) = f(x)$. Do you recognize this function?

EXAMPLE 8 Intervals of Convergence for $f(x)$, $f'(x)$, and $\int f(x)\, dx$

Consider the function given by

$$f(x) = \sum_{n=1}^{\infty} \frac{x^n}{n} = x + \frac{x^2}{2} + \frac{x^3}{3} + \cdots.$$

Find the intervals of convergence for each of the following.

a. $\int f(x)\, dx$ **b.** $f(x)$ **c.** $f'(x)$

Solution By Theorem 9.21, you have

$$f'(x) = \sum_{n=1}^{\infty} x^{n-1}$$
$$= 1 + x + x^2 + x^3 + \cdots$$

and

$$\int f(x)\, dx = C + \sum_{n=1}^{\infty} \frac{x^{n+1}}{n(n+1)}$$
$$= C + \frac{x^2}{1 \cdot 2} + \frac{x^3}{2 \cdot 3} + \frac{x^4}{3 \cdot 4} + \cdots.$$

By the Ratio Test, you can show that each series has a radius of convergence of $R = 1$. Considering the interval $(-1, 1)$, you have the following.

a. For $\int f(x)\, dx$, the series

$$\sum_{n=1}^{\infty} \frac{x^{n+1}}{n(n+1)} \qquad \text{Interval of convergence: } [-1, 1]$$

converges for $x = \pm 1$, and its interval of convergence is $[-1, 1]$. See Figure 9.21(a).

b. For $f(x)$, the series

$$\sum_{n=1}^{\infty} \frac{x^n}{n} \qquad \text{Interval of convergence: } [-1, 1)$$

converges for $x = -1$ and diverges for $x = 1$. So, its interval of convergence is $[-1, 1)$. See Figure 9.21(b).

c. For $f'(x)$, the series

$$\sum_{n=1}^{\infty} x^{n-1} \qquad \text{Interval of convergence: } (-1, 1)$$

diverges for $x = \pm 1$, and its interval of convergence is $(-1, 1)$. See Figure 9.21(c).

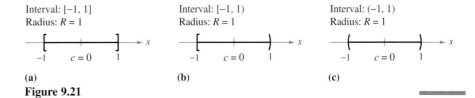

Figure 9.21

From Example 8, it appears that of the three series, the one for the derivative, $f'(x)$, is the least likely to converge at the endpoints. In fact, it can be shown that if the series for $f'(x)$ converges at the endpoints $x = c \pm R$, the series for $f(x)$ will also converge there.

Exercises for Section 9.8

In Exercises 1–4, state where the power series is centered.

1. $\sum_{n=0}^{\infty} n x^n$

2. $\sum_{n=1}^{\infty} \frac{(-1)^n 1 \cdot 3 \cdots (2n-1)}{2^n n!} x^n$

3. $\sum_{n=1}^{\infty} \frac{(x-2)^n}{n^3}$

4. $\sum_{n=0}^{\infty} \frac{(-1)^n (x-\pi)^{2n}}{(2n)!}$

In Exercises 5–10, find the radius of convergence of the power series.

5. $\sum_{n=0}^{\infty} (-1)^n \frac{x^n}{n+1}$

6. $\sum_{n=0}^{\infty} (2x)^n$

7. $\sum_{n=1}^{\infty} \frac{(2x)^n}{n^2}$

8. $\sum_{n=0}^{\infty} \frac{(-1)^n x^n}{2^n}$

9. $\sum_{n=0}^{\infty} \frac{(2x)^{2n}}{(2n)!}$

10. $\sum_{n=0}^{\infty} \frac{(2n)! x^{2n}}{n!}$

In Exercises 11–34, find the interval of convergence of the power series. (Be sure to include a check for convergence at the endpoints of the interval.)

11. $\sum_{n=0}^{\infty} \left(\frac{x}{2}\right)^n$

12. $\sum_{n=0}^{\infty} \left(\frac{x}{5}\right)^n$

13. $\sum_{n=1}^{\infty} \frac{(-1)^n x^n}{n}$

14. $\sum_{n=0}^{\infty} (-1)^{n+1}(n+1)x^n$

15. $\sum_{n=0}^{\infty} \frac{x^n}{n!}$

16. $\sum_{n=0}^{\infty} \frac{(3x)^n}{(2n)!}$

17. $\sum_{n=0}^{\infty} (2n)! \left(\frac{x}{2}\right)^n$

18. $\sum_{n=0}^{\infty} \frac{(-1)^n x^n}{(n+1)(n+2)}$

19. $\sum_{n=1}^{\infty} \frac{(-1)^{n+1} x^n}{4^n}$

20. $\sum_{n=0}^{\infty} \frac{(-1)^n n!(x-4)^n}{3^n}$

21. $\sum_{n=1}^{\infty} \frac{(-1)^{n+1}(x-5)^n}{n 5^n}$

22. $\sum_{n=0}^{\infty} \frac{(x-2)^{n+1}}{(n+1)4^{n+1}}$

23. $\sum_{n=0}^{\infty} \frac{(-1)^{n+1}(x-1)^{n+1}}{n+1}$

24. $\sum_{n=1}^{\infty} \frac{(-1)^{n+1}(x-2)^n}{n 2^n}$

25. $\sum_{n=1}^{\infty} \frac{(x-3)^{n-1}}{3^{n-1}}$

26. $\sum_{n=0}^{\infty} \frac{(-1)^n x^{2n+1}}{2n+1}$

27. $\sum_{n=1}^{\infty} \frac{n}{n+1}(-2x)^{n-1}$

28. $\sum_{n=0}^{\infty} \frac{(-1)^n x^{2n}}{n!}$

29. $\sum_{n=0}^{\infty} \frac{x^{2n+1}}{(2n+1)!}$

30. $\sum_{n=1}^{\infty} \frac{n! x^n}{(2n)!}$

31. $\sum_{n=1}^{\infty} \frac{2 \cdot 3 \cdot 4 \cdots (n+1) x^n}{n!}$

32. $\sum_{n=1}^{\infty} \left[\frac{2 \cdot 4 \cdot 6 \cdots 2n}{3 \cdot 5 \cdot 7 \cdots (2n+1)}\right] x^{2n+1}$

33. $\sum_{n=1}^{\infty} \frac{(-1)^{n+1} 3 \cdot 7 \cdot 11 \cdots (4n-1)(x-3)^n}{4^n}$

34. $\sum_{n=1}^{\infty} \frac{n!(x+1)^n}{1 \cdot 3 \cdot 5 \cdots (2n-1)}$

In Exercises 35 and 36, find the radius of convergence of the power series, where $c > 0$ and k is a positive integer.

35. $\sum_{n=1}^{\infty} \frac{(x-c)^{n-1}}{c^{n-1}}$

36. $\sum_{n=0}^{\infty} \frac{(n!)^k x^n}{(kn)!}$

In Exercises 37–40, find the interval of convergence of the power series. (Be sure to include a check for convergence at the endpoints of the interval.)

37. $\sum_{n=0}^{\infty} \left(\frac{x}{k}\right)^n, \quad k > 0$

38. $\sum_{n=1}^{\infty} \frac{(-1)^{n+1}(x-c)^n}{nc^n}$

39. $\sum_{n=1}^{\infty} \frac{k(k+1)(k+2) \cdots (k+n-1)x^n}{n!}, \quad k \geq 1$

40. $\sum_{n=1}^{\infty} \frac{n!(x-c)^n}{1 \cdot 3 \cdot 5 \cdots (2n-1)}$

In Exercises 41–44, write an equivalent series with the index of summation beginning at $n = 1$.

41. $\sum_{n=0}^{\infty} \frac{x^n}{n!}$

42. $\sum_{n=0}^{\infty} (-1)^{n+1}(n+1)x^n$

43. $\sum_{n=0}^{\infty} \frac{x^{2n+1}}{(2n+1)!}$

44. $\sum_{n=0}^{\infty} \frac{(-1)^n x^{2n+1}}{2n+1}$

In Exercises 45–48, find the intervals of convergence of (a) $f(x)$, (b) $f'(x)$, (c) $f''(x)$, and (d) $\int f(x)\,dx$. Include a check for convergence at the endpoints of the interval.

45. $f(x) = \sum_{n=0}^{\infty} \left(\frac{x}{2}\right)^n$

46. $f(x) = \sum_{n=1}^{\infty} \frac{(-1)^{n+1}(x-5)^n}{n 5^n}$

47. $f(x) = \sum_{n=0}^{\infty} \frac{(-1)^{n+1}(x-1)^{n+1}}{n+1}$

48. $f(x) = \sum_{n=1}^{\infty} \frac{(-1)^{n+1}(x-2)^n}{n}$

Writing In Exercises 49–52, match the graph of the first 10 terms of the sequence of partial sums of the series

$$g(x) = \sum_{n=0}^{\infty} \left(\frac{x}{3}\right)^n$$

with the indicated value of the function. [The graphs are labeled (a), (b), (c), and (d).] Explain how you made your choice.

(a)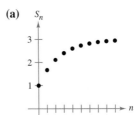

(b)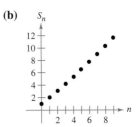

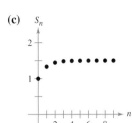
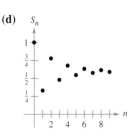

49. $g(1)$ **50.** $g(2)$
51. $g(3.1)$ **52.** $g(-2)$

Writing In Exercises 53–56, match the graph of the first 10 terms of the sequence of partial sums of the series

$$g(x) = \sum_{n=0}^{\infty} (2x)^n$$

with the indicated value of the function. [The graphs are labeled (a), (b), (c), and (d).] Explain how you made your choice.

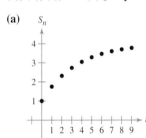
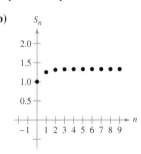
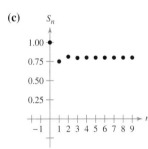
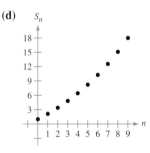

53. $g\left(\dfrac{1}{8}\right)$ **54.** $g\left(-\dfrac{1}{8}\right)$
55. $g\left(\dfrac{9}{16}\right)$ **56.** $g\left(\dfrac{3}{8}\right)$

Writing About Concepts

57. Define a power series centered at c.

58. Describe the radius of convergence of a power series. Describe the interval of convergence of a power series.

59. Describe the three basic forms of the domain of a power series.

Writing About Concepts (continued)

60. Describe how to differentiate and integrate a power series with a radius of convergence R. Will the series resulting from the operations of differentiation and integration have a different radius of convergence? Explain.

61. Give examples that show that the convergence of a power series at an endpoint of its interval of convergence may be either conditional or absolute. Explain your reasoning.

62. Write a power series that has the indicated interval of convergence. Explain your reasoning.
  (a) $(-2, 2)$ (b) $(-1, 1]$ (c) $(-1, 0)$ (d) $[-2, 6]$

63. Let $f(x) = \sum_{n=0}^{\infty} \dfrac{(-1)^n x^{2n+1}}{(2n+1)!}$ and $g(x) = \sum_{n=0}^{\infty} \dfrac{(-1)^n x^{2n}}{(2n)!}$.

  (a) Find the intervals of convergence of f and g.
  (b) Show that $f'(x) = g(x)$.
  (c) Show that $g'(x) = -f(x)$.
  (d) Identify the functions f and g.

64. Let $f(x) = \sum_{n=0}^{\infty} \dfrac{x^n}{n!}$.

  (a) Find the interval of convergence of f.
  (b) Show that $f'(x) = f(x)$.
  (c) Show that $f(0) = 1$.
  (d) Identify the function f.

In Exercises 65–70, show that the function represented by the power series is a solution of the differential equation.

65. $y = \sum_{n=0}^{\infty} \dfrac{(-1)^n x^{2n+1}}{(2n+1)!}$, $y'' + y = 0$

66. $y = \sum_{n=0}^{\infty} \dfrac{(-1)^n x^{2n}}{(2n)!}$, $y'' + y = 0$

67. $y = \sum_{n=0}^{\infty} \dfrac{x^{2n+1}}{(2n+1)!}$, $y'' - y = 0$

68. $y = \sum_{n=0}^{\infty} \dfrac{x^{2n}}{(2n)!}$, $y'' - y = 0$

69. $y = \sum_{n=0}^{\infty} \dfrac{x^{2n}}{2^n n!}$, $y'' - xy' - y = 0$

70. $y = 1 + \sum_{n=1}^{\infty} \dfrac{(-1)^n x^{4n}}{2^{2n} n! \cdot 3 \cdot 7 \cdot 11 \cdots (4n-1)}$, $y'' + x^2 y = 0$

71. *Bessel Function* The Bessel function of order 0 is

$$J_0(x) = \sum_{k=0}^{\infty} \dfrac{(-1)^k x^{2k}}{2^{2k}(k!)^2}.$$

  (a) Show that the series converges for all x.
  (b) Show that the series is a solution of the differential equation $x^2 J_0'' + x J_0' + x^2 J_0 = 0$.
  (c) Use a graphing utility to graph the polynomial composed of the first four terms of J_0.
  (d) Approximate $\int_0^1 J_0 \, dx$ accurate to two decimal places.

72. Bessel Function The Bessel function of order 1 is
$$J_1(x) = x \sum_{k=0}^{\infty} \frac{(-1)^k x^{2k}}{2^{2k+1} k!(k+1)!}.$$

(a) Show that the series converges for all x.

(b) Show that the series is a solution of the differential equation $x^2 J_1'' + x J_1' + (x^2 - 1) J_1 = 0$.

(c) Use a graphing utility to graph the polynomial composed of the first four terms of J_1.

(d) Show that $J_0'(x) = -J_1(x)$.

In Exercises 73–76, the series represents a well-known function. Use a computer algebra system to graph the partial sum S_{10} and identify the function from the graph.

73. $f(x) = \sum_{n=0}^{\infty} (-1)^n \frac{x^{2n}}{(2n)!}$

74. $f(x) = \sum_{n=0}^{\infty} (-1)^n \frac{x^{2n+1}}{(2n+1)!}$

75. $f(x) = \sum_{n=0}^{\infty} (-1)^n x^n$, $-1 < x < 1$

76. $f(x) = \sum_{n=0}^{\infty} (-1)^n \frac{x^{2n+1}}{2n+1}$, $-1 \le x \le 1$

77. Investigation In Exercise 11 you found that the interval of convergence of the geometric series $\sum_{n=0}^{\infty} \left(\frac{x}{2}\right)^n$ is $(-2, 2)$.

(a) Find the sum of the series when $x = \frac{3}{4}$. Use a graphing utility to graph the first six terms of the sequence of partial sums and the horizontal line representing the sum of the series.

(b) Repeat part (a) for $x = -\frac{3}{4}$.

(c) Write a short paragraph comparing the rate of convergence of the partial sums with the sum of the series in parts (a) and (b). How do the plots of the partial sums differ as they converge toward the sum of the series?

(d) Given any positive real number M, there exists a positive integer N such that the partial sum
$$\sum_{n=0}^{N} \left(\frac{3}{2}\right)^n > M.$$

Use a graphing utility to complete the table.

M	10	100	1000	10,000
N				

78. Investigation The interval of convergence of the series $\sum_{n=0}^{\infty} (3x)^n$ is $\left(-\frac{1}{3}, \frac{1}{3}\right)$.

(a) Find the sum of the series when $x = \frac{1}{6}$. Use a graphing utility to graph the first six terms of the sequence of partial sums and the horizontal line representing the sum of the series.

(b) Repeat part (a) for $x = -\frac{1}{6}$.

(c) Write a short paragraph comparing the rate of convergence of the partial sums with the sum of the series in parts (a) and (b). How do the plots of the partial sums differ as they converge toward the sum of the series?

(d) Given any positive real number M, there exists a positive integer N such that the partial sum
$$\sum_{n=0}^{N} \left(3 \cdot \frac{2}{3}\right)^n > M.$$

Use a graphing utility to complete the table.

M	10	100	1000	10,000
N				

True or False? In Exercises 79–82, determine whether the statement is true or false. If it is false, explain why or give an example that shows it is false.

79. If the power series $\sum_{n=0}^{\infty} a_n x^n$ converges for $x = 2$, then it also converges for $x = -2$.

80. If the power series $\sum_{n=0}^{\infty} a_n x^n$ converges for $x = 2$, then it also converges for $x = -1$.

81. If the interval of convergence for $\sum_{n=0}^{\infty} a_n x^n$ is $(-1, 1)$, then the interval of convergence for $\sum_{n=0}^{\infty} a_n (x-1)^n$ is $(0, 2)$.

82. If $f(x) = \sum_{n=0}^{\infty} a_n x^n$ converges for $|x| < 2$, then
$$\int_0^1 f(x)\, dx = \sum_{n=0}^{\infty} \frac{a_n}{n+1}.$$

83. Prove that the power series
$$\sum_{n=0}^{\infty} \frac{(n+p)!}{n!(n+q)!} x^n$$
has a radius of convergence of $R = \infty$ if p and q are positive integers.

84. Let $g(x) = 1 + 2x + x^2 + 2x^3 + x^4 + \cdots$, where the coefficients are $c_{2n} = 1$ and $c_{2n+1} = 2$ for $n \ge 0$.

(a) Find the interval of convergence of the series.

(b) Find an explicit formula for $g(x)$.

85. Let $f(x) = \sum_{n=0}^{\infty} c_n x^n$, where $c_{n+3} = c_n$ for $n \ge 0$.

(a) Find the interval of convergence of the series.

(b) Find an explicit formula for $f(x)$.

86. Prove that if the power series $\sum_{n=0}^{\infty} c_n x^n$ has a radius of convergence of R, then $\sum_{n=0}^{\infty} c_n x^{2n}$ has a radius of convergence of $\sqrt{R}$.

87. For $n > 0$, let $R > 0$ and $c_n > 0$. Prove that if the interval of convergence of the series $\sum_{n=0}^{\infty} c_n (x - x_0)^n$ is $[x_0 - R, x_0 + R]$, then the series converges conditionally at $x_0 + R$.

Section 9.9

Representation of Functions by Power Series

- Find a geometric power series that represents a function.
- Construct a power series using series operations.

Geometric Power Series

In this section and the next, you will study several techniques for finding a power series that represents a given function.

Consider the function given by $f(x) = 1/(1-x)$. The form of f closely resembles the sum of a geometric series

$$\sum_{n=0}^{\infty} ar^n = \frac{a}{1-r}, \quad |r| < 1.$$

In other words, if you let $a = 1$ and $r = x$, a power series representation for $1/(1-x)$, centered at 0, is

$$\frac{1}{1-x} = \sum_{n=0}^{\infty} x^n$$
$$= 1 + x + x^2 + x^3 + \cdots, \quad |x| < 1.$$

Of course, this series represents $f(x) = 1/(1-x)$ only on the interval $(-1, 1)$, whereas f is defined for all $x \neq 1$, as shown in Figure 9.22. To represent f in another interval, you must develop a different series. For instance, to obtain the power series centered at -1, you could write

$$\frac{1}{1-x} = \frac{1}{2-(x+1)} = \frac{1/2}{1-[(x+1)/2]} = \frac{a}{1-r}$$

which implies that $a = \frac{1}{2}$ and $r = (x+1)/2$. So, for $|x+1| < 2$, you have

$$\frac{1}{1-x} = \sum_{n=0}^{\infty} \left(\frac{1}{2}\right)\left(\frac{x+1}{2}\right)^n$$
$$= \frac{1}{2}\left[1 + \frac{(x+1)}{2} + \frac{(x+1)^2}{4} + \frac{(x+1)^3}{8} + \cdots\right], \quad |x+1| < 2$$

which converges on the interval $(-3, 1)$.

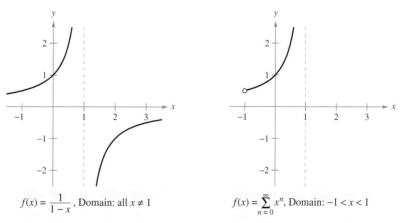

Figure 9.22

JOSEPH FOURIER (1768–1830)

Some of the early work in representing functions by power series was done by the French mathematician Joseph Fourier. Fourier's work is important in the history of calculus, partly because it forced eighteenth century mathematicians to question the then-prevailing narrow concept of a function. Both Cauchy and Dirichlet were motivated by Fourier's work with series, and in 1837 Dirichlet published the general definition of a function that is used today.

EXAMPLE 1 Finding a Geometric Power Series Centered at 0

Find a power series for $f(x) = \dfrac{4}{x+2}$, centered at 0.

Solution Writing $f(x)$ in the form $a/(1-r)$ produces

$$\frac{4}{2+x} = \frac{2}{1-(-x/2)} = \frac{a}{1-r}$$

which implies that $a = 2$ and $r = -x/2$. So, the power series for $f(x)$ is

$$\frac{4}{x+2} = \sum_{n=0}^{\infty} ar^n$$
$$= \sum_{n=0}^{\infty} 2\left(-\frac{x}{2}\right)^n$$
$$= 2\left(1 - \frac{x}{2} + \frac{x^2}{4} - \frac{x^3}{8} + \cdots\right).$$

This power series converges when

$$\left|-\frac{x}{2}\right| < 1$$

which implies that the interval of convergence is $(-2, 2)$.

Long Division

$$\begin{array}{r}
2 - x + \tfrac{1}{2}x^2 - \tfrac{1}{4}x^3 + \cdots \\
2 + x \overline{)\, 4 } \\
\underline{4 + 2x} \\
-2x \\
\underline{-2x - x^2} \\
x^2 \\
\underline{x^2 + \tfrac{1}{2}x^3} \\
-\tfrac{1}{2}x^3 \\
\underline{-\tfrac{1}{2}x^3 - \tfrac{1}{4}x^4}
\end{array}$$

Another way to determine a power series for a rational function such as the one in Example 1 is to use long division. For instance, by dividing $2 + x$ into 4, you obtain the result shown at the left.

EXAMPLE 2 Finding a Geometric Power Series Centered at 1

Find a power series for $f(x) = \dfrac{1}{x}$, centered at 1.

Solution Writing $f(x)$ in the form $a/(1-r)$ produces

$$\frac{1}{x} = \frac{1}{1-(-x+1)} = \frac{a}{1-r}$$

which implies that $a = 1$ and $r = 1 - x = -(x - 1)$. So, the power series for $f(x)$ is

$$\frac{1}{x} = \sum_{n=0}^{\infty} ar^n$$
$$= \sum_{n=0}^{\infty} [-(x-1)]^n$$
$$= \sum_{n=0}^{\infty} (-1)^n (x-1)^n$$
$$= 1 - (x-1) + (x-1)^2 - (x-1)^3 + \cdots.$$

This power series converges when

$$|x - 1| < 1$$

which implies that the interval of convergence is $(0, 2)$.

Operations with Power Series

The versatility of geometric power series will be shown later in this section, following a discussion of power series operations. These operations, used with differentiation and integration, provide a means of developing power series for a variety of elementary functions. (For simplicity, the following properties are stated for a series centered at 0.)

Operations with Power Series

Let $f(x) = \Sigma\, a_n x^n$ and $g(x) = \Sigma\, b_n x^n$.

1. $f(kx) = \displaystyle\sum_{n=0}^{\infty} a_n k^n x^n$

2. $f(x^N) = \displaystyle\sum_{n=0}^{\infty} a_n x^{nN}$

3. $f(x) \pm g(x) = \displaystyle\sum_{n=0}^{\infty} (a_n \pm b_n) x^n$

The operations described above can change the interval of convergence for the resulting series. For example, in the following addition, the interval of convergence for the sum is the *intersection* of the intervals of convergence of the two original series.

$$\underbrace{\sum_{n=0}^{\infty} x^n}_{(-1,\,1)} + \underbrace{\sum_{n=0}^{\infty} \left(\frac{x}{2}\right)^n}_{(-2,\,2)} = \underbrace{\sum_{n=0}^{\infty} \left(1 + \frac{1}{2^n}\right) x^n}_{(-1,\,1)}$$

EXAMPLE 3 Adding Two Power Series

Find a power series, centered at 0, for $f(x) = \dfrac{3x - 1}{x^2 - 1}$.

Solution Using partial fractions, you can write $f(x)$ as

$$\frac{3x - 1}{x^2 - 1} = \frac{2}{x + 1} + \frac{1}{x - 1}.$$

By adding the two geometric power series

$$\frac{2}{x + 1} = \frac{2}{1 - (-x)} = \sum_{n=0}^{\infty} 2(-1)^n x^n, \quad |x| < 1$$

and

$$\frac{1}{x - 1} = \frac{-1}{1 - x} = -\sum_{n=0}^{\infty} x^n, \quad |x| < 1$$

you obtain the following power series.

$$\frac{3x - 1}{x^2 - 1} = \sum_{n=0}^{\infty} [2(-1)^n - 1] x^n = 1 - 3x + x^2 - 3x^3 + x^4 - \cdots$$

The interval of convergence for this power series is $(-1, 1)$.

EXAMPLE 4 Finding a Power Series by Integration

Find a power series for $f(x) = \ln x$, centered at 1.

Solution From Example 2, you know that

$$\frac{1}{x} = \sum_{n=0}^{\infty} (-1)^n (x-1)^n. \qquad \text{Interval of convergence: } (0, 2)$$

Integrating this series produces

$$\ln x = \int \frac{1}{x} dx + C$$

$$= C + \sum_{n=0}^{\infty} (-1)^n \frac{(x-1)^{n+1}}{n+1}.$$

By letting $x = 1$, you can conclude that $C = 0$. Therefore,

$$\ln x = \sum_{n=0}^{\infty} (-1)^n \frac{(x-1)^{n+1}}{n+1}$$

$$= \frac{(x-1)}{1} - \frac{(x-1)^2}{2} + \frac{(x-1)^3}{3} - \frac{(x-1)^4}{4} + \cdots. \qquad \text{Interval of convergence: } (0, 2]$$

Note that the series converges at $x = 2$. This is consistent with the observation in the preceding section that integration of a power series may alter the convergence at the endpoints of the interval of convergence.

TECHNOLOGY In Section 9.7, the fourth-degree Taylor polynomial for the natural logarithmic function

$$\ln x \approx (x-1) - \frac{(x-1)^2}{2} + \frac{(x-1)^3}{3} - \frac{(x-1)^4}{4}$$

was used to approximate $\ln(1.1)$.

$$\ln(1.1) \approx (0.1) - \frac{1}{2}(0.1)^2 + \frac{1}{3}(0.1)^3 - \frac{1}{4}(0.1)^4$$

$$\approx 0.0953083$$

You now know from Example 4 that this polynomial represents the first four terms of the power series for $\ln x$. Moreover, using the Alternating Series Remainder, you can determine that the error in this approximation is less than

$$|R_4| \leq |a_5|$$

$$= \frac{1}{5}(0.1)^5$$

$$= 0.000002.$$

During the seventeenth and eighteenth centuries, mathematical tables for logarithms and values of other transcendental functions were computed in this manner. Such numerical techniques are far from outdated, because it is precisely by such means that many modern calculating devices are programmed to evaluate transcendental functions.

EXAMPLE 5 Finding a Power Series by Integration

Find a power series for $g(x) = \arctan x$, centered at 0.

Solution Because $D_x[\arctan x] = 1/(1 + x^2)$, you can use the series

$$f(x) = \frac{1}{1 + x} = \sum_{n=0}^{\infty} (-1)^n x^n.$$ Interval of convergence: $(-1, 1)$

Substituting x^2 for x produces

$$f(x^2) = \frac{1}{1 + x^2} = \sum_{n=0}^{\infty} (-1)^n x^{2n}.$$

Finally, by integrating, you obtain

$$\arctan x = \int \frac{1}{1 + x^2} dx + C$$

$$= C + \sum_{n=0}^{\infty} (-1)^n \frac{x^{2n+1}}{2n + 1}$$

$$= \sum_{n=0}^{\infty} (-1)^n \frac{x^{2n+1}}{2n + 1}$$ Let $x = 0$, then $C = 0$.

$$= x - \frac{x^3}{3} + \frac{x^5}{5} - \frac{x^7}{7} + \cdots.$$ Interval of convergence: $(-1, 1)$

It can be shown that the power series developed for $\arctan x$ in Example 5 also converges (to $\arctan x$) for $x = \pm 1$. For instance, when $x = 1$, you can write

$$\arctan 1 = 1 - \frac{1}{3} + \frac{1}{5} - \frac{1}{7} + \cdots$$

$$= \frac{\pi}{4}.$$

However, this series (developed by James Gregory in 1671) does not give us a practical way of approximating π because it converges so slowly that hundreds of terms would have to be used to obtain reasonable accuracy. Example 6 shows how to use *two* different arctangent series to obtain a very good approximation of π using only a few terms. This approximation was developed by John Machin in 1706.

EXAMPLE 6 Approximating π with a Series

Use the trigonometric identity

$$4 \arctan \frac{1}{5} - \arctan \frac{1}{239} = \frac{\pi}{4}$$

to approximate the number π [see Exercise 50(b)].

Solution By using only five terms from each of the series for $\arctan(1/5)$ and $\arctan(1/239)$, you obtain

$$4\left(4 \arctan \frac{1}{5} - \arctan \frac{1}{239}\right) \approx 3.1415926$$

which agrees with the exact value of π with an error of less than 0.0000001.

SRINIVASA RAMANUJAN (1887–1920)

Series that can be used to approximate π have interested mathematicians for the past 300 years. An amazing series for approximating $1/\pi$ was discovered by the Indian mathematician Srinivasa Ramanujan in 1914 (see Exercise 64). Each successive term of Ramanujan's series adds roughly eight more correct digits to the value of $1/\pi$. For more information about Ramanujan's work, see the article "Ramanujan and Pi" by Jonathan M. Borwein and Peter B. Borwein in *Scientific American*.

Exercises for Section 9.9

See www.CalcChat.com for worked-out solutions to odd-numbered exercises.

In Exercises 1–4, find a geometric power series for the function, centered at 0, (a) by the technique shown in Examples 1 and 2 and (b) by long division.

1. $f(x) = \dfrac{1}{2 - x}$
2. $f(x) = \dfrac{4}{5 - x}$
3. $f(x) = \dfrac{1}{2 + x}$
4. $f(x) = \dfrac{1}{1 + x}$

In Exercises 5–16, find a power series for the function, centered at c, and determine the interval of convergence.

5. $f(x) = \dfrac{1}{2 - x}, \quad c = 5$
6. $f(x) = \dfrac{4}{5 - x}, \quad c = -2$
7. $f(x) = \dfrac{3}{2x - 1}, \quad c = 0$
8. $f(x) = \dfrac{3}{2x - 1}, \quad c = 2$
9. $g(x) = \dfrac{1}{2x - 5}, \quad c = -3$
10. $h(x) = \dfrac{1}{2x - 5}, \quad c = 0$
11. $f(x) = \dfrac{3}{x + 2}, \quad c = 0$
12. $f(x) = \dfrac{4}{3x + 2}, \quad c = 2$
13. $g(x) = \dfrac{3x}{x^2 + x - 2}, \quad c = 0$
14. $g(x) = \dfrac{4x - 7}{2x^2 + 3x - 2}, \quad c = 0$
15. $f(x) = \dfrac{2}{1 - x^2}, \quad c = 0$
16. $f(x) = \dfrac{4}{4 + x^2}, \quad c = 0$

In Exercises 17–26, use the power series

$$\dfrac{1}{1 + x} = \sum_{n=0}^{\infty} (-1)^n x^n$$

to determine a power series, centered at 0, for the function. Identify the interval of convergence.

17. $h(x) = \dfrac{-2}{x^2 - 1} = \dfrac{1}{1 + x} + \dfrac{1}{1 - x}$
18. $h(x) = \dfrac{x}{x^2 - 1} = \dfrac{1}{2(1 + x)} - \dfrac{1}{2(1 - x)}$
19. $f(x) = -\dfrac{1}{(x + 1)^2} = \dfrac{d}{dx}\left[\dfrac{1}{x + 1}\right]$
20. $f(x) = \dfrac{2}{(x + 1)^3} = \dfrac{d^2}{dx^2}\left[\dfrac{1}{x + 1}\right]$
21. $f(x) = \ln(x + 1) = \displaystyle\int \dfrac{1}{x + 1}\,dx$
22. $f(x) = \ln(1 - x^2) = \displaystyle\int \dfrac{1}{1 + x}\,dx - \displaystyle\int \dfrac{1}{1 - x}\,dx$
23. $g(x) = \dfrac{1}{x^2 + 1}$
24. $f(x) = \ln(x^2 + 1)$
25. $h(x) = \dfrac{1}{4x^2 + 1}$
26. $f(x) = \arctan 2x$

Graphical and Numerical Analysis In Exercises 27 and 28, let

$$S_n = x - \dfrac{x^2}{2} + \dfrac{x^3}{3} - \dfrac{x^4}{4} + \cdots \pm \dfrac{x^n}{n}.$$

Use a graphing utility to confirm the inequality graphically. Then complete the table to confirm the inequality numerically.

x	0.0	0.2	0.4	0.6	0.8	1.0
S_n						
$\ln(x + 1)$						
S_{n+1}						

27. $S_2 \leq \ln(x + 1) \leq S_3$
28. $S_4 \leq \ln(x + 1) \leq S_5$

In Exercises 29 and 30, (a) graph several partial sums of the series, (b) find the sum of the series and its radius of convergence, (c) use 50 terms of the series to approximate the sum when $x = 0.5$, and (d) determine what the approximation represents and how good the approximation is.

29. $\displaystyle\sum_{n=1}^{\infty} \dfrac{(-1)^{n+1}(x - 1)^n}{n}$
30. $\displaystyle\sum_{n=0}^{\infty} \dfrac{(-1)^n x^{2n+1}}{(2n + 1)!}$

In Exercises 31–34, match the polynomial approximation of the function $f(x) = \arctan x$ with the correct graph. [The graphs are labeled (a), (b), (c), and (d).]

(a)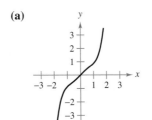

(b)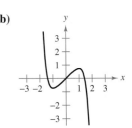

(c)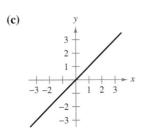

(d)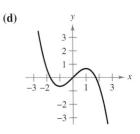

31. $g(x) = x$
32. $g(x) = x - \dfrac{x^3}{3}$
33. $g(x) = x - \dfrac{x^3}{3} + \dfrac{x^5}{5}$
34. $g(x) = x - \dfrac{x^3}{3} + \dfrac{x^5}{5} - \dfrac{x^7}{7}$

In Exercises 35–38, use the series for $f(x) = \arctan x$ to approximate the value, using $R_N \leq 0.001$.

35. $\arctan \dfrac{1}{4}$

36. $\displaystyle\int_0^{3/4} \arctan x^2 \, dx$

37. $\displaystyle\int_0^{1/2} \dfrac{\arctan x^2}{x} \, dx$

38. $\displaystyle\int_0^{1/2} x^2 \arctan x \, dx$

In Exercises 39–42, use the power series

$$\dfrac{1}{1-x} = \sum_{n=0}^{\infty} x^n, \quad |x| < 1.$$

Find the series representation of the function and determine its interval of convergence.

39. $f(x) = \dfrac{1}{(1-x)^2}$

40. $f(x) = \dfrac{x}{(1-x)^2}$

41. $f(x) = \dfrac{1+x}{(1-x)^2}$

42. $f(x) = \dfrac{x(1+x)}{(1-x)^2}$

43. **Probability** A fair coin is tossed repeatedly. The probability that the first head occurs on the nth toss is $P(n) = \left(\tfrac{1}{2}\right)^n$. When this game is repeated many times, the average number of tosses required until the first head occurs is

$$E(n) = \sum_{n=1}^{\infty} nP(n).$$

(This value is called the *expected value of* n.) Use the results of Exercises 39–42 to find $E(n)$. Is the answer what you expected? Why or why not?

44. Use the results of Exercises 39–42 to find the sum of each series.

(a) $\dfrac{1}{3} \displaystyle\sum_{n=1}^{\infty} n\left(\dfrac{2}{3}\right)^n$

(b) $\dfrac{1}{10} \displaystyle\sum_{n=1}^{\infty} n\left(\dfrac{9}{10}\right)^n$

Writing In Exercises 45–48, explain how to use the geometric series

$$g(x) = \dfrac{1}{1-x} = \sum_{n=0}^{\infty} x^n, \quad |x| < 1$$

to find the series for the function. Do not find the series.

45. $f(x) = \dfrac{1}{1+x}$

46. $f(x) = \dfrac{1}{1-x^2}$

47. $f(x) = \dfrac{5}{1+x}$

48. $f(x) = \ln(1-x)$

49. Prove that $\arctan x + \arctan y = \arctan \dfrac{x+y}{1-xy}$ for $xy \neq 1$ provided the value of the left side of the equation is between $-\pi/2$ and $\pi/2$.

50. Use the result of Exercise 49 to verify each identity.

(a) $\arctan \dfrac{120}{119} - \arctan \dfrac{1}{239} = \dfrac{\pi}{4}$

(b) $4 \arctan \dfrac{1}{5} - \arctan \dfrac{1}{239} = \dfrac{\pi}{4}$

[*Hint:* Use Exercise 49 twice to find $4 \arctan \tfrac{1}{5}$. Then use part (a).]

In Exercises 51 and 52, (a) verify the given equation and (b) use the equation and the series for the arctangent to approximate π to two-decimal-place accuracy.

51. $2 \arctan \dfrac{1}{2} - \arctan \dfrac{1}{7} = \dfrac{\pi}{4}$

52. $\arctan \dfrac{1}{2} + \arctan \dfrac{1}{3} = \dfrac{\pi}{4}$

In Exercises 53–58, find the sum of the convergent series by using a well-known function. Identify the function and explain how you obtained the sum.

53. $\displaystyle\sum_{n=1}^{\infty} (-1)^{n+1} \dfrac{1}{2^n n}$

54. $\displaystyle\sum_{n=1}^{\infty} (-1)^{n+1} \dfrac{1}{3^n n}$

55. $\displaystyle\sum_{n=1}^{\infty} (-1)^{n+1} \dfrac{2^n}{5^n n}$

56. $\displaystyle\sum_{n=0}^{\infty} (-1)^n \dfrac{1}{2n+1}$

57. $\displaystyle\sum_{n=0}^{\infty} (-1)^n \dfrac{1}{2^{2n+1}(2n+1)}$

58. $\displaystyle\sum_{n=1}^{\infty} (-1)^{n+1} \dfrac{1}{3^{2n-1}(2n-1)}$

Writing About Concepts

59. Use the results of Exercises 31–34 to make a geometric argument for why the series approximations of $f(x) = \arctan x$ have only odd powers of x.

60. Use the results of Exercises 31–34 to make a conjecture about the degrees of series approximations of $f(x) = \arctan x$ that have relative extrema.

61. One of the series in Exercises 53–58 converges to its sum at a much lower rate than the other five series. Which is it? Explain why this series converges so slowly. Use a graphing utility to illustrate the rate of convergence.

62. The radius of convergence of the power series $\displaystyle\sum_{n=0}^{\infty} a_n x^n$ is 3. What is the radius of convergence of the series $\displaystyle\sum_{n=1}^{\infty} n a_n x^{n-1}$? Explain.

63. The power series $\displaystyle\sum_{n=0}^{\infty} a_n x^n$ converges for $|x+1| < 4$. What can you conclude about the series $\displaystyle\sum_{n=0}^{\infty} a_n \dfrac{x^{n+1}}{n+1}$? Explain.

 64. Use a graphing utility to show that

$$\dfrac{\sqrt{8}}{9801} \sum_{n=0}^{\infty} \dfrac{(4n)!(1103 + 26{,}390n)}{(n!)^4 396^{4n}} = \dfrac{1}{\pi}.$$

(*Note:* This series was discovered by the Indian mathematician Srinivasa Ramanujan in 1914.)

In Exercises 65 and 66, find the sum of the series.

65. $\displaystyle\sum_{n=0}^{\infty} \dfrac{(-1)^n}{3^n(2n+1)}$

66. $\displaystyle\sum_{n=0}^{\infty} \dfrac{(-1)^n \pi^{2n+1}}{3^{2n+1}(2n+1)!}$

Section 9.10 Taylor and Maclaurin Series

- Find a Taylor or Maclaurin series for a function.
- Find a binomial series.
- Use a basic list of Taylor series to find other Taylor series.

Taylor Series and Maclaurin Series

In Section 9.9, you derived power series for several functions using geometric series with term-by-term differentiation or integration. In this section you will study a *general* procedure for deriving the power series for a function that has derivatives of all orders. The following theorem gives the form that *every* convergent power series must take.

> **THEOREM 9.22   The Form of a Convergent Power Series**
>
> If f is represented by a power series $f(x) = \Sigma a_n(x - c)^n$ for all x in an open interval I containing c, then $a_n = f^{(n)}(c)/n!$ and
>
> $$f(x) = f(c) + f'(c)(x - c) + \frac{f''(c)}{2!}(x - c)^2 + \cdots + \frac{f^{(n)}(c)}{n!}(x - c)^n + \cdots.$$

Proof   Suppose the power series $\Sigma a_n(x - c)^n$ has a radius of convergence R. Then, by Theorem 9.21, you know that the nth derivative of f exists for $|x - c| < R$, and by successive differentiation you obtain the following.

$$f^{(0)}(x) = a_0 + a_1(x - c) + a_2(x - c)^2 + a_3(x - c)^3 + a_4(x - c)^4 + \cdots$$
$$f^{(1)}(x) = a_1 + 2a_2(x - c) + 3a_3(x - c)^2 + 4a_4(x - c)^3 + \cdots$$
$$f^{(2)}(x) = 2a_2 + 3!a_3(x - c) + 4 \cdot 3a_4(x - c)^2 + \cdots$$
$$f^{(3)}(x) = 3!a_3 + 4!a_4(x - c) + \cdots$$
$$\vdots$$
$$f^{(n)}(x) = n!a_n + (n + 1)!a_{n+1}(x - c) + \cdots$$

Evaluating each of these derivatives at $x = c$ yields

$$f^{(0)}(c) = 0!a_0$$
$$f^{(1)}(c) = 1!a_1$$
$$f^{(2)}(c) = 2!a_2$$
$$f^{(3)}(c) = 3!a_3$$

and, in general, $f^{(n)}(c) = n!a_n$. By solving for a_n, you find that the coefficients of the power series representation of $f(x)$ are

$$a_n = \frac{f^{(n)}(c)}{n!}.$$

Notice that the coefficients of the power series in Theorem 9.22 are precisely the coefficients of the Taylor polynomials for $f(x)$ at c as defined in Section 9.7. For this reason, the series is called the **Taylor series** for $f(x)$ at c.

COLIN MACLAURIN (1698–1746)

The development of power series to represent functions is credited to the combined work of many seventeenth and eighteenth century mathematicians. Gregory, Newton, John and James Bernoulli, Leibniz, Euler, Lagrange, Wallis, and Fourier all contributed to this work. However, the two names that are most commonly associated with power series are Brook Taylor (1685–1731) and Colin Maclaurin.

NOTE   Be sure you understand Theorem 9.22. The theorem says that *if a power series converges to* $f(x)$, the series must be a Taylor series. The theorem does *not* say that every series formed with the Taylor coefficients $a_n = f^{(n)}(c)/n!$ will converge to $f(x)$.

> **Definitions of Taylor and Maclaurin Series**
>
> If a function f has derivatives of all orders at $x = c$, then the series
>
> $$\sum_{n=0}^{\infty} \frac{f^{(n)}(c)}{n!}(x - c)^n = f(c) + f'(c)(x - c) + \cdots + \frac{f^{(n)}(c)}{n!}(x - c)^n + \cdots$$
>
> is called the **Taylor series for $f(x)$ at c**. Moreover, if $c = 0$, then the series is the **Maclaurin series for f**.

If you know the pattern for the coefficients of the Taylor polynomials for a function, you can extend the pattern easily to form the corresponding Taylor series. For instance, in Example 4 in Section 9.7, you found the fourth Taylor polynomial for $\ln x$, centered at 1, to be

$$P_4(x) = (x - 1) - \frac{1}{2}(x - 1)^2 + \frac{1}{3}(x - 1)^3 - \frac{1}{4}(x - 1)^4.$$

From this pattern, you can obtain the Taylor series for $\ln x$ centered at $c = 1$,

$$(x - 1) - \frac{1}{2}(x - 1)^2 + \cdots + \frac{(-1)^{n+1}}{n}(x - 1)^n + \cdots.$$

EXAMPLE 1 Forming a Power Series

Use the function $f(x) = \sin x$ to form the Maclaurin series

$$\sum_{n=0}^{\infty} \frac{f^{(n)}(0)}{n!} x^n = f(0) + f'(0)x + \frac{f''(0)}{2!}x^2 + \frac{f^{(3)}(0)}{3!}x^3 + \frac{f^{(4)}(0)}{4!}x^4 + \cdots$$

and determine the interval of convergence.

Solution Successive differentiation of $f(x)$ yields

$$f(x) = \sin x \qquad f(0) = \sin 0 = 0$$
$$f'(x) = \cos x \qquad f'(0) = \cos 0 = 1$$
$$f''(x) = -\sin x \qquad f''(0) = -\sin 0 = 0$$
$$f^{(3)}(x) = -\cos x \qquad f^{(3)}(0) = -\cos 0 = -1$$
$$f^{(4)}(x) = \sin x \qquad f^{(4)}(0) = \sin 0 = 0$$
$$f^{(5)}(x) = \cos x \qquad f^{(5)}(0) = \cos 0 = 1$$

and so on. The pattern repeats after the third derivative. So, the power series is as follows.

$$\sum_{n=0}^{\infty} \frac{f^{(n)}(0)}{n!} x^n = f(0) + f'(0)x + \frac{f''(0)}{2!}x^2 + \frac{f^{(3)}(0)}{3!}x^3 + \frac{f^{(4)}(0)}{4!}x^4 + \cdots$$

$$\sum_{n=0}^{\infty} \frac{(-1)^n x^{2n+1}}{(2n+1)!} = 0 + (1)x + \frac{0}{2!}x^2 + \frac{(-1)}{3!}x^3 + \frac{0}{4!}x^4 + \frac{1}{5!}x^5 + \frac{0}{6!}x^6$$

$$+ \frac{(-1)}{7!}x^7 + \cdots$$

$$= x - \frac{x^3}{3!} + \frac{x^5}{5!} - \frac{x^7}{7!} + \cdots$$

By the Ratio Test, you can conclude that this series converges for all x.

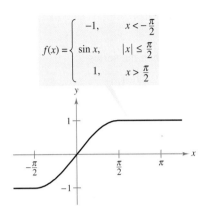

Figure 9.23

Notice that in Example 1 you cannot conclude that the power series converges to $\sin x$ for all x. You can simply conclude that the power series converges to some function, but you are not sure what function it is. This is a subtle, but important, point in dealing with Taylor or Maclaurin series. To persuade yourself that the series

$$f(c) + f'(c)(x - c) + \frac{f''(c)}{2!}(x - c)^2 + \cdots + \frac{f^{(n)}(c)}{n!}(x - c)^n + \cdots$$

might converge to a function other than f, remember that the derivatives are being evaluated at a single point. It can easily happen that another function will agree with the values of $f^{(n)}(x)$ when $x = c$ and disagree at other x-values. For instance, if you formed the power series (centered at 0) for the function shown in Figure 9.23, you would obtain the same series as in Example 1. You know that the series converges for all x, and yet it obviously cannot converge to both $f(x)$ and $\sin x$ for all x.

Let f have derivatives of all orders in an open interval I centered at c. The Taylor series for f may fail to converge for some x in I. Or, even if it is convergent, it may fail to have $f(x)$ as its sum. Nevertheless, Theorem 9.19 tells us that for each n,

$$f(x) = f(c) + f'(c)(x - c) + \frac{f''(c)}{2!}(x - c)^2 + \cdots + \frac{f^{(n)}(c)}{n!}(x - c)^n + R_n(x),$$

where

$$R_n(x) = \frac{f^{(n+1)}(z)}{(n + 1)!}(x - c)^{n+1}.$$

Note that in this remainder formula the particular value of z that makes the remainder formula true depends on the values of x and n. If $R_n \to 0$, then the following theorem tells us that the Taylor series for f actually converges to $f(x)$ for all x in I.

THEOREM 9.23 Convergence of Taylor Series

If $\lim_{n \to \infty} R_n = 0$ for all x in the interval I, then the Taylor series for f converges and equals $f(x)$,

$$f(x) = \sum_{n=0}^{\infty} \frac{f^{(n)}(c)}{n!}(x - c)^n.$$

Proof For a Taylor series, the nth partial sum coincides with the nth Taylor polynomial. That is, $S_n(x) = P_n(x)$. Moreover, because

$$P_n(x) = f(x) - R_n(x)$$

it follows that

$$\lim_{n \to \infty} S_n(x) = \lim_{n \to \infty} P_n(x)$$
$$= \lim_{n \to \infty} [f(x) - R_n(x)]$$
$$= f(x) - \lim_{n \to \infty} R_n(x).$$

So, for a given x, the Taylor series (the sequence of partial sums) converges to $f(x)$ if and only if $R_n(x) \to 0$ as $n \to \infty$.

NOTE Stated another way, Theorem 9.23 says that a power series formed with Taylor coefficients $a_n = f^{(n)}(c)/n!$ converges to the function from which it was derived at precisely those values for which the remainder approaches 0 as $n \to \infty$.

In Example 1, you derived the power series from the sine function and you also concluded that the series converges to some function on the entire real line. In Example 2, you will see that the series actually converges to sin x. The key observation is that although the value of z is not known, it is possible to obtain an upper bound for $|f^{(n+1)}(z)|$.

EXAMPLE 2 A Convergent Maclaurin Series

Show that the Maclaurin series for $f(x) = \sin x$ converges to $\sin x$ for all x.

Solution Using the result in Example 1, you need to show that

$$\sin x = x - \frac{x^3}{3!} + \frac{x^5}{5!} - \frac{x^7}{7!} + \cdots + \frac{(-1)^n x^{2n+1}}{(2n+1)!} + \cdots$$

is true for all x. Because

$$f^{(n+1)}(x) = \pm \sin x$$

or

$$f^{(n+1)}(x) = \pm \cos x$$

you know that $|f^{(n+1)}(z)| \leq 1$ for every real number z. Therefore, for any fixed x, you can apply Taylor's Theorem (Theorem 9.19) to conclude that

$$0 \leq |R_n(x)| = \left| \frac{f^{(n+1)}(z)}{(n+1)!} x^{n+1} \right| \leq \frac{|x|^{n+1}}{(n+1)!}.$$

From the discussion in Section 9.1 regarding the relative rates of convergence of exponential and factorial sequences, it follows that for a fixed x

$$\lim_{n \to \infty} \frac{|x|^{n+1}}{(n+1)!} = 0.$$

Finally, by the Squeeze Theorem, it follows that for all x, $R_n(x) \to 0$ as $n \to \infty$. So, by Theorem 9.23, the Maclaurin series for $\sin x$ converges to $\sin x$ for all x.

Figure 9.24 visually illustrates the convergence of the Maclaurin series for $\sin x$ by comparing the graphs of the Maclaurin polynomials $P_1(x)$, $P_3(x)$, $P_5(x)$, and $P_7(x)$ with the graph of the sine function. Notice that as the degree of the polynomial increases, its graph more closely resembles that of the sine function.

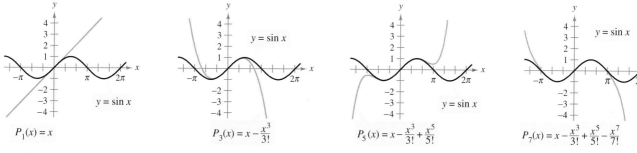

As n increases, the graph of P_n more closely resembles the sine function.
Figure 9.24

The guidelines for finding a Taylor series for $f(x)$ at c are summarized below.

> **Guidelines for Finding a Taylor Series**
>
> 1. Differentiate $f(x)$ several times and evaluate each derivative at c.
>
> $$f(c), f'(c), f''(c), f'''(c), \cdots, f^{(n)}(c), \cdots$$
>
> Try to recognize a pattern in these numbers.
>
> 2. Use the sequence developed in the first step to form the Taylor coefficients $a_n = f^{(n)}(c)/n!$, and determine the interval of convergence for the resulting power series
>
> $$f(c) + f'(c)(x - c) + \frac{f''(c)}{2!}(x - c)^2 + \cdots + \frac{f^{(n)}(c)}{n!}(x - c)^n + \cdots.$$
>
> 3. Within this interval of convergence, determine whether or not the series converges to $f(x)$.

The direct determination of Taylor or Maclaurin coefficients using successive differentiation can be difficult, and the next example illustrates a shortcut for finding the coefficients indirectly—using the coefficients of a known Taylor or Maclaurin series.

EXAMPLE 3 Maclaurin Series for a Composite Function

Find the Maclaurin series for $f(x) = \sin x^2$.

Solution To find the coefficients for this Maclaurin series directly, you must calculate successive derivatives of $f(x) = \sin x^2$. By calculating just the first two,

$$f'(x) = 2x \cos x^2 \quad \text{and} \quad f''(x) = -4x^2 \sin x^2 + 2 \cos x^2$$

you can see that this task would be quite cumbersome. Fortunately, there is an alternative. First consider the Maclaurin series for $\sin x$ found in Example 1.

$$g(x) = \sin x$$
$$= x - \frac{x^3}{3!} + \frac{x^5}{5!} - \frac{x^7}{7!} + \cdots$$

Now, because $\sin x^2 = g(x^2)$, you can substitute x^2 for x in the series for $\sin x$ to obtain

$$\sin x^2 = g(x^2)$$
$$= x^2 - \frac{x^6}{3!} + \frac{x^{10}}{5!} - \frac{x^{14}}{7!} + \cdots.$$

Be sure to understand the point illustrated in Example 3. Because direct computation of Taylor or Maclaurin coefficients can be tedious, the most practical way to find a Taylor or Maclaurin series is to develop power series for a *basic list* of elementary functions. From this list, you can determine power series for other functions by the operations of addition, subtraction, multiplication, division, differentiation, integration, or composition with known power series.

Binomial Series

Before presenting the basic list for elementary functions, you wll develop one more series—for a function of the form $f(x) = (1 + x)^k$. This produces the **binomial series.**

EXAMPLE 4 Binomial Series

Find the Maclaurin series for $f(x) = (1 + x)^k$ and determine its radius of convergence. Assume that k is not a positive integer.

Solution By successive differentiation, you have

$$f(x) = (1 + x)^k \qquad\qquad f(0) = 1$$
$$f'(x) = k(1 + x)^{k-1} \qquad\qquad f'(0) = k$$
$$f''(x) = k(k - 1)(1 + x)^{k-2} \qquad\qquad f''(0) = k(k - 1)$$
$$f'''(x) = k(k - 1)(k - 2)(1 + x)^{k-3} \qquad\qquad f'''(0) = k(k - 1)(k - 2)$$
$$\vdots \qquad\qquad \vdots$$
$$f^{(n)}(x) = k \cdots (k - n + 1)(1 + x)^{k-n} \qquad f^{(n)}(0) = k(k - 1) \cdots (k - n + 1)$$

which produces the series

$$1 + kx + \frac{k(k - 1)x^2}{2} + \cdots + \frac{k(k - 1) \cdots (k - n + 1)x^n}{n!} + \cdots.$$

Because $a_{n+1}/a_n \to 1$, you can apply the Ratio Test to conclude that the radius of convergence is $R = 1$. So, the series converges to some function in the interval $(-1, 1)$.

Note that Example 4 shows that the Taylor series for $(1 + x)^k$ converges to some function in the interval $(-1, 1)$. However, the example does not show that the series actually converges to $(1 + x)^k$. To do this, you could show that the remainder $R_n(x)$ converges to 0, as illustrated in Example 2.

EXAMPLE 5 Finding a Binomial Series

Find the power series for $f(x) = \sqrt[3]{1 + x}$.

Solution Using the binomial series

$$(1 + x)^k = 1 + kx + \frac{k(k - 1)x^2}{2!} + \frac{k(k - 1)(k - 2)x^3}{3!} + \cdots$$

let $k = \frac{1}{3}$ and write

$$(1 + x)^{1/3} = 1 + \frac{x}{3} - \frac{2x^2}{3^2 2!} + \frac{2 \cdot 5 x^3}{3^3 3!} - \frac{2 \cdot 5 \cdot 8 x^4}{3^4 4!} + \cdots$$

which converges for $-1 \le x \le 1$.

TECHNOLOGY Use a graphing utility to confirm the result in Example 5. When you graph the functions

$$f(x) = (1 + x)^{1/3} \quad \text{and} \quad P_4(x) = 1 + \frac{x}{3} - \frac{x^2}{9} + \frac{5x^3}{81} - \frac{10x^4}{243}$$

in the same viewing window, you should obtain the result shown in Figure 9.25.

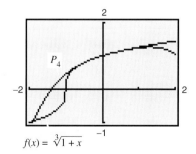

$f(x) = \sqrt[3]{1 + x}$

Figure 9.25

Deriving Taylor Series from a Basic List

The following list provides the power series for several elementary functions with the corresponding intervals of convergence.

Power Series for Elementary Functions

Function	Interval of Convergence
$\dfrac{1}{x} = 1 - (x-1) + (x-1)^2 - (x-1)^3 + (x-1)^4 - \cdots + (-1)^n (x-1)^n + \cdots$	$0 < x < 2$
$\dfrac{1}{1+x} = 1 - x + x^2 - x^3 + x^4 - x^5 + \cdots + (-1)^n x^n + \cdots$	$-1 < x < 1$
$\ln x = (x-1) - \dfrac{(x-1)^2}{2} + \dfrac{(x-1)^3}{3} - \dfrac{(x-1)^4}{4} + \cdots + \dfrac{(-1)^{n-1}(x-1)^n}{n} + \cdots$	$0 < x \leq 2$
$e^x = 1 + x + \dfrac{x^2}{2!} + \dfrac{x^3}{3!} + \dfrac{x^4}{4!} + \dfrac{x^5}{5!} + \cdots + \dfrac{x^n}{n!} + \cdots$	$-\infty < x < \infty$
$\sin x = x - \dfrac{x^3}{3!} + \dfrac{x^5}{5!} - \dfrac{x^7}{7!} + \dfrac{x^9}{9!} - \cdots + \dfrac{(-1)^n x^{2n+1}}{(2n+1)!} + \cdots$	$-\infty < x < \infty$
$\cos x = 1 - \dfrac{x^2}{2!} + \dfrac{x^4}{4!} - \dfrac{x^6}{6!} + \dfrac{x^8}{8!} - \cdots + \dfrac{(-1)^n x^{2n}}{(2n)!} + \cdots$	$-\infty < x < \infty$
$\arctan x = x - \dfrac{x^3}{3} + \dfrac{x^5}{5} - \dfrac{x^7}{7} + \dfrac{x^9}{9} - \cdots + \dfrac{(-1)^n x^{2n+1}}{2n+1} + \cdots$	$-1 \leq x \leq 1$
$\arcsin x = x + \dfrac{x^3}{2 \cdot 3} + \dfrac{1 \cdot 3 x^5}{2 \cdot 4 \cdot 5} + \dfrac{1 \cdot 3 \cdot 5 x^7}{2 \cdot 4 \cdot 6 \cdot 7} + \cdots + \dfrac{(2n)! x^{2n+1}}{(2^n n!)^2 (2n+1)} + \cdots$	$-1 \leq x \leq 1$
$(1+x)^k = 1 + kx + \dfrac{k(k-1)x^2}{2!} + \dfrac{k(k-1)(k-2)x^3}{3!} + \dfrac{k(k-1)(k-2)(k-3)x^4}{4!} + \cdots$	$-1 < x < 1$*

*The convergence at $x = \pm 1$ depends on the value of k.

NOTE The binomial series is valid for noninteger values of k. Moreover, if k happens to be a positive integer, the binomial series reduces to a simple binomial expansion.

EXAMPLE 6 Deriving a Power Series from a Basic List

Find the power series for $f(x) = \cos \sqrt{x}$.

Solution Using the power series

$$\cos x = 1 - \frac{x^2}{2!} + \frac{x^4}{4!} - \frac{x^6}{6!} + \frac{x^8}{8!} - \cdots$$

you can replace x by $\sqrt{x}$ to obtain the series

$$\cos \sqrt{x} = 1 - \frac{x}{2!} + \frac{x^2}{4!} - \frac{x^3}{6!} + \frac{x^4}{8!} - \cdots.$$

This series converges for all x in the domain of $\cos \sqrt{x}$—that is, for $x \geq 0$.

Power series can be multiplied and divided like polynomials. After finding the first few terms of the product (or quotient), you may be able to recognize a pattern.

EXAMPLE 7 Multiplication and Division of Power Series

Find the first three nonzero terms in each of the Maclaurin series.

a. $e^x \arctan x$ **b.** $\tan x$

Solution

a. Using the Maclaurin series for e^x and $\arctan x$ in the table, you have

$$e^x \arctan x = \left(1 + \frac{x}{1!} + \frac{x^2}{2!} + \frac{x^3}{3!} + \frac{x^4}{4!} + \cdots\right)\left(x - \frac{x^3}{3} + \frac{x^5}{5} - \cdots\right).$$

Multiply these expressions and collect like terms as you would for multiplying polynomials.

$$\begin{array}{l}
1 + x + \frac{1}{2}x^2 + \frac{1}{6}x^3 + \frac{1}{24}x^4 + \cdots \\
\qquad x \qquad\quad - \frac{1}{3}x^3 \qquad + \frac{1}{5}x^5 - \cdots \\
\hline
x + x^2 + \frac{1}{2}x^3 + \frac{1}{6}x^4 + \frac{1}{24}x^5 + \cdots \\
\qquad\qquad - \frac{1}{3}x^3 - \frac{1}{3}x^4 - \frac{1}{6}x^5 - \cdots \\
\qquad\qquad\qquad\qquad\qquad\quad + \frac{1}{5}x^5 + \cdots \\
\hline
x + x^2 + \frac{1}{6}x^3 - \frac{1}{6}x^4 + \frac{3}{40}x^5 + \cdots
\end{array}$$

So, $e^x \arctan x = x + x^2 + \frac{1}{6}x^3 + \cdots$.

b. Using the Maclaurin series for $\sin x$ and $\cos x$ in the table, you have

$$\tan x = \frac{\sin x}{\cos x} = \frac{x - \dfrac{x^3}{3!} + \dfrac{x^5}{5!} - \cdots}{1 - \dfrac{x^2}{2!} + \dfrac{x^4}{4!} - \cdots}.$$

Divide using long division.

$$\begin{array}{r}
x + \frac{1}{3}x^3 + \frac{2}{15}x^5 + \cdots \\
1 - \frac{1}{2}x^2 + \frac{1}{24}x^4 - \cdots \overline{\smash{\big)}\, x - \frac{1}{6}x^3 + \frac{1}{120}x^5 - \cdots} \\
\underline{x - \frac{1}{2}x^3 + \frac{1}{24}x^5 - \cdots} \\
\frac{1}{3}x^3 - \frac{1}{30}x^5 + \cdots \\
\underline{\frac{1}{3}x^3 - \frac{1}{6}x^5 + \cdots} \\
\frac{2}{15}x^5 + \cdots
\end{array}$$

So, $\tan x = x + \frac{1}{3}x^3 + \frac{2}{15}x^5 + \cdots$.

EXAMPLE 8 A Power Series for $\sin^2 x$

Find the power series for $f(x) = \sin^2 x$.

Solution Consider rewriting $\sin^2 x$ as follows.

$$\sin^2 x = \frac{1 - \cos 2x}{2} = \frac{1}{2} - \frac{\cos 2x}{2}$$

Now, use the series for $\cos x$.

$$\cos x = 1 - \frac{x^2}{2!} + \frac{x^4}{4!} - \frac{x^6}{6!} + \frac{x^8}{8!} - \cdots$$

$$\cos 2x = 1 - \frac{2^2}{2!}x^2 + \frac{2^4}{4!}x^4 - \frac{2^6}{6!}x^6 + \frac{2^8}{8!}x^8 - \cdots$$

$$-\frac{1}{2}\cos 2x = -\frac{1}{2} + \frac{2}{2!}x^2 - \frac{2^3}{4!}x^4 + \frac{2^5}{6!}x^6 - \frac{2^7}{8!}x^8 + \cdots$$

$$\sin^2 x = \frac{1}{2} - \frac{1}{2}\cos 2x = \frac{1}{2} - \frac{1}{2} + \frac{2}{2!}x^2 - \frac{2^3}{4!}x^4 + \frac{2^5}{6!}x^6 - \frac{2^7}{8!}x^8 + \cdots$$

$$= \frac{2}{2!}x^2 - \frac{2^3}{4!}x^4 + \frac{2^5}{6!}x^6 - \frac{2^7}{8!}x^8 + \cdots$$

This series converges for $-\infty < x < \infty$.

As mentioned in the preceding section, power series can be used to obtain tables of values of transcendental functions. They are also useful for estimating the values of definite integrals for which antiderivatives cannot be found. The next example demonstrates this use.

EXAMPLE 9 Power Series Approximation of a Definite Integral

Use a power series to approximate

$$\int_0^1 e^{-x^2}\, dx$$

with an error of less than 0.01.

Solution Replacing x with $-x^2$ in the series for e^x produces the following.

$$e^{-x^2} = 1 - x^2 + \frac{x^4}{2!} - \frac{x^6}{3!} + \frac{x^8}{4!} - \cdots$$

$$\int_0^1 e^{-x^2}\, dx = \left[x - \frac{x^3}{3} + \frac{x^5}{5 \cdot 2!} - \frac{x^7}{7 \cdot 3!} + \frac{x^9}{9 \cdot 4!} - \cdots \right]_0^1$$

$$= 1 - \frac{1}{3} + \frac{1}{10} - \frac{1}{42} + \frac{1}{216} - \cdots$$

Summing the first four terms, you have

$$\int_0^1 e^{-x^2}\, dx \approx 0.74$$

which, by the Alternating Series Test, has an error of less than $\frac{1}{216} \approx 0.005$.

Exercises for Section 9.10

In Exercises 1–10, use the definition to find the Taylor series (centered at c) for the function.

1. $f(x) = e^{2x}$, $c = 0$
2. $f(x) = e^{3x}$, $c = 0$
3. $f(x) = \cos x$, $c = \dfrac{\pi}{4}$
4. $f(x) = \sin x$, $c = \dfrac{\pi}{4}$
5. $f(x) = \ln x$, $c = 1$
6. $f(x) = e^x$, $c = 1$
7. $f(x) = \sin 2x$, $c = 0$
8. $f(x) = \ln(x^2 + 1)$, $c = 0$
9. $f(x) = \sec x$, $c = 0$ (first three nonzero terms)
10. $f(x) = \tan x$, $c = 0$ (first three nonzero terms)

In Exercises 11–14, prove that the Maclaurin series for the function converges to the function for all x.

11. $f(x) = \cos x$
12. $f(x) = e^{-2x}$
13. $f(x) = \sinh x$
14. $f(x) = \cosh x$

In Exercises 15–20, use the binomial series to find the Maclaurin series for the function.

15. $f(x) = \dfrac{1}{(1 + x)^2}$
16. $f(x) = \dfrac{1}{\sqrt{1 - x}}$
17. $f(x) = \dfrac{1}{\sqrt{4 + x^2}}$
18. $f(x) = \sqrt[4]{1 + x}$
19. $f(x) = \sqrt{1 + x^2}$
20. $f(x) = \sqrt{1 + x^3}$

In Exercises 21–30, find the Maclaurin series for the function. (Use the table of power series for elementary functions.)

21. $f(x) = e^{x^2/2}$
22. $g(x) = e^{-3x}$
23. $g(x) = \sin 3x$
24. $f(x) = \cos 4x$
25. $f(x) = \cos x^{3/2}$
26. $g(x) = 2 \sin x^3$
27. $f(x) = \tfrac{1}{2}(e^x - e^{-x}) = \sinh x$
28. $f(x) = e^x + e^{-x} = 2 \cosh x$
29. $f(x) = \cos^2 x$
30. $f(x) = \sinh^{-1} x = \ln\left(x + \sqrt{x^2 + 1}\right)$

$\left(\text{Hint: Integrate the series for } \dfrac{1}{\sqrt{x^2 + 1}}.\right)$

In Exercises 31–34, find the Maclaurin series for the function. (See Example 7.)

31. $f(x) = x \sin x$
32. $h(x) = x \cos x$
33. $g(x) = \begin{cases} \dfrac{\sin x}{x}, & x \neq 0 \\ 1, & x = 0 \end{cases}$
34. $f(x) = \begin{cases} \dfrac{\arcsin x}{x}, & x \neq 0 \\ 1, & x = 0 \end{cases}$

In Exercises 35 and 36, use a power series and the fact that $i^2 = -1$ to verify the formula.

35. $g(x) = \dfrac{1}{2i}(e^{ix} - e^{-ix}) = \sin x$
36. $g(x) = \tfrac{1}{2}(e^{ix} + e^{-ix}) = \cos x$

In Exercises 37–42, find the first four nonzero terms of the Maclaurin series for the function by multiplying or dividing the appropriate power series. Use the table of power series for elementary functions on page 682. Use a graphing utility to graph the function and its corresponding polynomial approximation.

37. $f(x) = e^x \sin x$
38. $g(x) = e^x \cos x$
39. $h(x) = \cos x \ln(1 + x)$
40. $f(x) = e^x \ln(1 + x)$
41. $g(x) = \dfrac{\sin x}{1 + x}$
42. $f(x) = \dfrac{e^x}{1 + x}$

In Exercises 43–46, match the polynomial with its graph. [The graphs are labeled (a), (b), (c), and (d).] Factor a common factor from each polynomial and identify the function approximated by the remaining Taylor polynomial.

(a)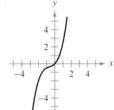

(b)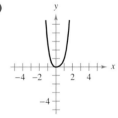

(c)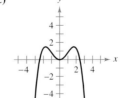

(d)

43. $y = x^2 - \dfrac{x^4}{3!}$
44. $y = x - \dfrac{x^3}{2!} + \dfrac{x^5}{4!}$
45. $y = x + x^2 + \dfrac{x^3}{2!}$
46. $y = x^2 - x^3 + x^4$

In Exercises 47 and 48, find a Maclaurin series for $f(x)$.

47. $f(x) = \int_0^x (e^{-t^2} - 1)\, dt$

48. $f(x) = \int_0^x \sqrt{1 + t^3}\, dt$

In Exercises 49–52, verify the sum. Then use a graphing utility to approximate the sum with an error of less than 0.0001.

49. $\sum_{n=1}^{\infty} (-1)^{n+1} \dfrac{1}{n} = \ln 2$

50. $\sum_{n=0}^{\infty} (-1)^n \left[\dfrac{1}{(2n+1)!} \right] = \sin 1$

51. $\sum_{n=0}^{\infty} \dfrac{2^n}{n!} = e^2$

52. $\sum_{n=1}^{\infty} (-1)^{n-1} \left(\dfrac{1}{n!} \right) = \dfrac{e-1}{e}$

In Exercises 53 and 54, use the series representation of the function f to find $\lim_{x \to 0} f(x)$ (if it exists).

53. $f(x) = \dfrac{1 - \cos x}{x}$

54. $f(x) = \dfrac{\sin x}{x}$

In Exercises 55–58, use a power series to approximate the value of the integral with an error of less than 0.0001. (In Exercises 55 and 56, assume that the integrand is defined as 1 when $x = 0$.)

55. $\int_0^1 \dfrac{\sin x}{x}\, dx$

56. $\int_0^{1/2} \dfrac{\arctan x}{x}\, dx$

57. $\int_{0.1}^{0.3} \sqrt{1 + x^3}\, dx$

58. $\int_0^{1/4} x \ln(x+1)\, dx$

Area In Exercises 59 and 60, use a power series to approximate the area of the region. Use a graphing utility to verify the result.

59. $\int_0^{\pi/2} \sqrt{x} \cos x\, dx$

60. $\int_{0.5}^{1} \cos \sqrt{x}\, dx$

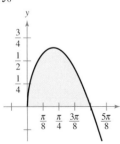
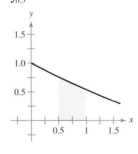

Probability In Exercises 61 and 62, approximate the normal probability with an error of less than 0.0001, where the probability is given by

$$P(a < x < b) = \dfrac{1}{\sqrt{2\pi}} \int_a^b e^{-x^2/2}\, dx.$$

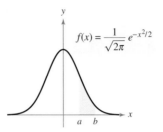

61. $P(0 < x < 1)$

62. $P(1 < x < 2)$

In Exercises 63–66, use a computer algebra system to find the fifth-degree Taylor polynomial (centered at c) for the function. Graph the function and the polynomial. Use the graph to determine the largest interval on which the polynomial is a reasonable approximation of the function.

63. $f(x) = x \cos 2x, \quad c = 0$

64. $f(x) = \sin \dfrac{x}{2} \ln(1+x), \quad c = 0$

65. $g(x) = \sqrt{x} \ln x, \quad c = 1$

66. $h(x) = \sqrt[3]{x} \arctan x, \quad c = 1$

Writing About Concepts

67. State the guidelines for finding a Taylor series.

68. If f is an even function, what must be true about the coefficients a_n in the Maclaurin series

$$f(x) = \sum_{n=0}^{\infty} a_n x^n?$$

Explain your reasoning.

69. Explain how to use the series

$$g(x) = e^x = \sum_{n=0}^{\infty} \dfrac{x^n}{n!}$$

to find the series for each function. Do not find the series.

(a) $f(x) = e^{-x}$

(b) $f(x) = e^{3x}$

(c) $f(x) = xe^x$

(d) $f(x) = e^{2x} + e^{-2x}$

70. Define the binomial series. What is its radius of convergence?

71. Projectile Motion A projectile fired from the ground follows the trajectory given by

$$y = \left(\tan\theta - \frac{g}{kv_0\cos\theta}\right)x - \frac{g}{k^2}\ln\left(1 - \frac{kx}{v_0\cos\theta}\right)$$

where v_0 is the initial speed, θ is the angle of projection, g is the acceleration due to gravity, and k is the drag factor caused by air resistance. Using the power series representation

$$\ln(1+x) = x - \frac{x^2}{2} + \frac{x^3}{3} - \frac{x^4}{4} + \cdots, \quad -1 < x < 1$$

verify that the trajectory can be rewritten as

$$y = (\tan\theta)x + \frac{gx^2}{2v_0^2\cos^2\theta} + \frac{kgx^3}{3v_0^3\cos^3\theta} + \frac{k^2 gx^4}{4v_0^4\cos^4\theta} + \cdots.$$

72. Projectile Motion Use the result of Exercise 71 to determine the series for the path of a projectile launched from ground level at an angle of $\theta = 60°$, with an initial speed of $v_0 = 64$ feet per second and a drag factor of $k = \frac{1}{16}$.

73. Investigation Consider the function f defined by

$$f(x) = \begin{cases} e^{-1/x^2}, & x \neq 0 \\ 0, & x = 0. \end{cases}$$

(a) Sketch a graph of the function.

(b) Use the alternative form of the definition of the derivative (Section 2.1) and L'Hôpital's Rule to show that $f'(0) = 0$. [By continuing this process, it can be shown that $f^{(n)}(0) = 0$ for $n \geq 1$.]

(c) Using the result in part (b), find the Maclaurin series for f. Does the series converge to f?

74. Investigation

(a) Find the power series centered at 0 for the function

$$f(x) = \frac{\ln(x^2 + 1)}{x^2}.$$

(b) Use a graphing utility to graph f and the eighth-degree Taylor polynomial $P_8(x)$ for f.

(c) Complete the table, where

$$F(x) = \int_0^x \frac{\ln(t^2 + 1)}{t^2}\,dt \quad \text{and} \quad G(x) = \int_0^x P_8(t)\,dt.$$

x	0.25	0.50	0.75	1.00	1.50	2.00
$F(x)$						
$G(x)$						

(d) Describe the relationship between the graphs of f and P_8 and the results given in the table in part (c).

75. Prove that $\lim\limits_{n\to\infty} \dfrac{x^n}{n!} = 0$ for any real x.

76. Find the Maclaurin series for

$$f(x) = \ln\frac{1+x}{1-x}$$

and determine its radius of convergence. Use the first four terms of the series to approximate $\ln 3$.

In Exercises 77–80, evaluate the binomial coefficient using the formula

$$\binom{k}{n} = \frac{k(k-1)(k-2)(k-3)\cdots(k-n+1)}{n!}$$

where k is a real number, n is a positive integer, and

$$\binom{k}{0} = 1.$$

77. $\dbinom{5}{3}$

78. $\dbinom{-2}{2}$

79. $\dbinom{0.5}{4}$

80. $\dbinom{-1/3}{5}$

81. Write the power series for $(1 + x)^k$ in terms of binomial coefficients.

82. Prove that e is irrational. $\Big[$Hint: Assume that $e = p/q$ is rational (p and q are integers) and consider

$$e = 1 + 1 + \frac{1}{2!} + \cdots + \frac{1}{n!} + \cdots.\Big]$$

83. Show that the Maclaurin series of the function

$$g(x) = \frac{x}{1-x-x^2}$$

is

$$\sum_{n=1}^{\infty} F_n x^n$$

where F_n is the nth Fibonacci number with $F_1 = F_2 = 1$ and $F_n = F_{n-2} + F_{n-1}$, for $n \geq 3$. (*Hint:* Write

$$\frac{x}{1-x-x^2} = a_0 + a_1 x + a_2 x^2 + \cdots$$

and multiply each side of this equation by $1 - x - x^2$.)

Putnam Exam Challenge

84. Assume that $|f(x)| \leq 1$ and $|f''(x)| \leq 1$ for all x on an interval of length at least 2. Show that $|f'(x)| \leq 2$ on the interval.

This problem was composed by the Committee on the Putnam Prize Competition. © The Mathematical Association of America. All rights reserved.

Review Exercises for Chapter 9

See www.CalcChat.com for worked-out solutions to odd-numbered exercises.

In Exercises 1 and 2, write an expression for the nth term of the sequence.

1. $1, \dfrac{1}{2}, \dfrac{1}{6}, \dfrac{1}{24}, \dfrac{1}{120}, \ldots$

2. $\dfrac{1}{2}, \dfrac{2}{5}, \dfrac{3}{10}, \dfrac{4}{17}, \ldots$

In Exercises 3–6, match the sequence with its graph. [The graphs are labeled (a), (b), (c), and (d).]

(a)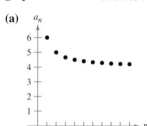

(b)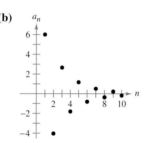

(c)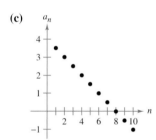

(d)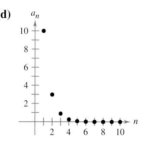

3. $a_n = 4 + \dfrac{2}{n}$

4. $a_n = 4 - \dfrac{1}{2}n$

5. $a_n = 10(0.3)^{n-1}$

6. $a_n = 6\left(-\dfrac{2}{3}\right)^{n-1}$

In Exercises 7 and 8, use a graphing utility to graph the first 10 terms of the sequence. Use the graph to make an inference about the convergence or divergence of the sequence. Verify your inference analytically and, if the sequence converges, find its limit.

7. $a_n = \dfrac{5n+2}{n}$

8. $a_n = \sin \dfrac{n\pi}{2}$

In Exercises 9–16, determine the convergence or divergence of the sequence with the given nth term. If the sequence converges, find its limit. (b and c are positive real numbers.)

9. $a_n = \dfrac{n+1}{n^2}$

10. $a_n = \dfrac{1}{\sqrt{n}}$

11. $a_n = \dfrac{n^3}{n^2+1}$

12. $a_n = \dfrac{n}{\ln n}$

13. $a_n = \sqrt{n+1} - \sqrt{n}$

14. $a_n = \left(1 + \dfrac{1}{2n}\right)^n$

15. $a_n = \dfrac{\sin\sqrt{n}}{\sqrt{n}}$

16. $a_n = (b^n + c^n)^{1/n}$

17. **Compound Interest** A deposit of $5000 is made in an account that earns 5% interest compounded quarterly. The balance in the account after n quarters is
$$A_n = 5000\left(1 + \dfrac{0.05}{4}\right)^n, \quad n = 1, 2, 3, \ldots.$$

   (a) Compute the first eight terms of the sequence $\{A_n\}$.

   (b) Find the balance in the account after 10 years by computing the 40th term of the sequence.

18. **Depreciation** A company buys a machine for $120,000. During the next 5 years the machine will depreciate at a rate of 30% per year. (That is, at the end of each year, the depreciated value will be 70% of what it was at the beginning of the year.)

   (a) Find a formula for the nth term of the sequence that gives the value V of the machine t full years after it was purchased.

   (b) Find the depreciated value of the machine at the end of 5 full years.

Numerical, Graphical, and Analytic Analysis In Exercises 19–22, (a) use a graphing utility to find the indicated partial sum S_k and complete the table, and (b) use a graphing utility to graph the first 10 terms of the sequence of partial sums.

k	5	10	15	20	25
S_k					

19. $\displaystyle\sum_{n=1}^{\infty} \left(\dfrac{3}{2}\right)^{n-1}$

20. $\displaystyle\sum_{n=1}^{\infty} \dfrac{(-1)^{n+1}}{2n}$

21. $\displaystyle\sum_{n=1}^{\infty} \dfrac{(-1)^{n+1}}{(2n)!}$

22. $\displaystyle\sum_{n=1}^{\infty} \dfrac{1}{n(n+1)}$

In Exercises 23–26, determine the convergence or divergence of the series.

23. $\displaystyle\sum_{n=0}^{\infty} (0.82)^n$

24. $\displaystyle\sum_{n=0}^{\infty} (1.82)^n$

25. $\displaystyle\sum_{n=1}^{\infty} \dfrac{(-1)^n n}{\ln n}$

26. $\displaystyle\sum_{n=0}^{\infty} \dfrac{2n+1}{3n+2}$

In Exercises 27–30, find the sum of the convergent series.

27. $\displaystyle\sum_{n=0}^{\infty} \left(\dfrac{2}{3}\right)^n$

28. $\displaystyle\sum_{n=0}^{\infty} \dfrac{2^{n+2}}{3^n}$

29. $\displaystyle\sum_{n=0}^{\infty} \left(\dfrac{1}{2^n} - \dfrac{1}{3^n}\right)$

30. $\displaystyle\sum_{n=0}^{\infty} \left[\left(\dfrac{2}{3}\right)^n - \dfrac{1}{(n+1)(n+2)}\right]$

In Exercises 31 and 32, (a) write the repeating decimal as a geometric series and (b) write its sum as the ratio of two integers.

31. $0.\overline{09}$ **32.** $0.\overline{923076}$

33. *Distance* A ball is dropped from a height of 8 meters. Each time it drops h meters, it rebounds $0.7h$ meters. Find the total distance traveled by the ball.

34. *Salary* You accept a job that pays a salary of $32,000 the first year. During the next 39 years, you will receive a 5.5% raise each year. What would be your total compensation over the 40-year period?

35. *Compound Interest* A deposit of $200 is made at the end of each month for 2 years in an account that pays 6% interest, compounded continuously. Determine the balance in the account at the end of 2 years.

36. *Compound Interest* A deposit of $100 is made at the end of each month for 10 years in an account that pays 3.5%, compounded monthly. Determine the balance in the account at the end of 10 years.

In Exercises 37–40, determine the convergence or divergence of the series.

37. $\sum_{n=1}^{\infty} \dfrac{\ln n}{n^4}$ **38.** $\sum_{n=1}^{\infty} \dfrac{1}{\sqrt[4]{n^3}}$

39. $\sum_{n=1}^{\infty} \left(\dfrac{1}{n^2} - \dfrac{1}{n}\right)$ **40.** $\sum_{n=1}^{\infty} \left(\dfrac{1}{n^2} - \dfrac{1}{2^n}\right)$

In Exercises 41–44, determine the convergence or divergence of the series.

41. $\sum_{n=1}^{\infty} \dfrac{1}{\sqrt{n^3 + 2n}}$ **42.** $\sum_{n=1}^{\infty} \dfrac{n+1}{n(n+2)}$

43. $\sum_{n=1}^{\infty} \dfrac{1 \cdot 3 \cdot 5 \cdots (2n-1)}{2 \cdot 4 \cdot 6 \cdots (2n)}$ **44.** $\sum_{n=1}^{\infty} \dfrac{1}{3^n - 5}$

In Exercises 45–48, determine the convergence or divergence of the series.

45. $\sum_{n=2}^{\infty} \dfrac{(-1)^n n}{n^2 - 3}$ **46.** $\sum_{n=1}^{\infty} \dfrac{(-1)^n \sqrt{n}}{n+1}$

47. $\sum_{n=4}^{\infty} \dfrac{(-1)^n n}{n - 3}$ **48.** $\sum_{n=2}^{\infty} \dfrac{(-1)^n \ln n^3}{n}$

In Exercises 49–52, determine the convergence or divergence of the series.

49. $\sum_{n=1}^{\infty} \dfrac{n}{e^{n^2}}$

50. $\sum_{n=1}^{\infty} \dfrac{n!}{e^n}$

51. $\sum_{n=1}^{\infty} \dfrac{2^n}{n^3}$

52. $\sum_{n=1}^{\infty} \dfrac{1 \cdot 3 \cdot 5 \cdots (2n-1)}{2 \cdot 5 \cdot 8 \cdots (3n-1)}$

 Numerical, Graphical, and Analytic Analysis In Exercises 53 and 54, (a) verify that the series converges, (b) use a graphing utility to find the indicated partial sum S_n and complete the table, (c) use a graphing utility to graph the first 10 terms of the sequence of partial sums, and (d) use the table to estimate the sum of the series.

n	5	10	15	20	25
S_n					

53. $\sum_{n=1}^{\infty} n\left(\dfrac{3}{5}\right)^n$ **54.** $\sum_{n=1}^{\infty} \dfrac{(-1)^{n-1} n}{n^3 + 5}$

55. *Writing* Use a graphing utility to complete the table for (a) $p = 2$ and (b) $p = 5$. Write a short paragraph describing and comparing the entries in the table.

N	5	10	20	30	40
$\sum_{n=1}^{N} \dfrac{1}{n^p}$					
$\int_{N}^{\infty} \dfrac{1}{x^p} dx$					

56. *Writing* You are told that the terms of a positive series appear to approach zero very slowly as n approaches infinity. (In fact, $a_{75} = 0.7$.) If you are given no other information, can you conclude that the series diverges? Support your answer with an example.

In Exercises 57 and 58, find the third-degree Taylor polynomial centered at c.

57. $f(x) = e^{-x/2}$, $c = 0$

58. $f(x) = \tan x$, $c = -\dfrac{\pi}{4}$

In Exercises 59–62, use a Taylor polynomial to approximate the function with an error of less than 0.001.

59. $\sin 95°$ **60.** $\cos(0.75)$

61. $\ln(1.75)$ **62.** $e^{-0.25}$

63. A Taylor polynomial centered at 0 will be used to approximate the cosine function. Find the degree of the polynomial required to obtain the desired accuracy over each interval.

Maximum Error	Interval
(a) 0.001	$[-0.5, 0.5]$
(b) 0.001	$[-1, 1]$
(c) 0.0001	$[-0.5, 0.5]$
(d) 0.0001	$[-2, 2]$

 64. Use a graphing utility to graph the cosine function and the Taylor polynomials in Exercise 63.

In Exercises 65–70, find the interval of convergence of the power series. (Be sure to include a check for convergence at the endpoints of the interval.)

65. $\sum_{n=0}^{\infty} \left(\frac{x}{10}\right)^n$

66. $\sum_{n=0}^{\infty} (2x)^n$

67. $\sum_{n=0}^{\infty} \frac{(-1)^n (x-2)^n}{(n+1)^2}$

68. $\sum_{n=1}^{\infty} \frac{3^n (x-2)^n}{n}$

69. $\sum_{n=0}^{\infty} n!(x-2)^n$

70. $\sum_{n=0}^{\infty} \frac{(x-2)^n}{2^n}$

In Exercises 71 and 72, show that the function represented by the power series is a solution of the differential equation.

71. $y = \sum_{n=0}^{\infty} (-1)^n \frac{x^{2n}}{4^n (n!)^2}$

$x^2 y'' + xy' + x^2 y = 0$

72. $y = \sum_{n=0}^{\infty} \frac{(-3)^n x^{2n}}{2^n n!}$

$y'' + 3xy' + 3y = 0$

In Exercises 73 and 74, find a geometric power series centered at 0 for the function.

73. $g(x) = \dfrac{2}{3-x}$

74. $h(x) = \dfrac{3}{2+x}$

75. Find a power series for the derivative of the function in Exercise 73.

76. Find a power series for the integral of the function in Exercise 74.

In Exercises 77 and 78, find a function represented by the series and give the domain of the function.

77. $1 + \dfrac{2}{3}x + \dfrac{4}{9}x^2 + \dfrac{8}{27}x^3 + \cdots$

78. $8 - 2(x-3) + \dfrac{1}{2}(x-3)^2 - \dfrac{1}{8}(x-3)^3 + \cdots$

In Exercises 79–86, find a power series for the function centered at c.

79. $f(x) = \sin x$, $c = \dfrac{3\pi}{4}$

80. $f(x) = \cos x$, $c = -\dfrac{\pi}{4}$

81. $f(x) = 3^x$, $c = 0$

82. $f(x) = \csc x$, $c = \dfrac{\pi}{2}$

(first three terms)

83. $f(x) = \dfrac{1}{x}$, $c = -1$

84. $f(x) = \sqrt{x}$, $c = 4$

85. $g(x) = \sqrt[5]{1+x}$, $c = 0$

86. $h(x) = \dfrac{1}{(1+x)^3}$, $c = 0$

In Exercises 87–92, find the sum of the convergent series by using a well-known function. Identify the function and explain how you obtained the sum.

87. $\sum_{n=1}^{\infty} (-1)^{n+1} \dfrac{1}{4^n n}$

88. $\sum_{n=1}^{\infty} (-1)^{n+1} \dfrac{1}{5^n n}$

89. $\sum_{n=0}^{\infty} \dfrac{1}{2^n n!}$

90. $\sum_{n=0}^{\infty} \dfrac{2^n}{3^n n!}$

91. $\sum_{n=0}^{\infty} (-1)^n \dfrac{2^{2n}}{3^{2n}(2n)!}$

92. $\sum_{n=0}^{\infty} (-1)^n \dfrac{1}{3^{2n+1}(2n+1)!}$

93. *Writing* One of the series in Exercises 41 and 49 converges to its sum at a much lower rate than the other series. Which is it? Explain why this series converges so slowly. Use a graphing utility to illustrate the rate of convergence.

94. Use the binomial series to find the Maclaurin series for

$$f(x) = \dfrac{1}{\sqrt{1+x^3}}.$$

95. *Forming Maclaurin Series* Determine the first four terms of the Maclaurin series for e^{2x}

(a) by using the definition of the Maclaurin series and the formula for the coefficient of the nth term, $a_n = f^{(n)}(0)/n!$.

(b) by replacing x by $2x$ in the series for e^x.

(c) by multiplying the series for e^x by itself, because $e^{2x} = e^x \cdot e^x$.

96. *Forming Maclaurin Series* Follow the pattern of Exercise 95 to find the first four terms of the series for $\sin 2x$. (*Hint:* $\sin 2x = 2 \sin x \cos x$.)

In Exercises 97–100, find the series representation of the function defined by the integral.

97. $\displaystyle\int_0^x \dfrac{\sin t}{t}\, dt$

98. $\displaystyle\int_0^x \cos \dfrac{\sqrt{t}}{2}\, dt$

99. $\displaystyle\int_0^x \dfrac{\ln(t+1)}{t}\, dt$

100. $\displaystyle\int_0^x \dfrac{e^t - 1}{t}\, dt$

In Exercises 101 and 102, use a power series to find the limit (if it exists). Verify the result by using L'Hôpital's Rule.

101. $\displaystyle\lim_{x \to 0^+} \dfrac{\arctan x}{\sqrt{x}}$

102. $\displaystyle\lim_{x \to 0} \dfrac{\arcsin x}{x}$

P.S. Problem Solving

1. The Cantor set (Georg Cantor, 1845–1918) is a subset of the unit interval $[0, 1]$. To construct the Cantor set, first remove the middle third $\left(\frac{1}{3}, \frac{2}{3}\right)$ of the interval, leaving two line segments. For the second step, remove the middle third of each of the two remaining segments, leaving four line segments. Continue this procedure indefinitely, as shown in the figure. The Cantor set consists of all numbers in the unit interval $[0, 1]$ that still remain.

   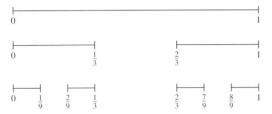

   (a) Find the total length of all the line segments that are removed.
   (b) Write down three numbers that are in the Cantor set.
   (c) Let C_n denote the total length of the remaining line segments after n steps. Find $\lim\limits_{n \to \infty} C_n$.

   GEORG CANTOR (1845–1918)
   Cantor was a German mathematician known for his work on the development of set theory, which is the basis of modern mathematical analysis. This theory extends to the concept of infinite (or transfinite) numbers.

2. It can be shown that
   $$\sum_{n=1}^{\infty} \frac{1}{n^2} = \frac{\pi^2}{6}$$ [see Example 3(b), Section 9.3].
   Use this fact to show that $\sum_{n=1}^{\infty} \frac{1}{(2n - 1)^2} = \frac{\pi^2}{8}$.

3. Let T be an equilateral triangle with sides of length 1. Let a_n be the number of circles that can be packed tightly in n rows inside the triangle. For example, $a_1 = 1$, $a_2 = 3$, and $a_3 = 6$, as shown in the figure. Let A_n be the combined area of the a_n circles. Find $\lim\limits_{n \to \infty} A_n$.

4. Identical blocks of unit length are stacked on top of each other at the edge of a table. The center of gravity of the top block must lie over the block below it, the center of gravity of the top two blocks must lie over the block below them, and so on (see figure).

   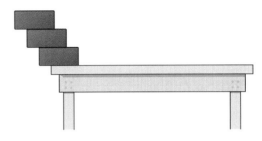

   (a) If there are three blocks, show that it is possible to stack them so that the left edge of the top block extends $\frac{11}{12}$ unit beyond the edge of the table.
   (b) Is it possible to stack the blocks so that the right edge of the top block extends beyond the edge of the table?
   (c) How far beyond the table can the blocks be stacked?

5. (a) Consider the power series
   $$\sum_{n=0}^{\infty} a_n x^n = 1 + 2x + 3x^2 + x^3 + 2x^4 + 3x^5 + x^6 + \cdots$$
   in which the coefficients $a_n = 1, 2, 3, 1, 2, 3, 1, \ldots$ are periodic of period $p = 3$. Find the radius of convergence and the sum of this power series.

   (b) Consider a power series
   $$\sum_{n=0}^{\infty} a_n x^n$$
   in which the coefficients are periodic, $(a_{n+p} = a_p)$ and $a_n > 0$. Find the radius of convergence and the sum of this power series.

6. For what values of the positive constants a and b does the following series converge absolutely? For what values does it converge conditionally?
   $$a - \frac{b}{2} + \frac{a}{3} - \frac{b}{4} + \frac{a}{5} - \frac{b}{6} + \frac{a}{7} - \frac{b}{8} + \cdots$$

7. (a) Find a power series for the function
   $$f(x) = xe^x$$
   centered at 0. Use this representation to find the sum of the infinite series
   $$\sum_{n=1}^{\infty} \frac{1}{n!(n + 2)}.$$

   (b) Differentiate the power series for $f(x) = xe^x$. Use the result to find the sum of the infinite series
   $$\sum_{n=0}^{\infty} \frac{n + 1}{n!}.$$

8. Find $f^{(12)}(0)$ if $f(x) = e^{x^2}$. (*Hint:* Do not calculate 12 derivatives.)

9. The graph of the function
$$f(x) = \begin{cases} 1, & x = 0 \\ \dfrac{\sin x}{x}, & x > 0 \end{cases}$$
is shown below. Use the Alternating Series Test to show that the improper integral $\int_1^\infty f(x)\,dx$ converges.

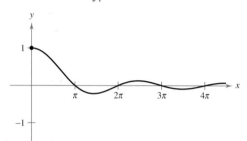

10. (a) Prove that $\displaystyle\int_2^\infty \frac{1}{x(\ln x)^p}\,dx$ converges if and only if $p > 1$.

    (b) Determine the convergence or divergence of the series
    $$\sum_{n=4}^\infty \frac{1}{n\ln(n^2)}.$$

11. (a) Consider the following sequence of numbers defined recursively.
    $$a_1 = 3$$
    $$a_2 = \sqrt{3}$$
    $$a_3 = \sqrt{3 + \sqrt{3}}$$
    $$\vdots$$
    $$a_{n+1} = \sqrt{3 + a_n}$$
    Write the decimal approximations for the first six terms of this sequence. Prove that the sequence converges and find its limit.

    (b) Consider the following sequence defined recursively by $a_1 = \sqrt{a}$ and $a_{n+1} = \sqrt{a + a_n}$, where $a > 2$.
    $$\sqrt{a},\ \sqrt{a + \sqrt{a}},\ \sqrt{a + \sqrt{a + \sqrt{a}}},\ \ldots$$
    Prove that this sequence converges and find its limit.

12. Let $\{a_n\}$ be a sequence of positive numbers satisfying $\lim_{n\to\infty}(a_n)^{1/n} = L < \dfrac{1}{r}$, $r > 0$. Prove that the series $\sum_{n=1}^\infty a_n r^n$ converges.

13. Consider the infinite series $\displaystyle\sum_{n=1}^\infty \frac{1}{2^{n+(-1)^n}}$.

    (a) Find the first five terms of the sequence of partial sums.

    (b) Show that the Ratio Test is inconclusive for this series.

    (c) Use the Root Test to test for the convergence or divergence of this series.

14. Derive each identity using the appropriate geometric series.

    (a) $\dfrac{1}{0.99} = 1.01010101\ldots$ (b) $\dfrac{1}{0.98} = 1.0204081632\ldots$

15. Consider an idealized population with the characteristic that each member of the population produces one offspring at the end of every time period. Each member has a life span of three time periods and the population begins with 10 newborn members. The following table shows the population during the first five time periods.

	Time Period				
Age Bracket	1	2	3	4	5
0–1	10	10	20	40	70
1–2		10	10	20	40
2–3			10	10	20
Total	10	20	40	70	130

The sequence for the total population has the property that
$$S_n = S_{n-1} + S_{n-2} + S_{n-3},\qquad n > 3.$$
Find the total population during each of the next five time periods.

16. Imagine you are stacking an infinite number of spheres of decreasing radii on top of each other, as shown in the figure. The radii of the spheres are 1 meter, $1/\sqrt{2}$ meter, $1/\sqrt{3}$ meter, etc. The spheres are made of a material that weighs 1 newton per cubic meter.

    (a) How high is this infinite stack of spheres?

    (b) What is the total surface area of all the spheres in the stack?

    (c) Show that the weight of the stack is finite.

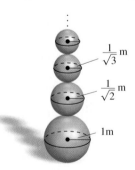

17. (a) Determine the convergence or divergence of the series
    $$\sum_{n=1}^\infty \frac{1}{2n}.$$

    (b) Determine the convergence or divergence of the series
    $$\sum_{n=1}^\infty \left(\sin\frac{1}{2n} - \sin\frac{1}{2n+1}\right).$$

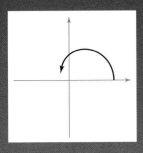

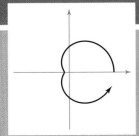

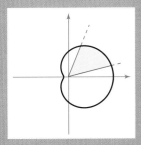

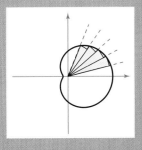

Conics, Parametric Equations, and Polar Coordinates

During the 2002 Winter Olympic Games, the Olympic rings were lighted high on a mountainside in Salt Lake City. The volunteers who installed the lights for the display took care to minimize the environmental impact of the project. How can you calculate the area enclosed by the display? Explain.

Harry Howe/Getty Images

In the polar coordinate system, graphing an equation involves tracing a curve about a fixed point called the pole. Consider a region bounded by a curve and by the rays that contain the endpoints of an interval on the curve. You can use sectors of circles to approximate the area of such a region. In Chapter 10, you will see how the limit process can be used to find this area.

693

Section 10.1

Conics and Calculus

- Understand the definition of a conic section.
- Analyze and write equations of parabolas using properties of parabolas.
- Analyze and write equations of ellipses using properties of ellipses.
- Analyze and write equations of hyperbolas using properties of hyperbolas.

Conic Sections

Each **conic section** (or simply **conic**) can be described as the intersection of a plane and a double-napped cone. Notice in Figure 10.1 that for the four basic conics, the intersecting plane does not pass through the vertex of the cone. When the plane passes through the vertex, the resulting figure is a **degenerate conic,** as shown in Figure 10.2.

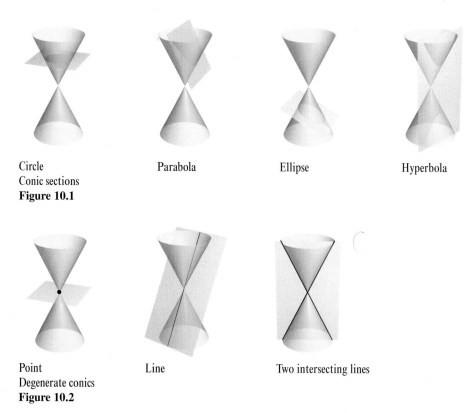

Circle   Parabola   Ellipse   Hyperbola
Conic sections
Figure 10.1

Point   Line   Two intersecting lines
Degenerate conics
Figure 10.2

HYPATIA (370–415 A.D.)

The Greeks discovered conic sections sometime between 600 and 300 B.C. By the beginning of the Alexandrian period, enough was known about conics for Apollonius (262–190 B.C.) to produce an eight-volume work on the subject. Later, toward the end of the Alexandrian period, Hypatia wrote a textbook entitled *On the Conics of Apollonius*. Her death marked the end of major mathematical discoveries in Europe for several hundred years.

The early Greeks were largely concerned with the geometric properties of conics. It was not until 1900 years later, in the early seventeenth century, that the broader applicability of conics became apparent. Conics then played a prominent role in the development of calculus.

There are several ways to study conics. You could begin as the Greeks did by defining the conics in terms of the intersections of planes and cones, or you could define them algebraically in terms of the general second-degree equation

$$Ax^2 + Bxy + Cy^2 + Dx + Ey + F = 0.$$ General second-degree equation

However, a third approach, in which each of the conics is defined as a **locus** (collection) of points satisfying a certain geometric property, works best. For example, a circle can be defined as the collection of all points (x, y) that are equidistant from a fixed point (h, k). This locus definition easily produces the standard equation of a circle

$$(x - h)^2 + (y - k)^2 = r^2.$$ Standard equation of a circle

FOR FURTHER INFORMATION To learn more about the mathematical activities of Hypatia, see the article "Hypatia and Her Mathematics" by Michael A. B. Deakin in *The American Mathematical Monthly*. To view this article, go to the website *www.matharticles.com*.

Parabolas

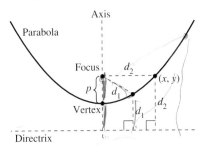

Figure 10.3

A **parabola** is the set of all points (x, y) that are equidistant from a fixed line called the **directrix** and a fixed point called the **focus** not on the line. The midpoint between the focus and the directrix is the **vertex,** and the line passing through the focus and the vertex is the **axis** of the parabola. Note in Figure 10.3 that a parabola is symmetric with respect to its axis.

THEOREM 10.1 Standard Equation of a Parabola

The **standard form** of the equation of a parabola with vertex (h, k) and directrix $y = k - p$ is

$$(x - h)^2 = 4p(y - k). \qquad \text{Vertical axis}$$

For directrix $x = h - p$, the equation is

$$(y - k)^2 = 4p(x - h). \qquad \text{Horizontal axis}$$

The focus lies on the axis p units (*directed distance*) from the vertex. The coordinates of the focus are as follows.

$(h, k + p)$ Vertical axis

$(h + p, k)$ Horizontal axis

EXAMPLE 1 Finding the Focus of a Parabola

Find the focus of the parabola given by $y = -\frac{1}{2}x^2 - x + \frac{1}{2}$.

Solution To find the focus, convert to standard form by completing the square.

$$y = \tfrac{1}{2} - x - \tfrac{1}{2}x^2 \qquad \text{Rewrite original equation.}$$
$$y = \tfrac{1}{2}(1 - 2x - x^2) \qquad \text{Factor out } \tfrac{1}{2}.$$
$$2y = 1 - 2x - x^2 \qquad \text{Multiply each side by 2.}$$
$$2y = 1 - (x^2 + 2x) \qquad \text{Group terms.}$$
$$2y = 2 - (x^2 + 2x + 1) \qquad \text{Add and subtract 1 on right side.}$$
$$x^2 + 2x + 1 = -2y + 2$$
$$(x + 1)^2 = -2(y - 1) \qquad \text{Write in standard form.}$$

Comparing this equation with $(x - h)^2 = 4p(y - k)$, you can conclude that

$$h = -1, \quad k = 1, \quad \text{and} \quad p = -\tfrac{1}{2}.$$

Because p is negative, the parabola opens downward, as shown in Figure 10.4. So, the focus of the parabola is p units from the vertex, or

$$(h, k + p) = \left(-1, \tfrac{1}{2}\right). \qquad \text{Focus}$$

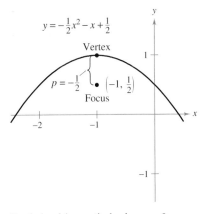

Parabola with a vertical axis, $p < 0$
Figure 10.4

A line segment that passes through the focus of a parabola and has endpoints on the parabola is called a **focal chord.** The specific focal chord perpendicular to the axis of the parabola is the **latus rectum.** The next example shows how to determine the length of the latus rectum and the length of the corresponding intercepted arc.

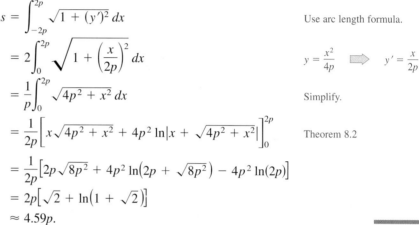

EXAMPLE 2 Focal Chord Length and Arc Length

Find the length of the latus rectum of the parabola given by $x^2 = 4py$. Then find the length of the parabolic arc intercepted by the latus rectum.

Solution Because the latus rectum passes through the focus $(0, p)$ and is perpendicular to the y-axis, the coordinates of its endpoints are $(-x, p)$ and (x, p). Substituting p for y in the equation of the parabola produces

$$x^2 = 4p(p) \quad \Longrightarrow \quad x = \pm 2p.$$

So, the endpoints of the latus rectum are $(-2p, p)$ and $(2p, p)$, and you can conclude that its length is $4p$, as shown in Figure 10.5. In contrast, the length of the intercepted arc is

$$\begin{aligned}
s &= \int_{-2p}^{2p} \sqrt{1 + (y')^2}\, dx && \text{Use arc length formula.}\\
&= 2\int_{0}^{2p} \sqrt{1 + \left(\frac{x}{2p}\right)^2}\, dx && y = \frac{x^2}{4p} \;\Longrightarrow\; y' = \frac{x}{2p}\\
&= \frac{1}{p}\int_{0}^{2p} \sqrt{4p^2 + x^2}\, dx && \text{Simplify.}\\
&= \frac{1}{2p}\left[x\sqrt{4p^2 + x^2} + 4p^2 \ln\left|x + \sqrt{4p^2 + x^2}\right|\right]_{0}^{2p} && \text{Theorem 8.2}\\
&= \frac{1}{2p}\left[2p\sqrt{8p^2} + 4p^2 \ln\left(2p + \sqrt{8p^2}\right) - 4p^2 \ln(2p)\right]\\
&= 2p\left[\sqrt{2} + \ln\left(1 + \sqrt{2}\right)\right]\\
&\approx 4.59p.
\end{aligned}$$

Length of latus rectum: $4p$
Figure 10.5

One widely used property of a parabola is its reflective property. In physics, a surface is called **reflective** if the tangent line at any point on the surface makes equal angles with an incoming ray and the resulting outgoing ray. The angle corresponding to the incoming ray is the **angle of incidence**, and the angle corresponding to the outgoing ray is the **angle of reflection**. One example of a reflective surface is a flat mirror.

Another type of reflective surface is that formed by revolving a parabola about its axis. A special property of parabolic reflectors is that they allow us to direct all incoming rays parallel to the axis through the focus of the parabola—this is the principle behind the design of the parabolic mirrors used in reflecting telescopes. Conversely, all light rays emanating from the focus of a parabolic reflector used in a flashlight are parallel, as shown in Figure 10.6.

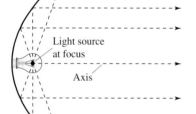

Parabolic reflector: light is reflected in parallel rays.
Figure 10.6

THEOREM 10.2 Reflective Property of a Parabola

Let P be a point on a parabola. The tangent line to the parabola at the point P makes equal angles with the following two lines.

1. The line passing through P and the focus
2. The line passing through P parallel to the axis of the parabola

indicates that in the HM mathSpace® *CD-ROM and the online* Eduspace® *system for this text, you will find an Open Exploration, which further explores this example using the computer algebra systems* Maple, Mathcad, Mathematica, *and* Derive.

Ellipses

More than a thousand years after the close of the Alexandrian period of Greek mathematics, Western civilization finally began a Renaissance of mathematical and scientific discovery. One of the principal figures in this rebirth was the Polish astronomer Nicolaus Copernicus. In his work *On the Revolutions of the Heavenly Spheres*, Copernicus claimed that all of the planets, including Earth, revolved about the sun in circular orbits. Although some of Copernicus's claims were invalid, the controversy set off by his heliocentric theory motivated astronomers to search for a mathematical model to explain the observed movements of the sun and planets. The first to find an accurate model was the German astronomer Johannes Kepler (1571–1630). Kepler discovered that the planets move about the sun in elliptical orbits, with the sun not as the center but as a focal point of the orbit.

The use of ellipses to explain the movements of the planets is only one of many practical and aesthetic uses. As with parabolas, you will begin your study of this second type of conic by defining it as a locus of points. Now, however, *two* focal points are used rather than one.

An **ellipse** is the set of all points (x, y) the sum of whose distances from two distinct fixed points called **foci** is constant. (See Figure 10.7.) The line through the foci intersects the ellipse at two points, called the **vertices.** The chord joining the vertices is the **major axis,** and its midpoint is the **center** of the ellipse. The chord perpendicular to the major axis at the center is the **minor axis** of the ellipse. (See Figure 10.8.)

NICOLAUS COPERNICUS (1473–1543)

Copernicus began to study planetary motion when asked to revise the calendar. At that time, the exact length of the year could not be accurately predicted using the theory that Earth was the center of the universe.

FOR FURTHER INFORMATION To learn about how an ellipse may be "exploded" into a parabola, see the article "Exploding the Ellipse" by Arnold Good in *Mathematics Teacher*. To view this article, go to the website *www.matharticles.com*.

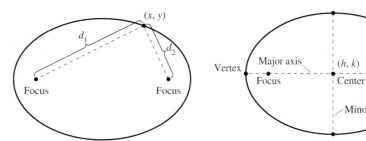

Figure 10.7

Figure 10.8

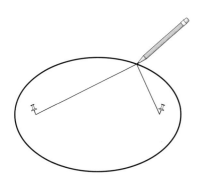

Figure 10.9

THEOREM 10.3 Standard Equation of an Ellipse

The standard form of the equation of an ellipse with center (h, k) and major and minor axes of lengths $2a$ and $2b$, where $a > b$, is

$$\frac{(x-h)^2}{a^2} + \frac{(y-k)^2}{b^2} = 1 \qquad \text{Major axis is horizontal.}$$

or

$$\frac{(x-h)^2}{b^2} + \frac{(y-k)^2}{a^2} = 1. \qquad \text{Major axis is vertical.}$$

The foci lie on the major axis, c units from the center, with $c^2 = a^2 - b^2$.

NOTE You can visualize the definition of an ellipse by imagining two thumbtacks placed at the foci, as shown in Figure 10.9. If the ends of a fixed length of string are fastened to the thumbtacks and the string is drawn taut with a pencil, the path traced by the pencil will be an ellipse.

EXAMPLE 3 Completing the Square

Find the center, vertices, and foci of the ellipse given by

$$4x^2 + y^2 - 8x + 4y - 8 = 0.$$

Solution By completing the square, you can write the original equation in standard form.

$$4x^2 + y^2 - 8x + 4y - 8 = 0 \qquad \text{Write original equation.}$$
$$4x^2 - 8x + y^2 + 4y = 8$$
$$4(x^2 - 2x + 1) + (y^2 + 4y + 4) = 8 + 4 + 4$$
$$4(x - 1)^2 + (y + 2)^2 = 16$$
$$\frac{(x-1)^2}{4} + \frac{(y+2)^2}{16} = 1 \qquad \text{Write in standard form.}$$

So, the major axis is parallel to the y-axis, where $h = 1$, $k = -2$, $a = 4$, $b = 2$, and $c = \sqrt{16 - 4} = 2\sqrt{3}$. So, you obtain the following.

Center:	$(1, -2)$	(h, k)
Vertices:	$(1, -6)$ and $(1, 2)$	$(h, k \pm a)$
Foci:	$(1, -2 - 2\sqrt{3})$ and $(1, -2 + 2\sqrt{3})$	$(h, k \pm c)$

The graph of the ellipse is shown in Figure 10.10.

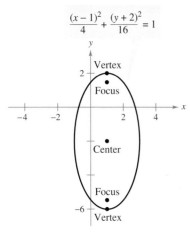

Ellipse with a vertical major axis
Figure 10.10

NOTE If the constant term $F = -8$ in the equation in Example 3 had been greater than or equal to 8, you would have obtained one of the following degenerate cases.

1. $F = 8$, single point, $(1, -2)$: $\dfrac{(x-1)^2}{4} + \dfrac{(y+2)^2}{16} = 0$

2. $F > 8$, no solution points: $\dfrac{(x-1)^2}{4} + \dfrac{(y+2)^2}{16} < 0$

EXAMPLE 4 The Orbit of the Moon

The moon orbits Earth in an elliptical path with the center of Earth at one focus, as shown in Figure 10.11. The major and minor axes of the orbit have lengths of 768,800 kilometers and 767,640 kilometers, respectively. Find the greatest and least distances (the apogee and perigee) from Earth's center to the moon's center.

Solution Begin by solving for a and b.

$2a = 768{,}800$	Length of major axis
$a = 384{,}400$	Solve for a.
$2b = 767{,}640$	Length of minor axis
$b = 383{,}820$	Solve for b.

Now, using these values, you can solve for c as follows.

$$c = \sqrt{a^2 - b^2} \approx 21{,}108$$

The greatest distance between the center of Earth and the center of the moon is $a + c \approx 405{,}508$ kilometers, and the least distance is $a - c \approx 363{,}292$ kilometers.

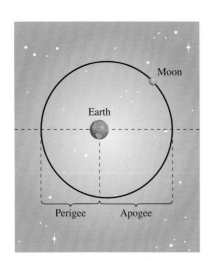

Figure 10.11

FOR FURTHER INFORMATION For more information on some uses of the reflective properties of conics, see the article "Parabolic Mirrors, Elliptic and Hyperbolic Lenses" by Mohsen Maesumi in *The American Mathematical Monthly*. Also see the article "The Geometry of Microwave Antennas" by William R. Parzynski in *Mathematics Teacher*.

Theorem 10.2 presented a reflective property of parabolas. Ellipses have a similar reflective property. You are asked to prove the following theorem in Exercise 110.

> **THEOREM 10.4   Reflective Property of an Ellipse**
>
> Let P be a point on an ellipse. The tangent line to the ellipse at point P makes equal angles with the lines through P and the foci.

One of the reasons that astronomers had difficulty in detecting that the orbits of the planets are ellipses is that the foci of the planetary orbits are relatively close to the center of the sun, making the orbits nearly circular. To measure the ovalness of an ellipse, you can use the concept of **eccentricity.**

> **Definition of Eccentricity of an Ellipse**
>
> The **eccentricity** e of an ellipse is given by the ratio
>
> $$e = \frac{c}{a}.$$

To see how this ratio is used to describe the shape of an ellipse, note that because the foci of an ellipse are located along the major axis between the vertices and the center, it follows that

$$0 < c < a.$$

For an ellipse that is nearly circular, the foci are close to the center and the ratio c/a is small, and for an elongated ellipse, the foci are close to the vertices and the ratio is close to 1, as shown in Figure 10.12. Note that $0 < e < 1$ for every ellipse.

The orbit of the moon has an eccentricity of $e = 0.0549$, and the eccentricities of the nine planetary orbits are as follows.

Mercury:	$e = 0.2056$	Saturn:	$e = 0.0542$
Venus:	$e = 0.0068$	Uranus:	$e = 0.0472$
Earth:	$e = 0.0167$	Neptune:	$e = 0.0086$
Mars:	$e = 0.0934$	Pluto:	$e = 0.2488$
Jupiter:	$e = 0.0484$		

You can use integration to show that the area of an ellipse is $A = \pi ab$. For instance, the area of the ellipse

$$\frac{x^2}{a^2} + \frac{y^2}{b^2} = 1$$

is given by

$$A = 4\int_0^a \frac{b}{a}\sqrt{a^2 - x^2}\,dx$$

$$= \frac{4b}{a}\int_0^{\pi/2} a^2 \cos^2\theta\,d\theta. \qquad \text{Trigonometric substitution } x = a\sin\theta.$$

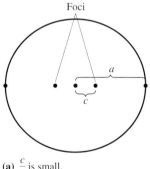

(a) $\frac{c}{a}$ is small.

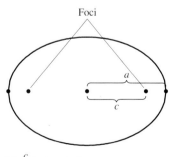

(b) $\frac{c}{a}$ is close to 1.

Eccentricity is the ratio $\frac{c}{a}$.
Figure 10.12

However, it is not so simple to find the *circumference* of an ellipse. The next example shows how to use eccentricity to set up an "elliptic integral" for the circumference of an ellipse.

EXAMPLE 5 Finding the Circumference of an Ellipse

Show that the circumference of the ellipse $(x^2/a^2) + (y^2/b^2) = 1$ is

$$4a \int_0^{\pi/2} \sqrt{1 - e^2 \sin^2 \theta}\, d\theta. \qquad e = \frac{c}{a}$$

Solution Because the given ellipse is symmetric with respect to both the x-axis and the y-axis, you know that its circumference C is four times the arc length of $y = (b/a)\sqrt{a^2 - x^2}$ in the first quadrant. The function y is differentiable for all x in the interval $[0, a]$ except at $x = a$. So, the circumference is given by the improper integral

$$C = \lim_{d \to a} 4 \int_0^d \sqrt{1 + (y')^2}\, dx = 4 \int_0^a \sqrt{1 + (y')^2}\, dx = 4 \int_0^a \sqrt{1 + \frac{b^2 x^2}{a^2(a^2 - x^2)}}\, dx.$$

Using the trigonometric substitution $x = a \sin \theta$, you obtain

$$C = 4 \int_0^{\pi/2} \sqrt{1 + \frac{b^2 \sin^2 \theta}{a^2 \cos^2 \theta}}\, (a \cos \theta)\, d\theta$$

$$= 4 \int_0^{\pi/2} \sqrt{a^2 \cos^2 \theta + b^2 \sin^2 \theta}\, d\theta$$

$$= 4 \int_0^{\pi/2} \sqrt{a^2(1 - \sin^2 \theta) + b^2 \sin^2 \theta}\, d\theta$$

$$= 4 \int_0^{\pi/2} \sqrt{a^2 - (a^2 - b^2)\sin^2 \theta}\, d\theta.$$

Because $e^2 = c^2/a^2 = (a^2 - b^2)/a^2$, you can rewrite this integral as

$$C = 4a \int_0^{\pi/2} \sqrt{1 - e^2 \sin^2 \theta}\, d\theta.$$

A great deal of time has been devoted to the study of elliptic integrals. Such integrals generally do not have elementary antiderivatives. To find the circumference of an ellipse, you must usually resort to an approximation technique.

AREA AND CIRCUMFERENCE OF AN ELLIPSE

In his work with elliptic orbits in the early 1600's, Johannes Kepler successfully developed a formula for the area of an ellipse, $A = \pi a b$. He was less successful in developing a formula for the circumference of an ellipse, however; the best he could do was to give the approximate formula $C = \pi(a + b)$.

EXAMPLE 6 Approximating the Value of an Elliptic Integral

Use the elliptic integral in Example 5 to approximate the circumference of the ellipse

$$\frac{x^2}{25} + \frac{y^2}{16} = 1.$$

Solution Because $e^2 = c^2/a^2 = (a^2 - b^2)/a^2 = 9/25$, you have

$$C = (4)(5) \int_0^{\pi/2} \sqrt{1 - \frac{9 \sin^2 \theta}{25}}\, d\theta.$$

Applying Simpson's Rule with $n = 4$ produces

$$C \approx 20 \left(\frac{\pi}{6}\right)\left(\frac{1}{4}\right)[1 + 4(0.9733) + 2(0.9055) + 4(0.8323) + 0.8]$$

$$\approx 28.36.$$

So, the ellipse has a circumference of about 28.36 units, as shown in Figure 10.13.

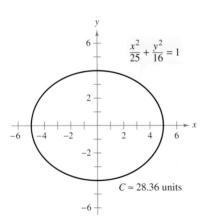

$C \approx 28.36$ units

Figure 10.13

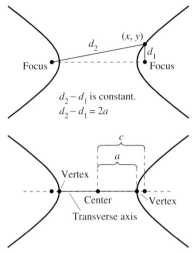

Figure 10.14

Hyperbolas

The definition of a hyperbola is similar to that of an ellipse. For an ellipse, the *sum* of the distances between the foci and a point on the ellipse is fixed, whereas for a hyperbola, the absolute value of the *difference* between these distances is fixed.

A **hyperbola** is the set of all points (x, y) for which the absolute value of the difference between the distances from two distinct fixed points called **foci** is constant. (See Figure 10.14.) The line through the two foci intersects a hyperbola at two points called the **vertices.** The line segment connecting the vertices is the **transverse axis,** and the midpoint of the transverse axis is the **center** of the hyperbola. One distinguishing feature of a hyperbola is that its graph has two separate *branches*.

> **THEOREM 10.5 Standard Equation of a Hyperbola**
>
> The standard form of the equation of a hyperbola with center at (h, k) is
>
> $$\frac{(x-h)^2}{a^2} - \frac{(y-k)^2}{b^2} = 1 \qquad \text{Transverse axis is horizontal.}$$
>
> or
>
> $$\frac{(y-k)^2}{a^2} - \frac{(x-h)^2}{b^2} = 1. \qquad \text{Transverse axis is vertical.}$$
>
> The vertices are a units from the center, and the foci are c units from the center, where, $c^2 = a^2 + b^2$.

NOTE The constants a, b, and c do not have the same relationship for hyperbolas as they do for ellipses. For hyperbolas, $c^2 = a^2 + b^2$, but for ellipses, $c^2 = a^2 - b^2$.

An important aid in sketching the graph of a hyperbola is the determination of its **asymptotes,** as shown in Figure 10.15. Each hyperbola has two asymptotes that intersect at the center of the hyperbola. The asymptotes pass through the vertices of a rectangle of dimensions $2a$ by $2b$, with its center at (h, k). The line segment of length $2b$ joining $(h, k + b)$ and $(h, k - b)$ is referred to as the **conjugate axis** of the hyperbola.

> **THEOREM 10.6 Asymptotes of a Hyperbola**
>
> For a *horizontal* transverse axis, the equations of the asymptotes are
>
> $$y = k + \frac{b}{a}(x - h) \qquad \text{and} \qquad y = k - \frac{b}{a}(x - h).$$
>
> For a *vertical* transverse axis, the equations of the asymptotes are
>
> $$y = k + \frac{a}{b}(x - h) \qquad \text{and} \qquad y = k - \frac{a}{b}(x - h).$$

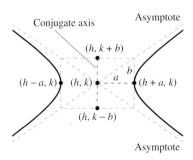

Figure 10.15

In Figure 10.15 you can see that the asymptotes coincide with the diagonals of the rectangle with dimensions $2a$ and $2b$, centered at (h, k). This provides you with a quick means of sketching the asymptotes, which in turn aids in sketching the hyperbola.

EXAMPLE 7 Using Asymptotes to Sketch a Hyperbola

Sketch the graph of the hyperbola whose equation is $4x^2 - y^2 = 16$.

Solution Begin by rewriting the equation in standard form.

$$\frac{x^2}{4} - \frac{y^2}{16} = 1$$

The transverse axis is horizontal and the vertices occur at $(-2, 0)$ and $(2, 0)$. The ends of the conjugate axis occur at $(0, -4)$ and $(0, 4)$. Using these four points, you can sketch the rectangle shown in Figure 10.16(a). By drawing the asymptotes through the corners of this rectangle, you can complete the sketch as shown in Figure 10.16(b).

TECHNOLOGY You can use a graphing utility to verify the graph obtained in Example 7 by solving the original equation for y and graphing the following equations.

$$y_1 = \sqrt{4x^2 - 16}$$
$$y_2 = -\sqrt{4x^2 - 16}$$

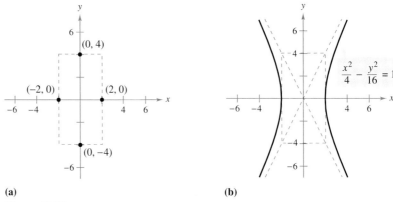

Figure 10.16

Definition of Eccentricity of a Hyperbola

The **eccentricity** e of a hyperbola is given by the ratio

$$e = \frac{c}{a}.$$

As with an ellipse, the **eccentricity** of a hyperbola is $e = c/a$. Because $c > a$ for hyperbolas, it follows that $e > 1$ for hyperbolas. If the eccentricity is large, the branches of the hyperbola are nearly flat. If the eccentricity is close to 1, the branches of the hyperbola are more pointed, as shown in Figure 10.17.

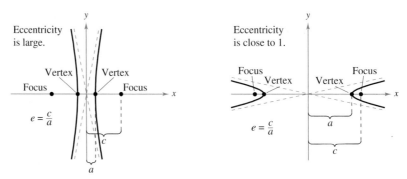

Figure 10.17

The following application was developed during World War II. It shows how the properties of hyperbolas can be used in radar and other detection systems.

EXAMPLE 8 A Hyperbolic Detection System

Two microphones, 1 mile apart, record an explosion. Microphone A receives the sound 2 seconds before microphone B. Where was the explosion?

Solution Assuming that sound travels at 1100 feet per second, you know that the explosion took place 2200 feet farther from B than from A, as shown in Figure 10.18. The locus of all points that are 2200 feet closer to A than to B is one branch of the hyperbola $(x^2/a^2) - (y^2/b^2) = 1$, where

$$c = \frac{1 \text{ mile}}{2} = \frac{5280 \text{ ft}}{2} = 2640 \text{ feet}$$

and

$$a = \frac{2200 \text{ ft}}{2} = 1100 \text{ feet}.$$

Because $c^2 = a^2 + b^2$, it follows that

$$b^2 = c^2 - a^2$$
$$= 5{,}759{,}600$$

and you can conclude that the explosion occurred somewhere on the right branch of the hyperbola given by

$$\frac{x^2}{1{,}210{,}000} - \frac{y^2}{5{,}759{,}600} = 1.$$

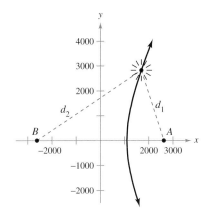

$2c = 5280$
$d_2 - d_1 = 2a = 2200$
Figure 10.18

In Example 8, you were able to determine only the hyperbola on which the explosion occurred, but not the exact location of the explosion. If, however, you had received the sound at a third position C, then two other hyperbolas would be determined. The exact location of the explosion would be the point at which these three hyperbolas intersect.

Another interesting application of conics involves the orbits of comets in our solar system. Of the 610 comets identified prior to 1970, 245 have elliptical orbits, 295 have parabolic orbits, and 70 have hyperbolic orbits. The center of the sun is a focus of each orbit, and each orbit has a vertex at the point at which the comet is closest to the sun. Undoubtedly, many comets with parabolic or hyperbolic orbits have not been identified—such comets pass through our solar system only once. Only comets with elliptical orbits such as Halley's comet remain in our solar system.

The type of orbit for a comet can be determined as follows.

1. Ellipse: $v < \sqrt{2GM/p}$
2. Parabola: $v = \sqrt{2GM/p}$
3. Hyperbola: $v > \sqrt{2GM/p}$

In these three formulas, p is the distance between one vertex and one focus of the comet's orbit (in meters), v is the velocity of the comet at the vertex (in meters per second), $M \approx 1.989 \times 10^{30}$ kilograms is the mass of the sun, and $G \approx 6.67 \times 10^{-8}$ cubic meters per kilogram-second squared is the gravitational constant.

CAROLINE HERSCHEL (1750–1848)

The first woman to be credited with detecting a new comet was the English astronomer Caroline Herschel. During her life, Caroline Herschel discovered a total of eight new comets.

Exercises for Section 10.1

In Exercises 1–8, match the equation with its graph. [The graphs are labeled (a), (b), (c), (d), (e), (f), (g), and (h).]

(a)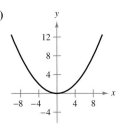

(b)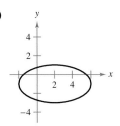

(c)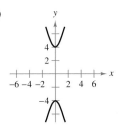

(d)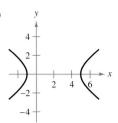

(e)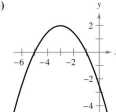

(f)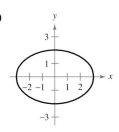

(g)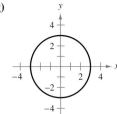

(h)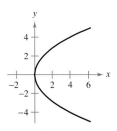

1. $y^2 = 4x$
2. $x^2 = 8y$
3. $(x + 3)^2 = -2(y - 2)$
4. $\dfrac{(x - 2)^2}{16} + \dfrac{(y + 1)^2}{4} = 1$
5. $\dfrac{x^2}{9} + \dfrac{y^2}{4} = 1$
6. $\dfrac{x^2}{9} + \dfrac{y^2}{9} = 1$
7. $\dfrac{y^2}{16} - \dfrac{x^2}{1} = 1$
8. $\dfrac{(x - 2)^2}{9} - \dfrac{y^2}{4} = 1$

In Exercises 9–16, find the vertex, focus, and directrix of the parabola, and sketch its graph.

9. $y^2 = -6x$
10. $x^2 + 8y = 0$
11. $(x + 3) + (y - 2)^2 = 0$
12. $(x - 1)^2 + 8(y + 2) = 0$
13. $y^2 - 4y - 4x = 0$
14. $y^2 + 6y + 8x + 25 = 0$
15. $x^2 + 4x + 4y - 4 = 0$
16. $y^2 + 4y + 8x - 12 = 0$

In Exercises 17–20, find the vertex, focus, and directrix of the parabola. Then use a graphing utility to graph the parabola.

17. $y^2 + x + y = 0$
18. $y = -\tfrac{1}{6}(x^2 - 8x + 6)$
19. $y^2 - 4x - 4 = 0$
20. $x^2 - 2x + 8y + 9 = 0$

In Exercises 21–28, find an equation of the parabola.

21. Vertex: $(3, 2)$
    Focus: $(1, 2)$
22. Vertex: $(-1, 2)$
    Focus: $(-1, 0)$
23. Vertex: $(0, 4)$
    Directrix: $y = -2$
24. Focus: $(2, 2)$
    Directrix: $x = -2$

25.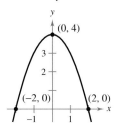

26.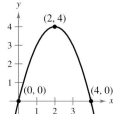

27. Axis is parallel to y-axis; graph passes through $(0, 3)$, $(3, 4)$, and $(4, 11)$.
28. Directrix: $y = -2$; endpoints of latus rectum are $(0, 2)$ and $(8, 2)$.

In Exercises 29–34, find the center, foci, vertices, and eccentricity of the ellipse, and sketch its graph.

29. $x^2 + 4y^2 = 4$
30. $5x^2 + 7y^2 = 70$
31. $\dfrac{(x - 1)^2}{9} + \dfrac{(y - 5)^2}{25} = 1$
32. $(x + 2)^2 + \dfrac{(y + 4)^2}{1/4} = 1$
33. $9x^2 + 4y^2 + 36x - 24y + 36 = 0$
34. $16x^2 + 25y^2 - 64x + 150y + 279 = 0$

In Exercises 35–38, find the center, foci, and vertices of the ellipse. Use a graphing utility to graph the ellipse.

35. $12x^2 + 20y^2 - 12x + 40y - 37 = 0$
36. $36x^2 + 9y^2 + 48x - 36y + 43 = 0$
37. $x^2 + 2y^2 - 3x + 4y + 0.25 = 0$
38. $2x^2 + y^2 + 4.8x - 6.4y + 3.12 = 0$

In Exercises 39–44, find an equation of the ellipse.

39. Center: $(0, 0)$
    Focus: $(2, 0)$
    Vertex: $(3, 0)$
40. Vertices: $(0, 2)$, $(4, 2)$
    Eccentricity: $\tfrac{1}{2}$
41. Vertices: $(3, 1)$, $(3, 9)$
    Minor axis length: 6
42. Foci: $(0, \pm 5)$
    Major axis length: 14

43. Center: (0, 0)
Major axis: horizontal
Points on the ellipse:
(3, 1), (4, 0)

44. Center: (1, 2)
Major axis: vertical
Points on the ellipse:
(1, 6), (3, 2)

In Exercises 45–52, find the center, foci, and vertices of the hyperbola, and sketch its graph using asymptotes as an aid.

45. $y^2 - \dfrac{x^2}{4} = 1$

46. $\dfrac{x^2}{25} - \dfrac{y^2}{9} = 1$

47. $\dfrac{(x-1)^2}{4} - \dfrac{(y+2)^2}{1} = 1$

48. $\dfrac{(y+1)^2}{144} - \dfrac{(x-4)^2}{25} = 1$

49. $9x^2 - y^2 - 36x - 6y + 18 = 0$
50. $y^2 - 9x^2 + 36x - 72 = 0$
51. $x^2 - 9y^2 + 2x - 54y - 80 = 0$
52. $9x^2 - 4y^2 + 54x + 8y + 78 = 0$

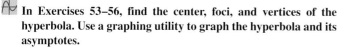
In Exercises 53–56, find the center, foci, and vertices of the hyperbola. Use a graphing utility to graph the hyperbola and its asymptotes.

53. $9y^2 - x^2 + 2x + 54y + 62 = 0$
54. $9x^2 - y^2 + 54x + 10y + 55 = 0$
55. $3x^2 - 2y^2 - 6x - 12y - 27 = 0$
56. $3y^2 - x^2 + 6x - 12y = 0$

In Exercises 57–64, find an equation of the hyperbola.

57. Vertices: $(\pm 1, 0)$
Asymptotes: $y = \pm 3x$

58. Vertices: $(0, \pm 3)$
Asymptotes: $y = \pm 3x$

59. Vertices: $(2, \pm 3)$
Point on graph: $(0, 5)$

60. Vertices: $(2, \pm 3)$
Foci: $(2, \pm 5)$

61. Center: $(0, 0)$
Vertex: $(0, 2)$
Focus: $(0, 4)$

62. Center: $(0, 0)$
Vertex: $(3, 0)$
Focus: $(5, 0)$

63. Vertices: $(0, 2), (6, 2)$
Asymptotes: $y = \dfrac{2}{3}x$
$y = 4 - \dfrac{2}{3}x$

64. Focus: $(10, 0)$
Asymptotes: $y = \pm \dfrac{3}{4}x$

In Exercises 65 and 66, find equations for (a) the tangent lines and (b) the normal lines to the hyperbola for the given value of x.

65. $\dfrac{x^2}{9} - y^2 = 1$, $x = 6$

66. $\dfrac{y^2}{4} - \dfrac{x^2}{2} = 1$, $x = 4$

In Exercises 67–76, classify the graph of the equation as a circle, a parabola, an ellipse, or a hyperbola.

67. $x^2 + 4y^2 - 6x + 16y + 21 = 0$
68. $4x^2 - y^2 - 4x - 3 = 0$
69. $y^2 - 4y - 4x = 0$
70. $25x^2 - 10x - 200y - 119 = 0$

71. $4x^2 + 4y^2 - 16y + 15 = 0$
72. $y^2 - 4y = x + 5$
73. $9x^2 + 9y^2 - 36x + 6y + 34 = 0$
74. $2x(x - y) = y(3 - y - 2x)$
75. $3(x - 1)^2 = 6 + 2(y + 1)^2$
76. $9(x + 3)^2 = 36 - 4(y - 2)^2$

Writing About Concepts

77. (a) Give the definition of a parabola.
  (b) Give the standard forms of a parabola with vertex at (h, k).
  (c) In your own words, state the reflective property of a parabola.

78. (a) Give the definition of an ellipse.
  (b) Give the standard forms of an ellipse with center at (h, k).

79. (a) Give the definition of a hyperbola.
  (b) Give the standard forms of a hyperbola with center at (h, k).
  (c) Write equations for the asymptotes of a hyperbola.

80. Define the eccentricity of an ellipse. In your own words, describe how changes in the eccentricity affect the ellipse.

81. *Solar Collector* A solar collector for heating water is constructed with a sheet of stainless steel that is formed into the shape of a parabola (see figure). The water will flow through a pipe that is located at the focus of the parabola. At what distance from the vertex is the pipe?

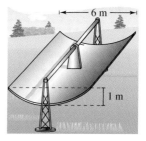
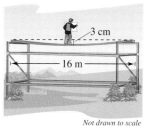

Not drawn to scale

Figure for 81   **Figure for 82**

82. *Beam Deflection* A simply supported beam that is 16 meters long has a load concentrated at the center (see figure). The deflection of the beam at its center is 3 centimeters. Assume that the shape of the deflected beam is parabolic.

(a) Find an equation of the parabola. (Assume that the origin is at the center of the beam.)

(b) How far from the center of the beam is the deflection 1 centimeter?

83. Find an equation of the tangent line to the parabola $y = ax^2$ at $x = x_0$. Prove that the x-intercept of this tangent line is $(x_0/2, 0)$.

84. (a) Prove that any two distinct tangent lines to a parabola intersect.

(b) Demonstrate the result of part (a) by finding the point of intersection of the tangent lines to the parabola $x^2 - 4x - 4y = 0$ at the points $(0, 0)$ and $(6, 3)$.

85. (a) Prove that if any two tangent lines to a parabola intersect at right angles, their point of intersection must lie on the directrix.

(b) Demonstrate the result of part (a) by proving that the tangent lines to the parabola $x^2 - 4x - 4y + 8 = 0$ at the points $(-2, 5)$ and $\left(3, \frac{5}{4}\right)$ intersect at right angles, and that the point of intersection lies on the directrix.

86. Find the point on the graph of $x^2 = 8y$ that is closest to the focus of the parabola.

87. Radio and Television Reception In mountainous areas, reception of radio and television is sometimes poor. Consider an idealized case where a hill is represented by the graph of the parabola $y = x - x^2$, a transmitter is located at the point $(-1, 1)$, and a receiver is located on the other side of the hill at the point $(x_0, 0)$. What is the closest the receiver can be to the hill so that the reception is unobstructed?

88. Modeling Data The table shows the average amounts of time A (in minutes) women spent watching television each day for the years 1996 to 2002. (*Source: Nielsen Media Research*)

Year	1996	1997	1998	1999	2000	2001	2002
A	274	273	273	280	286	291	298

(a) Use the regression capabilities of a graphing utility to find a model of the form $A = at^2 + bt + c$ for the data. Let t represent the year, with $t = 6$ corresponding to 1996.

(b) Use a graphing utility to plot the data and graph the model.

(c) Find dA/dt and sketch its graph for $6 \leq t \leq 12$. What information about the average amount of time women spent watching television is given by the graph of the derivative?

89. Architecture A church window is bounded above by a parabola and below by the arc of a circle (see figure). Find the surface area of the window.

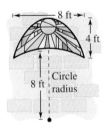

Figure for 89

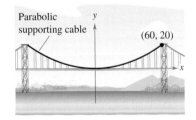

Figure for 91

90. Arc Length Find the arc length of the parabola $4x - y^2 = 0$ over the interval $0 \leq y \leq 4$.

91. Bridge Design A cable of a suspension bridge is suspended (in the shape of a parabola) between two towers that are 120 meters apart and 20 meters above the roadway (see figure). The cables touch the roadway midway between the towers.

(a) Find an equation for the parabolic shape of each cable.

(b) Find the length of the parabolic supporting cable.

92. Surface Area A satellite-signal receiving dish is formed by revolving the parabola given by $x^2 = 20y$ about the y-axis. The radius of the dish is r feet. Verify that the surface area of the dish is given by

$$2\pi \int_0^r x \sqrt{1 + \left(\frac{x}{10}\right)^2}\, dx = \frac{\pi}{15}[(100 + r^2)^{3/2} - 1000].$$

93. Investigation Sketch the graphs of $x^2 = 4py$ for $p = \frac{1}{4}, \frac{1}{2}, 1, \frac{3}{2}$, and 2 on the same coordinate axes. Discuss the change in the graphs as p increases.

94. Area Find a formula for the area of the shaded region in the figure.

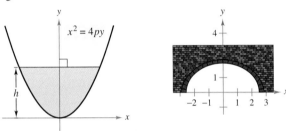

Figure for 94   Figure for 96

95. Writing On page 697, it was noted that an ellipse can be drawn using two thumbtacks, a string of fixed length (greater than the distance between the tacks), and a pencil. If the ends of the string are fastened at the tacks and the string is drawn taut with a pencil, the path traced by the pencil will be an ellipse.

(a) What is the length of the string in terms of a?

(b) Explain why the path is an ellipse.

96. Construction of a Semielliptical Arch A fireplace arch is to be constructed in the shape of a semiellipse. The opening is to have a height of 2 feet at the center and a width of 5 feet along the base (see figure). The contractor draws the outline of the ellipse by the method shown in Exercise 95. Where should the tacks be placed and what should be the length of the piece of string?

97. Sketch the ellipse that consists of all points (x, y) such that the sum of the distances between (x, y) and two fixed points is 16 units, and the foci are located at the centers of the two sets of concentric circles in the figure. To print an enlarged copy of the graph, go to the website *www.mathgraphs.com*.

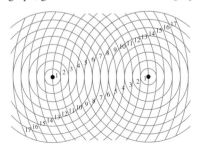

98. Orbit of Earth Earth moves in an elliptical orbit with the sun at one of the foci. The length of half of the major axis is 149,598,000 kilometers, and the eccentricity is 0.0167. Find the minimum distance (*perihelion*) and the maximum distance (*aphelion*) of Earth from the sun.

99. Satellite Orbit The *apogee* (the point in orbit farthest from Earth) and the *perigee* (the point in orbit closest to Earth) of an elliptical orbit of an Earth satellite are given by A and P. Show that the eccentricity of the orbit is

$$e = \frac{A - P}{A + P}.$$

100. Explorer 18 On November 27, 1963, the United States launched Explorer 18. Its low and high points above the surface of Earth were 119 miles and 123,000 miles. Find the eccentricity of its elliptical orbit.

101. Halley's Comet Probably the most famous of all comets, Halley's comet, has an elliptical orbit with the sun at the focus. Its maximum distance from the sun is approximately 35.29 AU (astronomical unit $\approx 92.956 \times 10^6$ miles), and its minimum distance is approximately 0.59 AU. Find the eccentricity of the orbit.

102. The equation of an ellipse with its center at the origin can be written as

$$\frac{x^2}{a^2} + \frac{y^2}{a^2(1-e^2)} = 1.$$

Show that as $e \to 0$, with a remaining fixed, the ellipse approaches a circle.

103. Consider a particle traveling clockwise on the elliptical path

$$\frac{x^2}{100} + \frac{y^2}{25} = 1.$$

The particle leaves the orbit at the point $(-8, 3)$ and travels in a straight line tangent to the ellipse. At what point will the particle cross the y-axis?

104. Volume The water tank on a fire truck is 16 feet long, and its cross sections are ellipses. Find the volume of water in the partially filled tank as shown in the figure.

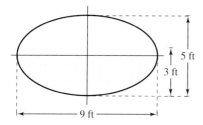

In Exercises 105 and 106, determine the points at which dy/dx is zero or does not exist to locate the endpoints of the major and minor axes of the ellipse.

105. $16x^2 + 9y^2 + 96x + 36y + 36 = 0$

106. $9x^2 + 4y^2 + 36x - 24y + 36 = 0$

Area and Volume In Exercises 107 and 108, find (a) the area of the region bounded by the ellipse, (b) the volume and surface area of the solid generated by revolving the region about its major axis (prolate spheroid), and (c) the volume and surface area of the solid generated by revolving the region about its minor axis (oblate spheroid).

107. $\dfrac{x^2}{4} + \dfrac{y^2}{1} = 1$

108. $\dfrac{x^2}{16} + \dfrac{y^2}{9} = 1$

 109. Arc Length Use the integration capabilities of a graphing utility to approximate to two-decimal-place accuracy the elliptical integral representing the circumference of the ellipse

$$\frac{x^2}{25} + \frac{y^2}{49} = 1.$$

110. Prove that the tangent line to an ellipse at a point P makes equal angles with lines through P and the foci (see figure). [*Hint*: (1) Find the slope of the tangent line at P, (2) find the slopes of the lines through P and each focus, and (3) use the formula for the tangent of the angle between two lines.]

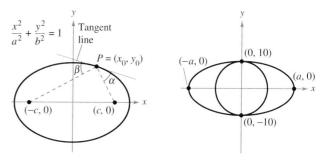

Figure for 110   **Figure for 111**

111. Geometry The area of the ellipse in the figure is twice the area of the circle. What is the length of the major axis?

 112. Conjecture

(a) Show that the equation of an ellipse can be written as

$$\frac{(x-h)^2}{a^2} + \frac{(y-k)^2}{a^2(1-e^2)} = 1.$$

(b) Use a graphing utility to graph the ellipse

$$\frac{(x-2)^2}{4} + \frac{(y-3)^2}{4(1-e^2)} = 1$$

for $e = 0.95$, $e = 0.75$, $e = 0.5$, $e = 0.25$, and $e = 0$.

(c) Use the results of part (b) to make a conjecture about the change in the shape of the ellipse as e approaches 0.

113. Find an equation of the hyperbola such that for any point on the hyperbola, the difference between its distances from the points $(2, 2)$ and $(10, 2)$ is 6.

114. Find an equation of the hyperbola such that for any point on the hyperbola, the difference between its distances from the points $(-3, 0)$ and $(-3, 3)$ is 2.

115. Sketch the hyperbola that consists of all points (x, y) such that the difference of the distances between (x, y) and two fixed points is 10 units, and the foci are located at the centers of the two sets of concentric circles in the figure. To print an enlarged copy of the graph, go to the website *www.mathgraphs.com*.

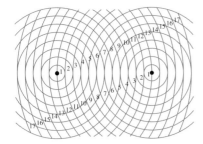

116. Consider a hyperbola centered at the origin with a horizontal transverse axis. Use the definition of a hyperbola to derive its standard form:

$$\frac{x^2}{a^2} - \frac{y^2}{b^2} = 1.$$

117. *Sound Location* A rifle positioned at point $(-c, 0)$ is fired at a target positioned at point $(c, 0)$. A person hears the sound of the rifle and the sound of the bullet hitting the target at the same time. Prove that the person is positioned on one branch of the hyperbola given by

$$\frac{x^2}{c^2 v_s^2 / v_m^2} - \frac{y^2}{c^2(v_m^2 - v_s^2)/v_m^2} = 1$$

where v_m is the muzzle velocity of the rifle and v_s is the speed of sound, which is about 1100 feet per second.

118. *Navigation* LORAN (long distance radio navigation) for aircraft and ships uses synchronized pulses transmitted by widely separated transmitting stations. These pulses travel at the speed of light (186,000 miles per second). The difference in the times of arrival of these pulses at an aircraft or ship is constant on a hyperbola having the transmitting stations as foci. Assume that two stations, 300 miles apart, are positioned on the rectangular coordinate system at $(-150, 0)$ and $(150, 0)$ and that a ship is traveling on a path with coordinates $(x, 75)$ (see figure). Find the x-coordinate of the position of the ship if the time difference between the pulses from the transmitting stations is 1000 microseconds (0.001 second).

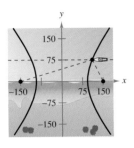

Figure for 118   Figure for 119

119. *Hyperbolic Mirror* A hyperbolic mirror (used in some telescopes) has the property that a light ray directed at the focus will be reflected to the other focus. The mirror in the figure has the equation $(x^2/36) - (y^2/64) = 1$. At which point on the mirror will light from the point $(0, 10)$ be reflected to the other focus?

120. Show that the equation of the tangent line to $\dfrac{x^2}{a^2} - \dfrac{y^2}{b^2} = 1$ at the point (x_0, y_0) is $(x_0/a^2)x - (y_0/b^2)y = 1$.

121. Show that the graphs of the equations intersect at right angles:

$$\frac{x^2}{a^2} + \frac{2y^2}{b^2} = 1 \quad \text{and} \quad \frac{x^2}{a^2 - b^2} - \frac{2y^2}{b^2} = 1.$$

122. Prove that the graph of the equation

$$Ax^2 + Cy^2 + Dx + Ey + F = 0$$

is one of the following (except in degenerate cases).

Conic	Condition
(a) Circle	$A = C$
(b) Parabola	$A = 0$ or $C = 0$ (but not both)
(c) Ellipse	$AC > 0$
(d) Hyperbola	$AC < 0$

True or False? In Exercises 123–128, determine whether the statement is true or false. If it is false, explain why or give an example that shows it is false.

123. It is possible for a parabola to intersect its directrix.

124. The point on a parabola closest to its focus is its vertex.

125. If C is the circumference of the ellipse

$$\frac{x^2}{a^2} + \frac{y^2}{b^2} = 1, \quad b < a$$

then $2\pi b \leq C \leq 2\pi a$.

126. If $D \neq 0$ or $E \neq 0$, then the graph of $y^2 - x^2 + Dx + Ey = 0$ is a hyperbola.

127. If the asymptotes of the hyperbola $(x^2/a^2) - (y^2/b^2) = 1$ intersect at right angles, then $a = b$.

128. Every tangent line to a hyperbola intersects the hyperbola only at the point of tangency.

Putnam Exam Challenge

129. For a point P on an ellipse, let d be the distance from the center of the ellipse to the line tangent to the ellipse at P. Prove that $(PF_1)(PF_2)d^2$ is constant as P varies on the ellipse, where PF_1 and PF_2 are the distances from P to the foci F_1 and F_2 of the ellipse.

130. Find the minimum value of $(u - v)^2 + \left(\sqrt{2 - u^2} - \dfrac{9}{v}\right)^2$ for $0 < u < \sqrt{2}$ and $v > 0$.

These problems were composed by the Committee on the Putnam Prize Competition.
© The Mathematical Association of America. All rights reserved.

Section 10.2 Plane Curves and Parametric Equations

- Sketch the graph of a curve given by a set of parametric equations.
- Eliminate the parameter in a set of parametric equations.
- Find a set of parametric equations to represent a curve.
- Understand two classic calculus problems, the tautochrone and brachistochrone problems.

Plane Curves and Parametric Equations

Until now, you have been representing a graph by a single equation involving *two* variables. In this section you will study situations in which *three* variables are used to represent a curve in the plane.

Consider the path followed by an object that is propelled into the air at an angle of 45°. If the initial velocity of the object is 48 feet per second, the object travels the parabolic path given by

$$y = -\frac{x^2}{72} + x \qquad \text{Rectangular equation}$$

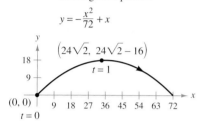

Rectangular equation:
$y = -\frac{x^2}{72} + x$

Parametric equations:
$x = 24\sqrt{2}\,t$
$y = -16t^2 + 24\sqrt{2}\,t$

Curvilinear motion: two variables for position, one variable for time
Figure 10.19

as shown in Figure 10.19. However, this equation does not tell the whole story. Although it does tell you *where* the object has been, it doesn't tell you *when* the object was at a given point (x, y). To determine this time, you can introduce a third variable t, called a **parameter**. By writing both x and y as functions of t, you obtain the **parametric equations**

$$x = 24\sqrt{2}\,t \qquad \text{Parametric equation for } x$$

and

$$y = -16t^2 + 24\sqrt{2}\,t. \qquad \text{Parametric equation for } y$$

From this set of equations, you can determine that at time $t = 0$, the object is at the point $(0, 0)$. Similarly, at time $t = 1$, the object is at the point $(24\sqrt{2}, 24\sqrt{2} - 16)$, and so on. (You will learn a method for determining this particular set of parametric equations—the equations of motion—later, in Section 12.3.)

For this particular motion problem, x and y are continuous functions of t, and the resulting path is called a **plane curve**.

Definition of a Plane Curve

If f and g are continuous functions of t on an interval I, then the equations

$$x = f(t) \quad \text{and} \quad y = g(t)$$

are called **parametric equations** and t is called the **parameter**. The set of points (x, y) obtained as t varies over the interval I is called the **graph** of the parametric equations. Taken together, the parametric equations and the graph are called a **plane curve**, denoted by C.

NOTE At times it is important to distinguish between a graph (the set of points) and a curve (the points together with their defining parametric equations). When it is important, we will make the distinction explicit. When it is not important, we will use C to represent the graph or the curve.

When sketching (by hand) a curve represented by a set of parametric equations, you can plot points in the *xy*-plane. Each set of coordinates (x, y) is determined from a value chosen for the parameter t. By plotting the resulting points in order of increasing values of t, the curve is traced out in a specific direction. This is called the **orientation** of the curve.

EXAMPLE 1 Sketching a Curve

Sketch the curve described by the parametric equations

$$x = t^2 - 4 \quad \text{and} \quad y = \frac{t}{2}, \quad -2 \leq t \leq 3.$$

Solution For values of t on the given interval, the parametric equations yield the points (x, y) shown in the table.

t	-2	-1	0	1	2	3
x	0	-3	-4	-3	0	5
y	-1	$-\frac{1}{2}$	0	$\frac{1}{2}$	1	$\frac{3}{2}$

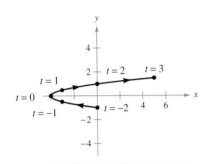

Parametric equations:
$x = t^2 - 4$ and $y = \frac{t}{2}, -2 \leq t \leq 3$

Figure 10.20

By plotting these points in order of increasing t and using the continuity of f and g, you obtain the curve C shown in Figure 10.20. Note that the arrows on the curve indicate its orientation as t increases from -2 to 3.

NOTE From the Vertical Line Test, you can see that the graph shown in Figure 10.20 does not define y as a function of x. This points out one benefit of parametric equations—they can be used to represent graphs that are more general than graphs of functions.

It often happens that two different sets of parametric equations have the same graph. For example, the set of parametric equations

$$x = 4t^2 - 4 \quad \text{and} \quad y = t, \quad -1 \leq t \leq \frac{3}{2}$$

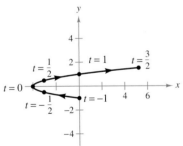

Parametric equations:
$x = 4t^2 - 4$ and $y = t, -1 \leq t \leq \frac{3}{2}$

Figure 10.21

has the same graph as the set given in Example 1. However, comparing the values of t in Figures 10.20 and 10.21, you can see that the second graph is traced out more *rapidly* (considering t as time) than the first graph. So, in applications, different parametric representations can be used to represent various *speeds* at which objects travel along a given path.

> **TECHNOLOGY** Most graphing utilities have a *parametric* graphing mode. If you have access to such a utility, use it to confirm the graphs shown in Figures 10.20 and 10.21. Does the curve given by
>
> $$x = 4t^2 - 8t \quad \text{and} \quad y = 1 - t, \quad -\tfrac{1}{2} \leq t \leq 2$$
>
> represent the same graph as that shown in Figures 10.20 and 10.21? What do you notice about the *orientation* of this curve?

Eliminating the Parameter

Finding a rectangular equation that represents the graph of a set of parametric equations is called **eliminating the parameter.** For instance, you can eliminate the parameter from the set of parametric equations in Example 1 as follows.

Parametric equations	⇒	Solve for t in one equation.	⇒	Substitute into second equation.	⇒	Rectangular equation
$x = t^2 - 4$ $y = t/2$		$t = 2y$		$x = (2y)^2 - 4$		$x = 4y^2 - 4$

Once you have eliminated the parameter, you can recognize that the equation $x = 4y^2 - 4$ represents a parabola with a horizontal axis and vertex at $(-4, 0)$, as shown in Figure 10.20.

The range of x and y implied by the parametric equations may be altered by the change to rectangular form. In such instances the domain of the rectangular equation must be adjusted so that its graph matches the graph of the parametric equations. Such a situation is demonstrated in the next example.

EXAMPLE 2 Adjusting the Domain After Eliminating the Parameter

Sketch the curve represented by the equations

$$x = \frac{1}{\sqrt{t+1}} \quad \text{and} \quad y = \frac{t}{t+1}, \quad t > -1$$

by eliminating the parameter and adjusting the domain of the resulting rectangular equation.

Solution Begin by solving one of the parametric equations for t. For instance, you can solve the first equation for t as follows.

$$x = \frac{1}{\sqrt{t+1}} \quad \text{Parametric equation for } x$$

$$x^2 = \frac{1}{t+1} \quad \text{Square each side.}$$

$$t + 1 = \frac{1}{x^2}$$

$$t = \frac{1}{x^2} - 1 = \frac{1 - x^2}{x^2} \quad \text{Solve for } t.$$

Now, substituting into the parametric equation for y produces

$$y = \frac{t}{t+1} \quad \text{Parametric equation for } y$$

$$y = \frac{(1 - x^2)/x^2}{[(1 - x^2)/x^2] + 1} \quad \text{Substitute } (1 - x^2)/x^2 \text{ for } t.$$

$$y = 1 - x^2. \quad \text{Simplify.}$$

The rectangular equation, $y = 1 - x^2$, is defined for all values of x, but from the parametric equation for x you can see that the curve is defined only when $t > -1$. This implies that you should restrict the domain of x to positive values, as shown in Figure 10.22.

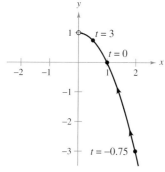

Parametric equations:
$x = \frac{1}{\sqrt{t+1}}, \quad y = \frac{t}{t+1}, \quad t > -1$

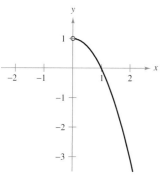

Rectangular equation:
$y = 1 - x^2, \quad x > 0$

Figure 10.22

It is not necessary for the parameter in a set of parametric equations to represent time. The next example uses an *angle* as the parameter.

EXAMPLE 3 Using Trigonometry to Eliminate a Parameter

Sketch the curve represented by

$$x = 3\cos\theta \quad \text{and} \quad y = 4\sin\theta, \quad 0 \le \theta \le 2\pi$$

by eliminating the parameter and finding the corresponding rectangular equation.

Solution Begin by solving for $\cos\theta$ and $\sin\theta$ in the given equations.

$$\cos\theta = \frac{x}{3} \quad \text{and} \quad \sin\theta = \frac{y}{4} \qquad \text{Solve for } \cos\theta \text{ and } \sin\theta.$$

Next, make use of the identity $\sin^2\theta + \cos^2\theta = 1$ to form an equation involving only x and y.

$$\cos^2\theta + \sin^2\theta = 1 \qquad \text{Trigonometric identity}$$

$$\left(\frac{x}{3}\right)^2 + \left(\frac{y}{4}\right)^2 = 1 \qquad \text{Substitute.}$$

$$\frac{x^2}{9} + \frac{y^2}{16} = 1 \qquad \text{Rectangular equation}$$

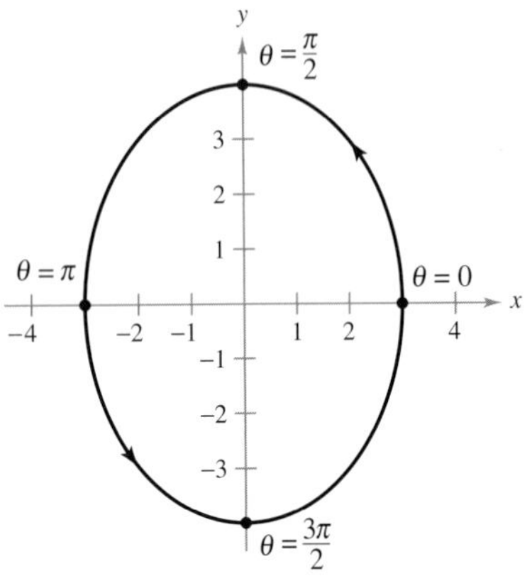

From this rectangular equation you can see that the graph is an ellipse centered at $(0, 0)$, with vertices at $(0, 4)$ and $(0, -4)$ and minor axis of length $2b = 6$, as shown in Figure 10.23. Note that the ellipse is traced out *counterclockwise* as θ varies from 0 to 2π.

Parametric equations:
$x = 3\cos\theta, \ y = 4\sin\theta$

Rectangular equation:
$\dfrac{x^2}{9} + \dfrac{y^2}{16} = 1$

Figure 10.23

Using the technique shown in Example 3, you can conclude that the graph of the parametric equations

$$x = h + a\cos\theta \quad \text{and} \quad y = k + b\sin\theta, \quad 0 \le \theta \le 2\pi$$

is the ellipse (traced counterclockwise) given by

$$\frac{(x-h)^2}{a^2} + \frac{(y-k)^2}{b^2} = 1.$$

The graph of the parametric equations

$$x = h + a\sin\theta \quad \text{and} \quad y = k + b\cos\theta, \quad 0 \le \theta \le 2\pi$$

is also the ellipse (traced clockwise) given by

$$\frac{(x-h)^2}{a^2} + \frac{(y-k)^2}{b^2} = 1.$$

Use a graphing utility in *parametric* mode to graph several ellipses.

In Examples 2 and 3, it is important to realize that eliminating the parameter is primarily an *aid to curve sketching*. If the parametric equations represent the path of a moving object, the graph alone is not sufficient to describe the object's motion. You still need the parametric equations to tell you the *position*, *direction*, and *speed* at a given time.

Finding Parametric Equations

The first three examples in this section illustrate techniques for sketching the graph represented by a set of parametric equations. You will now investigate the reverse problem. How can you determine a set of parametric equations for a given graph or a given physical description? From the discussion following Example 1, you know that such a representation is not unique. This is demonstrated further in the following example, which finds two different parametric representations for a given graph.

EXAMPLE 4 Finding Parametric Equations for a Given Graph

Find a set of parametric equations to represent the graph of $y = 1 - x^2$, using each of the following parameters.

a. $t = x$ **b.** The slope $m = \dfrac{dy}{dx}$ at the point (x, y)

Solution

a. Letting $x = t$ produces the parametric equations

$$x = t \quad \text{and} \quad y = 1 - x^2 = 1 - t^2.$$

b. To write x and y in terms of the parameter m, you can proceed as follows.

$$m = \frac{dy}{dx} = -2x \qquad \text{Differentiate } y = 1 - x^2.$$

$$x = -\frac{m}{2} \qquad \text{Solve for } x.$$

This produces a parametric equation for x. To obtain a parametric equation for y, substitute $-m/2$ for x in the original equation.

$$y = 1 - x^2 \qquad \text{Write original rectangular equation.}$$

$$y = 1 - \left(-\frac{m}{2}\right)^2 \qquad \text{Substitute } -m/2 \text{ for } x.$$

$$y = 1 - \frac{m^2}{4} \qquad \text{Simplify.}$$

So, the parametric equations are

$$x = -\frac{m}{2} \quad \text{and} \quad y = 1 - \frac{m^2}{4}.$$

In Figure 10.24, note that the resulting curve has a right-to-left orientation as determined by the direction of increasing values of slope m. For part (a), the curve would have the opposite orientation.

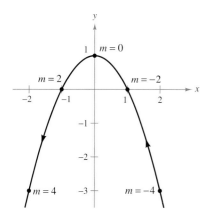

Rectangular equation: $y = 1 - x^2$
Parametric equations:
$x = -\dfrac{m}{2}, y = 1 - \dfrac{m^2}{4}$

Figure 10.24

TECHNOLOGY To be efficient at using a graphing utility, it is important that you develop skill in representing a graph by a set of parametric equations. The reason for this is that many graphing utilities have only three graphing modes—(1) functions, (2) parametric equations, and (3) polar equations. Most graphing utilities are not programmed to graph a general equation. For instance, suppose you want to graph the hyperbola $x^2 - y^2 = 1$. To graph the hyperbola in *function* mode, you need two equations: $y = \sqrt{x^2 - 1}$ and $y = -\sqrt{x^2 - 1}$. In *parametric* mode, you can represent the graph by $x = \sec t$ and $y = \tan t$.

CYCLOIDS

Galileo first called attention to the cycloid, once recommending that it be used for the arches of bridges. Pascal once spent 8 days attempting to solve many of the problems of cycloids, such as finding the area under one arch, and the volume of the solid of revolution formed by revolving the curve about a line. The cycloid has so many interesting properties and has caused so many quarrels among mathematicians that it has been called "the Helen of geometry" and "the apple of discord."

FOR FURTHER INFORMATION For more information on cycloids, see the article "The Geometry of Rolling Curves" by John Bloom and Lee Whitt in *The American Mathematical Monthly*. To view this article, go to the website *www.matharticles.com*.

EXAMPLE 5 Parametric Equations for a Cycloid

Determine the curve traced by a point P on the circumference of a circle of radius a rolling along a straight line in a plane. Such a curve is called a **cycloid**.

Solution Let the parameter θ be the measure of the circle's rotation, and let the point $P = (x, y)$ begin at the origin. When $\theta = 0$, P is at the origin. When $\theta = \pi$, P is at a maximum point $(\pi a, 2a)$. When $\theta = 2\pi$, P is back on the x-axis at $(2\pi a, 0)$. From Figure 10.25, you can see that $\angle APC = 180° - \theta$. So,

$$\sin \theta = \sin(180° - \theta) = \sin(\angle APC) = \frac{AC}{a} = \frac{BD}{a}$$

$$\cos \theta = -\cos(180° - \theta) = -\cos(\angle APC) = \frac{AP}{-a}$$

which implies that

$$AP = -a \cos \theta \quad \text{and} \quad BD = a \sin \theta.$$

Because the circle rolls along the x-axis, you know that $OD = \overset{\frown}{PD} = a\theta$. Furthermore, because $BA = DC = a$, you have

$$x = OD - BD = a\theta - a \sin \theta$$
$$y = BA + AP = a - a \cos \theta.$$

So, the parametric equations are

$$x = a(\theta - \sin \theta) \quad \text{and} \quad y = a(1 - \cos \theta).$$

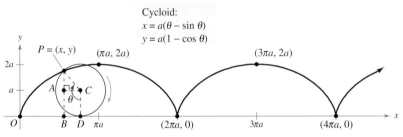

Figure 10.25

The cycloid in Figure 10.25 has sharp corners at the values $x = 2n\pi a$. Notice that the derivatives $x'(\theta)$ and $y'(\theta)$ are both zero at the points for which $\theta = 2n\pi$.

$$x(\theta) = a(\theta - \sin \theta) \qquad y(\theta) = a(1 - \cos \theta)$$
$$x'(\theta) = a - a \cos \theta \qquad y'(\theta) = a \sin \theta$$
$$x'(2n\pi) = 0 \qquad y'(2n\pi) = 0$$

Between these points, the cycloid is called **smooth**.

TECHNOLOGY Some graphing utilities allow you to simulate the motion of an object that is moving in the plane or in space. If you have access to such a utility, use it to trace out the path of the cycloid shown in Figure 10.25.

Definition of a Smooth Curve

A curve C represented by $x = f(t)$ and $y = g(t)$ on an interval I is called **smooth** if f' and g' are continuous on I and not simultaneously 0, except possibly at the endpoints of I. The curve C is called **piecewise smooth** if it is smooth on each subinterval of some partition of I.

The Tautochrone and Brachistochrone Problems

The type of curve described in Example 5 is related to one of the most famous pairs of problems in the history of calculus. The first problem (called the **tautochrone problem**) began with Galileo's discovery that the time required to complete a full swing of a given pendulum is *approximately* the same whether it makes a large movement at high speed or a small movement at lower speed (see Figure 10.26). Late in his life, Galileo (1564–1642) realized that he could use this principle to construct a clock. However, he was not able to conquer the mechanics of actual construction. Christian Huygens (1629–1695) was the first to design and construct a working model. In his work with pendulums, Huygens realized that a pendulum does not take exactly the same time to complete swings of varying lengths. (This doesn't affect a pendulum clock, because the length of the circular arc is kept constant by giving the pendulum a slight boost each time it passes its lowest point.) But, in studying the problem, Huygens discovered that a ball rolling back and forth on an inverted cycloid does complete each cycle in exactly the same time.

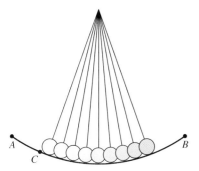

The time required to complete a full swing of the pendulum when starting from point C is only approximately the same as when starting from point A.
Figure 10.26

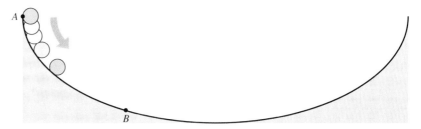

An inverted cycloid is the path down which a ball will roll in the shortest time.
Figure 10.27

The second problem, which was posed by John Bernoulli in 1696, is called the **brachistochrone problem**—in Greek, *brachys* means short and *chronos* means time. The problem was to determine the path down which a particle will slide from point A to point B in the *shortest time*. Several mathematicians took up the challenge, and the following year the problem was solved by Newton, Leibniz, L'Hôpital, John Bernoulli, and James Bernoulli. As it turns out, the solution is not a straight line from A to B, but an inverted cycloid passing through the points A and B, as shown in Figure 10.27. The amazing part of the solution is that a particle starting at rest at *any* other point C of the cycloid between A and B will take exactly the same time to reach B, as shown in Figure 10.28.

JAMES BERNOULLI (1654–1705)

James Bernoulli, also called Jacques, was the older brother of John. He was one of several accomplished mathematicians of the Swiss Bernoulli family. James's mathematical accomplishments have given him a prominent place in the early development of calculus.

A ball starting at point C takes the same time to reach point B as one that starts at point A.
Figure 10.28

FOR FURTHER INFORMATION To see a proof of the famous brachistochrone problem, see the article "A New Minimization Proof for the Brachistochrone" by Gary Lawlor in *The American Mathematical Monthly*. To view this article, go to the website *www.matharticles.com*.

Exercises for Section 10.2

See www.CalcChat.com for worked-out solutions to odd-numbered exercises.

1. Consider the parametric equations $x = \sqrt{t}$ and $y = 1 - t$.

(a) Complete the table.

t	0	1	2	3	4
x					
y					

(b) Plot the points (x, y) generated in the table, and sketch a graph of the parametric equations. Indicate the orientation of the graph.

(c) Use a graphing utility to confirm your graph in part (b).

(d) Find the rectangular equation by eliminating the parameter, and sketch its graph. Compare the graph in part (b) with the graph of the rectangular equation.

2. Consider the parametric equations $x = 4\cos^2\theta$ and $y = 2\sin\theta$.

(a) Complete the table.

θ	$-\dfrac{\pi}{2}$	$-\dfrac{\pi}{4}$	0	$\dfrac{\pi}{4}$	$\dfrac{\pi}{2}$
x					
y					

(b) Plot the points (x, y) generated in the table, and sketch a graph of the parametric equations. Indicate the orientation of the graph.

(c) Use a graphing utility to confirm your graph in part (b).

(d) Find the rectangular equation by eliminating the parameter, and sketch its graph. Compare the graph in part (b) with the graph of the rectangular equation.

(e) If values of θ were selected from the interval $[\pi/2, 3\pi/2]$ for the table in part (a), would the graph in part (b) be different? Explain.

In Exercises 3–20, sketch the curve represented by the parametric equations (indicate the orientation of the curve), and write the corresponding rectangular equation by eliminating the parameter.

3. $x = 3t - 1$, $y = 2t + 1$
4. $x = 3 - 2t$, $y = 2 + 3t$
5. $x = t + 1$, $y = t^2$
6. $x = 2t^2$, $y = t^4 + 1$
7. $x = t^3$, $y = \dfrac{t^2}{2}$
8. $x = t^2 + t$, $y = t^2 - t$
9. $x = \sqrt{t}$, $y = t - 2$
10. $x = \sqrt[4]{t}$, $y = 3 - t$
11. $x = t - 1$, $y = \dfrac{t}{t-1}$
12. $x = 1 + \dfrac{1}{t}$, $y = t - 1$
13. $x = 2t$, $y = |t - 2|$
14. $x = |t - 1|$, $y = t + 2$
15. $x = e^t$, $y = e^{3t} + 1$
16. $x = e^{-t}$, $y = e^{2t} - 1$
17. $x = \sec\theta$, $y = \cos\theta$, $0 \le \theta < \pi/2$, $\pi/2 < \theta \le \pi$
18. $x = \tan^2\theta$, $y = \sec^2\theta$
19. $x = 3\cos\theta$, $y = 3\sin\theta$
20. $x = 2\cos\theta$, $y = 6\sin\theta$

 In Exercises 21–32, use a graphing utility to graph the curve represented by the parametric equations (indicate the orientation of the curve). Eliminate the parameter and write the corresponding rectangular equation.

21. $x = 4\sin 2\theta$, $y = 2\cos 2\theta$
22. $x = \cos\theta$, $y = 2\sin 2\theta$
23. $x = 4 + 2\cos\theta$
     $y = -1 + \sin\theta$
24. $x = 4 + 2\cos\theta$
     $y = -1 + 2\sin\theta$
25. $x = 4 + 2\cos\theta$
     $y = -1 + 4\sin\theta$
26. $x = \sec\theta$
     $y = \tan\theta$
27. $x = 4\sec\theta$, $y = 3\tan\theta$
28. $x = \cos^3\theta$, $y = \sin^3\theta$
29. $x = t^3$, $y = 3\ln t$
30. $x = \ln 2t$, $y = t^2$
31. $x = e^{-t}$, $y = e^{3t}$
32. $x = e^{2t}$, $y = e^t$

Comparing Plane Curves In Exercises 33–36, determine any differences between the curves of the parametric equations. Are the graphs the same? Are the orientations the same? Are the curves smooth?

33. (a) $x = t$
       $y = 2t + 1$
   (b) $x = \cos\theta$
       $y = 2\cos\theta + 1$
   (c) $x = e^{-t}$
       $y = 2e^{-t} + 1$
   (d) $x = e^t$
       $y = 2e^t + 1$

34. (a) $x = 2\cos\theta$
       $y = 2\sin\theta$
   (b) $x = \sqrt{4t^2 - 1}/|t|$
       $y = 1/t$
   (c) $x = \sqrt{t}$
       $y = \sqrt{4 - t}$
   (d) $x = -\sqrt{4 - e^{2t}}$
       $y = e^t$

35. (a) $x = \cos\theta$
       $y = 2\sin^2\theta$
       $0 < \theta < \pi$
   (b) $x = \cos(-\theta)$
       $y = 2\sin^2(-\theta)$
       $0 < \theta < \pi$

36. (a) $x = t + 1$, $y = t^3$
   (b) $x = -t + 1$, $y = (-t)^3$

37. *Conjecture*

(a) Use a graphing utility to graph the curves represented by the two sets of parametric equations.

   $x = 4\cos t$ $x = 4\cos(-t)$
   $y = 3\sin t$ $y = 3\sin(-t)$

(b) Describe the change in the graph when the sign of the parameter is changed.

(c) Make a conjecture about the change in the graph of parametric equations when the sign of the parameter is changed.

(d) Test your conjecture with another set of parametric equations.

38. *Writing* Review Exercises 33–36 and write a short paragraph describing how the graphs of curves represented by different sets of parametric equations can differ even though eliminating the parameter from each yields the same rectangular equation.

In Exercises 39–42, eliminate the parameter and obtain the standard form of the rectangular equation.

39. Line through (x_1, y_1) and (x_2, y_2):
    $x = x_1 + t(x_2 - x_1)$, $y = y_1 + t(y_2 - y_1)$
40. Circle: $x = h + r\cos\theta$, $y = k + r\sin\theta$
41. Ellipse: $x = h + a\cos\theta$, $y = k + b\sin\theta$
42. Hyperbola: $x = h + a\sec\theta$, $y = k + b\tan\theta$

In Exercises 43–50, use the results of Exercises 39–42 to find a set of parametric equations for the line or conic.

43. Line: passes through $(0, 0)$ and $(5, -2)$
44. Line: passes through $(1, 4)$ and $(5, -2)$
45. Circle: center: $(2, 1)$; radius: 4
46. Circle: center: $(-3, 1)$; radius: 3
47. Ellipse: vertices: $(\pm 5, 0)$; foci: $(\pm 4, 0)$
48. Ellipse: vertices: $(4, 7), (4, -3)$; foci: $(4, 5), (4, -1)$
49. Hyperbola: vertices: $(\pm 4, 0)$; foci: $(\pm 5, 0)$
50. Hyperbola: vertices: $(0, \pm 1)$; foci: $(0, \pm 2)$

In Exercises 51–54, find two different sets of parametric equations for the rectangular equation.

51. $y = 3x - 2$
52. $y = \dfrac{2}{x - 1}$
53. $y = x^3$
54. $y = x^2$

In Exercises 55–62, use a graphing utility to graph the curve represented by the parametric equations. Indicate the direction of the curve. Identify any points at which the curve is not smooth.

55. Cycloid: $x = 2(\theta - \sin\theta)$, $y = 2(1 - \cos\theta)$
56. Cycloid: $x = \theta + \sin\theta$, $y = 1 - \cos\theta$
57. Prolate cycloid: $x = \theta - \tfrac{3}{2}\sin\theta$, $y = 1 - \tfrac{3}{2}\cos\theta$
58. Prolate cycloid: $x = 2\theta - 4\sin\theta$, $y = 2 - 4\cos\theta$
59. Hypocycloid: $x = 3\cos^3\theta$, $y = 3\sin^3\theta$
60. Curtate cycloid: $x = 2\theta - \sin\theta$, $y = 2 - \cos\theta$
61. Witch of Agnesi: $x = 2\cot\theta$, $y = 2\sin^2\theta$
62. Folium of Descartes: $x = \dfrac{3t}{1 + t^3}$, $y = \dfrac{3t^2}{1 + t^3}$

Writing About Concepts

63. State the definition of a plane curve given by parametric equations.
64. Explain the process of sketching a plane curve given by parametric equations. What is meant by the orientation of the curve?
65. State the definition of a smooth curve.

Writing About Concepts (continued)

66. Match each set of parametric equations with the correct graph. [The graphs are labeled (a), (b), (c), (d), (e), and (f).] Explain your reasoning.

    (a)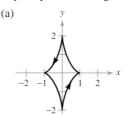
    (b)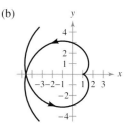
    (c)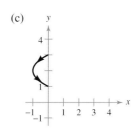
    (d)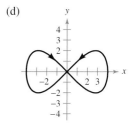
    (e)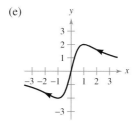
    (f)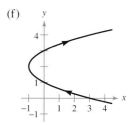

    (i) $x = t^2 - 1$, $y = t + 2$
    (ii) $x = \sin^2\theta - 1$, $y = \sin\theta + 2$
    (iii) Lissajous curve: $x = 4\cos\theta$, $y = 2\sin 2\theta$
    (iv) Evolute of ellipse: $x = \cos^3\theta$, $y = 2\sin^3\theta$
    (v) Involute of circle: $x = \cos\theta + \theta\sin\theta$,
        $y = \sin\theta - \theta\cos\theta$
    (vi) Serpentine curve: $x = \cot\theta$, $y = 4\sin\theta\cos\theta$

67. **Curtate Cycloid** A wheel of radius a rolls along a line without slipping. The curve traced by a point P that is b units from the center ($b < a$) is called a **curtate cycloid** (see figure). Use the angle θ to find a set of parametric equations for this curve.

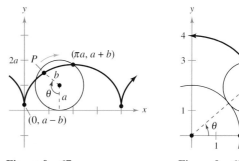

Figure for 67   Figure for 68

68. Epicycloid A circle of radius 1 rolls around the outside of a circle of radius 2 without slipping. The curve traced by a point on the circumference of the smaller circle is called an epicycloid (see figure on previous page). Use the angle θ to find a set of parametric equations for this curve.

True or False? **In Exercises 69 and 70, determine whether the statement is true or false. If it is false, explain why or give an example that shows it is false.**

69. The graph of the parametric equations $x = t^2$ and $y = t^2$ is the line $y = x$.

70. If y is a function of t and x is a function of t, then y is a function of x.

Projectile Motion **In Exercises 71 and 72, consider a projectile launched at a height h feet above the ground and at an angle θ with the horizontal. If the initial velocity is v_0 feet per second, the path of the projectile is modeled by the parametric equations $x = (v_0 \cos \theta)t$ and $y = h + (v_0 \sin \theta)t - 16t^2$.**

71. The center field fence in a ballpark is 10 feet high and 400 feet from home plate. The ball is hit 3 feet above the ground. It leaves the bat at an angle of θ degrees with the horizontal at a speed of 100 miles per hour (see figure).

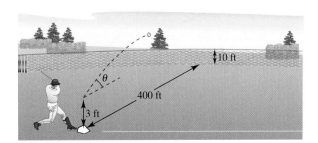

(a) Write a set of parametric equations for the path of the ball.

(b) Use a graphing utility to graph the path of the ball when $\theta = 15°$. Is the hit a home run?

(c) Use a graphing utility to graph the path of the ball when $\theta = 23°$. Is the hit a home run?

(d) Find the minimum angle at which the ball must leave the bat in order for the hit to be a home run.

72. A rectangular equation for the path of a projectile is $y = 5 + x - 0.005x^2$.

(a) Eliminate the parameter t from the position function for the motion of a projectile to show that the rectangular equation is
$$y = -\frac{16 \sec^2 \theta}{v_0^2} x^2 + (\tan \theta) x + h.$$

(b) Use the result of part (a) to find h, v_0, and θ. Find the parametric equations of the path.

(c) Use a graphing utility to graph the rectangular equation for the path of the projectile. Confirm your answer in part (b) by sketching the curve represented by the parametric equations.

(d) Use a graphing utility to approximate the maximum height of the projectile and its range.

Section Project: Cycloids

In Greek, the word *cycloid* means *wheel*, the word *hypocycloid* means *under the wheel*, and the word *epicycloid* means *upon the wheel*. Match the hypocycloid or epicycloid with its graph. [The graphs are labeled (a), (b), (c), (d), (e), and (f).]

Hypocycloid, $H(A, B)$

Path traced by a fixed point on a circle of radius B as it rolls around the inside of a circle of radius A

$$x = (A - B) \cos t + B \cos\left(\frac{A - B}{B}\right)t$$

$$y = (A - B) \sin t - B \sin\left(\frac{A - B}{B}\right)t$$

Epicycloid, $E(A, B)$

Path traced by a fixed point on a circle of radius B as it rolls around the *outside* of a circle of radius A

$$x = (A + B) \cos t - B \cos\left(\frac{A + B}{B}\right)t$$

$$y = (A + B) \sin t - B \sin\left(\frac{A + B}{B}\right)t$$

  I. $H(8, 3)$ II. $E(8, 3)$
III. $H(8, 7)$ IV. $E(24, 3)$
 V. $H(24, 7)$ VI. $E(24, 7)$

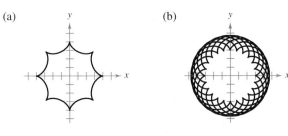

(a)   (b)

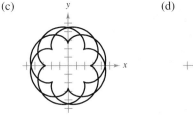

(c)   (d)

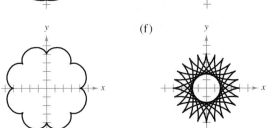

(e)   (f)

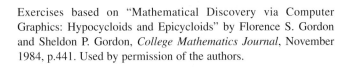

Exercises based on "Mathematical Discovery via Computer Graphics: Hypocycloids and Epicycloids" by Florence S. Gordon and Sheldon P. Gordon, *College Mathematics Journal*, November 1984, p.441. Used by permission of the authors.

Section 10.3
Parametric Equations and Calculus

- Find the slope of a tangent line to a curve given by a set of parametric equations.
- Find the arc length of a curve given by a set of parametric equations.
- Find the area of a surface of revolution (parametric form).

Slope and Tangent Lines

Now that you can represent a graph in the plane by a set of parametric equations, it is natural to ask how to use calculus to study plane curves. To begin, let's take another look at the projectile represented by the parametric equations

$$x = 24\sqrt{2}\,t \quad \text{and} \quad y = -16t^2 + 24\sqrt{2}\,t$$

as shown in Figure 10.29. From Section 10.2, you know that these equations enable you to locate the position of the projectile at a given time. You also know that the object is initially projected at an angle of $45°$. But how can you find the angle θ representing the object's direction at some other time t? The following theorem answers this question by giving a formula for the slope of the tangent line as a function of t.

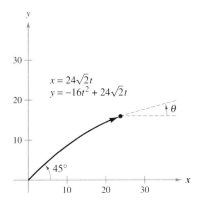

At time t, the angle of elevation of the projectile is θ, the slope of the tangent line at that point.
Figure 10.29

THEOREM 10.7 Parametric Form of the Derivative

If a smooth curve C is given by the equations $x = f(t)$ and $y = g(t)$, then the slope of C at (x, y) is

$$\frac{dy}{dx} = \frac{dy/dt}{dx/dt}, \quad \frac{dx}{dt} \neq 0.$$

Proof In Figure 10.30, consider $\Delta t > 0$ and let

$$\Delta y = g(t + \Delta t) - g(t) \quad \text{and} \quad \Delta x = f(t + \Delta t) - f(t).$$

Because $\Delta x \to 0$ as $\Delta t \to 0$, you can write

$$\frac{dy}{dx} = \lim_{\Delta x \to 0} \frac{\Delta y}{\Delta x}$$

$$= \lim_{\Delta t \to 0} \frac{g(t + \Delta t) - g(t)}{f(t + \Delta t) - f(t)}.$$

Dividing both the numerator and denominator by Δt, you can use the differentiability of f and g to conclude that

$$\frac{dy}{dx} = \lim_{\Delta t \to 0} \frac{[g(t + \Delta t) - g(t)]/\Delta t}{[f(t + \Delta t) - f(t)]/\Delta t}$$

$$= \frac{\displaystyle\lim_{\Delta t \to 0} \frac{g(t + \Delta t) - g(t)}{\Delta t}}{\displaystyle\lim_{\Delta t \to 0} \frac{f(t + \Delta t) - f(t)}{\Delta t}}$$

$$= \frac{g'(t)}{f'(t)}$$

$$= \frac{dy/dt}{dx/dt}.$$

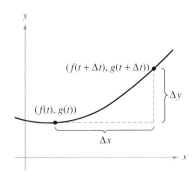

The slope of the secant line through the points $(f(t), g(t))$ and $(f(t + \Delta t), g(t + \Delta t))$ is $\Delta y/\Delta x$.
Figure 10.30

STUDY TIP The curve traced out in Example 1 is a circle. Use the formula

$$\frac{dy}{dx} = -\tan t$$

to find the slopes at the points $(1, 0)$ and $(0, 1)$.

EXAMPLE 1 Differentiation and Parametric Form

Find dy/dx for the curve given by $x = \sin t$ and $y = \cos t$.

Solution

$$\frac{dy}{dx} = \frac{dy/dt}{dx/dt} = \frac{-\sin t}{\cos t} = -\tan t$$

Because dy/dx is a function of t, you can use Theorem 10.7 repeatedly to find *higher-order* derivatives. For instance,

$$\frac{d^2y}{dx^2} = \frac{d}{dx}\left[\frac{dy}{dx}\right] = \frac{\frac{d}{dt}\left[\frac{dy}{dx}\right]}{dx/dt} \qquad \text{Second derivative}$$

$$\frac{d^3y}{dx^3} = \frac{d}{dx}\left[\frac{d^2y}{dx^2}\right] = \frac{\frac{d}{dt}\left[\frac{d^2y}{dx^2}\right]}{dx/dt}. \qquad \text{Third derivative}$$

EXAMPLE 2 Finding Slope and Concavity

For the curve given by

$$x = \sqrt{t} \quad \text{and} \quad y = \frac{1}{4}(t^2 - 4), \quad t \geq 0$$

find the slope and concavity at the point $(2, 3)$.

Solution Because

$$\frac{dy}{dx} = \frac{dy/dt}{dx/dt} = \frac{(1/2)t}{(1/2)t^{-1/2}} = t^{3/2} \qquad \text{Parametric form of first derivative}$$

you can find the second derivative to be

$$\frac{d^2y}{dx^2} = \frac{\frac{d}{dt}[dy/dx]}{dx/dt} = \frac{\frac{d}{dt}[t^{3/2}]}{dx/dt} = \frac{(3/2)t^{1/2}}{(1/2)t^{-1/2}} = 3t. \qquad \text{Parametric form of second derivative}$$

At $(x, y) = (2, 3)$, it follows that $t = 4$, and the slope is

$$\frac{dy}{dx} = (4)^{3/2} = 8.$$

Moreover, when $t = 4$, the second derivative is

$$\frac{d^2y}{dx^2} = 3(4) = 12 > 0$$

and you can conclude that the graph is concave upward at $(2, 3)$, as shown in Figure 10.31.

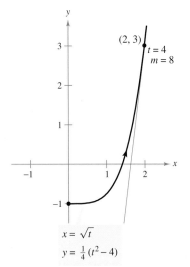

The graph is concave upward at $(2, 3)$, when $t = 4$.
Figure 10.31

Because the parametric equations $x = f(t)$ and $y = g(t)$ need not define y as a function of x, it is possible for a plane curve to loop around and cross itself. At such points the curve may have more than one tangent line, as shown in the next example.

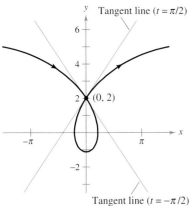

$x = 2t - \pi \sin t$
$y = 2 - \pi \cos t$

This prolate cycloid has two tangent lines at the point $(0, 2)$.
Figure 10.32

EXAMPLE 3 A Curve with Two Tangent Lines at a Point

The **prolate cycloid** given by

$$x = 2t - \pi \sin t \quad \text{and} \quad y = 2 - \pi \cos t$$

crosses itself at the point $(0, 2)$, as shown in Figure 10.32. Find the equations of both tangent lines at this point.

Solution Because $x = 0$ and $y = 2$ when $t = \pm \pi/2$, and

$$\frac{dy}{dx} = \frac{dy/dt}{dx/dt} = \frac{\pi \sin t}{2 - \pi \cos t}$$

you have $dy/dx = -\pi/2$ when $t = -\pi/2$ and $dy/dx = \pi/2$ when $t = \pi/2$. So, the two tangent lines at $(0, 2)$ are

$$y - 2 = -\left(\frac{\pi}{2}\right)x \quad \text{Tangent line when } t = -\frac{\pi}{2}$$

$$y - 2 = \left(\frac{\pi}{2}\right)x. \quad \text{Tangent line when } t = \frac{\pi}{2}$$

If $dy/dt = 0$ and $dx/dt \neq 0$ when $t = t_0$, the curve represented by $x = f(t)$ and $y = g(t)$ has a horizontal tangent at $(f(t_0), g(t_0))$. For instance, in Example 3, the given curve has a horizontal tangent at the point $(0, 2 - \pi)$ (when $t = 0$). Similarly, if $dx/dt = 0$ and $dy/dt \neq 0$ when $t = t_0$, the curve represented by $x = f(t)$ and $y = g(t)$ has a vertical tangent at $(f(t_0), g(t_0))$.

Arc Length

You have seen how parametric equations can be used to describe the path of a particle moving in the plane. You will now develop a formula for determining the *distance* traveled by the particle along its path.

Recall from Section 7.4 that the formula for the arc length of a curve C given by $y = h(x)$ over the interval $[x_0, x_1]$ is

$$s = \int_{x_0}^{x_1} \sqrt{1 + [h'(x)]^2}\, dx$$

$$= \int_{x_0}^{x_1} \sqrt{1 + \left(\frac{dy}{dx}\right)^2}\, dx.$$

If C is represented by the parametric equations $x = f(t)$ and $y = g(t)$, $a \leq t \leq b$, and if $dx/dt = f'(t) > 0$, you can write

$$s = \int_{x_0}^{x_1} \sqrt{1 + \left(\frac{dy}{dx}\right)^2}\, dx = \int_{x_0}^{x_1} \sqrt{1 + \left(\frac{dy/dt}{dx/dt}\right)^2}\, dx$$

$$= \int_a^b \sqrt{\frac{(dx/dt)^2 + (dy/dt)^2}{(dx/dt)^2}}\, \frac{dx}{dt}\, dt$$

$$= \int_a^b \sqrt{\left(\frac{dx}{dt}\right)^2 + \left(\frac{dy}{dt}\right)^2}\, dt$$

$$= \int_a^b \sqrt{[f'(t)]^2 + [g'(t)]^2}\, dt.$$

NOTE When applying the arc length formula to a curve, be sure that the curve is traced out only once on the interval of integration. For instance, the circle given by $x = \cos t$ and $y = \sin t$ is traced out once on the interval $0 \leq t \leq 2\pi$, but is traced out twice on the interval $0 \leq t \leq 4\pi$.

THEOREM 10.8 Arc Length in Parametric Form

If a smooth curve C is given by $x = f(t)$ and $y = g(t)$ such that C does not intersect itself on the interval $a \leq t \leq b$ (except possibly at the endpoints), then the arc length of C over the interval is given by

$$s = \int_a^b \sqrt{\left(\frac{dx}{dt}\right)^2 + \left(\frac{dy}{dt}\right)^2}\, dt = \int_a^b \sqrt{[f'(t)]^2 + [g'(t)]^2}\, dt.$$

In the preceding section you saw that if a circle rolls along a line, a point on its circumference will trace a path called a cycloid. If the circle rolls around the circumference of another circle, the path of the point is an **epicycloid.** The next example shows how to find the arc length of an epicycloid.

ARCH OF A CYCLOID

The arc length of an arch of a cycloid was first calculated in 1658 by British architect and mathematician Christopher Wren, famous for rebuilding many buildings and churches in London, including St. Paul's Cathedral.

EXAMPLE 4 Finding Arc Length

A circle of radius 1 rolls around the circumference of a larger circle of radius 4, as shown in Figure 10.33. The epicycloid traced by a point on the circumference of the smaller circle is given by

$$x = 5 \cos t - \cos 5t$$

and

$$y = 5 \sin t - \sin 5t.$$

Find the distance traveled by the point in one complete trip about the larger circle.

Solution Before applying Theorem 10.8, note in Figure 10.33 that the curve has sharp points when $t = 0$ and $t = \pi/2$. Between these two points, dx/dt and dy/dt are not simultaneously 0. So, the portion of the curve generated from $t = 0$ to $t = \pi/2$ is smooth. To find the total distance traveled by the point, you can find the arc length of that portion lying in the first quadrant and multiply by 4.

$$\begin{aligned}
s &= 4\int_0^{\pi/2} \sqrt{\left(\frac{dx}{dt}\right)^2 + \left(\frac{dy}{dt}\right)^2}\, dt && \text{Parametric form for arc length} \\
&= 4\int_0^{\pi/2} \sqrt{(-5\sin t + 5\sin 5t)^2 + (5\cos t - 5\cos 5t)^2}\, dt \\
&= 20\int_0^{\pi/2} \sqrt{2 - 2\sin t \sin 5t - 2\cos t \cos 5t}\, dt \\
&= 20\int_0^{\pi/2} \sqrt{2 - 2\cos 4t}\, dt \\
&= 20\int_0^{\pi/2} \sqrt{4\sin^2 2t}\, dt && \text{Trigonometric identity} \\
&= 40\int_0^{\pi/2} \sin 2t\, dt \\
&= -20\Big[\cos 2t\Big]_0^{\pi/2} \\
&= 40
\end{aligned}$$

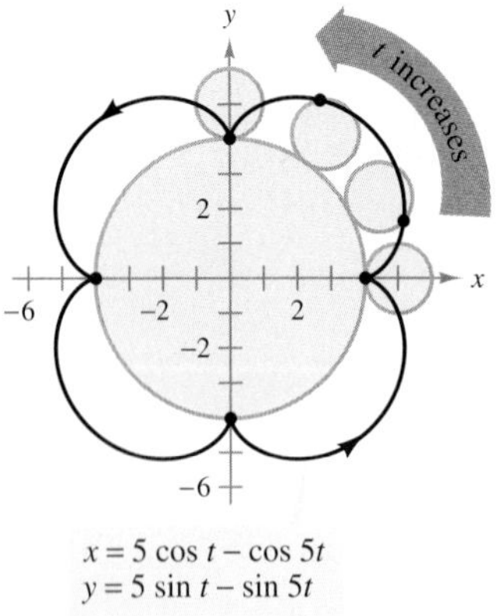

$x = 5\cos t - \cos 5t$
$y = 5\sin t - \sin 5t$

An epicycloid is traced by a point on the smaller circle as it rolls around the larger circle.
Figure 10.33

For the epicycloid shown in Figure 10.33, an arc length of 40 seems about right because the circumference of a circle of radius 6 is $2\pi r = 12\pi \approx 37.7$.

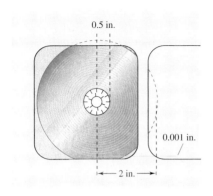

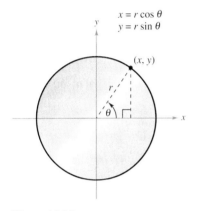

Figure 10.34

NOTE The graph of $r = a\theta$ is called the **spiral of Archimedes**. The graph of $r = \theta/2000\pi$ (in Example 5) is of this form.

EXAMPLE 5 **Length of a Recording Tape**

A recording tape 0.001 inch thick is wound around a reel whose inner radius is 0.5 inch and whose outer radius is 2 inches, as shown in Figure 10.34. How much tape is required to fill the reel?

Solution To create a model for this problem, assume that as the tape is wound around the reel its distance r from the center increases linearly at a rate of 0.001 inch per revolution, or

$$r = (0.001)\frac{\theta}{2\pi} = \frac{\theta}{2000\pi}, \qquad 1000\pi \le \theta \le 4000\pi$$

where θ is measured in radians. You can determine the coordinates of the point (x, y) corresponding to a given radius to be

$$x = r \cos \theta$$

and

$$y = r \sin \theta.$$

Substituting for r, you obtain the parametric equations

$$x = \left(\frac{\theta}{2000\pi}\right)\cos\theta \quad \text{and} \quad y = \left(\frac{\theta}{2000\pi}\right)\sin\theta.$$

You can use the arc length formula to determine the total length of the tape to be

$$\begin{aligned}
s &= \int_{1000\pi}^{4000\pi} \sqrt{\left(\frac{dx}{d\theta}\right)^2 + \left(\frac{dy}{d\theta}\right)^2}\, d\theta \\
&= \frac{1}{2000\pi}\int_{1000\pi}^{4000\pi} \sqrt{(-\theta\sin\theta + \cos\theta)^2 + (\theta\cos\theta + \sin\theta)^2}\, d\theta \\
&= \frac{1}{2000\pi}\int_{1000\pi}^{4000\pi} \sqrt{\theta^2 + 1}\, d\theta \\
&= \frac{1}{2000\pi}\left(\frac{1}{2}\right)\left[\theta\sqrt{\theta^2+1} + \ln\left|\theta + \sqrt{\theta^2+1}\right|\right]_{1000\pi}^{4000\pi} \quad \text{Integration tables (Appendix B), Formula 26} \\
&\approx 11{,}781 \text{ inches} \\
&\approx 982 \text{ feet}
\end{aligned}$$

FOR FURTHER INFORMATION For more information on the mathematics of recording tape, see "Tape Counters" by Richard L. Roth in *The American Mathematical Monthly*. To view this article, go to the website *www.matharticles.com*.

The length of the tape in Example 5 can be approximated by adding the circumferences of circular pieces of tape. The smallest circle has a radius of 0.501 and the largest has a radius of 2.

$$\begin{aligned}
s &\approx 2\pi(0.501) + 2\pi(0.502) + 2\pi(0.503) + \cdots + 2\pi(2.000) \\
&= \sum_{i=1}^{1500} 2\pi(0.5 + 0.001i) \\
&= 2\pi[1500(0.5) + 0.001(1500)(1501)/2] \\
&\approx 11{,}786 \text{ inches}
\end{aligned}$$

Area of a Surface of Revolution

You can use the formula for the area of a surface of revolution in rectangular form to develop a formula for surface area in parametric form.

> **THEOREM 10.9 Area of a Surface of Revolution**
>
> If a smooth curve C given by $x = f(t)$ and $y = g(t)$ does not cross itself on an interval $a \leq t \leq b$, then the area S of the surface of revolution formed by revolving C about the coordinate axes is given by the following.
>
> 1. $S = 2\pi \int_a^b g(t) \sqrt{\left(\dfrac{dx}{dt}\right)^2 + \left(\dfrac{dy}{dt}\right)^2}\, dt$ Revolution about the x-axis: $g(t) \geq 0$
>
> 2. $S = 2\pi \int_a^b f(t) \sqrt{\left(\dfrac{dx}{dt}\right)^2 + \left(\dfrac{dy}{dt}\right)^2}\, dt$ Revolution about the y-axis: $f(t) \geq 0$

These formulas are easy to remember if you think of the differential of arc length as

$$ds = \sqrt{\left(\dfrac{dx}{dt}\right)^2 + \left(\dfrac{dy}{dt}\right)^2}\, dt.$$

Then the formulas are written as follows.

1. $S = 2\pi \int_a^b g(t)\, ds$ 2. $S = 2\pi \int_a^b f(t)\, ds$

EXAMPLE 6 Finding the Area of a Surface of Revolution

Let C be the arc of the circle

$$x^2 + y^2 = 9$$

from $(3, 0)$ to $(3/2, 3\sqrt{3}/2)$, as shown in Figure 10.35. Find the area of the surface formed by revolving C about the x-axis.

Solution You can represent C parametrically by the equations

$$x = 3\cos t \quad \text{and} \quad y = 3\sin t, \quad 0 \leq t \leq \pi/3.$$

(Note that you can determine the interval for t by observing that $t = 0$ when $x = 3$ and $t = \pi/3$ when $x = 3/2$.) On this interval, C is smooth and y is nonnegative, and you can apply Theorem 10.9 to obtain a surface area of

$$S = 2\pi \int_0^{\pi/3} (3\sin t)\sqrt{(-3\sin t)^2 + (3\cos t)^2}\, dt \quad \text{Formula for area of a surface of revolution}$$

$$= 6\pi \int_0^{\pi/3} \sin t \sqrt{9(\sin^2 t + \cos^2 t)}\, dt$$

$$= 6\pi \int_0^{\pi/3} 3\sin t\, dt \quad \text{Trigonometric identity}$$

$$= -18\pi \Big[\cos t\Big]_0^{\pi/3}$$

$$= -18\pi \left(\dfrac{1}{2} - 1\right)$$

$$= 9\pi.$$

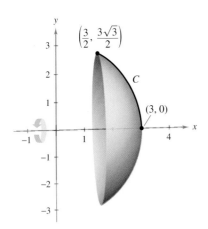

This surface of revolution has a surface area of 9π.
Figure 10.35

Exercises for Section 10.3

In Exercises 1–4, find dy/dx.

1. $x = t^2$, $y = 5 - 4t$
2. $x = \sqrt[3]{t}$, $y = 4 - t$
3. $x = \sin^2 \theta$, $y = \cos^2 \theta$
4. $x = 2e^\theta$, $y = e^{-\theta/2}$

In Exercises 5–14, find dy/dx and d^2y/dx^2, and find the slope and concavity (if possible) at the given value of the parameter.

Parametric Equations	Point
5. $x = 2t$, $y = 3t - 1$	$t = 3$
6. $x = \sqrt{t}$, $y = 3t - 1$	$t = 1$
7. $x = t + 1$, $y = t^2 + 3t$	$t = -1$
8. $x = t^2 + 3t + 2$, $y = 2t$	$t = 0$
9. $x = 2\cos\theta$, $y = 2\sin\theta$	$\theta = \dfrac{\pi}{4}$
10. $x = \cos\theta$, $y = 3\sin\theta$	$\theta = 0$
11. $x = 2 + \sec\theta$, $y = 1 + 2\tan\theta$	$\theta = \dfrac{\pi}{6}$
12. $x = \sqrt{t}$, $y = \sqrt{t-1}$	$t = 2$
13. $x = \cos^3\theta$, $y = \sin^3\theta$	$\theta = \dfrac{\pi}{4}$
14. $x = \theta - \sin\theta$, $y = 1 - \cos\theta$	$\theta = \pi$

In Exercises 15 and 16, find an equation of the tangent line at each given point on the curve.

15. $x = 2\cot\theta$
    $y = 2\sin^2\theta$

16. $x = 2 - 3\cos\theta$
    $y = 3 + 2\sin\theta$

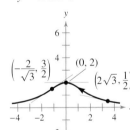
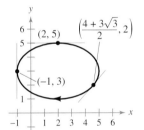

In Exercises 17–20, (a) use a graphing utility to graph the curve represented by the parametric equations, (b) use a graphing utility to find dx/dt, dy/dt, and dy/dx at the given value of the parameter, (c) find an equation of the tangent line to the curve at the given value of the parameter, and (d) use a graphing utility to graph the curve and the tangent line from part (c).

Parametric Equations	Parameter
17. $x = 2t$, $y = t^2 - 1$	$t = 2$
18. $x = t - 1$, $y = \dfrac{1}{t} + 1$	$t = 1$
19. $x = t^2 - t + 2$, $y = t^3 - 3t$	$t = -1$
20. $x = 4\cos\theta$, $y = 3\sin\theta$	$\theta = \dfrac{3\pi}{4}$

In Exercises 21–24, find the equations of the tangent lines at the point where the curve crosses itself.

21. $x = 2\sin 2t$, $y = 3\sin t$
22. $x = 2 - \pi\cos t$, $y = 2t - \pi\sin t$
23. $x = t^2 - t$, $y = t^3 - 3t - 1$
24. $x = t^3 - 6t$, $y = t^2$

In Exercises 25 and 26, find all points (if any) of horizontal and vertical tangency to the portion of the curve shown.

25. Involute of a circle:
    $x = \cos\theta + \theta\sin\theta$
    $y = \sin\theta - \theta\cos\theta$

26. $x = 2\theta$
    $y = 2(1 - \cos\theta)$

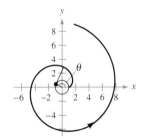
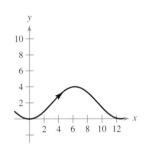

In Exercises 27–36, find all points (if any) of horizontal and vertical tangency to the curve. Use a graphing utility to confirm your results.

27. $x = 1 - t$, $y = t^2$
28. $x = t + 1$, $y = t^2 + 3t$
29. $x = 1 - t$, $y = t^3 - 3t$
30. $x = t^2 - t + 2$, $y = t^3 - 3t$
31. $x = 3\cos\theta$, $y = 3\sin\theta$
32. $x = \cos\theta$, $y = 2\sin 2\theta$
33. $x = 4 + 2\cos\theta$, $y = -1 + \sin\theta$
34. $x = 4\cos^2\theta$, $y = 2\sin\theta$
35. $x = \sec\theta$, $y = \tan\theta$
36. $x = \cos^2\theta$, $y = \cos\theta$

In Exercises 37–42, determine the t intervals on which the curve is concave downward or concave upward.

37. $x = t^2$, $y = t^3 - t$
38. $x = 2 + t^2$, $y = t^2 + t^3$
39. $x = 2t + \ln t$, $y = 2t - \ln t$
40. $x = t^2$, $y = \ln t$
41. $x = \sin t$, $y = \cos t$, $0 < t < \pi$
42. $x = 2\cos t$, $y = \sin t$, $0 < t < 2\pi$

Arc Length In Exercises 43–46, write an integral that represents the arc length of the curve on the given interval. Do not evaluate the integral.

Parametric Equations	Interval
43. $x = 2t - t^2$, $y = 2t^{3/2}$	$1 \leq t \leq 2$
44. $x = \ln t$, $y = t + 1$	$1 \leq t \leq 6$
45. $x = e^t + 2$, $y = 2t + 1$	$-2 \leq t \leq 2$
46. $x = t + \sin t$, $y = t - \cos t$	$0 \leq t \leq \pi$

Arc Length In Exercises 47–52, find the arc length of the curve on the given interval.

Parametric Equations	Interval
47. $x = t^2$, $y = 2t$	$0 \leq t \leq 2$
48. $x = t^2 + 1$, $y = 4t^3 + 3$	$-1 \leq t \leq 0$
49. $x = e^{-t} \cos t$, $y = e^{-t} \sin t$	$0 \leq t \leq \frac{\pi}{2}$
50. $x = \arcsin t$, $y = \ln\sqrt{1 - t^2}$	$0 \leq t \leq \frac{1}{2}$
51. $x = \sqrt{t}$, $y = 3t - 1$	$0 \leq t \leq 1$
52. $x = t$, $y = \frac{t^5}{10} + \frac{1}{6t^3}$	$1 \leq t \leq 2$

Arc Length In Exercises 53–56, find the arc length of the curve on the interval $[0, 2\pi]$.

53. Hypocycloid perimeter: $x = a \cos^3 \theta$, $y = a \sin^3 \theta$
54. Circle circumference: $x = a \cos \theta$, $y = a \sin \theta$
55. Cycloid arch: $x = a(\theta - \sin \theta)$, $y = a(1 - \cos \theta)$
56. Involute of a circle: $x = \cos \theta + \theta \sin \theta$, $y = \sin \theta - \theta \cos \theta$

57. Path of a Projectile The path of a projectile is modeled by the parametric equations

$$x = (90 \cos 30°)t \quad \text{and} \quad y = (90 \sin 30°)t - 16t^2$$

where x and y are measured in feet.

(a) Use a graphing utility to graph the path of the projectile.

(b) Use a graphing utility to approximate the range of the projectile.

(c) Use the integration capabilities of a graphing utility to approximate the arc length of the path. Compare this result with the range of the projectile.

58. Path of a Projectile If the projectile in Exercise 57 is launched at an angle θ with the horizontal, its parametric equations are

$$x = (90 \cos \theta)t \quad \text{and} \quad y = (90 \sin \theta)t - 16t^2.$$

Use a graphing utility to find the angle that maximizes the range of the projectile. What angle maximizes the arc length of the trajectory?

59. Folium of Descartes Consider the parametric equations

$$x = \frac{4t}{1 + t^3} \quad \text{and} \quad y = \frac{4t^2}{1 + t^3}.$$

(a) Use a graphing utility to graph the curve represented by the parametric equations.

(b) Use a graphing utility to find the points of horizontal tangency to the curve.

(c) Use the integration capabilities of a graphing utility to approximate the arc length of the closed loop. (*Hint:* Use symmetry and integrate over the interval $0 \leq t \leq 1$.)

60. Witch of Agnesi Consider the parametric equations

$$x = 4 \cot \theta \quad \text{and} \quad y = 4 \sin^2 \theta, \quad -\frac{\pi}{2} \leq \theta \leq \frac{\pi}{2}.$$

(a) Use a graphing utility to graph the curve represented by the parametric equations.

(b) Use a graphing utility to find the points of horizontal tangency to the curve.

(c) Use the integration capabilities of a graphing utility to approximate the arc length over the interval $\pi/4 \leq \theta \leq \pi/2$.

61. Writing

(a) Use a graphing utility to graph each set of parametric equations.

$x = t - \sin t$	$x = 2t - \sin(2t)$
$y = 1 - \cos t$	$y = 1 - \cos(2t)$
$0 \leq t \leq 2\pi$	$0 \leq t \leq \pi$

(b) Compare the graphs of the two sets of parametric equations in part (a). If the curve represents the motion of a particle and t is time, what can you infer about the average speeds of the particle on the paths represented by the two sets of parametric equations?

(c) Without graphing the curve, determine the time required for a particle to traverse the same path as in parts (a) and (b) if the path is modeled by

$$x = \tfrac{1}{2}t - \sin(\tfrac{1}{2}t) \quad \text{and} \quad y = 1 - \cos(\tfrac{1}{2}t).$$

62. Writing

(a) Each set of parametric equations represents the motion of a particle. Use a graphing utility to graph each set.

First Particle	Second Particle
$x = 3 \cos t$	$x = 4 \sin t$
$y = 4 \sin t$	$y = 3 \cos t$
$0 \leq t \leq 2\pi$	$0 \leq t \leq 2\pi$

(b) Determine the number of points of intersection.

(c) Will the particles ever be at the same place at the same time? If so, identify the points.

(d) Explain what happens if the motion of the second particle is represented by

$$x = 2 + 3 \sin t, \quad y = 2 - 4 \cos t, \quad 0 \leq t \leq 2\pi.$$

Surface Area In Exercises 63–66, write an integral that represents the area of the surface generated by revolving the curve about the *x*-axis. Use a graphing utility to approximate the integral.

Parametric Equations	Interval
63. $x = 4t$, $y = t + 1$	$0 \leq t \leq 2$
64. $x = \frac{1}{4}t^2$, $y = t + 2$	$0 \leq t \leq 4$
65. $x = \cos^2 \theta$, $y = \cos \theta$	$0 \leq \theta \leq \frac{\pi}{2}$
66. $x = \theta + \sin \theta$, $y = \theta + \cos \theta$	$0 \leq \theta \leq \frac{\pi}{2}$

Surface Area In Exercises 67–72, find the area of the surface generated by revolving the curve about each given axis.

67. $x = t$, $y = 2t$, $0 \leq t \leq 4$,   (a) *x*-axis   (b) *y*-axis
68. $x = t$, $y = 4 - 2t$, $0 \leq t \leq 2$,   (a) *x*-axis   (b) *y*-axis
69. $x = 4 \cos \theta$, $y = 4 \sin \theta$, $0 \leq \theta \leq \frac{\pi}{2}$,   *y*-axis
70. $x = \frac{1}{3}t^3$, $y = t + 1$, $1 \leq t \leq 2$,   *y*-axis
71. $x = a \cos^3 \theta$, $y = a \sin^3 \theta$, $0 \leq \theta \leq \pi$,   *x*-axis
72. $x = a \cos \theta$, $y = b \sin \theta$, $0 \leq \theta \leq 2\pi$,
  (a) *x*-axis   (b) *y*-axis

Writing About Concepts

73. Give the parametric form of the derivative.
74. Mentally determine dy/dx.
  (a) $x = t$, $y = 4$ (b) $x = t$, $y = 4t - 3$
75. Sketch a graph of a curve defined by the parametric equations $x = g(t)$ and $y = f(t)$ such that $dx/dt > 0$ and $dy/dt < 0$ for all real numbers *t*.
76. Sketch a graph of a curve defined by the parametric equations $x = g(t)$ and $y = f(t)$ such that $dx/dt < 0$ and $dy/dt < 0$ for all real numbers *t*.
77. Give the integral formula for arc length in parametric form.
78. Give the integral formulas for the areas of the surfaces of revolution formed when a smooth curve *C* is revolved about (a) the *x*-axis and (b) the *y*-axis.

79. Use integration by substitution to show that if *y* is a continuous function of *x* on the interval $a \leq x \leq b$, where $x = f(t)$ and $y = g(t)$, then
$$\int_a^b y \, dx = \int_{t_1}^{t_2} g(t) f'(t) \, dt$$
where $f(t_1) = a$, $f(t_2) = b$, and both *g* and f' are continuous on $[t_1, t_2]$.

80. *Surface Area* A portion of a sphere of radius *r* is removed by cutting out a circular cone with its vertex at the center of the sphere. The vertex of the cone forms an angle of 2θ. Find the surface area removed from the sphere.

Area In Exercises 81 and 82, find the area of the region. (Use the result of Exercise 79.)

81. $x = 2 \sin^2 \theta$
  $y = 2 \sin^2 \theta \tan \theta$
  $0 \leq \theta < \frac{\pi}{2}$

82. $x = 2 \cot \theta$
  $y = 2 \sin^2 \theta$
  $0 < \theta < \pi$

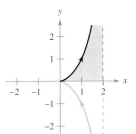
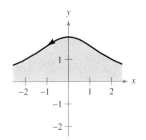

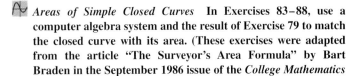 *Areas of Simple Closed Curves* In Exercises 83–88, use a computer algebra system and the result of Exercise 79 to match the closed curve with its area. (These exercises were adapted from the article "The Surveyor's Area Formula" by Bart Braden in the September 1986 issue of the *College Mathematics Journal*, by permission of the author.)

(a) $\frac{8}{3}ab$ (b) $\frac{3}{8}\pi a^2$ (c) $2\pi a^2$
(d) πab (e) $2\pi ab$ (f) $6\pi a^2$

83. Ellipse: $(0 \leq t \leq 2\pi)$
  $x = b \cos t$
  $y = a \sin t$

84. Astroid: $(0 \leq t \leq 2\pi)$
  $x = a \cos^3 t$
  $y = a \sin^3 t$

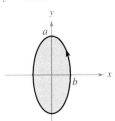
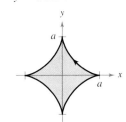

85. Cardioid: $(0 \leq t \leq 2\pi)$
  $x = 2a \cos t - a \cos 2t$
  $y = 2a \sin t - a \sin 2t$

86. Deltoid: $(0 \leq t \leq 2\pi)$
  $x = 2a \cos t + a \cos 2t$
  $y = 2a \sin t - a \sin 2t$

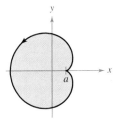
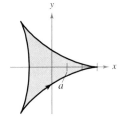

87. Hourglass: $(0 \le t \le 2\pi)$

$x = a \sin 2t$

$y = b \sin t$

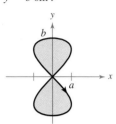

88. Teardrop: $(0 \le t \le 2\pi)$

$x = 2a \cos t - a \sin 2t$

$y = b \sin t$

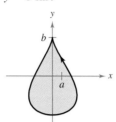

Centroid In Exercises 89 and 90, find the centroid of the region bounded by the graph of the parametric equations and the coordinate axes. (Use the result of Exercise 79.)

89. $x = \sqrt{t},\ y = 4 - t$

90. $x = \sqrt{4 - t},\ y = \sqrt{t}$

Volume In Exercises 91 and 92, find the volume of the solid formed by revolving the region bounded by the graphs of the given equations about the x-axis. (Use the result of Exercise 79.)

91. $x = 3 \cos \theta,\ y = 3 \sin \theta$

92. $x = \cos \theta,\ y = 3 \sin \theta,\ a > 0$

93. Cycloid Use the parametric equations

$$x = a(\theta - \sin \theta) \quad \text{and} \quad y = a(1 - \cos \theta),\ a > 0$$

to answer the following.

(a) Find dy/dx and d^2y/dx^2.

(b) Find the equations of the tangent line at the point where $\theta = \pi/6$.

(c) Find all points (if any) of horizontal tangency.

(d) Determine where the curve is concave upward or concave downward.

(e) Find the length of one arc of the curve.

94. Use the parametric equations

$$x = t^2\sqrt{3} \quad \text{and} \quad y = 3t - \frac{1}{3}t^3$$

to answer the following.

(a) Use a graphing utility to graph the curve on the interval $-3 \le t \le 3$.

(b) Find dy/dx and d^2y/dx^2.

(c) Find the equation of the tangent line at the point $\left(\sqrt{3},\ \frac{8}{3}\right)$.

(d) Find the length of the curve.

(e) Find the surface area generated by revolving the curve about the x-axis.

95. Involute of a Circle The involute of a circle is described by the endpoint P of a string that is held taut as it is unwound from a spool that does not turn (see figure). Show that a parametric representation of the involute is

$$x = r(\cos \theta + \theta \sin \theta) \quad \text{and} \quad y = r(\sin \theta - \theta \cos \theta).$$

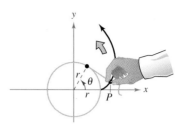

Figure for 95

96. Involute of a Circle The figure shows a piece of string tied to a circle with a radius of one unit. The string is just long enough to reach the opposite side of the circle. Find the area that is covered when the string is unwound counterclockwise.

97. (a) Use a graphing utility to graph the curve given by

$$x = \frac{1 - t^2}{1 + t^2},\quad y = \frac{2t}{1 + t^2},\quad -20 \le t \le 20.$$

(b) Describe the graph and confirm your result analytically.

(c) Discuss the speed at which the curve is traced as t increases from -20 to 20.

98. Tractrix A person moves from the origin along the positive y-axis pulling a weight at the end of a 12-meter rope. Initially, the weight is located at the point $(12, 0)$.

(a) In Exercise 86 of Section 8.7, it was shown that the path of the weight is modeled by the rectangular equation

$$y = -12 \ln\left(\frac{12 - \sqrt{144 - x^2}}{x}\right) - \sqrt{144 - x^2}$$

where $0 < x \le 12$. Use a graphing utility to graph the rectangular equation.

(b) Use a graphing utility to graph the parametric equations

$$x = 12\, \text{sech}\, \frac{t}{12} \quad \text{and} \quad y = t - 12 \tanh \frac{t}{12}$$

where $t \ge 0$. How does this graph compare with the graph in part (a)? Which graph (if either) do you think is a better representation of the path?

(c) Use the parametric equations for the tractrix to verify that the distance from the y-intercept of the tangent line to the point of tangency is independent of the location of the point of tangency.

True or False? In Exercises 99 and 100, determine whether the statement is true or false. If it is false, explain why or give an example that shows it is false.

99. If $x = f(t)$ and $y = g(t)$, then $d^2y/dx^2 = g''(t)/f''(t)$.

100. The curve given by $x = t^3,\ y = t^2$ has a horizontal tangent at the origin because $dy/dt = 0$ when $t = 0$.

Section 10.4 Polar Coordinates and Polar Graphs

- Understand the polar coordinate system.
- Rewrite rectangular coordinates and equations in polar form and vice versa.
- Sketch the graph of an equation given in polar form.
- Find the slope of a tangent line to a polar graph.
- Identify several types of special polar graphs.

Polar Coordinates

So far, you have been representing graphs as collections of points (x, y) on the rectangular coordinate system. The corresponding equations for these graphs have been in either rectangular or parametric form. In this section you will study a coordinate system called the **polar coordinate system.**

To form the polar coordinate system in the plane, fix a point O, called the **pole** (or **origin**), and construct from O an initial ray called the **polar axis,** as shown in Figure 10.36. Then each point P in the plane can be assigned **polar coordinates** (r, θ), as follows.

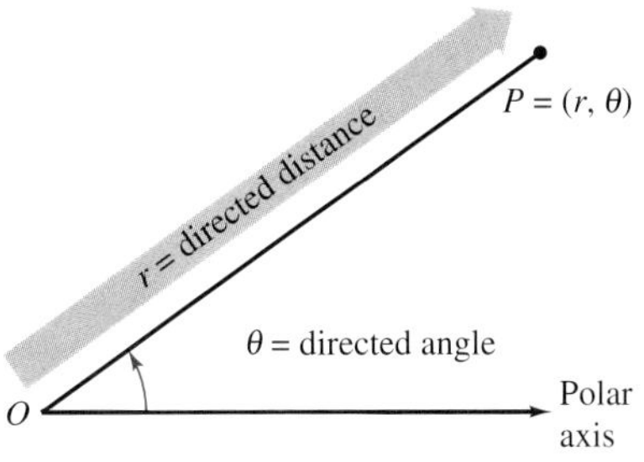

$r = $ *directed distance* from O to P

$\theta = $ *directed angle*, counterclockwise from polar axis to segment $\overline{OP}$

Polar coordinates
Figure 10.36

Figure 10.37 shows three points on the polar coordinate system. Notice that in this system, it is convenient to locate points with respect to a grid of concentric circles intersected by **radial lines** through the pole.

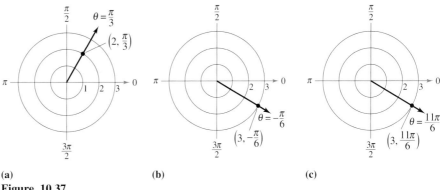

(a)   (b)   (c)
Figure 10.37

With rectangular coordinates, each point (x, y) has a unique representation. This is not true with polar coordinates. For instance, the coordinates (r, θ) and $(r, 2\pi + \theta)$ represent the same point [see parts (b) and (c) in Figure 10.37]. Also, because r is a *directed distance*, the coordinates (r, θ) and $(-r, \theta + \pi)$ represent the same point. In general, the point (r, θ) can be written as

$$(r, \theta) = (r, \theta + 2n\pi)$$

or

$$(r, \theta) = (-r, \theta + (2n + 1)\pi)$$

where n is any integer. Moreover, the pole is represented by $(0, \theta)$, where θ is any angle.

POLAR COORDINATES

The mathematician credited with first using polar coordinates was James Bernoulli, who introduced them in 1691. However, there is some evidence that it may have been Isaac Newton who first used them.

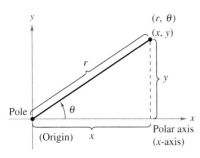

Relating polar and rectangular coordinates
Figure 10.38

Coordinate Conversion

To establish the relationship between polar and rectangular coordinates, let the polar axis coincide with the positive x-axis and the pole with the origin, as shown in Figure 10.38. Because (x, y) lies on a circle of radius r, it follows that $r^2 = x^2 + y^2$. Moreover, for $r > 0$, the definition of the trigonometric functions implies that

$$\tan \theta = \frac{y}{x}, \quad \cos \theta = \frac{x}{r}, \quad \text{and} \quad \sin \theta = \frac{y}{r}.$$

If $r < 0$, you can show that the same relationships hold.

THEOREM 10.10 Coordinate Conversion

The polar coordinates (r, θ) of a point are related to the rectangular coordinates (x, y) of the point as follows.

1. $x = r \cos \theta$
   $y = r \sin \theta$

2. $\tan \theta = \frac{y}{x}$
   $r^2 = x^2 + y^2$

EXAMPLE 1 Polar-to-Rectangular Conversion

a. For the point $(r, \theta) = (2, \pi)$,
$$x = r \cos \theta = 2 \cos \pi = -2 \quad \text{and} \quad y = r \sin \theta = 2 \sin \pi = 0.$$
So, the rectangular coordinates are $(x, y) = (-2, 0)$.

b. For the point $(r, \theta) = (\sqrt{3}, \pi/6)$,
$$x = \sqrt{3} \cos \frac{\pi}{6} = \frac{3}{2} \quad \text{and} \quad y = \sqrt{3} \sin \frac{\pi}{6} = \frac{\sqrt{3}}{2}.$$
So, the rectangular coordinates are $(x, y) = (3/2, \sqrt{3}/2)$. See Figure 10.39.

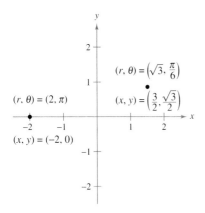

To convert from polar to rectangular coordinates, let $x = r \cos \theta$ and $y = r \sin \theta$.
Figure 10.39

EXAMPLE 2 Rectangular-to-Polar Conversion

a. For the second quadrant point $(x, y) = (-1, 1)$,
$$\tan \theta = \frac{y}{x} = -1 \quad \Longrightarrow \quad \theta = \frac{3\pi}{4}.$$

Because θ was chosen to be in the same quadrant as (x, y), you should use a positive value of r.
$$r = \sqrt{x^2 + y^2}$$
$$= \sqrt{(-1)^2 + (1)^2}$$
$$= \sqrt{2}$$

This implies that *one* set of polar coordinates is $(r, \theta) = (\sqrt{2}, 3\pi/4)$.

b. Because the point $(x, y) = (0, 2)$ lies on the positive y-axis, choose $\theta = \pi/2$ and $r = 2$, and one set of polar coordinates is $(r, \theta) = (2, \pi/2)$.

See Figure 10.40.

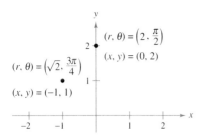

To convert from rectangular to polar coordinates, let $\tan \theta = y/x$ and $r = \sqrt{x^2 + y^2}$.
Figure 10.40

Polar Graphs

One way to sketch the graph of a polar equation is to convert to rectangular coordinates and then sketch the graph of the rectangular equation.

EXAMPLE 3 Graphing Polar Equations

Describe the graph of each polar equation. Confirm each description by converting to a rectangular equation.

a. $r = 2$ **b.** $\theta = \dfrac{\pi}{3}$ **c.** $r = \sec\theta$

Solution

a. The graph of the polar equation $r = 2$ consists of all points that are two units from the pole. In other words, this graph is a circle centered at the origin with a radius of 2. [See Figure 10.41(a).] You can confirm this by using the relationship $r^2 = x^2 + y^2$ to obtain the rectangular equation

$$x^2 + y^2 = 2^2. \quad \text{Rectangular equation}$$

b. The graph of the polar equation $\theta = \pi/3$ consists of all points on the line that makes an angle of $\pi/3$ with the positive x-axis. [See Figure 10.41(b).] You can confirm this by using the relationship $\tan\theta = y/x$ to obtain the rectangular equation

$$y = \sqrt{3}\,x. \quad \text{Rectangular equation}$$

c. The graph of the polar equation $r = \sec\theta$ is not evident by simple inspection, so you can begin by converting to rectangular form using the relationship $r\cos\theta = x$.

$$r = \sec\theta \quad \text{Polar equation}$$
$$r\cos\theta = 1$$
$$x = 1 \quad \text{Rectangular equation}$$

From the rectangular equation, you can see that the graph is a vertical line. [See Figure 10.41(c).]

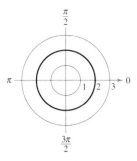

(a) Circle: $r = 2$

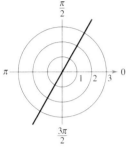

(b) Radial line: $\theta = \dfrac{\pi}{3}$

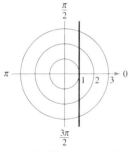

(c) Vertical line: $r = \sec\theta$
Figure 10.41

TECHNOLOGY Sketching the graphs of complicated polar equations *by hand* can be tedious. With technology, however, the task is not difficult. If your graphing utility has a *polar* mode, use it to graph the equations in the exercise set. If your graphing utility doesn't have a *polar* mode, but does have a *parametric* mode, you can graph $r = f(\theta)$ by writing the equation as

$$x = f(\theta)\cos\theta$$
$$y = f(\theta)\sin\theta.$$

For instance, the graph of $r = \tfrac{1}{2}\theta$ shown in Figure 10.42 was produced with a graphing calculator in *parametric* mode. This equation was graphed using the parametric equations

$$x = \dfrac{1}{2}\theta\cos\theta$$
$$y = \dfrac{1}{2}\theta\sin\theta$$

with the values of θ varying from -4π to 4π. This curve is of the form $r = a\theta$ and is called a **spiral of Archimedes**.

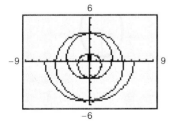

Spiral of Archimedes
Figure 10.42

EXAMPLE 4 Sketching a Polar Graph

Sketch the graph of $r = 2 \cos 3\theta$.

Solution Begin by writing the polar equation in parametric form.

$$x = 2 \cos 3\theta \cos \theta \quad \text{and} \quad y = 2 \cos 3\theta \sin \theta$$

After some experimentation, you will find that the entire curve, which is called a **rose curve,** can be sketched by letting θ vary from 0 to π, as shown in Figure 10.43. If you try duplicating this graph with a graphing utility, you will find that by letting θ vary from 0 to 2π, you will actually trace the entire curve *twice*.

NOTE One way to sketch the graph of $r = 2 \cos 3\theta$ by hand is to make a table of values.

θ	0	$\dfrac{\pi}{6}$	$\dfrac{\pi}{3}$	$\dfrac{\pi}{2}$	$\dfrac{2\pi}{3}$
r	2	0	-2	0	2

By extending the table and plotting the points, you will obtain the curve shown in Example 4.

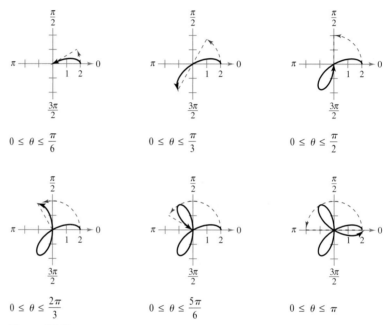

Figure 10.43

Use a graphing utility to experiment with other rose curves (they are of the form $r = a \cos n\theta$ or $r = a \sin n\theta$). For instance, Figure 10.44 shows the graphs of two other rose curves.

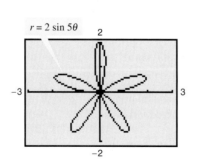
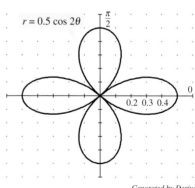

Rose curves
Figure 10.44

Slope and Tangent Lines

To find the slope of a tangent line to a polar graph, consider a differentiable function given by $r = f(\theta)$. To find the slope in polar form, use the parametric equations

$$x = r \cos \theta = f(\theta) \cos \theta \quad \text{and} \quad y = r \sin \theta = f(\theta) \sin \theta.$$

Using the parametric form of dy/dx given in Theorem 10.7, you have

$$\frac{dy}{dx} = \frac{dy/d\theta}{dx/d\theta}$$

$$= \frac{f(\theta) \cos \theta + f'(\theta) \sin \theta}{-f(\theta) \sin \theta + f'(\theta) \cos \theta}$$

which establishes the following theorem.

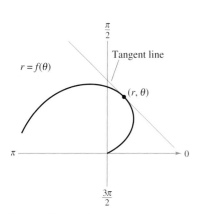

Tangent line to polar curve
Figure 10.45

THEOREM 10.11 Slope in Polar Form

If f is a differentiable function of θ, then the *slope* of the tangent line to the graph of $r = f(\theta)$ at the point (r, θ) is

$$\frac{dy}{dx} = \frac{dy/d\theta}{dx/d\theta} = \frac{f(\theta) \cos \theta + f'(\theta) \sin \theta}{-f(\theta) \sin \theta + f'(\theta) \cos \theta}$$

provided that $dx/d\theta \neq 0$ at (r, θ). (See Figure 10.45.)

From Theorem 10.11, you can make the following observations.

1. Solutions to $\dfrac{dy}{d\theta} = 0$ yield horizontal tangents, provided that $\dfrac{dx}{d\theta} \neq 0$.

2. Solutions to $\dfrac{dx}{d\theta} = 0$ yield vertical tangents, provided that $\dfrac{dy}{d\theta} \neq 0$.

If $dy/d\theta$ and $dx/d\theta$ are *simultaneously* 0, no conclusion can be drawn about tangent lines.

EXAMPLE 5 Finding Horizontal and Vertical Tangent Lines

Find the horizontal and vertical tangent lines of $r = \sin \theta$, $0 \leq \theta \leq \pi$.

Solution Begin by writing the equation in parametric form.

$$x = r \cos \theta = \sin \theta \cos \theta$$

and

$$y = r \sin \theta = \sin \theta \sin \theta = \sin^2 \theta$$

Next, differentiate x and y with respect to θ and set each derivative equal to 0.

$$\frac{dx}{d\theta} = \cos^2 \theta - \sin^2 \theta = \cos 2\theta = 0 \quad \Longrightarrow \quad \theta = \frac{\pi}{4}, \frac{3\pi}{4}$$

$$\frac{dy}{d\theta} = 2 \sin \theta \cos \theta = \sin 2\theta = 0 \quad \Longrightarrow \quad \theta = 0, \frac{\pi}{2}$$

So, the graph has vertical tangent lines at $\left(\sqrt{2}/2, \pi/4\right)$ and $\left(\sqrt{2}/2, 3\pi/4\right)$, and it has horizontal tangent lines at $(0, 0)$ and $(1, \pi/2)$, as shown in Figure 10.46.

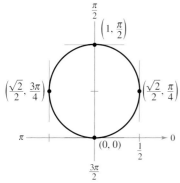

Horizontal and vertical tangent lines of $r = \sin \theta$
Figure 10.46

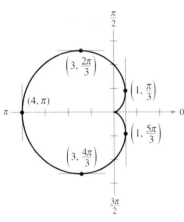

Horizontal and vertical tangent lines of $r = 2(1 - \cos \theta)$
Figure 10.47

EXAMPLE 6 Finding Horizontal and Vertical Tangent Lines

Find the horizontal and vertical tangents to the graph of $r = 2(1 - \cos \theta)$.

Solution Using $y = r \sin \theta$, differentiate and set $dy/d\theta$ equal to 0.

$$y = r \sin \theta = 2(1 - \cos \theta) \sin \theta$$

$$\frac{dy}{d\theta} = 2[(1 - \cos \theta)(\cos \theta) + \sin \theta(\sin \theta)]$$

$$= -2(2 \cos \theta + 1)(\cos \theta - 1) = 0$$

So, $\cos \theta = -\frac{1}{2}$ and $\cos \theta = 1$, and you can conclude that $dy/d\theta = 0$ when $\theta = 2\pi/3, 4\pi/3$, and 0. Similarly, using $x = r \cos \theta$, you have

$$x = r \cos \theta = 2 \cos \theta - 2 \cos^2 \theta$$

$$\frac{dx}{d\theta} = -2 \sin \theta + 4 \cos \theta \sin \theta = 2 \sin \theta(2 \cos \theta - 1) = 0.$$

So, $\sin \theta = 0$ or $\cos \theta = \frac{1}{2}$, and you can conclude that $dx/d\theta = 0$ when $\theta = 0$, π, $\pi/3$, and $5\pi/3$. From these results, and from the graph shown in Figure 10.47, you can conclude that the graph has horizontal tangents at $(3, 2\pi/3)$ and $(3, 4\pi/3)$, and has vertical tangents at $(1, \pi/3)$, $(1, 5\pi/3)$, and $(4, \pi)$. This graph is called a **cardioid**. Note that both derivatives $(dy/d\theta$ and $dx/d\theta)$ are 0 when $\theta = 0$. Using this information alone, you don't know whether the graph has a horizontal or vertical tangent line at the pole. From Figure 10.47, however, you can see that the graph has a cusp at the pole.

Theorem 10.11 has an important consequence. Suppose the graph of $r = f(\theta)$ passes through the pole when $\theta = \alpha$ and $f'(\alpha) \neq 0$. Then the formula for dy/dx simplifies as follows.

$$\frac{dy}{dx} = \frac{f'(\alpha) \sin \alpha + f(\alpha) \cos \alpha}{f'(\alpha) \cos \alpha - f(\alpha) \sin \alpha} = \frac{f'(\alpha) \sin \alpha + 0}{f'(\alpha) \cos \alpha - 0} = \frac{\sin \alpha}{\cos \alpha} = \tan \alpha$$

So, the line $\theta = \alpha$ is tangent to the graph at the pole, $(0, \alpha)$.

THEOREM 10.12 Tangent Lines at the Pole

If $f(\alpha) = 0$ and $f'(\alpha) \neq 0$, then the line $\theta = \alpha$ is tangent at the pole to the graph of $r = f(\theta)$.

Theorem 10.12 is useful because it states that the zeros of $r = f(\theta)$ can be used to find the tangent lines at the pole. Note that because a polar curve can cross the pole more than once, it can have more than one tangent line at the pole. For example, the rose curve

$$f(\theta) = 2 \cos 3\theta$$

has three tangent lines at the pole, as shown in Figure 10.48. For this curve, $f(\theta) = 2 \cos 3\theta$ is 0 when θ is $\pi/6$, $\pi/2$, and $5\pi/6$. Moreover, the derivative $f'(\theta) = -6 \sin 3\theta$ is not 0 for these values of θ.

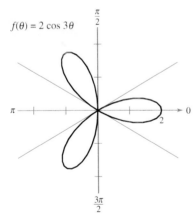

$f(\theta) = 2 \cos 3\theta$

This rose curve has three tangent lines ($\theta = \pi/6$, $\theta = \pi/2$, and $\theta = 5\pi/6$) at the pole.
Figure 10.48

Special Polar Graphs

Several important types of graphs have equations that are simpler in polar form than in rectangular form. For example, the polar equation of a circle having a radius of a and centered at the origin is simply $r = a$. Later in the text you will come to appreciate this benefit. For now, several other types of graphs that have simpler equations in polar form are shown below. (Conics are considered in Section 10.6.)

Limaçons

$r = a \pm b \cos \theta$

$r = a \pm b \sin \theta$

$(a > 0, b > 0)$

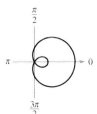
$\dfrac{a}{b} < 1$
Limaçon with inner loop

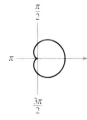
$\dfrac{a}{b} = 1$
Cardioid (heart-shaped)

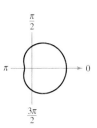
$1 < \dfrac{a}{b} < 2$
Dimpled limaçon

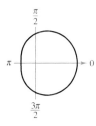
$\dfrac{a}{b} \geq 2$
Convex limaçon

Rose Curves

n petals if n is odd

$2n$ petals if n is even

$(n \geq 2)$

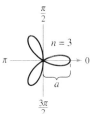
$r = a \cos n\theta$
Rose curve

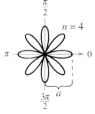
$r = a \cos n\theta$
Rose curve

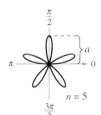
$r = a \sin n\theta$
Rose curve

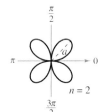
$r = a \sin n\theta$
Rose curve

Circles and Lemniscates

$r = a \cos \theta$
Circle

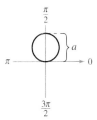
$r = a \sin \theta$
Circle

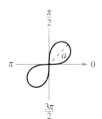
$r^2 = a^2 \sin 2\theta$
Lemniscate

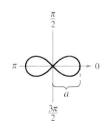
$r^2 = a^2 \cos 2\theta$
Lemniscate

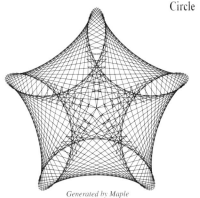
Generated by Maple

TECHNOLOGY The rose curves described above are of the form $r = a \cos n\theta$ or $r = a \sin n\theta$, where n is a positive integer that is greater than or equal to 2. Use a graphing utility to graph $r = a \cos n\theta$ or $r = a \sin n\theta$ for some noninteger values of n. Are these graphs also rose curves? For example, try sketching the graph of $r = \cos \frac{2}{3}\theta$, $0 \leq \theta \leq 6\pi$.

FOR FURTHER INFORMATION For more information on rose curves and related curves, see the article "A Rose is a Rose . . ." by Peter M. Maurer in *The American Mathematical Monthly*. The computer-generated graph at the left is the result of an algorithm that Maurer calls "The Rose." To view this article, go to the website *www.matharticles.com*.

Exercises for Section 10.4

In Exercises 1–6, plot the point in polar coordinates and find the corresponding rectangular coordinates for the point.

1. $(4, 3\pi/6)$
2. $(-2, 7\pi/4)$
3. $(-4, -\pi/3)$
4. $(0, -7\pi/6)$
5. $(\sqrt{2}, 2.36)$
6. $(-3, -1.57)$

In Exercises 7–10, use the *angle* feature of a graphing utility to find the rectangular coordinates for the point given in polar coordinates. Plot the point.

7. $(5, 3\pi/4)$
8. $(-2, 11\pi/6)$
9. $(-3.5, 2.5)$
10. $(8.25, 1.3)$

In Exercises 11–16, the rectangular coordinates of a point are given. Plot the point and find *two* sets of polar coordinates for the point for $0 \le \theta < 2\pi$.

11. $(1, 1)$
12. $(0, -5)$
13. $(-3, 4)$
14. $(4, -2)$
15. $(\sqrt{3}, -1)$
16. $(3, -\sqrt{3})$

In Exercises 17–20, use the *angle* feature of a graphing utility to find one set of polar coordinates for the point given in rectangular coordinates.

17. $(3, -2)$
18. $(3\sqrt{2}, 3\sqrt{2})$
19. $(\frac{5}{2}, \frac{4}{3})$
20. $(0, -5)$

21. Plot the point $(4, 3.5)$ if the point is given in (a) rectangular coordinates and (b) polar coordinates.

22. *Graphical Reasoning*

    (a) Set the window format of a graphing utility to rectangular coordinates and locate the cursor at any position off the axes. Move the cursor horizontally and vertically. Describe any changes in the displayed coordinates of the points.

    (b) Set the window format of a graphing utility to polar coordinates and locate the cursor at any position off the axes. Move the cursor horizontally and vertically. Describe any changes in the displayed coordinates of the points.

    (c) Why are the results in parts (a) and (b) different?

In Exercises 23–26, match the graph with its polar equation. [The graphs are labeled (a), (b), (c), and (d).]

(a)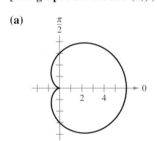
(b)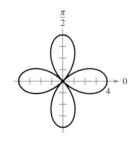
(c)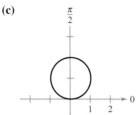
(d)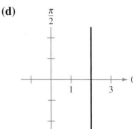

23. $r = 2 \sin \theta$
24. $r = 4 \cos 2\theta$
25. $r = 3(1 + \cos \theta)$
26. $r = 2 \sec \theta$

In Exercises 27–34, convert the rectangular equation to polar form and sketch its graph.

27. $x^2 + y^2 = a^2$
28. $x^2 + y^2 - 2ax = 0$
29. $y = 4$
30. $x = 10$
31. $3x - y + 2 = 0$
32. $xy = 4$
33. $y^2 = 9x$
34. $(x^2 + y^2)^2 - 9(x^2 - y^2) = 0$

In Exercises 35–42, convert the polar equation to rectangular form and sketch its graph.

35. $r = 3$
36. $r = -2$
37. $r = \sin \theta$
38. $r = 5 \cos \theta$
39. $r = \theta$
40. $\theta = \dfrac{5\pi}{6}$
41. $r = 3 \sec \theta$
42. $r = 2 \csc \theta$

In Exercises 43–52, use a graphing utility to graph the polar equation. Find an interval for θ over which the graph is traced *only once*.

43. $r = 3 - 4 \cos \theta$
44. $r = 5(1 - 2 \sin \theta)$
45. $r = 2 + \sin \theta$
46. $r = 4 + 3 \cos \theta$
47. $r = \dfrac{2}{1 + \cos \theta}$
48. $r = \dfrac{2}{4 - 3 \sin \theta}$
49. $r = 2 \cos\left(\dfrac{3\theta}{2}\right)$
50. $r = 3 \sin\left(\dfrac{5\theta}{2}\right)$
51. $r^2 = 4 \sin 2\theta$
52. $r^2 = \dfrac{1}{\theta}$

53. Convert the equation
$$r = 2(h \cos \theta + k \sin \theta)$$
to rectangular form and verify that it is the equation of a circle. Find the radius and the rectangular coordinates of the center of the circle.

54. *Distance Formula*

(a) Verify that the Distance Formula for the distance between the two points (r_1, θ_1) and (r_2, θ_2) in polar coordinates is
$$d = \sqrt{r_1^2 + r_2^2 - 2r_1 r_2 \cos(\theta_1 - \theta_2)}.$$

(b) Describe the positions of the points relative to each other if $\theta_1 = \theta_2$. Simplify the Distance Formula for this case. Is the simplification what you expected? Explain.

(c) Simplify the Distance Formula if $\theta_1 - \theta_2 = 90°$. Is the simplification what you expected? Explain.

(d) Choose two points on the polar coordinate system and find the distance between them. Then choose different polar representations of the same two points and apply the Distance Formula again. Discuss the result.

In Exercises 55–58, use the result of Exercise 54 to approximate the distance between the two points in polar coordinates.

55. $\left(4, \frac{2\pi}{3}\right)$, $\left(2, \frac{\pi}{6}\right)$ **56.** $\left(10, \frac{7\pi}{6}\right)$, $(3, \pi)$

57. $(2, 0.5)$, $(7, 1.2)$ **58.** $(4, 2.5)$, $(12, 1)$

In Exercises 59 and 60, find dy/dx and the slopes of the tangent lines shown on the graph of the polar equation.

59. $r = 2 + 3 \sin \theta$ **60.** $r = 2(1 - \sin \theta)$

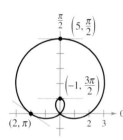 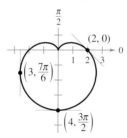

 In Exercises 61–64, use a graphing utility to (a) graph the polar equation, (b) draw the tangent line at the given value of θ, and (c) find dy/dx at the given value of θ. (Hint: Let the increment between the values of θ equal $\pi/24$.)

61. $r = 3(1 - \cos \theta)$, $\theta = \frac{\pi}{2}$ **62.** $r = 3 - 2 \cos \theta$, $\theta = 0$

63. $r = 3 \sin \theta$, $\theta = \frac{\pi}{3}$ **64.** $r = 4$, $\theta = \frac{\pi}{4}$

In Exercises 65 and 66, find the points of horizontal and vertical tangency (if any) to the polar curve.

65. $r = 1 - \sin \theta$ **66.** $r = a \sin \theta$

In Exercises 67 and 68, find the points of horizontal tangency (if any) to the polar curve.

67. $r = 2 \csc \theta + 3$ **68.** $r = a \sin \theta \cos^2 \theta$

 In Exercises 69–72, use a graphing utility to graph the polar equation and find all points of horizontal tangency.

69. $r = 4 \sin \theta \cos^2 \theta$ **70.** $r = 3 \cos 2\theta \sec \theta$

71. $r = 2 \csc \theta + 5$ **72.** $r = 2 \cos(3\theta - 2)$

In Exercises 73–80, sketch a graph of the polar equation and find the tangents at the pole.

73. $r = 3 \sin \theta$ **74.** $r = 3 \cos \theta$

75. $r = 2(1 - \sin \theta)$ **76.** $r = 3(1 - \cos \theta)$

77. $r = 2 \cos 3\theta$ **78.** $r = -\sin 5\theta$

79. $r = 3 \sin 2\theta$ **80.** $r = 3 \cos 2\theta$

In Exercises 81–92, sketch a graph of the polar equation.

81. $r = 5$ **82.** $r = 2$

83. $r = 4(1 + \cos \theta)$ **84.** $r = 1 + \sin \theta$

85. $r = 3 - 2 \cos \theta$ **86.** $r = 5 - 4 \sin \theta$

87. $r = 3 \csc \theta$ **88.** $r = \dfrac{6}{2 \sin \theta - 3 \cos \theta}$

89. $r = 2\theta$ **90.** $r = \dfrac{1}{\theta}$

91. $r^2 = 4 \cos 2\theta$ **92.** $r^2 = 4 \sin \theta$

 In Exercises 93–96, use a graphing utility to graph the equation and show that the given line is an asymptote of the graph.

Name of Graph	Polar Equation	Asymptote
93. Conchoid	$r = 2 - \sec \theta$	$x = -1$
94. Conchoid	$r = 2 + \csc \theta$	$y = 1$
95. Hyperbolic spiral	$r = 2/\theta$	$y = 2$
96. Strophoid	$r = 2 \cos 2\theta \sec \theta$	$x = -2$

Writing About Concepts

97. Describe the differences between the rectangular coordinate system and the polar coordinate system.

98. Give the equations for the coordinate conversion from rectangular to polar coordinates and vice versa.

99. For constants a and b, describe the graphs of the equations $r = a$ and $\theta = b$ in polar coordinates.

100. How are the slopes of tangent lines determined in polar coordinates? What are tangent lines at the pole and how are they determined?

101. Sketch the graph of $r = 4 \sin \theta$ over each interval.

(a) $0 \leq \theta \leq \dfrac{\pi}{2}$ (b) $\dfrac{\pi}{2} \leq \theta \leq \pi$ (c) $-\dfrac{\pi}{2} \leq \theta \leq \dfrac{\pi}{2}$

 102. *Think About It* Use a graphing utility to graph the polar equation $r = 6[1 + \cos(\theta - \phi)]$ for (a) $\phi = 0$, (b) $\phi = \pi/4$, and (c) $\phi = \pi/2$. Use the graphs to describe the effect of the angle ϕ. Write the equation as a function of $\sin \theta$ for part (c).

103. Verify that if the curve whose polar equation is $r = f(\theta)$ is rotated about the pole through an angle ϕ, then an equation for the rotated curve is $r = f(\theta - \phi)$.

104. The polar form of an equation for a curve is $r = f(\sin \theta)$. Show that the form becomes

(a) $r = f(-\cos \theta)$ if the curve is rotated counterclockwise $\pi/2$ radians about the pole.

(b) $r = f(-\sin \theta)$ if the curve is rotated counterclockwise π radians about the pole.

(c) $r = f(\cos \theta)$ if the curve is rotated counterclockwise $3\pi/2$ radians about the pole.

In Exercises 105–108, use the results of Exercises 103 and 104.

105. Write an equation for the limaçon $r = 2 - \sin \theta$ after it has been rotated by the given amount. Use a graphing utility to graph the rotated limaçon.

(a) $\dfrac{\pi}{4}$ (b) $\dfrac{\pi}{2}$ (c) π (d) $\dfrac{3\pi}{2}$

106. Write an equation for the rose curve $r = 2 \sin 2\theta$ after it has been rotated by the given amount. Verify the results by using a graphing utility to graph the rotated rose curve.

(a) $\dfrac{\pi}{6}$ (b) $\dfrac{\pi}{2}$ (c) $\dfrac{2\pi}{3}$ (d) π

107. Sketch the graph of each equation.

(a) $r = 1 - \sin \theta$ (b) $r = 1 - \sin\left(\theta - \dfrac{\pi}{4}\right)$

108. Prove that the tangent of the angle ψ ($0 \le \psi \le \pi/2$) between the radial line and the tangent line at the point (r, θ) on the graph of $r = f(\theta)$ (see figure) is given by $\tan \psi = |r/(dr/d\theta)|$.

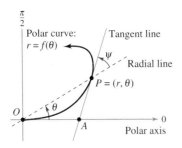

In Exercises 109–114, use the result of Exercise 108 to find the angle ψ between the radial and tangent lines to the graph for the indicated value of θ. Use a graphing utility to graph the polar equation, the radial line, and the tangent line for the indicated value of θ. Identify the angle ψ.

Polar Equation	Value of θ
109. $r = 2(1 - \cos \theta)$	$\theta = \pi$
110. $r = 3(1 - \cos \theta)$	$\theta = 3\pi/4$
111. $r = 2 \cos 3\theta$	$\theta = \pi/4$
112. $r = 4 \sin 2\theta$	$\theta = \pi/6$
113. $r = \dfrac{6}{1 - \cos \theta}$	$\theta = 2\pi/3$
114. $r = 5$	$\theta = \pi/6$

True or False? **In Exercises 115–118, determine whether the statement is true or false. If it is false, explain why or give an example that shows it is false.**

115. If (r_1, θ_1) and (r_2, θ_2) represent the same point on the polar coordinate system, then $|r_1| = |r_2|$.

116. If (r, θ_1) and (r, θ_2) represent the same point on the polar coordinate system, then $\theta_1 = \theta_2 + 2\pi n$ for some integer n.

117. If $x > 0$, then the point (x, y) on the rectangular coordinate system can be represented by (r, θ) on the polar coordinate system, where $r = \sqrt{x^2 + y^2}$ and $\theta = \arctan(y/x)$.

118. The polar equations $r = \sin 2\theta$ and $r = -\sin 2\theta$ have the same graph.

Section Project: Anamorphic Art

Use the anamorphic transformations

$$r = y + 16 \quad \text{and} \quad \theta = -\dfrac{\pi}{8}x, \quad -\dfrac{3\pi}{4} \le \theta \le \dfrac{3\pi}{4}$$

to sketch the transformed polar image of the rectangular graph. When the reflection (in a cylindrical mirror centered at the pole) of each polar image is viewed from the polar axis, the viewer will see the original rectangular image.

(a) $y = 3$ (b) $x = 2$

(c) $y = x + 5$ (d) $x^2 + (y - 5)^2 = 5^2$

This example of anamorphic art is from the Museum of Science and Industry in Manchester, England. When the reflection of the transformed "polar painting" is viewed in the mirror, the viewer sees faces.

FOR FURTHER INFORMATION For more information on anamorphic art, see the article "Anamorphisms" by Philip Hickin in the *Mathematical Gazette*.

Section 10.5 Area and Arc Length in Polar Coordinates

- Find the area of a region bounded by a polar graph.
- Find the points of intersection of two polar graphs.
- Find the arc length of a polar graph.
- Find the area of a surface of revolution (polar form).

Area of a Polar Region

The development of a formula for the area of a polar region parallels that for the area of a region on the rectangular coordinate system, but uses sectors of a circle instead of rectangles as the basic element of area. In Figure 10.49, note that the area of a circular sector of radius r is given by $\frac{1}{2}\theta r^2$, provided θ is measured in radians.

Consider the function given by $r = f(\theta)$, where f is continuous and nonnegative in the interval given by $\alpha \leq \theta \leq \beta$. The region bounded by the graph of f and the radial lines $\theta = \alpha$ and $\theta = \beta$ is shown in Figure 10.50(a). To find the area of this region, partition the interval $[\alpha, \beta]$ into n equal subintervals

$$\alpha = \theta_0 < \theta_1 < \theta_2 < \cdots < \theta_{n-1} < \theta_n = \beta.$$

Then, approximate the area of the region by the sum of the areas of the n sectors, as shown in Figure 10.50(b).

$$\text{Radius of } i\text{th sector} = f(\theta_i)$$

$$\text{Central angle of } i\text{th sector} = \frac{\beta - \alpha}{n} = \Delta\theta$$

$$A \approx \sum_{i=1}^{n} \left(\frac{1}{2}\right) \Delta\theta [f(\theta_i)]^2$$

Taking the limit as $n \to \infty$ produces

$$A = \lim_{n \to \infty} \frac{1}{2} \sum_{i=1}^{n} [f(\theta_i)]^2 \Delta\theta$$

$$= \frac{1}{2} \int_{\alpha}^{\beta} [f(\theta)]^2 \, d\theta$$

which leads to the following theorem.

The area of a sector of a circle is $A = \frac{1}{2}\theta r^2$.
Figure 10.49

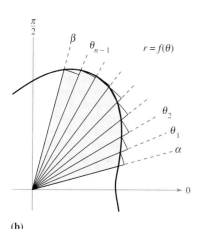

Figure 10.50

THEOREM 10.13 Area in Polar Coordinates

If f is continuous and nonnegative on the interval $[\alpha, \beta]$, $0 < \beta - \alpha \leq 2\pi$, then the area of the region bounded by the graph of $r = f(\theta)$ between the radial lines $\theta = \alpha$ and $\theta = \beta$ is given by

$$A = \frac{1}{2} \int_{\alpha}^{\beta} [f(\theta)]^2 \, d\theta$$

$$= \frac{1}{2} \int_{\alpha}^{\beta} r^2 \, d\theta. \qquad 0 < \beta - \alpha \leq 2\pi$$

NOTE You can use the same formula to find the area of a region bounded by the graph of a continuous *nonpositive* function. However, the formula is not necessarily valid if f takes on both positive *and* negative values in the interval $[\alpha, \beta]$.

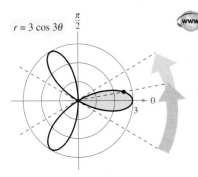

$r = 3 \cos 3\theta$

The area of one petal of the rose curve that lies between the radial lines $\theta = -\pi/6$ and $\theta = \pi/6$ is $3\pi/4$.
Figure 10.51

EXAMPLE 1 Finding the Area of a Polar Region

Find the area of one petal of the rose curve given by $r = 3 \cos 3\theta$.

Solution In Figure 10.51, you can see that the right petal is traced as θ increases from $-\pi/6$ to $\pi/6$. So, the area is

$$A = \frac{1}{2}\int_\alpha^\beta r^2\, d\theta = \frac{1}{2}\int_{-\pi/6}^{\pi/6} (3\cos 3\theta)^2\, d\theta \quad \text{Formula for area in polar coordinates}$$

$$= \frac{9}{2}\int_{-\pi/6}^{\pi/6} \frac{1 + \cos 6\theta}{2}\, d\theta \quad \text{Trigonometric identity}$$

$$= \frac{9}{4}\left[\theta + \frac{\sin 6\theta}{6}\right]_{-\pi/6}^{\pi/6}$$

$$= \frac{9}{4}\left(\frac{\pi}{6} + \frac{\pi}{6}\right)$$

$$= \frac{3\pi}{4}.$$

NOTE To find the area of the region lying inside all three petals of the rose curve in Example 1, you could not simply integrate between 0 and 2π. In doing this you would obtain $9\pi/2$, which is twice the area of the three petals. The duplication occurs because the rose curve is traced twice as θ increases from 0 to 2π.

EXAMPLE 2 Finding the Area Bounded by a Single Curve

Find the area of the region lying between the inner and outer loops of the limaçon $r = 1 - 2\sin\theta$.

Solution In Figure 10.52, note that the inner loop is traced as θ increases from $\pi/6$ to $5\pi/6$. So, the area inside the *inner loop* is

$$A_1 = \frac{1}{2}\int_\alpha^\beta r^2\, d\theta = \frac{1}{2}\int_{\pi/6}^{5\pi/6} (1 - 2\sin\theta)^2\, d\theta \quad \text{Formula for area in polar coordinates}$$

$$= \frac{1}{2}\int_{\pi/6}^{5\pi/6} (1 - 4\sin\theta + 4\sin^2\theta)\, d\theta$$

$$= \frac{1}{2}\int_{\pi/6}^{5\pi/6} \left[1 - 4\sin\theta + 4\left(\frac{1 - \cos 2\theta}{2}\right)\right] d\theta \quad \text{Trigonometric identity}$$

$$= \frac{1}{2}\int_{\pi/6}^{5\pi/6} (3 - 4\sin\theta - 2\cos 2\theta)\, d\theta \quad \text{Simplify.}$$

$$= \frac{1}{2}\left[3\theta + 4\cos\theta - \sin 2\theta\right]_{\pi/6}^{5\pi/6}$$

$$= \frac{1}{2}(2\pi - 3\sqrt{3})$$

$$= \pi - \frac{3\sqrt{3}}{2}.$$

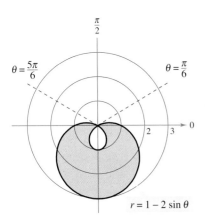

$r = 1 - 2\sin\theta$

The area between the inner and outer loops is approximately 8.34.
Figure 10.52

In a similar way, you can integrate from $5\pi/6$ to $13\pi/6$ to find that the area of the region lying inside the outer loop is $A_2 = 2\pi + (3\sqrt{3}/2)$. The area of the region lying between the two loops is the difference of A_2 and A_1.

$$A = A_2 - A_1 = \left(2\pi + \frac{3\sqrt{3}}{2}\right) - \left(\pi - \frac{3\sqrt{3}}{2}\right) = \pi + 3\sqrt{3} \approx 8.34$$

Points of Intersection of Polar Graphs

Because a point may be represented in different ways in polar coordinates, care must be taken in determining the points of intersection of two polar graphs. For example, consider the points of intersection of the graphs of

$$r = 1 - 2\cos\theta \quad \text{and} \quad r = 1$$

as shown in Figure 10.53. If, as with rectangular equations, you attempted to find the points of intersection by solving the two equations simultaneously, you would obtain

$$r = 1 - 2\cos\theta \quad \text{First equation}$$
$$1 = 1 - 2\cos\theta \quad \text{Substitute } r = 1 \text{ from 2nd equation into 1st equation.}$$
$$\cos\theta = 0 \quad \text{Simplify.}$$
$$\theta = \frac{\pi}{2}, \frac{3\pi}{2}. \quad \text{Solve for } \theta.$$

The corresponding points of intersection are $(1, \pi/2)$ and $(1, 3\pi/2)$. However, from Figure 10.53 you can see that there is a *third* point of intersection that did not show up when the two polar equations were solved simultaneously. (This is one reason why you should sketch a graph when finding the area of a polar region.) The reason the third point was not found is that it does not occur with the same coordinates in the two graphs. On the graph of $r = 1$, the point occurs with coordinates $(1, \pi)$, but on the graph of $r = 1 - 2\cos\theta$, the point occurs with coordinates $(-1, 0)$.

You can compare the problem of finding points of intersection of two polar graphs with that of finding collision points of two satellites in intersecting orbits about Earth, as shown in Figure 10.54. The satellites will not collide as long as they reach the points of intersection at different times (θ-values). Collisions will occur only at the points of intersection that are "simultaneous points"—those reached at the same time (θ-value).

NOTE Because the pole can be represented by $(0, \theta)$, where θ is *any* angle, you should check separately for the pole when finding points of intersection.

FOR FURTHER INFORMATION For more information on using technology to find points of intersection, see the article "Finding Points of Intersection of Polar-Coordinate Graphs" by Warren W. Esty in *Mathematics Teacher*. To view this article, go to the website *www.matharticles.com*.

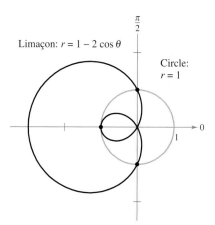

Three points of intersection: $(1, \pi/2)$, $(-1, 0), (1, 3\pi/2)$
Figure 10.53

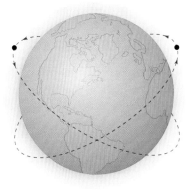

The paths of satellites can cross without causing a collision.
Figure 10.54

EXAMPLE 3 Finding the Area of a Region Between Two Curves

Find the area of the region common to the two regions bounded by the following curves.

$$r = -6\cos\theta \qquad \text{Circle}$$
$$r = 2 - 2\cos\theta \qquad \text{Cardioid}$$

Solution Because both curves are symmetric with respect to the *x*-axis, you can work with the upper half-plane, as shown in Figure 10.55. The gray shaded region lies between the circle and the radial line $\theta = 2\pi/3$. Because the circle has coordinates $(0, \pi/2)$ at the pole, you can integrate between $\pi/2$ and $2\pi/3$ to obtain the area of this region. The region that is shaded red is bounded by the radial lines $\theta = 2\pi/3$ and $\theta = \pi$ and the cardioid. So, you can find the area of this second region by integrating between $2\pi/3$ and π. The sum of these two integrals gives the area of the common region lying *above* the radial line $\theta = \pi$.

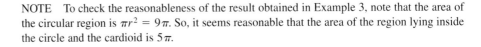

$$\frac{A}{2} = \frac{1}{2}\int_{\pi/2}^{2\pi/3}(-6\cos\theta)^2\,d\theta + \frac{1}{2}\int_{2\pi/3}^{\pi}(2 - 2\cos\theta)^2\,d\theta$$

$$= 18\int_{\pi/2}^{2\pi/3}\cos^2\theta\,d\theta + \frac{1}{2}\int_{2\pi/3}^{\pi}(4 - 8\cos\theta + 4\cos^2\theta)\,d\theta$$

$$= 9\int_{\pi/2}^{2\pi/3}(1 + \cos 2\theta)\,d\theta + \int_{2\pi/3}^{\pi}(3 - 4\cos\theta + \cos 2\theta)\,d\theta$$

$$= 9\left[\theta + \frac{\sin 2\theta}{2}\right]_{\pi/2}^{2\pi/3} + \left[3\theta - 4\sin\theta + \frac{\sin 2\theta}{2}\right]_{2\pi/3}^{\pi}$$

$$= 9\left(\frac{2\pi}{3} - \frac{\sqrt{3}}{4} - \frac{\pi}{2}\right) + \left(3\pi - 2\pi + 2\sqrt{3} + \frac{\sqrt{3}}{4}\right)$$

$$= \frac{5\pi}{2}$$

$$\approx 7.85$$

Finally, multiplying by 2, you can conclude that the total area is 5π.

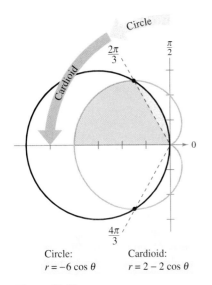

Figure 10.55

NOTE To check the reasonableness of the result obtained in Example 3, note that the area of the circular region is $\pi r^2 = 9\pi$. So, it seems reasonable that the area of the region lying inside the circle and the cardioid is 5π.

To see the benefit of polar coordinates for finding the area in Example 3, consider the following integral, which gives the comparable area in rectangular coordinates.

$$\frac{A}{2} = \int_{-4}^{-3/2}\sqrt{2\sqrt{1 - 2x} - x^2 - 2x + 2}\,dx + \int_{-3/2}^{0}\sqrt{-x^2 - 6x}\,dx$$

Use the integration capabilities of a graphing utility to show that you obtain the same area as that found in Example 3.

Arc Length in Polar Form

NOTE When applying the arc length formula to a polar curve, be sure that the curve is traced out only once on the interval of integration. For instance, the rose curve given by $r = \cos 3\theta$ is traced out once on the interval $0 \leq \theta \leq \pi$, but is traced out twice on the interval $0 \leq \theta \leq 2\pi$.

The formula for the length of a polar arc can be obtained from the arc length formula for a curve described by parametric equations. (See Exercise 77.)

THEOREM 10.14 Arc Length of a Polar Curve

Let f be a function whose derivative is continuous on an interval $\alpha \leq \theta \leq \beta$. The length of the graph of $r = f(\theta)$ from $\theta = \alpha$ to $\theta = \beta$ is

$$s = \int_\alpha^\beta \sqrt{[f(\theta)]^2 + [f'(\theta)]^2}\, d\theta = \int_\alpha^\beta \sqrt{r^2 + \left(\frac{dr}{d\theta}\right)^2}\, d\theta.$$

EXAMPLE 4 Finding the Length of a Polar Curve

Find the length of the arc from $\theta = 0$ to $\theta = 2\pi$ for the cardioid

$$r = f(\theta) = 2 - 2\cos\theta$$

as shown in Figure 10.56.

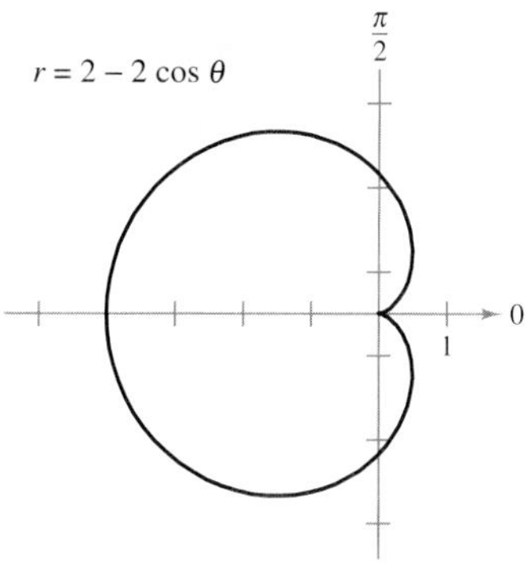

Solution Because $f'(\theta) = 2\sin\theta$, you can find the arc length as follows.

$$\begin{aligned}
s &= \int_\alpha^\beta \sqrt{[f(\theta)]^2 + [f'(\theta)]^2}\, d\theta && \text{Formula for arc length of a polar curve}\\
&= \int_0^{2\pi} \sqrt{(2 - 2\cos\theta)^2 + (2\sin\theta)^2}\, d\theta \\
&= 2\sqrt{2}\int_0^{2\pi} \sqrt{1 - \cos\theta}\, d\theta && \text{Simplify.}\\
&= 2\sqrt{2}\int_0^{2\pi} \sqrt{2\sin^2\frac{\theta}{2}}\, d\theta && \text{Trigonometric identity}\\
&= 4\int_0^{2\pi} \sin\frac{\theta}{2}\, d\theta && \sin\frac{\theta}{2} \geq 0 \text{ for } 0 \leq \theta \leq 2\pi\\
&= 8\left[-\cos\frac{\theta}{2}\right]_0^{2\pi}\\
&= 8(1 + 1)\\
&= 16
\end{aligned}$$

In the fifth step of the solution, it is legitimate to write

$$\sqrt{2\sin^2(\theta/2)} = \sqrt{2}\sin(\theta/2)$$

rather than

$$\sqrt{2\sin^2(\theta/2)} = \sqrt{2}\,|\sin(\theta/2)|$$

because $\sin(\theta/2) \geq 0$ for $0 \leq \theta \leq 2\pi$.

NOTE Using Figure 10.56, you can determine the reasonableness of this answer by comparing it with the circumference of a circle. For example, a circle of radius $\frac{5}{2}$ has a circumference of $5\pi \approx 15.7$.

Figure 10.56

$r = 2 - 2\cos\theta$

Area of a Surface of Revolution

The polar coordinate versions of the formulas for the area of a surface of revolution can be obtained from the parametric versions given in Theorem 10.9, using the equations $x = r \cos \theta$ and $y = r \sin \theta$.

> **THEOREM 10.15 Area of a Surface of Revolution**
>
> Let f be a function whose derivative is continuous on an interval $\alpha \leq \theta \leq \beta$. The area of the surface formed by revolving the graph of $r = f(\theta)$ from $\theta = \alpha$ to $\theta = \beta$ about the indicated line is as follows.
>
> 1. $S = 2\pi \displaystyle\int_\alpha^\beta f(\theta) \sin \theta \sqrt{[f(\theta)]^2 + [f'(\theta)]^2} \, d\theta$ About the polar axis
>
> 2. $S = 2\pi \displaystyle\int_\alpha^\beta f(\theta) \cos \theta \sqrt{[f(\theta)]^2 + [f'(\theta)]^2} \, d\theta$ About the line $\theta = \dfrac{\pi}{2}$

NOTE When using Theorem 10.15, check to see that the graph of $r = f(\theta)$ is traced only once on the interval $\alpha \leq \theta \leq \beta$. For example, the circle given by $r = \cos \theta$ is traced only once on the interval $0 \leq \theta \leq \pi$.

EXAMPLE 5 Finding the Area of a Surface of Revolution

Find the area of the surface formed by revolving the circle $r = f(\theta) = \cos \theta$ about the line $\theta = \pi/2$, as shown in Figure 10.57.

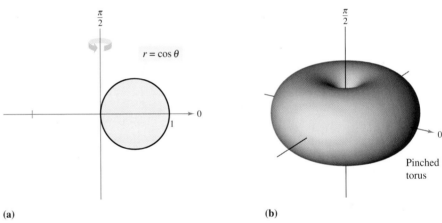

(a)   (b)

Pinched torus

Figure 10.57

Solution You can use the second formula given in Theorem 10.15 with $f'(\theta) = -\sin \theta$. Because the circle is traced once as θ increases from 0 to π, you have

$$S = 2\pi \int_\alpha^\beta f(\theta) \cos \theta \sqrt{[f(\theta)]^2 + [f'(\theta)]^2} \, d\theta \quad \text{Formula for area of a surface of revolution}$$

$$= 2\pi \int_0^\pi \cos \theta (\cos \theta) \sqrt{\cos^2 \theta + \sin^2 \theta} \, d\theta$$

$$= 2\pi \int_0^\pi \cos^2 \theta \, d\theta \quad \text{Trigonometric identity}$$

$$= \pi \int_0^\pi (1 + \cos 2\theta) \, d\theta \quad \text{Trigonometric identity}$$

$$= \pi \left[\theta + \frac{\sin 2\theta}{2} \right]_0^\pi = \pi^2.$$

Exercises for Section 10.5

In Exercises 1–4, write an integral that represents the area of the shaded region shown in the figure. Do not evaluate the integral.

1. $r = 2 \sin \theta$
2. $r = \cos 2\theta$

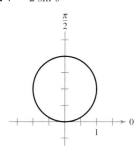
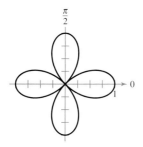

3. $r = 1 - \sin \theta$
4. $r = 1 - \cos 2\theta$

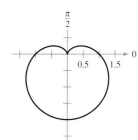
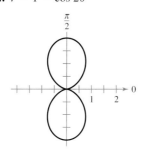

In Exercises 5 and 6, find the area of the region bounded by the graph of the polar equation using (a) a geometric formula and (b) integration.

5. $r = 8 \sin \theta$
6. $r = 3 \cos \theta$

In Exercises 7–12, find the area of the region.

7. One petal of $r = 2 \cos 3\theta$
8. One petal of $r = 6 \sin 2\theta$
9. One petal of $r = \cos 2\theta$
10. One petal of $r = \cos 5\theta$
11. Interior of $r = 1 - \sin \theta$
12. Interior of $r = 1 - \sin \theta$ (above the polar axis)

 In Exercises 13–16, use a graphing utility to graph the polar equation and find the area of the given region.

13. Inner loop of $r = 1 + 2 \cos \theta$
14. Inner loop of $r = 4 - 6 \sin \theta$
15. Between the loops of $r = 1 + 2 \cos \theta$
16. Between the loops of $r = 2(1 + 2 \sin \theta)$

In Exercises 17–26, find the points of intersection of the graphs of the equations.

17. $r = 1 + \cos \theta$
    $r = 1 - \cos \theta$

18. $r = 3(1 + \sin \theta)$
    $r = 3(1 - \sin \theta)$

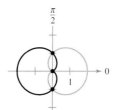
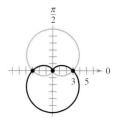

19. $r = 1 + \cos \theta$
    $r = 1 - \sin \theta$

20. $r = 2 - 3 \cos \theta$
    $r = \cos \theta$

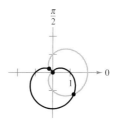
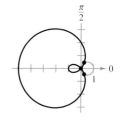

21. $r = 4 - 5 \sin \theta$
    $r = 3 \sin \theta$

22. $r = 1 + \cos \theta$
    $r = 3 \cos \theta$

23. $r = \dfrac{\theta}{2}$
    $r = 2$

24. $\theta = \dfrac{\pi}{4}$
    $r = 2$

25. $r = 4 \sin 2\theta$
    $r = 2$

26. $r = 3 + \sin \theta$
    $r = 2 \csc \theta$

 In Exercises 27 and 28, use a graphing utility to approximate the points of intersection of the graphs of the polar equations. Confirm your results analytically.

27. $r = 2 + 3 \cos \theta$
    $r = \dfrac{\sec \theta}{2}$

28. $r = 3(1 - \cos \theta)$
    $r = \dfrac{6}{1 - \cos \theta}$

 Writing **In Exercises 29 and 30, use a graphing utility to find the points of intersection of the graphs of the polar equations. Watch the graphs as they are traced in the viewing window. Explain why the pole is not a point of intersection obtained by solving the equations simultaneously.**

29. $r = \cos \theta$
    $r = 2 - 3 \sin \theta$

30. $r = 4 \sin \theta$
    $r = 2(1 + \sin \theta)$

In Exercises 31–36, use a graphing utility to graph the polar equations and find the area of the given region.

31. Common interior of $r = 4 \sin 2\theta$ and $r = 2$
32. Common interior of $r = 3(1 + \sin \theta)$ and $r = 3(1 - \sin \theta)$
33. Common interior of $r = 3 - 2 \sin \theta$ and $r = -3 + 2 \sin \theta$
34. Common interior of $r = 5 - 3 \sin \theta$ and $r = 5 - 3 \cos \theta$
35. Common interior of $r = 4 \sin \theta$ and $r = 2$
36. Inside $r = 3 \sin \theta$ and outside $r = 2 - \sin \theta$

In Exercises 37–40, find the area of the region.

37. Inside $r = a(1 + \cos \theta)$ and outside $r = a \cos \theta$
38. Inside $r = 2a \cos \theta$ and outside $r = a$
39. Common interior of $r = a(1 + \cos \theta)$ and $r = a \sin \theta$
40. Common interior of $r = a \cos \theta$ and $r = a \sin \theta$ where $a > 0$

41. **Antenna Radiation** The radiation from a transmitting antenna is not uniform in all directions. The intensity from a particular antenna is modeled by
$$r = a \cos^2 \theta.$$
   (a) Convert the polar equation to rectangular form.
   (b) Use a graphing utility to graph the model for $a = 4$ and $a = 6$.
   (c) Find the area of the geographical region between the two curves in part (b).

42. **Area** The area inside one or more of the three interlocking circles
$$r = 2a \cos \theta, \quad r = 2a \sin \theta, \quad \text{and} \quad r = a$$
is divided into seven regions. Find the area of each region.

43. **Conjecture** Find the area of the region enclosed by
$$r = a \cos(n\theta)$$
for $n = 1, 2, 3, \ldots$. Use the results to make a conjecture about the area enclosed by the function if n is even and if n is odd.

44. **Area** Sketch the strophoid
$$r = \sec \theta - 2 \cos \theta, \quad -\frac{\pi}{2} < \theta < \frac{\pi}{2}.$$
Convert this equation to rectangular coordinates. Find the area enclosed by the loop.

In Exercises 45–48, find the length of the curve over the given interval.

Polar Equation	Interval
45. $r = a$	$0 \leq \theta \leq 2\pi$
46. $r = 2a \cos \theta$	$-\frac{\pi}{2} \leq \theta \leq \frac{\pi}{2}$
47. $r = 1 + \sin \theta$	$0 \leq \theta \leq 2\pi$
48. $r = 8(1 + \cos \theta)$	$0 \leq \theta \leq 2\pi$

In Exercises 49–54, use a graphing utility to graph the polar equation over the given interval. Use the integration capabilities of the graphing utility to approximate the length of the curve accurate to two decimal places.

49. $r = 2\theta, \quad 0 \leq \theta \leq \frac{\pi}{2}$
50. $r = \sec \theta, \quad 0 \leq \theta \leq \frac{\pi}{3}$
51. $r = \frac{1}{\theta}, \quad \pi \leq \theta \leq 2\pi$
52. $r = e^{\theta}, \quad 0 \leq \theta \leq \pi$
53. $r = \sin(3 \cos \theta), \quad 0 \leq \theta \leq \pi$
54. $r = 2 \sin(2 \cos \theta), \quad 0 \leq \theta \leq \pi$

In Exercises 55–58, find the area of the surface formed by revolving the curve about the given line.

Polar Equation	Interval	Axis of Revolution
55. $r = 6 \cos \theta$	$0 \leq \theta \leq \frac{\pi}{2}$	Polar axis
56. $r = a \cos \theta$	$0 \leq \theta \leq \frac{\pi}{2}$	$\theta = \frac{\pi}{2}$
57. $r = e^{a\theta}$	$0 \leq \theta \leq \frac{\pi}{2}$	$\theta = \frac{\pi}{2}$
58. $r = a(1 + \cos \theta)$	$0 \leq \theta \leq \pi$	Polar axis

In Exercises 59 and 60, use the integration capabilities of a graphing utility to approximate to two decimal places the area of the surface formed by revolving the curve about the polar axis.

59. $r = 4 \cos 2\theta, \quad 0 \leq \theta \leq \frac{\pi}{4}$
60. $r = \theta, \quad 0 \leq \theta \leq \pi$

Writing About Concepts

61. Give the integral formulas for area and arc length in polar coordinates.
62. Explain why finding points of intersection of polar graphs may require further analysis beyond solving two equations simultaneously.
63. Which integral yields the arc length of $r = 3(1 - \cos 2\theta)$? State why the other integrals are incorrect.

   (a) $3 \int_0^{2\pi} \sqrt{(1 - \cos 2\theta)^2 + 4 \sin^2 2\theta} \, d\theta$

   (b) $12 \int_0^{\pi/4} \sqrt{(1 - \cos 2\theta)^2 + 4 \sin^2 2\theta} \, d\theta$

   (c) $3 \int_0^{\pi} \sqrt{(1 - \cos 2\theta)^2 + 4 \sin^2 2\theta} \, d\theta$

   (d) $6 \int_0^{\pi/2} \sqrt{(1 - \cos 2\theta)^2 + 4 \sin^2 2\theta} \, d\theta$

64. Give the integral formulas for the area of the surface of revolution formed when the graph of $r = f(\theta)$ is revolved about (a) the x-axis and (b) the y-axis.

65. Surface Area of a Torus Find the surface area of the torus generated by revolving the circle given by $r = 2$ about the line $r = 5 \sec \theta$.

66. Surface Area of a Torus Find the surface area of the torus generated by revolving the circle given by $r = a$ about the line $r = b \sec \theta$, where $0 < a < b$.

67. Approximating Area Consider the circle $r = 8 \cos \theta$.

(a) Find the area of the circle.

(b) Complete the table giving the areas A of the sectors of the circle between $\theta = 0$ and the values of θ in the table.

θ	0.2	0.4	0.6	0.8	1.0	1.2	1.4
A							

(c) Use the table in part (b) to approximate the values of θ for which the sector of the circle composes $\frac{1}{4}$, $\frac{1}{2}$, and $\frac{3}{4}$ of the total area of the circle.

(d) Use a graphing utility to approximate, to two decimal places, the angles θ for which the sector of the circle composes $\frac{1}{4}$, $\frac{1}{2}$, and $\frac{3}{4}$ of the total area of the circle.

(e) Do the results of part (d) depend on the radius of the circle? Explain.

68. Approximate Area Consider the circle $r = 3 \sin \theta$.

(a) Find the area of the circle.

(b) Complete the table giving the areas A of the sectors of the circle between $\theta = 0$ and the values of θ in the table.

θ	0.2	0.4	0.6	0.8	1.0	1.2	1.4
A							

(c) Use the table in part (b) to approximate the values of θ for which the sector of the circle composes $\frac{1}{8}$, $\frac{1}{4}$, and $\frac{1}{2}$ of the total area of the circle.

(d) Use a graphing utility to approximate, to two decimal places, the angles θ for which the sector of the circle composes $\frac{1}{8}$, $\frac{1}{4}$, and $\frac{1}{2}$ of the total area of the circle.

69. What conic section does the following polar equation represent?

$r = a \sin \theta + b \cos \theta$

70. Area Find the area of the circle given by $r = \sin \theta + \cos \theta$. Check your result by converting the polar equation to rectangular form, then using the formula for the area of a circle.

71. Spiral of Archimedes The curve represented by the equation $r = a\theta$, where a is a constant, is called the spiral of Archimedes.

(a) Use a graphing utility to graph $r = \theta$, where $\theta \geq 0$. What happens to the graph of $r = a\theta$ as a increases? What happens if $\theta \leq 0$?

(b) Determine the points on the spiral $r = a\theta$ ($a > 0$, $\theta \geq 0$), where the curve crosses the polar axis.

(c) Find the length of $r = \theta$ over the interval $0 \leq \theta \leq 2\pi$.

(d) Find the area under the curve $r = \theta$ for $0 \leq \theta \leq 2\pi$.

72. Logarithmic Spiral The curve represented by the equation $r = ae^{b\theta}$, where a and b are constants, is called a **logarithmic spiral**. The figure below shows the graph of $r = e^{\theta/6}$, $-2\pi \leq \theta \leq 2\pi$. Find the area of the shaded region.

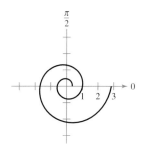

73. The larger circle in the figure below is the graph of $r = 1$. Find the polar equation of the smaller circle such that the shaded regions are equal.

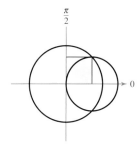

74. Folium of Descartes A curve called the **folium of Descartes** can be represented by the parametric equations

$$x = \frac{3t}{1 + t^3} \quad \text{and} \quad y = \frac{3t^2}{1 + t^3}.$$

(a) Convert the parametric equations to polar form.

(b) Sketch the graph of the polar equation from part (a).

(c) Use a graphing utility to approximate the area enclosed by the loop of the curve.

True or False? In Exercises 75 and 76, determine whether the statement is true or false. If it is false, explain why or give an example that shows it is false.

75. If $f(\theta) > 0$ for all θ and $g(\theta) < 0$ for all θ, then the graphs of $r = f(\theta)$ and $r = g(\theta)$ do not intersect.

76. If $f(\theta) = g(\theta)$ for $\theta = 0$, $\pi/2$, and $3\pi/2$, then the graphs of $r = f(\theta)$ and $r = g(\theta)$ have at least four points of intersection.

77. Use the formula for the arc length of a curve in parametric form to derive the formula for the arc length of a polar curve.

Section 10.6 Polar Equations of Conics and Kepler's Laws

- Analyze and write polar equations of conics.
- Understand and use Kepler's Laws of planetary motion.

EXPLORATION

Graphing Conics Set a graphing utility to *polar* mode and enter polar equations of the form

$$r = \frac{a}{1 \pm b \cos \theta}$$

or

$$r = \frac{a}{1 \pm b \sin \theta}.$$

As long as $a \neq 0$, the graph should be a conic. Describe the values of a and b that produce parabolas. What values produce ellipses? What values produce hyperbolas?

Polar Equations of Conics

In this chapter you have seen that the rectangular equations of ellipses and hyperbolas take simple forms when the origin lies at their *centers*. As it happens, there are many important applications of conics in which it is more convenient to use one of the foci as the reference point (the origin) for the coordinate system. For example, the sun lies at a focus of Earth's orbit. Similarly, the light source of a parabolic reflector lies at its focus. In this section you will see that polar equations of conics take simple forms if one of the foci lies at the pole.

The following theorem uses the concept of *eccentricity*, as defined in Section 10.1, to classify the three basic types of conics. A proof of this theorem is given in Appendix A.

THEOREM 10.16 Classification of Conics by Eccentricity

Let F be a fixed point (*focus*) and D be a fixed line (*directrix*) in the plane. Let P be another point in the plane and let e (*eccentricity*) be the ratio of the distance between P and F to the distance between P and D. The collection of all points P with a given eccentricity is a conic.

1. The conic is an ellipse if $0 < e < 1$.
2. The conic is a parabola if $e = 1$.
3. The conic is a hyperbola if $e > 1$.

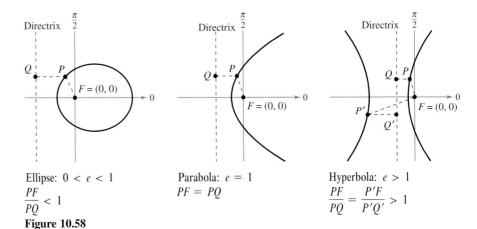

Ellipse: $0 < e < 1$
$\dfrac{PF}{PQ} < 1$

Parabola: $e = 1$
$PF = PQ$

Hyperbola: $e > 1$
$\dfrac{PF}{PQ} = \dfrac{P'F}{P'Q'} > 1$

Figure 10.58

In Figure 10.58, note that for each type of conic the pole corresponds to the fixed point (focus) given in the definition. The benefit of this location can be seen in the proof of the following theorem.

THEOREM 10.17 Polar Equations of Conics

The graph of a polar equation of the form

$$r = \frac{ed}{1 \pm e \cos \theta} \quad \text{or} \quad r = \frac{ed}{1 \pm e \sin \theta}$$

is a conic, where $e > 0$ is the eccentricity and $|d|$ is the distance between the focus at the pole and its corresponding directrix.

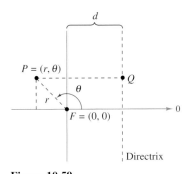

Figure 10.59

Proof The following is a proof for $r = ed/(1 + e \cos \theta)$ with $d > 0$. In Figure 10.59, consider a vertical directrix d units to the right of the focus $F = (0, 0)$. If $P = (r, \theta)$ is a point on the graph of $r = ed/(1 + e \cos \theta)$, the distance between P and the directrix can be shown to be

$$PQ = |d - x| = |d - r \cos \theta| = \left| \frac{r(1 + e \cos \theta)}{e} - r \cos \theta \right| = \left| \frac{r}{e} \right|.$$

Because the distance between P and the pole is simply $PF = |r|$, the ratio of PF to PQ is $PF/PQ = |r|/|r/e| = |e| = e$ and, by Theorem 10.16, the graph of the equation must be a conic. The proofs of the other cases are similar.

The four types of equations indicated in Theorem 10.17 can be classified as follows, where $d > 0$.

a. Horizontal directrix above the pole: $\quad r = \dfrac{ed}{1 + e \sin \theta}$

b. Horizontal directrix below the pole: $\quad r = \dfrac{ed}{1 - e \sin \theta}$

c. Vertical directrix to the right of the pole: $r = \dfrac{ed}{1 + e \cos \theta}$

d. Vertical directrix to the left of the pole: $\quad r = \dfrac{ed}{1 - e \cos \theta}$

Figure 10.60 illustrates these four possibilities for a parabola.

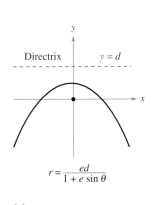
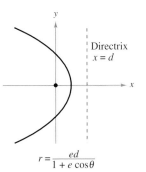

(a) (b) (c) (d)

The four types of polar equations for a parabola
Figure 10.60

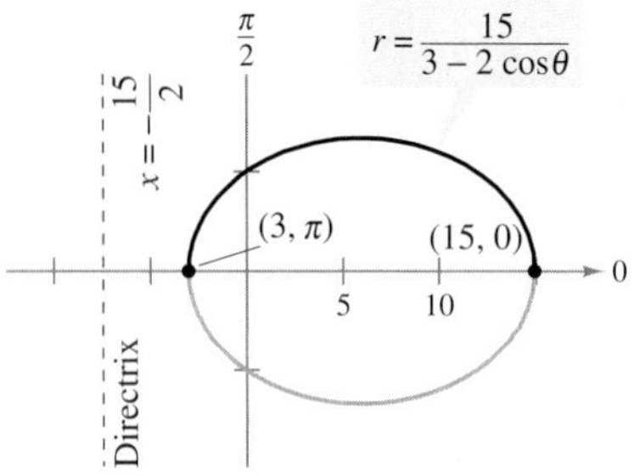

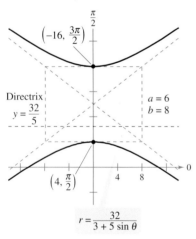

The graph of the conic is an ellipse with $e = \frac{2}{3}$.
Figure 10.61

EXAMPLE 1 Determining a Conic from Its Equation

Sketch the graph of the conic given by $r = \dfrac{15}{3 - 2\cos\theta}$.

Solution To determine the type of conic, rewrite the equation as

$$r = \frac{15}{3 - 2\cos\theta} \qquad \text{Write original equation.}$$

$$= \frac{5}{1 - (2/3)\cos\theta}. \qquad \text{Divide numerator and denominator by 3.}$$

So, the graph is an ellipse with $e = \frac{2}{3}$. You can sketch the upper half of the ellipse by plotting points from $\theta = 0$ to $\theta = \pi$, as shown in Figure 10.61. Then, using symmetry with respect to the polar axis, you can sketch the lower half.

For the ellipse in Figure 10.61, the major axis is horizontal and the vertices lie at $(15, 0)$ and $(3, \pi)$. So, the length of the *major* axis is $2a = 18$. To find the length of the minor axis, you can use the equations $e = c/a$ and $b^2 = a^2 - c^2$ to conclude

$$b^2 = a^2 - c^2 = a^2 - (ea)^2 = a^2(1 - e^2). \qquad \text{Ellipse}$$

Because $e = \frac{2}{3}$, you have

$$b^2 = 9^2\left[1 - \left(\tfrac{2}{3}\right)^2\right] = 45$$

which implies that $b = \sqrt{45} = 3\sqrt{5}$. So, the length of the minor axis is $2b = 6\sqrt{5}$. A similar analysis for hyperbolas yields

$$b^2 = c^2 - a^2 = (ea)^2 - a^2 = a^2(e^2 - 1). \qquad \text{Hyperbola}$$

EXAMPLE 2 Sketching a Conic from Its Polar Equation

Sketch the graph of the polar equation $r = \dfrac{32}{3 + 5\sin\theta}$.

Solution Dividing the numerator and denominator by 3 produces

$$r = \frac{32/3}{1 + (5/3)\sin\theta}.$$

Because $e = \frac{5}{3} > 1$, the graph is a hyperbola. Because $d = \frac{32}{5}$, the directrix is the line $y = \frac{32}{5}$. The transverse axis of the hyperbola lies on the line $\theta = \pi/2$, and the vertices occur at

$$(r, \theta) = \left(4, \frac{\pi}{2}\right) \quad \text{and} \quad (r, \theta) = \left(-16, \frac{3\pi}{2}\right).$$

Because the length of the transverse axis is 12, you can see that $a = 6$. To find b, write

$$b^2 = a^2(e^2 - 1) = 6^2\left[\left(\tfrac{5}{3}\right)^2 - 1\right] = 64.$$

Therefore, $b = 8$. Finally, you can use a and b to determine the asymptotes of the hyperbola and obtain the sketch shown in Figure 10.62.

The graph of the conic is a hyperbola with $e = \frac{5}{3}$.
Figure 10.62

JOHANNES KEPLER (1571–1630)

Kepler formulated his three laws from the extensive data recorded by Danish astronomer Tycho Brahe, and from direct observation of the orbit of Mars.

Kepler's Laws

Kepler's Laws, named after the German astronomer Johannes Kepler, can be used to describe the orbits of the planets about the sun.

1. Each planet moves in an elliptical orbit with the sun as a focus.
2. A ray from the sun to the planet sweeps out equal areas of the ellipse in equal times.
3. The square of the period is proportional to the cube of the mean distance between the planet and the sun.*

Although Kepler derived these laws empirically, they were later validated by Newton. In fact, Newton was able to show that each law can be deduced from a set of universal laws of motion and gravitation that govern the movement of all heavenly bodies, including comets and satellites. This is shown in the next example, involving the comet named after the English mathematician and physicist Edmund Halley (1656–1742).

EXAMPLE 3   Halley's Comet

Halley's comet has an elliptical orbit with the sun at one focus and has an eccentricity of $e \approx 0.967$. The length of the major axis of the orbit is approximately 35.88 astronomical units. (An astronomical unit is defined to be the mean distance between Earth and the sun, 93 million miles.) Find a polar equation for the orbit. How close does Halley's comet come to the sun?

Solution   Using a vertical axis, you can choose an equation of the form

$$r = \frac{ed}{(1 + e \sin \theta)}.$$

Because the vertices of the ellipse occur when $\theta = \pi/2$ and $\theta = 3\pi/2$, you can determine the length of the major axis to be the sum of the r-values of the vertices, as shown in Figure 10.63. That is,

$$2a = \frac{0.967d}{1 + 0.967} + \frac{0.967d}{1 - 0.967}$$

$$35.88 \approx 27.79d. \qquad 2a \approx 35.88$$

So, $d \approx 1.204$ and $ed \approx (0.967)(1.204) \approx 1.164$. Using this value in the equation produces

$$r = \frac{1.164}{1 + 0.967 \sin \theta}$$

where r is measured in astronomical units. To find the closest point to the sun (the focus), you can write $c = ea \approx (0.967)(17.94) \approx 17.35$. Because c is the distance between the focus and the center, the closest point is

$$a - c \approx 17.94 - 17.35$$
$$\approx 0.59 \text{ AU}$$
$$\approx 55{,}000{,}000 \text{ miles}$$

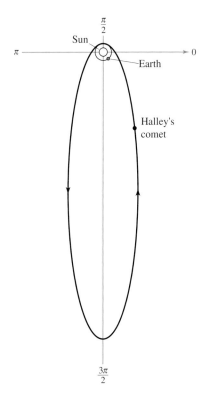

Figure 10.63

---

*If Earth is used as a reference with a period of 1 year and a distance of 1 astronomical unit, the proportionality constant is 1. For example, because Mars has a mean distance to the sun of $D = 1.524$ AU, its period P is given by $D^3 = P^2$. So, the period for Mars is $P = 1.88$.

Kepler's Second Law states that as a planet moves about the sun, a ray from the sun to the planet sweeps out equal areas in equal times. This law can also be applied to comets or asteroids with elliptical orbits. For example, Figure 10.64 shows the orbit of the asteroid Apollo about the sun. Applying Kepler's Second Law to this asteroid, you know that the closer it is to the sun, the greater its velocity, because a short ray must be moving quickly to sweep out as much area as a long ray.

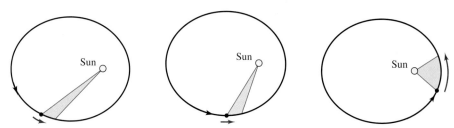

A ray from the sun to the asteroid sweeps out equal areas in equal times.
Figure 10.64

EXAMPLE 4 The Asteroid Apollo

The asteroid Apollo has a period of 661 Earth days, and its orbit is approximated by the ellipse

$$r = \frac{1}{1 + (5/9)\cos\theta} = \frac{9}{9 + 5\cos\theta}$$

where r is measured in astronomical units. How long does it take Apollo to move from the position given by $\theta = -\pi/2$ to $\theta = \pi/2$, as shown in Figure 10.65?

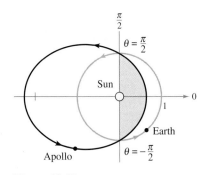

Figure 10.65

Solution Begin by finding the area swept out as θ increases from $-\pi/2$ to $\pi/2$.

$$A = \frac{1}{2}\int_\alpha^\beta r^2\, d\theta \qquad \text{Formula for area of a polar graph}$$

$$= \frac{1}{2}\int_{-\pi/2}^{\pi/2} \left(\frac{9}{9 + 5\cos\theta}\right)^2 d\theta$$

Using the substitution $u = \tan(\theta/2)$, as discussed in Section 8.6, you obtain

$$A = \frac{81}{112}\left[\frac{-5\sin\theta}{9 + 5\cos\theta} + \frac{18}{\sqrt{56}}\arctan\frac{\sqrt{56}\tan(\theta/2)}{14}\right]_{-\pi/2}^{\pi/2} \approx 0.90429.$$

Because the major axis of the ellipse has length $2a = 81/28$ and the eccentricity is $e = 5/9$, you can determine that $b = a\sqrt{1 - e^2} = 9/\sqrt{56}$. So, the area of the ellipse is

$$\text{Area of ellipse} = \pi ab = \pi\left(\frac{81}{56}\right)\left(\frac{9}{\sqrt{56}}\right) \approx 5.46507.$$

Because the time required to complete the orbit is 661 days, you can apply Kepler's Second Law to conclude that the time t required to move from the position $\theta = -\pi/2$ to $\theta = \pi/2$ is given by

$$\frac{t}{661} = \frac{\text{area of elliptical segment}}{\text{area of ellipse}} \approx \frac{0.90429}{5.46507}$$

which implies that $t \approx 109$ days.

Exercises for Section 10.6

Graphical Reasoning In Exercises 1–4, use a graphing utility to graph the polar equation when (a) $e = 1$, (b) $e = 0.5$, and (c) $e = 1.5$. Identify the conic.

1. $r = \dfrac{2e}{1 + e \cos \theta}$
2. $r = \dfrac{2e}{1 - e \cos \theta}$
3. $r = \dfrac{2e}{1 - e \sin \theta}$
4. $r = \dfrac{2e}{1 + e \sin \theta}$

5. **Writing** Consider the polar equation
$$r = \dfrac{4}{1 + e \sin \theta}.$$

   (a) Use a graphing utility to graph the equation for $e = 0.1$, $e = 0.25$, $e = 0.5$, $e = 0.75$, and $e = 0.9$. Identify the conic and discuss the change in its shape as $e \to 1^-$ and $e \to 0^+$.

   (b) Use a graphing utility to graph the equation for $e = 1$. Identify the conic.

   (c) Use a graphing utility to graph the equation for $e = 1.1$, $e = 1.5$, and $e = 2$. Identify the conic and discuss the change in its shape as $e \to 1^+$ and $e \to \infty$.

6. Consider the polar equation
$$r = \dfrac{4}{1 - 0.4 \cos \theta}.$$

   (a) Identify the conic without graphing the equation.

   (b) Without graphing the following polar equations, describe how each differs from the polar equation above.
   $$r = \dfrac{4}{1 + 0.4 \cos \theta}, \quad r = \dfrac{4}{1 - 0.4 \sin \theta}$$

   (c) Verify the results of part (b) graphically.

In Exercises 7–12, match the polar equation with the correct graph. [The graphs are labeled (a), (b), (c), (d), (e), and (f).]

(a) 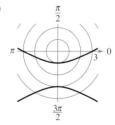   (b)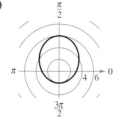

(c)    (d)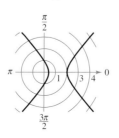

(e) 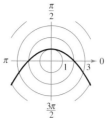   (f)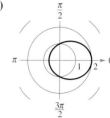

7. $r = \dfrac{6}{1 - \cos \theta}$
8. $r = \dfrac{2}{2 - \cos \theta}$
9. $r = \dfrac{3}{1 - 2 \sin \theta}$
10. $r = \dfrac{2}{1 + \sin \theta}$
11. $r = \dfrac{6}{2 - \sin \theta}$
12. $r = \dfrac{2}{2 + 3 \cos \theta}$

In Exercises 13–22, find the eccentricity and the distance from the pole to the directrix of the conic. Then sketch and identify the graph. Use a graphing utility to confirm your results.

13. $r = \dfrac{-1}{1 - \sin \theta}$
14. $r = \dfrac{6}{1 + \cos \theta}$
15. $r = \dfrac{6}{2 + \cos \theta}$
16. $r = \dfrac{5}{5 + 3 \sin \theta}$
17. $r(2 + \sin \theta) = 4$
18. $r(3 - 2 \cos \theta) = 6$
19. $r = \dfrac{5}{-1 + 2 \cos \theta}$
20. $r = \dfrac{-6}{3 + 7 \sin \theta}$
21. $r = \dfrac{3}{2 + 6 \sin \theta}$
22. $r = \dfrac{4}{1 + 2 \cos \theta}$

In Exercises 23–26, use a graphing utility to graph the polar equation. Identify the graph.

23. $r = \dfrac{3}{-4 + 2 \sin \theta}$
24. $r = \dfrac{-3}{2 + 4 \sin \theta}$
25. $r = \dfrac{-1}{1 - \cos \theta}$
26. $r = \dfrac{2}{2 + 3 \sin \theta}$

In Exercises 27–30, use a graphing utility to graph the conic. Describe how the graph differs from that in the indicated exercise.

27. $r = \dfrac{-1}{1 - \sin(\theta - \pi/4)}$ (See Exercise 13.)

28. $r = \dfrac{6}{1 + \cos(\theta - \pi/3)}$ (See Exercise 14.)

29. $r = \dfrac{6}{2 + \cos(\theta + \pi/6)}$ (See Exercise 15.)

30. $r = \dfrac{-6}{3 + 7\sin(\theta + 2\pi/3)}$ (See Exercise 20.)

31. Write the equation for the ellipse rotated $\pi/4$ radian clockwise from the ellipse
$$r = \dfrac{5}{5 + 3\cos\theta}.$$

32. Write the equation for the parabola rotated $\pi/6$ radian counterclockwise from the parabola
$$r = \dfrac{2}{1 + \sin\theta}.$$

In Exercises 33–44, find a polar equation for the conic with its focus at the pole. (For convenience, the equation for the directrix is given in rectangular form.)

Conic	Eccentricity	Directrix
33. Parabola	$e = 1$	$x = -1$
34. Parabola	$e = 1$	$y = 1$
35. Ellipse	$e = \tfrac{1}{2}$	$y = 1$
36. Ellipse	$e = \tfrac{3}{4}$	$y = -2$
37. Hyperbola	$e = 2$	$x = 1$
38. Hyperbola	$e = \tfrac{3}{2}$	$x = -1$

Conic	Vertex or Vertices
39. Parabola	$\left(1, -\tfrac{\pi}{2}\right)$
40. Parabola	$(5, \pi)$
41. Ellipse	$(2, 0), (8, \pi)$
42. Ellipse	$\left(2, \tfrac{\pi}{2}\right), \left(4, \tfrac{3\pi}{2}\right)$
43. Hyperbola	$\left(1, \tfrac{3\pi}{2}\right), \left(9, \tfrac{3\pi}{2}\right)$
44. Hyperbola	$(2, 0), (10, 0)$

Writing About Concepts

45. Classify the conics by their eccentricities.

46. Explain how the graph of each conic differs from the graph of $r = \dfrac{4}{1 + \sin\theta}$.

(a) $r = \dfrac{4}{1 - \cos\theta}$ (b) $r = \dfrac{4}{1 - \sin\theta}$

(c) $r = \dfrac{4}{1 + \cos\theta}$ (d) $r = \dfrac{4}{1 - \sin(\theta - \pi/4)}$

47. Identify each conic.

(a) $r = \dfrac{5}{1 - 2\cos\theta}$ (b) $r = \dfrac{5}{10 - \sin\theta}$

(c) $r = \dfrac{5}{3 - 3\cos\theta}$ (d) $r = \dfrac{5}{1 - 3\sin(\theta - \pi/4)}$

48. Describe what happens to the distance between the directrix and the center of an ellipse if the foci remain fixed and e approaches 0.

49. Show that the polar equation for $\dfrac{x^2}{a^2} + \dfrac{y^2}{b^2} = 1$ is
$$r^2 = \dfrac{b^2}{1 - e^2\cos^2\theta}. \quad \text{Ellipse}$$

50. Show that the polar equation for $\dfrac{x^2}{a^2} - \dfrac{y^2}{b^2} = 1$ is
$$r^2 = \dfrac{-b^2}{1 - e^2\cos^2\theta}. \quad \text{Hyperbola}$$

In Exercises 51–54, use the results of Exercises 49 and 50 to write the polar form of the equation of the conic.

51. Ellipse: focus at $(4, 0)$; vertices at $(5, 0), (5, \pi)$

52. Hyperbola: focus at $(5, 0)$; vertices at $(4, 0), (4, \pi)$

53. $\dfrac{x^2}{9} - \dfrac{y^2}{16} = 1$

54. $\dfrac{x^2}{4} + y^2 = 1$

 In Exercises 55 and 56, use the integration capabilities of a graphing utility to approximate to two decimal places the area of the region bounded by the graph of the polar equation.

55. $r = \dfrac{3}{2 - \cos\theta}$

56. $r = \dfrac{2}{3 - 2\sin\theta}$

57. *Explorer 18* On November 27, 1963, the United States launched Explorer 18. Its low and high points above the surface of Earth were approximately 119 miles and 123,000 miles (see figure). The center of Earth is the focus of the orbit. Find the polar equation for the orbit and find the distance between the surface of Earth and the satellite when $\theta = 60°$. (Assume that the radius of Earth is 4000 miles.)

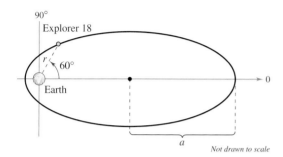

58. *Planetary Motion* The planets travel in elliptical orbits with the sun as a focus, as shown in the figure.

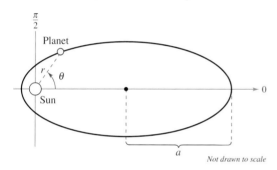

(a) Show that the polar equation of the orbit is given by

$$r = \frac{(1 - e^2)a}{1 - e \cos \theta}$$

where e is the eccentricity.

(b) Show that the minimum distance (*perihelion*) from the sun to the planet is $r = a(1 - e)$ and the maximum distance (*aphelion*) is $r = a(1 + e)$.

In Exercises 59–62, use Exercise 58 to find the polar equation of the elliptical orbit of the planet, and the perihelion and aphelion distances.

59. Earth $a = 1.496 \times 10^8$ kilometers
                   $e = 0.0167$

60. Saturn $a = 1.427 \times 10^9$ kilometers
                   $e = 0.0542$

61. Pluto $a = 5.906 \times 10^9$ kilometers
                   $e = 0.2488$

62. Mercury $a = 5.791 \times 10^7$ kilometers
                   $e = 0.2056$

 63. *Planetary Motion* In Exercise 61, the polar equation for the elliptical orbit of Pluto was found. Use the equation and a computer algebra system to perform each of the following.

(a) Approximate the area swept out by a ray from the sun to the planet as θ increases from 0 to $\pi/9$. Use this result to determine the number of years for the planet to move through this arc if the period of one revolution around the sun is 248 years.

(b) By trial and error, approximate the angle α such that the area swept out by a ray from the sun to the planet as θ increases from π to α equals the area found in part (a) (see figure). Does the ray sweep through a larger or smaller angle than in part (a) to generate the same area? Why is this the case?

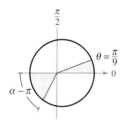

(c) Approximate the distances the planet traveled in parts (a) and (b). Use these distances to approximate the average number of kilometers per year the planet traveled in the two cases.

64. *Comet Hale-Bopp* The comet Hale-Bopp has an elliptical orbit with the sun at one focus and has an eccentricity of $e \approx 0.995$. The length of the major axis of the orbit is approximately 250 astronomical units.

(a) Find the length of its minor axis.

(b) Find a polar equation for the orbit.

(c) Find the perihelion and aphelion distances.

In Exercises 65 and 66, let r_0 represent the distance from the focus to the nearest vertex, and let r_1 represent the distance from the focus to the farthest vertex.

65. Show that the eccentricity of an ellipse can be written as

$$e = \frac{r_1 - r_0}{r_1 + r_0}. \text{ Then show that } \frac{r_1}{r_0} = \frac{1 + e}{1 - e}.$$

66. Show that the eccentricity of a hyperbola can be written as

$$e = \frac{r_1 + r_0}{r_1 - r_0}. \text{ Then show that } \frac{r_1}{r_0} = \frac{e + 1}{e - 1}.$$

In Exercises 67 and 68, show that the graphs of the given equations intersect at right angles.

67. $r = \dfrac{ed}{1 + \sin \theta}$ and $r = \dfrac{ed}{1 - \sin \theta}$

68. $r = \dfrac{c}{1 + \cos \theta}$ and $r = \dfrac{d}{1 - \cos \theta}$

Review Exercises for Chapter 10

See www.CalcChat.com for worked-out solutions to odd-numbered exercises.

In Exercises 1–6, match the equation with the correct graph. [The graphs are labeled (a), (b), (c), (d), (e), and (f).]

(a)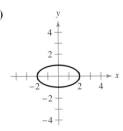

(b)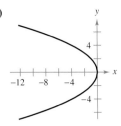

(c)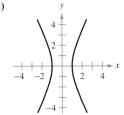

(d)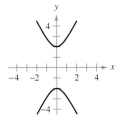

(e)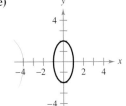

(f)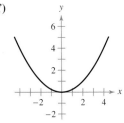

1. $4x^2 + y^2 = 4$
2. $4x^2 - y^2 = 4$
3. $y^2 = -4x$
4. $y^2 - 4x^2 = 4$
5. $x^2 + 4y^2 = 4$
6. $x^2 = 4y$

In Exercises 7–12, analyze the equation and sketch its graph. Use a graphing utility to confirm your results.

7. $16x^2 + 16y^2 - 16x + 24y - 3 = 0$
8. $y^2 - 12y - 8x + 20 = 0$
9. $3x^2 - 2y^2 + 24x + 12y + 24 = 0$
10. $4x^2 + y^2 - 16x + 15 = 0$
11. $3x^2 + 2y^2 - 12x + 12y + 29 = 0$
12. $4x^2 - 4y^2 - 4x + 8y - 11 = 0$

In Exercises 13 and 14, find an equation of the parabola.

13. Vertex: $(0, 2)$; directrix: $x = -3$
14. Vertex: $(4, 2)$; focus: $(4, 0)$

In Exercises 15 and 16, find an equation of the ellipse.

15. Vertices: $(-3, 0), (7, 0)$; foci: $(0, 0), (4, 0)$
16. Center: $(0, 0)$; solution points: $(1, 2), (2, 0)$

In Exercises 17 and 18, find an equation of the hyperbola.

17. Vertices: $(\pm 4, 0)$; foci: $(\pm 6, 0)$
18. Foci: $(0, \pm 8)$; asymptotes: $y = \pm 4x$

In Exercises 19 and 20, use a graphing utility to approximate the perimeter of the ellipse.

19. $\dfrac{x^2}{9} + \dfrac{y^2}{4} = 1$

20. $\dfrac{x^2}{4} + \dfrac{y^2}{25} = 1$

21. A line is tangent to the parabola $y = x^2 - 2x + 2$ and perpendicular to the line $y = x - 2$. Find the equation of the line.

22. A line is tangent to the parabola $3x^2 + y = x - 6$ and perpendicular to the line $2x + y = 5$. Find the equation of the line.

23. **Satellite Antenna** A cross section of a large parabolic antenna is modeled by the graph of

$$y = \frac{x^2}{200}, \quad -100 \le x \le 100.$$

The receiving and transmitting equipment is positioned at the focus.

(a) Find the coordinates of the focus.

(b) Find the surface area of the antenna.

24. **Fire Truck** Consider a fire truck with a water tank 16 feet long whose vertical cross sections are ellipses modeled by the equation

$$\frac{x^2}{16} + \frac{y^2}{9} = 1.$$

(a) Find the volume of the tank.

(b) Find the force on the end of the tank when it is full of water. (The density of water is 62.4 pounds per cubic foot.)

(c) Find the depth of the water in the tank if it is $\frac{3}{4}$ full (by volume) and the truck is on level ground.

(d) Approximate the tank's surface area.

In Exercises 25–30, sketch the curve represented by the parametric equations (indicate the orientation of the curve), and write the corresponding rectangular equation by eliminating the parameter.

25. $x = 1 + 4t, \ y = 2 - 3t$
26. $x = t + 4, \ y = t^2$
27. $x = 6\cos\theta, \ y = 6\sin\theta$
28. $x = 3 + 3\cos\theta, \ y = 2 + 5\sin\theta$
29. $x = 2 + \sec\theta, \ y = 3 + \tan\theta$
30. $x = 5\sin^3\theta, \ y = 5\cos^3\theta$

In Exercises 31–34, find a parametric representation of the line or conic.

31. Line: passes through $(-2, 6)$ and $(3, 2)$
32. Circle: center at $(5, 3)$; radius 2
33. Ellipse: center at $(-3, 4)$; horizontal major axis of length 8 and minor axis of length 6
34. Hyperbola: vertices at $(0, \pm 4)$; foci at $(0, \pm 5)$

 35. *Rotary Engine* The rotary engine was developed by Felix Wankel in the 1950s. It features a rotor, which is a modified equilateral triangle. The rotor moves in a chamber that, in two dimensions, is an epitrochoid. Use a graphing utility to graph the chamber modeled by the parametric equations.

$$x = \cos 3\theta + 5 \cos \theta$$

and

$$y = \sin 3\theta + 5 \sin \theta.$$

36. *Serpentine Curve* Consider the parametric equations $x = 2 \cot \theta$ and $y = 4 \sin \theta \cos \theta$, $0 < \theta < \pi$.

   (a) Use a graphing utility to graph the curve.

   (b) Eliminate the parameter to show that the rectangular equation of the serpentine curve is $(4 + x^2)y = 8x$.

In Exercises 37–46, (a) find dy/dx and all points of horizontal tangency, (b) eliminate the parameter where possible, and (c) sketch the curve represented by the parametric equations.

37. $x = 1 + 4t$, $y = 2 - 3t$
38. $x = t + 4$, $y = t^2$
39. $x = \dfrac{1}{t}$, $y = 2t + 3$
40. $x = \dfrac{1}{t}$, $y = t^2$
41. $x = \dfrac{1}{2t + 1}$
    $y = \dfrac{1}{t^2 - 2t}$
42. $x = 2t - 1$
    $y = \dfrac{1}{t^2 - 2t}$
43. $x = 3 + 2 \cos \theta$
    $y = 2 + 5 \sin \theta$
44. $x = 6 \cos \theta$
    $y = 6 \sin \theta$
45. $x = \cos^3 \theta$
    $y = 4 \sin^3 \theta$
46. $x = e^t$
    $y = e^{-t}$

In Exercises 47–50, find all points (if any) of horizontal and vertical tangency to the curve. Use a graphing utility to confirm your results.

47. $x = 4 - t$, $y = t^2$
48. $x = t + 2$, $y = t^3 - 2t$
49. $x = 2 + 2 \sin \theta$, $y = 1 + \cos \theta$
50. $x = 2 - 2 \cos \theta$, $y = 2 \sin 2\theta$

 In Exercises 51 and 52, (a) use a graphing utility to graph the curve represented by the parametric equations, (b) use a graphing utility to find $dx/d\theta$, $dy/d\theta$, and dy/dx for $\theta = \pi/6$, and (c) use a graphing utility to graph the tangent line to the curve when $\theta = \pi/6$.

51. $x = \cot \theta$
    $y = \sin 2\theta$

52. $x = 2\theta - \sin \theta$
    $y = 2 - \cos \theta$

Arc Length **In Exercises 53 and 54, find the arc length of the curve on the given interval.**

53. $x = r(\cos \theta + \theta \sin \theta)$
    $y = r(\sin \theta - \theta \cos \theta)$
    $0 \leq \theta \leq \pi$

54. $x = 6 \cos \theta$
    $y = 6 \sin \theta$
    $0 \leq \theta \leq \pi$

Surface Area **In Exercises 55 and 56, find the area of the surface generated by revolving the curve about (a) the x-axis and (b) the y-axis.**

55. $x = t$, $y = 3t$, $0 \leq t \leq 2$

56. $x = 2 \cos \theta$, $y = 2 \sin \theta$, $0 \leq \theta \leq \dfrac{\pi}{2}$

Area **In Exercises 57 and 58, find the area of the region.**

57. $x = 3 \sin \theta$
    $y = 2 \cos \theta$
    $-\dfrac{\pi}{2} \leq \theta \leq \dfrac{\pi}{2}$

58. $x = 2 \cos \theta$
    $y = \sin \theta$
    $0 \leq \theta \leq \pi$

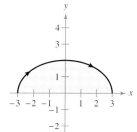
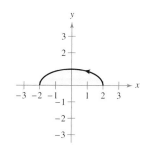

In Exercises 59–62, plot the point in polar coordinates and find the corresponding rectangular coordinates of the point.

59. $\left(3, \dfrac{\pi}{2}\right)$
60. $\left(-4, \dfrac{11\pi}{6}\right)$
61. $(\sqrt{3}, 1.56)$
62. $(-2, -2.45)$

In Exercises 63 and 64, the rectangular coordinates of a point are given. Plot the point and find *two* sets of polar coordinates of the point for $0 \leq \theta < 2\pi$.

63. $(4, -4)$
64. $(-1, 3)$

In Exercises 65–72, convert the polar equation to rectangular form.

65. $r = 3 \cos \theta$
66. $r = 10$
67. $r = -2(1 + \cos \theta)$
68. $r = \dfrac{1}{2 - \cos \theta}$
69. $r^2 = \cos 2\theta$
70. $r = 4 \sec\left(\theta - \dfrac{\pi}{3}\right)$
71. $r = 4 \cos 2\theta \sec \theta$
72. $\theta = \dfrac{3\pi}{4}$

In Exercises 73–76, convert the rectangular equation to polar form.

73. $(x^2 + y^2)^2 = ax^2y$
74. $x^2 + y^2 - 4x = 0$
75. $x^2 + y^2 = a^2\left(\arctan \dfrac{y}{x}\right)^2$
76. $(x^2 + y^2)\left(\arctan \dfrac{y}{x}\right)^2 = a^2$

In Exercises 77–88, sketch a graph of the polar equation.

77. $r = 4$
78. $\theta = \dfrac{\pi}{12}$
79. $r = -\sec \theta$
80. $r = 3 \csc \theta$
81. $r = -2(1 + \cos \theta)$
82. $r = 3 - 4 \cos \theta$
83. $r = 4 - 3 \cos \theta$
84. $r = 2\theta$
85. $r = -3 \cos 2\theta$
86. $r = \cos 5\theta$
87. $r^2 = 4 \sin^2 2\theta$
88. $r^2 = \cos 2\theta$

 In Exercises 89–92, use a graphing utility to graph the polar equation.

89. $r = \dfrac{3}{\cos(\theta - \pi/4)}$
90. $r = 2 \sin \theta \cos^2 \theta$
91. $r = 4 \cos 2\theta \sec \theta$
92. $r = 4(\sec \theta - \cos \theta)$

 In Exercises 93 and 94, (a) find the tangents at the pole, (b) find all points of vertical and horizontal tangency, and (c) use a graphing utility to graph the polar equation and draw a tangent line to the graph for $\theta = \pi/6$.

93. $r = 1 - 2 \cos \theta$
94. $r^2 = 4 \sin 2\theta$

95. Find the angle between the circle $r = 3 \sin \theta$ and the limaçon $r = 4 - 5 \sin \theta$ at the point of intersection $(3/2, \pi/6)$.

96. **True or False?** There is a unique polar coordinate representation for each point in the plane. Explain.

In Exercises 97 and 98, show that the graphs of the polar equations are orthogonal at the points of intersection. Use a graphing utility to confirm your results graphically.

97. $r = 1 + \cos \theta$
    $r = 1 - \cos \theta$
98. $r = a \sin \theta$
    $r = a \cos \theta$

In Exercises 99–102, find the area of the region.

99. Interior of $r = 2 + \cos \theta$
100. Interior of $r = 5(1 - \sin \theta)$
101. Interior of $r^2 = 4 \sin 2\theta$
102. Common interior of $r = 4 \cos \theta$ and $r = 2$

 In Exercises 103–106, use a graphing utility to graph the polar equation. Set up an integral for finding the area of the given region and use the integration capabilities of a graphing utility to approximate the integral accurate to two decimal places.

103. Interior of $r = \sin \theta \cos^2 \theta$
104. Interior of $r = 4 \sin 3\theta$
105. Common interior of $r = 3$ and $r^2 = 18 \sin 2\theta$
106. Region bounded by the polar axis and $r = e^\theta$ for $0 \leq \theta \leq \pi$

In Exercises 107 and 108, find the length of the curve over the given interval.

Polar Equation	Interval
107. $r = a(1 - \cos \theta)$	$0 \leq \theta \leq \pi$
108. $r = a \cos 2\theta$	$-\dfrac{\pi}{2} \leq \theta \leq \dfrac{\pi}{2}$

 In Exercises 109 and 110, write an integral that represents the area of the surface formed by revolving the curve about the given line. Use a graphing utility to approximate the integral.

Polar Equation	Interval	Axis of Revolution
109. $r = 1 + 4 \cos \theta$	$0 \leq \theta \leq \dfrac{\pi}{2}$	Polar axis
110. $r = 2 \sin \theta$	$0 \leq \theta \leq \dfrac{\pi}{2}$	$\theta = \dfrac{\pi}{2}$

In Exercises 111–116, sketch and identify the graph. Use a graphing utility to confirm your results.

111. $r = \dfrac{2}{1 - \sin \theta}$
112. $r = \dfrac{2}{1 + \cos \theta}$
113. $r = \dfrac{6}{3 + 2 \cos \theta}$
114. $r = \dfrac{4}{5 - 3 \sin \theta}$
115. $r = \dfrac{4}{2 - 3 \sin \theta}$
116. $r = \dfrac{8}{2 - 5 \cos \theta}$

In Exercises 117–122, find a polar equation for the line or conic with its focus at the pole.

117. Circle
    Center: $(5, \pi/2)$
    Solution point: $(0, 0)$
118. Line
    Solution point: $(0, 0)$
    Slope: $\sqrt{3}$
119. Parabola
    Vertex: $(2, \pi)$
120. Parabola
    Vertex: $(2, \pi/2)$
121. Ellipse
    Vertices: $(5, 0), (1, \pi)$
122. Hyperbola
    Vertices: $(1, 0), (7, 0)$

P.S. Problem Solving

1. Consider the parabola $x^2 = 4y$ and the focal chord $y = \frac{3}{4}x + 1$.
   (a) Sketch the graph of the parabola and the focal chord.
   (b) Show that the tangent lines to the parabola at the endpoints of the focal chord intersect at right angles.
   (c) Show that the tangent lines to the parabola at the endpoints of the focal chord intersect on the directrix of the parabola.

2. Consider the parabola $x^2 = 4py$ and one of its focal chords.
   (a) Show that the tangent lines to the parabola at the endpoints of the focal chord intersect at right angles.
   (b) Show that the tangent lines to the parabola at the endpoints of the focal chord intersect on the directrix of the parabola.

3. Prove Theorem 10.2, Reflective Property of a Parabola, as shown in the figure.

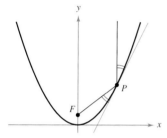

4. Consider the hyperbola
   $$\frac{x^2}{a^2} - \frac{y^2}{b^2} = 1$$
   with foci F_1 and F_2, as shown in the figure. Let T be the tangent line at a point M on the hyperbola. Show that incoming rays of light aimed at one focus are reflected by a hyperbolic mirror toward the other focus.

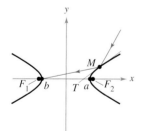

Figure for 4   **Figure for 5**

5. Consider a circle of radius a tangent to the y-axis and the line $x = 2a$, as shown in the figure. Let A be the point where the segment OB intersects the circle. The **cissoid of Diocles** consists of all points P such that $OP = AB$.
   (a) Find a polar equation of the cissoid.
   (b) Find a set of parametric equations for the cissoid that does not contain trigonometric functions.
   (c) Find a rectangular equation of the cissoid.

6. Consider the region bounded by the ellipse $x^2/a^2 + y^2/b^2 = 1$, with eccentricity $e = c/a$.
   (a) Show that the area of the region is πab.
   (b) Show that the solid (oblate spheroid) generated by revolving the region about the minor axis of the ellipse has a volume $V = 4\pi^2 b/3$ and a surface area of
   $$S = 2\pi a^2 + \pi\left(\frac{b^2}{e}\right)\ln\left(\frac{1+e}{1-e}\right).$$
   (c) Show that the solid (prolate spheroid) generated by revolving the region about the major axis of the ellipse has a volume of $V = 4\pi ab^2/3$ and a surface area of
   $$S = 2\pi b^2 + 2\pi\left(\frac{ab}{e}\right)\arcsin e.$$

7. The curve given by the parametric equations
   $$x(t) = \frac{1-t^2}{1+t^2} \text{ and } y(t) = \frac{t(1-t^2)}{1+t^2}$$
   is called a **strophoid**.
   (a) Find a rectangular equation of the strophoid.
   (b) Find a polar equation of the strophoid.
   (c) Sketch a graph of the strophoid.
   (d) Find the equations of the two tangent lines at the origin.
   (e) Find the points on the graph where the tangent lines are horizontal.

8. Find a rectangular equation of the portion of the cycloid given by the parametric equations $x = a(\theta - \sin\theta)$ and $y = a(1 - \cos\theta)$, $0 \leq \theta \leq \pi$, as shown in the figure.

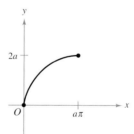

9. Consider the **cornu spiral** given by
   $$x(t) = \int_0^t \cos\left(\frac{\pi u^2}{2}\right)du \text{ and } y(t) = \int_0^t \sin\left(\frac{\pi u^2}{2}\right)du.$$
   (a) Use a graphing utility to graph the spiral over the interval $-\pi \leq t \leq \pi$.
   (b) Show that the cornu spiral is symmetric with respect to the origin.
   (c) Find the length of the cornu spiral from $t = 0$ to $t = a$. What is the length of the spiral from $t = -\pi$ to $t = \pi$?

10. A particle is moving along the path described by the parametric equations $x = 1/t$ and $y = \sin t/t$, for $1 \leq t < \infty$, as shown in the figure. Find the length of this path.

11. Let a and b be positive constants. Find the area of the region in the first quadrant bounded by the graph of the polar equation
$$r = \frac{ab}{(a \sin \theta + b \cos \theta)}, \quad 0 \leq \theta \leq \frac{\pi}{2}.$$

12. Consider the right triangle shown in the figure.

    (a) Show that the area of the triangle is $A(\alpha) = \dfrac{1}{2}\displaystyle\int_0^\alpha \sec^2 \theta \, d\theta$.

    (b) Show that $\tan \alpha = \displaystyle\int_0^\alpha \sec^2 \theta \, d\theta$.

    (c) Use part (b) to derive the formula for the derivative of the tangent function.

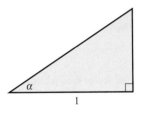 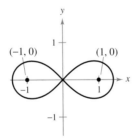

Figure for 12   **Figure for 13**

13. Determine the polar equation of the set of all points (r, θ), the product of whose distances from the points $(1, 0)$ and $(-1, 0)$ is equal to 1, as shown in the figure.

14. Four dogs are located at the corners of a square with sides of length d. The dogs all move counterclockwise at the same speed directly toward the next dog, as shown in the figure. Find the polar equation of a dog's path as it spirals toward the center of the square.

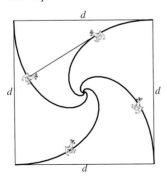

15. An air traffic controller spots two planes at the same altitude flying toward each other (see figure). Their flight paths are 20° and 315°. One plane is 150 miles from point P with a speed of 375 miles per hour. The other is 190 miles from point P with a speed of 450 miles per hour.

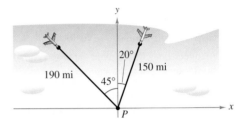

    (a) Find parametric equations for the path of each plane where t is the time in hours, with $t = 0$ corresponding to the time at which the air traffic controller spots the planes.

    (b) Use the result of part (a) to write the distance between the planes as a function of t.

    (c) Use a graphing utility to graph the function in part (b). When will the distance between the planes be minimum? If the planes must keep a separation of at least 3 miles, is the requirement met?

16. Use a graphing utility to graph the curve shown below. The curve is given by
$$r = e^{\cos \theta} - 2 \cos 4\theta + \sin^5 \frac{\theta}{12}.$$

Over what interval must θ vary to produce the curve?

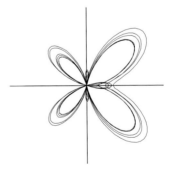

FOR FURTHER INFORMATION For more information on this curve, see the article "A Study in Step Size" by Temple H. Fay in *Mathematics Magazine*. To view this article, go to the website www.matharticles.com.

17. Use a graphing utility to graph the polar equation $r = \cos 5\theta + n \cos \theta$, for $0 \leq \theta < \pi$ and for the integers $n = -5$ to $n = 5$. What values of n produce the "heart" portion of the curve? What values of n produce the "bell" portion? (This curve, created by Michael W. Chamberlin, appeared in *The College Mathematics Journal*.)

Vectors and the Geometry of Space

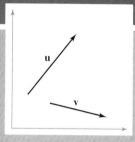

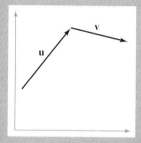

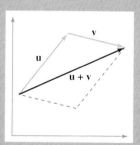

Vectors indicate quantities that involve both magnitude and direction. In Chapter 11, you will study operations of vectors in the plane and in space. You will also learn how to represent vector operations geometrically. For example, the graphs shown above represent vector addition in the plane.

To make a hot air balloon rise, a pilot increases the flow of propane gas to a burner, which in turn increases the temperature of the air inside the balloon. To slow the balloon's ascent and eventually force it to descend, the pilot decreases the temperature of the air inside the balloon by opening the parachute valve at the top of the balloon. How do you think pilots control the horizontal speed and direction of the hot air balloon? Explain.

Reuters/Corbis

Section 11.1 Vectors in the Plane

- Write the component form of a vector.
- Perform vector operations and interpret the results geometrically.
- Write a vector as a linear combination of standard unit vectors.
- Use vectors to solve problems involving force or velocity.

Component Form of a Vector

Many quantities in geometry and physics, such as area, volume, temperature, mass, and time, can be characterized by a single real number scaled to appropriate units of measure. These are called **scalar quantities,** and the real number associated with each is called a **scalar.**

Other quantities, such as force, velocity, and acceleration, involve both magnitude and direction and cannot be characterized completely by a single real number. A **directed line segment** is used to represent such a quantity, as shown in Figure 11.1. The directed line segment $\overrightarrow{PQ}$ has **initial point** P and **terminal point** Q, and its **length** (or **magnitude**) is denoted by $\|\overrightarrow{PQ}\|$. Directed line segments that have the same length and direction are **equivalent,** as shown in Figure 11.2. The set of all directed line segments that are equivalent to a given directed line segment $\overrightarrow{PQ}$ is a **vector in the plane** and is denoted by $\mathbf{v} = \overrightarrow{PQ}$. In typeset material, vectors are usually denoted by lowercase, boldface letters such as $\mathbf{u}$, $\mathbf{v}$, and $\mathbf{w}$. When written by hand, however, vectors are often denoted by letters with arrows above them, such as $\vec{u}$, $\vec{v}$, and $\vec{w}$.

Be sure you see that a vector in the plane can be represented by many different directed line segments—all pointing in the same direction and all of the same length.

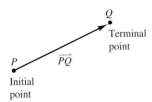

A directed line segment
Figure 11.1

Equivalent directed line segments
Figure 11.2

EXAMPLE 1   Vector Representation by Directed Line Segments

Let $\mathbf{v}$ be represented by the directed line segment from $(0, 0)$ to $(3, 2)$, and let $\mathbf{u}$ be represented by the directed line segment from $(1, 2)$ to $(4, 4)$. Show that $\mathbf{v}$ and $\mathbf{u}$ are equivalent.

Solution   Let $P(0, 0)$ and $Q(3, 2)$ be the initial and terminal points of $\mathbf{v}$, and let $R(1, 2)$ and $S(4, 4)$ be the initial and terminal points of $\mathbf{u}$, as shown in Figure 11.3. You can use the Distance Formula to show that $\overrightarrow{PQ}$ and $\overrightarrow{RS}$ have the *same length*.

$$\|\overrightarrow{PQ}\| = \sqrt{(3-0)^2 + (2-0)^2} = \sqrt{13} \qquad \text{Length of } \overrightarrow{PQ}$$
$$\|\overrightarrow{RS}\| = \sqrt{(4-1)^2 + (4-2)^2} = \sqrt{13} \qquad \text{Length of } \overrightarrow{RS}$$

Both line segments have the *same direction*, because they both are directed toward the upper right on lines having the same slope.

$$\text{Slope of } \overrightarrow{PQ} = \frac{2-0}{3-0} = \frac{2}{3}$$

and

$$\text{Slope of } \overrightarrow{RS} = \frac{4-2}{4-1} = \frac{2}{3}$$

Because $\overrightarrow{PQ}$ and $\overrightarrow{RS}$ have the same length and direction, you can conclude that the two vectors are equivalent. That is, $\mathbf{v}$ and $\mathbf{u}$ are equivalent.

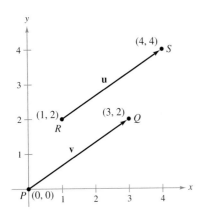

The vectors $\mathbf{u}$ and $\mathbf{v}$ are equivalent.
Figure 11.3

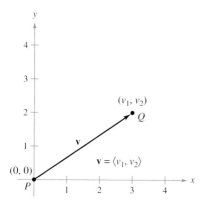

The standard position of a vector
Figure 11.4

The directed line segment whose initial point is the origin is often the most convenient representative of a set of equivalent directed line segments such as those shown in Figure 11.3. This representation of **v** is said to be in **standard position.** A directed line segment whose initial point is the origin can be uniquely represented by the coordinates of its terminal point $Q(v_1, v_2)$, as shown in Figure 11.4.

Definition of Component Form of a Vector in the Plane

If **v** is a vector in the plane whose initial point is the origin and whose terminal point is (v_1, v_2), then the **component form of v** is given by

$$\mathbf{v} = \langle v_1, v_2 \rangle.$$

The coordinates v_1 and v_2 are called the **components of v.** If both the initial point and the terminal point lie at the origin, then **v** is called the **zero vector** and is denoted by $\mathbf{0} = \langle 0, 0 \rangle$.

This definition implies that two vectors $\mathbf{u} = \langle u_1, u_2 \rangle$ and $\mathbf{v} = \langle v_1, v_2 \rangle$ are **equal** if and only if $u_1 = v_1$ and $u_2 = v_2$.

The following procedures can be used to convert directed line segments to component form or vice versa.

NOTE It is important to understand that a vector represents a *set* of directed line segments (each having the same length and direction). In practice, however, it is common not to distinguish between a vector and one of its representatives.

1. If $P(p_1, p_2)$ and $Q(q_1, q_2)$ are the initial and terminal points of a directed line segment, the component form of the vector **v** represented by $\overrightarrow{PQ}$ is $\langle v_1, v_2 \rangle = \langle q_1 - p_1, q_2 - p_2 \rangle$. Moreover, the **length** (or **magnitude**) of **v** is

$$\|\mathbf{v}\| = \sqrt{(q_1 - p_1)^2 + (q_2 - p_2)^2} \qquad \text{Length of a vector}$$
$$= \sqrt{v_1^2 + v_2^2}.$$

2. If $\mathbf{v} = \langle v_1, v_2 \rangle$, **v** can be represented by the directed line segment, in standard position, from $P(0, 0)$ to $Q(v_1, v_2)$.

The length of **v** is also called the **norm of v.** If $\|\mathbf{v}\| = 1$, **v** is a **unit vector.** Moreover, $\|\mathbf{v}\| = 0$ if and only if **v** is the zero vector **0**.

EXAMPLE 2 Finding the Component Form and Length of a Vector

Find the component form and length of the vector **v** that has initial point $(3, -7)$ and terminal point $(-2, 5)$.

Solution Let $P(3, -7) = (p_1, p_2)$ and $Q(-2, 5) = (q_1, q_2)$. Then the components of $\mathbf{v} = \langle v_1, v_2 \rangle$ are

$$v_1 = q_1 - p_1 = -2 - 3 = -5$$
$$v_2 = q_2 - p_2 = 5 - (-7) = 12.$$

So, as shown in Figure 11.5, $\mathbf{v} = \langle -5, 12 \rangle$, and the length of **v** is

$$\|\mathbf{v}\| = \sqrt{(-5)^2 + 12^2}$$
$$= \sqrt{169}$$
$$= 13.$$

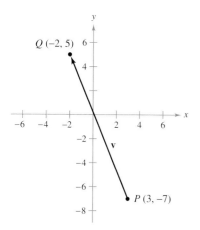

Component form of **v**: $\mathbf{v} = \langle -5, 12 \rangle$
Figure 11.5

Vector Operations

> **Definitions of Vector Addition and Scalar Multiplication**
>
> Let $\mathbf{u} = \langle u_1, u_2 \rangle$ and $\mathbf{v} = \langle v_1, v_2 \rangle$ be vectors and let c be a scalar.
>
> 1. The **vector sum** of $\mathbf{u}$ and $\mathbf{v}$ is the vector $\mathbf{u} + \mathbf{v} = \langle u_1 + v_1, u_2 + v_2 \rangle$.
> 2. The **scalar multiple** of c and $\mathbf{u}$ is the vector $c\mathbf{u} = \langle cu_1, cu_2 \rangle$.
> 3. The **negative** of $\mathbf{v}$ is the vector
> $$-\mathbf{v} = (-1)\mathbf{v} = \langle -v_1, -v_2 \rangle.$$
> 4. The **difference** of $\mathbf{u}$ and $\mathbf{v}$ is
> $$\mathbf{u} - \mathbf{v} = \mathbf{u} + (-\mathbf{v}) = \langle u_1 - v_1, u_2 - v_2 \rangle.$$

The scalar multiplication of $\mathbf{v}$
Figure 11.6

Geometrically, the scalar multiple of a vector $\mathbf{v}$ and a scalar c is the vector that is $|c|$ times as long as $\mathbf{v}$, as shown in Figure 11.6. If c is positive, $c\mathbf{v}$ has the same direction as $\mathbf{v}$. If c is negative, $c\mathbf{v}$ has the opposite direction.

The sum of two vectors can be represented geometrically by positioning the vectors (without changing their magnitudes or directions) so that the initial point of one coincides with the terminal point of the other, as shown in Figure 11.7. The vector $\mathbf{u} + \mathbf{v}$, called the **resultant vector,** is the diagonal of a parallelogram having $\mathbf{u}$ and $\mathbf{v}$ as its adjacent sides.

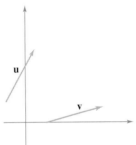 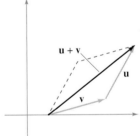

To find $\mathbf{u} + \mathbf{v}$, (1) move the initial point of $\mathbf{v}$ to the terminal point of $\mathbf{u}$, or (2) move the initial point of $\mathbf{u}$ to the terminal point of $\mathbf{v}$.

Figure 11.7

ISAAC WILLIAM ROWAN HAMILTON (1805–1865)

Some of the earliest work with vectors was done by the Irish mathematician William Rowan Hamilton. Hamilton spent many years developing a system of vector-like quantities called *quaternions*. Although Hamilton was convinced of the benefits of quaternions, the operations he defined did not produce good models for physical phenomena. It wasn't until the latter half of the nineteenth century that the Scottish physicist James Maxwell (1831–1879) restructured Hamilton's quaternions in a form useful for representing physical quantities such as force, velocity, and acceleration.

Figure 11.8 shows the equivalence of the geometric and algebraic definitions of vector addition and scalar multiplication, and presents (at far right) a geometric interpretation of $\mathbf{u} - \mathbf{v}$.

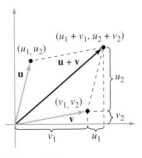 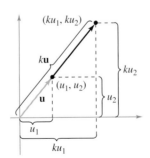 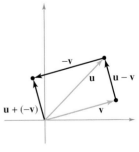

Vector addition   Scalar multiplication   Vector subtraction
Figure 11.8

EXAMPLE 3 Vector Operations

Given $\mathbf{v} = \langle -2, 5 \rangle$ and $\mathbf{w} = \langle 3, 4 \rangle$, find each of the vectors.

a. $\frac{1}{2}\mathbf{v}$ **b.** $\mathbf{w} - \mathbf{v}$ **c.** $\mathbf{v} + 2\mathbf{w}$

Solution

a. $\frac{1}{2}\mathbf{v} = \langle \frac{1}{2}(-2), \frac{1}{2}(5) \rangle = \langle -1, \frac{5}{2} \rangle$

b. $\mathbf{w} - \mathbf{v} = \langle w_1 - v_1, w_2 - v_2 \rangle = \langle 3 - (-2), 4 - 5 \rangle = \langle 5, -1 \rangle$

c. Using $2\mathbf{w} = \langle 6, 8 \rangle$, you have

$$\mathbf{v} + 2\mathbf{w} = \langle -2, 5 \rangle + \langle 6, 8 \rangle$$
$$= \langle -2 + 6, 5 + 8 \rangle$$
$$= \langle 4, 13 \rangle.$$

Vector addition and scalar multiplication share many properties of ordinary arithmetic, as shown in the following theorem.

THEOREM 11.1 Properties of Vector Operations

Let $\mathbf{u}$, $\mathbf{v}$, and $\mathbf{w}$ be vectors in the plane, and let c and d be scalars.

1. $\mathbf{u} + \mathbf{v} = \mathbf{v} + \mathbf{u}$ — Commutative Property
2. $(\mathbf{u} + \mathbf{v}) + \mathbf{w} = \mathbf{u} + (\mathbf{v} + \mathbf{w})$ — Associative Property
3. $\mathbf{u} + \mathbf{0} = \mathbf{u}$ — Additive Identity Property
4. $\mathbf{u} + (-\mathbf{u}) = \mathbf{0}$ — Additive Inverse Property
5. $c(d\mathbf{u}) = (cd)\mathbf{u}$
6. $(c + d)\mathbf{u} = c\mathbf{u} + d\mathbf{u}$ — Distributive Property
7. $c(\mathbf{u} + \mathbf{v}) = c\mathbf{u} + c\mathbf{v}$ — Distributive Property
8. $1(\mathbf{u}) = \mathbf{u}, 0(\mathbf{u}) = \mathbf{0}$

Proof The proof of the *Associative Property* of vector addition uses the Associative Property of addition of real numbers.

$$(\mathbf{u} + \mathbf{v}) + \mathbf{w} = [\langle u_1, u_2 \rangle + \langle v_1, v_2 \rangle] + \langle w_1, w_2 \rangle$$
$$= \langle u_1 + v_1, u_2 + v_2 \rangle + \langle w_1, w_2 \rangle$$
$$= \langle (u_1 + v_1) + w_1, (u_2 + v_2) + w_2 \rangle$$
$$= \langle u_1 + (v_1 + w_1), u_2 + (v_2 + w_2) \rangle$$
$$= \langle u_1, u_2 \rangle + \langle v_1 + w_1, v_2 + w_2 \rangle = \mathbf{u} + (\mathbf{v} + \mathbf{w})$$

Similarly, the proof of the *Distributive Property* of vectors depends on the Distributive Property of real numbers.

$$(c + d)\mathbf{u} = (c + d)\langle u_1, u_2 \rangle$$
$$= \langle (c + d)u_1, (c + d)u_2 \rangle$$
$$= \langle cu_1 + du_1, cu_2 + du_2 \rangle$$
$$= \langle cu_1, cu_2 \rangle + \langle du_1, du_2 \rangle = c\mathbf{u} + d\mathbf{u}$$

The other properties can be proved in a similar manner.

Any set of vectors (with an accompanying set of scalars) that satisfies the eight properties given in Theorem 11.1 is a **vector space**.* The eight properties are the *vector space axioms*. So, this theorem states that the set of vectors in the plane (with the set of real numbers) forms a vector space.

THEOREM 11.2 Length of a Scalar Multiple

Let **v** be a vector and let c be a scalar. Then

$$\|c\mathbf{v}\| = |c|\,\|\mathbf{v}\|. \qquad |c| \text{ is the absolute value of } c.$$

Proof Because $c\mathbf{v} = \langle cv_1, cv_2 \rangle$, it follows that

$$\begin{aligned}
\|c\mathbf{v}\| = \|\langle cv_1, cv_2 \rangle\| &= \sqrt{(cv_1)^2 + (cv_2)^2} \\
&= \sqrt{c^2 v_1^2 + c^2 v_2^2} \\
&= \sqrt{c^2(v_1^2 + v_2^2)} \\
&= |c|\sqrt{v_1^2 + v_2^2} \\
&= |c|\,\|\mathbf{v}\|.
\end{aligned}$$

In many applications of vectors, it is useful to find a unit vector that has the same direction as a given vector. The following theorem gives a procedure for doing this.

THEOREM 11.3 Unit Vector in the Direction of v

If **v** is a nonzero vector in the plane, then the vector

$$\mathbf{u} = \frac{\mathbf{v}}{\|\mathbf{v}\|} = \frac{1}{\|\mathbf{v}\|}\mathbf{v}$$

has length 1 and the same direction as **v**.

Proof Because $1/\|\mathbf{v}\|$ is positive and $\mathbf{u} = (1/\|\mathbf{v}\|)\mathbf{v}$, you can conclude that **u** has the same direction as **v**. To see that $\|\mathbf{u}\| = 1$, note that

$$\begin{aligned}
\|\mathbf{u}\| &= \left\|\left(\frac{1}{\|\mathbf{v}\|}\right)\mathbf{v}\right\| \\
&= \left|\frac{1}{\|\mathbf{v}\|}\right|\,\|\mathbf{v}\| \\
&= \frac{1}{\|\mathbf{v}\|}\,\|\mathbf{v}\| \\
&= 1.
\end{aligned}$$

So, **u** has length 1 and the same direction as **v**.

In Theorem 11.3, **u** is called a **unit vector in the direction of v**. The process of multiplying **v** by $1/\|\mathbf{v}\|$ to get a unit vector is called **normalization of v**.

EMMY NOETHER (1882–1935)

One person who contributed to our knowledge of axiomatic systems was the German mathematician Emmy Noether. Noether is generally recognized as the leading woman mathematician in recent history.

FOR FURTHER INFORMATION For more information on Emmy Noether, see the article "Emmy Noether, Greatest Woman Mathematician" by Clark Kimberling in *The Mathematics Teacher*. To view this article, go to the website *www.matharticles.com*.

---

* For more information about vector spaces, see *Elementary Linear Algebra*, Fifth Edition, by Larson, Edwards, and Falvo (Boston: Houghton Mifflin Company, 2004).

EXAMPLE 4   Finding a Unit Vector

Find a unit vector in the direction of $\mathbf{v} = \langle -2, 5 \rangle$ and verify that it has length 1.

Solution   From Theorem 11.3, the unit vector in the direction of $\mathbf{v}$ is

$$\frac{\mathbf{v}}{\|\mathbf{v}\|} = \frac{\langle -2, 5 \rangle}{\sqrt{(-2)^2 + (5)^2}} = \frac{1}{\sqrt{29}} \langle -2, 5 \rangle = \left\langle \frac{-2}{\sqrt{29}}, \frac{5}{\sqrt{29}} \right\rangle.$$

This vector has length 1, because

$$\sqrt{\left(\frac{-2}{\sqrt{29}}\right)^2 + \left(\frac{5}{\sqrt{29}}\right)^2} = \sqrt{\frac{4}{29} + \frac{25}{29}} = \sqrt{\frac{29}{29}} = 1.$$

Generally, the length of the sum of two vectors is not equal to the sum of their lengths. To see this, consider the vectors $\mathbf{u}$ and $\mathbf{v}$ as shown in Figure 11.9. By considering $\mathbf{u}$ and $\mathbf{v}$ as two sides of a triangle, you can see that the length of the third side is $\|\mathbf{u} + \mathbf{v}\|$, and you have

$$\|\mathbf{u} + \mathbf{v}\| \leq \|\mathbf{u}\| + \|\mathbf{v}\|.$$

Equality occurs only if the vectors $\mathbf{u}$ and $\mathbf{v}$ have the *same direction*. This result is called the **triangle inequality** for vectors. (You are asked to prove this in Exercise 89, Section 11.3.)

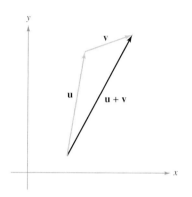

Triangle inequality
Figure 11.9

Standard Unit Vectors

The unit vectors $\langle 1, 0 \rangle$ and $\langle 0, 1 \rangle$ are called the **standard unit vectors** in the plane and are denoted by

$$\mathbf{i} = \langle 1, 0 \rangle \quad \text{and} \quad \mathbf{j} = \langle 0, 1 \rangle \qquad \text{Standard unit vectors}$$

as shown in Figure 11.10. These vectors can be used to represent any vector uniquely, as follows.

$$\mathbf{v} = \langle v_1, v_2 \rangle = \langle v_1, 0 \rangle + \langle 0, v_2 \rangle = v_1 \langle 1, 0 \rangle + v_2 \langle 0, 1 \rangle = v_1 \mathbf{i} + v_2 \mathbf{j}$$

The vector $\mathbf{v} = v_1 \mathbf{i} + v_2 \mathbf{j}$ is called a **linear combination** of $\mathbf{i}$ and $\mathbf{j}$. The scalars v_1 and v_2 are called the **horizontal** and **vertical components of v.**

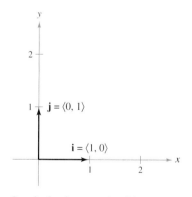

Standard unit vectors **i** and **j**
Figure 11.10

EXAMPLE 5   Writing a Linear Combination of Unit Vectors

Let $\mathbf{u}$ be the vector with initial point $(2, -5)$ and terminal point $(-1, 3)$, and let $\mathbf{v} = 2\mathbf{i} - \mathbf{j}$. Write each vector as a linear combination of $\mathbf{i}$ and $\mathbf{j}$.

a. $\mathbf{u}$ b. $\mathbf{w} = 2\mathbf{u} - 3\mathbf{v}$

Solution

a. $\mathbf{u} = \langle q_1 - p_1, q_2 - p_2 \rangle$
$= \langle -1 - 2, 3 - (-5) \rangle$
$= \langle -3, 8 \rangle = -3\mathbf{i} + 8\mathbf{j}$

b. $\mathbf{w} = 2\mathbf{u} - 3\mathbf{v} = 2(-3\mathbf{i} + 8\mathbf{j}) - 3(2\mathbf{i} - \mathbf{j})$
$= -6\mathbf{i} + 16\mathbf{j} - 6\mathbf{i} + 3\mathbf{j}$
$= -12\mathbf{i} + 19\mathbf{j}$

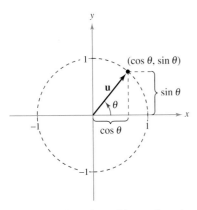

The angle θ from the positive x-axis to the vector **u**
Figure 11.11

If **u** is a unit vector and θ is the angle (measured counterclockwise) from the positive x-axis to **u**, then the terminal point of **u** lies on the unit circle, and you have

$$\mathbf{u} = \langle \cos\theta, \sin\theta \rangle = \cos\theta\mathbf{i} + \sin\theta\mathbf{j} \qquad \text{Unit vector}$$

as shown in Figure 11.11. Moreover, it follows that any other nonzero vector **v** making an angle θ with the positive x-axis has the same direction as **u**, and you can write

$$\mathbf{v} = \|\mathbf{v}\|\langle \cos\theta, \sin\theta \rangle = \|\mathbf{v}\|\cos\theta\mathbf{i} + \|\mathbf{v}\|\sin\theta\mathbf{j}.$$

EXAMPLE 6 Writing a Vector of Given Magnitude and Direction

The vector **v** has a magnitude of 3 and makes an angle of $30° = \pi/6$ with the positive x-axis. Write **v** as a linear combination of the unit vectors **i** and **j**.

Solution Because the angle between **v** and the positive x-axis is $\theta = \pi/6$, you can write the following.

$$\begin{aligned}\mathbf{v} &= \|\mathbf{v}\|\cos\theta\mathbf{i} + \|\mathbf{v}\|\sin\theta\mathbf{j}\\ &= 3\cos\frac{\pi}{6}\mathbf{i} + 3\sin\frac{\pi}{6}\mathbf{j}\\ &= \frac{3\sqrt{3}}{2}\mathbf{i} + \frac{3}{2}\mathbf{j}\end{aligned}$$

Applications of Vectors

Vectors have many applications in physics and engineering. One example is force. A vector can be used to represent force because force has both magnitude and direction. If two or more forces are acting on an object, then the **resultant force** on the object is the vector sum of the vector forces.

EXAMPLE 7 Finding the Resultant Force

Two tugboats are pushing an ocean liner, as shown in Figure 11.12. Each boat is exerting a force of 400 pounds. What is the resultant force on the ocean liner?

Solution Using Figure 11.12, you can represent the forces exerted by the first and second tugboats as

$$\begin{aligned}\mathbf{F}_1 &= 400\langle \cos 20°, \sin 20° \rangle\\ &= 400\cos(20°)\mathbf{i} + 400\sin(20°)\mathbf{j}\\ \mathbf{F}_2 &= 400\langle \cos(-20°), \sin(-20°) \rangle\\ &= 400\cos(20°)\mathbf{i} - 400\sin(20°)\mathbf{j}.\end{aligned}$$

The resultant force on the ocean liner is

$$\begin{aligned}\mathbf{F} &= \mathbf{F}_1 + \mathbf{F}_2\\ &= [400\cos(20°)\mathbf{i} + 400\sin(20°)\mathbf{j}] + [400\cos(20°)\mathbf{i} - 400\sin(20°)\mathbf{j}]\\ &= 800\cos(20°)\mathbf{i}\\ &\approx 752\mathbf{i}.\end{aligned}$$

So, the resultant force on the ocean liner is approximately 752 pounds in the direction of the positive x-axis.

The resultant force on the ocean liner that is exerted by the two tugboats.
Figure 11.12

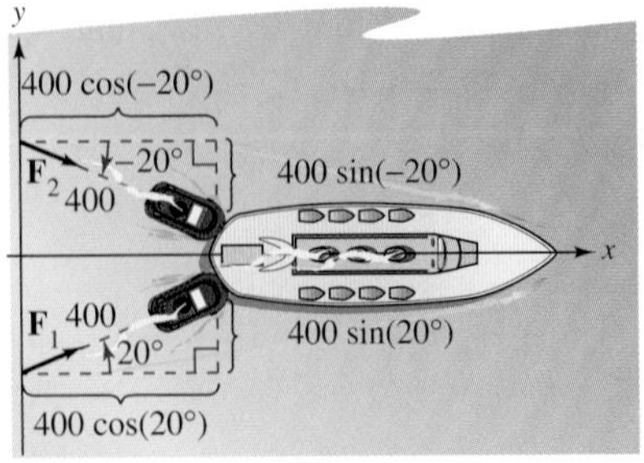

In surveying and navigation, a bearing is a direction that measures the acute angle that a path or line of sight makes with a fixed north-south line. In air navigation, bearings are measured in degrees clockwise from north.

EXAMPLE 8 Finding a Velocity

An airplane is traveling at a fixed altitude with a negligible wind factor. The airplane is traveling at a speed of 500 miles per hour with a bearing of 330°, as shown in Figure 11.13(a). As the airplane reaches a certain point, it encounters wind with a velocity of 70 miles per hour in the direction N 45° E (45° east of north), as shown in Figure 11.13(b). What are the resultant speed and direction of the airplane?

Solution Using Figure 11.13(a), represent the velocity of the airplane (alone) as

$$\mathbf{v}_1 = 500\cos(120°)\mathbf{i} + 500\sin(120°)\mathbf{j}.$$

The velocity of the wind is represented by the vector

$$\mathbf{v}_2 = 70\cos(45°)\mathbf{i} + 70\sin(45°)\mathbf{j}.$$

The resultant velocity of the airplane (in the wind) is

$$\mathbf{v} = \mathbf{v}_1 + \mathbf{v}_2 = 500\cos(120°)\mathbf{i} + 500\sin(120°)\mathbf{j} + 70\cos(45°)\mathbf{i} + 70\sin(45°)\mathbf{j}$$
$$\approx -200.5\mathbf{i} + 482.5\mathbf{j}.$$

To find the resultant speed and direction, write $\mathbf{v} = \|\mathbf{v}\|(\cos\theta\,\mathbf{i} + \sin\theta\,\mathbf{j})$. Because $\|\mathbf{v}\| \approx \sqrt{(-200.5)^2 + (482.5)^2} \approx 522.5$, you can write

$$\mathbf{v} \approx 522.5\left(\frac{-200.5}{522.5}\mathbf{i} + \frac{482.5}{522.5}\mathbf{j}\right) \approx 522.5\,[\cos(112.6°)\mathbf{i} + \sin(112.6°)\mathbf{j}].$$

The new speed of the airplane, as altered by the wind, is approximately 522.5 miles per hour in a path that makes an angle of 112.6° with the positive x-axis.

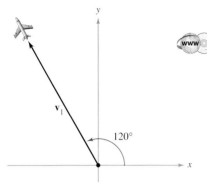

(a) Direction without wind

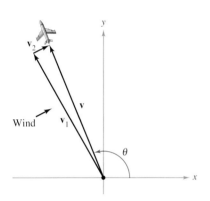

(b) Direction with wind
Figure 11.13

Exercises for Section 11.1

See www.CalcChat.com for worked-out solutions to odd-numbered exercises.

In Exercises 1–4, (a) find the component form of the vector **v** and (b) sketch the vector with its initial point at the origin.

1.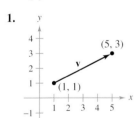

2.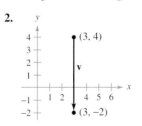

3.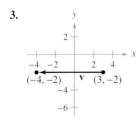

4.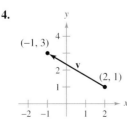

In Exercises 5–8, find the vectors **u** and **v** whose initial and terminal points are given. Show that **u** and **v** are equivalent.

5. **u**: $(3, 2)$, $(5, 6)$
   v: $(-1, 4)$, $(1, 8)$

6. **u**: $(-4, 0)$, $(1, 8)$
   v: $(2, -1)$, $(7, 7)$

7. **u**: $(0, 3)$, $(6, -2)$
   v: $(3, 10)$, $(9, 5)$

8. **u**: $(-4, -1)$, $(11, -4)$
   v: $(10, 13)$, $(25, 10)$

In Exercises 9–16, the initial and terminal points of a vector **v** are given. (a) Sketch the given directed line segment, (b) write the vector in component form, and (c) sketch the vector with its initial point at the origin.

Initial Point	Terminal Point		Initial Point	Terminal Point
9. $(1, 2)$	$(5, 5)$		10. $(2, -6)$	$(3, 6)$
11. $(10, 2)$	$(6, -1)$		12. $(0, -4)$	$(-5, -1)$

indicates that in the HM mathSpace® CD-ROM and the online Eduspace® system for this text, you will find an Open Exploration, which further explores this example using the computer algebra systems Maple, Mathcad, Mathematica, and Derive.

	Initial Point	Terminal Point		Initial Point	Terminal Point
13.	(6, 2)	(6, 6)	14.	(7, −1)	(−3, −1)
15.	$(\frac{3}{2}, \frac{4}{3})$	$(\frac{1}{2}, 3)$	16.	(0.12, 0.60)	(0.84, 1.25)

In Exercises 17 and 18, sketch each scalar multiple of v.

17. $\mathbf{v} = \langle 2, 3 \rangle$
    (a) $2\mathbf{v}$ (b) $-3\mathbf{v}$ (c) $\frac{7}{2}\mathbf{v}$ (d) $\frac{2}{3}\mathbf{v}$

18. $\mathbf{v} = \langle -1, 5 \rangle$
    (a) $4\mathbf{v}$ (b) $-\frac{1}{2}\mathbf{v}$ (c) $0\mathbf{v}$ (d) $-6\mathbf{v}$

In Exercises 19–22, use the figure to sketch a graph of the vector. To print an enlarged copy of the graph, go to the website www.mathgraphs.com.

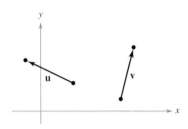

19. $-\mathbf{u}$
20. $2\mathbf{u}$
21. $\mathbf{u} - \mathbf{v}$
22. $\mathbf{u} + 2\mathbf{v}$

In Exercises 23 and 24, find (a) $\frac{2}{3}\mathbf{u}$, (b) $\mathbf{v} - \mathbf{u}$, and (c) $2\mathbf{u} + 5\mathbf{v}$.

23. $\mathbf{u} = \langle 4, 9 \rangle$
    $\mathbf{v} = \langle 2, -5 \rangle$

24. $\mathbf{u} = \langle -3, -8 \rangle$
    $\mathbf{v} = \langle 8, 25 \rangle$

In Exercises 25–28, find the vector v where $\mathbf{u} = \langle 2, -1 \rangle$ and $\mathbf{w} = \langle 1, 2 \rangle$. Illustrate the vector operations geometrically.

25. $\mathbf{v} = \frac{3}{2}\mathbf{u}$
26. $\mathbf{v} = \mathbf{u} + \mathbf{w}$
27. $\mathbf{v} = \mathbf{u} + 2\mathbf{w}$
28. $\mathbf{v} = 5\mathbf{u} - 3\mathbf{w}$

In Exercises 29 and 30, the vector v and its initial point are given. Find the terminal point.

29. $\mathbf{v} = \langle -1, 3 \rangle$; Initial point: (4, 2)
30. $\mathbf{v} = \langle 4, -9 \rangle$; Initial point: (3, 2)

In Exercises 31–36, find the magnitude of v.

31. $\mathbf{v} = \langle 4, 3 \rangle$
32. $\mathbf{v} = \langle 12, -5 \rangle$
33. $\mathbf{v} = 6\mathbf{i} - 5\mathbf{j}$
34. $\mathbf{v} = -10\mathbf{i} + 3\mathbf{j}$
35. $\mathbf{v} = 4\mathbf{j}$
36. $\mathbf{v} = \mathbf{i} - \mathbf{j}$

In Exercises 37–40, find the unit vector in the direction of u and verify that it has length 1.

37. $\mathbf{u} = \langle 3, 12 \rangle$
38. $\mathbf{u} = \langle 5, 15 \rangle$
39. $\mathbf{u} = \langle \frac{3}{2}, \frac{5}{2} \rangle$
40. $\mathbf{u} = \langle -6.2, 3.4 \rangle$

In Exercises 41–44, find the following.

(a) $\|\mathbf{u}\|$ (b) $\|\mathbf{v}\|$ (c) $\|\mathbf{u} + \mathbf{v}\|$

(d) $\left\| \frac{\mathbf{u}}{\|\mathbf{u}\|} \right\|$ (e) $\left\| \frac{\mathbf{v}}{\|\mathbf{v}\|} \right\|$ (f) $\left\| \frac{\mathbf{u} + \mathbf{v}}{\|\mathbf{u} + \mathbf{v}\|} \right\|$

41. $\mathbf{u} = \langle 1, -1 \rangle$
    $\mathbf{v} = \langle -1, 2 \rangle$

42. $\mathbf{u} = \langle 0, 1 \rangle$
    $\mathbf{v} = \langle 3, -3 \rangle$

43. $\mathbf{u} = \langle 1, \frac{1}{2} \rangle$
    $\mathbf{v} = \langle 2, 3 \rangle$

44. $\mathbf{u} = \langle 2, -4 \rangle$
    $\mathbf{v} = \langle 5, 5 \rangle$

In Exercises 45 and 46, sketch a graph of u, v, and u + v. Then demonstrate the triangle inequality using the vectors u and v.

45. $\mathbf{u} = \langle 2, 1 \rangle$, $\mathbf{v} = \langle 5, 4 \rangle$
46. $\mathbf{u} = \langle -3, 2 \rangle$, $\mathbf{v} = \langle 1, -2 \rangle$

In Exercises 47–50, find the vector v with the given magnitude and the same direction as u.

	Magnitude	Direction
47.	$\|\mathbf{v}\| = 4$	$\mathbf{u} = \langle 1, 1 \rangle$
48.	$\|\mathbf{v}\| = 4$	$\mathbf{u} = \langle -1, 1 \rangle$
49.	$\|\mathbf{v}\| = 2$	$\mathbf{u} = \langle \sqrt{3}, 3 \rangle$
50.	$\|\mathbf{v}\| = 3$	$\mathbf{u} = \langle 0, 3 \rangle$

In Exercises 51–54, find the component form of v given its magnitude and the angle it makes with the positive x-axis.

51. $\|\mathbf{v}\| = 3$, $\theta = 0°$
52. $\|\mathbf{v}\| = 5$, $\theta = 120°$
53. $\|\mathbf{v}\| = 2$, $\theta = 150°$
54. $\|\mathbf{v}\| = 1$, $\theta = 3.5°$

In Exercises 55–58, find the component form of u + v given the lengths of u and v and the angles that u and v make with the positive x-axis.

55. $\|\mathbf{u}\| = 1$, $\theta_\mathbf{u} = 0°$
    $\|\mathbf{v}\| = 3$, $\theta_\mathbf{v} = 45°$

56. $\|\mathbf{u}\| = 4$, $\theta_\mathbf{u} = 0°$
    $\|\mathbf{v}\| = 2$, $\theta_\mathbf{v} = 60°$

57. $\|\mathbf{u}\| = 2$, $\theta_\mathbf{u} = 4$
    $\|\mathbf{v}\| = 1$, $\theta_\mathbf{v} = 2$

58. $\|\mathbf{u}\| = 5$, $\theta_\mathbf{u} = -0.5$
    $\|\mathbf{v}\| = 5$, $\theta_\mathbf{v} = 0.5$

Writing About Concepts

59. In your own words, state the difference between a scalar and a vector. Give examples of each.
60. Give geometric descriptions of the operations of addition of vectors and multiplication of a vector by a scalar.
61. Identify the quantity as a scalar or as a vector. Explain your reasoning.
    (a) The muzzle velocity of a gun
    (b) The price of a company's stock
62. Identify the quantity as a scalar or as a vector. Explain your reasoning.
    (a) The air temperature in a room
    (b) The weight of a car

In Exercises 63–68, find a and b such that $\mathbf{v} = a\mathbf{u} + b\mathbf{w}$, where $\mathbf{u} = \langle 1, 2 \rangle$ and $\mathbf{w} = \langle 1, -1 \rangle$.

63. $\mathbf{v} = \langle 2, 1 \rangle$
64. $\mathbf{v} = \langle 0, 3 \rangle$
65. $\mathbf{v} = \langle 3, 0 \rangle$
66. $\mathbf{v} = \langle 3, 3 \rangle$
67. $\mathbf{v} = \langle 1, 1 \rangle$
68. $\mathbf{v} = \langle -1, 7 \rangle$

In Exercises 69–74, find a unit vector (a) parallel to and (b) normal to the graph of $f(x)$ at the given point. Then sketch a graph of the vectors and the function.

Function	Point
69. $f(x) = x^2$	$(3, 9)$
70. $f(x) = -x^2 + 5$	$(1, 4)$
71. $f(x) = x^3$	$(1, 1)$
72. $f(x) = x^3$	$(-2, -8)$
73. $f(x) = \sqrt{25 - x^2}$	$(3, 4)$
74. $f(x) = \tan x$	$\left(\dfrac{\pi}{4}, 1\right)$

In Exercises 75 and 76, find the component form of $\mathbf{v}$ given the magnitudes of $\mathbf{u}$ and $\mathbf{u} + \mathbf{v}$ and the angles that $\mathbf{u}$ and $\mathbf{u} + \mathbf{v}$ make with the positive x-axis.

75. $\|\mathbf{u}\| = 1, \theta = 45°$
    $\|\mathbf{u} + \mathbf{v}\| = \sqrt{2}, \theta = 90°$

76. $\|\mathbf{u}\| = 4, \theta = 30°$
    $\|\mathbf{u} + \mathbf{v}\| = 6, \theta = 120°$

77. Programming You are given the magnitudes of $\mathbf{u}$ and $\mathbf{v}$ and the angles $\mathbf{u}$ and $\mathbf{v}$ make with the positive x-axis. Write a program for a graphing utility in which the output is the following.
(a) $\mathbf{u} + \mathbf{v}$
(b) $\|\mathbf{u} + \mathbf{v}\|$
(c) The angle $\mathbf{u} + \mathbf{v}$ makes with the positive x-axis

78. Programming Use the program you wrote in Exercise 77 to find the magnitude and direction of the resultant of the vectors shown.

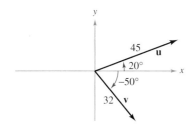

In Exercises 79 and 80, use a graphing utility to find the magnitude and direction of the resultant of the vectors.

79.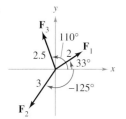

80.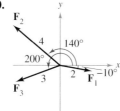

81. Numerical and Graphical Analysis Forces with magnitudes of 180 newtons and 275 newtons act on a hook (see figure). The angle between the two forces is θ degrees.
(a) If $\theta = 30°$, find the direction and magnitude of the resultant force.
(b) Write the magnitude M and direction α of the resultant force as functions of θ, where $0° \leq \theta \leq 180°$.
(c) Use a graphing utility to complete the table.

θ	0°	30°	60°	90°	120°	150°	180°
M							
α							

(d) Use a graphing utility to graph the two functions M and α.
(e) Explain why one of the functions decreases for increasing values of θ whereas the other does not.

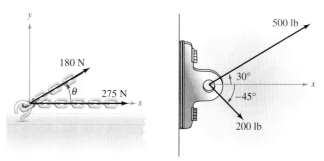

Figure for 81   **Figure for 82**

82. Resultant Force Forces with magnitudes of 500 pounds and 200 pounds act on a machine part at angles of 30° and −45°, respectively, with the x-axis (see figure). Find the direction and magnitude of the resultant force.

83. Resultant Force Three forces with magnitudes of 75 pounds, 100 pounds, and 125 pounds act on an object at angles of 30°, 45°, and 120°, respectively, with the positive x-axis. Find the direction and magnitude of the resultant force.

84. Resultant Force Three forces with magnitudes of 400 newtons, 280 newtons, and 350 newtons act on an object at angles of −30°, 45°, and 135°, respectively, with the positive x-axis. Find the direction and magnitude of the resultant force.

85. Think About It Consider two forces of equal magnitude acting on a point.
(a) If the magnitude of the resultant is the sum of the magnitudes of the two forces, make a conjecture about the angle between the forces.
(b) If the resultant of the forces is $\mathbf{0}$, make a conjecture about the angle between the forces.
(c) Can the magnitude of the resultant be greater than the sum of the magnitudes of the two forces? Explain.

86. **Graphical Reasoning** Consider two forces $\mathbf{F}_1 = \langle 20, 0 \rangle$ and $\mathbf{F}_2 = 10 \langle \cos \theta, \sin \theta \rangle$.
    (a) Find $\|\mathbf{F}_1 + \mathbf{F}_2\|$.
    (b) Determine the magnitude of the resultant as a function of θ. Use a graphing utility to graph the function for $0 \le \theta < 2\pi$.
    (c) Use the graph in part (b) to determine the range of the function. What is its maximum and for what value of θ does it occur? What is its minimum and for what value of θ does it occur?
    (d) Explain why the magnitude of the resultant is never 0.

87. Three vertices of a parallelogram are $(1, 2)$, $(3, 1)$, and $(8, 4)$. Find the three possible fourth vertices (see figure).

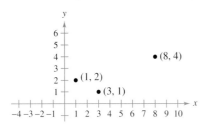

88. Use vectors to find the points of trisection of the line segment with endpoints $(1, 2)$ and $(7, 5)$.

Cable Tension In Exercises 89 and 90, use the figure to determine the tension in each cable supporting the given load.

89.

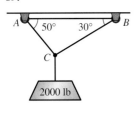

90.

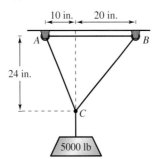

91. **Projectile Motion** A gun with a muzzle velocity of 1200 feet per second is fired at an angle of 6° above the horizontal. Find the vertical and horizontal components of the velocity.

92. **Shared Load** To carry a 100-pound cylindrical weight, two workers lift on the ends of short ropes tied to an eyelet on the top center of the cylinder. One rope makes a 20° angle away from the vertical and the other makes a 30° angle (see figure).
    (a) Find each rope's tension if the resultant force is vertical.
    (b) Find the vertical component of each worker's force.

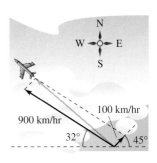

Figure for 92   **Figure for 93**

93. **Navigation** A plane is flying in the direction 302°. Its speed with respect to the air is 900 kilometers per hour. The wind at the plane's altitude is from the southwest at 100 kilometers per hour (see figure). What is the true direction of the plane, and what is its speed with respect to the ground?

94. **Navigation** A plane flies at a constant groundspeed of 400 miles per hour due east and encounters a 50-mile-per-hour wind from the northwest. Find the airspeed and compass direction that will allow the plane to maintain its groundspeed and eastward direction.

True or False? In Exercises 95–100, determine whether the statement is true or false. If it is false, explain why or give an example that shows it is false.

95. If $\mathbf{u}$ and $\mathbf{v}$ have the same magnitude and direction, then $\mathbf{u}$ and $\mathbf{v}$ are equivalent.
96. If $\mathbf{u}$ is a unit vector in the direction of $\mathbf{v}$, then $\mathbf{v} = \|\mathbf{v}\| \mathbf{u}$.
97. If $\mathbf{u} = a\mathbf{i} + b\mathbf{j}$ is a unit vector, then $a^2 + b^2 = 1$.
98. If $\mathbf{v} = a\mathbf{i} + b\mathbf{j} = \mathbf{0}$, then $a = -b$.
99. If $a = b$, then $\|a\mathbf{i} + b\mathbf{j}\| = \sqrt{2}a$.
100. If $\mathbf{u}$ and $\mathbf{v}$ have the same magnitude but opposite directions, then $\mathbf{u} + \mathbf{v} = \mathbf{0}$.
101. Prove that $\mathbf{u} = (\cos \theta)\mathbf{i} - (\sin \theta)\mathbf{j}$ and $\mathbf{v} = (\sin \theta)\mathbf{i} + (\cos \theta)\mathbf{j}$ are unit vectors for any angle θ.
102. **Geometry** Using vectors, prove that the line segment joining the midpoints of two sides of a triangle is parallel to, and one-half the length of, the third side.
103. **Geometry** Using vectors, prove that the diagonals of a parallelogram bisect each other.
104. Prove that the vector $\mathbf{w} = \|\mathbf{u}\|\mathbf{v} + \|\mathbf{v}\|\mathbf{u}$ bisects the angle between $\mathbf{u}$ and $\mathbf{v}$.
105. Consider the vector $\mathbf{u} = \langle x, y \rangle$. Describe the set of all points (x, y) such that $\|\mathbf{u}\| = 5$.

Putnam Exam Challenge

106. A coast artillery gun can fire at any angle of elevation between 0° and 90° in a fixed vertical plane. If air resistance is neglected and the muzzle velocity is constant ($= v_0$), determine the set H of points in the plane and above the horizontal which can be hit.

This problem was composed by the Committee on the Putnam Prize Competition.
© The Mathematical Association of America. All rights reserved.

Section 11.2
Space Coordinates and Vectors in Space

- Understand the three-dimensional rectangular coordinate system.
- Analyze vectors in space.
- Use three-dimensional vectors to solve real-life problems.

Coordinates in Space

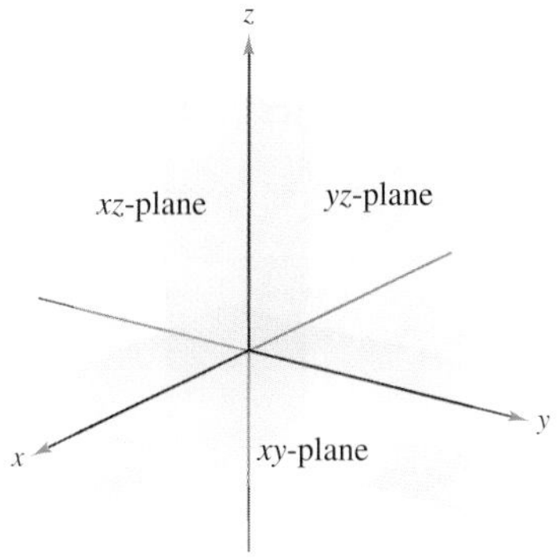

Up to this point in the text, you have been primarily concerned with the two-dimensional coordinate system. Much of the remaining part of your study of calculus will involve the three-dimensional coordinate system.

Before extending the concept of a vector to three dimensions, you must be able to identify points in the **three-dimensional coordinate system.** You can construct this system by passing a z-axis perpendicular to both the x- and y-axes at the origin. Figure 11.14 shows the positive portion of each coordinate axis. Taken as pairs, the axes determine three **coordinate planes**: the **xy-plane**, the **xz-plane**, and the **yz-plane.** These three coordinate planes separate three-space into eight **octants.** The first octant is the one for which all three coordinates are positive. In this three-dimensional system, a point P in space is determined by an ordered triple (x, y, z) where x, y, and z are as follows.

x = directed distance from yz-plane to P

y = directed distance from xz-plane to P

z = directed distance from xy-plane to P

Several points are shown in Figure 11.15.

The three-dimensional coordinate system
Figure 11.14

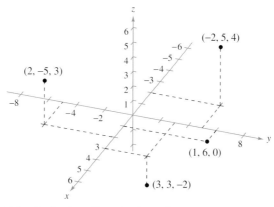

Points in the three-dimensional coordinate system are represented by ordered triples.
Figure 11.15

A three-dimensional coordinate system can have either a **left-handed** or a **right-handed** orientation. To determine the orientation of a system, imagine that you are standing at the origin, with your arms pointing in the direction of the positive x- and y-axes, and with the z-axis pointing up, as shown in Figure 11.16. The system is right-handed or left-handed depending on which hand points along the x-axis. In this text, you will work exclusively with the right-handed system.

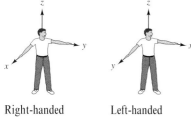

Right-handed system Left-handed system
Figure 11.16

NOTE The three-dimensional rotatable graphs that are available in the *HM mathSpace®* CD-ROM and the online *Eduspace®* system for this text will help you visualize points or objects in a three-dimensional coordinate system.

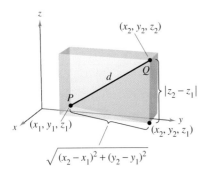

The distance between two points in space
Figure 11.17

Many of the formulas established for the two-dimensional coordinate system can be extended to three dimensions. For example, to find the distance between two points in space, you can use the Pythagorean Theorem twice, as shown in Figure 11.17. By doing this, you will obtain the formula for the distance between the points (x_1, y_1, z_1) and (x_2, y_2, z_2).

$$d = \sqrt{(x_2 - x_1)^2 + (y_2 - y_1)^2 + (z_2 - z_1)^2} \quad \text{Distance Formula}$$

EXAMPLE 1 Finding the Distance Between Two Points in Space

The distance between the points $(2, -1, 3)$ and $(1, 0, -2)$ is

$$\begin{aligned} d &= \sqrt{(1-2)^2 + (0+1)^2 + (-2-3)^2} \quad \text{Distance Formula} \\ &= \sqrt{1 + 1 + 25} \\ &= \sqrt{27} \\ &= 3\sqrt{3}. \end{aligned}$$

A **sphere** with center at (x_0, y_0, z_0) and radius r is defined to be the set of all points (x, y, z) such that the distance between (x, y, z) and (x_0, y_0, z_0) is r. You can use the Distance Formula to find the **standard equation of a sphere** of radius r, centered at (x_0, y_0, z_0). If (x, y, z) is an arbitrary point on the sphere, the equation of the sphere is

$$(x - x_0)^2 + (y - y_0)^2 + (z - z_0)^2 = r^2 \quad \text{Equation of sphere}$$

as shown in Figure 11.18. Moreover, the midpoint of the line segment joining the points (x_1, y_1, z_1) and (x_2, y_2, z_2) has coordinates

$$\left(\frac{x_1 + x_2}{2}, \frac{y_1 + y_2}{2}, \frac{z_1 + z_2}{2}\right). \quad \text{Midpoint Rule}$$

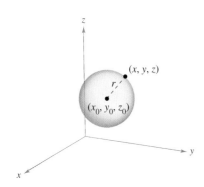

Figure 11.18

EXAMPLE 2 Finding the Equation of a Sphere

Find the standard equation of the sphere that has the points $(5, -2, 3)$ and $(0, 4, -3)$ as endpoints of a diameter.

Solution By the Midpoint Rule, the center of the sphere is

$$\left(\frac{5 + 0}{2}, \frac{-2 + 4}{2}, \frac{3 - 3}{2}\right) = \left(\frac{5}{2}, 1, 0\right). \quad \text{Midpoint Rule}$$

By the Distance Formula, the radius is

$$r = \sqrt{\left(0 - \frac{5}{2}\right)^2 + (4 - 1)^2 + (-3 - 0)^2} = \sqrt{\frac{97}{4}} = \frac{\sqrt{97}}{2}.$$

Therefore, the standard equation of the sphere is

$$\left(x - \frac{5}{2}\right)^2 + (y - 1)^2 + z^2 = \frac{97}{4}. \quad \text{Equation of sphere}$$

Vectors in Space

In space, vectors are denoted by ordered triples $\mathbf{v} = \langle v_1, v_2, v_3 \rangle$. The **zero vector** is denoted by $\mathbf{0} = \langle 0, 0, 0 \rangle$. Using the unit vectors $\mathbf{i} = \langle 1, 0, 0 \rangle$, $\mathbf{j} = \langle 0, 1, 0 \rangle$, and $\mathbf{k} = \langle 0, 0, 1 \rangle$ in the direction of the positive z-axis, the **standard unit vector notation** for $\mathbf{v}$ is

$$\mathbf{v} = v_1 \mathbf{i} + v_2 \mathbf{j} + v_3 \mathbf{k}$$

as shown in Figure 11.19. If $\mathbf{v}$ is represented by the directed line segment from $P(p_1, p_2, p_3)$ to $Q(q_1, q_2, q_3)$, as shown in Figure 11.20, the component form of $\mathbf{v}$ is given by subtracting the coordinates of the initial point from the coordinates of the terminal point, as follows.

$$\mathbf{v} = \langle v_1, v_2, v_3 \rangle = \langle q_1 - p_1, q_2 - p_2, q_3 - p_3 \rangle$$

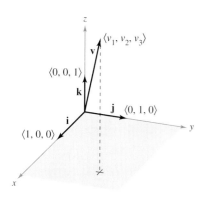

The standard unit vectors in space
Figure 11.19

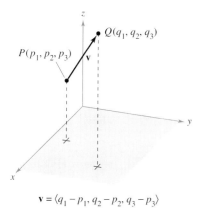

$\mathbf{v} = \langle q_1 - p_1, q_2 - p_2, q_3 - p_3 \rangle$

Figure 11.20

Vectors in Space

Let $\mathbf{u} = \langle u_1, u_2, u_3 \rangle$ and $\mathbf{v} = \langle v_1, v_2, v_3 \rangle$ be vectors in space and let c be a scalar.

1. *Equality of Vectors:* $\mathbf{u} = \mathbf{v}$ if and only if $u_1 = v_1$, $u_2 = v_2$, and $u_3 = v_3$.
2. *Component Form:* If $\mathbf{v}$ is represented by the directed line segment from $P(p_1, p_2, p_3)$ to $Q(q_1, q_2, q_3)$, then
$$\mathbf{v} = \langle v_1, v_2, v_3 \rangle = \langle q_1 - p_1, q_2 - p_2, q_3 - p_3 \rangle.$$
3. *Length:* $\|\mathbf{v}\| = \sqrt{v_1^2 + v_2^2 + v_3^2}$
4. *Unit Vector in the Direction of* $\mathbf{v}$: $\dfrac{\mathbf{v}}{\|\mathbf{v}\|} = \left(\dfrac{1}{\|\mathbf{v}\|}\right)\langle v_1, v_2, v_3 \rangle$, $\mathbf{v} \neq \mathbf{0}$
5. *Vector Addition:* $\mathbf{v} + \mathbf{u} = \langle v_1 + u_1, v_2 + u_2, v_3 + u_3 \rangle$
6. *Scalar Multiplication:* $c\mathbf{v} = \langle cv_1, cv_2, cv_3 \rangle$

NOTE The properties of vector addition and scalar multiplication given in Theorem 11.1 are also valid for vectors in space.

EXAMPLE 3 Finding the Component Form of a Vector in Space

Find the component form and magnitude of the vector $\mathbf{v}$ having initial point $(-2, 3, 1)$ and terminal point $(0, -4, 4)$. Then find a unit vector in the direction of $\mathbf{v}$.

Solution The component form of $\mathbf{v}$ is

$$\mathbf{v} = \langle q_1 - p_1, q_2 - p_2, q_3 - p_3 \rangle = \langle 0 - (-2), -4 - 3, 4 - 1 \rangle$$
$$= \langle 2, -7, 3 \rangle$$

which implies that its magnitude is

$$\|\mathbf{v}\| = \sqrt{2^2 + (-7)^2 + 3^2} = \sqrt{62}.$$

The unit vector in the direction of $\mathbf{v}$ is

$$\mathbf{u} = \frac{\mathbf{v}}{\|\mathbf{v}\|} = \frac{1}{\sqrt{62}}\langle 2, -7, 3 \rangle.$$

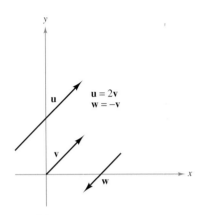

Parallel vectors
Figure 11.21

Recall from the definition of scalar multiplication that positive scalar multiples of a nonzero vector **v** have the same direction as **v**, whereas negative multiples have the direction opposite of **v**. In general, two nonzero vectors **u** and **v** are **parallel** if there is some scalar c such that $\mathbf{u} = c\mathbf{v}$.

> **Definition of Parallel Vectors**
>
> Two nonzero vectors **u** and **v** are **parallel** if there is some scalar c such that $\mathbf{u} = c\mathbf{v}$.

For example, in Figure 11.21, the vectors **u**, **v**, and **w** are parallel because $\mathbf{u} = 2\mathbf{v}$ and $\mathbf{w} = -\mathbf{v}$.

EXAMPLE 4 Parallel Vectors

Vector **w** has initial point $(2, -1, 3)$ and terminal point $(-4, 7, 5)$. Which of the following vectors is parallel to **w**?

a. $\mathbf{u} = \langle 3, -4, -1 \rangle$

b. $\mathbf{v} = \langle 12, -16, 4 \rangle$

Solution Begin by writing **w** in component form.

$$\mathbf{w} = \langle -4 - 2, 7 - (-1), 5 - 3 \rangle = \langle -6, 8, 2 \rangle$$

a. Because $\mathbf{u} = \langle 3, -4, -1 \rangle = -\frac{1}{2}\langle -6, 8, 2 \rangle = -\frac{1}{2}\mathbf{w}$, you can conclude that **u** is parallel to **w**.

b. In this case, you want to find a scalar c such that

$$\langle 12, -16, 4 \rangle = c\langle -6, 8, 2 \rangle.$$
$$12 = -6c \rightarrow c = -2$$
$$-16 = 8c \rightarrow c = -2$$
$$4 = 2c \rightarrow c = 2$$

Because there is no c for which the equation has a solution, the vectors are not parallel.

EXAMPLE 5 Using Vectors to Determine Collinear Points

Determine whether the points $P(1, -2, 3)$, $Q(2, 1, 0)$, and $R(4, 7, -6)$ are collinear.

Solution The component forms of $\overrightarrow{PQ}$ and $\overrightarrow{PR}$ are

$$\overrightarrow{PQ} = \langle 2 - 1, 1 - (-2), 0 - 3 \rangle = \langle 1, 3, -3 \rangle$$

and

$$\overrightarrow{PR} = \langle 4 - 1, 7 - (-2), -6 - 3 \rangle = \langle 3, 9, -9 \rangle.$$

These two vectors have a common initial point. So, P, Q, and R lie on the same line if and only if $\overrightarrow{PQ}$ and $\overrightarrow{PR}$ are parallel—which they are because $\overrightarrow{PR} = 3\overrightarrow{PQ}$, as shown in Figure 11.22.

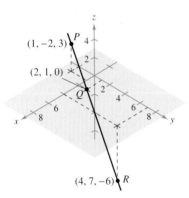

The points P, Q, and R lie on the same line.
Figure 11.22

EXAMPLE 6 Standard Unit Vector Notation

a. Write the vector $\mathbf{v} = 4\mathbf{i} - 5\mathbf{k}$ in component form.

b. Find the terminal point of the vector $\mathbf{v} = 7\mathbf{i} - \mathbf{j} + 3\mathbf{k}$, given that the initial point is $P(-2, 3, 5)$.

Solution

a. Because $\mathbf{j}$ is missing, its component is 0 and
$$\mathbf{v} = 4\mathbf{i} - 5\mathbf{k} = \langle 4, 0, -5 \rangle.$$

b. You need to find $Q(q_1, q_2, q_3)$ such that $\mathbf{v} = \overrightarrow{PQ} = 7\mathbf{i} - \mathbf{j} + 3\mathbf{k}$. This implies that $q_1 - (-2) = 7$, $q_2 - 3 = -1$, and $q_3 - 5 = 3$. The solution of these three equations is $q_1 = 5$, $q_2 = 2$, and $q_3 = 8$. Therefore, Q is $(5, 2, 8)$.

Application

EXAMPLE 7 Measuring Force

A television camera weighing 120 pounds is supported by a tripod, as shown in Figure 11.23. Represent the force exerted on each leg of the tripod as a vector.

Solution Let the vectors $\mathbf{F}_1$, $\mathbf{F}_2$, and $\mathbf{F}_3$ represent the forces exerted on the three legs. From Figure 11.23, you can determine the directions of $\mathbf{F}_1$, $\mathbf{F}_2$, and $\mathbf{F}_3$ to be as follows.

$$\overrightarrow{PQ_1} = \langle 0 - 0, -1 - 0, 0 - 4 \rangle = \langle 0, -1, -4 \rangle$$

$$\overrightarrow{PQ_2} = \left\langle \frac{\sqrt{3}}{2} - 0, \frac{1}{2} - 0, 0 - 4 \right\rangle = \left\langle \frac{\sqrt{3}}{2}, \frac{1}{2}, -4 \right\rangle$$

$$\overrightarrow{PQ_3} = \left\langle -\frac{\sqrt{3}}{2} - 0, \frac{1}{2} - 0, 0 - 4 \right\rangle = \left\langle -\frac{\sqrt{3}}{2}, \frac{1}{2}, -4 \right\rangle$$

Because each leg has the same length, and the total force is distributed equally among the three legs, you know that $\|\mathbf{F}_1\| = \|\mathbf{F}_2\| = \|\mathbf{F}_3\|$. So, there exists a constant c such that

$$\mathbf{F}_1 = c\langle 0, -1, -4 \rangle, \quad \mathbf{F}_2 = c\left\langle \frac{\sqrt{3}}{2}, \frac{1}{2}, -4 \right\rangle, \quad \text{and} \quad \mathbf{F}_3 = c\left\langle -\frac{\sqrt{3}}{2}, \frac{1}{2}, -4 \right\rangle.$$

Let the total force exerted by the object be given by $\mathbf{F} = -120\mathbf{k}$. Then, using the fact that

$$\mathbf{F} = \mathbf{F}_1 + \mathbf{F}_2 + \mathbf{F}_3$$

you can conclude that $\mathbf{F}_1$, $\mathbf{F}_2$, and $\mathbf{F}_3$ all have a vertical component of -40. This implies that $c(-4) = -40$ and $c = 10$. Therefore, the forces exerted on the legs can be represented by

$$\mathbf{F}_1 = \langle 0, -10, -40 \rangle$$
$$\mathbf{F}_2 = \langle 5\sqrt{3}, 5, -40 \rangle$$
$$\mathbf{F}_3 = \langle -5\sqrt{3}, 5, -40 \rangle.$$

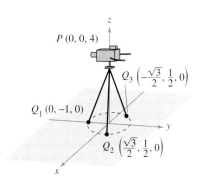

Figure 11.23

Exercises for Section 11.2

In Exercises 1–4, plot the points on the same three-dimensional coordinate system.

1. (a) $(2, 1, 3)$ (b) $(-1, 2, 1)$
2. (a) $(3, -2, 5)$ (b) $(\frac{3}{2}, 4, -2)$
3. (a) $(5, -2, 2)$ (b) $(5, -2, -2)$
4. (a) $(0, 4, -5)$ (b) $(4, 0, 5)$

In Exercises 5 and 6, approximate the coordinates of the points.

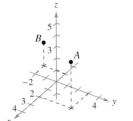

In Exercises 7–10, find the coordinates of the point.

7. The point is located three units behind the yz-plane, four units to the right of the xz-plane, and five units above the xy-plane.
8. The point is located seven units in front of the yz-plane, two units to the left of the xz-plane, and one unit below the xy-plane.
9. The point is located on the x-axis, 10 units in front of the yz-plane.
10. The point is located in the yz-plane, three units to the right of the xz-plane, and two units above the xy-plane.
11. *Think About It* What is the z-coordinate of any point in the xy-plane?
12. *Think About It* What is the x-coordinate of any point in the yz-plane?

In Exercises 13–24, determine the location of a point (x, y, z) that satisfies the condition(s).

13. $z = 6$
14. $y = 2$
15. $x = 4$
16. $z = -3$
17. $y < 0$
18. $x < 0$
19. $|y| \leq 3$
20. $|x| > 4$
21. $xy > 0, \quad z = -3$
22. $xy < 0, \quad z = 4$
23. $xyz < 0$
24. $xyz > 0$

In Exercises 25–28, find the distance between the points.

25. $(0, 0, 0), \quad (5, 2, 6)$
26. $(-2, 3, 2), \quad (2, -5, -2)$
27. $(1, -2, 4), \quad (6, -2, -2)$
28. $(2, 2, 3), \quad (4, -5, 6)$

In Exercises 29–32, find the lengths of the sides of the triangle with the indicated vertices, and determine whether the triangle is a right triangle, an isosceles triangle, or neither.

29. $(0, 0, 0), (2, 2, 1), (2, -4, 4)$
30. $(5, 3, 4), (7, 1, 3), (3, 5, 3)$
31. $(1, -3, -2), (5, -1, 2), (-1, 1, 2)$
32. $(5, 0, 0), (0, 2, 0), (0, 0, -3)$

33. *Think About It* The triangle in Exercise 29 is translated five units upward along the z-axis. Determine the coordinates of the translated triangle.
34. *Think About It* The triangle in Exercise 30 is translated three units to the right along the y-axis. Determine the coordinates of the translated triangle.

In Exercises 35 and 36, find the coordinates of the midpoint of the line segment joining the points.

35. $(5, -9, 7), (-2, 3, 3)$
36. $(4, 0, -6), (8, 8, 20)$

In Exercises 37–40, find the standard equation of the sphere.

37. Center: $(0, 2, 5)$ 38. Center: $(4, -1, 1)$
    Radius: 2   Radius: 5
39. Endpoints of a diameter: $(2, 0, 0), (0, 6, 0)$
40. Center: $(-3, 2, 4)$, tangent to the yz-plane

In Exercises 41–44, complete the square to write the equation of the sphere in standard form. Find the center and radius.

41. $x^2 + y^2 + z^2 - 2x + 6y + 8z + 1 = 0$
42. $x^2 + y^2 + z^2 + 9x - 2y + 10z + 19 = 0$
43. $9x^2 + 9y^2 + 9z^2 - 6x + 18y + 1 = 0$
44. $4x^2 + 4y^2 + 4z^2 - 4x - 32y + 8z + 33 = 0$

In Exercises 45–48, describe the solid satisfying the condition.

45. $x^2 + y^2 + z^2 \leq 36$ 46. $x^2 + y^2 + z^2 > 4$
47. $x^2 + y^2 + z^2 < 4x - 6y + 8z - 13$
48. $x^2 + y^2 + z^2 > -4x + 6y - 8z - 13$

In Exercises 49–52, (a) find the component form of the vector v and (b) sketch the vector with its initial point at the origin.

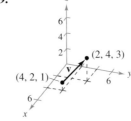

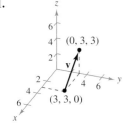

51.
52.

In Exercises 53–56, find the component form and magnitude of the vector u with the given initial and terminal points. Then find a unit vector in the direction of u.

	Initial Point	Terminal Point
53.	$(3, 2, 0)$	$(4, 1, 6)$
54.	$(4, -5, 2)$	$(-1, 7, -3)$
55.	$(-4, 3, 1)$	$(-5, 3, 0)$
56.	$(1, -2, 4)$	$(2, 4, -2)$

In Exercises 57 and 58, the initial and terminal points of a vector v are given. (a) Sketch the directed line segment, (b) find the component form of the vector, and (c) sketch the vector with its initial point at the origin.

57. Initial point: $(-1, 2, 3)$
Terminal point: $(3, 3, 4)$

58. Initial point: $(2, -1, -2)$
Terminal point: $(-4, 3, 7)$

In Exercises 59 and 60, the vector v and its initial point are given. Find the terminal point.

59. $\mathbf{v} = \langle 3, -5, 6 \rangle$
Initial point: $(0, 6, 2)$

60. $\mathbf{v} = \langle 1, -\frac{2}{3}, \frac{1}{2} \rangle$
Initial point: $(0, 2, \frac{5}{2})$

In Exercises 61 and 62, find each scalar multiple of v and sketch its graph.

61. $\mathbf{v} = \langle 1, 2, 2 \rangle$
(a) $2\mathbf{v}$ (b) $-\mathbf{v}$
(c) $\frac{3}{2}\mathbf{v}$ (d) $0\mathbf{v}$

62. $\mathbf{v} = \langle 2, -2, 1 \rangle$
(a) $-\mathbf{v}$ (b) $2\mathbf{v}$
(c) $\frac{1}{2}\mathbf{v}$ (d) $\frac{5}{2}\mathbf{v}$

In Exercises 63–68, find the vector z, given that $\mathbf{u} = \langle 1, 2, 3 \rangle$, $\mathbf{v} = \langle 2, 2, -1 \rangle$, and $\mathbf{w} = \langle 4, 0, -4 \rangle$.

63. $\mathbf{z} = \mathbf{u} - \mathbf{v}$
64. $\mathbf{z} = \mathbf{u} - \mathbf{v} + 2\mathbf{w}$
65. $\mathbf{z} = 2\mathbf{u} + 4\mathbf{v} - \mathbf{w}$
66. $\mathbf{z} = 5\mathbf{u} - 3\mathbf{v} - \frac{1}{2}\mathbf{w}$
67. $2\mathbf{z} - 3\mathbf{u} = \mathbf{w}$
68. $2\mathbf{u} + \mathbf{v} - \mathbf{w} + 3\mathbf{z} = \mathbf{0}$

In Exercises 69–72, determine which of the vectors is (are) parallel to z. Use a graphing utility to confirm your results.

69. $\mathbf{z} = \langle 3, 2, -5 \rangle$
(a) $\langle -6, -4, 10 \rangle$
(b) $\langle 2, \frac{4}{3}, -\frac{10}{3} \rangle$
(c) $\langle 6, 4, 10 \rangle$
(d) $\langle 1, -4, 2 \rangle$

70. $\mathbf{z} = \frac{1}{2}\mathbf{i} - \frac{2}{3}\mathbf{j} + \frac{3}{4}\mathbf{k}$
(a) $6\mathbf{i} - 4\mathbf{j} + 9\mathbf{k}$
(b) $-\mathbf{i} + \frac{4}{3}\mathbf{j} - \frac{3}{2}\mathbf{k}$
(c) $12\mathbf{i} + 9\mathbf{k}$
(d) $\frac{3}{4}\mathbf{i} - \mathbf{j} + \frac{9}{8}\mathbf{k}$

71. $\mathbf{z}$ has initial point $(1, -1, 3)$ and terminal point $(-2, 3, 5)$.
(a) $-6\mathbf{i} + 8\mathbf{j} + 4\mathbf{k}$ (b) $4\mathbf{j} + 2\mathbf{k}$

72. $\mathbf{z}$ has initial point $(5, 4, 1)$ and terminal point $(-2, -4, 4)$.
(a) $\langle 7, 6, 2 \rangle$ (b) $\langle 14, 16, -6 \rangle$

In Exercises 73–76, use vectors to determine whether the points are collinear.

73. $(0, -2, -5), (3, 4, 4), (2, 2, 1)$
74. $(4, -2, 7), (-2, 0, 3), (7, -3, 9)$
75. $(1, 2, 4), (2, 5, 0), (0, 1, 5)$
76. $(0, 0, 0), (1, 3, -2), (2, -6, 4)$

In Exercises 77 and 78, use vectors to show that the points form the vertices of a parallelogram.

77. $(2, 9, 1), (3, 11, 4), (0, 10, 2), (1, 12, 5)$
78. $(1, 1, 3), (9, -1, -2), (11, 2, -9), (3, 4, -4)$

In Exercises 79–84, find the magnitude of v.

79. $\mathbf{v} = \langle 0, 0, 0 \rangle$
80. $\mathbf{v} = \langle 1, 0, 3 \rangle$
81. $\mathbf{v} = \mathbf{i} - 2\mathbf{j} - 3\mathbf{k}$
82. $\mathbf{v} = -4\mathbf{i} + 3\mathbf{j} + 7\mathbf{k}$
83. Initial point of $\mathbf{v}$: $(1, -3, 4)$
Terminal point of $\mathbf{v}$: $(1, 0, -1)$
84. Initial point of $\mathbf{v}$: $(0, -1, 0)$
Terminal point of $\mathbf{v}$: $(1, 2, -2)$

In Exercises 85–88, find a unit vector (a) in the direction of u and (b) in the direction opposite of u.

85. $\mathbf{u} = \langle 2, -1, 2 \rangle$
86. $\mathbf{u} = \langle 6, 0, 8 \rangle$
87. $\mathbf{u} = \langle 3, 2, -5 \rangle$
88. $\mathbf{u} = \langle 8, 0, 0 \rangle$

89. *Programming* You are given the component forms of the vectors **u** and **v**. Write a program for a graphing utility in which the output is (a) the component form of $\mathbf{u} + \mathbf{v}$, (b) $\|\mathbf{u} + \mathbf{v}\|$, (c) $\|\mathbf{u}\|$, and (d) $\|\mathbf{v}\|$.

90. *Programming* Run the program you wrote in Exercise 89 for the vectors $\mathbf{u} = \langle -1, 3, 4 \rangle$ and $\mathbf{v} = \langle 5, 4.5, -6 \rangle$.

In Exercises 91 and 92, determine the values of c that satisfy the equation. Let $\mathbf{u} = \mathbf{i} + 2\mathbf{j} + 3\mathbf{k}$ and $\mathbf{v} = 2\mathbf{i} + 2\mathbf{j} - \mathbf{k}$.

91. $\|c\mathbf{v}\| = 5$
92. $\|c\mathbf{u}\| = 3$

In Exercises 93–96, find the vector v with the given magnitude and direction u.

	Magnitude	Direction
93.	10	$\mathbf{u} = \langle 0, 3, 3 \rangle$
94.	3	$\mathbf{u} = \langle 1, 1, 1 \rangle$
95.	$\frac{3}{2}$	$\mathbf{u} = \langle 2, -2, 1 \rangle$
96.	$\sqrt{5}$	$\mathbf{u} = \langle -4, 6, 2 \rangle$

In Exercises 97 and 98, sketch the vector v and write its component form.

97. **v** lies in the yz-plane, has magnitude 2, and makes an angle of 30° with the positive y-axis.

98. **v** lies in the xz-plane, has magnitude 5, and makes an angle of 45° with the positive z-axis.

In Exercises 99 and 100, use vectors to find the point that lies two-thirds of the way from P to Q.

99. $P(4, 3, 0)$, $Q(1, -3, 3)$ **100.** $P(1, 2, 5)$, $Q(6, 8, 2)$

101. Let $\mathbf{u} = \mathbf{i} + \mathbf{j}$, $\mathbf{v} = \mathbf{j} + \mathbf{k}$, and $\mathbf{w} = a\mathbf{u} + b\mathbf{v}$.

(a) Sketch **u** and **v**.

(b) If $\mathbf{w} = \mathbf{0}$, show that a and b must both be zero.

(c) Find a and b such that $\mathbf{w} = \mathbf{i} + 2\mathbf{j} + \mathbf{k}$.

(d) Show that no choice of a and b yields $\mathbf{w} = \mathbf{i} + 2\mathbf{j} + 3\mathbf{k}$.

102. Writing The initial and terminal points of the vector **v** are (x_1, y_1, z_1) and (x, y, z). Describe the set of all points (x, y, z) such that $\|\mathbf{v}\| = 4$.

Writing About Concepts

103. A point in the three-dimensional coordinate system has coordinates (x_0, y_0, z_0). Describe what each coordinate measures.

104. Give the formula for the distance between the points (x_1, y_1, z_1) and (x_2, y_2, z_2).

105. Give the standard equation of a sphere of radius r, centered at (x_0, y_0, z_0).

106. State the definition of parallel vectors.

107. Let A, B, and C be vertices of a triangle. Find $\overrightarrow{AB} + \overrightarrow{BC} + \overrightarrow{CA}$.

108. Let $\mathbf{r} = \langle x, y, z \rangle$ and $\mathbf{r}_0 = \langle 1, 1, 1 \rangle$. Describe the set of all points (x, y, z) such that $\|\mathbf{r} - \mathbf{r}_0\| = 2$.

109. Numerical, Graphical, and Analytic Analysis The lights in an auditorium are 24-pound discs of radius 18 inches. Each disc is supported by three equally spaced cables that are L inches long (see figure).

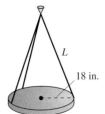

(a) Write the tension T in each cable as a function of L. Determine the domain of the function.

(b) Use a graphing utility and the function in part (a) to complete the table.

L	20	25	30	35	40	45	50
T							

(c) Use a graphing utility to graph the function in part (a). Determine the asymptotes of the graph.

(d) Confirm the asymptotes of the graph in part (c) analytically.

(e) Determine the minimum length of each cable if a cable is designed to carry a maximum load of 10 pounds.

110. Think About It Suppose the length of each cable in Exercise 109 has a fixed length $L = a$, and the radius of each disc is r_0 inches. Make a conjecture about the limit $\lim_{r_0 \to a^-} T$ and give a reason for your answer.

111. Diagonal of a Cube Find the component form of the unit vector **v** in the direction of the diagonal of the cube shown in the figure.

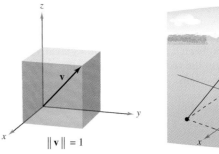

Figure for 111   **Figure for 112**

112. Tower Guy Wire The guy wire to a 100-foot tower has a tension of 550 pounds. Using the distances shown in the figure, write the component form of the vector **F** representing the tension in the wire.

113. Load Supports Find the tension in each of the supporting cables in the figure if the weight of the crate is 500 newtons.

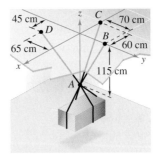 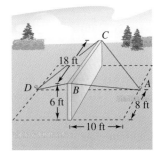

Figure for 113   **Figure for 114**

114. Construction A precast concrete wall is temporarily kept in its vertical position by ropes (see figure). Find the total force exerted on the pin at position A. The tensions in AB and AC are 420 pounds and 650 pounds.

115. Write an equation whose graph consists of the set of points $P(x, y, z)$ that are twice as far from $A(0, -1, 1)$ as from $B(1, 2, 0)$.

Section 11.3 The Dot Product of Two Vectors

- Use properties of the dot product of two vectors.
- Find the angle between two vectors using the dot product.
- Find the direction cosines of a vector in space.
- Find the projection of a vector onto another vector.
- Use vectors to find the work done by a constant force.

EXPLORATION

Interpreting a Dot Product Several vectors are shown below on the unit circle. Find the dot products of several pairs of vectors. Then find the angle between each pair that you used. Make a conjecture about the relationship between the dot product of two vectors and the angle between the vectors.

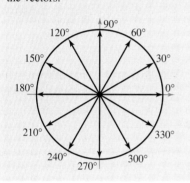

The Dot Product

So far you have studied two operations with vectors—vector addition and multiplication by a scalar—each of which yields another vector. In this section you will study a third vector operation, called the **dot product.** This product yields a scalar, rather than a vector.

Definition of Dot Product

The **dot product** of $\mathbf{u} = \langle u_1, u_2 \rangle$ and $\mathbf{v} = \langle v_1, v_2 \rangle$ is

$$\mathbf{u} \cdot \mathbf{v} = u_1 v_1 + u_2 v_2.$$

The **dot product** of $\mathbf{u} = \langle u_1, u_2, u_3 \rangle$ and $\mathbf{v} = \langle v_1, v_2, v_3 \rangle$ is

$$\mathbf{u} \cdot \mathbf{v} = u_1 v_1 + u_2 v_2 + u_3 v_3.$$

NOTE Because the dot product of two vectors yields a scalar, it is also called the **inner product** (or **scalar product**) of the two vectors.

THEOREM 11.4 Properties of the Dot Product

Let $\mathbf{u}$, $\mathbf{v}$, and $\mathbf{w}$ be vectors in the plane or in space and let c be a scalar.

1. $\mathbf{u} \cdot \mathbf{v} = \mathbf{v} \cdot \mathbf{u}$ Commutative Property
2. $\mathbf{u} \cdot (\mathbf{v} + \mathbf{w}) = \mathbf{u} \cdot \mathbf{v} + \mathbf{u} \cdot \mathbf{w}$ Distributive Property
3. $c(\mathbf{u} \cdot \mathbf{v}) = c\mathbf{u} \cdot \mathbf{v} = \mathbf{u} \cdot c\mathbf{v}$
4. $\mathbf{0} \cdot \mathbf{v} = 0$
5. $\mathbf{v} \cdot \mathbf{v} = \|\mathbf{v}\|^2$

Proof To prove the first property, let $\mathbf{u} = \langle u_1, u_2, u_3 \rangle$ and $\mathbf{v} = \langle v_1, v_2, v_3 \rangle$. Then

$$\mathbf{u} \cdot \mathbf{v} = u_1 v_1 + u_2 v_2 + u_3 v_3$$
$$= v_1 u_1 + v_2 u_2 + v_3 u_3$$
$$= \mathbf{v} \cdot \mathbf{u}.$$

For the fifth property, let $\mathbf{v} = \langle v_1, v_2, v_3 \rangle$. Then

$$\mathbf{v} \cdot \mathbf{v} = v_1^2 + v_2^2 + v_3^2$$
$$= \left(\sqrt{v_1^2 + v_2^2 + v_3^2}\right)^2$$
$$= \|\mathbf{v}\|^2.$$

Proofs of the other properties are left to you.

EXAMPLE 1 Finding Dot Products

Given $\mathbf{u} = \langle 2, -2 \rangle$, $\mathbf{v} = \langle 5, 8 \rangle$, and $\mathbf{w} = \langle -4, 3 \rangle$, find each of the following.

a. $\mathbf{u} \cdot \mathbf{v}$ **b.** $(\mathbf{u} \cdot \mathbf{v})\mathbf{w}$

c. $\mathbf{u} \cdot (2\mathbf{v})$ **d.** $\|\mathbf{w}\|^2$

Solution

a. $\mathbf{u} \cdot \mathbf{v} = \langle 2, -2 \rangle \cdot \langle 5, 8 \rangle = 2(5) + (-2)(8) = -6$

b. $(\mathbf{u} \cdot \mathbf{v})\mathbf{w} = -6\langle -4, 3 \rangle = \langle 24, -18 \rangle$

c. $\mathbf{u} \cdot (2\mathbf{v}) = 2(\mathbf{u} \cdot \mathbf{v}) = 2(-6) = -12$ Theorem 11.4

d. $\|\mathbf{w}\|^2 = \mathbf{w} \cdot \mathbf{w}$ Theorem 11.4

$\quad\quad = \langle -4, 3 \rangle \cdot \langle -4, 3 \rangle$ Substitute $\langle -4, 3 \rangle$ for $\mathbf{w}$.

$\quad\quad = (-4)(-4) + (3)(3)$ Definition of dot product

$\quad\quad = 25$ Simplify.

Notice that the result of part (b) is a *vector* quantity, whereas the results of the other three parts are *scalar* quantities.

Angle Between Two Vectors

The **angle between two nonzero vectors** is the angle θ, $0 \leq \theta \leq \pi$, between their respective standard position vectors, as shown in Figure 11.24. The next theorem shows how to find this angle using the dot product. (Note that the angle between the zero vector and another vector is not defined here.)

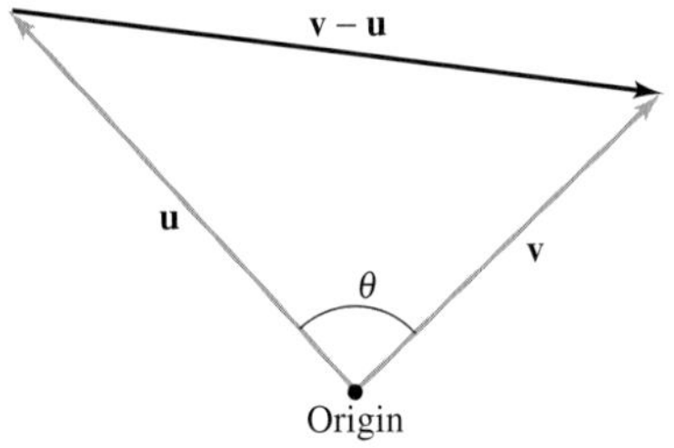

The angle between two vectors
Figure 11.24

THEOREM 11.5 Angle Between Two Vectors

If θ is the angle between two nonzero vectors $\mathbf{u}$ and $\mathbf{v}$, then

$$\cos \theta = \frac{\mathbf{u} \cdot \mathbf{v}}{\|\mathbf{u}\| \|\mathbf{v}\|}.$$

Proof Consider the triangle determined by vectors $\mathbf{u}$, $\mathbf{v}$, and $\mathbf{v} - \mathbf{u}$, as shown in Figure 11.24. By the Law of Cosines, you can write

$$\|\mathbf{v} - \mathbf{u}\|^2 = \|\mathbf{u}\|^2 + \|\mathbf{v}\|^2 - 2\|\mathbf{u}\| \|\mathbf{v}\| \cos \theta.$$

Using the properties of the dot product, the left side can be rewritten as

$$\|\mathbf{v} - \mathbf{u}\|^2 = (\mathbf{v} - \mathbf{u}) \cdot (\mathbf{v} - \mathbf{u})$$
$$= (\mathbf{v} - \mathbf{u}) \cdot \mathbf{v} - (\mathbf{v} - \mathbf{u}) \cdot \mathbf{u}$$
$$= \mathbf{v} \cdot \mathbf{v} - \mathbf{u} \cdot \mathbf{v} - \mathbf{v} \cdot \mathbf{u} + \mathbf{u} \cdot \mathbf{u}$$
$$= \|\mathbf{v}\|^2 - 2\mathbf{u} \cdot \mathbf{v} + \|\mathbf{u}\|^2$$

and substitution back into the Law of Cosines yields

$$\|\mathbf{v}\|^2 - 2\mathbf{u} \cdot \mathbf{v} + \|\mathbf{u}\|^2 = \|\mathbf{u}\|^2 + \|\mathbf{v}\|^2 - 2\|\mathbf{u}\| \|\mathbf{v}\| \cos \theta$$

$$-2\mathbf{u} \cdot \mathbf{v} = -2\|\mathbf{u}\| \|\mathbf{v}\| \cos \theta$$

$$\cos \theta = \frac{\mathbf{u} \cdot \mathbf{v}}{\|\mathbf{u}\| \|\mathbf{v}\|}.$$

If the angle between two vectors is known, rewriting Theorem 11.5 in the form

$$\mathbf{u} \cdot \mathbf{v} = \|\mathbf{u}\| \|\mathbf{v}\| \cos \theta \qquad \text{Alternative form of dot product}$$

produces an alternative way to calculate the dot product. From this form, you can see that because $\|\mathbf{u}\|$ and $\|\mathbf{v}\|$ are always positive, $\mathbf{u} \cdot \mathbf{v}$ and $\cos \theta$ will always have the same sign. Figure 11.25 shows the possible orientations of two vectors.

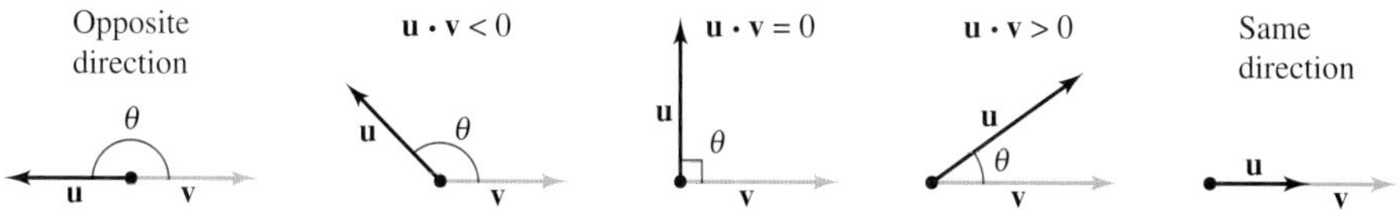

Opposite direction	$\mathbf{u} \cdot \mathbf{v} < 0$	$\mathbf{u} \cdot \mathbf{v} = 0$	$\mathbf{u} \cdot \mathbf{v} > 0$	Same direction
$\theta = \pi$	$\pi/2 < \theta < \pi$	$\theta = \pi/2$	$0 < \theta < \pi/2$	$\theta = 0$
$\cos \theta = -1$	$-1 < \cos \theta < 0$	$\cos \theta = 0$	$0 < \cos \theta < 1$	$\cos \theta = 1$

Figure 11.25

From Theorem 11.5, you can see that two nonzero vectors meet at a right angle if and only if their dot product is zero. Two such vectors are said to be **orthogonal.**

Definition of Orthogonal Vectors

The vectors $\mathbf{u}$ and $\mathbf{v}$ are orthogonal if $\mathbf{u} \cdot \mathbf{v} = 0$.

NOTE   The terms "perpendicular," "orthogonal," and "normal" all mean essentially the same thing—meeting at right angles. However, it is common to say that two vectors are *orthogonal*, two lines or planes are *perpendicular*, and a vector is *normal* to a given line or plane.

From this definition, it follows that the zero vector is orthogonal to every vector $\mathbf{u}$, because $\mathbf{0} \cdot \mathbf{u} = 0$. Moreover, for $0 \leq \theta \leq \pi$, you know that $\cos \theta = 0$ if and only if $\theta = \pi/2$. So, you can use Theorem 11.5 to conclude that two *nonzero* vectors are orthogonal if and only if the angle between them is $\pi/2$.

EXAMPLE 2   Finding the Angle Between Two Vectors

For $\mathbf{u} = \langle 3, -1, 2 \rangle$, $\mathbf{v} = \langle -4, 0, 2 \rangle$, $\mathbf{w} = \langle 1, -1, -2 \rangle$, and $\mathbf{z} = \langle 2, 0, -1 \rangle$, find the angle between each pair of vectors.

a. $\mathbf{u}$ and $\mathbf{v}$ **b.** $\mathbf{u}$ and $\mathbf{w}$ **c.** $\mathbf{v}$ and $\mathbf{z}$

Solution

a. $\cos \theta = \dfrac{\mathbf{u} \cdot \mathbf{v}}{\|\mathbf{u}\| \|\mathbf{v}\|} = \dfrac{-12 + 0 + 4}{\sqrt{14} \sqrt{20}} = \dfrac{-8}{2\sqrt{14}\sqrt{5}} = \dfrac{-4}{\sqrt{70}}$

Because $\mathbf{u} \cdot \mathbf{v} < 0$, $\theta = \arccos \dfrac{-4}{\sqrt{70}} \approx 2.069$ radians.

b. $\cos \theta = \dfrac{\mathbf{u} \cdot \mathbf{w}}{\|\mathbf{u}\| \|\mathbf{w}\|} = \dfrac{3 + 1 - 4}{\sqrt{14} \sqrt{6}} = \dfrac{0}{\sqrt{84}} = 0$

Because $\mathbf{u} \cdot \mathbf{w} = 0$, $\mathbf{u}$ and $\mathbf{w}$ are *orthogonal*. So, $\theta = \pi/2$.

c. $\cos \theta = \dfrac{\mathbf{v} \cdot \mathbf{z}}{\|\mathbf{v}\| \|\mathbf{z}\|} = \dfrac{-8 + 0 - 2}{\sqrt{20} \sqrt{5}} = \dfrac{-10}{\sqrt{100}} = -1$

Consequently, $\theta = \pi$. Note that $\mathbf{v}$ and $\mathbf{z}$ are parallel, with $\mathbf{v} = -2\mathbf{z}$.

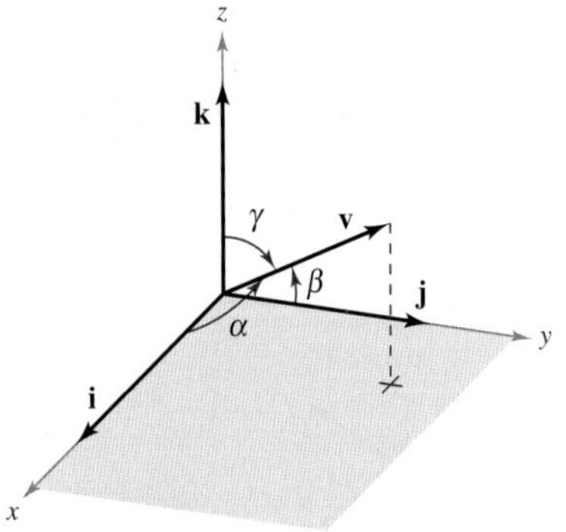

Direction Cosines

For a vector in the plane, you have seen that it is convenient to measure direction in terms of the angle, measured counterclockwise, *from* the positive x-axis *to* the vector. In space it is more convenient to measure direction in terms of the angles *between* the nonzero vector **v** and the three unit vectors **i**, **j**, and **k**, as shown in Figure 11.26. The angles α, β, and γ are the **direction angles of v,** and $\cos \alpha$, $\cos \beta$, and $\cos \gamma$ are the **direction cosines of v.** Because

$$\mathbf{v} \cdot \mathbf{i} = \|\mathbf{v}\| \|\mathbf{i}\| \cos \alpha = \|\mathbf{v}\| \cos \alpha$$

and

$$\mathbf{v} \cdot \mathbf{i} = \langle v_1, v_2, v_3 \rangle \cdot \langle 1, 0, 0 \rangle = v_1$$

it follows that $\cos \alpha = v_1/\|\mathbf{v}\|$. By similar reasoning with the unit vectors **j** and **k**, you have

$\cos \alpha = \dfrac{v_1}{\|\mathbf{v}\|}$ α is the angle between **v** and **i**.

$\cos \beta = \dfrac{v_2}{\|\mathbf{v}\|}$ β is the angle between **v** and **j**.

$\cos \gamma = \dfrac{v_3}{\|\mathbf{v}\|}.$ γ is the angle between **v** and **k**.

Consequently, any nonzero vector **v** in space has the normalized form

$$\frac{\mathbf{v}}{\|\mathbf{v}\|} = \frac{v_1}{\|\mathbf{v}\|}\mathbf{i} + \frac{v_2}{\|\mathbf{v}\|}\mathbf{j} + \frac{v_3}{\|\mathbf{v}\|}\mathbf{k} = \cos \alpha \mathbf{i} + \cos \beta \mathbf{j} + \cos \gamma \mathbf{k}$$

and because $\mathbf{v}/\|\mathbf{v}\|$ is a unit vector, it follows that

$$\cos^2 \alpha + \cos^2 \beta + \cos^2 \gamma = 1.$$

Direction angles
Figure 11.26

EXAMPLE 3   Finding Direction Angles

Find the direction cosines and angles for the vector $\mathbf{v} = 2\mathbf{i} + 3\mathbf{j} + 4\mathbf{k}$, and show that $\cos^2 \alpha + \cos^2 \beta + \cos^2 \gamma = 1$.

Solution   Because $\|\mathbf{v}\| = \sqrt{2^2 + 3^2 + 4^2} = \sqrt{29}$, you can write the following.

$\cos \alpha = \dfrac{v_1}{\|\mathbf{v}\|} = \dfrac{2}{\sqrt{29}}$ ⇒ $\alpha \approx 68.2°$ Angle between **v** and **i**

$\cos \beta = \dfrac{v_2}{\|\mathbf{v}\|} = \dfrac{3}{\sqrt{29}}$ ⇒ $\beta \approx 56.1°$ Angle between **v** and **j**

$\cos \gamma = \dfrac{v_3}{\|\mathbf{v}\|} = \dfrac{4}{\sqrt{29}}$ ⇒ $\gamma \approx 42.0°$ Angle between **v** and **k**

Furthermore, the sum of the squares of the direction cosines is

$$\cos^2 \alpha + \cos^2 \beta + \cos^2 \gamma = \frac{4}{29} + \frac{9}{29} + \frac{16}{29}$$

$$= \frac{29}{29}$$

$$= 1.$$

See Figure 11.27.

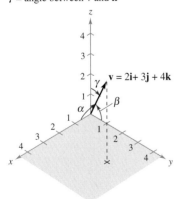

$\alpha =$ angle between **v** and **i**
$\beta =$ angle between **v** and **j**
$\gamma =$ angle between **v** and **k**

The direction angles of **v**
Figure 11.27

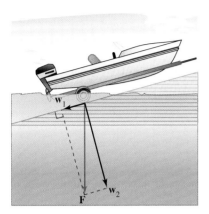

The force due to gravity pulls the boat against the ramp and down the ramp.
Figure 11.28

Projections and Vector Components

You have already seen applications in which two vectors are added to produce a resultant vector. Many applications in physics and engineering pose the reverse problem—decomposing a given vector into the sum of two **vector components**. The following physical example enables you to see the usefulness of this procedure.

Consider a boat on an inclined ramp, as shown in Figure 11.28. The force **F** due to gravity pulls the boat *down* the ramp and *against* the ramp. These two forces, $\mathbf{w}_1$ and $\mathbf{w}_2$, are orthogonal—they are called the vector components of **F**.

$$\mathbf{F} = \mathbf{w}_1 + \mathbf{w}_2 \qquad \text{Vector components of } \mathbf{F}$$

The forces $\mathbf{w}_1$ and $\mathbf{w}_2$ help you analyze the effect of gravity on the boat. For example, $\mathbf{w}_1$ indicates the force necessary to keep the boat from rolling down the ramp, whereas $\mathbf{w}_2$ indicates the force that the tires must withstand.

Definition of Projection and Vector Components

Let **u** and **v** be nonzero vectors. Moreover, let $\mathbf{u} = \mathbf{w}_1 + \mathbf{w}_2$, where $\mathbf{w}_1$ is parallel to **v** and $\mathbf{w}_2$ is orthogonal to **v**, as shown in Figure 11.29.

1. $\mathbf{w}_1$ is called the **projection of u onto v** or the **vector component of u along v**, and is denoted by $\mathbf{w}_1 = \text{proj}_\mathbf{v}\mathbf{u}$.
2. $\mathbf{w}_2 = \mathbf{u} - \mathbf{w}_1$ is called the **vector component of u orthogonal to v.**

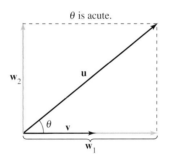
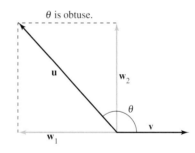

$\mathbf{w}_1 = \text{proj}_\mathbf{v}\mathbf{u} = $ projection of **u** onto **v** = vector component of **u** along **v**
$\mathbf{w}_2 = $ vector component of **u** orthogonal to **v**
Figure 11.29

EXAMPLE 4 Finding a Vector Component of u Orthogonal to v

Find the vector component of $\mathbf{u} = \langle 7, 4 \rangle$ that is orthogonal to $\mathbf{v} = \langle 2, 3 \rangle$, given that $\mathbf{w}_1 = \text{proj}_\mathbf{v}\mathbf{u} = \langle 4, 6 \rangle$ and

$$\mathbf{u} = \langle 7, 4 \rangle = \mathbf{w}_1 + \mathbf{w}_2.$$

Solution   Because $\mathbf{u} = \mathbf{w}_1 + \mathbf{w}_2$, where $\mathbf{w}_1$ is parallel to **v**, it follows that $\mathbf{w}_2$ is the vector component of **u** orthogonal to **v**. So, you have

$$\begin{aligned}\mathbf{w}_2 &= \mathbf{u} - \mathbf{w}_1 \\ &= \langle 7, 4 \rangle - \langle 4, 6 \rangle \\ &= \langle 3, -2 \rangle.\end{aligned}$$

Check to see that $\mathbf{w}_2$ is orthogonal to **v**, as shown in Figure 11.30.

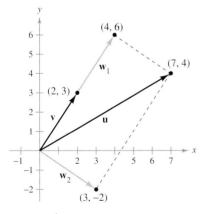

$\mathbf{u} = \mathbf{w}_1 + \mathbf{w}_2$
Figure 11.30

NOTE Note the distinction between the terms "component" and "vector component." For example, using the standard unit vectors with $\mathbf{u} = u_1\mathbf{i} + u_2\mathbf{j}$, u_1 is the *component* of $\mathbf{u}$ in the direction of $\mathbf{i}$ and $u_1\mathbf{i}$ is the *vector component* in the direction of $\mathbf{i}$.

From Example 4, you can see that it is easy to find the vector component $\mathbf{w}_2$ once you have found the projection, $\mathbf{w}_1$, of $\mathbf{u}$ onto $\mathbf{v}$. To find this projection, use the dot product given in the theorem below, which you will prove in Exercise 90.

THEOREM 11.6 Projection Using the Dot Product

If $\mathbf{u}$ and $\mathbf{v}$ are nonzero vectors, then the projection of $\mathbf{u}$ onto $\mathbf{v}$ is given by

$$\operatorname{proj}_{\mathbf{v}} \mathbf{u} = \left(\frac{\mathbf{u} \cdot \mathbf{v}}{\|\mathbf{v}\|^2}\right)\mathbf{v}.$$

The projection of $\mathbf{u}$ onto $\mathbf{v}$ can be written as a scalar multiple of a unit vector in the direction of $\mathbf{v}$. That is,

$$\left(\frac{\mathbf{u} \cdot \mathbf{v}}{\|\mathbf{v}\|^2}\right)\mathbf{v} = \left(\frac{\mathbf{u} \cdot \mathbf{v}}{\|\mathbf{v}\|}\right)\frac{\mathbf{v}}{\|\mathbf{v}\|} = (k)\frac{\mathbf{v}}{\|\mathbf{v}\|} \quad \Longrightarrow \quad k = \frac{\mathbf{u} \cdot \mathbf{v}}{\|\mathbf{v}\|} = \|\mathbf{u}\|\cos\theta.$$

The scalar k is called the **component of u in the direction of v**.

EXAMPLE 5 Decomposing a Vector into Vector Components

Find the projection of $\mathbf{u}$ onto $\mathbf{v}$ and the vector component of $\mathbf{u}$ orthogonal to $\mathbf{v}$ for the vectors $\mathbf{u} = 3\mathbf{i} - 5\mathbf{j} + 2\mathbf{k}$ and $\mathbf{v} = 7\mathbf{i} + \mathbf{j} - 2\mathbf{k}$ shown in Figure 11.31.

Solution The projection of $\mathbf{u}$ onto $\mathbf{v}$ is

$$\mathbf{w}_1 = \left(\frac{\mathbf{u} \cdot \mathbf{v}}{\|\mathbf{v}\|^2}\right)\mathbf{v} = \left(\frac{12}{54}\right)(7\mathbf{i} + \mathbf{j} - 2\mathbf{k}) = \frac{14}{9}\mathbf{i} + \frac{2}{9}\mathbf{j} - \frac{4}{9}\mathbf{k}.$$

The vector component of $\mathbf{u}$ orthogonal to $\mathbf{v}$ is the vector

$$\mathbf{w}_2 = \mathbf{u} - \mathbf{w}_1 = (3\mathbf{i} - 5\mathbf{j} + 2\mathbf{k}) - \left(\frac{14}{9}\mathbf{i} + \frac{2}{9}\mathbf{j} - \frac{4}{9}\mathbf{k}\right) = \frac{13}{9}\mathbf{i} - \frac{47}{9}\mathbf{j} + \frac{22}{9}\mathbf{k}.$$

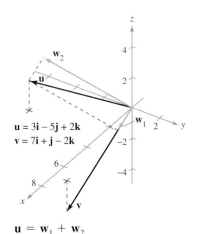

$\mathbf{u} = 3\mathbf{i} - 5\mathbf{j} + 2\mathbf{k}$
$\mathbf{v} = 7\mathbf{i} + \mathbf{j} - 2\mathbf{k}$

$\mathbf{u} = \mathbf{w}_1 + \mathbf{w}_2$
Figure 11.31

EXAMPLE 6 Finding a Force

A 600-pound boat sits on a ramp inclined at 30°, as shown in Figure 11.32. What force is required to keep the boat from rolling down the ramp?

Solution Because the force due to gravity is vertical and downward, you can represent the gravitational force by the vector $\mathbf{F} = -600\mathbf{j}$. To find the force required to keep the boat from rolling down the ramp, project $\mathbf{F}$ onto a unit vector $\mathbf{v}$ in the direction of the ramp, as follows.

$$\mathbf{v} = \cos 30°\mathbf{i} + \sin 30°\mathbf{j} = \frac{\sqrt{3}}{2}\mathbf{i} + \frac{1}{2}\mathbf{j} \qquad \text{Unit vector along ramp}$$

Therefore, the projection of $\mathbf{F}$ onto $\mathbf{v}$ is given by

$$\mathbf{w}_1 = \operatorname{proj}_{\mathbf{v}} \mathbf{F} = \left(\frac{\mathbf{F} \cdot \mathbf{v}}{\|\mathbf{v}\|^2}\right)\mathbf{v} = (\mathbf{F} \cdot \mathbf{v})\mathbf{v} = (-600)\left(\frac{1}{2}\right)\mathbf{v} = -300\left(\frac{\sqrt{3}}{2}\mathbf{i} + \frac{1}{2}\mathbf{j}\right).$$

The magnitude of this force is 300, and therefore a force of 300 pounds is required to keep the boat from rolling down the ramp.

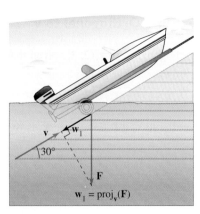

Figure 11.32

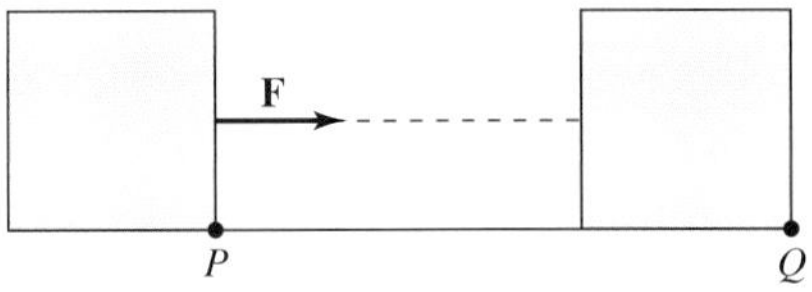

Work

The work W done by the constant force **F** acting along the line of motion of an object is given by

$$W = (\text{magnitude of force})(\text{distance}) = \|\mathbf{F}\|\|\overrightarrow{PQ}\|$$

as shown in Figure 11.33(a). If the constant force **F** is not directed along the line of motion, you can see from Figure 11.33(b) that the work W done by the force is

$$W = \|\text{proj}_{\overrightarrow{PQ}}\mathbf{F}\|\|\overrightarrow{PQ}\| = (\cos\theta)\|\mathbf{F}\|\|\overrightarrow{PQ}\| = \mathbf{F}\cdot\overrightarrow{PQ}.$$

This notion of work is summarized in the following definition.

Definition of Work

The work W done by a constant force **F** as its point of application moves along the vector $\overrightarrow{PQ}$ is given by either of the following.

1. $W = \|\text{proj}_{\overrightarrow{PQ}}\mathbf{F}\|\|\overrightarrow{PQ}\|$ Projection form
2. $W = \mathbf{F}\cdot\overrightarrow{PQ}$ Dot product form

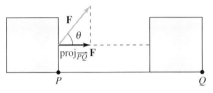

(a) Force acts along the line of motion.

Work = $\|\text{proj}_{\overrightarrow{PQ}}\mathbf{F}\|\|\overrightarrow{PQ}\|$

(b) Force acts at angle θ with the line of motion.
Figure 11.33

EXAMPLE 7   Finding Work

To close a sliding door, a person pulls on a rope with a constant force of 50 pounds at a constant angle of 60°, as shown in Figure 11.34. Find the work done in moving the door 12 feet to its closed position.

Solution   Using a projection, you can calculate the work as follows.

$$\begin{aligned} W &= \|\text{proj}_{\overrightarrow{PQ}}\mathbf{F}\|\|\overrightarrow{PQ}\| && \text{Projection form for work} \\ &= \cos(60°)\|\mathbf{F}\|\|\overrightarrow{PQ}\| \\ &= \frac{1}{2}(50)(12) \\ &= 300 \text{ foot-pounds} \end{aligned}$$

Figure 11.34

Exercises for Section 11.3

See www.CalcChat.com for worked-out solutions to odd-numbered exercises.

In Exercises 1–8, find (a) $\mathbf{u}\cdot\mathbf{v}$, (b) $\mathbf{u}\cdot\mathbf{u}$, (c) $\|\mathbf{u}\|^2$, (d) $(\mathbf{u}\cdot\mathbf{v})\mathbf{v}$, and (e) $\mathbf{u}\cdot(2\mathbf{v})$.

1. $\mathbf{u} = \langle 3, 4\rangle$, $\mathbf{v} = \langle 2, -3\rangle$
2. $\mathbf{u} = \langle 4, 10\rangle$, $\mathbf{v} = \langle -2, 3\rangle$
3. $\mathbf{u} = \langle 5, -1\rangle$, $\mathbf{v} = \langle -3, 2\rangle$
4. $\mathbf{u} = \langle -4, 8\rangle$, $\mathbf{v} = \langle 6, 3\rangle$
5. $\mathbf{u} = \langle 2, -3, 4\rangle$, $\mathbf{v} = \langle 0, 6, 5\rangle$
6. $\mathbf{u} = \mathbf{i}$, $\mathbf{v} = \mathbf{i}$
7. $\mathbf{u} = 2\mathbf{i} - \mathbf{j} + \mathbf{k}$
   $\mathbf{v} = \mathbf{i} - \mathbf{k}$
8. $\mathbf{u} = 2\mathbf{i} + \mathbf{j} - 2\mathbf{k}$
   $\mathbf{v} = \mathbf{i} - 3\mathbf{j} + 2\mathbf{k}$

In Exercises 9 and 10, find $\mathbf{u}\cdot\mathbf{v}$.

9. $\|\mathbf{u}\| = 8$, $\|\mathbf{v}\| = 5$, and the angle between **u** and **v** is $\pi/3$.
10. $\|\mathbf{u}\| = 40$, $\|\mathbf{v}\| = 25$, and the angle between **u** and **v** is $5\pi/6$.

In Exercises 11–18, find the angle θ between the vectors.

11. $\mathbf{u} = \langle 1, 1\rangle$, $\mathbf{v} = \langle 2, -2\rangle$
12. $\mathbf{u} = \langle 3, 1\rangle$, $\mathbf{v} = \langle 2, -1\rangle$
13. $\mathbf{u} = 3\mathbf{i} + \mathbf{j}$, $\mathbf{v} = -2\mathbf{i} + 4\mathbf{j}$
14. $\mathbf{u} = \cos\left(\dfrac{\pi}{6}\right)\mathbf{i} + \sin\left(\dfrac{\pi}{6}\right)\mathbf{j}$
    $\mathbf{v} = \cos\left(\dfrac{3\pi}{4}\right)\mathbf{i} + \sin\left(\dfrac{3\pi}{4}\right)\mathbf{j}$
15. $\mathbf{u} = \langle 1, 1, 1\rangle$
    $\mathbf{v} = \langle 2, 1, -1\rangle$
16. $\mathbf{u} = 3\mathbf{i} + 2\mathbf{j} + \mathbf{k}$
    $\mathbf{v} = 2\mathbf{i} - 3\mathbf{j}$
17. $\mathbf{u} = 3\mathbf{i} + 4\mathbf{j}$
    $\mathbf{v} = -2\mathbf{j} + 3\mathbf{k}$
18. $\mathbf{u} = 2\mathbf{i} - 3\mathbf{j} + \mathbf{k}$
    $\mathbf{v} = \mathbf{i} - 2\mathbf{j} + \mathbf{k}$

In Exercises 19–26, determine whether u and v are orthogonal, parallel, or neither.

19. $\mathbf{u} = \langle 4, 0\rangle$, $\mathbf{v} = \langle 1, 1\rangle$
20. $\mathbf{u} = \langle 2, 18\rangle$, $\mathbf{v} = \langle \frac{3}{2}, -\frac{1}{6}\rangle$

21. $\mathbf{u} = \langle 4, 3 \rangle$
$\mathbf{v} = \langle \frac{1}{2}, -\frac{2}{3} \rangle$

22. $\mathbf{u} = -\frac{1}{3}(\mathbf{i} - 2\mathbf{j})$
$\mathbf{v} = 2\mathbf{i} - 4\mathbf{j}$

23. $\mathbf{u} = \mathbf{j} + 6\mathbf{k}$
$\mathbf{v} = \mathbf{i} - 2\mathbf{j} - \mathbf{k}$

24. $\mathbf{u} = -2\mathbf{i} + 3\mathbf{j} - \mathbf{k}$
$\mathbf{v} = 2\mathbf{i} + \mathbf{j} - \mathbf{k}$

25. $\mathbf{u} = \langle 2, -3, 1 \rangle$
$\mathbf{v} = \langle -1, -1, -1 \rangle$

26. $\mathbf{u} = \langle \cos\theta, \sin\theta, -1 \rangle$
$\mathbf{v} = \langle \sin\theta, -\cos\theta, 0 \rangle$

In Exercises 27–30, the vertices of a triangle are given. Determine whether the triangle is an acute triangle, an obtuse triangle, or a right triangle. Explain your reasoning.

27. $(1, 2, 0), (0, 0, 0), (-2, 1, 0)$
28. $(-3, 0, 0), (0, 0, 0), (1, 2, 3)$
29. $(2, -3, 4), (0, 1, 2), (-1, 2, 0)$
30. $(2, -7, 3), (-1, 5, 8), (4, 6, -1)$

In Exercises 31–34, find the direction cosines of **u** and demonstrate that the sum of the squares of the direction cosines is 1.

31. $\mathbf{u} = \mathbf{i} + 2\mathbf{j} + 2\mathbf{k}$
32. $\mathbf{u} = 5\mathbf{i} + 3\mathbf{j} - \mathbf{k}$
33. $\mathbf{u} = \langle 0, 6, -4 \rangle$
34. $\mathbf{u} = \langle a, b, c \rangle$

In Exercises 35–38, find the direction angles of the vector.

35. $\mathbf{u} = 3\mathbf{i} + 2\mathbf{j} - 2\mathbf{k}$
36. $\mathbf{u} = -4\mathbf{i} + 3\mathbf{j} + 5\mathbf{k}$
37. $\mathbf{u} = \langle -1, 5, 2 \rangle$
38. $\mathbf{u} = \langle -2, 6, 1 \rangle$

In Exercises 39 and 40, use a graphing utility to find the magnitude and direction angles of the resultant of forces $\mathbf{F}_1$ and $\mathbf{F}_2$ with initial points at the origin. The magnitude and terminal point of each vector are given.

Vector	Magnitude	Terminal Point
39. $\mathbf{F}_1$	50 lb	$(10, 5, 3)$
$\mathbf{F}_2$	80 lb	$(12, 7, -5)$
40. $\mathbf{F}_1$	300 N	$(-20, -10, 5)$
$\mathbf{F}_2$	100 N	$(5, 15, 0)$

41. *Load-Supporting Cables* A load is supported by three cables, as shown in the figure. Find the direction angles of the load-supporting cable *OA*.

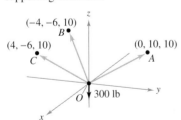

42. *Load-Supporting Cables* The tension in the cable *OA* in Exercise 41 is 200 newtons. Determine the weight of the load.

In Exercises 43–46, find the component of **u** that is orthogonal to **v**, given $\mathbf{w}_1 = \text{proj}_{\mathbf{v}}\mathbf{u}$.

43. $\mathbf{u} = \langle 6, 7 \rangle$, $\mathbf{v} = \langle 1, 4 \rangle$, $\text{proj}_{\mathbf{v}}\mathbf{u} = \langle 2, 8 \rangle$
44. $\mathbf{u} = \langle 9, 7 \rangle$, $\mathbf{v} = \langle 1, 3 \rangle$, $\text{proj}_{\mathbf{v}}\mathbf{u} = \langle 3, 9 \rangle$
45. $\mathbf{u} = \langle 0, 3, 3 \rangle$, $\mathbf{v} = \langle -1, 1, 1 \rangle$, $\text{proj}_{\mathbf{v}}\mathbf{u} = \langle -2, 2, 2 \rangle$
46. $\mathbf{u} = \langle 8, 2, 0 \rangle$, $\mathbf{v} = \langle 2, 1, -1 \rangle$, $\text{proj}_{\mathbf{v}}\mathbf{u} = \langle 6, 3, -3 \rangle$

In Exercises 47–50, (a) find the projection of **u** onto **v**, and (b) find the vector component of **u** orthogonal to **v**.

47. $\mathbf{u} = \langle 2, 3 \rangle$, $\mathbf{v} = \langle 5, 1 \rangle$
48. $\mathbf{u} = \langle 2, -3 \rangle$, $\mathbf{v} = \langle 3, 2 \rangle$
49. $\mathbf{u} = \langle 2, 1, 2 \rangle$
$\mathbf{v} = \langle 0, 3, 4 \rangle$
50. $\mathbf{u} = \langle 1, 0, 4 \rangle$
$\mathbf{v} = \langle 3, 0, 2 \rangle$

Writing About Concepts

51. Define the dot product of vectors **u** and **v**.

52. State the definition of orthogonal vectors. If vectors are neither parallel nor orthogonal, how do you find the angle between them? Explain.

53. What is known about θ, the angle between two nonzero vectors **u** and **v**, if
(a) $\mathbf{u} \cdot \mathbf{v} = 0$? (b) $\mathbf{u} \cdot \mathbf{v} > 0$? (c) $\mathbf{u} \cdot \mathbf{v} < 0$?

54. Determine which of the following are defined for nonzero vectors **u**, **v**, and **w**. Explain your reasoning.
(a) $\mathbf{u} \cdot (\mathbf{v} + \mathbf{w})$ (b) $(\mathbf{u} \cdot \mathbf{v})\mathbf{w}$
(c) $\mathbf{u} \cdot \mathbf{v} + \mathbf{w}$ (d) $\|\mathbf{u}\| \cdot (\mathbf{v} + \mathbf{w})$

55. Describe direction cosines and direction angles of a vector **v**.

56. Give a geometric description of the projection of **u** onto **v**.

57. What can be said about the vectors **u** and **v** if (a) the projection of **u** onto **v** equals **u** and (b) the projection of **u** onto **v** equals **0**?

58. If the projection of **u** onto **v** has the same magnitude as the projection of **v** onto **u**, can you conclude that $\|\mathbf{u}\| = \|\mathbf{v}\|$? Explain.

59. *Revenue* The vector $\mathbf{u} = \langle 3240, 1450, 2235 \rangle$ gives the numbers of hamburgers, chicken sandwiches, and cheeseburgers, respectively, sold at a fast-food restaurant in one week. The vector $\mathbf{v} = \langle 1.35, 2.65, 1.85 \rangle$ gives the prices (in dollars) per unit for the three food items. Find the dot product $\mathbf{u} \cdot \mathbf{v}$, and explain what information it gives.

60. *Revenue* Repeat Exercise 59 after increasing prices by 4%. Identify the vector operation used to increase prices by 4%.

61. *Programming* Given vectors **u** and **v** in component form, write a program for a graphing utility in which the output is (a) $\|\mathbf{u}\|$, (b) $\|\mathbf{v}\|$, and (c) the angle between **u** and **v**.

62. *Programming* Use the program you wrote in Exercise 61 to find the angle between the vectors $\mathbf{u} = \langle 8, -4, 2 \rangle$ and $\mathbf{v} = \langle 2, 5, 2 \rangle$.

63. *Programming* Given vectors **u** and **v** in component form, write a program for a graphing utility in which the output is the component form of the projection of **u** onto **v**.

64. *Programming* Use the program you wrote in Exercise 63 to find the projection of **u** onto **v** for $\mathbf{u} = \langle 5, 6, 2 \rangle$ and $\mathbf{v} = \langle -1, 3, 4 \rangle$.

Think About It In Exercises 65 and 66, use the figure to determine mentally the projection of **u** onto **v**. (The coordinates of the terminal points of the vectors in standard position are given.) Verify your results analytically.

65.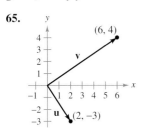

66.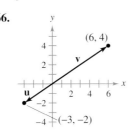

In Exercises 67–70, find two vectors in opposite directions that are orthogonal to the vector **u**. (The answers are not unique.)

67. $\mathbf{u} = \frac{1}{2}\mathbf{i} - \frac{2}{3}\mathbf{j}$
68. $\mathbf{u} = -8\mathbf{i} + 3\mathbf{j}$
69. $\mathbf{u} = \langle 3, 1, -2 \rangle$
70. $\mathbf{u} = \langle 0, -3, 6 \rangle$

71. *Braking Load* A 48,000-pound truck is parked on a 10° slope (see figure). Assume the only force to overcome is that due to gravity. Find (a) the force required to keep the truck from rolling down the hill and (b) the force perpendicular to the hill.

Figure for 71

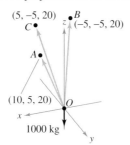

Figure for 72

72. *Load-Supporting Cables* Find the magnitude of the projection of the load-supporting cable OA onto the positive z-axis as shown in the figure.

73. *Work* An object is pulled 10 feet across a floor, using a force of 85 pounds. The direction of the force is 60° above the horizontal (see figure). Find the work done.

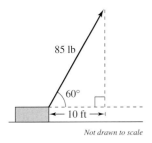

Figure for 73

Figure for 74

74. *Work* A toy wagon is pulled by exerting a force of 25 pounds on a handle that makes a 20° angle with the horizontal (see figure in left column). Find the work done in pulling the wagon 50 feet.

True or False? In Exercises 75 and 76, determine whether the statement is true or false. If it is false, explain why or give an example that shows it is false.

75. If $\mathbf{u} \cdot \mathbf{v} = \mathbf{u} \cdot \mathbf{w}$ and $\mathbf{u} \neq \mathbf{0}$, then $\mathbf{v} = \mathbf{w}$.
76. If **u** and **v** are orthogonal to **w**, then $\mathbf{u} + \mathbf{v}$ is orthogonal to **w**.
77. Find the angle between a cube's diagonal and one of its edges.
78. Find the angle between the diagonal of a cube and the diagonal of one of its sides.

In Exercises 79–82, (a) find the unit tangent vectors to each curve at their points of intersection and (b) find the angles $(0 \leq \theta \leq 90°)$ between the curves at their points of intersection.

79. $y = x^2$, $y = x^{1/3}$
80. $y = x^3$, $y = x^{1/3}$
81. $y = 1 - x^2$, $y = x^2 - 1$
82. $(y + 1)^2 = x$, $y = x^3 - 1$

83. Use vectors to prove that the diagonals of a rhombus are perpendicular.
84. Use vectors to prove that a parallelogram is a rectangle if and only if its diagonals are equal in length.

85. *Bond Angle* Consider a regular tetrahedron with vertices $(0, 0, 0)$, $(k, k, 0)$, $(k, 0, k)$, and $(0, k, k)$, where k is a positive real number.

(a) Sketch the graph of the tetrahedron.
(b) Find the length of each edge.
(c) Find the angle between any two edges.
(d) Find the angle between the line segments from the centroid $(k/2, k/2, k/2)$ to two vertices. This is the bond angle for a molecule such as CH_4 or $PbCl_4$, where the structure of the molecule is a tetrahedron.

86. Consider the vectors
$$\mathbf{u} = \langle \cos \alpha, \sin \alpha, 0 \rangle$$
and
$$\mathbf{v} = \langle \cos \beta, \sin \beta, 0 \rangle$$
where $\alpha > \beta$. Find the dot product of the vectors and use the result to prove the identity
$$\cos(\alpha - \beta) = \cos \alpha \cos \beta + \sin \alpha \sin \beta.$$

87. Prove that $\|\mathbf{u} - \mathbf{v}\|^2 = \|\mathbf{u}\|^2 + \|\mathbf{v}\|^2 - 2\mathbf{u} \cdot \mathbf{v}$.
88. Prove the **Cauchy-Schwarz Inequality** $|\mathbf{u} \cdot \mathbf{v}| \leq \|\mathbf{u}\| \|\mathbf{v}\|$.
89. Prove the triangle inequality $\|\mathbf{u} + \mathbf{v}\| \leq \|\mathbf{u}\| + \|\mathbf{v}\|$.
90. Prove Theorem 11.6.

Section 11.4 The Cross Product of Two Vectors in Space

- Find the cross product of two vectors in space.
- Use the triple scalar product of three vectors in space.

The Cross Product

Many applications in physics, engineering, and geometry involve finding a vector in space that is orthogonal to two given vectors. In this section you will study a product that will yield such a vector. It is called the **cross product**, and it is most conveniently defined and calculated using the standard unit vector form. Because the cross product yields a vector, it is also called the **vector product.**

Definition of Cross Product of Two Vectors in Space

Let $\mathbf{u} = u_1\mathbf{i} + u_2\mathbf{j} + u_3\mathbf{k}$ and $\mathbf{v} = v_1\mathbf{i} + v_2\mathbf{j} + v_3\mathbf{k}$ be vectors in space. The **cross product** of $\mathbf{u}$ and $\mathbf{v}$ is the vector

$$\mathbf{u} \times \mathbf{v} = (u_2v_3 - u_3v_2)\mathbf{i} - (u_1v_3 - u_3v_1)\mathbf{j} + (u_1v_2 - u_2v_1)\mathbf{k}.$$

NOTE Be sure you see that this definition applies only to three-dimensional vectors. The cross product is not defined for two-dimensional vectors.

A convenient way to calculate $\mathbf{u} \times \mathbf{v}$ is to use the following *determinant form* with cofactor expansion. (This 3×3 determinant form is used simply to help remember the formula for the cross product—it is technically not a determinant because the entries of the corresponding matrix are not all real numbers.)

$$\mathbf{u} \times \mathbf{v} = \begin{vmatrix} \mathbf{i} & \mathbf{j} & \mathbf{k} \\ u_1 & u_2 & u_3 \\ v_1 & v_2 & v_3 \end{vmatrix} \quad \begin{matrix} \longleftarrow \text{Put ``}\mathbf{u}\text{'' in Row 2.} \\ \longleftarrow \text{Put ``}\mathbf{v}\text{'' in Row 3.} \end{matrix}$$

$$= \begin{vmatrix} \mathbf{i} & \mathbf{j} & \mathbf{k} \\ u_1 & u_2 & u_3 \\ v_1 & v_2 & v_3 \end{vmatrix}\mathbf{i} - \begin{vmatrix} \mathbf{i} & \mathbf{j} & \mathbf{k} \\ u_1 & u_2 & u_3 \\ v_1 & v_2 & v_3 \end{vmatrix}\mathbf{j} + \begin{vmatrix} \mathbf{i} & \mathbf{j} & \mathbf{k} \\ u_1 & u_2 & u_3 \\ v_1 & v_2 & v_3 \end{vmatrix}\mathbf{k}$$

$$= \begin{vmatrix} u_2 & u_3 \\ v_2 & v_3 \end{vmatrix}\mathbf{i} - \begin{vmatrix} u_1 & u_3 \\ v_1 & v_3 \end{vmatrix}\mathbf{j} + \begin{vmatrix} u_1 & u_2 \\ v_1 & v_2 \end{vmatrix}\mathbf{k}$$

$$= (u_2v_3 - u_3v_2)\mathbf{i} - (u_1v_3 - u_3v_1)\mathbf{j} + (u_1v_2 - u_2v_1)\mathbf{k}$$

Note the minus sign in front of the **j**-component. Each of the three 2×2 determinants can be evaluated by using the following diagonal pattern.

$$\begin{vmatrix} a & b \\ c & d \end{vmatrix} = ad - bc$$

Here are a couple of examples.

$$\begin{vmatrix} 2 & 4 \\ 3 & -1 \end{vmatrix} = (2)(-1) - (4)(3) = -2 - 12 = -14$$

$$\begin{vmatrix} 4 & 0 \\ -6 & 3 \end{vmatrix} = (4)(3) - (0)(-6) = 12$$

EXPLORATION

Geometric Property of the Cross Product Three pairs of vectors are shown below. Use the definition to find the cross product of each pair. Sketch all three vectors in a three-dimensional system. Describe any relationships among the three vectors. Use your description to write a conjecture about $\mathbf{u}$, $\mathbf{v}$, and $\mathbf{u} \times \mathbf{v}$.

a. $\mathbf{u} = \langle 3, 0, 3 \rangle$, $\mathbf{v} = \langle 3, 0, -3 \rangle$

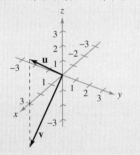

b. $\mathbf{u} = \langle 0, 3, 3 \rangle$, $\mathbf{v} = \langle 0, -3, 3 \rangle$

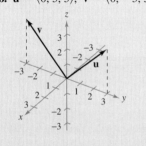

c. $\mathbf{u} = \langle 3, 3, 0 \rangle$, $\mathbf{v} = \langle 3, -3, 0 \rangle$

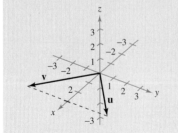

NOTATION FOR DOT AND CROSS PRODUCTS

The notation for the dot product and cross product of vectors was first introduced by the American physicist Josiah Willard Gibbs (1839–1903). In the early 1880s, Gibbs built a system to represent physical quantities called "vector analysis." The system was a departure from Hamilton's theory of quaternions.

EXAMPLE 1 Finding the Cross Product

Given $\mathbf{u} = \mathbf{i} - 2\mathbf{j} + \mathbf{k}$ and $\mathbf{v} = 3\mathbf{i} + \mathbf{j} - 2\mathbf{k}$, find each of the following.

a. $\mathbf{u} \times \mathbf{v}$ **b.** $\mathbf{v} \times \mathbf{u}$ **c.** $\mathbf{v} \times \mathbf{v}$

Solution

a. $\mathbf{u} \times \mathbf{v} = \begin{vmatrix} \mathbf{i} & \mathbf{j} & \mathbf{k} \\ 1 & -2 & 1 \\ 3 & 1 & -2 \end{vmatrix} = \begin{vmatrix} -2 & 1 \\ 1 & -2 \end{vmatrix}\mathbf{i} - \begin{vmatrix} 1 & 1 \\ 3 & -2 \end{vmatrix}\mathbf{j} + \begin{vmatrix} 1 & -2 \\ 3 & 1 \end{vmatrix}\mathbf{k}$

$= (4 - 1)\mathbf{i} - (-2 - 3)\mathbf{j} + (1 + 6)\mathbf{k}$

$= 3\mathbf{i} + 5\mathbf{j} + 7\mathbf{k}$

b. $\mathbf{v} \times \mathbf{u} = \begin{vmatrix} \mathbf{i} & \mathbf{j} & \mathbf{k} \\ 3 & 1 & -2 \\ 1 & -2 & 1 \end{vmatrix} = \begin{vmatrix} 1 & -2 \\ -2 & 1 \end{vmatrix}\mathbf{i} - \begin{vmatrix} 3 & -2 \\ 1 & 1 \end{vmatrix}\mathbf{j} + \begin{vmatrix} 3 & 1 \\ 1 & -2 \end{vmatrix}\mathbf{k}$

$= (1 - 4)\mathbf{i} - (3 + 2)\mathbf{j} + (-6 - 1)\mathbf{k}$

$= -3\mathbf{i} - 5\mathbf{j} - 7\mathbf{k}$

Note that this result is the negative of that in part (a).

c. $\mathbf{v} \times \mathbf{v} = \begin{vmatrix} \mathbf{i} & \mathbf{j} & \mathbf{k} \\ 3 & 1 & -2 \\ 3 & 1 & -2 \end{vmatrix} = \mathbf{0}$

The results obtained in Example 1 suggest some interesting *algebraic* properties of the cross product. For instance, $\mathbf{u} \times \mathbf{v} = -(\mathbf{v} \times \mathbf{u})$, and $\mathbf{v} \times \mathbf{v} = \mathbf{0}$. These properties, and several others, are summarized in the following theorem.

THEOREM 11.7 Algebraic Properties of the Cross Product

Let $\mathbf{u}$, $\mathbf{v}$, and $\mathbf{w}$ be vectors in space, and let c be a scalar.

1. $\mathbf{u} \times \mathbf{v} = -(\mathbf{v} \times \mathbf{u})$
2. $\mathbf{u} \times (\mathbf{v} + \mathbf{w}) = (\mathbf{u} \times \mathbf{v}) + (\mathbf{u} \times \mathbf{w})$
3. $c(\mathbf{u} \times \mathbf{v}) = (c\mathbf{u}) \times \mathbf{v} = \mathbf{u} \times (c\mathbf{v})$
4. $\mathbf{u} \times \mathbf{0} = \mathbf{0} \times \mathbf{u} = \mathbf{0}$
5. $\mathbf{u} \times \mathbf{u} = \mathbf{0}$
6. $\mathbf{u} \cdot (\mathbf{v} \times \mathbf{w}) = (\mathbf{u} \times \mathbf{v}) \cdot \mathbf{w}$

Proof To prove Property 1, let $\mathbf{u} = u_1\mathbf{i} + u_2\mathbf{j} + u_3\mathbf{k}$ and $\mathbf{v} = v_1\mathbf{i} + v_2\mathbf{j} + v_3\mathbf{k}$. Then,

$\mathbf{u} \times \mathbf{v} = (u_2v_3 - u_3v_2)\mathbf{i} - (u_1v_3 - u_3v_1)\mathbf{j} + (u_1v_2 - u_2v_1)\mathbf{k}$

and

$\mathbf{v} \times \mathbf{u} = (v_2u_3 - v_3u_2)\mathbf{i} - (v_1u_3 - v_3u_1)\mathbf{j} + (v_1u_2 - v_2u_1)\mathbf{k}$

which implies that $\mathbf{u} \times \mathbf{v} = -(\mathbf{v} \times \mathbf{u})$. Proofs of Properties 2, 3, 5, and 6 are left as exercises (see Exercises 57–60).

Note that Property 1 of Theorem 11.7 indicates that the cross product is *not commutative*. In particular, this property indicates that the vectors $\mathbf{u} \times \mathbf{v}$ and $\mathbf{v} \times \mathbf{u}$ have equal lengths but opposite directions. The following theorem lists some other *geometric* properties of the cross product of two vectors.

NOTE   It follows from Properties 1 and 2 in Theorem 11.8 that if $\mathbf{n}$ is a unit vector orthogonal to both $\mathbf{u}$ and $\mathbf{v}$, then

$$\mathbf{u} \times \mathbf{v} = \pm(\|\mathbf{u}\|\,\|\mathbf{v}\| \sin \theta)\mathbf{n}.$$

THEOREM 11.8   Geometric Properties of the Cross Product

Let $\mathbf{u}$ and $\mathbf{v}$ be nonzero vectors in space, and let θ be the angle between $\mathbf{u}$ and $\mathbf{v}$.

1. $\mathbf{u} \times \mathbf{v}$ is orthogonal to both $\mathbf{u}$ and $\mathbf{v}$.
2. $\|\mathbf{u} \times \mathbf{v}\| = \|\mathbf{u}\|\,\|\mathbf{v}\| \sin \theta$
3. $\mathbf{u} \times \mathbf{v} = \mathbf{0}$ if and only if $\mathbf{u}$ and $\mathbf{v}$ are scalar multiples of each other.
4. $\|\mathbf{u} \times \mathbf{v}\|$ = area of parallelogram having $\mathbf{u}$ and $\mathbf{v}$ as adjacent sides.

Proof   To prove Property 2, note because $\cos \theta = (\mathbf{u} \cdot \mathbf{v})/(\|\mathbf{u}\|\,\|\mathbf{v}\|)$, it follows that

$$\begin{aligned}
\|\mathbf{u}\|\,\|\mathbf{v}\| \sin \theta &= \|\mathbf{u}\|\,\|\mathbf{v}\|\sqrt{1 - \cos^2 \theta} \\
&= \|\mathbf{u}\|\,\|\mathbf{v}\|\sqrt{1 - \frac{(\mathbf{u} \cdot \mathbf{v})^2}{\|\mathbf{u}\|^2\|\mathbf{v}\|^2}} \\
&= \sqrt{\|\mathbf{u}\|^2\|\mathbf{v}\|^2 - (\mathbf{u} \cdot \mathbf{v})^2} \\
&= \sqrt{(u_1^2 + u_2^2 + u_3^2)(v_1^2 + v_2^2 + v_3^2) - (u_1v_1 + u_2v_2 + u_3v_3)^2} \\
&= \sqrt{(u_2v_3 - u_3v_2)^2 + (u_1v_3 - u_3v_1)^2 + (u_1v_2 - u_2v_1)^2} \\
&= \|\mathbf{u} \times \mathbf{v}\|.
\end{aligned}$$

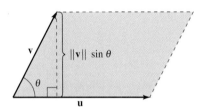

The vectors $\mathbf{u}$ and $\mathbf{v}$ form adjacent sides of a parallelogram.
Figure 11.35

To prove Property 4, refer to Figure 11.35, which is a parallelogram having $\mathbf{v}$ and $\mathbf{u}$ as adjacent sides. Because the height of the parallelogram is $\|\mathbf{v}\| \sin \theta$, the area is

$$\begin{aligned}
\text{Area} &= (\text{base})(\text{height}) \\
&= \|\mathbf{u}\|\,\|\mathbf{v}\| \sin \theta \\
&= \|\mathbf{u} \times \mathbf{v}\|.
\end{aligned}$$

Proofs of Properties 1 and 3 are left as exercises (see Exercises 61 and 62).

Both $\mathbf{u} \times \mathbf{v}$ and $\mathbf{v} \times \mathbf{u}$ are perpendicular to the plane determined by $\mathbf{u}$ and $\mathbf{v}$. One way to remember the orientations of the vectors $\mathbf{u}$, $\mathbf{v}$, and $\mathbf{u} \times \mathbf{v}$ is to compare them with the unit vectors $\mathbf{i}$, $\mathbf{j}$, and $\mathbf{k} = \mathbf{i} \times \mathbf{j}$, as shown in Figure 11.36. The three vectors $\mathbf{u}$, $\mathbf{v}$, and $\mathbf{u} \times \mathbf{v}$ form a *right-handed system*, whereas the three vectors $\mathbf{u}$, $\mathbf{v}$, and $\mathbf{v} \times \mathbf{u}$ form a *left-handed system*.

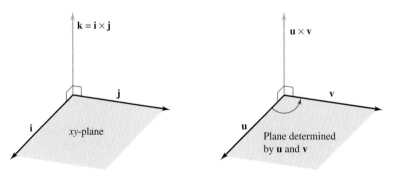

Right-handed systems
Figure 11.36

EXAMPLE 2 Using the Cross Product

Find a unit vector that is orthogonal to both

$$\mathbf{u} = \mathbf{i} - 4\mathbf{j} + \mathbf{k} \quad \text{and} \quad \mathbf{v} = 2\mathbf{i} + 3\mathbf{j}.$$

Solution The cross product $\mathbf{u} \times \mathbf{v}$, as shown in Figure 11.37, is orthogonal to both $\mathbf{u}$ and $\mathbf{v}$.

$$\mathbf{u} \times \mathbf{v} = \begin{vmatrix} \mathbf{i} & \mathbf{j} & \mathbf{k} \\ 1 & -4 & 1 \\ 2 & 3 & 0 \end{vmatrix} \quad \text{Cross product}$$

$$= -3\mathbf{i} + 2\mathbf{j} + 11\mathbf{k}$$

Because

$$\|\mathbf{u} \times \mathbf{v}\| = \sqrt{(-3)^2 + 2^2 + 11^2} = \sqrt{134}$$

a unit vector orthogonal to both $\mathbf{u}$ and $\mathbf{v}$ is

$$\frac{\mathbf{u} \times \mathbf{v}}{\|\mathbf{u} \times \mathbf{v}\|} = -\frac{3}{\sqrt{134}}\mathbf{i} + \frac{2}{\sqrt{134}}\mathbf{j} + \frac{11}{\sqrt{134}}\mathbf{k}.$$

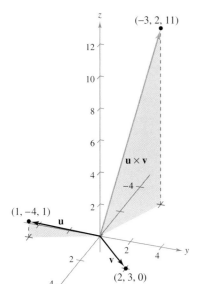

The vector $\mathbf{u} \times \mathbf{v}$ is orthogonal to both $\mathbf{u}$ and $\mathbf{v}$.
Figure 11.37

NOTE In Example 2, note that you could have used the cross product $\mathbf{v} \times \mathbf{u}$ to form a unit vector that is orthogonal to both $\mathbf{u}$ and $\mathbf{v}$. With that choice, you would have obtained the negative of the unit vector found in the example.

EXAMPLE 3 Geometric Application of the Cross Product

Show that the quadrilateral with vertices at the following points is a parallelogram, and find its area.

$$A = (5, 2, 0) \qquad B = (2, 6, 1)$$
$$C = (2, 4, 7) \qquad D = (5, 0, 6)$$

Solution From Figure 11.38 you can see that the sides of the quadrilateral correspond to the following four vectors.

$$\overrightarrow{AB} = -3\mathbf{i} + 4\mathbf{j} + \mathbf{k} \qquad \overrightarrow{CD} = 3\mathbf{i} - 4\mathbf{j} - \mathbf{k} = -\overrightarrow{AB}$$
$$\overrightarrow{AD} = 0\mathbf{i} - 2\mathbf{j} + 6\mathbf{k} \qquad \overrightarrow{CB} = 0\mathbf{i} + 2\mathbf{j} - 6\mathbf{k} = -\overrightarrow{AD}$$

So, $\overrightarrow{AB}$ is parallel to $\overrightarrow{CD}$ and $\overrightarrow{AD}$ is parallel to $\overrightarrow{CB}$, and you can conclude that the quadrilateral is a parallelogram with $\overrightarrow{AB}$ and $\overrightarrow{AD}$ as adjacent sides. Moreover, because

$$\overrightarrow{AB} \times \overrightarrow{AD} = \begin{vmatrix} \mathbf{i} & \mathbf{j} & \mathbf{k} \\ -3 & 4 & 1 \\ 0 & -2 & 6 \end{vmatrix} \quad \text{Cross product}$$

$$= 26\mathbf{i} + 18\mathbf{j} + 6\mathbf{k}$$

the area of the parallelogram is

$$\|\overrightarrow{AB} \times \overrightarrow{AD}\| = \sqrt{1036} \approx 32.19.$$

Is the parallelogram a rectangle? You can determine whether it is by finding the angle between the vectors $\overrightarrow{AB}$ and $\overrightarrow{AD}$.

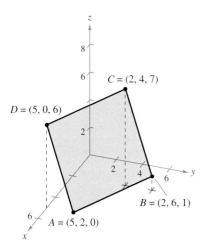

The area of the parallelogram is approximately 32.19.
Figure 11.38

In physics, the cross product can be used to measure **torque**—the **moment M of a force F about a point P,** as shown in Figure 11.39. If the point of application of the force is Q, the moment of **F** about P is given by

$$\mathbf{M} = \overrightarrow{PQ} \times \mathbf{F}. \qquad \text{Moment of } \mathbf{F} \text{ about } P$$

The magnitude of the moment **M** measures the tendency of the vector $\overrightarrow{PQ}$ to rotate counterclockwise (using the right-hand rule) about an axis directed along the vector **M**.

EXAMPLE 4 An Application of the Cross Product

A vertical force of 50 pounds is applied to the end of a one-foot lever that is attached to an axle at point P, as shown in Figure 11.40. Find the moment of this force about the point P when $\theta = 60°$.

Solution If you represent the 50-pound force as $\mathbf{F} = -50\mathbf{k}$ and the lever as

$$\overrightarrow{PQ} = \cos(60°)\mathbf{j} + \sin(60°)\mathbf{k} = \frac{1}{2}\mathbf{j} + \frac{\sqrt{3}}{2}\mathbf{k}$$

the moment of **F** about P is given by

$$\mathbf{M} = \overrightarrow{PQ} \times \mathbf{F} = \begin{vmatrix} \mathbf{i} & \mathbf{j} & \mathbf{k} \\ 0 & \frac{1}{2} & \frac{\sqrt{3}}{2} \\ 0 & 0 & -50 \end{vmatrix} = -25\mathbf{i}. \qquad \text{Moment of } \mathbf{F} \text{ about } P$$

The magnitude of this moment is 25 foot-pounds.

NOTE In Example 4, note that the moment (the tendency of the lever to rotate about its axle) is dependent on the angle θ. When $\theta = \pi/2$, the moment is **0**. The moment is greatest when $\theta = 0$.

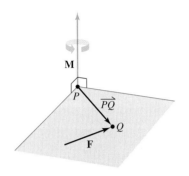

The moment of **F** about P
Figure 11.39

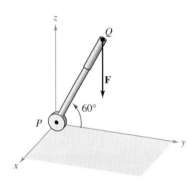

A vertical force of 50 pounds is applied at point Q.
Figure 11.40

The Triple Scalar Product

For vectors **u**, **v**, and **w** in space, the dot product of **u** and $\mathbf{v} \times \mathbf{w}$

$$\mathbf{u} \cdot (\mathbf{v} \times \mathbf{w})$$

is called the **triple scalar product,** as defined in Theorem 11.9. The proof of this theorem is left as an exercise (see Exercise 56).

FOR FURTHER INFORMATION To see how the cross product is used to model the torque of the robot arm of a space shuttle, see the article "The Long Arm of Calculus" by Ethan Berkove and Rich Marchand in *The College Mathematics Journal.* To view this article, go to the website *www.matharticles.com.*

THEOREM 11.9 The Triple Scalar Product

For $\mathbf{u} = u_1\mathbf{i} + u_2\mathbf{j} + u_3\mathbf{k}$, $\mathbf{v} = v_1\mathbf{i} + v_2\mathbf{j} + v_3\mathbf{k}$, and $\mathbf{w} = w_1\mathbf{i} + w_2\mathbf{j} + w_3\mathbf{k}$, the triple scalar product is given by

$$\mathbf{u} \cdot (\mathbf{v} \times \mathbf{w}) = \begin{vmatrix} u_1 & u_2 & u_3 \\ v_1 & v_2 & v_3 \\ w_1 & w_2 & w_3 \end{vmatrix}.$$

NOTE The value of a determinant is multiplied by -1 if two rows are interchanged. After two such interchanges, the value of the determinant will be unchanged. So, the following triple scalar products are equivalent.

$$\mathbf{u} \cdot (\mathbf{v} \times \mathbf{w}) = \mathbf{v} \cdot (\mathbf{w} \times \mathbf{u}) = \mathbf{w} \cdot (\mathbf{u} \times \mathbf{v})$$

SECTION 11.4 The Cross Product of Two Vectors in Space 795

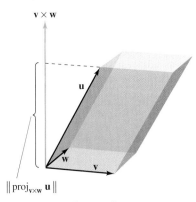

Area of base = $\|\mathbf{v} \times \mathbf{w}\|$
Volume of parallelepiped = $|\mathbf{u} \cdot (\mathbf{v} \times \mathbf{w})|$
Figure 11.41

If the vectors **u**, **v**, and **w** do not lie in the same plane, the triple scalar product $\mathbf{u} \cdot (\mathbf{v} \times \mathbf{w})$ can be used to determine the volume of the parallelepiped (a polyhedron, all of whose faces are parallelograms) with **u**, **v**, and **w** as adjacent edges, as shown in Figure 11.41. This is established in the following theorem.

THEOREM 11.10 Geometric Property of Triple Scalar Product

The volume V of a parallelepiped with vectors **u**, **v**, and **w** as adjacent edges is given by
$$V = |\mathbf{u} \cdot (\mathbf{v} \times \mathbf{w})|.$$

Proof In Figure 11.41, note that
$$\|\mathbf{v} \times \mathbf{w}\| = \text{area of base}$$
and
$$\|\text{proj}_{\mathbf{v} \times \mathbf{w}} \mathbf{u}\| = \text{height of parallelepiped}.$$

Therefore, the volume is
$$\begin{aligned} V = (\text{height})(\text{area of base}) &= \|\text{proj}_{\mathbf{v} \times \mathbf{w}} \mathbf{u}\| \|\mathbf{v} \times \mathbf{w}\| \\ &= \left|\frac{\mathbf{u} \cdot (\mathbf{v} \times \mathbf{w})}{\|\mathbf{v} \times \mathbf{w}\|}\right| \|\mathbf{v} \times \mathbf{w}\| \\ &= |\mathbf{u} \cdot (\mathbf{v} \times \mathbf{w})|. \end{aligned}$$

EXAMPLE 5 Volume by the Triple Scalar Product

Find the volume of the parallelepiped shown in Figure 11.42 having $\mathbf{u} = 3\mathbf{i} - 5\mathbf{j} + \mathbf{k}$, $\mathbf{v} = 2\mathbf{j} - 2\mathbf{k}$, and $\mathbf{w} = 3\mathbf{i} + \mathbf{j} + \mathbf{k}$ as adjacent edges.

Solution By Theorem 11.10, you have

$$\begin{aligned} V &= |\mathbf{u} \cdot (\mathbf{v} \times \mathbf{w})| \quad \text{Triple scalar product} \\ &= \begin{vmatrix} 3 & -5 & 1 \\ 0 & 2 & -2 \\ 3 & 1 & 1 \end{vmatrix} \\ &= 3\begin{vmatrix} 2 & -2 \\ 1 & 1 \end{vmatrix} - (-5)\begin{vmatrix} 0 & -2 \\ 3 & 1 \end{vmatrix} + (1)\begin{vmatrix} 0 & 2 \\ 3 & 1 \end{vmatrix} \\ &= 3(4) + 5(6) + 1(-6) \\ &= 36. \end{aligned}$$

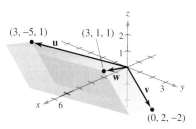

The parallelepiped has a volume of 36.
Figure 11.42

A natural consequence of Theorem 11.10 is that the volume of the parallelepiped is 0 if and only if the three vectors are coplanar. That is, if the vectors $\mathbf{u} = \langle u_1, u_2, u_3 \rangle$, $\mathbf{v} = \langle v_1, v_2, v_3 \rangle$, and $\mathbf{w} = \langle w_1, w_2, w_3 \rangle$ have the same initial point, they lie in the same plane if and only if

$$\mathbf{u} \cdot (\mathbf{v} \times \mathbf{w}) = \begin{vmatrix} u_1 & u_2 & u_3 \\ v_1 & v_2 & v_3 \\ w_1 & w_2 & w_3 \end{vmatrix} = 0.$$

Exercises for Section 11.4

In Exercises 1–6, find the cross product of the unit vectors and sketch your result.

1. $\mathbf{j} \times \mathbf{i}$
2. $\mathbf{i} \times \mathbf{j}$
3. $\mathbf{j} \times \mathbf{k}$
4. $\mathbf{k} \times \mathbf{j}$
5. $\mathbf{i} \times \mathbf{k}$
6. $\mathbf{k} \times \mathbf{i}$

In Exercises 7–10, find (a) $\mathbf{u} \times \mathbf{v}$, (b) $\mathbf{v} \times \mathbf{u}$, and (c) $\mathbf{v} \times \mathbf{v}$.

7. $\mathbf{u} = -2\mathbf{i} + 3\mathbf{j} + 4\mathbf{k}$
   $\mathbf{v} = 3\mathbf{i} + 7\mathbf{j} + 2\mathbf{k}$
8. $\mathbf{u} = 3\mathbf{i} + 5\mathbf{k}$
   $\mathbf{v} = 2\mathbf{i} + 3\mathbf{j} - 2\mathbf{k}$
9. $\mathbf{u} = \langle 7, 3, 2 \rangle$
   $\mathbf{v} = \langle 1, -1, 5 \rangle$
10. $\mathbf{u} = \langle 3, -2, -2 \rangle$
    $\mathbf{v} = \langle 1, 5, 1 \rangle$

In Exercises 11–16, find $\mathbf{u} \times \mathbf{v}$ and show that it is orthogonal to both $\mathbf{u}$ and $\mathbf{v}$.

11. $\mathbf{u} = \langle 2, -3, 1 \rangle$
    $\mathbf{v} = \langle 1, -2, 1 \rangle$
12. $\mathbf{u} = \langle -1, 1, 2 \rangle$
    $\mathbf{v} = \langle 0, 1, 0 \rangle$
13. $\mathbf{u} = \langle 12, -3, 0 \rangle$
    $\mathbf{v} = \langle -2, 5, 0 \rangle$
14. $\mathbf{u} = \langle -10, 0, 6 \rangle$
    $\mathbf{v} = \langle 7, 0, 0 \rangle$
15. $\mathbf{u} = \mathbf{i} + \mathbf{j} + \mathbf{k}$
    $\mathbf{v} = 2\mathbf{i} + \mathbf{j} - \mathbf{k}$
16. $\mathbf{u} = \mathbf{i} + 6\mathbf{j}$
    $\mathbf{v} = -2\mathbf{i} + \mathbf{j} + \mathbf{k}$

Think About It In Exercises 17–20, use the vectors $\mathbf{u}$ and $\mathbf{v}$ shown in the figure to sketch a vector in the direction of the indicated cross product in a right-handed system.

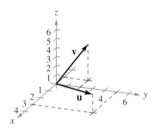

17. $\mathbf{u} \times \mathbf{v}$
18. $\mathbf{v} \times \mathbf{u}$
19. $(-\mathbf{v}) \times \mathbf{u}$
20. $\mathbf{u} \times (\mathbf{u} \times \mathbf{v})$

In Exercises 21–24, use a computer algebra system to find $\mathbf{u} \times \mathbf{v}$ and a unit vector orthogonal to $\mathbf{u}$ and $\mathbf{v}$.

21. $\mathbf{u} = \langle 4, -3.5, 7 \rangle$
    $\mathbf{v} = \langle -1, 8, 4 \rangle$
22. $\mathbf{u} = \langle -8, -6, 4 \rangle$
    $\mathbf{v} = \langle 10, -12, -2 \rangle$
23. $\mathbf{u} = -3\mathbf{i} + 2\mathbf{j} - 5\mathbf{k}$
    $\mathbf{v} = \frac{1}{2}\mathbf{i} - \frac{3}{4}\mathbf{j} + \frac{1}{10}\mathbf{k}$
24. $\mathbf{u} = \frac{2}{3}\mathbf{k}$
    $\mathbf{v} = \frac{1}{2}\mathbf{i} + 6\mathbf{k}$

25. *Programming* Given the vectors $\mathbf{u}$ and $\mathbf{v}$ in component form, write a program for a graphing utility in which the output is $\mathbf{u} \times \mathbf{v}$ and $\|\mathbf{u} \times \mathbf{v}\|$.

26. *Programming* Use the program you wrote in Exercise 25 to find $\mathbf{u} \times \mathbf{v}$ and $\|\mathbf{u} \times \mathbf{v}\|$ for $\mathbf{u} = \langle -2, 6, 10 \rangle$ and $\mathbf{v} = \langle 3, 8, 5 \rangle$.

Area In Exercises 27–30, find the area of the parallelogram that has the given vectors as adjacent sides. Use a computer algebra system or a graphing utility to verify your result.

27. $\mathbf{u} = \mathbf{j}$
    $\mathbf{v} = \mathbf{j} + \mathbf{k}$
28. $\mathbf{u} = \mathbf{i} + \mathbf{j} + \mathbf{k}$
    $\mathbf{v} = \mathbf{j} + \mathbf{k}$
29. $\mathbf{u} = \langle 3, 2, -1 \rangle$
    $\mathbf{v} = \langle 1, 2, 3 \rangle$
30. $\mathbf{u} = \langle 2, -1, 0 \rangle$
    $\mathbf{v} = \langle -1, 2, 0 \rangle$

Area In Exercises 31 and 32, verify that the points are the vertices of a parallelogram, and find its area.

31. $(1, 1, 1)$, $(2, 3, 4)$, $(6, 5, 2)$, $(7, 7, 5)$
32. $(2, -3, 1)$, $(6, 5, -1)$, $(3, -6, 4)$, $(7, 2, 2)$

Area In Exercises 33–36, find the area of the triangle with the given vertices. $\left(\text{Hint: } \frac{1}{2}\|\mathbf{u} \times \mathbf{v}\| \text{ is the area of the triangle having } \mathbf{u} \text{ and } \mathbf{v} \text{ as adjacent sides.} \right)$

33. $(0, 0, 0)$, $(1, 2, 3)$, $(-3, 0, 0)$
34. $(2, -3, 4)$, $(0, 1, 2)$, $(-1, 2, 0)$
35. $(2, -7, 3)$, $(-1, 5, 8)$, $(4, 6, -1)$
36. $(1, 2, 0)$, $(-2, 1, 0)$, $(0, 0, 0)$

37. *Torque* A child applies the brakes on a bicycle by applying a downward force of 20 pounds on the pedal when the crank makes a 40° angle with the horizontal (see figure). The crank is 6 inches in length. Find the torque at P.

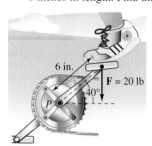

Figure for 37   Figure for 38

38. *Torque* Both the magnitude and the direction of the force on a crankshaft change as the crankshaft rotates. Find the torque on the crankshaft using the position and data shown in the figure.

39. *Optimization* A force of 60 pounds acts on the pipe wrench shown in the figure on the next page.

   (a) Find the magnitude of the moment about O by evaluating $\|\overrightarrow{OA} \times \mathbf{F}\|$. Use a graphing utility to graph the resulting function of θ.

   (b) Use the result of part (a) to determine the magnitude of the moment when $\theta = 45°$.

   (c) Use the result of part (a) to determine the angle θ when the magnitude of the moment is maximum. Is the answer what you expected? Why or why not?

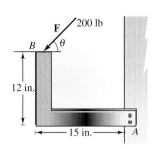

Figure for 39 **Figure for 40**

40. *Optimization* A force of 200 pounds acts on the bracket shown in the figure.

(a) Determine the vector $\vec{AB}$ and the vector **F** representing the force. (**F** will be in terms of θ.)

(b) Find the magnitude of the moment about A by evaluating $\|\vec{AB} \times \mathbf{F}\|$.

(c) Use the result of part (b) to determine the magnitude of the moment when $\theta = 30°$.

(d) Use the result of part (b) to determine the angle θ when the magnitude of the moment is maximum. At that angle, what is the relationship between the vectors **F** and $\vec{AB}$? Is it what you expected? Why or why not?

(e) Use a graphing utility to graph the function for the magnitude of the moment about A for $0° \leq \theta \leq 180°$. Find the zero of the function in the given domain. Interpret the meaning of the zero in the context of the problem.

In Exercises 41–44, find $\mathbf{u} \cdot (\mathbf{v} \times \mathbf{w})$.

41. $\mathbf{u} = \mathbf{i}$
    $\mathbf{v} = \mathbf{j}$
    $\mathbf{w} = \mathbf{k}$

42. $\mathbf{u} = \langle 1, 1, 1 \rangle$
    $\mathbf{v} = \langle 2, 1, 0 \rangle$
    $\mathbf{w} = \langle 0, 0, 1 \rangle$

43. $\mathbf{u} = \langle 2, 0, 1 \rangle$
    $\mathbf{v} = \langle 0, 3, 0 \rangle$
    $\mathbf{w} = \langle 0, 0, 1 \rangle$

44. $\mathbf{u} = \langle 2, 0, 0 \rangle$
    $\mathbf{v} = \langle 1, 1, 1 \rangle$
    $\mathbf{w} = \langle 0, 2, 2 \rangle$

Volume In Exercises 45 and 46, use the triple scalar product to find the volume of the parallelepiped having adjacent edges **u**, **v**, and **w**.

45. $\mathbf{u} = \mathbf{i} + \mathbf{j}$
    $\mathbf{v} = \mathbf{j} + \mathbf{k}$
    $\mathbf{w} = \mathbf{i} + \mathbf{k}$

46. $\mathbf{u} = \langle 1, 3, 1 \rangle$
    $\mathbf{v} = \langle 0, 6, 6 \rangle$
    $\mathbf{w} = \langle -4, 0, -4 \rangle$

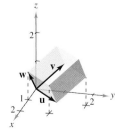
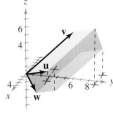

Volume In Exercises 47 and 48, find the volume of the parallelepiped with the given vertices (see figures).

47. $(0, 0, 0), (3, 0, 0), (0, 5, 1), (3, 5, 1)$
    $(2, 0, 5), (5, 0, 5), (2, 5, 6), (5, 5, 6)$

48. $(0, 0, 0), (1, 1, 0), (1, 0, 2), (0, 1, 1)$
    $(2, 1, 2), (1, 1, 3), (1, 2, 1), (2, 2, 3)$

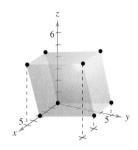
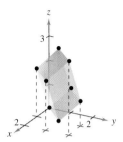

Figure for 47 **Figure for 48**

Writing About Concepts

49. Define the cross product of vectors **u** and **v**.

50. State the geometric properties of the cross product.

51. If the magnitudes of two vectors are doubled, how will the magnitude of the cross product of the vectors change? Explain.

52. The vertices of a triangle in space are $(x_1, y_1, z_1), (x_2, y_2, z_2)$, and (x_3, y_3, z_3). Explain how to find a vector perpendicular to the triangle.

True or False? In Exercises 53–55, determine whether the statement is true or false. If it is false, explain why or give an example that shows it is false.

53. It is possible to find the cross product of two vectors in a two-dimensional coordinate system.

54. If $\mathbf{u} \neq \mathbf{0}$ and $\mathbf{u} \times \mathbf{v} = \mathbf{u} \times \mathbf{w}$, then $\mathbf{v} = \mathbf{w}$.

55. If $\mathbf{u} \neq \mathbf{0}$, $\mathbf{u} \cdot \mathbf{v} = \mathbf{u} \cdot \mathbf{w}$, and $\mathbf{u} \times \mathbf{v} = \mathbf{u} \times \mathbf{w}$, then $\mathbf{v} = \mathbf{w}$.

56. Prove Theorem 11.9.

In Exercises 57–62, prove the property of the cross product.

57. $\mathbf{u} \times (\mathbf{v} + \mathbf{w}) = (\mathbf{u} \times \mathbf{v}) + (\mathbf{u} \times \mathbf{w})$

58. $c(\mathbf{u} \times \mathbf{v}) = (c\mathbf{u}) \times \mathbf{v} = \mathbf{u} \times (c\mathbf{v})$

59. $\mathbf{u} \times \mathbf{u} = \mathbf{0}$

60. $\mathbf{u} \cdot (\mathbf{v} \times \mathbf{w}) = (\mathbf{u} \times \mathbf{v}) \cdot \mathbf{w}$

61. $\mathbf{u} \times \mathbf{v}$ is orthogonal to both **u** and **v**.

62. $\mathbf{u} \times \mathbf{v} = \mathbf{0}$ if and only if **u** and **v** are scalar multiples of each other.

63. Prove $\|\mathbf{u} \times \mathbf{v}\| = \|\mathbf{u}\| \, \|\mathbf{v}\|$ if **u** and **v** are orthogonal.

64. Prove $\mathbf{u} \times (\mathbf{v} \times \mathbf{w}) = (\mathbf{u} \cdot \mathbf{w})\mathbf{v} - (\mathbf{u} \cdot \mathbf{v})\mathbf{w}$.

Section 11.5 Lines and Planes in Space

- Write a set of parametric equations for a line in space.
- Write a linear equation to represent a plane in space.
- Sketch the plane given by a linear equation.
- Find the distances between points, planes, and lines in space.

Lines in Space

In the plane, *slope* is used to determine an equation of a line. In space, it is more convenient to use *vectors* to determine the equation of a line.

In Figure 11.43, consider the line L through the point $P(x_1, y_1, z_1)$ and parallel to the vector $\mathbf{v} = \langle a, b, c \rangle$. The vector $\mathbf{v}$ is a **direction vector** for the line L, and a, b, and c are **direction numbers**. One way of describing the line L is to say that it consists of all points $Q(x, y, z)$ for which the vector $\overrightarrow{PQ}$ is parallel to $\mathbf{v}$. This means that $\overrightarrow{PQ}$ is a scalar multiple of $\mathbf{v}$, and you can write $\overrightarrow{PQ} = t\mathbf{v}$, where t is a scalar (a real number).

$$\overrightarrow{PQ} = \langle x - x_1, y - y_1, z - z_1 \rangle = \langle at, bt, ct \rangle = t\mathbf{v}$$

By equating corresponding components, you can obtain **parametric equations** of a line in space.

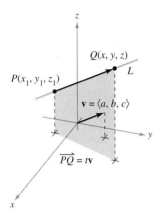

Line L and its direction vector $\mathbf{v}$
Figure 11.43

> **THEOREM 11.11 Parametric Equations of a Line in Space**
>
> A line L parallel to the vector $\mathbf{v} = \langle a, b, c \rangle$ and passing through the point $P(x_1, y_1, z_1)$ is represented by the **parametric equations**
>
> $$x = x_1 + at, \quad y = y_1 + bt, \quad \text{and} \quad z = z_1 + ct.$$

If the direction numbers a, b, and c are all nonzero, you can eliminate the parameter t to obtain **symmetric equations** of the line.

$$\frac{x - x_1}{a} = \frac{y - y_1}{b} = \frac{z - z_1}{c} \qquad \text{Symmetric equations}$$

EXAMPLE 1 Finding Parametric and Symmetric Equations

Find parametric and symmetric equations of the line L that passes through the point $(1, -2, 4)$ and is parallel to $\mathbf{v} = \langle 2, 4, -4 \rangle$.

Solution To find a set of parametric equations of the line, use the coordinates $x_1 = 1$, $y_1 = -2$, and $z_1 = 4$ and direction numbers $a = 2$, $b = 4$, and $c = -4$ (see Figure 11.44).

$$x = 1 + 2t, \quad y = -2 + 4t, \quad z = 4 - 4t \qquad \text{Parametric equations}$$

Because a, b, and c are all nonzero, a set of symmetric equations is

$$\frac{x - 1}{2} = \frac{y + 2}{4} = \frac{z - 4}{-4}. \qquad \text{Symmetric equations}$$

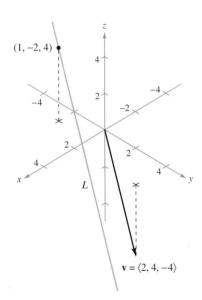

The vector $\mathbf{v}$ is parallel to the line L.
Figure 11.44

Neither parametric equations nor symmetric equations of a given line are unique. For instance, in Example 1, by letting $t = 1$ in the parametric equations you would obtain the point $(3, 2, 0)$. Using this point with the direction numbers $a = 2$, $b = 4$, and $c = -4$ would produce a different set of parametric equations

$$x = 3 + 2t, \quad y = 2 + 4t, \quad \text{and} \quad z = -4t.$$

EXAMPLE 2 Parametric Equations of a Line Through Two Points

Find a set of parametric equations of the line that passes through the points $(-2, 1, 0)$ and $(1, 3, 5)$.

Solution Begin by using the points $P(-2, 1, 0)$ and $Q(1, 3, 5)$ to find a direction vector for the line passing through P and Q, given by

$$\mathbf{v} = \overrightarrow{PQ} = \langle 1 - (-2), 3 - 1, 5 - 0 \rangle = \langle 3, 2, 5 \rangle = \langle a, b, c \rangle.$$

Using the direction numbers $a = 3$, $b = 2$, and $c = 5$ with the point $P(-2, 1, 0)$, you can obtain the parametric equations

$$x = -2 + 3t, \quad y = 1 + 2t, \quad \text{and} \quad z = 5t.$$

NOTE As t varies over all real numbers, the parametric equations in Example 2 determine the points (x, y, z) on the line. In particular, note that $t = 0$ and $t = 1$ give the original points $(-2, 1, 0)$ and $(1, 3, 5)$.

Planes in Space

You have seen how an equation of a line in space can be obtained from a point on the line and a vector *parallel* to it. You will now see that an equation of a plane in space can be obtained from a point in the plane and a vector *normal* (perpendicular) to the plane.

Consider the plane containing the point $P(x_1, y_1, z_1)$ having a nonzero normal vector $\mathbf{n} = \langle a, b, c \rangle$, as shown in Figure 11.45. This plane consists of all points $Q(x, y, z)$ for which vector $\overrightarrow{PQ}$ is orthogonal to $\mathbf{n}$. Using the dot product, you can write the following.

$$\mathbf{n} \cdot \overrightarrow{PQ} = 0$$
$$\langle a, b, c \rangle \cdot \langle x - x_1, y - y_1, z - z_1 \rangle = 0$$
$$a(x - x_1) + b(y - y_1) + c(z - z_1) = 0$$

The third equation of the plane is said to be in **standard form**.

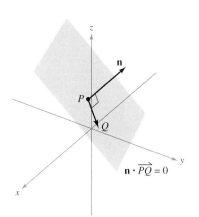

The normal vector $\mathbf{n}$ is orthogonal to each vector $\overrightarrow{PQ}$ in the plane.
Figure 11.45

THEOREM 11.12 Standard Equation of a Plane in Space

The plane containing the point (x_1, y_1, z_1) and having a normal vector $\mathbf{n} = \langle a, b, c \rangle$ can be represented, in **standard form,** by the equation

$$a(x - x_1) + b(y - y_1) + c(z - z_1) = 0.$$

By regrouping terms, you obtain the **general form** of the equation of a plane in space.

$$ax + by + cz + d = 0 \qquad \text{General form of equation of plane}$$

Given the general form of the equation of a plane, it is easy to find a normal vector to the plane. Simply use the coefficients of x, y, and z and write $\mathbf{n} = \langle a, b, c \rangle$.

EXAMPLE 3 Finding an Equation of a Plane in Three-Space

Find the general equation of the plane containing the points $(2, 1, 1)$, $(0, 4, 1)$, and $(-2, 1, 4)$.

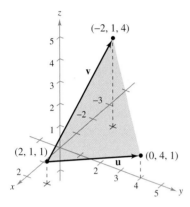

A plane determined by $\mathbf{u}$ and $\mathbf{v}$
Figure 11.46

Solution To apply Theorem 11.12 you need a point in the plane and a vector that is normal to the plane. There are three choices for the point, but no normal vector is given. To obtain a normal vector, use the cross product of vectors $\mathbf{u}$ and $\mathbf{v}$ extending from the point $(2, 1, 1)$ to the points $(0, 4, 1)$ and $(-2, 1, 4)$, as shown in Figure 11.46. The component forms of $\mathbf{u}$ and $\mathbf{v}$ are

$$\mathbf{u} = \langle 0 - 2, 4 - 1, 1 - 1 \rangle = \langle -2, 3, 0 \rangle$$
$$\mathbf{v} = \langle -2 - 2, 1 - 1, 4 - 1 \rangle = \langle -4, 0, 3 \rangle$$

and it follows that

$$\mathbf{n} = \mathbf{u} \times \mathbf{v}$$
$$= \begin{vmatrix} \mathbf{i} & \mathbf{j} & \mathbf{k} \\ -2 & 3 & 0 \\ -4 & 0 & 3 \end{vmatrix}$$
$$= 9\mathbf{i} + 6\mathbf{j} + 12\mathbf{k}$$
$$= \langle a, b, c \rangle$$

is normal to the given plane. Using the direction numbers for $\mathbf{n}$ and the point $(x_1, y_1, z_1) = (2, 1, 1)$, you can determine an equation of the plane to be

$$a(x - x_1) + b(y - y_1) + c(z - z_1) = 0$$
$$9(x - 2) + 6(y - 1) + 12(z - 1) = 0 \qquad \text{Standard form}$$
$$9x + 6y + 12z - 36 = 0 \qquad \text{General form}$$
$$3x + 2y + 4z - 12 = 0. \qquad \text{Simplified general form}$$

NOTE In Example 3, check to see that each of the three original points satisfies the equation $3x + 2y + 4z - 12 = 0$.

Two distinct planes in three-space either are parallel or intersect in a line. If they intersect, you can determine the angle $(0 \leq \theta \leq \pi/2)$ between them from the angle between their normal vectors, as shown in Figure 11.47. Specifically, if vectors $\mathbf{n}_1$ and $\mathbf{n}_2$ are normal to two intersecting planes, the angle θ between the normal vectors is equal to the angle between the two planes and is given by

$$\cos \theta = \frac{|\mathbf{n}_1 \cdot \mathbf{n}_2|}{\|\mathbf{n}_1\| \|\mathbf{n}_2\|}. \qquad \text{Angle between two planes}$$

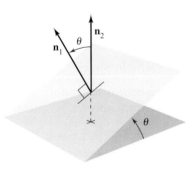

The angle θ between two planes
Figure 11.47

Consequently, two planes with normal vectors $\mathbf{n}_1$ and $\mathbf{n}_2$ are

1. *perpendicular* if $\mathbf{n}_1 \cdot \mathbf{n}_2 = 0$.
2. *parallel* if $\mathbf{n}_1$ is a scalar multiple of $\mathbf{n}_2$.

EXAMPLE 4 Finding the Line of Intersection of Two Planes

Find the angle between the two planes given by

$$x - 2y + z = 0 \qquad \text{Equation of plane 1}$$
$$2x + 3y - 2z = 0 \qquad \text{Equation of plane 2}$$

and find parametric equations of their line of intersection (see Figure 11.48).

Solution Normal vectors for the planes are $\mathbf{n}_1 = \langle 1, -2, 1 \rangle$ and $\mathbf{n}_2 = \langle 2, 3, -2 \rangle$. Consequently, the angle between the two planes is determined as follows.

$$\cos \theta = \frac{|\mathbf{n}_1 \cdot \mathbf{n}_2|}{\|\mathbf{n}_1\| \|\mathbf{n}_2\|} \qquad \text{Cosine of angle between } \mathbf{n}_1 \text{ and } \mathbf{n}_2$$

$$= \frac{|-6|}{\sqrt{6} \sqrt{17}}$$

$$= \frac{6}{\sqrt{102}}$$

$$\approx 0.59409$$

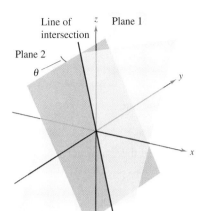

Figure 11.48

This implies that the angle between the two planes is $\theta \approx 53.55°$. You can find the line of intersection of the two planes by simultaneously solving the two linear equations representing the planes. One way to do this is to multiply the first equation by -2 and add the result to the second equation.

$$\begin{aligned} x - 2y + z &= 0 \\ 2x + 3y - 2z &= 0 \end{aligned} \quad\Longrightarrow\quad \begin{aligned} -2x + 4y - 2z &= 0 \\ \underline{2x + 3y - 2z = 0} \\ 7y - 4z &= 0 \end{aligned} \quad\Longrightarrow\quad y = \frac{4z}{7}$$

Substituting $y = 4z/7$ back into one of the original equations, you can determine that $x = z/7$. Finally, by letting $t = z/7$, you obtain the parametric equations

$$x = t, \quad y = 4t, \quad \text{and} \quad z = 7t \qquad \text{Line of intersection}$$

which indicate that 1, 4, and 7 are direction numbers for the line of intersection.

Note that the direction numbers in Example 4 can be obtained from the cross product of the two normal vectors as follows.

$$\mathbf{n}_1 \times \mathbf{n}_2 = \begin{vmatrix} \mathbf{i} & \mathbf{j} & \mathbf{k} \\ 1 & -2 & 1 \\ 2 & 3 & -2 \end{vmatrix}$$

$$= \begin{vmatrix} -2 & 1 \\ 3 & -2 \end{vmatrix} \mathbf{i} - \begin{vmatrix} 1 & 1 \\ 2 & -2 \end{vmatrix} \mathbf{j} + \begin{vmatrix} 1 & -2 \\ 2 & 3 \end{vmatrix} \mathbf{k}$$

$$= \mathbf{i} + 4\mathbf{j} + 7\mathbf{k}$$

This means that the line of intersection of the two planes is parallel to the cross product of their normal vectors.

NOTE The three-dimensional rotatable graphs that are available in the *HM mathSpace®* CD-ROM and the online *Eduspace®* system for this text can help you visualize surfaces such as those shown in Figure 11.48. If you have access to these graphs, you should use them to help your spatial intuition when studying this section and other sections in the text that deal with vectors, curves, or surfaces in space.

Sketching Planes in Space

If a plane in space intersects one of the coordinate planes, the line of intersection is called the **trace** of the given plane in the coordinate plane. To sketch a plane in space, it is helpful to find its points of intersection with the coordinate axes and its traces in the coordinate planes. For example, consider the plane given by

$$3x + 2y + 4z = 12. \qquad \text{Equation of plane}$$

You can find the xy-trace by letting $z = 0$ and sketching the line

$$3x + 2y = 12 \qquad xy\text{-trace}$$

in the xy-plane. This line intersects the x-axis at $(4, 0, 0)$ and the y-axis at $(0, 6, 0)$. In Figure 11.49, this process is continued by finding the yz-trace and the xz-trace, and then shading the triangular region lying in the first octant.

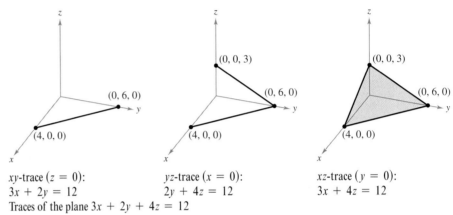

xy-trace ($z = 0$):
$3x + 2y = 12$

yz-trace ($x = 0$):
$2y + 4z = 12$

xz-trace ($y = 0$):
$3x + 4z = 12$

Traces of the plane $3x + 2y + 4z = 12$
Figure 11.49

If an equation of a plane has a missing variable, such as $2x + z = 1$, the plane must be *parallel to the axis* represented by the missing variable, as shown in Figure 11.50. If two variables are missing from an equation of a plane, it is *parallel to the coordinate plane* represented by the missing variables, as shown in Figure 11.51.

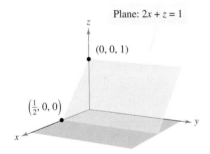

Plane $2x + z = 1$ is parallel to the y-axis.
Figure 11.50

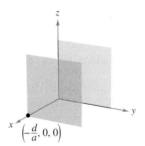

Plane $ax + d = 0$ is parallel to the yz-plane.

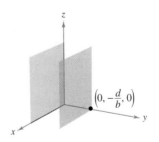

Plane $by + d = 0$ is parallel to the xz-plane.

Figure 11.51

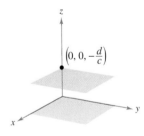

Plane $cz + d = 0$ is parallel to the xy-plane.

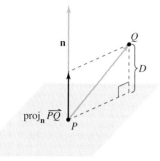

$D = \|\text{proj}_{\mathbf{n}} \overrightarrow{PQ}\|$

The distance between a point and a plane
Figure 11.52

Distances Between Points, Planes, and Lines

This section is concluded with the following discussion of two basic types of problems involving distance in space.

1. Finding the distance between a point and a plane
2. Finding the distance between a point and a line

The solutions of these problems illustrate the versatility and usefulness of vectors in coordinate geometry: the first problem uses the *dot product* of two vectors, and the second problem uses the *cross product*.

The distance D between a point Q and a plane is the length of the shortest line segment connecting Q to the plane, as shown in Figure 11.52. If P is *any* point in the plane, you can find this distance by projecting the vector $\overrightarrow{PQ}$ onto the normal vector $\mathbf{n}$. The length of this projection is the desired distance.

THEOREM 11.13 Distance Between a Point and a Plane

The distance between a plane and a point Q (not in the plane) is

$$D = \|\text{proj}_{\mathbf{n}} \overrightarrow{PQ}\| = \frac{|\overrightarrow{PQ} \cdot \mathbf{n}|}{\|\mathbf{n}\|}$$

where P is a point in the plane and $\mathbf{n}$ is normal to the plane.

To find a point in the plane given by $ax + by + cz + d = 0$ $(a \neq 0)$, let $y = 0$ and $z = 0$. Then, from the equation $ax + d = 0$, you can conclude that the point $(-d/a, 0, 0)$ lies in the plane.

EXAMPLE 5 Finding the Distance Between a Point and a Plane

Find the distance between the point $Q(1, 5, -4)$ and the plane given by

$$3x - y + 2z = 6.$$

Solution You know that $\mathbf{n} = \langle 3, -1, 2 \rangle$ is normal to the given plane. To find a point in the plane, let $y = 0$ and $z = 0$, and obtain the point $P(2, 0, 0)$. The vector from P to Q is given by

$$\overrightarrow{PQ} = \langle 1 - 2, 5 - 0, -4 - 0 \rangle$$
$$= \langle -1, 5, -4 \rangle.$$

Using the Distance Formula given in Theorem 11.13 produces

$$D = \frac{|\overrightarrow{PQ} \cdot \mathbf{n}|}{\|\mathbf{n}\|} = \frac{|\langle -1, 5, -4 \rangle \cdot \langle 3, -1, 2 \rangle|}{\sqrt{9 + 1 + 4}} \quad \text{Distance between a point and a plane}$$
$$= \frac{|-3 - 5 - 8|}{\sqrt{14}}$$
$$= \frac{16}{\sqrt{14}}.$$

NOTE The choice of the point P in Example 5 is arbitrary. Try choosing a different point in the plane to verify that you obtain the same distance.

From Theorem 11.13, you can determine that the distance between the point $Q(x_0, y_0, z_0)$ and the plane given by $ax + by + cz + d = 0$ is

$$D = \frac{|a(x_0 - x_1) + b(y_0 - y_1) + c(z_0 - z_1)|}{\sqrt{a^2 + b^2 + c^2}}$$

or

$$D = \frac{|ax_0 + by_0 + cz_0 + d|}{\sqrt{a^2 + b^2 + c^2}}$$ Distance between a point and a plane

where $P(x_1, y_1, z_1)$ is a point in the plane and $d = -(ax_1 + by_1 + cz_1)$.

EXAMPLE 6   **Finding the Distance Between Two Parallel Planes**

Find the distance between the two parallel planes given by

$$3x - y + 2z - 6 = 0 \quad \text{and} \quad 6x - 2y + 4z + 4 = 0.$$

Solution The two planes are shown in Figure 11.53. To find the distance between the planes, choose a point in the first plane, say $(x_0, y_0, z_0) = (2, 0, 0)$. Then, from the second plane, you can determine that $a = 6$, $b = -2$, $c = 4$, and $d = 4$, and conclude that the distance is

$$\begin{aligned} D &= \frac{|ax_0 + by_0 + cz_0 + d|}{\sqrt{a^2 + b^2 + c^2}} \quad\quad \text{Distance between a point and a plane} \\ &= \frac{|6(2) + (-2)(0) + (4)(0) + 4|}{\sqrt{6^2 + (-2)^2 + 4^2}} \\ &= \frac{16}{\sqrt{56}} = \frac{8}{\sqrt{14}} \approx 2.14. \end{aligned}$$

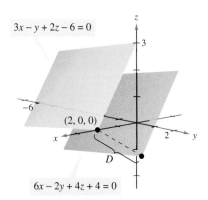

The distance between the parallel planes is approximately 2.14.
Figure 11.53

The formula for the distance between a point and a line in space resembles that for the distance between a point and a plane—except that you replace the dot product with the length of the cross product and the normal vector **n** with a direction vector for the line.

THEOREM 11.14   **Distance Between a Point and a Line in Space**

The distance between a point Q and a line in space is given by

$$D = \frac{\|\overrightarrow{PQ} \times \mathbf{u}\|}{\|\mathbf{u}\|}$$

where **u** is a direction vector for the line and P is a point on the line.

Proof In Figure 11.54, let D be the distance between the point Q and the given line. Then $D = \|\overrightarrow{PQ}\| \sin \theta$, where θ is the angle between **u** and $\overrightarrow{PQ}$. By Theorem 11.8, you have

$$\|\mathbf{u}\| \|\overrightarrow{PQ}\| \sin \theta = \|\mathbf{u} \times \overrightarrow{PQ}\| = \|\overrightarrow{PQ} \times \mathbf{u}\|.$$

Consequently,

$$D = \|\overrightarrow{PQ}\| \sin \theta = \frac{\|\overrightarrow{PQ} \times \mathbf{u}\|}{\|\mathbf{u}\|}.$$

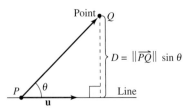

The distance between a point and a line
Figure 11.54

EXAMPLE 7 Finding the Distance Between a Point and a Line

Find the distance between the point $Q(3, -1, 4)$ and the line given by

$$x = -2 + 3t, \quad y = -2t, \quad \text{and} \quad z = 1 + 4t.$$

Solution Using the direction numbers 3, -2, and 4, you know that a direction vector for the line is

$$\mathbf{u} = \langle 3, -2, 4 \rangle. \qquad \text{Direction vector for line}$$

To find a point on the line, let $t = 0$ and obtain

$$P = (-2, 0, 1). \qquad \text{Point on the line}$$

So,

$$\overrightarrow{PQ} = \langle 3 - (-2), -1 - 0, 4 - 1 \rangle = \langle 5, -1, 3 \rangle$$

and you can form the cross product

$$\overrightarrow{PQ} \times \mathbf{u} = \begin{vmatrix} \mathbf{i} & \mathbf{j} & \mathbf{k} \\ 5 & -1 & 3 \\ 3 & -2 & 4 \end{vmatrix} = 2\mathbf{i} - 11\mathbf{j} - 7\mathbf{k} = \langle 2, -11, -7 \rangle.$$

Finally, using Theorem 11.14, you can find the distance to be

$$D = \frac{\|\overrightarrow{PQ} \times \mathbf{u}\|}{\|\mathbf{u}\|}$$

$$= \frac{\sqrt{174}}{\sqrt{29}}$$

$$= \sqrt{6} \approx 2.45. \qquad \text{See Figure 11.55.}$$

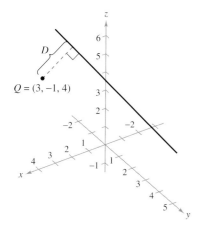

The distance between the point Q and the line is $\sqrt{6} \approx 2.45$.
Figure 11.55

Exercises for Section 11.5

See www.CalcChat.com for worked-out solutions to odd-numbered exercises.

In Exercises 1 and 2, the figure shows the graph of a line given by the parametric equations. (a) Draw an arrow on the line to indicate its orientation. To print an enlarged copy of the graph, go to the website *www.mathgraphs.com*. (b) Find the coordinates of two points, P and Q, on the line. Determine the vector $\overrightarrow{PQ}$. What is the relationship between the components of the vector and the coefficients of t in the parametric equations? Why is this true? (c) Determine the coordinates of any points of intersection with the coordinate planes. If the line does not intersect a coordinate plane, explain why.

1. $x = 1 + 3t$
$y = 2 - t$
$z = 2 + 5t$

2. $x = 2 - 3t$
$y = 2$
$z = 1 - t$

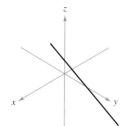
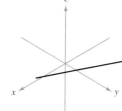

In Exercises 3–8, find sets of (a) parametric equations and (b) symmetric equations of the line through the point parallel to the given vector or line. (For each line, write the direction numbers as integers.)

Point	Parallel to
3. $(0, 0, 0)$	$\mathbf{v} = \langle 1, 2, 3 \rangle$
4. $(0, 0, 0)$	$\mathbf{v} = \langle -2, \frac{5}{2}, 1 \rangle$
5. $(-2, 0, 3)$	$\mathbf{v} = 2\mathbf{i} + 4\mathbf{j} - 2\mathbf{k}$
6. $(-3, 0, 2)$	$\mathbf{v} = 6\mathbf{j} + 3\mathbf{k}$
7. $(1, 0, 1)$	$x = 3 + 3t, y = 5 - 2t, z = -7 + t$
8. $(-3, 5, 4)$	$\dfrac{x-1}{3} = \dfrac{y+1}{-2} = z - 3$

In Exercises 9–12, find sets of (a) parametric equations and (b) symmetric equations of the line through the two points. (For each line, write the direction numbers as integers.)

9. $(5, -3, -2), \left(-\frac{2}{3}, \frac{2}{3}, 1\right)$
10. $(2, 0, 2), (1, 4, -3)$
11. $(2, 3, 0), (10, 8, 12)$
12. $(0, 0, 25), (10, 10, 0)$

In Exercises 13–20, find a set of parametric equations of the line.

13. The line passes through the point $(2, 3, 4)$ and is parallel to the xz-plane and the yz-plane.

14. The line passes through the point $(-4, 5, 2)$ and is parallel to the xy-plane and the yz-plane.

15. The line passes through the point $(2, 3, 4)$ and is perpendicular to the plane given by $3x + 2y - z = 6$.

16. The line passes through the point $(-4, 5, 2)$ and is perpendicular to the plane given by $-x + 2y + z = 5$.

17. The line passes through the point $(5, -3, -4)$ and is parallel to $\mathbf{v} = \langle 2, -1, 3 \rangle$.

18. The line passes through the point $(-1, 4, -3)$ and is parallel to $\mathbf{v} = 5\mathbf{i} - \mathbf{j}$.

19. The line passes through the point $(2, 1, 2)$ and is parallel to the line $x = -t, y = 1 + t, z = -2 + t$.

20. The line passes through the point $(-6, 0, 8)$ and is parallel to the line $x = 5 - 2t, y = -4 + 2t, z = 0$.

In Exercises 21–24, find the coordinates of a point P on the line and a vector $\mathbf{v}$ parallel to the line.

21. $x = 3 - t, \quad y = -1 + 2t, \quad z = -2$
22. $x = 4t, \quad y = 5 - t, \quad z = 4 + 3t$
23. $\dfrac{x - 7}{4} = \dfrac{y + 6}{2} = z + 2$
24. $\dfrac{x + 3}{5} = \dfrac{y}{8} = \dfrac{z - 3}{6}$

In Exercises 25 and 26, determine if any of the lines are parallel or identical.

25. L_1: $x = 6 - 3t, \quad y = -2 + 2t, \quad z = 5 + 4t$
    L_2: $x = 6t, \quad y = 2 - 4t, \quad z = 13 - 8t$
    L_3: $x = 10 - 6t, \quad y = 3 + 4t, \quad z = 7 + 8t$
    L_4: $x = -4 + 6t, \quad y = 3 + 4t, \quad z = 5 - 6t$

26. L_1: $\dfrac{x - 8}{4} = \dfrac{y + 5}{-2} = \dfrac{z + 9}{3}$
    L_2: $\dfrac{x + 7}{2} = \dfrac{y - 4}{1} = \dfrac{z + 6}{5}$
    L_3: $\dfrac{x + 4}{-8} = \dfrac{y - 1}{4} = \dfrac{z + 18}{-6}$
    L_4: $\dfrac{x - 2}{-2} = \dfrac{y + 3}{1} = \dfrac{z - 4}{1.5}$

In Exercises 27–30, determine whether the lines intersect, and if so, find the point of intersection and the cosine of the angle of intersection.

27. $x = 4t + 2, \quad y = 3, \quad z = -t + 1$
    $x = 2s + 2, \quad y = 2s + 3, \quad z = s + 1$

28. $x = -3t + 1, y = 4t + 1, z = 2t + 4$
    $x = 3s + 1, y = 2s + 4, z = -s + 1$

29. $\dfrac{x}{3} = \dfrac{y - 2}{-1} = z + 1, \quad \dfrac{x - 1}{4} = y + 2 = \dfrac{z + 3}{-3}$

30. $\dfrac{x - 2}{-3} = \dfrac{y - 2}{6} = z - 3, \quad \dfrac{x - 3}{2} = y + 5 = \dfrac{z + 2}{4}$

In Exercises 31 and 32, use a computer algebra system to graph the pair of intersecting lines and find the point of intersection.

31. $x = 2t + 3, y = 5t - 2, z = -t + 1$
    $x = -2s + 7, y = s + 8, z = 2s - 1$

32. $x = 2t - 1, y = -4t + 10, z = t$
    $x = -5s - 12, y = 3s + 11, z = -2s - 4$

Cross Product In Exercises 33 and 34, (a) find the coordinates of three points $P, Q,$ and R in the plane, and determine the vectors $\overrightarrow{PQ}$ and $\overrightarrow{PR}$. (b) Find $\overrightarrow{PQ} \times \overrightarrow{PR}$. What is the relationship between the components of the cross product and the coefficients of the equation of the plane? Why is this true?

33. $4x - 3y - 6z = 6$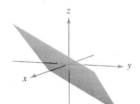

34. $2x + 3y + 4z = 4$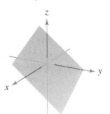

In Exercises 35–40, find an equation of the plane passing through the point perpendicular to the given vector or line.

Point	Perpendicular to
35. $(2, 1, 2)$	$\mathbf{n} = \mathbf{i}$
36. $(1, 0, -3)$	$\mathbf{n} = \mathbf{k}$
37. $(3, 2, 2)$	$\mathbf{n} = 2\mathbf{i} + 3\mathbf{j} - \mathbf{k}$
38. $(0, 0, 0)$	$\mathbf{n} = -3\mathbf{i} + 2\mathbf{k}$
39. $(0, 0, 6)$	$x = 1 - t, y = 2 + t, z = 4 - 2t$
40. $(3, 2, 2)$	$\dfrac{x - 1}{4} = y + 2 = \dfrac{z + 3}{-3}$

In Exercises 41–52, find an equation of the plane.

41. The plane passes through $(0, 0, 0), (1, 2, 3),$ and $(-2, 3, 3)$.
42. The plane passes through $(2, 3, -2), (3, 4, 2),$ and $(1, -1, 0)$.
43. The plane passes through $(1, 2, 3), (3, 2, 1),$ and $(-1, -2, 2)$.
44. The plane passes through the point $(1, 2, 3)$ and is parallel to the yz-plane.
45. The plane passes through the point $(1, 2, 3)$ and is parallel to the xy-plane.
46. The plane contains the y-axis and makes an angle of $\pi/6$ with the positive x-axis.

47. The plane contains the lines given by

$$\frac{x-1}{-2} = y - 4 = z \quad \text{and} \quad \frac{x-2}{-3} = \frac{y-1}{4} = \frac{z-2}{-1}.$$

48. The plane passes through the point $(2, 2, 1)$ and contains the line given by

$$\frac{x}{2} = \frac{y-4}{-1} = z.$$

49. The plane passes through the points $(2, 2, 1)$ and $(-1, 1, -1)$ and is perpendicular to the plane $2x - 3y + z = 3$.

50. The plane passes through the points $(3, 2, 1)$ and $(3, 1, -5)$ and is perpendicular to the plane $6x + 7y + 2z = 10$.

51. The plane passes through the points $(1, -2, -1)$ and $(2, 5, 6)$ and is parallel to the x-axis.

52. The plane passes through the points $(4, 2, 1)$ and $(-3, 5, 7)$ and is parallel to the z-axis.

In Exercises 53 and 54, sketch a graph of the line and find the points (if any) where the line intersects the xy-, xz-, and yz-planes.

53. $x = 1 - 2t, \quad y = -2 + 3t, \quad z = -4 + t$

54. $\dfrac{x-2}{3} = y + 1 = \dfrac{z-3}{2}$

In Exercises 55 and 56, find an equation of the plane that contains all the points that are equidistant from the given points.

55. $(2, 2, 0), \quad (0, 2, 2)$

56. $(-3, 1, 2), \quad (6, -2, 4)$

In Exercises 57–62, determine whether the planes are parallel, orthogonal, or neither. If they are neither parallel nor orthogonal, find the angle of intersection.

57. $5x - 3y + z = 4$
$x + 4y + 7z = 1$

58. $3x + y - 4z = 3$
$-9x - 3y + 12z = 4$

59. $x - 3y + 6z = 4$
$5x + y - z = 4$

60. $3x + 2y - z = 7$
$x - 4y + 2z = 0$

61. $x - 5y - z = 1$
$5x - 25y - 5z = -3$

62. $2x - z = 1$
$4x + y + 8z = 10$

In Exercises 63–70, label any intercepts and sketch a graph of the plane.

63. $4x + 2y + 6z = 12$

64. $3x + 6y + 2z = 6$

65. $2x - y + 3z = 4$

66. $2x - y + z = 4$

67. $y + z = 5$

68. $x + 2y = 4$

69. $x = 5$

70. $z = 8$

In Exercises 71–74, use a computer algebra system to graph the plane.

71. $2x + y - z = 6$

72. $x - 3z = 3$

73. $-5x + 4y - 6z = -8$

74. $2.1x - 4.7y - z = -3$

In Exercises 75 and 76, determine if any of the planes are parallel or identical.

75. $P_1: 3x - 2y + 5z = 10$
$P_2: -6x + 4y - 10z = 5$
$P_3: -3x + 2y + 5z = 8$
$P_4: 75x - 50y + 125z = 250$

76. $P_1: -60x + 90y + 30z = 27$
$P_2: 6x - 9y - 3z = 2$
$P_3: -20x + 30y + 10z = 9$
$P_4: 12x - 18y + 6z = 5$

In Exercises 77–80, describe the family of planes represented by the equation, where c is any real number.

77. $x + y + z = c$

78. $x + y = c$

79. $cy + z = 0$

80. $x + cz = 0$

In Exercises 81 and 82, find a set of parametric equations for the line of intersection of the planes.

81. $3x + 2y - z = 7$
$x - 4y + 2z = 0$

82. $6x - 3y + z = 5$
$-x + y + 5z = 5$

In Exercises 83–86, find the point(s) of intersection (if any) of the plane and the line. Also determine whether the line lies in the plane.

83. $2x - 2y + z = 12, \quad x - \dfrac{1}{2} = \dfrac{y + (3/2)}{-1} = \dfrac{z + 1}{2}$

84. $2x + 3y = -5, \quad \dfrac{x-1}{4} = \dfrac{y}{2} = \dfrac{z-3}{6}$

85. $2x + 3y = 10, \quad \dfrac{x-1}{3} = \dfrac{y+1}{-2} = z - 3$

86. $5x + 3y = 17, \quad \dfrac{x-4}{2} = \dfrac{y+1}{-3} = \dfrac{z+2}{5}$

In Exercises 87–90, find the distance between the point and the plane.

87. $(0, 0, 0)$
$2x + 3y + z = 12$

88. $(0, 0, 0)$
$8x - 4y + z = 8$

89. $(2, 8, 4)$
$2x + y + z = 5$

90. $(3, 2, 1)$
$x - y + 2z = 4$

In Exercises 91–94, verify that the two planes are parallel, and find the distance between the planes.

91. $x - 3y + 4z = 10$
    $x - 3y + 4z = 6$

92. $4x - 4y + 9z = 7$
    $4x - 4y + 9z = 18$

93. $-3x + 6y + 7z = 1$
    $6x - 12y - 14z = 25$

94. $2x - 4z = 4$
    $2x - 4z = 10$

In Exercises 95–98, find the distance between the point and the line given by the set of parametric equations.

95. $(1, 5, -2)$; $x = 4t - 2$, $y = 3$, $z = -t + 1$
96. $(1, -2, 4)$; $x = 2t$, $y = t - 3$, $z = 2t + 2$
97. $(-2, 1, 3)$; $x = 1 - t$, $y = 2 + t$, $z = -2t$
98. $(4, -1, 5)$; $x = 3$, $y = 1 + 3t$, $z = 1 + t$

In Exercises 99 and 100, verify that the lines are parallel, and find the distance between them.

99. L_1: $x = 2 - t$, $y = 3 + 2t$, $z = 4 + t$
    L_2: $x = 3t$, $y = 1 - 6t$, $z = 4 - 3t$

100. L_1: $x = 3 + 6t$, $y = -2 + 9t$, $z = 1 - 12t$
     L_2: $x = -1 + 4t$, $y = 3 + 6t$, $z = -8t$

Writing About Concepts

101. Give the parametric equations and the symmetric equations of a line in space. Describe what is required to find these equations.

102. Give the standard equation of a plane in space. Describe what is required to find this equation.

103. Describe a method of finding the line of intersection of two planes.

104. Describe each surface given by the equations $x = a$, $y = b$, and $z = c$.

105. Describe a method for determining when two planes
    $a_1x + b_1y + c_1z + d_1 = 0$
    and
    $a_2x + b_2y + c_2z + d_2 = 0$
    are (a) parallel and (b) perpendicular. Explain your reasoning.

106. Let L_1 and L_2 be nonparallel lines that do not intersect. Is it possible to find a nonzero vector $\mathbf{v}$ such that $\mathbf{v}$ is perpendicular to both L_1 and L_2? Explain your reasoning.

107. Find an equation of the plane with x-intercept $(a, 0, 0)$, y-intercept $(0, b, 0)$, and z-intercept $(0, 0, c)$. (Assume a, b, and c are nonzero.)

108. (a) Describe and find an equation for the surface generated by all points (x, y, z) that are four units from the point $(3, -2, 5)$.
     (b) Describe and find an equation for the surface generated by all points (x, y, z) that are four units from the plane
     $4x - 3y + z = 10$.

109. *Modeling Data* Per capita consumptions (in gallons) of different types of plain milk in the United States from 1994 to 2000 are shown in the table. Consumption of light and skim milks, reduced-fat milk, and whole milk are represented by the variables x, y, and z, respectively. *(Source: U.S. Department of Agriculture)*

Year	1994	1995	1996	1997	1998	1999	2000
x	5.8	6.2	6.4	6.6	6.5	6.3	6.1
y	8.7	8.2	8.0	7.7	7.4	7.3	7.1
z	8.8	8.4	8.4	8.2	7.8	7.9	7.8

A model for the data is given by $0.04x - 0.64y + z = 3.4$.

(a) Complete a fourth row in the table using the model to approximate z for the given values of x and y. Compare the approximations with the actual values of z.

(b) According to this model, any increases in consumption of two types of milk will have what effect on the consumption of the third type?

110. *Mechanical Design* A chute at the top of a grain elevator of a combine funnels the grain into a bin (see figure). Find the angle between two adjacent sides.

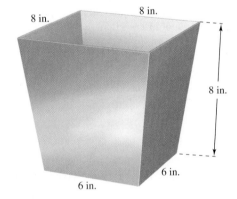

111. *Distance* Two insects are crawling along different lines in three-space. At time t (in minutes), the first insect is at the point (x, y, z) on the line

$$x = 6 + t, \quad y = 8 - t, \quad z = 3 + t.$$

Also, at time t, the second insect is at the point (x, y, z) on the line

$$x = 1 + t, \quad y = 2 + t, \quad z = 2t.$$

Assume distances are given in inches.

(a) Find the distance between the two insects at time $t = 0$.

(b) Use a graphing utility to graph the distance between the insects from $t = 0$ to $t = 10$.

(c) Using the graph from part (b), what can you conclude about the distance between the insects?

(d) How close do the insects get?

112. Find the standard equation of the sphere with center $(-3, 2, 4)$ that is tangent to the plane given by $2x + 4y - 3z = 8$.

113. Find the point of intersection of the plane $3x - y + 4z = 7$ and the line through $(5, 4, -3)$ that is perpendicular to this plane.

114. Show that the plane $2x - y - 3z = 4$ is parallel to the line $x = -2 + 2t$, $y = -1 + 4t$, $z = 4$, and find the distance between them.

115. Find the point of intersection of the line through $(1, -3, 1)$ and $(3, -4, 2)$, and the plane given by $x - y + z = 2$.

116. Find a set of parametric equations for the line passing through the point $(1, 0, 2)$ that is parallel to the plane given by $x + y + z = 5$, and perpendicular to the line $x = t$, $y = 1 + t, z = 1 + t$.

True or False? In Exercises 117–120, determine whether the statement is true or false. If it is false, explain why or give an example that shows it is false.

117. If $\mathbf{v} = a_1\mathbf{i} + b_1\mathbf{j} + c_1\mathbf{k}$ is any vector in the plane given by $a_2x + b_2y + c_2z + d_2 = 0$, then $a_1a_2 + b_1b_2 + c_1c_2 = 0$.

118. Every pair of lines in space are either intersecting or parallel.

119. Two planes in space are either intersecting or parallel.

120. If two lines L_1 and L_2 are parallel to a plane P, then L_1 and L_2 are parallel.

Section Project: Distances in Space

You have learned two distance formulas in this section—the distance between a point and a plane, and the distance between a point and a line. In this project you will study a third distance problem—the distance between two skew lines. Two lines in space are *skew* if they are neither parallel nor intersecting (see figure).

(a) Consider the following two lines in space.

$L_1: x = 4 + 5t, \ y = 5 + 5t, \ z = 1 - 4t$
$L_2: x = 4 + s, \ y = -6 + 8s, \ z = 7 - 3s$

(i) Show that these lines are not parallel.

(ii) Show that these lines do not intersect, and therefore are skew lines.

(iii) Show that the two lines lie in parallel planes.

(iv) Find the distance between the parallel planes from part (iii). This is the distance between the original skew lines.

(b) Use the procedure in part (a) to find the distance between the lines.

$L_1: x = 2t, \ y = 4t, \ z = 6t$
$L_2: x = 1 - s, \ y = 4 + s, \ z = -1 + s$

(c) Use the procedure in part (a) to find the distance between the lines.

$L_1: x = 3t, \ y = 2 - t, \ z = -1 + t$
$L_2: x = 1 + 4s, \ y = -2 + s, \ z = -3 - 3s$

(d) Develop a formula for finding the distance between the skew lines.

$L_1: x = x_1 + a_1t, \ y = y_1 + b_1t, \ z = z_1 + c_1t$
$L_2: x = x_2 + a_2s, \ y = y_2 + b_2s, \ z = z_2 + c_2s$

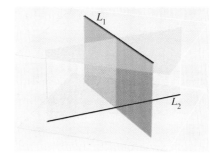

Section 11.6 Surfaces in Space

- Recognize and write equations for cylindrical surfaces.
- Recognize and write equations for quadric surfaces.
- Recognize and write equations for surfaces of revolution.

Cylindrical Surfaces

The first five sections of this chapter contained the vector portion of the preliminary work necessary to study vector calculus and the calculus of space. In this and the next section, you will study surfaces in space and alternative coordinate systems for space. You have already studied two special types of surfaces.

1. **Spheres:** $(x - x_0)^2 + (y - y_0)^2 + (z - z_0)^2 = r^2$ Section 11.2
2. **Planes:** $ax + by + cz + d = 0$ Section 11.5

A third type of surface in space is called a **cylindrical surface,** or simply a **cylinder.** To define a cylinder, consider the familiar right circular cylinder shown in Figure 11.56. You can imagine that this cylinder is generated by a vertical line moving around the circle $x^2 + y^2 = a^2$ in the xy-plane. This circle is called a **generating curve** for the cylinder, as indicated in the following definition.

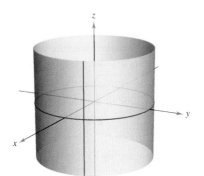

Right circular cylinder:
$x^2 + y^2 = a^2$

Rulings are parallel to z-axis.
Figure 11.56

Definition of a Cylinder

Let C be a curve in a plane and let L be a line not in a parallel plane. The set of all lines parallel to L and intersecting C is called a **cylinder.** C is called the **generating curve** (or **directrix**) of the cylinder, and the parallel lines are called **rulings.**

NOTE Without loss of generality, you can assume that C lies in one of the three coordinate planes. Moreover, this text restricts the discussion to *right* cylinders—cylinders whose rulings are perpendicular to the coordinate plane containing C, as shown in Figure 11.57.

For the right circular cylinder shown in Figure 11.56, the equation of the generating curve is

$$x^2 + y^2 = a^2.$$ Equation of generating curve in xy-plane

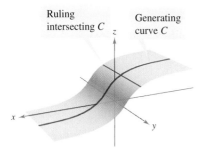

Cylinder: Rulings intersect C and are parallel to the given line.
Figure 11.57

To find an equation for the cylinder, note that you can generate any one of the rulings by fixing the values of x and y and then allowing z to take on all real values. In this sense, the value of z is arbitrary and is, therefore, not included in the equation. In other words, the equation of this cylinder is simply the equation of its generating curve.

$$x^2 + y^2 = a^2$$ Equation of cylinder in space

Equations of Cylinders

The equation of a cylinder whose rulings are parallel to one of the coordinate axes contains only the variables corresponding to the other two axes.

EXAMPLE 1 Sketching a Cylinder

Sketch the surface represented by each equation.

a. $z = y^2$ **b.** $z = \sin x, \quad 0 \leq x \leq 2\pi$

Solution

a. The graph is a cylinder whose generating curve, $z = y^2$, is a parabola in the yz-plane. The rulings of the cylinder are parallel to the x-axis, as shown in Figure 11.58(a).

b. The graph is a cylinder generated by the sine curve in the xz-plane. The rulings are parallel to the y-axis, as shown in Figure 11.58(b).

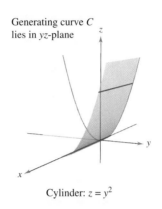

(a) Rulings are parallel to x-axis.   (b) Rulings are parallel to y-axis.
Figure 11.58

Quadric Surfaces

STUDY TIP In the table on pages 812 and 813, only one of several orientations of each quadric surface is shown. If the surface is oriented along a different axis, its standard equation will change accordingly, as illustrated in Examples 2 and 3. The fact that the two types of paraboloids have one variable raised to the first power can be helpful in classifying quadric surfaces. The other four types of basic quadric surfaces have equations that are of *second degree* in all three variables.

The fourth basic type of surface in space is a **quadric surface.** Quadric surfaces are the three-dimensional analogs of conic sections.

Quadric Surface

The equation of a **quadric surface** in space is a second-degree equation of the form

$$Ax^2 + By^2 + Cz^2 + Dxy + Exz + Fyz + Gx + Hy + Iz + J = 0.$$

There are six basic types of quadric surfaces: **ellipsoid, hyperboloid of one sheet, hyperboloid of two sheets, elliptic cone, elliptic paraboloid,** and **hyperbolic paraboloid.**

The intersection of a surface with a plane is called the **trace of the surface** in the plane. To visualize a surface in space, it is helpful to determine its traces in some well-chosen planes. The traces of quadric surfaces are conics. These traces, together with the **standard form** of the equation of each quadric surface, are shown in the table on pages 812 and 813.

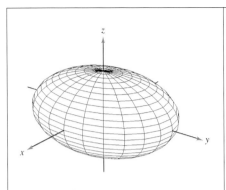

Ellipsoid

$$\frac{x^2}{a^2} + \frac{y^2}{b^2} + \frac{z^2}{c^2} = 1$$

Trace	Plane
Ellipse	Parallel to xy-plane
Ellipse	Parallel to xz-plane
Ellipse	Parallel to yz-plane

The surface is a sphere if $a = b = c \neq 0$.

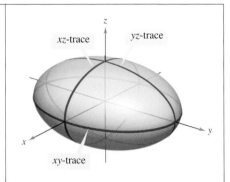

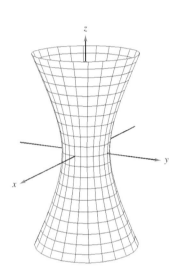

Hyperboloid of One Sheet

$$\frac{x^2}{a^2} + \frac{y^2}{b^2} - \frac{z^2}{c^2} = 1$$

Trace	Plane
Ellipse	Parallel to xy-plane
Hyperbola	Parallel to xz-plane
Hyperbola	Parallel to yz-plane

The axis of the hyperboloid corresponds to the variable whose coefficient is negative.

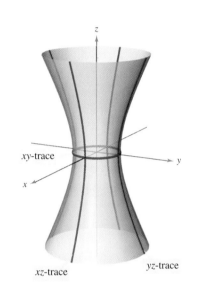

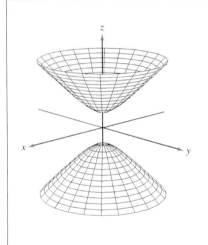

Hyperboloid of Two Sheets

$$\frac{z^2}{c^2} - \frac{x^2}{a^2} - \frac{y^2}{b^2} = 1$$

Trace	Plane
Ellipse	Parallel to xy-plane
Hyperbola	Parallel to xz-plane
Hyperbola	Parallel to yz-plane

The axis of the hyperboloid corresponds to the variable whose coefficient is positive. There is no trace in the coordinate plane perpendicular to this axis.

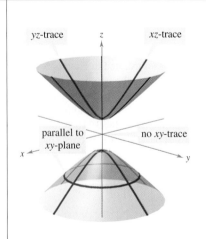

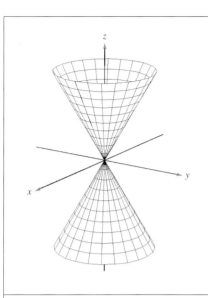

Elliptic Cone

$$\frac{x^2}{a^2} + \frac{y^2}{b^2} - \frac{z^2}{c^2} = 0$$

Trace	Plane
Ellipse	Parallel to xy-plane
Hyperbola	Parallel to xz-plane
Hyperbola	Parallel to yz-plane

The axis of the cone corresponds to the variable whose coefficient is negative. The traces in the coordinate planes parallel to this axis are intersecting lines.

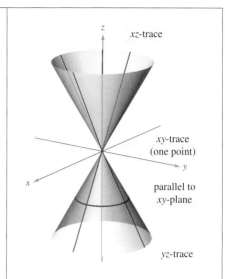

Elliptic Paraboloid

$$z = \frac{x^2}{a^2} + \frac{y^2}{b^2}$$

Trace	Plane
Ellipse	Parallel to xy-plane
Parabola	Parallel to xz-plane
Parabola	Parallel to yz-plane

The axis of the paraboloid corresponds to the variable raised to the first power.

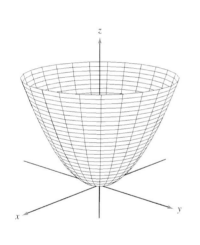

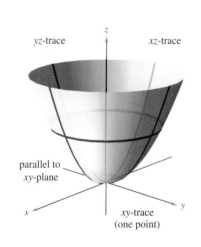

Hyperbolic Paraboloid

$$z = \frac{y^2}{b^2} - \frac{x^2}{a^2}$$

Trace	Plane
Hyperbola	Parallel to xy-plane
Parabola	Parallel to xz-plane
Parabola	Parallel to yz-plane

The axis of the paraboloid corresponds to the variable raised to the first power.

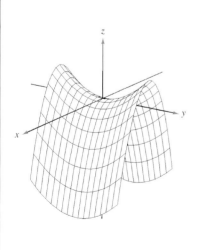

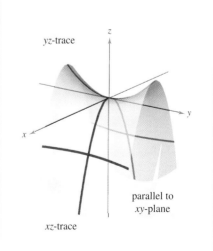

To classify a quadric surface, begin by writing the surface in standard form. Then, determine several traces taken in the coordinate planes *or* taken in planes that are parallel to the coordinate planes.

EXAMPLE 2 Sketching a Quadric Surface

Classify and sketch the surface given by $4x^2 - 3y^2 + 12z^2 + 12 = 0$.

Solution Begin by writing the equation in standard form.

$$4x^2 - 3y^2 + 12z^2 + 12 = 0 \qquad \text{Write original equation.}$$

$$\frac{x^2}{-3} + \frac{y^2}{4} - z^2 - 1 = 0 \qquad \text{Divide by } -12.$$

$$\frac{y^2}{4} - \frac{x^2}{3} - \frac{z^2}{1} = 1 \qquad \text{Standard form}$$

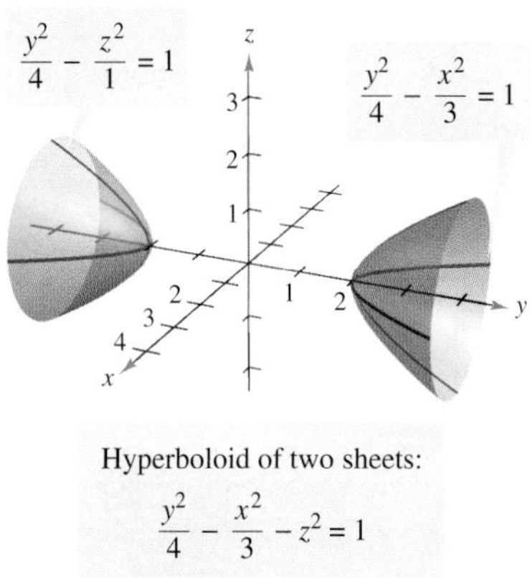

From the table on pages 812 and 813, you can conclude that the surface is a hyperboloid of two sheets with the y-axis as its axis. To sketch the graph of this surface, it helps to find the traces in the coordinate planes.

xy-trace $(z = 0)$: $\quad \dfrac{y^2}{4} - \dfrac{x^2}{3} = 1 \qquad$ Hyperbola

xz-trace $(y = 0)$: $\quad \dfrac{x^2}{3} + \dfrac{z^2}{1} = -1 \qquad$ No trace

yz-trace $(x = 0)$: $\quad \dfrac{y^2}{4} - \dfrac{z^2}{1} = 1 \qquad$ Hyperbola

The graph is shown in Figure 11.59.

Hyperboloid of two sheets:
$$\frac{y^2}{4} - \frac{x^2}{3} - z^2 = 1$$

Figure 11.59

EXAMPLE 3 Sketching a Quadric Surface

Classify and sketch the surface given by $x - y^2 - 4z^2 = 0$.

Solution Because x is raised only to the first power, the surface is a paraboloid. The axis of the paraboloid is the x-axis. In the standard form, the equation is

$$x = y^2 + 4z^2. \qquad \text{Standard form}$$

Some convenient traces are as follows.

xy-trace $(z = 0)$: $\qquad x = y^2 \qquad$ Parabola

xz-trace $(y = 0)$: $\qquad x = 4z^2 \qquad$ Parabola

parallel to yz-plane $(x = 4)$: $\quad \dfrac{y^2}{4} + \dfrac{z^2}{1} = 1 \quad$ Ellipse

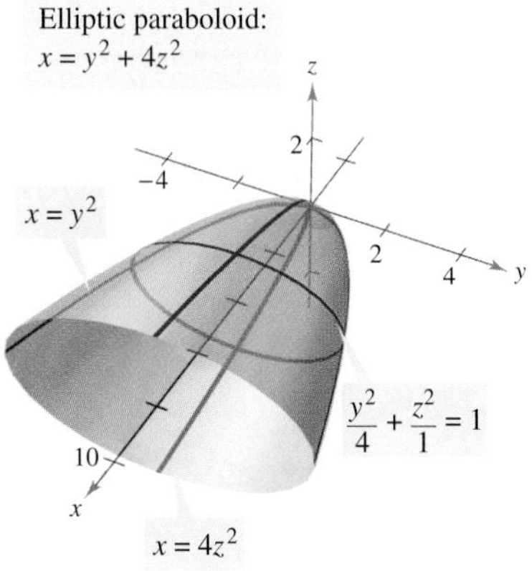

The surface is an *elliptic* paraboloid, as shown in Figure 11.60.

Elliptic paraboloid:
$x = y^2 + 4z^2$

Figure 11.60

Some second-degree equations in x, y, and z do not represent any of the basic types of quadric surfaces. Here are two examples.

$$x^2 + y^2 + z^2 = 0 \qquad \text{Single point}$$

$$x^2 + y^2 = 1 \qquad \text{Right circular cylinder}$$

For a quadric surface not centered at the origin, you can form the standard equation by completing the square, as demonstrated in Example 4.

EXAMPLE 4 A Quadric Surface Not Centered at the Origin

Classify and sketch the surface given by

$$x^2 + 2y^2 + z^2 - 4x + 4y - 2z + 3 = 0.$$

Solution Completing the square for each variable produces the following.

$$(x^2 - 4x + \quad) + 2(y^2 + 2y + \quad) + (z^2 - 2z + \quad) = -3$$
$$(x^2 - 4x + 4) + 2(y^2 + 2y + 1) + (z^2 - 2z + 1) = -3 + 4 + 2 + 1$$
$$(x - 2)^2 + 2(y + 1)^2 + (z - 1)^2 = 4$$
$$\frac{(x - 2)^2}{4} + \frac{(y + 1)^2}{2} + \frac{(z - 1)^2}{4} = 1$$

From this equation, you can see that the quadric surface is an ellipsoid that is centered at $(2, -1, 1)$. Its graph is shown in Figure 11.61.

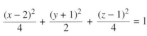
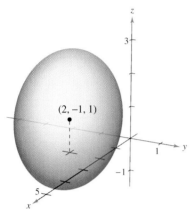

An ellipsoid centered at $(2, -1, 1)$
Figure 11.61

TECHNOLOGY

A computer algebra system can help you visualize a surface in space.* Most of these computer algebra systems create three-dimensional illusions by sketching several traces of the surface and then applying a "hidden-line" routine that blocks out portions of the surface that lie behind other portions of the surface. Two examples of figures that were generated by *Mathematica* are shown below.

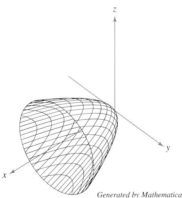

Generated by Mathematica

Elliptic paraboloid

$$x = \frac{y^2}{2} + \frac{z^2}{2}$$

Generated by Mathematica

Hyperbolic paraboloid

$$z = \frac{y^2}{16} - \frac{x^2}{16}$$

Using a graphing utility to graph a surface in space requires practice. For one thing, you must know enough about the surface to be able to specify a *viewing window* that gives a representative view of the surface. Also, you can often improve the view of a surface by rotating the axes. For instance, note that the elliptic paraboloid in the figure is seen from a line of sight that is "higher" than the line of sight used to view the hyperbolic paraboloid.

---

*Some 3-D graphing utilities require surfaces to be entered with parametric equations. For a discussion of this technique, see Section 15.5.

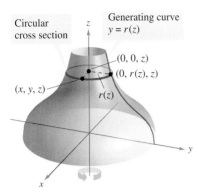

Figure 11.62

Surfaces of Revolution

The fifth special type of surface you will study is called a **surface of revolution**. In Section 7.4, you studied a method for finding the *area* of such a surface. You will now look at a procedure for finding its *equation*. Consider the graph of the **radius function**

$$y = r(z) \qquad \text{Generating curve}$$

in the yz-plane. If this graph is revolved about the z-axis, it forms a surface of revolution, as shown in Figure 11.62. The trace of the surface in the plane $z = z_0$ is a circle whose radius is $r(z_0)$ and whose equation is

$$x^2 + y^2 = [r(z_0)]^2. \qquad \text{Circular trace in plane: } z = z_0$$

Replacing z_0 with z produces an equation that is valid for all values of z. In a similar manner, you can obtain equations for surfaces of revolution for the other two axes, and the results are summarized as follows.

Surface of Revolution

If the graph of a radius function r is revolved about one of the coordinate axes, the equation of the resulting surface of revolution has one of the following forms.

1. Revolved about the x-axis: $y^2 + z^2 = [r(x)]^2$
2. Revolved about the y-axis: $x^2 + z^2 = [r(y)]^2$
3. Revolved about the z-axis: $x^2 + y^2 = [r(z)]^2$

EXAMPLE 5 **Finding an Equation for a Surface of Revolution**

a. An equation for the surface of revolution formed by revolving the graph of

$$y = \frac{1}{z} \qquad \text{Radius function}$$

about the z-axis is

$$x^2 + y^2 = [r(z)]^2 \qquad \text{Revolved about the } z\text{-axis}$$

$$x^2 + y^2 = \left(\frac{1}{z}\right)^2. \qquad \text{Substitute } 1/z \text{ for } r(z).$$

b. To find an equation for the surface formed by revolving the graph of $9x^2 = y^3$ about the y-axis, solve for x in terms of y to obtain

$$x = \tfrac{1}{3}y^{3/2} = r(y). \qquad \text{Radius function}$$

So, the equation for this surface is

$$x^2 + z^2 = [r(y)]^2 \qquad \text{Revolved about the } y\text{-axis}$$
$$x^2 + z^2 = \left(\tfrac{1}{3}y^{3/2}\right)^2 \qquad \text{Substitute } \tfrac{1}{3}y^{3/2} \text{ for } r(y).$$
$$x^2 + z^2 = \tfrac{1}{9}y^3. \qquad \text{Equation of surface}$$

The graph is shown in Figure 11.63.

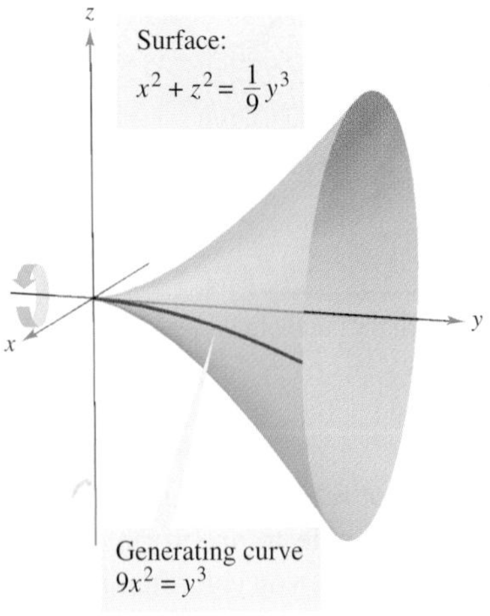

Figure 11.63

The generating curve for a surface of revolution is not unique. For instance, the surface

$$x^2 + z^2 = e^{-2y}$$

can be formed by revolving either the graph of $x = e^{-y}$ about the y-axis or the graph of $z = e^{-y}$ about the y-axis, as shown in Figure 11.64.

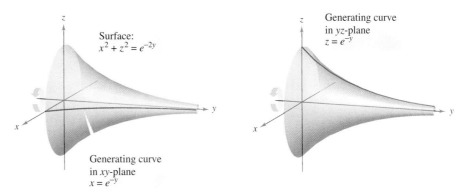

Figure 11.64

EXAMPLE 6 Finding a Generating Curve for a Surface of Revolution

Find a generating curve and the axis of revolution for the surface given by

$$x^2 + 3y^2 + z^2 = 9.$$

Solution You now know that the equation has one of the following forms.

$$x^2 + y^2 = [r(z)]^2 \qquad \text{Revolved about } z\text{-axis}$$
$$y^2 + z^2 = [r(x)]^2 \qquad \text{Revolved about } x\text{-axis}$$
$$x^2 + z^2 = [r(y)]^2 \qquad \text{Revolved about } y\text{-axis}$$

Because the coefficients of x^2 and z^2 are equal, you should choose the third form and write

$$x^2 + z^2 = 9 - 3y^2.$$

The y-axis is the axis of revolution. You can choose a generating curve from either of the following traces.

$$x^2 = 9 - 3y^2 \qquad \text{Trace in } xy\text{-plane}$$
$$z^2 = 9 - 3y^2 \qquad \text{Trace in } yz\text{-plane}$$

For example, using the first trace, the generating curve is the semiellipse given by

$$x = \sqrt{9 - 3y^2}. \qquad \text{Generating curve}$$

The graph of this surface is shown in Figure 11.65.

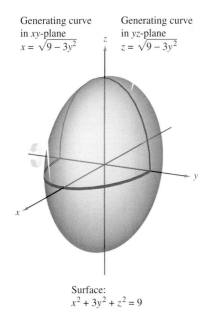

Figure 11.65

Exercises for Section 11.6

In Exercises 1–6, match the equation with its graph. [The graphs are labeled (a), (b), (c), (d), (e), and (f).]

(a)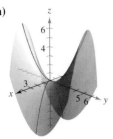

(b)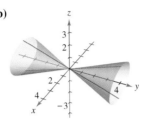

(c)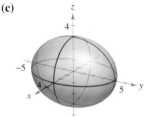

(d)

(e)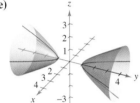

(f)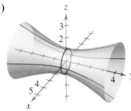

1. $\dfrac{x^2}{9} + \dfrac{y^2}{16} + \dfrac{z^2}{9} = 1$
2. $15x^2 - 4y^2 + 15z^2 = -4$
3. $4x^2 - y^2 + 4z^2 = 4$
4. $y^2 = 4x^2 + 9z^2$
5. $4x^2 - 4y + z^2 = 0$
6. $4x^2 - y^2 + 4z = 0$

In Exercises 7–16, describe and sketch the surface.

7. $z = 3$
8. $x = 4$
9. $y^2 + z^2 = 9$
10. $x^2 + z^2 = 25$
11. $x^2 - y = 0$
12. $y^2 + z = 4$
13. $4x^2 + y^2 = 4$
14. $y^2 - z^2 = 4$
15. $z - \sin y = 0$
16. $z - e^y = 0$

17. **Think About It** The four figures are graphs of the quadric surface $z = x^2 + y^2$. Match each of the four graphs with the point in space from which the paraboloid is viewed. The four points are $(0, 0, 20)$, $(0, 20, 0)$, $(20, 0, 0)$, and $(10, 10, 20)$.

(a)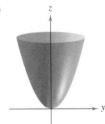

(b)

(c)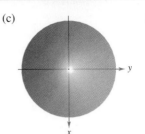

(d)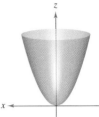

Figures for 17

18. Use a computer algebra system to graph a view of the cylinder $y^2 + z^2 = 4$ from each point.
    (a) $(10, 0, 0)$
    (b) $(0, 10, 0)$
    (c) $(10, 10, 10)$

In Exercises 19–30, identify and sketch the quadric surface. Use a computer algebra system to confirm your sketch.

19. $x^2 + \dfrac{y^2}{4} + z^2 = 1$
20. $\dfrac{x^2}{16} + \dfrac{y^2}{25} + \dfrac{z^2}{25} = 1$
21. $16x^2 - y^2 + 16z^2 = 4$
22. $z^2 - x^2 - \dfrac{y^2}{4} = 1$
23. $x^2 - y + z^2 = 0$
24. $z = x^2 + 4y^2$
25. $x^2 - y^2 + z = 0$
26. $3z = -y^2 + x^2$
27. $z^2 = x^2 + \dfrac{y^2}{4}$
28. $x^2 = 2y^2 + 2z^2$
29. $16x^2 + 9y^2 + 16z^2 - 32x - 36y + 36 = 0$
30. $9x^2 + y^2 - 9z^2 - 54x - 4y - 54z + 4 = 0$

In Exercises 31–40, use a computer algebra system to graph the surface. (*Hint:* It may be necessary to solve for z and acquire two equations to graph the surface.)

31. $z = 2 \sin x$
32. $z = x^2 + 0.5y^2$
33. $z^2 = x^2 + 4y^2$
34. $4y = x^2 + z^2$
35. $x^2 + y^2 = \left(\dfrac{2}{z}\right)^2$
36. $x^2 + y^2 = e^{-z}$
37. $z = 4 - \sqrt{|xy|}$
38. $z = \dfrac{-x}{8 + x^2 + y^2}$
39. $4x^2 - y^2 + 4z^2 = -16$
40. $9x^2 + 4y^2 - 8z^2 = 72$

In Exercises 41–44, sketch the region bounded by the graphs of the equations.

41. $z = 2\sqrt{x^2 + y^2}$, $z = 2$
42. $z = \sqrt{4 - x^2}$, $y = \sqrt{4 - x^2}$, $x = 0$, $y = 0$, $z = 0$
43. $x^2 + y^2 = 1$, $x + z = 2$, $z = 0$
44. $z = \sqrt{4 - x^2 - y^2}$, $y = 2z$, $z = 0$

In Exercises 45–50, find an equation for the surface of revolution generated by revolving the curve in the indicated coordinate plane about the given axis.

Equation of Curve	Coordinate Plane	Axis of Revolution
45. $z^2 = 4y$	yz-plane	y-axis
46. $z = 3y$	yz-plane	y-axis
47. $z = 2y$	yz-plane	z-axis
48. $2z = \sqrt{4 - x^2}$	xz-plane	x-axis
49. $xy = 2$	xy-plane	x-axis
50. $z = \ln y$	yz-plane	z-axis

In Exercises 51 and 52, find an equation of a generating curve given the equation of its surface of revolution.

51. $x^2 + y^2 - 2z = 0$ 52. $x^2 + z^2 = \cos^2 y$

Writing About Concepts

53. State the definition of a cylinder.
54. What is meant by the trace of a surface? How do you find a trace?
55. Identify the six quadric surfaces and give the standard form of each.
56. What does the equation $z = x^2$ represent in the xz-plane? What does it represent in three-space?

In Exercises 57 and 58, use the shell method to find the volume of the solid below the surface of revolution and above the xy-plane.

57. The curve $z = 4x - x^2$ in the xz-plane is revolved about the z-axis.
58. The curve $z = \sin y$ ($0 \le y \le \pi$) in the yz-plane is revolved about the z-axis.

In Exercises 59 and 60, analyze the trace when the surface $z = \frac{1}{2}x^2 + \frac{1}{4}y^2$ is intersected by the indicated planes.

59. Find the lengths of the major and minor axes and the coordinates of the foci of the ellipse generated when the surface is intersected by the planes given by
    (a) $z = 2$ and   (b) $z = 8$.
60. Find the coordinates of the focus of the parabola formed when the surface is intersected by the planes given by
    (a) $y = 4$ and   (b) $x = 2$.

In Exercises 61 and 62, find an equation of the surface satisfying the conditions, and identify the surface.

61. The set of all points equidistant from the point $(0, 2, 0)$ and the plane $y = -2$

62. The set of all points equidistant from the point $(0, 0, 4)$ and the xy-plane

63. *Geography* Because of the forces caused by its rotation, Earth is an oblate ellipsoid rather than a sphere. The equatorial radius is 3963 miles and the polar radius is 3950 miles. Find an equation of the ellipsoid. (Assume that the center of Earth is at the origin and that the trace formed by the plane $z = 0$ corresponds to the equator.)

64. *Machine Design* The top of a rubber bushing designed to absorb vibrations in an automobile is the surface of revolution generated by revolving the curve $z = \frac{1}{2}y^2 + 1$ ($0 \le y \le 2$) in the yz-plane about the z-axis.
    (a) Find an equation for the surface of revolution.
    (b) All measurements are in centimeters and the bushing is set on the xy-plane. Use the shell method to find its volume.
    (c) The bushing has a hole of diameter 1 centimeter through its center and parallel to the axis of revolution. Find the volume of the rubber bushing.

65. Determine the intersection of the hyperbolic paraboloid $z = y^2/b^2 - x^2/a^2$ with the plane $bx + ay - z = 0$. (Assume $a, b > 0$.)

66. Explain why the curve of intersection of the surfaces $x^2 + 3y^2 - 2z^2 + 2y = 4$ and $2x^2 + 6y^2 - 4z^2 - 3x = 2$ lies in a plane.

True or False? In Exercises 67 and 68, determine whether the statement is true or false. If it is false, explain why or give an example that shows it is false.

67. A sphere is an ellipsoid.
68. The generating curve for a surface of revolution is unique.

69. *Think About It* Three types of classic "topological" surfaces are shown below. The sphere and torus have both an "inside" and an "outside." Does the Klein bottle have both an inside and an outside? Explain.

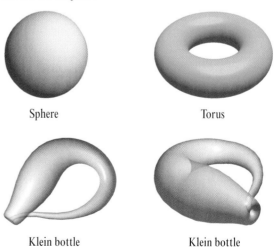

Sphere   Torus

Klein bottle   Klein bottle

Section 11.7 Cylindrical and Spherical Coordinates

- Use cylindrical coordinates to represent surfaces in space.
- Use spherical coordinates to represent surfaces in space.

Cylindrical Coordinates

You have already seen that some two-dimensional graphs are easier to represent in polar coordinates than in rectangular coordinates. A similar situation exists for surfaces in space. In this section, you will study two alternative space-coordinate systems. The first, the **cylindrical coordinate system,** is an extension of polar coordinates in the plane to three-dimensional space.

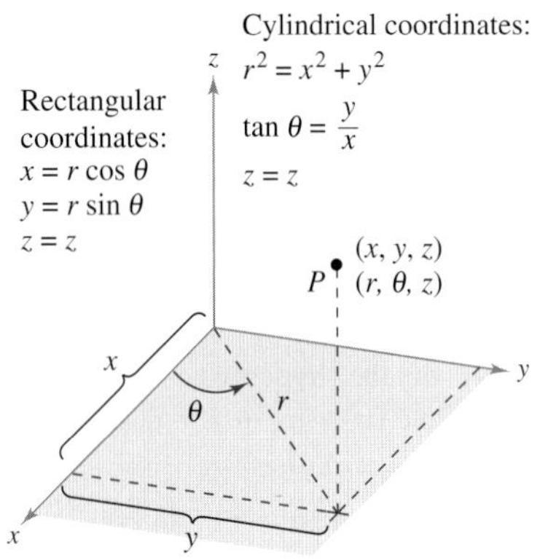

Rectangular coordinates:
$x = r\cos\theta$
$y = r\sin\theta$
$z = z$

Cylindrical coordinates:
$r^2 = x^2 + y^2$
$\tan\theta = \dfrac{y}{x}$
$z = z$

Figure 11.66

> **The Cylindrical Coordinate System**
>
> In a **cylindrical coordinate system,** a point P in space is represented by an ordered triple (r, θ, z).
>
> 1. (r, θ) is a polar representation of the projection of P in the xy-plane.
> 2. z is the directed distance from (r, θ) to P.

To convert from rectangular to cylindrical coordinates (or vice versa), use the following conversion guidelines for polar coordinates, as illustrated in Figure 11.66.

Cylindrical to rectangular:

$$x = r\cos\theta, \quad y = r\sin\theta, \quad z = z$$

Rectangular to cylindrical:

$$r^2 = x^2 + y^2, \quad \tan\theta = \frac{y}{x}, \quad z = z$$

The point $(0, 0, 0)$ is called the **pole.** Moreover, because the representation of a point in the polar coordinate system is not unique, it follows that the representation in the cylindrical coordinate system is also not unique.

EXAMPLE 1 Converting from Cylindrical to Rectangular Coordinates

Convert the point $(r, \theta, z) = \left(4, \dfrac{5\pi}{6}, 3\right)$ to rectangular coordinates.

Solution Using the cylindrical-to-rectangular conversion equations produces

$$x = 4\cos\frac{5\pi}{6} = 4\left(-\frac{\sqrt{3}}{2}\right) = -2\sqrt{3}$$

$$y = 4\sin\frac{5\pi}{6} = 4\left(\frac{1}{2}\right) = 2$$

$$z = 3.$$

So, in rectangular coordinates, the point is $(x, y, z) = \left(-2\sqrt{3}, 2, 3\right)$, as shown in Figure 11.67.

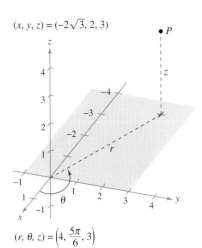

Figure 11.67

EXAMPLE 2 Converting from Rectangular to Cylindrical Coordinates

Convert the point $(x, y, z) = (1, \sqrt{3}, 2)$ to cylindrical coordinates.

Solution Use the rectangular-to-cylindrical conversion equations.

$$r = \pm\sqrt{1 + 3} = \pm 2$$

$$\tan \theta = \sqrt{3} \implies \theta = \arctan(\sqrt{3}) + n\pi = \frac{\pi}{3} + n\pi$$

$$z = 2$$

You have two choices for r and infinitely many choices for θ. As shown in Figure 11.68, two convenient representations of the point are

$$\left(2, \frac{\pi}{3}, 2\right) \qquad r > 0 \text{ and } \theta \text{ in Quadrant I}$$

$$\left(-2, \frac{4\pi}{3}, 2\right). \qquad r < 0 \text{ and } \theta \text{ in Quadrant III}$$

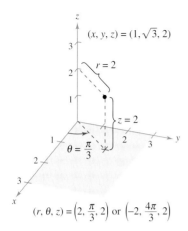

Figure 11.68

Cylindrical coordinates are especially convenient for representing cylindrical surfaces and surfaces of revolution with the z-axis as the axis of symmetry, as shown in Figure 11.69.

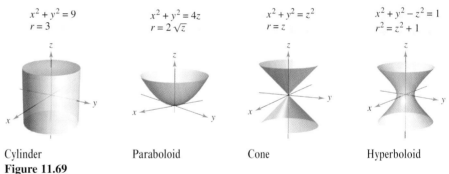

Cylinder   Paraboloid   Cone   Hyperboloid
Figure 11.69

Vertical planes containing the z-axis and horizontal planes also have simple cylindrical coordinate equations, as shown in Figure 11.70.

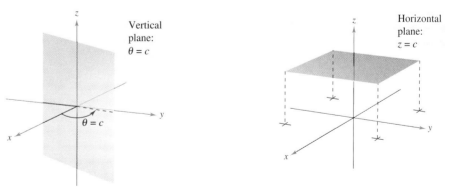

Figure 11.70

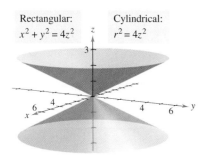

Rectangular:
$x^2 + y^2 = 4z^2$

Cylindrical:
$r^2 = 4z^2$

Figure 11.71

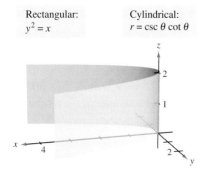

Rectangular:
$y^2 = x$

Cylindrical:
$r = \csc\theta \cot\theta$

Figure 11.72

EXAMPLE 3   Rectangular-to-Cylindrical Conversion

Find an equation in cylindrical coordinates for the surface represented by each rectangular equation.

a. $x^2 + y^2 = 4z^2$

b. $y^2 = x$

Solution

a. From the preceding section, you know that the graph $x^2 + y^2 = 4z^2$ is a "double-napped" cone with its axis along the z-axis, as shown in Figure 11.71. If you replace $x^2 + y^2$ with r^2, the equation in cylindrical coordinates is

$$x^2 + y^2 = 4z^2 \qquad \text{Rectangular equation}$$
$$r^2 = 4z^2. \qquad \text{Cylindrical equation}$$

b. The graph of the surface $y^2 = x$ is a parabolic cylinder with rulings parallel to the z-axis, as shown in Figure 11.72. By replacing y^2 with $r^2 \sin^2\theta$ and x with $r \cos\theta$, you obtain the following equation in cylindrical coordinates.

$$y^2 = x \qquad \text{Rectangular equation}$$
$$r^2 \sin^2\theta = r\cos\theta \qquad \text{Substitute } r\sin\theta \text{ for } y \text{ and } r\cos\theta \text{ for } x.$$
$$r(r\sin^2\theta - \cos\theta) = 0 \qquad \text{Collect terms and factor.}$$
$$r\sin^2\theta - \cos\theta = 0 \qquad \text{Divide each side by } r.$$
$$r = \frac{\cos\theta}{\sin^2\theta} \qquad \text{Solve for } r.$$
$$r = \csc\theta \cot\theta \qquad \text{Cylindrical equation}$$

Note that this equation includes a point for which $r = 0$, so nothing was lost by dividing each side by the factor r.

Converting from rectangular coordinates to cylindrical coordinates is more straightforward than converting from cylindrical coordinates to rectangular coordinates, as demonstrated in Example 4.

EXAMPLE 4   Cylindrical-to-Rectangular Conversion

Find an equation in rectangular coordinates for the surface represented by the cylindrical equation

$$r^2 \cos 2\theta + z^2 + 1 = 0.$$

Solution

$$r^2 \cos 2\theta + z^2 + 1 = 0 \qquad \text{Cylindrical equation}$$
$$r^2(\cos^2\theta - \sin^2\theta) + z^2 + 1 = 0 \qquad \text{Trigonometric identity}$$
$$r^2\cos^2\theta - r^2\sin^2\theta + z^2 = -1$$
$$x^2 - y^2 + z^2 = -1 \qquad \text{Replace } r\cos\theta \text{ with } x \text{ and } r\sin\theta \text{ with } y.$$
$$y^2 - x^2 - z^2 = 1 \qquad \text{Rectangular equation}$$

This is a hyperboloid of two sheets whose axis lies along the y-axis, as shown in Figure 11.73.

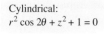

Cylindrical:
$r^2 \cos 2\theta + z^2 + 1 = 0$

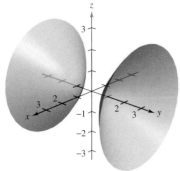

Rectangular:
$y^2 - x^2 - z^2 = 1$

Figure 11.73

Spherical Coordinates

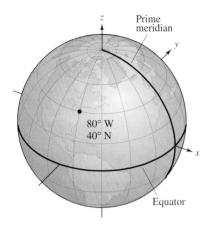

Figure 11.74

In the **spherical coordinate system,** each point is represented by an ordered triple: the first coordinate is a distance, and the second and third coordinates are angles. This system is similar to the latitude-longitude system used to identify points on the surface of Earth. For example, the point on the surface of Earth whose latitude is 40° North (of the equator) and whose longitude is 80° West (of the prime meridian) is shown in Figure 11.74. Assuming that the Earth is spherical and has a radius of 4000 miles, you would label this point as

$$(4000, -80°, 50°).$$

Radius — 80° clockwise from prime meridian — 50° down from North Pole

---

The Spherical Coordinate System

In a **spherical coordinate system,** a point P in space is represented by an ordered triple (ρ, θ, ϕ).

1. ρ is the distance between P and the origin, $\rho \geq 0$.
2. θ is the same angle used in cylindrical coordinates for $r \geq 0$.
3. ϕ is the angle *between* the positive z-axis and the line segment $\overrightarrow{OP}$, $0 \leq \phi \leq \pi$.

Note that the first and third coordinates, ρ and ϕ, are nonnegative. ρ is the lowercase Greek letter *rho*, and ϕ is the lowercase Greek letter *phi*.

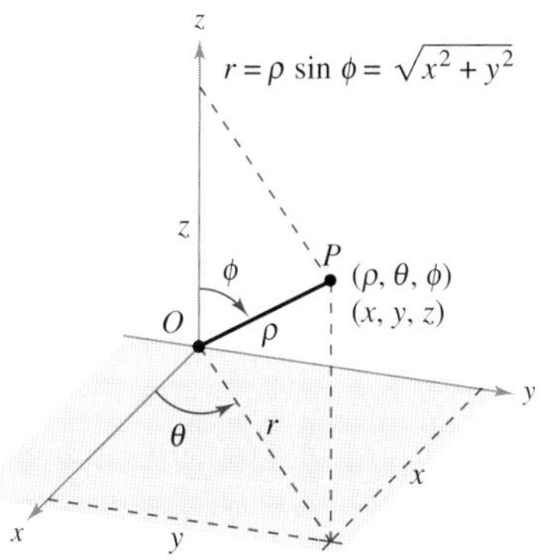

The relationship between rectangular and spherical coordinates is illustrated in Figure 11.75. To convert from one system to the other, use the following.

Spherical to rectangular:

$$x = \rho \sin \phi \cos \theta, \quad y = \rho \sin \phi \sin \theta, \quad z = \rho \cos \phi$$

Rectangular to spherical:

$$\rho^2 = x^2 + y^2 + z^2, \quad \tan \theta = \frac{y}{x}, \quad \phi = \arccos\left(\frac{z}{\sqrt{x^2 + y^2 + z^2}}\right)$$

To change coordinates between the cylindrical and spherical systems, use the following.

Spherical to cylindrical $(r \geq 0)$:

$$r^2 = \rho^2 \sin^2 \phi, \quad \theta = \theta, \quad z = \rho \cos \phi$$

Cylindrical to spherical $(r \geq 0)$:

$$\rho = \sqrt{r^2 + z^2}, \quad \theta = \theta, \quad \phi = \arccos\left(\frac{z}{\sqrt{r^2 + z^2}}\right)$$

Spherical coordinates
Figure 11.75

The spherical coordinate system is useful primarily for surfaces in space that have a *point* or *center* of symmetry. For example, Figure 11.76 shows three surfaces with simple spherical equations.

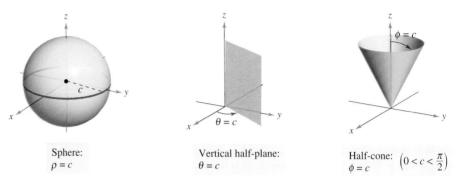

Sphere:
$\rho = c$

Vertical half-plane:
$\theta = c$

Half-cone: $\left(0 < c < \dfrac{\pi}{2}\right)$
$\phi = c$

Figure 11.76

EXAMPLE 5 Rectangular-to-Spherical Conversion

Find an equation in spherical coordinates for the surface represented by each rectangular equation.

a. Cone: $x^2 + y^2 = z^2$
b. Sphere: $x^2 + y^2 + z^2 - 4z = 0$

Solution

a. Making the appropriate replacements for x, y, and z in the given equation yields the following.

$$x^2 + y^2 = z^2$$
$$\rho^2 \sin^2 \phi \cos^2 \theta + \rho^2 \sin^2 \phi \sin^2 \theta = \rho^2 \cos^2 \phi$$
$$\rho^2 \sin^2 \phi (\cos^2 \theta + \sin^2 \theta) = \rho^2 \cos^2 \phi$$
$$\rho^2 \sin^2 \phi = \rho^2 \cos^2 \phi$$
$$\dfrac{\sin^2 \phi}{\cos^2 \phi} = 1 \qquad \rho \geq 0$$
$$\tan^2 \phi = 1 \qquad \phi = \pi/4 \text{ or } \phi = 3\pi/4$$

The equation $\phi = \pi/4$ represents the *upper* half-cone, and the equation $\phi = 3\pi/4$ represents the *lower* half-cone.

b. Because $\rho^2 = x^2 + y^2 + z^2$ and $z = \rho \cos \phi$, the given equation has the following spherical form.

$$\rho^2 - 4\rho \cos \phi = 0 \quad \Longrightarrow \quad \rho(\rho - 4 \cos \phi) = 0$$

Temporarily discarding the possibility that $\rho = 0$, you have the spherical equation

$$\rho - 4 \cos \phi = 0 \quad \text{or} \quad \rho = 4 \cos \phi.$$

Note that the solution set for this equation includes a point for which $\rho = 0$, so nothing is lost by discarding the factor ρ. The sphere represented by the equation $\rho = 4 \cos \phi$ is shown in Figure 11.77.

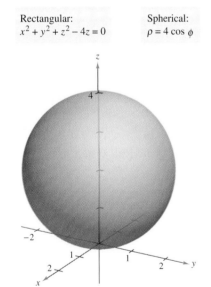

Rectangular:
$x^2 + y^2 + z^2 - 4z = 0$

Spherical:
$\rho = 4 \cos \phi$

Figure 11.77

Exercises for Section 11.7

See www.CalcChat.com for worked-out solutions to odd-numbered exercises.

In Exercises 1–6, convert the point from cylindrical coordinates to rectangular coordinates.

1. $(5, 0, 2)$
2. $(4, \pi/2, -2)$
3. $(2, \pi/3, 2)$
4. $(6, -\pi/4, 2)$
5. $(4, 7\pi/6, 3)$
6. $(1, 3\pi/2, 1)$

In Exercises 7–12, convert the point from rectangular coordinates to cylindrical coordinates.

7. $(0, 5, 1)$
8. $(2\sqrt{2}, -2\sqrt{2}, 4)$
9. $(1, \sqrt{3}, 4)$
10. $(2\sqrt{3}, -2, 6)$
11. $(2, -2, -4)$
12. $(-3, 2, -1)$

In Exercises 13–20, find an equation in cylindrical coordinates for the equation given in rectangular coordinates.

13. $z = 5$
14. $x = 4$
15. $x^2 + y^2 + z^2 = 10$
16. $z = x^2 + y^2 - 2$
17. $y = x^2$
18. $x^2 + y^2 = 8x$
19. $y^2 = 10 - z^2$
20. $x^2 + y^2 + z^2 - 3z = 0$

In Exercises 21–28, find an equation in rectangular coordinates for the equation given in cylindrical coordinates, and sketch its graph.

21. $r = 2$
22. $z = 2$
23. $\theta = \pi/6$
24. $r = \frac{1}{2}z$
25. $r = 2\sin\theta$
26. $r = 2\cos\theta$
27. $r^2 + z^2 = 4$
28. $z = r^2 \cos^2\theta$

In Exercises 29–34, convert the point from rectangular coordinates to spherical coordinates.

29. $(4, 0, 0)$
30. $(1, 1, 1)$
31. $(-2, 2\sqrt{3}, 4)$
32. $(2, 2, 4\sqrt{2})$
33. $(\sqrt{3}, 1, 2\sqrt{3})$
34. $(-4, 0, 0)$

In Exercises 35–40, convert the point from spherical coordinates to rectangular coordinates.

35. $(4, \pi/6, \pi/4)$
36. $(12, 3\pi/4, \pi/9)$
37. $(12, -\pi/4, 0)$
38. $(9, \pi/4, \pi)$
39. $(5, \pi/4, 3\pi/4)$
40. $(6, \pi, \pi/2)$

In Exercises 41–48, find an equation in spherical coordinates for the equation given in rectangular coordinates.

41. $y = 3$
42. $z = 2$
43. $x^2 + y^2 + z^2 = 36$
44. $x^2 + y^2 - 3z^2 = 0$
45. $x^2 + y^2 = 9$
46. $x = 10$
47. $x^2 + y^2 = 2z^2$
48. $x^2 + y^2 + z^2 - 9z = 0$

In Exercises 49–56, find an equation in rectangular coordinates for the equation given in spherical coordinates, and sketch its graph.

49. $\rho = 2$
50. $\theta = \frac{3\pi}{4}$
51. $\phi = \frac{\pi}{6}$
52. $\phi = \frac{\pi}{2}$
53. $\rho = 4\cos\phi$
54. $\rho = 2\sec\phi$
55. $\rho = \csc\phi$
56. $\rho = 4\csc\phi\sec\theta$

In Exercises 57–64, convert the point from cylindrical coordinates to spherical coordinates.

57. $(4, \pi/4, 0)$
58. $(3, -\pi/4, 0)$
59. $(4, \pi/2, 4)$
60. $(2, 2\pi/3, -2)$
61. $(4, -\pi/6, 6)$
62. $(-4, \pi/3, 4)$
63. $(12, \pi, 5)$
64. $(4, \pi/2, 3)$

In Exercises 65–72, convert the point from spherical coordinates to cylindrical coordinates.

65. $(10, \pi/6, \pi/2)$
66. $(4, \pi/18, \pi/2)$
67. $(36, \pi, \pi/2)$
68. $(18, \pi/3, \pi/3)$
69. $(6, -\pi/6, \pi/3)$
70. $(5, -5\pi/6, \pi)$
71. $(8, 7\pi/6, \pi/6)$
72. $(7, \pi/4, 3\pi/4)$

In Exercises 73–86, use a computer algebra system or graphing utility to convert the point from one system to another among the rectangular, cylindrical, and spherical coordinate systems.

	Rectangular	Cylindrical	Spherical
73.	$(4, 6, 3)$		
74.	$(6, -2, -3)$		
75.		$(5, \pi/9, 8)$	
76.		$(10, -0.75, 6)$	
77.			$(20, 2\pi/3, \pi/4)$
78.			$(7.5, 0.25, 1)$
79.	$(3, -2, 2)$		
80.	$(3\sqrt{2}, 3\sqrt{2}, -3)$		
81.	$(5/2, 4/3, -3/2)$		
82.	$(0, -5, 4)$		
83.		$(5, 3\pi/4, -5)$	
84.		$(-2, 11\pi/6, 3)$	
85.			$(-3.5, 2.5, 6)$
86.			$(8.25, 1.3, -4)$

In Exercises 87–92, match the equation (written in terms of cylindrical or spherical coordinates) with its graph. [The graphs are labeled (a), (b), (c), (d), (e), and (f).]

(a)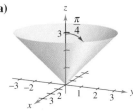

(b)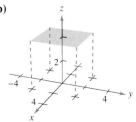

(c)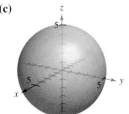

(d)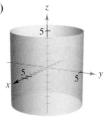

(e)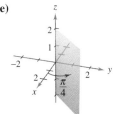

(f)

87. $r = 5$
88. $\theta = \dfrac{\pi}{4}$
89. $\rho = 5$
90. $\phi = \dfrac{\pi}{4}$
91. $r^2 = z$
92. $\rho = 4 \sec \phi$

Writing About Concepts

93. Give the equations for the coordinate conversion from rectangular to cylindrical coordinates and vice versa.
94. For constants a, b, and c, describe the graphs of the equations $r = a$, $\theta = b$, and $z = c$ in cylindrical coordinates.
95. Give the equations for the coordinate conversion from rectangular to spherical coordinates and vice versa.
96. For constants a, b, and c, describe the graphs of the equations $\rho = a$, $\theta = b$, and $\phi = c$ in spherical coordinates.

In Exercises 97–104, convert the rectangular equation to an equation in (a) cylindrical coordinates and (b) spherical coordinates.

97. $x^2 + y^2 + z^2 = 16$
98. $4(x^2 + y^2) = z^2$
99. $x^2 + y^2 + z^2 - 2z = 0$
100. $x^2 + y^2 = z$
101. $x^2 + y^2 = 4y$
102. $x^2 + y^2 = 16$
103. $x^2 - y^2 = 9$
104. $y = 4$

In Exercises 105–108, sketch the solid that has the given description in cylindrical coordinates.

105. $0 \leq \theta \leq \pi/2, 0 \leq r \leq 2, 0 \leq z \leq 4$
106. $-\pi/2 \leq \theta \leq \pi/2, 0 \leq r \leq 3, 0 \leq z \leq r \cos \theta$
107. $0 \leq \theta \leq 2\pi, 0 \leq r \leq a, r \leq z \leq a$
108. $0 \leq \theta \leq 2\pi, 2 \leq r \leq 4, z^2 \leq -r^2 + 6r - 8$

In Exercises 109–112, sketch the solid that has the given description in spherical coordinates.

109. $0 \leq \theta \leq 2\pi, 0 \leq \phi \leq \pi/6, 0 \leq \rho \leq a \sec \phi$
110. $0 \leq \theta \leq 2\pi, \pi/4 \leq \phi \leq \pi/2, 0 \leq \rho \leq 1$
111. $0 \leq \theta \leq \pi/2, 0 \leq \phi \leq \pi/2, 0 \leq \rho \leq 2$
112. $0 \leq \theta \leq \pi, 0 \leq \phi \leq \pi/2, 1 \leq \rho \leq 3$

Think About It In Exercises 113–118, find inequalities that describe the solid, and state the coordinate system used. Position the solid on the coordinate system such that the inequalities are as simple as possible.

113. A cube with each edge 10 centimeters long
114. A cylindrical shell 8 meters long with an inside diameter of 0.75 meter and an outside diameter of 1.25 meters
115. A spherical shell with inside and outside radii of 4 inches and 6 inches, respectively
116. The solid that remains after a hole 1 inch in diameter is drilled through the center of a sphere 6 inches in diameter
117. The solid inside both $x^2 + y^2 + z^2 = 9$ and $\left(x - \tfrac{3}{2}\right)^2 + y^2 = \tfrac{9}{4}$
118. The solid between the spheres $x^2 + y^2 + z^2 = 4$ and $x^2 + y^2 + z^2 = 9$, and inside the cone $z^2 = x^2 + y^2$

True or False? In Exercises 119–122, determine whether the statement is true or false. If it is false, explain why or give an example that shows it is false.

119. In spherical coordinates, the equation $\theta = c$ represents an entire plane.
120. The equations $\rho = 2$ and $x^2 + y^2 + z^2 = 4$ represent the same surface.
121. The cylindrical coordinates of a point (x, y, z) are unique.
122. The spherical coordinates of a point (x, y, z) are unique.
123. Identify the curve of intersection of the surfaces (in cylindrical coordinates) $z = \sin \theta$ and $r = 1$.
124. Identify the curve of intersection of the surfaces (in spherical coordinates) $\rho = 2 \sec \phi$ and $\rho = 4$.

Review Exercises for Chapter 11

See www.CalcChat.com for worked-out solutions to odd-numbered exercises.

In Exercises 1 and 2, let $\mathbf{u} = \overrightarrow{PQ}$ and $\mathbf{v} = \overrightarrow{PR}$, and find (a) the component forms of $\mathbf{u}$ and $\mathbf{v}$, (b) the magnitude of $\mathbf{v}$, and (c) $2\mathbf{u} + \mathbf{v}$.

1. $P = (1, 2), Q = (4, 1), R = (5, 4)$
2. $P = (-2, -1), Q = (5, -1), R = (2, 4)$

In Exercises 3 and 4, find the component form of $\mathbf{v}$ given its magnitude and the angle it makes with the positive x-axis.

3. $\|\mathbf{v}\| = 8$, $\theta = 120°$
4. $\|\mathbf{v}\| = \frac{1}{2}$, $\theta = 225°$

5. Find the coordinates of the point in the xy-plane four units to the right of the xz-plane and five units behind the yz-plane.

6. Find the coordinates of the point located on the y-axis and seven units to the left of the xz-plane.

In Exercises 7 and 8, determine the location of a point (x, y, z) that satisfies the condition.

7. $yz > 0$
8. $xy < 0$

In Exercises 9 and 10, find the standard equation of the sphere.

9. Center: $(3, -2, 6)$; Diameter: 15
10. Endpoints of a diameter: $(0, 0, 4), (4, 6, 0)$

In Exercises 11 and 12, complete the square to write the equation of the sphere in standard form. Find the center and radius.

11. $x^2 + y^2 + z^2 - 4x - 6y + 4 = 0$
12. $x^2 + y^2 + z^2 - 10x + 6y - 4z + 34 = 0$

In Exercises 13 and 14, the initial and terminal points of a vector are given. Sketch the directed line segment and find the component form of the vector.

13. Initial point: $(2, -1, 3)$
    Terminal point: $(4, 4, -7)$
14. Initial point: $(6, 2, 0)$
    Terminal point: $(3, -3, 8)$

In Exercises 15 and 16, use vectors to determine whether the points are collinear.

15. $(3, 4, -1), (-1, 6, 9), (5, 3, -6)$
16. $(5, -4, 7), (8, -5, 5), (11, 6, 3)$

17. Find a unit vector in the direction of $\mathbf{u} = \langle 2, 3, 5 \rangle$.
18. Find the vector $\mathbf{v}$ of magnitude 8 in the direction $\langle 6, -3, 2 \rangle$.

In Exercises 19 and 20, let $\mathbf{u} = \overrightarrow{PQ}$ and $\mathbf{v} = \overrightarrow{PR}$, and find (a) the component forms of $\mathbf{u}$ and $\mathbf{v}$, (b) $\mathbf{u} \cdot \mathbf{v}$, and (c) $\mathbf{v} \cdot \mathbf{v}$.

19. $P = (5, 0, 0), Q = (4, 4, 0), R = (2, 0, 6)$
20. $P = (2, -1, 3), Q = (0, 5, 1), R = (5, 5, 0)$

In Exercises 21 and 22, determine whether $\mathbf{u}$ and $\mathbf{v}$ are orthogonal, parallel, or neither.

21. $\mathbf{u} = \langle 7, -2, 3 \rangle$
    $\mathbf{v} = \langle -1, 4, 5 \rangle$
22. $\mathbf{u} = \langle -4, 3, -6 \rangle$
    $\mathbf{v} = \langle 16, -12, 24 \rangle$

In Exercises 23–26, find the angle θ between the vectors.

23. $\mathbf{u} = 5[\cos(3\pi/4)\mathbf{i} + \sin(3\pi/4)\mathbf{j}]$
    $\mathbf{v} = 2[\cos(2\pi/3)\mathbf{i} + \sin(2\pi/3)\mathbf{j}]$
24. $\mathbf{u} = \langle 4, -1, 5 \rangle$, $\mathbf{v} = \langle 3, 2, -2 \rangle$
25. $\mathbf{u} = \langle 10, -5, 15 \rangle$, $\mathbf{v} = \langle -2, 1, -3 \rangle$
26. $\mathbf{u} = \langle 1, 0, -3 \rangle$, $\mathbf{v} = \langle 2, -2, 1 \rangle$

27. Find two vectors in opposite directions that are orthogonal to the vector $\mathbf{u} = \langle 5, 6, -3 \rangle$.

28. *Work* An object is pulled 8 feet across a floor using a force of 75 pounds. The direction of the force is 30° above the horizontal. Find the work done.

In Exercises 29–32, let $\mathbf{u} = \langle 3, -2, 1 \rangle$, $\mathbf{v} = \langle 2, -4, -3 \rangle$, and $\mathbf{w} = \langle -1, 2, 2 \rangle$.

29. Show that $\mathbf{u} \cdot \mathbf{u} = \|\mathbf{u}\|^2$.
30. Find the angle between $\mathbf{u}$ and $\mathbf{v}$.
31. Determine the projection of $\mathbf{w}$ onto $\mathbf{u}$.
32. Find the work done in moving an object along the vector $\mathbf{u}$ if the applied force is $\mathbf{w}$.

In Exercises 33–38, let $\mathbf{u} = \langle 3, -2, 1 \rangle$, $\mathbf{v} = \langle 2, -4, -3 \rangle$, and $\mathbf{w} = \langle -1, 2, 2 \rangle$.

33. Determine a unit vector perpendicular to the plane containing $\mathbf{v}$ and $\mathbf{w}$.
34. Show that $\mathbf{u} \times \mathbf{v} = -(\mathbf{v} \times \mathbf{u})$.
35. Find the volume of the solid whose edges are $\mathbf{u}$, $\mathbf{v}$, and $\mathbf{w}$.
36. Show that $\mathbf{u} \times (\mathbf{v} + \mathbf{w}) = (\mathbf{u} \times \mathbf{v}) + (\mathbf{u} \times \mathbf{w})$.
37. Find the area of the parallelogram with adjacent sides $\mathbf{u}$ and $\mathbf{v}$.
38. Find the area of the triangle with adjacent sides $\mathbf{v}$ and $\mathbf{w}$.

39. *Torque* The specifications for a tractor state that the torque on a bolt with head size $\frac{7}{8}$ inch cannot exceed 200 foot-pounds. Determine the maximum force $\|\mathbf{F}\|$ that can be applied to the wrench in the figure.

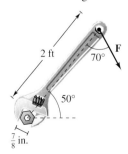

40. Volume Use the triple scalar product to find the volume of the parallelepiped having adjacent edges $\mathbf{u} = 2\mathbf{i} + \mathbf{j}$, $\mathbf{v} = 2\mathbf{j} + \mathbf{k}$, and $\mathbf{w} = -\mathbf{j} + 2\mathbf{k}$.

In Exercises 41 and 42, find sets of (a) parametric equations and (b) symmetric equations of the line through the two points. (For each line, write the direction numbers as integers.)

41. $(3, 0, 2)$, $(9, 11, 6)$ **42.** $(-1, 4, 3)$, $(8, 10, 5)$

In Exercises 43–46, find (a) a set of parametric equations and (b) a set of symmetric equations for the line.

43. The line passes through the point $(1, 2, 3)$ and is perpendicular to the xz-plane.

44. The line passes through the point $(1, 2, 3)$ and is parallel to the line given by $x = y = z$.

45. The intersection of the planes $3x - 3y - 7z = -4$ and $x - y + 2z = 3$

46. The line passes through the point $(0, 1, 4)$ and is perpendicular to $\mathbf{u} = \langle 2, -5, 1 \rangle$ and $\mathbf{v} = \langle -3, 1, 4 \rangle$.

In Exercises 47–50, find an equation of the plane.

47. The plane passes through $(-3, -4, 2)$, $(-3, 4, 1)$, and $(1, 1, -2)$.

48. The plane passes through the point $(-2, 3, 1)$ and is perpendicular to $\mathbf{n} = 3\mathbf{i} - \mathbf{j} + \mathbf{k}$.

49. The plane contains the lines given by

$$\frac{x-1}{-2} = y = z + 1$$

and

$$\frac{x+1}{-2} = y - 1 = z - 2.$$

50. The plane passes through the points $(5, 1, 3)$ and $(2, -2, 1)$ and is perpendicular to the plane $2x + y - z = 4$.

51. Find the distance between the point $(1, 0, 2)$ and the plane $2x - 3y + 6z = 6$.

52. Find the distance between the point $(3, -2, 4)$ and the plane $2x - 5y + z = 10$.

53. Find the distance between the planes $5x - 3y + z = 2$ and $5x - 3y + z = -3$.

54. Find the distance between the point $(-5, 1, 3)$ and the line given by $x = 1 + t$, $y = 3 - 2t$, and $z = 5 - t$.

In Exercises 55–64, describe and sketch the surface.

55. $x + 2y + 3z = 6$
56. $y = z^2$
57. $y = \frac{1}{2}z$
58. $y = \cos z$

59. $\dfrac{x^2}{16} + \dfrac{y^2}{9} + z^2 = 1$

60. $16x^2 + 16y^2 - 9z^2 = 0$

61. $\dfrac{x^2}{16} - \dfrac{y^2}{9} + z^2 = -1$

62. $\dfrac{x^2}{25} + \dfrac{y^2}{4} - \dfrac{z^2}{100} = 1$

63. $x^2 + z^2 = 4$

64. $y^2 + z^2 = 16$

65. Find an equation of a generating curve of the surface of revolution $y^2 + z^2 - 4x = 0$.

66. Find an equation for the surface of revolution generated by revolving the curve $z^2 = 2y$ in the yz-plane about the y-axis.

In Exercises 67 and 68, convert the point from rectangular coordinates to (a) cylindrical coordinates and (b) spherical coordinates.

67. $\left(-2\sqrt{2}, 2\sqrt{2}, 2\right)$ **68.** $\left(\dfrac{\sqrt{3}}{4}, \dfrac{3}{4}, \dfrac{3\sqrt{3}}{2}\right)$

In Exercises 69 and 70, convert the point from cylindrical coordinates to spherical coordinates.

69. $\left(100, -\dfrac{\pi}{6}, 50\right)$ **70.** $\left(81, -\dfrac{5\pi}{6}, 27\sqrt{3}\right)$

In Exercises 71 and 72, convert the point from spherical coordinates to cylindrical coordinates.

71. $\left(25, -\dfrac{\pi}{4}, \dfrac{3\pi}{4}\right)$

72. $\left(12, -\dfrac{\pi}{2}, \dfrac{2\pi}{3}\right)$

In Exercises 73 and 74, convert the rectangular equation to an equation in (a) cylindrical coordinates and (b) spherical coordinates.

73. $x^2 - y^2 = 2z$

74. $x^2 + y^2 + z^2 = 16$

In Exercises 75 and 76, find an equation in rectangular coordinates for the equation given in cylindrical coordinates, and sketch its graph.

75. $r = 4 \sin \theta$ **76.** $z = 4$

In Exercises 77 and 78, find an equation in rectangular coordinates for the equation given in spherical coordinates, and sketch its graph.

77. $\theta = \dfrac{\pi}{4}$ **78.** $\rho = 2 \cos \theta$

P.S. Problem Solving

1. Using vectors, prove the Law of Sines: If **a**, **b**, and **c** are the three sides of the triangle shown in the figure, then

   $$\frac{\sin A}{\|\mathbf{a}\|} = \frac{\sin B}{\|\mathbf{b}\|} = \frac{\sin C}{\|\mathbf{c}\|}.$$

   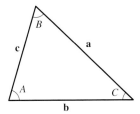

2. Consider the function $f(x) = \int_0^x \sqrt{t^4 + 1}\, dt$.

   (a) Use a graphing utility to graph the function on the interval $-2 \le x \le 2$.

   (b) Find a unit vector parallel to the graph of f at the point $(0, 0)$.

   (c) Find a unit vector perpendicular to the graph of f at the point $(0, 0)$.

   (d) Find the parametric equations of the tangent line to the graph of f at the point $(0, 0)$.

3. Using vectors, prove that the line segments joining the midpoints of the sides of a parallelogram form a parallelogram (see figure).

   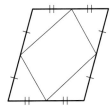

4. Using vectors, prove that the diagonals of a rhombus are perpendicular (see figure).

   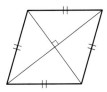

5. (a) Find the shortest distance between the point $Q(2, 0, 0)$ and the line determined by the points $P_1(0, 0, 1)$ and $P_2(0, 1, 2)$.

   (b) Find the shortest distance between the point $Q(2, 0, 0)$ and the line segment joining the points $P_1(0, 0, 1)$ and $P_2(0, 1, 2)$.

6. Let P_0 be a point in the plane with normal vector **n**. Describe the set of points P in the plane for which $(\mathbf{n} + \overrightarrow{PP_0})$ is orthogonal to $(\mathbf{n} - \overrightarrow{PP_0})$.

7. (a) Find the volume of the solid bounded below by the paraboloid $z = x^2 + y^2$ and above by the plane $z = 1$.

   (b) Find the volume of the solid bounded below by the elliptic paraboloid $z = \dfrac{x^2}{a^2} + \dfrac{y^2}{b^2}$ and above by the plane $z = k$, where $k > 0$.

   (c) Show that the volume of the solid in part (b) is equal to one-half the product of the area of the base times the altitude, as shown in the figure.

   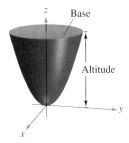

8. (a) Use the disk method to find the volume of the sphere $x^2 + y^2 + z^2 = r^2$.

   (b) Find the volume of the ellipsoid $\dfrac{x^2}{a^2} + \dfrac{y^2}{b^2} + \dfrac{z^2}{c^2} = 1$.

9. Sketch the graph of each equation given in spherical coordinates.

   (a) $\rho = 2 \sin \phi$

   (b) $\rho = 2 \cos \phi$

10. Sketch the graph of each equation given in cylindrical coordinates.

    (a) $r = 2 \cos \theta$

    (b) $z = r^2 \cos 2\theta$

11. Prove the following property of the cross product.

    $(\mathbf{u} \times \mathbf{v}) \times (\mathbf{w} \times \mathbf{z}) = (\mathbf{u} \times \mathbf{v} \cdot \mathbf{z})\mathbf{w} - (\mathbf{u} \times \mathbf{v} \cdot \mathbf{w})\mathbf{z}$

12. Consider the line given by the parametric equations

    $x = -t + 3, \quad y = \tfrac{1}{2}t + 1, \quad z = 2t - 1$

    and the point $(4, 3, s)$ for any real number s.

    (a) Write the distance between the point and the line as a function of s.

    (b) Use a graphing utility to graph the function in part (a). Use the graph to find the value of s such that the distance between the point and the line is minimum.

    (c) Use the *zoom* feature of a graphing utility to zoom out several times on the graph in part (b). Does it appear that the graph has slant asymptotes? Explain. If it appears to have slant asymptotes, find them.

13. A tetherball weighing 1 pound is pulled outward from the pole by a horizontal force **u** until the rope makes an angle of θ degrees with the pole (see figure).

(a) Determine the resulting tension in the rope and the magnitude of **u** when $\theta = 30°$.

(b) Write the tension T in the rope and the magnitude of **u** as functions of θ. Determine the domains of the functions.

(c) Use a graphing utility to complete the table.

θ	0°	10°	20°	30°	40°	50°	60°
T							
$\|\mathbf{u}\|$							

(d) Use a graphing utility to graph the two functions for $0° \leq \theta \leq 60°$.

(e) Compare T and $\|\mathbf{u}\|$ as θ increases.

(f) Find (if possible) $\lim_{\theta \to \pi/2^-} T$ and $\lim_{\theta \to \pi/2^-} \|\mathbf{u}\|$. Are the results what you expected? Explain.

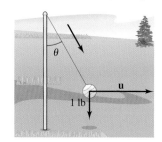
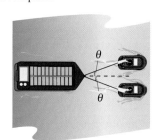

Figure for 13 **Figure for 14**

 14. A loaded barge is being towed by two tugboats, and the magnitude of the resultant is 6000 pounds directed along the axis of the barge (see figure). Each towline makes an angle of θ degrees with the axis of the barge.

(a) Find the tension in the towlines if $\theta = 20°$.

(b) Write the tension T of each line as a function of θ. Determine the domain of the function.

(c) Use a graphing utility to complete the table.

θ	10°	20°	30°	40°	50°	60°
T						

(d) Use a graphing utility to graph the tension function.

(e) Explain why the tension increases as θ increases.

15. Consider the vectors $\mathbf{u} = \langle \cos\alpha, \sin\alpha, 0 \rangle$ and $\mathbf{v} = \langle \cos\beta, \sin\beta, 0 \rangle$, where $\alpha > \beta$. Find the cross product of the vectors and use the result to prove the identity

$$\sin(\alpha - \beta) = \sin\alpha\cos\beta - \cos\alpha\sin\beta.$$

16. Los Angeles is located at 34.05° North latitude and 118.24° West longitude, and Rio de Janeiro, Brazil is located at 22.90° South latitude and 43.23° West longitude (see figure). Assume that Earth is spherical and has a radius of 4000 miles.

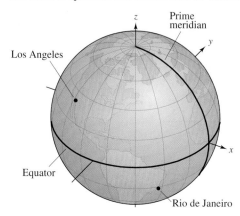

(a) Find the spherical coordinates for the location of each city.

(b) Find the rectangular coordinates for the location of each city.

(c) Find the angle (in radians) between the vectors from the center of Earth to each city.

(d) Find the great-circle distance s between the cities. (Hint: $s = r\theta$)

(e) Repeat parts (a)–(d) for the cities of Boston, located at 42.36° North latitude and 71.06° West longitude, and Honolulu, located at 21.31° North latitude and 157.86° West longitude.

17. Consider the plane that passes through the points P, R, and S. Show that the distance from a point Q to this plane is

$$\text{Distance} = \frac{|\mathbf{u} \cdot (\mathbf{v} \times \mathbf{w})|}{\|\mathbf{u} \times \mathbf{v}\|}$$

where $\mathbf{u} = \overrightarrow{PR}$, $\mathbf{v} = \overrightarrow{PS}$, and $\mathbf{w} = \overrightarrow{PQ}$.

18. Show that the distance between the parallel planes $ax + by + cz + d_1 = 0$ and $ax + by + cz + d_2 = 0$ is

$$\text{Distance} = \frac{|d_1 - d_2|}{\sqrt{a^2 + b^2 + c^2}}.$$

19. Show that the curve of intersection of the plane $z = 2y$ and the cylinder $x^2 + y^2 = 1$ is an ellipse.

20. Read the article "Tooth Tables: Solution of a Dental Problem by Vector Algebra" by Gary Hosler Meisters in *Mathematics Magazine*. (To view this article, go to the website www.matharticles.com.) Then write a paragraph explaining how vectors and vector algebra can be used in the construction of dental inlays.

Appendices

Appendix A   Proofs of Selected Theorems   A2

Appendix B   Integration Tables   A20

The remaining appendices are located on the website that accompanies this text at *college.hmco.com*.

Appendix C   Additional Topics in Differential Equations
- C.1   Exact First-Order Equations
- C.2   Second-Order Homogeneous Linear Equations
- C.3   Second-Order Nonhomogeneous Linear Equations
- C.4   Series Solutions of Differential Equations

Appendix D   Precalculus Review
- D.1   Real Numbers and the Real Number Line
- D.2   The Cartesian Plane
- D.3   Review of Trigonometric Functions

Appendix E   Rotation and the General Second-Degree Equation

Appendix F   Complex Numbers

Appendix G   Business and Economic Applications

A Proofs of Selected Theorems

> **THEOREM 1.2 Properties of Limits (Properties 2, 3, 4, and 5) (page 59)**
>
> Let b and c be real numbers, let n be a positive integer, and let f and g be functions with the following limits.
>
> $$\lim_{x \to c} f(x) = L \quad \text{and} \quad \lim_{x \to c} g(x) = K$$
>
> **2. Sum or difference:** $\displaystyle\lim_{x \to c} [f(x) \pm g(x)] = L \pm K$
>
> **3. Product:** $\displaystyle\lim_{x \to c} [f(x)g(x)] = LK$
>
> **4. Quotient:** $\displaystyle\lim_{x \to c} \frac{f(x)}{g(x)} = \frac{L}{K}, \quad$ provided $K \neq 0$
>
> **5. Power:** $\displaystyle\lim_{x \to c} [f(x)]^n = L^n$

Proof To prove Property 2, choose $\varepsilon > 0$. Because $\varepsilon/2 > 0$, you know that there exists $\delta_1 > 0$ such that $0 < |x - c| < \delta_1$ implies $|f(x) - L| < \varepsilon/2$. You also know that there exists $\delta_2 > 0$ such that $0 < |x - c| < \delta_2$ implies $|g(x) - K| < \varepsilon/2$. Let δ be the smaller of δ_1 and δ_2; then $0 < |x - c| < \delta$ implies that

$$|f(x) - L| < \frac{\varepsilon}{2} \quad \text{and} \quad |g(x) - K| < \frac{\varepsilon}{2}.$$

So, you can apply the triangle inequality to conclude that

$$|[f(x) + g(x)] - (L + K)| \leq |f(x) - L| + |g(x) - K| < \frac{\varepsilon}{2} + \frac{\varepsilon}{2} = \varepsilon$$

which implies that

$$\lim_{x \to c} [f(x) + g(x)] = L + K = \lim_{x \to c} f(x) + \lim_{x \to c} g(x).$$

The proof that

$$\lim_{x \to c} [f(x) - g(x)] = L - K$$

is similar.

To prove Property 3, given that

$$\lim_{x \to c} f(x) = L \quad \text{and} \quad \lim_{x \to c} g(x) = K$$

you can write

$$f(x)g(x) = [f(x) - L][g(x) - K] + [Lg(x) + Kf(x)] - LK.$$

Because the limit of $f(x)$ is L, and the limit of $g(x)$ is K, you have

$$\lim_{x \to c} [f(x) - L] = 0 \quad \text{and} \quad \lim_{x \to c} [g(x) - K] = 0.$$

Let $0 < \varepsilon < 1$. Then there exists $\delta > 0$ such that if $0 < |x - c| < \delta$, then
$$|f(x) - L - 0| < \varepsilon \quad \text{and} \quad |g(x) - K - 0| < \varepsilon$$
which implies that
$$|[f(x) - L][g(x) - K] - 0| = |f(x) - L||g(x) - K| < \varepsilon\varepsilon < \varepsilon.$$
So,
$$\lim_{x \to c} [f(x) - L][g(x) - K] = 0.$$
Furthermore, by Property 1, you have
$$\lim_{x \to c} Lg(x) = LK \quad \text{and} \quad \lim_{x \to c} Kf(x) = KL.$$
Finally, by Property 2, you obtain
$$\lim_{x \to c} f(x)g(x) = \lim_{x \to c} [f(x) - L][g(x) - K] + \lim_{x \to c} Lg(x) + \lim_{x \to c} Kf(x) - \lim_{x \to c} LK$$
$$= 0 + LK + KL - LK$$
$$= LK.$$

To prove Property 4, note that it is sufficient to prove that
$$\lim_{x \to c} \frac{1}{g(x)} = \frac{1}{K}.$$

Then you can use Property 3 to write
$$\lim_{x \to c} \frac{f(x)}{g(x)} = \lim_{x \to c} f(x) \frac{1}{g(x)} = \lim_{x \to c} f(x) \cdot \lim_{x \to c} \frac{1}{g(x)} = \frac{L}{K}.$$

Let $\varepsilon > 0$. Because $\lim_{x \to c} g(x) = K$, there exists $\delta_1 > 0$ such that if
$$0 < |x - c| < \delta_1, \text{ then } |g(x) - K| < \frac{|K|}{2}$$
which implies that
$$|K| = |g(x) + [|K| - g(x)]| \leq |g(x)| + ||K| - g(x)| < |g(x)| + \frac{|K|}{2}.$$

That is, for $0 < |x - c| < \delta_1$,
$$\frac{|K|}{2} < |g(x)| \quad \text{or} \quad \frac{1}{|g(x)|} < \frac{2}{|K|}.$$

Similarly, there exists a $\delta_2 > 0$ such that if $0 < |x - c| < \delta_2$, then
$$|g(x) - K| < \frac{|K|^2}{2} \varepsilon.$$

Let δ be the smaller of δ_1 and δ_2. For $0 < |x - c| < \delta$, you have
$$\left|\frac{1}{g(x)} - \frac{1}{K}\right| = \left|\frac{K - g(x)}{g(x)K}\right| = \frac{1}{|K|} \cdot \frac{1}{|g(x)|} |K - g(x)| < \frac{1}{|K|} \cdot \frac{2}{|K|} \frac{|K|^2}{2} \varepsilon = \varepsilon.$$

So, $\lim_{x \to c} \frac{1}{g(x)} = \frac{1}{K}$.

Finally, the proof of Property 5 can be obtained by a straightforward application of mathematical induction coupled with Property 3.

> **THEOREM 1.4 The Limit of a Function Involving a Radical (page 60)**
>
> Let n be a positive integer. The following limit is valid for all c if n is odd, and is valid for $c > 0$ if n is even.
>
> $$\lim_{x \to c} \sqrt[n]{x} = \sqrt[n]{c}.$$

Proof Consider the case for which $c > 0$ and n is any positive integer. For a given $\varepsilon > 0$, you need to find $\delta > 0$ such that

$$\left| \sqrt[n]{x} - \sqrt[n]{c} \right| < \varepsilon \quad \text{whenever} \quad 0 < |x - c| < \delta$$

which is the same as saying

$$-\varepsilon < \sqrt[n]{x} - \sqrt[n]{c} < \varepsilon \quad \text{whenever} \quad -\delta < x - c < \delta.$$

Assume $\varepsilon < \sqrt[n]{c}$, which implies that $0 < \sqrt[n]{c} - \varepsilon < \sqrt[n]{c}$. Now, let δ be the smaller of the two numbers.

$$c - \left(\sqrt[n]{c} - \varepsilon \right)^n \quad \text{and} \quad \left(\sqrt[n]{c} + \varepsilon \right)^n - c$$

Then you have

$$-\delta < x - c \qquad\qquad < \delta$$
$$-\left[c - \left(\sqrt[n]{c} - \varepsilon \right)^n \right] < x - c \qquad\qquad < \left(\sqrt[n]{c} + \varepsilon \right)^n - c$$
$$\left(\sqrt[n]{c} - \varepsilon \right)^n - c < x - c \qquad\qquad < \left(\sqrt[n]{c} + \varepsilon \right)^n - c$$
$$\left(\sqrt[n]{c} - \varepsilon \right)^n < x \qquad\qquad < \left(\sqrt[n]{c} + \varepsilon \right)^n$$
$$\sqrt[n]{c} - \varepsilon < \sqrt[n]{x} \qquad\qquad < \sqrt[n]{c} + \varepsilon$$
$$-\varepsilon < \sqrt[n]{x} - \sqrt[n]{c} < \varepsilon.$$

> **THEOREM 1.5 The Limit of a Composite Function (page 61)**
>
> If f and g are functions such that $\lim_{x \to c} g(x) = L$ and $\lim_{x \to L} f(x) = f(L)$, then
>
> $$\lim_{x \to c} f(g(x)) = f\left(\lim_{x \to c} g(x) \right) = f(L).$$

Proof For a given $\varepsilon > 0$, you must find $\delta > 0$ such that

$$|f(g(x)) - f(L)| < \varepsilon \quad \text{whenever} \quad 0 < |x - c| < \delta.$$

Because the limit of $f(x)$ as $x \to L$ is $f(L)$, you know there exists $\delta_1 > 0$ such that

$$|f(u) - f(L)| < \varepsilon \quad \text{whenever} \quad |u - L| < \delta_1.$$

Moreover, because the limit of $g(x)$ as $x \to c$ is L, you know there exists $\delta > 0$ such that

$$|g(x) - L| < \delta_1 \quad \text{whenever} \quad 0 < |x - c| < \delta.$$

Finally, letting $u = g(x)$, you have

$$|f(g(x)) - f(L)| < \varepsilon \quad \text{whenever} \quad 0 < |x - c| < \delta.$$

> **THEOREM 1.7 Functions That Agree at All But One Point (page 62)**
>
> Let c be a real number and let $f(x) = g(x)$ for all $x \neq c$ in an open interval containing c. If the limit of $g(x)$ as x approaches c exists, then the limit of $f(x)$ also exists and
>
> $$\lim_{x \to c} f(x) = \lim_{x \to c} g(x).$$

Proof Let L be the limit of $g(x)$ as $x \to c$. Then, for each $\varepsilon > 0$ there exists a $\delta > 0$ such that $f(x) = g(x)$ in the open intervals $(c - \delta, c)$ and $(c, c + \delta)$, and

$$|g(x) - L| < \varepsilon \quad \text{whenever} \quad 0 < |x - c| < \delta.$$

Because $f(x) = g(x)$ for all x in the open interval other than $x = c$, it follows that

$$|f(x) - L| < \varepsilon \quad \text{whenever} \quad 0 < |x - c| < \delta.$$

So, the limit of $f(x)$ as $x \to c$ is also L.

> **THEOREM 1.8 The Squeeze Theorem (page 65)**
>
> If $h(x) \leq f(x) \leq g(x)$ for all x in an open interval containing c, except possibly at c itself, and if
>
> $$\lim_{x \to c} h(x) = L = \lim_{x \to c} g(x)$$
>
> then $\lim_{x \to c} f(x)$ exists and is equal to L.

Proof For $\varepsilon > 0$ there exist $\delta_1 > 0$ and $\delta_2 > 0$ such that

$$|h(x) - L| < \varepsilon \quad \text{whenever} \quad 0 < |x - c| < \delta_1$$

and

$$|g(x) - L| < \varepsilon \quad \text{whenever} \quad 0 < |x - c| < \delta_2.$$

Because $h(x) \leq f(x) \leq g(x)$ for all x in an open interval containing c, except possibly at c itself, there exists $\delta_3 > 0$ such that $h(x) \leq f(x) \leq g(x)$ for $0 < |x - c| < \delta_3$. Let δ be the smallest of δ_1, δ_2, and δ_3. Then, if $0 < |x - c| < \delta$, it follows that $|h(x) - L| < \varepsilon$ and $|g(x) - L| < \varepsilon$, which implies that

$$-\varepsilon < h(x) - L < \varepsilon \quad \text{and} \quad -\varepsilon < g(x) - L < \varepsilon$$
$$L - \varepsilon < h(x) \quad \text{and} \quad g(x) < L + \varepsilon.$$

Now, because $h(x) \leq f(x) \leq g(x)$, it follows that $L - \varepsilon < f(x) < L + \varepsilon$, which implies that $|f(x) - L| < \varepsilon$. Therefore,

$$\lim_{x \to c} f(x) = L.$$

> **THEOREM 1.14 Vertical Asymptotes (page 85)**
>
> Let f and g be continuous on an open interval containing c. If $f(c) \neq 0$, $g(c) = 0$, and there exists an open interval containing c such that $g(x) \neq 0$ for all $x \neq c$ in the interval, then the graph of the function given by
>
> $$h(x) = \frac{f(x)}{g(x)}$$
>
> has a vertical asymptote at $x = c$.

Proof Consider the case for which $f(c) > 0$, and there exists $b > c$ such that $c < x < b$ implies $g(x) > 0$. Then for $M > 0$, choose δ_1 such that

$$0 < x - c < \delta_1 \quad \text{implies that} \quad \frac{f(c)}{2} < f(x) < \frac{3f(c)}{2}$$

and δ_2 such that

$$0 < x - c < \delta_2 \quad \text{implies that} \quad 0 < g(x) < \frac{f(c)}{2M}.$$

Now let δ be the smaller of δ_1 and δ_2. Then it follows that

$$0 < x - c < \delta \quad \text{implies that} \quad \frac{f(x)}{g(x)} > \frac{f(c)}{2}\left[\frac{2M}{f(c)}\right] = M.$$

So, it follows that

$$\lim_{x \to c^+} \frac{f(x)}{g(x)} = \infty$$

and the line $x = c$ is a vertical asymptote of the graph of h.

> **Alternative Form of the Derivative (page 101)**
>
> The derivative of f at c is given by
>
> $$f'(c) = \lim_{x \to c} \frac{f(x) - f(c)}{x - c}$$
>
> provided this limit exists.

Proof The derivative of f at c is given by

$$f'(c) = \lim_{\Delta x \to 0} \frac{f(c + \Delta x) - f(c)}{\Delta x}.$$

Let $x = c + \Delta x$. Then $x \to c$ as $\Delta x \to 0$. So, replacing $c + \Delta x$ by x, you have

$$f'(c) = \lim_{\Delta x \to 0} \frac{f(c + \Delta x) - f(c)}{\Delta x} = \lim_{x \to c} \frac{f(x) - f(c)}{x - c}.$$

> **THEOREM 2.10   The Chain Rule (page 131)**
>
> If $y = f(u)$ is a differentiable function of u, and $u = g(x)$ is a differentiable function of x, then $y = f(g(x))$ is a differentiable function of x and
>
> $$\frac{dy}{dx} = \frac{dy}{du} \cdot \frac{du}{dx}$$
>
> or, equivalently,
>
> $$\frac{d}{dx}[f(g(x))] = f'(g(x))g'(x).$$

Proof In Section 2.4, you let $h(x) = f(g(x))$ and used the alternative form of the derivative to show that $h'(c) = f'(g(c))g'(c)$, provided $g(x) \neq g(c)$ for values of x other than c. Now consider a more general proof. Begin by considering the derivative of f.

$$f'(x) = \lim_{\Delta x \to 0} \frac{f(x + \Delta x) - f(x)}{\Delta x} = \lim_{\Delta x \to 0} \frac{\Delta y}{\Delta x}$$

For a fixed value of x, define a function η such that

$$\eta(\Delta x) = \begin{cases} 0, & \Delta x = 0 \\ \frac{\Delta y}{\Delta x} - f'(x), & \Delta x \neq 0. \end{cases}$$

Because the limit of $\eta(\Delta x)$ as $\Delta x \to 0$ doesn't depend on the value of $\eta(0)$, you have

$$\lim_{\Delta x \to 0} \eta(\Delta x) = \lim_{\Delta x \to 0}\left[\frac{\Delta y}{\Delta x} - f'(x)\right] = 0$$

and you can conclude that η is continuous at 0. Moreover, because $\Delta y = 0$ when $\Delta x = 0$, the equation

$$\Delta y = \Delta x \eta(\Delta x) + \Delta x f'(x)$$

is valid whether Δx is zero or not. Now, by letting $\Delta u = g(x + \Delta x) - g(x)$, you can use the continuity of g to conclude that

$$\lim_{\Delta x \to 0} \Delta u = \lim_{\Delta x \to 0}[g(x + \Delta x) - g(x)] = 0$$

which implies that

$$\lim_{\Delta x \to 0} \eta(\Delta u) = 0.$$

Finally,

$$\Delta y = \Delta u \eta(\Delta u) + \Delta u f'(u) \to \frac{\Delta y}{\Delta x} = \frac{\Delta u}{\Delta x}\eta(\Delta u) + \frac{\Delta u}{\Delta x}f'(u), \quad \Delta x \neq 0$$

and taking the limit as $\Delta x \to 0$, you have

$$\frac{dy}{dx} = \frac{du}{dx}\left[\lim_{\Delta x \to 0} \eta(\Delta u)\right] + \frac{du}{dx}f'(u) = \frac{dy}{dx}(0) + \frac{du}{dx}f'(u)$$

$$= \frac{du}{dx}f'(u)$$

$$= \frac{du}{dx} \cdot \frac{dy}{du}.$$

Concavity Interpretation (page 190)

1. Let f be differentiable on an open interval I. If the graph of f is concave *upward* on I, then the graph of f lies *above* all of its tangent lines on I.
2. Let f be differentiable on an open interval I. If the graph of f is concave *downward* on I, then the graph of f lies *below* all of its tangent lines on I.

Proof Assume that f is concave upward on $I = (a, b)$. Then, f' is increasing on (a, b). Let c be a point in the interval $I = (a, b)$. The equation of the tangent line to the graph of f at c is given by

$$g(x) = f(c) + f'(c)(x - c).$$

If x is in the open interval (c, b), then the directed distance from point $(x, f(x))$ (on the graph of f) to the point $(x, g(x))$ (on the tangent line) is given by

$$\begin{aligned}d &= f(x) - [f(c) + f'(c)(x - c)] \\ &= f(x) - f(c) - f'(c)(x - c).\end{aligned}$$

Moreover, by the Mean Value Theorem there exists a number z in (c, x) such that

$$f'(z) = \frac{f(x) - f(c)}{x - c}.$$

So, you have

$$\begin{aligned}d &= f(x) - f(c) - f'(c)(x - c) \\ &= f'(z)(x - c) - f'(c)(x - c) \\ &= [f'(z) - f'(c)](x - c).\end{aligned}$$

The second factor $(x - c)$ is positive because $c < x$. Moreover, because f' is increasing, it follows that the first factor $[f'(z) - f'(c)]$ is also positive. Therefore, $d > 0$ and you can conclude that the graph of f lies above the tangent line at x. If x is in the open interval (a, c), a similar argument can be given. This proves the first statement. The proof of the second statement is similar.

THEOREM 3.10 Limits at Infinity (page 199)

If r is a positive rational number and c is any real number, then

$$\lim_{x \to \infty} \frac{c}{x^r} = 0.$$

Furthermore, if x^r is defined when $x < 0$, then $\displaystyle\lim_{x \to -\infty} \frac{c}{x^r} = 0$.

Proof Begin by proving that

$$\lim_{x \to \infty} \frac{1}{x} = 0.$$

For $\varepsilon > 0$, let $M = 1/\varepsilon$. Then, for $x > M$, you have

$$x > M = \frac{1}{\varepsilon} \implies \frac{1}{x} < \varepsilon \implies \left|\frac{1}{x} - 0\right| < \varepsilon.$$

So, by the definition of a limit at infinity, you can conclude that the limit of $1/x$ as $x \to \infty$ is 0. Now, using this result, and letting $r = m/n$, you can write the following.

$$\lim_{x \to \infty} \frac{c}{x^r} = \lim_{x \to \infty} \frac{c}{x^{m/n}}$$

$$= c \left[\lim_{x \to \infty} \left(\frac{1}{\sqrt[n]{x}} \right)^m \right]$$

$$= c \left(\lim_{x \to \infty} \sqrt[n]{\frac{1}{x}} \right)^m$$

$$= c \left(\sqrt[n]{\lim_{x \to \infty} \frac{1}{x}} \right)^m$$

$$= c \left(\sqrt[n]{0} \right)^m$$

$$= 0$$

The proof of the second part of the theorem is similar.

THEOREM 4.2   Summation Formulas (page 260)

1. $\sum_{i=1}^{n} c = cn$
2. $\sum_{i=1}^{n} i = \dfrac{n(n+1)}{2}$
3. $\sum_{i=1}^{n} i^2 = \dfrac{n(n+1)(2n+1)}{6}$
4. $\sum_{i=1}^{n} i^3 = \dfrac{n^2(n+1)^2}{4}$

Proof   The proof of Property 1 is straightforward. By adding c to itself n times, you obtain a sum of cn.

To prove Property 2, write the sum in increasing and decreasing order and add corresponding terms, as follows.

$$\sum_{i=1}^{n} i = 1 \;+\; 2 \;+\; 3 \;+\; \cdots \;+\; (n-1) \;+\; n$$
$$\downarrow \quad \downarrow \quad \downarrow \quad \quad \downarrow \quad \downarrow$$
$$\sum_{i=1}^{n} i = n \;+\; (n-1) \;+\; (n-2) \;+\; \cdots \;+\; 2 \;+\; 1$$
$$\downarrow \quad \downarrow \quad \downarrow \quad \quad \downarrow \quad \downarrow$$
$$2\sum_{i=1}^{n} i = \underbrace{(n+1) + (n+1) + (n+1) + \cdots + (n+1) + (n+1)}_{n \text{ terms}}$$

So,

$$\sum_{i=1}^{n} i = \frac{n(n+1)}{2}.$$

To prove Property 3, use mathematical induction. First, if $n = 1$, the result is true because

$$\sum_{i=1}^{1} i^2 = 1^2 = 1 = \frac{1(1+1)(2+1)}{6}.$$

Now, assuming the result is true for $n = k$, you can show that it is true for $n = k + 1$, as follows.

$$\sum_{i=1}^{k+1} i^2 = \sum_{i=1}^{k} i^2 + (k+1)^2$$

$$= \frac{k(k+1)(2k+1)}{6} + (k+1)^2$$

$$= \frac{k+1}{6}(2k^2 + k + 6k + 6)$$

$$= \frac{k+1}{6}[(2k+3)(k+2)]$$

$$= \frac{(k+1)(k+2)[2(k+1)+1]}{6}$$

Property 4 can be proved using a similar argument with mathematical induction.

THEOREM 4.8 Preservation of Inequality (page 278)

1. If f is integrable and nonnegative on the closed interval $[a, b]$, then

$$0 \leq \int_a^b f(x)\,dx.$$

2. If f and g are integrable on the closed interval $[a, b]$ and $f(x) \leq g(x)$ for every x in $[a, b]$, then

$$\int_a^b f(x)\,dx \leq \int_a^b g(x)\,dx.$$

Proof To prove Property 1, suppose, on the contrary, that

$$\int_a^b f(x)\,dx = I < 0.$$

Then, let $a = x_0 < x_1 < x_2 < \cdots < x_n = b$ be a partition of $[a, b]$, and let

$$R = \sum_{i=1}^{n} f(c_i)\,\Delta x_i$$

be a Riemann sum. Because $f(x) \geq 0$, it follows that $R \geq 0$. Now, for $\|\Delta\|$ sufficiently small, you have $|R - I| < -I/2$, which implies that

$$\sum_{i=1}^{n} f(c_i)\,\Delta x_i = R < I - \frac{I}{2} < 0$$

which is not possible. From this contradiction, you can conclude that

$$0 \leq \int_a^b f(x)\,dx.$$

To prove Property 2 of the theorem, note that $f(x) \le g(x)$ implies that $g(x) - f(x) \ge 0$. So, you can apply the result of Property 1 to conclude that

$$0 \le \int_a^b [g(x) - f(x)]\, dx$$

$$0 \le \int_a^b g(x)\, dx - \int_a^b f(x)\, dx$$

$$\int_a^b f(x)\, dx \le \int_a^b g(x)\, dx.$$

Properties of the Natural Logarithmic Function (page 323)

$$\lim_{x \to 0^+} \ln x = -\infty \quad \text{and} \quad \lim_{x \to \infty} \ln x = \infty$$

Proof To begin, show that $\ln 2 \ge \frac{1}{2}$. From the Mean Value Theorem for Integrals, you can write

$$\ln 2 = \int_1^2 \frac{1}{x}\, dx = \frac{1}{c}(2 - 1) = \frac{1}{c}$$

where c is in $[1, 2]$. This implies that

$$1 \le c \le 2$$

$$1 \ge \frac{1}{c} \ge \frac{1}{2}$$

$$1 \ge \ln 2 \ge \frac{1}{2}.$$

Now, let N be any positive (large) number. Because $\ln x$ is increasing, it follows that if $x > 2^{2N}$, then

$$\ln x > \ln 2^{2N} = 2N \ln 2.$$

However, because $\ln 2 \ge \frac{1}{2}$, it follows that

$$\ln x > 2N \ln 2 \ge 2N\left(\frac{1}{2}\right) = N.$$

This verifies the second limit. To verify the first limit, let $z = 1/x$. Then, $z \to \infty$ as $x \to 0^+$, and you can write

$$\lim_{x \to 0^+} \ln x = \lim_{x \to 0^+} \left(-\ln \frac{1}{x}\right)$$

$$= \lim_{z \to \infty} (-\ln z)$$

$$= -\lim_{z \to \infty} \ln z$$

$$= -\infty$$

> **THEOREM 5.8 Continuity and Differentiability of Inverse Functions (page 345)**
>
> Let f be a function whose domain is an interval I. If f has an inverse function, then the following statements are true.
>
> 1. If f is continuous on its domain, then f^{-1} is continuous on its domain.
> 2. If f is increasing on its domain, then f^{-1} is increasing on its domain.
> 3. If f is decreasing on its domain, then f^{-1} is decreasing on its domain.
> 4. If f is differentiable at c and $f'(c) \neq 0$, then f^{-1} is differentiable at $f(c)$.

Proof To prove Property 1, first show that if f is continuous on I and has an inverse function, then f is strictly monotonic on I. Suppose that f were not strictly monotonic. Then there would exist numbers x_1, x_2, x_3 in I such that $x_1 < x_2 < x_3$, but $f(x_2)$ is not between $f(x_1)$ and $f(x_3)$. Without loss of generality, assume $f(x_1) < f(x_3) < f(x_2)$. By the Intermediate Value Theorem, there exists a number x_0 between x_1 and x_2 such that $f(x_0) = f(x_3)$. So, f is not one-to-one and cannot have an inverse function. So, f must be strictly monotonic.

Because f is continuous, the Intermediate Value Theorem implies that the set of values of f

$$\{f(x) : x \in I\}$$

forms an interval J. Assume that a is an interior point of J. From the previous argument, $f^{-1}(a)$ is an interior point of I. Let $\varepsilon > 0$. There exists $0 < \varepsilon_1 < \varepsilon$ such that

$$I_1 = (f^{-1}(a) - \varepsilon_1, f^{-1}(a) + \varepsilon_1) \subseteq I.$$

Because f is strictly monotonic on I_1, the set of values $\{f(x) : x \in I_1\}$ forms an interval $J_1 \subseteq J$. Let $\delta > 0$ such that $(a - \delta, a + \delta) \subseteq J_1$. Finally, if

$$|y - a| < \delta, \text{ then } |f^{-1}(y) - f^{-1}(a)| < \varepsilon_1 < \varepsilon.$$

So, f^{-1} is continuous at a. A similar proof can be given if a is an endpoint.

To prove Property 2, let y_1 and y_2 be in the domain of f^{-1}, with $y_1 < y_2$. Then, there exist x_1 and x_2 in the domain of f such that

$$f(x_1) = y_1 < y_2 = f(x_2).$$

Because f is increasing, $f(x_1) < f(x_2)$ holds precisely when $x_1 < x_2$. Therefore,

$$f^{-1}(y_1) = x_1 < x_2 = f^{-1}(y_2)$$

which implies that f^{-1} is increasing. (Property 3 can be proved in a similar way.)

Finally, to prove Property 4, consider the limit

$$(f^{-1})'(a) = \lim_{y \to a} \frac{f^{-1}(y) - f^{-1}(a)}{y - a}$$

where a is in the domain of f^{-1} and $f^{-1}(a) = c$. Because f is differentiable at c, f is continuous at c, and so is f^{-1} at a. So, $y \to a$ implies that $x \to c$, and you have

$$(f^{-1})'(a) = \lim_{x \to c} \frac{x - c}{f(x) - f(c)}$$

$$= \lim_{x \to c} \frac{1}{\left(\dfrac{f(x) - f(c)}{x - c}\right)}$$

$$= \frac{1}{\lim\limits_{x \to c} \dfrac{f(x) - f(c)}{x - c}}$$

$$= \frac{1}{f'(c)}.$$

So, $(f^{-1})'(a)$ exists, and f^{-1} is differentiable at $f(c)$.

THEOREM 5.9   The Derivative of an Inverse Function (page 345)

Let f be a function that is differentiable on an interval I. If f has an inverse function g, then g is differentiable at any x for which $f'(g(x)) \neq 0$. Moreover,

$$g'(x) = \frac{1}{f'(g(x))}, \quad f'(g(x)) \neq 0.$$

Proof   From the proof of Theorem 5.8, letting $a = x$, you know that g is differentiable. Using the Chain Rule, differentiate both sides of the equation $x = f(g(x))$ to obtain

$$1 = f'(g(x)) \frac{d}{dx}[g(x)].$$

Because $f'(g(x)) \neq 0$, you can divide by this quantity to obtain

$$\frac{d}{dx}[g(x)] = \frac{1}{f'(g(x))}.$$

THEOREM 5.15   A Limit Involving e (page 364)

$$\lim_{x \to \infty} \left(1 + \frac{1}{x}\right)^x = \lim_{x \to \infty} \left(\frac{x + 1}{x}\right)^x = e$$

Proof   Let $y = \lim\limits_{x \to \infty} \left(1 + \dfrac{1}{x}\right)^x$. Taking the natural logarithm of each side, you have

$$\ln y = \ln\left[\lim_{x \to \infty} \left(1 + \frac{1}{x}\right)^x\right].$$

Because the natural logarithmic function is continuous, you can write

$$\ln y = \lim_{x \to \infty} \left[x \ln\left(1 + \frac{1}{x}\right) \right] = \lim_{x \to \infty} \left\{ \frac{\ln[1 + (1/x)]}{1/x} \right\}.$$

Letting $x = \frac{1}{t}$, you have

$$\ln y = \lim_{t \to 0^+} \frac{\ln(1 + t)}{t}$$

$$= \lim_{t \to 0^+} \frac{\ln(1 + t) - \ln 1}{t}$$

$$= \frac{d}{dx} \ln x \text{ at } x = 1$$

$$= \frac{1}{x} \text{ at } x = 1$$

$$= 1.$$

Finally, because $\ln y = 1$, you know that $y = e$, and you can conclude that

$$\lim_{x \to \infty} \left(1 + \frac{1}{x}\right)^x = e.$$

THEOREM 8.3 The Extended Mean Value Theorem (page 568)

If f and g are differentiable on an open interval (a, b) and continuous on $[a, b]$ such that $g'(x) \neq 0$ for any x in (a, b), then there exists a point c in (a, b) such that

$$\frac{f'(c)}{g'(c)} = \frac{f(b) - f(a)}{g(b) - g(a)}.$$

Proof You can assume that $g(a) \neq g(b)$, because otherwise, by Rolle's Theorem, it would follow that $g'(x) = 0$ for some x in (a, b). Now, define $h(x)$ to be

$$h(x) = f(x) - \left[\frac{f(b) - f(a)}{g(b) - g(a)}\right] g(x).$$

Then

$$h(a) = f(a) - \left[\frac{f(b) - f(a)}{g(b) - g(a)}\right] g(a) = \frac{f(a)g(b) - f(b)g(a)}{g(b) - g(a)}$$

and

$$h(b) = f(b) - \left[\frac{f(b) - f(a)}{g(b) - g(a)}\right] g(b) = \frac{f(a)g(b) - f(b)g(a)}{g(b) - g(a)}$$

and by Rolle's Theorem there exists a point c in (a, b) such that

$$h'(c) = f'(c) - \frac{f(b) - f(a)}{g(b) - g(a)} g'(c) = 0$$

which implies that

$$\frac{f'(c)}{g'(c)} = \frac{f(b) - f(a)}{g(b) - g(a)}.$$

> **THEOREM 8.4 L'Hôpital's Rule (page 568)**
>
> Let f and g be functions that are differentiable on an open interval (a, b) containing c, except possibly at c itself. Assume that $g'(x) \neq 0$ for all x in (a, b), except possibly at c itself. If the limit of $f(x)/g(x)$ as x approaches c produces the indeterminate form $0/0$, then
>
> $$\lim_{x \to c} \frac{f(x)}{g(x)} = \lim_{x \to c} \frac{f'(x)}{g'(x)}$$
>
> provided the limit on the right exists (or is infinite). This result also applies if the limit of $f(x)/g(x)$ as x approaches c produces any one of the indeterminate forms ∞/∞, $(-\infty)/\infty$, $\infty/(-\infty)$, or $(-\infty)/(-\infty)$.

You can use the Extended Mean Value Theorem to prove L'Hôpital's Rule. Of the several different cases of this rule, the proof of only one case is illustrated. The remaining cases where $x \to c^-$ and $x \to c$ are left for you to prove.

Proof Consider the case for which

$$\lim_{x \to c^+} f(x) = 0 \quad \text{and} \quad \lim_{x \to c^+} g(x) = 0.$$

Define the following new functions:

$$F(x) = \begin{cases} f(x), & x \neq c \\ 0, & x = c \end{cases} \quad \text{and} \quad G(x) = \begin{cases} g(x), & x \neq c \\ 0, & x = c \end{cases}.$$

For any x, $c < x < b$, F and G are differentiable on $(c, x]$ and continuous on $[c, x]$. You can apply the Extended Mean Value Theorem to conclude that there exists a number z in (c, x) such that

$$\frac{F'(z)}{G'(z)} = \frac{F(x) - F(c)}{G(x) - G(c)}$$

$$= \frac{F(x)}{G(x)}$$

$$= \frac{f'(z)}{g'(z)}$$

$$= \frac{f(x)}{g(x)}.$$

Finally, by letting x approach c from the right, $x \to c^+$, you have $z \to c^+$ because $c < z < x$, and

$$\lim_{x \to c^+} \frac{f(x)}{g(x)} = \lim_{x \to c^+} \frac{f'(z)}{g'(z)}$$

$$= \lim_{z \to c^+} \frac{f'(z)}{g'(z)}$$

$$= \lim_{x \to c^+} \frac{f'(x)}{g'(x)}.$$

> **THEOREM 9.19 Taylor's Theorem (page 654)**
>
> If a function f is differentiable through order $n + 1$ in an interval I containing c, then, for each x in I, there exists z between x and c such that
>
> $$f(x) = f(c) + f'(c)(x - c) + \frac{f''(c)}{2!}(x - c)^2 + \cdots + \frac{f^{(n)}(c)}{n!}(x - c)^n + R_n(x)$$
>
> where
>
> $$R_n(x) = \frac{f^{(n+1)}(z)}{(n + 1)!}(x - c)^{n+1}.$$

Proof To find $R_n(x)$, fix x in I ($x \neq c$) and write

$$R_n(x) = f(x) - P_n(x)$$

where $P_n(x)$ is the nth Taylor polynomial for $f(x)$. Then let g be a function of t defined by

$$g(t) = f(x) - f(t) - f'(t)(x - t) - \cdots - \frac{f^{(n)}(t)}{n!}(x - t)^n - R_n(x)\frac{(x - t)^{n+1}}{(x - c)^{n+1}}.$$

The reason for defining g in this way is that differentiation with respect to t has a telescoping effect. For example, you have

$$\frac{d}{dt}[-f(t) - f'(t)(x - t)] = -f'(t) + f'(t) - f''(t)(x - t)$$
$$= -f''(t)(x - t).$$

The result is that the derivative $g'(t)$ simplifies to

$$g'(t) = -\frac{f^{(n+1)}(t)}{n!}(x - t)^n + (n + 1)R_n(x)\frac{(x - t)^n}{(x - c)^{n+1}}$$

for all t between c and x. Moreover, for a fixed x,

$$g(c) = f(x) - [P_n(x) + R_n(x)] = f(x) - f(x) = 0$$

and

$$g(x) = f(x) - f(x) - 0 - \cdots - 0 = f(x) - f(x) = 0.$$

Therefore, g satisfies the conditions of Rolle's Theorem, and it follows that there is a number z between c and x such that $g'(z) = 0$. Substituting z for t in the equation for $g'(t)$ and then solving for $R_n(x)$, you obtain

$$g'(z) = -\frac{f^{(n+1)}(z)}{n!}(x - z)^n + (n + 1)R_n(x)\frac{(x - z)^n}{(x - c)^{n+1}} = 0$$

$$R_n(x) = \frac{f^{(n+1)}(z)}{(n + 1)!}(x - c)^{n+1}.$$

Finally, because $g(c) = 0$, you have

$$0 = f(x) - f(c) - f'(c)(x - c) - \cdots - \frac{f^{(n)}(c)}{n!}(x - c)^n - R_n(x)$$

$$f(x) = f(c) + f'(c)(x - c) + \cdots + \frac{f^{(n)}(c)}{n!}(x - c)^n + R_n(x).$$

> **THEOREM 9.20 Convergence of a Power Series (page 660)**
>
> For a power series centered at c, precisely one of the following is true.
>
> 1. The series converges only at c.
> 2. There exists a real number $R > 0$ such that the series converges absolutely for $|x - c| < R$, and diverges for $|x - c| > R$.
> 3. The series converges absolutely for all x.
>
> The number R is the **radius of convergence** of the power series. If the series converges only at c, the radius of convergence is $R = 0$, and if the series converges for all x, the radius of convergence is $R = \infty$. The set of all values of x for which the power series converges is the **interval of convergence** of the power series.

Proof In order to simplify the notation, the theorem for the power series $\sum a_n x^n$ centered at $x = 0$ will be proved. The proof for a power series centered at $x = c$ follows easily. A key step in this proof uses the completeness property of the set of real numbers: If a nonempty set S of real numbers has an upper bound, then it must have a least upper bound (see page 601).

It must be shown that if a power series $\sum a_n x^n$ converges at $x = d$, $d \neq 0$, then it converges for all b satisfying $|b| < |d|$. Because $\sum a_n x^n$ converges, $\lim\limits_{x \to \infty} a_n d^n = 0$. So, there exists $N > 0$ such that $a_n d^n < 1$ for all $n \geq N$. Then for $n \geq N$,

$$|a_n b^n| = \left|a_n b^n \frac{d^n}{d^n}\right| = |a_n d^n| \left|\frac{b^n}{d^n}\right| < \left|\frac{b^n}{d^n}\right|.$$

So, for $|b| < |d|$, $\left|\frac{b}{d}\right| < 1$, which implies that

$$\sum \left|\frac{b^n}{d^n}\right|$$

is a convergent geometric series. By the Comparison Test, the series $\sum a_n b^n$ converges.

Similarly, if the power series $\sum a_n x^n$ diverges at $x = b$, where $b \neq 0$, then it diverges for all d satisfying $|d| > |b|$. If $\sum a_n d^n$ converged, then the argument above would imply that $\sum a_n b^n$ converged as well.

Finally, to prove the theorem, suppose that neither case 1 nor case 3 is true. Then there exist points b and d such that $\sum a_n x^n$ converges at b and diverges at d. Let $S = \{x : \sum a_n x^n \text{ converges}\}$. S is nonempty because $b \in S$. If $x \in S$ then $|x| \leq |d|$, which shows that $|d|$ is an upper bound for the nonempty set S. By the completeness property, S has a least upper bound, R.

Now, if $|x| > R$, then $x \notin S$ so $\sum a_n x^n$ diverges. And if $|x| < R$, then $|x|$ is not an upper bound for S, so there exists b in S satisfying $|b| > |x|$. Since $b \in S$, $\sum a_n b^n$ converges, which implies that $\sum a_n x^n$ converges.

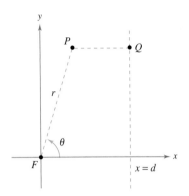

Figure A.1

> **THEOREM 10.16 Classification of Conics by Eccentricity (page 748)**
>
> Let F be a fixed point (*focus*) and let D be a fixed line (*directrix*) in the plane. Let P be another point in the plane and let e (*eccentricity*) be the ratio of the distance between P and F to the distance between P and D. The collection of all points P with a given eccentricity is a conic.
>
> 1. The conic is an ellipse if $0 < e < 1$.
> 2. The conic is a parabola if $e = 1$.
> 3. The conic is a hyperbola if $e > 1$.

Proof If $e = 1$, then, by definition, the conic must be a parabola. If $e \neq 1$, then you can consider the focus F to lie at the origin and the directrix $x = d$ to lie to the right of the origin, as shown in Figure A.1. For the point $P = (r, \theta) = (x, y)$, you have $|PF| = r$ and $|PQ| = d - r\cos\theta$. Given that $e = |PF|/|PQ|$, it follows that

$$|PF| = |PQ|e \quad \Longrightarrow \quad r = e(d - r\cos\theta).$$

By converting to rectangular coordinates and squaring each side, you obtain

$$x^2 + y^2 = e^2(d - x)^2 = e^2(d^2 - 2dx + x^2).$$

Completing the square produces

$$\left(x + \frac{e^2 d}{1 - e^2}\right)^2 + \frac{y^2}{1 - e^2} = \frac{e^2 d^2}{(1 - e^2)^2}.$$

If $e < 1$, this equation represents an ellipse. If $e > 1$, then $1 - e^2 < 0$, and the equation represents a hyperbola.

> **THEOREM 13.4 Sufficient Condition for Differentiability (page 917)**
>
> If f is a function of x and y, where f_x and f_y are continuous in an open region R, then f is differentiable on R.

Proof Let S be the surface defined by $z = f(x, y)$, where f, f_x, and f_y are continuous at (x, y). Let A, B, and C be points on surface S, as shown in Figure A.2. From this figure, you can see that the change in f from point A to point C is given by

$$\begin{aligned}\Delta z &= f(x + \Delta x, y + \Delta y) - f(x, y) \\ &= [f(x + \Delta x, y) - f(x, y)] + [f(x + \Delta x, y + \Delta y) - f(x + \Delta x, y)] \\ &= \Delta z_1 + \Delta z_2.\end{aligned}$$

Between A and B, y is fixed and x changes. So, by the Mean Value Theorem, there is a value x_1 between x and $x + \Delta x$ such that

$$\Delta z_1 = f(x + \Delta x, y) - f(x, y) = f_x(x_1, y)\, \Delta x.$$

Similarly, between B and C, x is fixed and y changes, and there is a value y_1 between y and $y + \Delta y$ such that

$$\Delta z_2 = f(x + \Delta x, y + \Delta y) - f(x + \Delta x, y) = f_y(x + \Delta x, y_1)\, \Delta y.$$

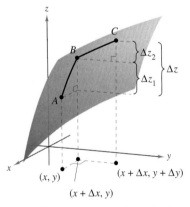

$\Delta z = f(x + \Delta x, y + \Delta y) - f(x, y)$

Figure A.2

By combining these two results, you can write

$$\Delta z = \Delta z_1 + \Delta z_2 = f_x(x_1, y)\Delta x + f_y(x + \Delta x, y_1)\Delta y.$$

If you define ε_1 and ε_2 as

$$\varepsilon_1 = f_x(x_1, y) - f_x(x, y) \quad \text{and} \quad \varepsilon_2 = f_y(x + \Delta x, y_1) - f_y(x, y)$$

it follows that

$$\Delta z = \Delta z_1 + \Delta z_2 = [\varepsilon_1 + f_x(x, y)]\Delta x + [\varepsilon_2 + f_y(x, y)]\Delta y$$
$$= [f_x(x, y)\Delta x + f_y(x, y)\Delta y] + \varepsilon_1 \Delta x + \varepsilon_2 \Delta y.$$

By the continuity of f_x and f_y and the fact that $x \le x_1 \le x + \Delta x$ and $y \le y_1 \le y + \Delta y$, it follows that $\varepsilon_1 \to 0$ and $\varepsilon_2 \to 0$ as $\Delta x \to 0$ and $\Delta y \to 0$. Therefore, by definition, f is differentiable.

THEOREM 13.6 Chain Rule: One Independent Variable (page 923)

Let $w = f(x, y)$, where f is a differentiable function of x and y. If $x = g(t)$ and $y = h(t)$, where g and h are differentiable functions of t, then w is a differentiable function of t, and

$$\frac{dw}{dt} = \frac{\partial w}{\partial x}\frac{dx}{dt} + \frac{\partial w}{\partial y}\frac{dy}{dt}.$$

Proof Because g and h are differentiable functions of t, you know that both Δx and Δy approach zero as Δt approaches zero. Moreover, because f is a differentiable function of x and y, you know that

$$\Delta w = \frac{\partial w}{\partial x}\Delta x + \frac{\partial w}{\partial y}\Delta y + \varepsilon_1 \Delta x + \varepsilon_2 \Delta y$$

where both ε_1 and $\varepsilon_2 \to 0$ as $(\Delta x, \Delta y) \to (0, 0)$. So, for $\Delta t \ne 0$

$$\frac{\Delta w}{\Delta t} = \frac{\partial w}{\partial x}\frac{\Delta x}{\Delta t} + \frac{\partial w}{\partial y}\frac{\Delta y}{\Delta t} + \varepsilon_1 \frac{\Delta x}{\Delta t} + \varepsilon_2 \frac{\Delta y}{\Delta t}$$

from which it follows that

$$\frac{dw}{dt} = \lim_{\Delta t \to 0}\frac{\Delta w}{\Delta t} = \frac{\partial w}{\partial x}\frac{dx}{dt} + \frac{\partial w}{\partial y}\frac{dy}{dt} + 0\left(\frac{dx}{dt}\right) + 0\left(\frac{dy}{dt}\right)$$
$$= \frac{\partial w}{\partial x}\frac{dx}{dt} + \frac{\partial w}{\partial y}\frac{dy}{dt}.$$

B Integration Tables

Forms Involving u^n

1. $\displaystyle\int u^n\,du = \frac{u^{n+1}}{n+1} + C,\ n \neq -1$

2. $\displaystyle\int \frac{1}{u}\,du = \ln|u| + C$

Forms Involving $a + bu$

3. $\displaystyle\int \frac{u}{a+bu}\,du = \frac{1}{b^2}\bigl(bu - a\ln|a+bu|\bigr) + C$

4. $\displaystyle\int \frac{u}{(a+bu)^2}\,du = \frac{1}{b^2}\left(\frac{a}{a+bu} + \ln|a+bu|\right) + C$

5. $\displaystyle\int \frac{u}{(a+bu)^n}\,du = \frac{1}{b^2}\left[\frac{-1}{(n-2)(a+bu)^{n-2}} + \frac{a}{(n-1)(a+bu)^{n-1}}\right] + C,\ n \neq 1, 2$

6. $\displaystyle\int \frac{u^2}{a+bu}\,du = \frac{1}{b^3}\left[-\frac{bu}{2}(2a - bu) + a^2\ln|a+bu|\right] + C$

7. $\displaystyle\int \frac{u^2}{(a+bu)^2}\,du = \frac{1}{b^3}\left(bu - \frac{a^2}{a+bu} - 2a\ln|a+bu|\right) + C$

8. $\displaystyle\int \frac{u^2}{(a+bu)^3}\,du = \frac{1}{b^3}\left[\frac{2a}{a+bu} - \frac{a^2}{2(a+bu)^2} + \ln|a+bu|\right] + C$

9. $\displaystyle\int \frac{u^2}{(a+bu)^n}\,du = \frac{1}{b^3}\left[\frac{-1}{(n-3)(a+bu)^{n-3}} + \frac{2a}{(n-2)(a+bu)^{n-2}} - \frac{a^2}{(n-1)(a+bu)^{n-1}}\right] + C,\ n \neq 1, 2, 3$

10. $\displaystyle\int \frac{1}{u(a+bu)}\,du = \frac{1}{a}\ln\left|\frac{u}{a+bu}\right| + C$

11. $\displaystyle\int \frac{1}{u(a+bu)^2}\,du = \frac{1}{a}\left(\frac{1}{a+bu} + \frac{1}{a}\ln\left|\frac{u}{a+bu}\right|\right) + C$

12. $\displaystyle\int \frac{1}{u^2(a+bu)}\,du = -\frac{1}{a}\left(\frac{1}{u} + \frac{b}{a}\ln\left|\frac{u}{a+bu}\right|\right) + C$

13. $\displaystyle\int \frac{1}{u^2(a+bu)^2}\,du = -\frac{1}{a^2}\left[\frac{a+2bu}{u(a+bu)} + \frac{2b}{a}\ln\left|\frac{u}{a+bu}\right|\right] + C$

Forms Involving $a + bu + cu^2$, $b^2 \neq 4ac$

14. $\displaystyle \int \frac{1}{a + bu + cu^2}\,du = \begin{cases} \dfrac{2}{\sqrt{4ac - b^2}} \arctan \dfrac{2cu + b}{\sqrt{4ac - b^2}} + C, & b^2 < 4ac \\[2mm] \dfrac{1}{\sqrt{b^2 - 4ac}} \ln \left| \dfrac{2cu + b - \sqrt{b^2 - 4ac}}{2cu + b + \sqrt{b^2 - 4ac}} \right| + C, & b^2 > 4ac \end{cases}$

15. $\displaystyle \int \frac{u}{a + bu + cu^2}\,du = \frac{1}{2c}\left(\ln|a + bu + cu^2| - b \int \frac{1}{a + bu + cu^2}\,du \right)$

Forms Involving $\sqrt{a + bu}$

16. $\displaystyle \int u^n \sqrt{a + bu}\,du = \frac{2}{b(2n + 3)}\left[u^n(a + bu)^{3/2} - na \int u^{n-1} \sqrt{a + bu}\,du \right]$

17. $\displaystyle \int \frac{1}{u\sqrt{a + bu}}\,du = \begin{cases} \dfrac{1}{\sqrt{a}} \ln \left| \dfrac{\sqrt{a + bu} - \sqrt{a}}{\sqrt{a + bu} + \sqrt{a}} \right| + C, & a > 0 \\[2mm] \dfrac{2}{\sqrt{-a}} \arctan \sqrt{\dfrac{a + bu}{-a}} + C, & a < 0 \end{cases}$

18. $\displaystyle \int \frac{1}{u^n \sqrt{a + bu}}\,du = \frac{-1}{a(n - 1)}\left[\frac{\sqrt{a + bu}}{u^{n-1}} + \frac{(2n - 3)b}{2} \int \frac{1}{u^{n-1} \sqrt{a + bu}}\,du \right],\ n \neq 1$

19. $\displaystyle \int \frac{\sqrt{a + bu}}{u}\,du = 2\sqrt{a + bu} + a \int \frac{1}{u\sqrt{a + bu}}\,du$

20. $\displaystyle \int \frac{\sqrt{a + bu}}{u^n}\,du = \frac{-1}{a(n - 1)}\left[\frac{(a + bu)^{3/2}}{u^{n-1}} + \frac{(2n - 5)b}{2} \int \frac{\sqrt{a + bu}}{u^{n-1}}\,du \right],\ n \neq 1$

21. $\displaystyle \int \frac{u}{\sqrt{a + bu}}\,du = \frac{-2(2a - bu)}{3b^2}\sqrt{a + bu} + C$

22. $\displaystyle \int \frac{u^n}{\sqrt{a + bu}}\,du = \frac{2}{(2n + 1)b}\left(u^n \sqrt{a + bu} - na \int \frac{u^{n-1}}{\sqrt{a + bu}}\,du \right)$

Forms Involving $a^2 \pm u^2$, $a > 0$

23. $\displaystyle \int \frac{1}{a^2 + u^2}\,du = \frac{1}{a} \arctan \frac{u}{a} + C$

24. $\displaystyle \int \frac{1}{u^2 - a^2}\,du = -\int \frac{1}{a^2 - u^2}\,du = \frac{1}{2a} \ln \left| \frac{u - a}{u + a} \right| + C$

25. $\displaystyle \int \frac{1}{(a^2 \pm u^2)^n}\,du = \frac{1}{2a^2(n - 1)}\left[\frac{u}{(a^2 \pm u^2)^{n-1}} + (2n - 3) \int \frac{1}{(a^2 \pm u^2)^{n-1}}\,du \right],\ n \neq 1$

Forms Involving $\sqrt{u^2 \pm a^2}$, $a > 0$

26. $\displaystyle \int \sqrt{u^2 \pm a^2}\,du = \frac{1}{2}\left(u\sqrt{u^2 \pm a^2} \pm a^2 \ln|u + \sqrt{u^2 \pm a^2}| \right) + C$

27. $\displaystyle \int u^2 \sqrt{u^2 \pm a^2}\,du = \frac{1}{8}\left[u(2u^2 \pm a^2)\sqrt{u^2 \pm a^2} - a^4 \ln|u + \sqrt{u^2 \pm a^2}| \right] + C$

28. $\displaystyle \int \frac{\sqrt{u^2 + a^2}}{u}\,du = \sqrt{u^2 + a^2} - a \ln \left| \frac{a + \sqrt{u^2 + a^2}}{u} \right| + C$

29. $\displaystyle\int \frac{\sqrt{u^2 - a^2}}{u}\, du = \sqrt{u^2 - a^2} - a\,\text{arcsec}\,\frac{|u|}{a} + C$

30. $\displaystyle\int \frac{\sqrt{u^2 \pm a^2}}{u^2}\, du = \frac{-\sqrt{u^2 \pm a^2}}{u} + \ln\left|u + \sqrt{u^2 \pm a^2}\right| + C$

31. $\displaystyle\int \frac{1}{\sqrt{u^2 \pm a^2}}\, du = \ln\left|u + \sqrt{u^2 \pm a^2}\right| + C$

32. $\displaystyle\int \frac{1}{u\sqrt{u^2 + a^2}}\, du = \frac{-1}{a}\ln\left|\frac{a + \sqrt{u^2 + a^2}}{u}\right| + C$

33. $\displaystyle\int \frac{1}{u\sqrt{u^2 - a^2}}\, du = \frac{1}{a}\,\text{arcsec}\,\frac{|u|}{a} + C$

34. $\displaystyle\int \frac{u^2}{\sqrt{u^2 \pm a^2}}\, du = \frac{1}{2}\left(u\sqrt{u^2 \pm a^2} \mp a^2 \ln\left|u + \sqrt{u^2 \pm a^2}\right|\right) + C$

35. $\displaystyle\int \frac{1}{u^2\sqrt{u^2 \pm a^2}}\, du = \mp\frac{\sqrt{u^2 \pm a^2}}{a^2 u} + C$

36. $\displaystyle\int \frac{1}{(u^2 \pm a^2)^{3/2}}\, du = \frac{\pm u}{a^2\sqrt{u^2 \pm a^2}} + C$

Forms Involving $\sqrt{a^2 - u^2}$, $a > 0$

37. $\displaystyle\int \sqrt{a^2 - u^2}\, du = \frac{1}{2}\left(u\sqrt{a^2 - u^2} + a^2 \arcsin\frac{u}{a}\right) + C$

38. $\displaystyle\int u^2\sqrt{a^2 - u^2}\, du = \frac{1}{8}\left[u(2u^2 - a^2)\sqrt{a^2 - u^2} + a^4 \arcsin\frac{u}{a}\right] + C$

39. $\displaystyle\int \frac{\sqrt{a^2 - u^2}}{u}\, du = \sqrt{a^2 - u^2} - a\ln\left|\frac{a + \sqrt{a^2 - u^2}}{u}\right| + C$

40. $\displaystyle\int \frac{\sqrt{a^2 - u^2}}{u^2}\, du = \frac{-\sqrt{a^2 - u^2}}{u} - \arcsin\frac{u}{a} + C$

41. $\displaystyle\int \frac{1}{\sqrt{a^2 - u^2}}\, du = \arcsin\frac{u}{a} + C$

42. $\displaystyle\int \frac{1}{u\sqrt{a^2 - u^2}}\, du = \frac{-1}{a}\ln\left|\frac{a + \sqrt{a^2 - u^2}}{u}\right| + C$

43. $\displaystyle\int \frac{u^2}{\sqrt{a^2 - u^2}}\, du = \frac{1}{2}\left(-u\sqrt{a^2 - u^2} + a^2 \arcsin\frac{u}{a}\right) + C$

44. $\displaystyle\int \frac{1}{u^2\sqrt{a^2 - u^2}}\, du = \frac{-\sqrt{a^2 - u^2}}{a^2 u} + C$

45. $\displaystyle\int \frac{1}{(a^2 - u^2)^{3/2}}\, du = \frac{u}{a^2\sqrt{a^2 - u^2}} + C$

Forms Involving sin u or cos u

46. $\int \sin u \, du = -\cos u + C$

47. $\int \cos u \, du = \sin u + C$

48. $\int \sin^2 u \, du = \frac{1}{2}(u - \sin u \cos u) + C$

49. $\int \cos^2 u \, du = \frac{1}{2}(u + \sin u \cos u) + C$

50. $\int \sin^n u \, du = -\frac{\sin^{n-1} u \cos u}{n} + \frac{n-1}{n} \int \sin^{n-2} u \, du$

51. $\int \cos^n u \, du = \frac{\cos^{n-1} u \sin u}{n} + \frac{n-1}{n} \int \cos^{n-2} u \, du$

52. $\int u \sin u \, du = \sin u - u \cos u + C$

53. $\int u \cos u \, du = \cos u + u \sin u + C$

54. $\int u^n \sin u \, du = -u^n \cos u + n \int u^{n-1} \cos u \, du$

55. $\int u^n \cos u \, du = u^n \sin u - n \int u^{n-1} \sin u \, du$

56. $\int \frac{1}{1 \pm \sin u} \, du = \tan u \mp \sec u + C$

57. $\int \frac{1}{1 \pm \cos u} \, du = -\cot u \pm \csc u + C$

58. $\int \frac{1}{\sin u \cos u} \, du = \ln|\tan u| + C$

Forms Involving tan u, cot u, sec u, csc u

59. $\int \tan u \, du = -\ln|\cos u| + C$

60. $\int \cot u \, du = \ln|\sin u| + C$

61. $\int \sec u \, du = \ln|\sec u + \tan u| + C$

62. $\int \csc u \, du = \ln|\csc u - \cot u| + C$ or $\int \csc u \, du = -\ln|\csc u + \cot u| + C$

63. $\int \tan^2 u \, du = -u + \tan u + C$

64. $\int \cot^2 u \, du = -u - \cot u + C$

65. $\int \sec^2 u \, du = \tan u + C$

66. $\int \csc^2 u \, du = -\cot u + C$

67. $\int \tan^n u \, du = \frac{\tan^{n-1} u}{n-1} - \int \tan^{n-2} u \, du, \; n \neq 1$

68. $\int \cot^n u \, du = -\frac{\cot^{n-1} u}{n-1} - \int (\cot^{n-2} u) \, du, \; n \neq 1$

69. $\int \sec^n u \, du = \frac{\sec^{n-2} u \tan u}{n-1} + \frac{n-2}{n-1} \int \sec^{n-2} u \, du, \; n \neq 1$

70. $\int \csc^n u \, du = -\frac{\csc^{n-2} u \cot u}{n-1} + \frac{n-2}{n-1} \int \csc^{n-2} u \, du, \; n \neq 1$

71. $\int \frac{1}{1 \pm \tan u} \, du = \frac{1}{2}(u \pm \ln|\cos u \pm \sin u|) + C$

72. $\int \frac{1}{1 \pm \cot u} \, du = \frac{1}{2}(u \mp \ln|\sin u \pm \cos u|) + C$

73. $\int \frac{1}{1 \pm \sec u} \, du = u + \cot u \mp \csc u + C$

74. $\int \frac{1}{1 \pm \csc u} \, du = u - \tan u \pm \sec u + C$

Forms Involving Inverse Trigonometric Functions

75. $\displaystyle\int \arcsin u \, du = u \arcsin u + \sqrt{1 - u^2} + C$

76. $\displaystyle\int \arccos u \, du = u \arccos u - \sqrt{1 - u^2} + C$

77. $\displaystyle\int \arctan u \, du = u \arctan u - \ln\sqrt{1 + u^2} + C$

78. $\displaystyle\int \text{arccot } u \, du = u \, \text{arccot } u + \ln\sqrt{1 + u^2} + C$

79. $\displaystyle\int \text{arcsec } u \, du = u \, \text{arcsec } u - \ln\left|u + \sqrt{u^2 - 1}\right| + C$

80. $\displaystyle\int \text{arccsc } u \, du = u \, \text{arccsc } u + \ln\left|u + \sqrt{u^2 - 1}\right| + C$

Forms Involving e^u

81. $\displaystyle\int e^u \, du = e^u + C$

82. $\displaystyle\int u e^u \, du = (u - 1)e^u + C$

83. $\displaystyle\int u^n e^u \, du = u^n e^u - n \int u^{n-1} e^u \, du$

84. $\displaystyle\int \frac{1}{1 + e^u} \, du = u - \ln(1 + e^u) + C$

85. $\displaystyle\int e^{au} \sin bu \, du = \frac{e^{au}}{a^2 + b^2}(a \sin bu - b \cos bu) + C$

86. $\displaystyle\int e^{au} \cos bu \, du = \frac{e^{au}}{a^2 + b^2}(a \cos bu + b \sin bu) + C$

Forms Involving ln u

87. $\displaystyle\int \ln u \, du = u(-1 + \ln u) + C$

88. $\displaystyle\int u \ln u \, du = \frac{u^2}{4}(-1 + 2 \ln u) + C$

89. $\displaystyle\int u^n \ln u \, du = \frac{u^{n+1}}{(n + 1)^2}[-1 + (n + 1) \ln u] + C, \; n \neq -1$

90. $\displaystyle\int (\ln u)^2 \, du = u\left[2 - 2 \ln u + (\ln u)^2\right] + C$

91. $\displaystyle\int (\ln u)^n \, du = u(\ln u)^n - n \int (\ln u)^{n-1} \, du$

Forms Involving Hyperbolic Functions

92. $\displaystyle\int \cosh u \, du = \sinh u + C$

93. $\displaystyle\int \sinh u \, du = \cosh u + C$

94. $\displaystyle\int \text{sech}^2 u \, du = \tanh u + C$

95. $\displaystyle\int \text{csch}^2 u \, du = -\coth u + C$

96. $\displaystyle\int \text{sech } u \tanh u \, du = -\text{sech } u + C$

97. $\displaystyle\int \text{csch } u \coth u \, du = -\text{csch } u + C$

Forms Involving Inverse Hyperbolic Functions (in logarithmic form)

98. $\displaystyle\int \frac{du}{\sqrt{u^2 \pm a^2}} = \ln\left(u + \sqrt{u^2 \pm a^2}\right) + C$

99. $\displaystyle\int \frac{du}{a^2 - u^2} = \frac{1}{2a} \ln\left|\frac{a + u}{a - u}\right| + C$

100. $\displaystyle\int \frac{du}{u\sqrt{a^2 \pm u^2}} = -\frac{1}{a} \ln \frac{a + \sqrt{a^2 \pm u^2}}{|u|} + C$

Answers to Odd-Numbered Exercises

Chapter 7
Section 7.1 (page 452)

1. $-\int_0^6 (x^2 - 6x)\, dx$ 3. $\int_0^3 (-2x^2 + 6x)\, dx$

5. $-6\int_0^1 (x^3 - x)\, dx$

7. 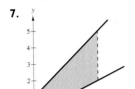   9.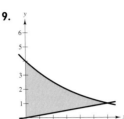

11. 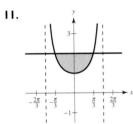   13. (a) $\frac{125}{6}$ (b) $\frac{125}{6}$ 15. d

17. 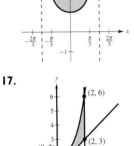   19.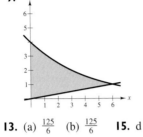

2     $\frac{32}{3}$

21. 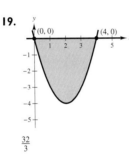   23.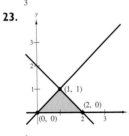

$\frac{9}{2}$ 1

25. 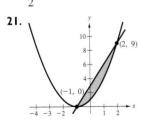   27.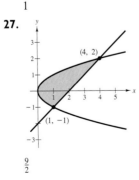

$\frac{3}{2}$ $\frac{9}{2}$

29. 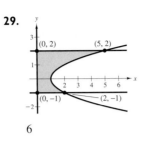   31.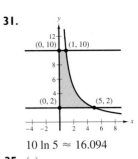

6         $10 \ln 5 \approx 16.094$

33. (a) 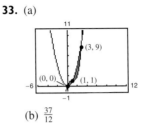   35. (a)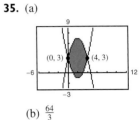

(b) $\frac{37}{12}$ (b) $\frac{64}{3}$

37. (a) 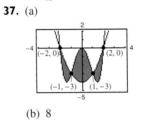   39. (a)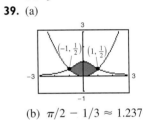

(b) 8   (b) $\pi/2 - 1/3 \approx 1.237$

41. (a) 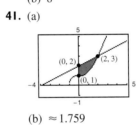   43.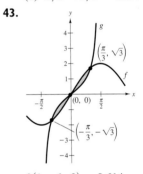

(b) ≈ 1.759

                $2(1 - \ln 2) \approx 0.614$

45. 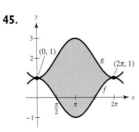   47.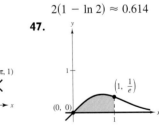

$4\pi \approx 12.566$ $(1/2)(1 - 1/e) \approx 0.316$

49. (a) 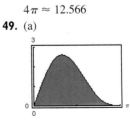   51. (a)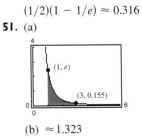

(b) 4   (b) ≈ 1.323

53. (a) 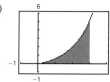 (b) Function is difficult to integrate.
(c) ≈ 4.772

55. (a) 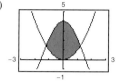 (b) Intersections are difficult to find.
(c) ≈ 6.304

57. $F(x) = \frac{1}{4}x^2 + x$
(a) $F(0) = 0$ (b) $F(2) = 3$
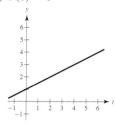 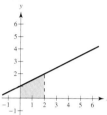
(c) $F(6) = 15$
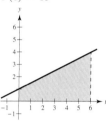

59. $F(\alpha) = (2/\pi)[\sin(\pi\alpha/2) + 1]$
(a) $F(-1) = 0$ (b) $F(0) = 2/\pi \approx 0.6366$
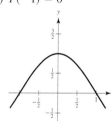
(c) $F(1/2) = (\sqrt{2} + 2)/\pi \approx 1.0868$
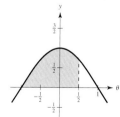

61. 14 **63.** 16
65. Answers will vary. Sample answers:
(a) ≈ 846 ft² (b) ≈ 848 ft²
67. $\int_{-2}^{1} [x^3 - (3x - 2)]\,dx = \frac{27}{4}$

69. $\int_{0}^{1} \left[\frac{1}{x^2 + 1} - \left(-\frac{1}{2}x + 1\right) \right] dx \approx 0.0354$

71. Answers will vary. Example: $x^4 - 2x^2 + 1 \leq 1 - x^2$ on $[-1, 1]$
$\int_{-1}^{1} [(1 - x^2) - (x^4 - 2x^2 + 1)]\,dx = \frac{4}{15}$

73. Offer 2 is better because the cumulative salary (area under the curve) is greater.

75. $b = 9(1 - 1/\sqrt[3]{4}) \approx 3.330$ **77.** $a = 4 - 2\sqrt{2} \approx 1.172$
79. Answers will vary. **81.** $1.625 billion
Sample answer: $\frac{1}{6}$

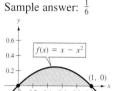

83. (a) $R = 270.32 e^{0.057t}$ (b) $E = 239.97 e^{0.040t}$
or or
$R = 270.32(1.058^t)$ $E = 239.97(1.042^t)$
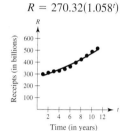 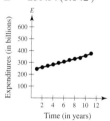

(c) Answer will vary. Sample answer: ≈ 931.6 billion
(d) No. The model for total receipts is increasing at a greater rate than the model for total expenditures. No.

85. Answers will vary. Sample answer: $193,156
87. (a) $k = 3.125$ (b) 13.02083
89. (a) 5.908 m² (b) 11.816 m³ (c) 59,082 lb
91. True **93.** $\sqrt{3}/2 + 7\pi/24 \approx 2.7823$
95. Putnam Problem A1, 1993

Section 7.2 (page 463)

1. $\pi \int_{0}^{1} (-x + 1)^2\,dx = \frac{\pi}{3}$ **3.** $\pi \int_{1}^{4} (\sqrt{x})^2\,dx = \frac{15\pi}{2}$

5. $\pi \int_{0}^{1} [(x^2)^2 - (x^3)^2]\,dx = \frac{2\pi}{35}$ **7.** $\pi \int_{0}^{4} (\sqrt{y})^2\,dy = 8\pi$

9. $\pi \int_{0}^{1} (y^{3/2})^2\,dy = \frac{\pi}{4}$

11. (a) 8π (b) $128\pi/5$ (c) $256\pi/15$ (d) $192\pi/5$
13. (a) $32\pi/3$ (b) $64\pi/3$ **15.** 18π
17. $\pi(16 \ln 2 - \frac{3}{4}) \approx 32.485$ **19.** $208\pi/3$ **21.** $384\pi/5$
23. $\pi \ln 4$ **25.** $3\pi/4$ **27.** $(\pi/2)(1 - 1/e^2) \approx 1.358$
29. $277\pi/3$ **31.** 8π **33.** $\pi^2/2 \approx 4.935$
35. $(\pi/2)(e^2 - 1) \approx 10.036$ **37.** 1.969 **39.** 15.4115
41. A sine curve from $[0, \pi/2]$ revolved about the x-axis.
43. (a) The area appears to be close to 1 and therefore the volume (Area² × π) is near 3.

45. The parabola $y = 4x - x^2$ is a horizontal translation of the parabola $y = 4 - x^2$. Therefore, their volumes are equal.
47. 18π **49.** Proof **51.** $\pi r^2 h[1 - h/H + h^2/(3H^2)]$
53. $\pi/30$ **55.** (a) 60π (b) 50π
57. One-fourth: 32.64 ft; Three-fourths: 67.36 ft
59. (a) ii; right circular cylinder of radius r and height h
(b) iv; ellipsoid whose underlying ellipse has the equation $(x/b)^2 + (y/a)^2 = 1$
(c) iii; sphere of radius r
(d) i; right circular cone of radius r and height h
(e) v; torus of cross-sectional radius r and other radius R
61. (a) $\frac{81}{10}$ (b) $\frac{9}{2}$
63. (a) $1/10$ (b) $\pi/80$ (c) $\sqrt{3}/40$ (d) $\pi/20$
65. $V = \frac{4}{3}\pi(R^2 - r^2)^{3/2}$ **67.** $\pi/3$ **69.** $2\pi/15$
71. $\pi/2$ **73.** $\pi/6$
75. (a) When $a = 1$: represents a square
When $a = 2$: represents a circle
(b) $A = 4\int_0^1 (1 - x^a)^{1/a} dx$

To approximate the volume of the solid, form n slices, each of whose area is approximated by the integral above. Then sum the volumes of these n slices.
77. (a) Proof (b) $V = 2\pi^2 r^2 R$

Section 7.3 (page 472)

1. $2\pi \int_0^2 x^2 dx = \frac{16\pi}{3}$ **3.** $2\pi \int_0^4 x\sqrt{x} dx = \frac{128\pi}{5}$
5. $2\pi \int_0^2 x^3 dx = 8\pi$ **7.** $2\pi \int_0^2 x(4x - 2x^2) dx = \frac{16\pi}{3}$
9. $2\pi \int_0^2 x(x^2 - 4x + 4) dx = \frac{8\pi}{3}$
11. $2\pi \int_0^1 x\left(\frac{1}{\sqrt{2\pi}}e^{-x^2/2}\right) dx = \sqrt{2\pi}\left(1 - \frac{1}{\sqrt{e}}\right) \approx 0.986$
13. $2\pi \int_0^2 y(2 - y) dy = \frac{8\pi}{3}$
15. $2\pi \left[\int_0^{1/2} y\, dy + \int_{1/2}^1 y\left(\frac{1}{y} - 1\right) dy\right] = \frac{\pi}{2}$
17. $2\pi \left[\int_0^8 y^{4/3} dy\right] = \frac{768\pi}{7}$ **19.** $2\pi \int_0^2 y(4 - 2y) dy = 16\pi/3$
21. 16π **23.** 64π
25. Shell method; it is much easier to put x in terms of y rather than vice versa.
27. (a) $128\pi/7$ (b) $64\pi/5$ (c) $96\pi/5$
29. (a) $\pi a^3/15$ (b) $\pi a^3/15$ (c) $4\pi a^3/15$
31. (a) The rectangles would be vertical.
(b) The rectangles would be horizontal.
33. Both integrals yield the volume of the solid generated by revolving the region bounded by the graphs of $y = \sqrt{x - 1}$, $y = 0$, and $x = 5$ about the x-axis.

35. (a)
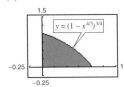
(b) 1.506

37. (a)
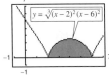
(b) 187.25

39. d **41.** Diameter $= 2\sqrt{4 - 2\sqrt{3}} \approx 1.464$ **43.** $4\pi^2$
45. (a) Proof (b) (i) $V = 2\pi$ (ii) $V = 6\pi^2$
47. (a) region bounded by $y = x^2$, $y = 0$, $x = 0$, $x = 2$
(b) revolved about the y-axis
49. (a) region bounded by $x = \sqrt{6 - y}$, $y = 0$, $x = 0$
(b) revolved about $y = -2$
51. Proof
53. (a) $R_1(n) = n/(n + 1)$ (b) $\lim_{n \to \infty} R_1(n) = 1$
(c) $R_2(n) = n/(n + 2)$ (d) $\lim_{n \to \infty} R_2(n) = 1$
(e) As $n \to \infty$, the graph approaches the line $x = 1$.
55. (a) $\approx 121{,}475$ ft^3 (b) $\approx 121{,}475$ ft^3
57. (a) $64\pi/3$ (b) $2048\pi/35$ (c) $8192\pi/105$ **59.** $c = 2$

Section 7.4 (page 483)

1. (a) and (b) 13 **3.** $\frac{2}{3}(2\sqrt{2} - 1) \approx 1.219$
5. $5\sqrt{5} - 2\sqrt{2} \approx 8.352$ **7.** $779/240 \approx 3.246$
9. $\ln[(\sqrt{2} + 1)/(\sqrt{2} - 1)] \approx 1.763$
11. $\frac{1}{2}(e^2 - 1/e^2) \approx 3.627$ **13.** $\frac{76}{3}$
15. (a)
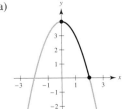
(b) $\int_0^2 \sqrt{1 + 4x^2} dx$
(c) ≈ 4.647

17. (a)
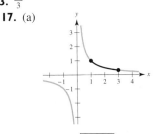
(b) $\int_1^3 \sqrt{1 + \frac{1}{x^4}} dx$
(c) ≈ 2.147

19. (a)
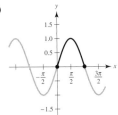
(b) $\int_0^\pi \sqrt{1 + \cos^2 x} dx$
(c) ≈ 3.820

21. (a)
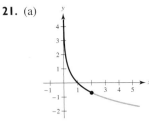
(b) $\int_0^2 \sqrt{1 + e^{-2y}} dy$
$= \int_{e^{-2}}^1 \sqrt{1 + \frac{1}{x^2}} dx$
(c) ≈ 2.221

23. (a)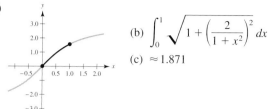
(b) $\int_0^1 \sqrt{1 + \left(\frac{2}{1+x^2}\right)^2}\, dx$
(c) ≈ 1.871

25. b   **27.** (a) 64.125 (b) 64.525 (c) 64.666 (d) 64.672

29. (a)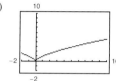
(b) No; $f'(0)$ is not defined.
(c) ≈ 10.5131

31. (a)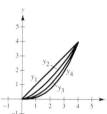
(b) y_1, y_2, y_3, y_4
(c) $s_1 \approx 5.657$; $s_2 \approx 5.759$; $s_3 \approx 5.916$; $s_4 \approx 6.063$

33. Fleeing object: $\frac{2}{3}$ unit
Pursuer: $\frac{1}{2}\int_0^1 \frac{x+1}{\sqrt{x}}\, dx = \frac{1}{2}\left[\frac{2}{3}x^{3/2} + 2x^{1/2}\right]_0^1$
$= \frac{4}{3} = 2\left(\frac{2}{3}\right)$

35. $20[\sinh 1 - \sinh(-1)] \approx 47.0$ m   **37.** $3 \arcsin \frac{2}{3} \approx 2.1892$

39. $2\pi \int_0^3 \frac{1}{3}x^3\sqrt{1 + x^4}\, dx = \frac{\pi}{9}(82\sqrt{82} - 1) \approx 258.85$

41. $2\pi \int_1^2 \left(\frac{x^3}{6} + \frac{1}{2x}\right)\left(\frac{x^2}{2} + \frac{1}{2x^2}\right) dx = \frac{47\pi}{16} \approx 9.23$

43. $2\pi \int_1^8 x\sqrt{1 + \frac{1}{9x^{4/3}}}\, dx = \frac{\pi}{27}(145\sqrt{145} - 10\sqrt{10}) \approx 199.48$

45. 14.424

47. A rectifiable curve is a curve with a finite arc length.

49. The integral formula for the area of a surface of revolution is derived from the formula for the lateral surface area of the frustum of a right circular cone. The formula is $S = 2\pi rL$, where $r = \frac{1}{2}(r_1 + r_2)$, which is the average radius of the frustum, and L is the length of a line segment on the frustum.

51. Proof   **53.** $6\pi(3 - \sqrt{5}) \approx 14.40$

55. Surface area $= \pi/27$ ft$^2 \approx 16.8$ in.^{2}
Amount of glass $= (\pi/27)(0.015/12)$ ft^3
≈ 0.00015 ft^3
≈ 0.25 in.^{3}

57. (a) Answers will vary. Sample answers: 5207.62 in.^{3}
(b) Answers will vary. Sample answers: 1168.64 in.^{2}
(c) $r = 0.0040y^3 - 0.142y^2 + 1.23y + 7.9$
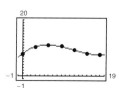
(d) 5279.64 in.^{3}; 1179.5 in.^{2}

59. (a) $\pi(1 - 1/b)$ (b) $2\pi \int_1^b \sqrt{x^4 + 1}/x^3\, dx$
(c) $\lim_{b\to\infty} V = \lim_{b\to\infty} \pi(1 - 1/b) = \pi$
(d) Since $\frac{\sqrt{x^4+1}}{x^3} > \frac{\sqrt{x^4}}{x^3} = \frac{1}{x} > 0$ on $[1, b]$,
we have $\int_1^b \frac{\sqrt{x^4+1}}{x^3}\, dx > \int_1^b \frac{1}{x}\, dx = \left[\ln x\right]_1^b = \ln b$
and $\lim_{b\to\infty} \ln b \to \infty$. Thus, $\lim_{b\to\infty} 2\pi \int_1^b \frac{\sqrt{x^4+1}}{x^3}\, dx = \infty$.

61. Answers will vary.   **63.** $192\pi/5$ **65.** Proof
67. Putnam Problem A1, 1939

Section 7.5   (page 493)

1. 1000 ft-lb   **3.** 448 N-m
5. If an object is moved a distance D in the direction of an applied constant force F, then the work W done by the force is defined as $W = FD$.
7. c, d, a, b; The area under the curves increases in this order.
9. 30.625 in.-lb ≈ 2.55 ft-lb   **11.** 8750 N-cm $= 87.5$ N-m
13. 160 in.-lb ≈ 13.3 ft-lb   **15.** 37.125 ft-lb
17. (a) 487.805 mile-tons $\approx 5.151 \times 10^9$ ft-lb
(b) 1395.349 mile-tons $\approx 1.473 \times 10^{10}$ ft-lb
19. (a) 2.93×10^4 mile-tons $\approx 3.10 \times 10^{11}$ ft-lb
(b) 3.38×10^4 mile-tons $\approx 3.57 \times 10^{11}$ ft-lb
21. (a) 2496 ft-lb (b) 9984 ft-lb   **23.** $470,400\pi$ N-m
25. 2995.2π ft-lb   **27.** $20,217.6\pi$ ft-lb   **29.** 2457π ft-lb
31. 337.5 ft-lb   **33.** 300 ft-lb   **35.** 168.75 ft-lb
37. 7987.5 ft-lb   **39.** $2000 \ln(3/2) \approx 810.93$ ft-lb   **41.** $3k/4$
43. 3249.4 ft-lb   **45.** 10,330.3 ft-lb

Section 7.6   (page 504)

1. $\bar{x} = -\frac{6}{7}$ **3.** $\bar{x} = 12$ **5.** (a) $\bar{x} = 17$ (b) $\bar{x} = -3$
7. $x = 6$ ft   **9.** $(\bar{x}, \bar{y}) = \left(\frac{10}{9}, -\frac{1}{9}\right)$ **11.** $(\bar{x}, \bar{y}) = \left(\frac{5}{8}, \frac{13}{16}\right)$
13. $M_x = 4\rho$, $M_y = 64\rho/5$, $(\bar{x}, \bar{y}) = (12/5, 3/4)$
15. $M_x = \rho/35$, $M_y = \rho/20$, $(\bar{x}, \bar{y}) = (3/5, 12/35)$
17. $M_x = 99\rho/5$, $M_y = 27\rho/4$, $(\bar{x}, \bar{y}) = (3/2, 22/5)$
19. $M_x = 192\rho/7$, $M_y = 96\rho$, $(\bar{x}, \bar{y}) = (5, 10/7)$
21. $M_x = 0$, $M_y = 256\rho/15$, $(\bar{x}, \bar{y}) = (8/5, 0)$
23. $M_x = 27\rho/4$, $M_y = -27\rho/10$, $(\bar{x}, \bar{y}) = (-3/5, 3/2)$
25. $A = \int_0^1 (x - x^2)\, dx = \frac{1}{6}$
$M_x = \int_0^1 \left(\frac{x + x^2}{2}\right)(x - x^2)\, dx = \frac{1}{15}$
$M_y = \int_0^1 x(x - x^2)\, dx = \frac{1}{12}$
27. $A = \int_0^3 (2x + 4)\, dx = 21$
$M_x = \int_0^3 \left(\frac{2x+4}{2}\right)(2x + 4)\, dx = 78$
$M_y = \int_0^3 x(2x + 4)\, dx = 36$

29.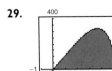
$(\bar{x}, \bar{y}) = (3.0, 126.0)$

31.
$(\bar{x}, \bar{y}) = (0, 16.2)$

33. $(\bar{x}, \bar{y}) = \left(\dfrac{b}{3}, \dfrac{c}{3}\right)$ **35.** $(\bar{x}, \bar{y}) = \left(\dfrac{(a+2b)c}{3(a+b)}, \dfrac{a^2+ab+b^2}{3(a+b)}\right)$

37. $(\bar{x}, \bar{y}) = (0, 4b/(3\pi))$

39. (a)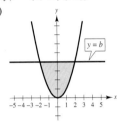
(b) $\bar{x} = 0$ by symmetry
(c) $M_y = \displaystyle\int_{-\sqrt{b}}^{\sqrt{b}} x(b - x^2)\, dx = 0$ because $x(b-x^2)$ is an odd function.
(d) $\bar{y} > b/2$ because the area is greater for $y > b/2$.
(e) $\bar{y} = (3/5)b$

41. (a) $(\bar{x}, \bar{y}) = (0, 12.98)$
(b) $y = (-1.02 \times 10^{-5})x^4 - 0.0019x^2 + 29.28$
(c) $(\bar{x}, \bar{y}) = (0, 12.85)$

43.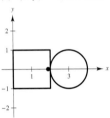
$(\bar{x}, \bar{y}) = \left(\dfrac{4+3\pi}{4+\pi}, 0\right)$

45.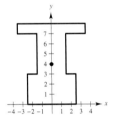
$(\bar{x}, \bar{y}) = \left(0, \dfrac{135}{34}\right)$

47. $(\bar{x}, \bar{y}) = \left(\dfrac{2+3\pi}{2+\pi}, 0\right)$ **49.** $160\pi^2 \approx 1579.14$

51. $128\pi/3 \approx 134.04$

53. The center of mass $(\bar{x}, \bar{y})$ is $\bar{x} = M_y/m$ and $\bar{y} = M_x/m$, where:
1. $m = m_1 + m_2 + \cdots + m_n$ is the total mass of the system.
2. $M_y = m_1 x_1 + m_2 x_2 + \cdots + m_n x_n$ is the moment about the y-axis.
3. $M_x = m_1 y_1 + m_2 y_2 + \cdots + m_n y_n$ is the moment about the x-axis.

55. (a) $\left(\dfrac{5}{6}, 2\dfrac{5}{18}\right)$; The plane region has been translated 2 units up.
(b) $\left(2\dfrac{5}{6}, \dfrac{5}{18}\right)$; The plane region has been translated 2 units to the right.
(c) $\left(\dfrac{5}{6}, -\dfrac{5}{18}\right)$; The plane region has been reflected across the x-axis.
(d) Not possible

57. $(\bar{x}, \bar{y}) = (0, 2r/\pi)$

59. $(\bar{x}, \bar{y}) = \left(\dfrac{n+1}{n+2}, \dfrac{n+1}{4n+2}\right)$; As $n \to \infty$, the region shrinks toward the line segments $y = 0$ for $0 \le x \le 1$ and $x = 1$ for $0 \le y \le 1$; $(\bar{x}, \bar{y}) \to \left(1, \dfrac{1}{4}\right)$.

Section 7.7 (page 511)

1. 936 lb   **3.** 748.8 lb   **5.** 1123.2 lb   **7.** 748.8 lb

9. 1064.96 lb   **11.** 117,600 N   **13.** 2,381,400 N
15. 2814 lb   **17.** 6753.6 lb   **19.** 94.5 lb   **21.** Proof
23. Proof   **25.** 960 lb   **27.** 3010.8 lb   **29.** 6448.7 lb
31. (a) $3\sqrt{2}/2 \approx 2.12$ ft
(b) The pressure increases with increasing depth.
33. The fluid force F of constant weight-density w (per unit of volume) against a submerged vertical plane region from $y = c$ to $y = d$ is
$$F = w \lim_{\|\Delta\| \to 0} \sum_{i=1}^{n} h(y_i) L(y_i) \Delta y = w \int_c^d h(y) L(y)\, dy$$
where $h(y)$ is the depth of the fluid at y and $L(y)$ is the horizontal length of the region at y.

Review Exercises for Chapter 7 (page 513)

1.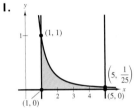
$4/5$

3.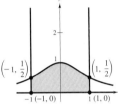
$\pi/2$

5.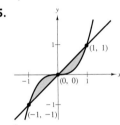

7.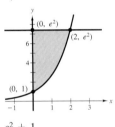
$e^2 + 1$

9.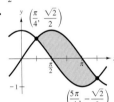
$2\sqrt{2}$

11.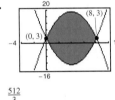
$\dfrac{512}{3}$

13.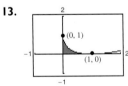
$\dfrac{1}{6}$

15. $\displaystyle\int_0^2 [0 - (y^2 - 2y)]\, dy = \int_{-1}^0 2\sqrt{x+1}\, dx = \dfrac{4}{3}$

17. $\displaystyle\int_0^2 \left[1 - \left(1 - \dfrac{x}{2}\right)\right] dx + \int_2^3 [1 - (x-2)]\, dx$
$= \displaystyle\int_0^1 [(y+2) - (2-2y)]\, dy = \dfrac{3}{2}$

19. Job 1. The salary for job 1 is greater than the salary for job 2 for all the years except the first and tenth years.

21. (a) $64\pi/3$ (b) $128\pi/3$ (c) $64\pi/3$ (d) $160\pi/3$

23. (a) 64π (b) 48π **25.** $\pi^2/4$
27. $(4\pi/3)(20 - 9 \ln 3) \approx 42.359$ **29.** $\frac{4}{15}$ **31.** 1.958 ft
33. $\frac{8}{15}(1 + 6\sqrt{3}) \approx 6.076$ **35.** 4018.2 ft
37. 15π **39.** 50 in.-lb $\approx$ 4.167 ft-lb
41. $104{,}000\pi$ ft-lb $\approx$ 163.4 foot-tons   **43.** 250 ft-lb
45. $a = 15/4$ **47.** $(\bar{x}, \bar{y}) = (a/5, a/5)$ **49.** $(\bar{x}, \bar{y}) = (0, 2a^2/5)$
51. $(\bar{x}, \bar{y}) = \left(\dfrac{2(9\pi + 49)}{3(\pi + 9)}, 0\right)$

53. Let D = surface of liquid; ρ = weight per cubic volume.
$$F = \rho \int_c^d (D - y)[f(y) - g(y)]\,dy$$
$$= \rho\left[\int_c^d D[f(y) - g(y)]\,dy - \int_c^d y[f(y) - g(y)]\,dy\right]$$
$$= \rho\left[\int_c^d [f(y) - g(y)]\,dy\right]\left[D - \dfrac{\int_c^d y[f(y) - g(y)]\,dy}{\int_c^d [f(y) - g(y)]\,dy}\right]$$
$$= \rho(\text{area})(D - \bar{y})$$
$$= \rho(\text{area})(\text{depth of centroid})$$

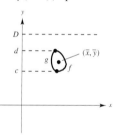

P.S. Problem Solving   (page 515)

1. 3   **3.** (a) $4\pi^2$ (b) $2\pi^2 r^2 R$ **5.** $\pi h^3/6$
7. (a) Area S is 16 times area R.
(b) Let point A be (a, a^3). The equation of the tangent line to the curve $y = x^3$ at A is $y = 3a^2 x - 2a^3$, and point B is $(-2a, -8a^3)$. Area R is
$$\int_{-2a}^{a} (x^3 - 3a^2 x + 2a^3)\,dx = \dfrac{27a^4}{4}.$$
Then, the equation of the tangent line to the curve $y = x^3$ at B is $y = 12a^2 x + 16a^3$, and point C is $(4a, 64a^3)$. Area S is
$$\int_{-2a}^{4a} (12a^2 x + 16a^3 - x^3)\,dx = 108a^4.$$
Therefore, area S is 16 times area R.

9. (a) $\dfrac{ds}{dx} = \sqrt{1 + [f'(x)]^2}$
(b) $ds = \sqrt{1 + [f'(x)]^2}\,dx$; $(ds)^2 = (dx)^2 + (dy)^2$
(c) $\displaystyle\int_1^x \sqrt{1 + \tfrac{9}{4}t}\,dt$
(d) $s(2) \approx 2.0858$. This is the arc length of the curve.

11.

(a) $\left(\dfrac{12}{7}, 0\right)$

(b) $\left(\dfrac{2b}{b + 1}, 0\right)$

(c) $(2, 0)$

13. (a) 12   (b) 7.5
15. Consumer surplus: 1600; Producer surplus: 400
17. Wall at shallow end: 9984 pounds
Wall at deep end: 39,936 pounds
Side wall: $19{,}968 + 26{,}624 = 46{,}592$ lb

Chapter 8

Section 8.1   (page 522)

1. b   **3.** c
5. $\displaystyle\int u^n\,du$ **7.** $\displaystyle\int \dfrac{du}{u}$
$u = 3x - 2, n = 4$ $u = 1 - 2\sqrt{x}$
9. $\displaystyle\int \dfrac{du}{\sqrt{a^2 - u^2}}$ **11.** $\displaystyle\int \sin u\,du$
$u = t, a = 1$ $u = t^2$
13. $\displaystyle\int e^u\,du$ **15.** $(x - 4)^6 + C$
$u = \sin x$
17. $-5/[4(z - 4)^4] + C$ **19.** $\tfrac{1}{2}v^2 - 1/[6(3v - 1)^2] + C$
21. $-\tfrac{1}{3}\ln|-t^3 + 9t + 1| + C$ **23.** $\tfrac{1}{2}x^2 + x + \ln|x - 1| + C$
25. $\ln(1 + e^x) + C$ **27.** $\tfrac{1}{15}x(12x^4 + 20x^2 + 15) + C$
29. $\sin(2\pi x^2)/(4\pi) + C$ **31.** $-(1/\pi)\csc \pi x + C$
33. $\tfrac{1}{5}e^{5x} + C$ **35.** $2\ln(1 + e^x) + C$ **37.** $(\ln x)^2 + C$
39. $-\ln(1 - \sin x) + C = \ln|\sec x(\sec x + \tan x)| + C$
41. $\csc\theta + \cot\theta + C = (1 + \cos\theta)/\sin\theta + C$
43. $-\tfrac{1}{2}\arcsin(2t - 1) + C$ **45.** $\tfrac{1}{2}\ln|\cos(2/t)| + C$
47. $3\arcsin[(x - 3)/3] + C$ **49.** $\tfrac{1}{4}\arctan[(2x + 1)/8] + C$
51. (a) 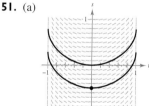   (b) $\tfrac{1}{2}\arcsin t^2 - \tfrac{1}{2}$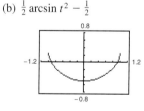

53. (a) 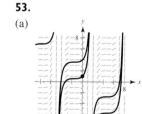   (b) $2\tan x + 2\sec x - x - 1 + C$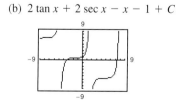

55. $y = 3e^{0.2x}$

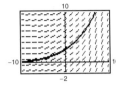

57. $y = \frac{1}{2}e^{2x} + 2e^x + x + C$

59. $y = \frac{1}{2}\arctan(\tan x/2) + C$ **61.** $\frac{1}{2}$
63. $\frac{1}{2}(1 - e^{-1}) \approx 0.316$ **65.** 4 **67.** $\pi/18$
69. $5\sqrt{5} \approx 11.18$ **71.** $\frac{3}{2}\ln(\frac{34}{9}) + \frac{2}{3}\arctan(\frac{5}{3}) \approx 2.68$
73. $\frac{4}{3} \approx 1.333$
75. $\frac{1}{3}\arctan[\frac{1}{3}(x + 2)] + C$ **77.** $\tan\theta - \sec\theta + C$
Graphs will vary. Graphs will vary.
Example: Example:

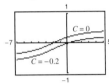 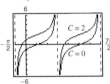

One graph is a vertical   One graph is a vertical
translation of the other. translation of the other.

79. Power rule: $\int u^n \, du = \frac{u^{n+1}}{n+1} + C;\ u = x^2 + 1,\ du = 2x,\ n = 3$

81. Log rule: $\int \frac{du}{u} = \ln|u| + C;\ u = x^2 + 1,\ du = 2x$

83. Using laws of logarithms, $y_1 = e^{x+C_1} = e^x \cdot e^{C_1}$, where e^{C_1} is a constant. Therefore, e^{C_1} can be replaced by C, resulting in $y_2 = Ce^x$.

85. $a = \sqrt{2},\ b = \frac{\pi}{4};\ -\frac{1}{\sqrt{2}}\ln\left|\csc\left(x + \frac{\pi}{4}\right) + \cot\left(x + \frac{\pi}{4}\right)\right| + C$

87.

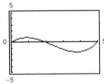

Negative; more area below the x-axis than above

89. a
91. (a)

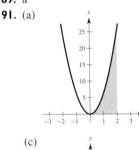

(b)

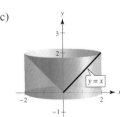

(c)

93. (a) $\pi(1 - e^{-1}) \approx 1.986$ (b) $b = \sqrt{\ln\left(\frac{3\pi}{3\pi - 4}\right)} \approx 0.743$

95. $(8\pi/3)(10\sqrt{10} - 1) \approx 256.545$
97. $\frac{1}{3}\arctan 3 \approx 0.416$ **99.** ≈ 1.0320
101. (a) $\frac{1}{3}\sin x(\cos^2 x + 2)$
  (b) $\frac{1}{15}\sin x(3\cos^4 x + 4\cos^2 x + 8)$
  (c) $\frac{1}{35}\sin x(5\cos^6 x + 6\cos^4 x + 8\cos^2 x + 16)$
  (d) Use the power reducing formula, $\int \cos^n x\, dx = \frac{\cos^{n-1}\sin^n x}{n} + \frac{n-1}{n}\int \cos^{n-2} x\, dx$ until your integral is $\int \cos x\, dx$, which you can then integrate.

103. Proof

Section 8.2 (page 531)

1. b **2.** d **3.** c **4.** a **5.** $u = x,\ dv = e^{2x}\,dx$
7. $u = (\ln x)^2,\ dv = dx$ **9.** $u = x,\ dv = \sec^2 x\, dx$
11. $-(1/4e^{2x})(2x + 1) + C$ **13.** $e^x(x^3 - 3x^2 + 6x - 6) + C$
15. $\frac{1}{3}e^{x^3} + C$ **17.** $\frac{1}{4}[2(t^2 - 1)\ln|t + 1| - t^2 + 2t] + C$
19. $\frac{1}{3}(\ln x)^3 + C$ **21.** $e^{2x}/[4(2x + 1)] + C$
23. $(x - 1)^2 e^x + C$ **25.** $\frac{2}{15}(x - 1)^{3/2}(3x + 2) + C$
27. $x \sin x + \cos x + C$
29. $(6x - x^3)\cos x + (3x^2 - 6)\sin x + C$
31. $-t \csc t - \ln|\csc t + \cot t| + C$
33. $x \arctan x - \frac{1}{2}\ln(1 + x^2) + C$
35. $\frac{1}{5}e^{2x}(2\sin x - \cos x) + C$ **37.** $y = \frac{1}{2}e^{x^2} + C$
39. $y = \frac{2}{405}(27t^2 - 24t + 32)\sqrt{2 + 3t} + C$ **41.** $\sin y = x^2 + C$
43. (a) 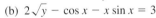 (b) $2\sqrt{y} - \cos x - x\sin x = 3$

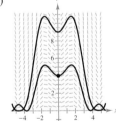 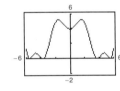

45. 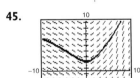 **47.** $4 - 12/e^2$ **49.** $\pi/2 - 1$
51. $(\pi - 3\sqrt{3} + 6)/6 \approx 0.658$

53. $\frac{1}{2}[e(\sin 1 - \cos 1) + 1] \approx 0.909$
55. $(24 \ln 2 - 7)/9 \approx 1.071$
57. $8 \arcsec 4 + \sqrt{3}/2 - \sqrt{15}/2 - 2\pi/3 \approx 7.380$
59. $(e^{2x}/4)(2x^2 - 2x + 1) + C$
61. $(3x^2 - 6)\sin x - (x^3 - 6x)\cos x + C$
63. $x \tan x + \ln|\cos x| + C$ **65.** $2(\sin\sqrt{x} - \sqrt{x}\cos\sqrt{x}) + C$
67. $\frac{128}{15}$ **69.** $\frac{1}{2}x[\cos(\ln x) + \sin(\ln x)] + C$ **71.** Product Rule
73. No **75.** Yes. Let $u = x^2$ and $dv = e^{2x}\,dx$.
77. Yes. Let $u = x,\ dv = (1\sqrt{x + 1})\,dx$. (Substitution also works. Let $u = \sqrt{x + 1}$).

79. (a) $-(e^{-4t}/128)(32t^3 + 24t^2 + 12t + 3) + C$
(b) Graphs will vary. Example: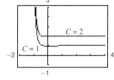
(c) One graph is a vertical translation of the other.

81. (a) $\frac{1}{13}(2e^{-\pi} + 3) \approx 0.2374$
(b) Graphs will vary. Example: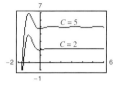
(c) One graph is a vertical translation of the other.

83. $\frac{2}{5}(2x - 3)^{3/2}(x + 1) + C$ **85.** $\frac{1}{3}\sqrt{4 + x^2}(x^2 - 8) + C$

87. $n = 0$: $x(\ln x - 1) + C$
$n = 1$: $\frac{1}{4}x^2(2 \ln x - 1) + C$
$n = 2$: $\frac{1}{9}x^3(3 \ln x - 1) + C$
$n = 3$: $\frac{1}{16}x^4(4 \ln x - 1) + C$
$n = 4$: $\frac{1}{25}x^5(5 \ln x - 1) + C$

$$\int x^n \ln x \, dx = \frac{x^{n+1}}{(n+1)^2}[(n+1)\ln x - 1] + C$$

89–93. Proofs
95. $\frac{1}{16}x^4(4 \ln x - 1) + C$ **97.** $\frac{1}{13}e^{2x}(2 \cos 3x + 3 \sin 3x) + C$

99. 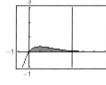 **101.**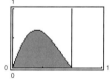

$1 - \dfrac{5}{e^4} \approx 0.908$ $\dfrac{\pi}{1 + \pi^2}\left(\dfrac{1}{e} + 1\right) \approx 0.395$

103. (a) 1 (b) $\pi(e - 2) \approx 2.257$ (c) $\frac{1}{2}\pi(e^2 + 1) \approx 13.177$
(d) $\left(\dfrac{e^2 + 1}{4}, \dfrac{e - 2}{2}\right) \approx (2.097, 0.359)$

105. In Example 6, we showed that the centroid of an equivalent region was $(1, \pi/8)$. By symmetry, the centroid of this region is $(1, \pi/8)$.

107. $[7/(10\pi)](1 - e^{-4\pi}) \approx 0.223$ **109.** $931,265
111. Proof   **113.** $b_n = [8h/(n\pi)^2]\sin(n\pi/2)$
115. Shell: $V = \pi\left[b^2 f(b) - a^2 f(a) - \int_a^b x^2 f'(x)\, dx\right]$

Disk: $V = \pi\left[b^2 f(b) - a^2 f(a) - \int_{f(a)}^{f(b)} [f^{-1}(y)]^2\, dy\right]$

Both methods yield the same volume because $x = f^{-1}(y)$, $f'(x)\, dx = dy$, if $y = f(a)$ then $x = a$, and if $y = f(b)$ then $x = b$.

117. (a) $y = \frac{1}{4}(3 \sin 2x - 6x \cos 2x)$ (b)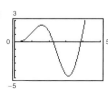

(c) You obtain the following points.

n	x^n	y^n
0	0	0
1	0.05	0.000250
2	0.10	0.001992
3	0.15	0.006689
4	0.20	0.015745
80	4.00	1.615019

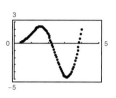

(d) You obtain the following points.

n	x^n	y^n
0	0	0
1	0.1	0.001992
2	0.2	0.015745
3	0.3	0.052081
4	0.4	0.119993
40	4.0	1.615019

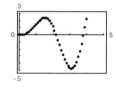

(e) Smaller step size, more data
119. The graph of $y = x \sin x$ is below the graph of $y = x$ on $[0, \pi/2]$.

Section 8.3 (page 540)

1. c **2.** a **3.** d **4.** b
5. $-\frac{1}{4}\cos^4 x + C$ **7.** $\frac{1}{12}\sin^6 2x + C$
9. $-\frac{1}{3}\cos^3 x + \frac{2}{5}\cos^5 x - \frac{1}{7}\cos^7 x + C$
11. $\frac{2}{3}\sin^{3/2}\theta - \frac{2}{7}\sin^{7/2}\theta + C$ **13.** $\frac{1}{12}(6x + \sin 6x) + C$
15. $\frac{1}{8}(\alpha - \frac{1}{4}\sin 4\alpha) + C$ or $\frac{1}{8}\alpha - \frac{1}{32}\sin 4\alpha + C$
17. $\frac{1}{8}(2x^2 - 2x \sin 2x - \cos 2x) + C$ **19.** $\frac{2}{3}$ **21.** $\frac{16}{35}$
23. $5\pi/32$ **25.** $\frac{1}{3}\ln|\sec 3x + \tan 3x| + C$
27. $\frac{1}{15}\tan 5x(3 + \tan^2 5x) + C$
29. $(\sec \pi x \tan \pi x + \ln|\sec \pi x + \tan \pi x|)/(2\pi) + C$
31. $\tan^4(x/4) - 2\tan^2(x/4) - 4\ln|\cos(x/4)| + C$
33. $\frac{1}{2}\tan^2 x + C$ **35.** $\frac{1}{3}\tan^3 x + C$ **37.** $\frac{1}{24}\sec^6 4x + C$
39. $\frac{1}{3}\sec^3 x + C$ **41.** $\ln|\sec x + \tan x| - \sin x + C$
43. $(12\pi\theta - 8 \sin 2\pi\theta + \sin 4\pi\theta)/(32\pi) + C$
45. $y = \frac{1}{9}\sec^3 3x - \frac{1}{3}\sec 3x + C$
47. (a) 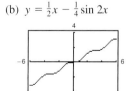 (b) $y = \frac{1}{2}x - \frac{1}{4}\sin 2x$

49. 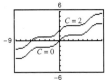   **51.** $-\frac{1}{10}(\cos 5x + 5\cos x) + C$

53. $\frac{1}{8}(2\sin 2\theta - \sin 4\theta) + C$ **55.** $\frac{1}{4}(\ln|\csc^2 2x| - \cot^2 2x) + C$
57. $-\cot\theta - \frac{1}{3}\cot^3\theta + C$ **59.** $\ln|\csc t - \cot t| + \cos t + C$
61. $\ln|\csc x - \cot x| + \cos x + C$ **63.** $t - 2\tan t + C$
65. π **67.** $\frac{1}{2}(1 - \ln 2)$ **69.** $\ln 2$ **71.** $\frac{4}{3}$
73. $\frac{1}{16}(6x + 8\sin x + \sin 2x) + C$
Graphs will vary. Example:

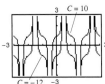

75. $\left[\sec^3 \pi x \tan \pi x + \frac{3}{2}(\sec \pi x \tan \pi x + \ln|\sec \pi x + \tan \pi x|)\right]/(4\pi) + C$
Graphs will vary. Example:

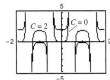

77. $(\sec^5 \pi x)/(5\pi) + C$ **79.** $3\sqrt{2}/10$ **81.** $3\pi/16$
Graphs will vary. Example:

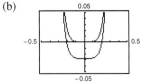

83. (a) Save one sine factor and convert the remaining factors to cosines. Then, expand and integrate.
(b) Save one cosine factor and convert the remaining factors to sines. Then, expand and integrate.
(c) Make repeated use of the power reducing formulas to convert the integrand to odd powers of the cosine. Then, proceed as in part (b).
85. (a) $\frac{1}{18}\tan^6 3x + \frac{1}{12}\tan^4 3x + C_1$, $\frac{1}{18}\sec^6 3x - \frac{1}{12}\sec^4 3x + C_2$
(b) (c) Proof

87. $\frac{1}{3}$ **89.** 1   **91.** $2\pi(1 - \pi/4) \approx 1.348$
93. (a) $\pi^2/2$ (b) $(\bar{x}, \bar{y}) = (\pi/2, \pi/8)$ **95–97.** Proofs
99. $-\frac{1}{15}\cos x(3\sin^4 x + 4\sin^2 x + 8) + C$
101. $\frac{5}{6\pi}\tan\frac{2\pi x}{5}\left(\sec^2\frac{2\pi x}{5} + 2\right) + C$

103. (a) $H(t) \approx 57.72 - 23.36\cos(\pi t/6) - 2.75\sin(\pi t/6)$
(b) $L(t) \approx 42.04 - 20.91\cos(\pi t/6) - 4.33\sin(\pi t/6)$
(c) The maximum difference is at $t \approx 4.9$ or early summer.

105. Proof

Section 8.4 (page 549)

1. b   **2.** d   **3.** a   **4.** c   **5.** $x/(25\sqrt{25-x^2}) + C$
7. $5\ln\left|(5 - \sqrt{25-x^2})/x\right| + \sqrt{25-x^2} + C$
9. $\ln\left|x + \sqrt{x^2-4}\right| + C$ **11.** $\frac{1}{15}(x^2-4)^{3/2}(3x^2+8) + C$
13. $\frac{1}{3}(1+x^2)^{3/2} + C$ **15.** $\frac{1}{2}[\arctan x + x/(1+x^2)] + C$
17. $\frac{1}{2}x\sqrt{4+9x^2} + \frac{2}{3}\ln\left|3x + \sqrt{4+9x^2}\right| + C$
19. $\frac{25}{4}\arcsin(2x/5) + \frac{1}{2}x\sqrt{25-4x^2} + C$
21. $\sqrt{x^2+9} + C$ **23.** $\arcsin(x/4) + C$
25. $4\arcsin(x/2) + x\sqrt{4-x^2} + C$ **27.** $\ln\left|x + \sqrt{x^2-9}\right| + C$
29. $-\frac{(1-x^2)^{3/2}}{3x^3} + C$ **31.** $-\frac{1}{3}\ln\left|\frac{\sqrt{4x^2+9}+3}{2x}\right| + C$
33. $5\sqrt{x^2+5}/(x+5) + C$ **35.** $\frac{1}{3}(1+e^{2x})^{3/2} + C$
37. $\frac{1}{2}\left(\arcsin e^x + e^x\sqrt{1-e^{2x}}\right) + C$
39. $\frac{1}{4}(x/(x^2+2) + (1/\sqrt{2})\arctan(x/\sqrt{2})) + C$
41. $x\,\text{arcsec}\,2x - \frac{1}{2}\ln\left|2x + \sqrt{4x^2-1}\right| + C$
43. $\arcsin[(x-2)/2] + C$
45. $\sqrt{x^2+4x+8} - 2\ln\left|\sqrt{x^2+4x+8} + (x+2)\right| + C$
47. (a) and (b) $\sqrt{3} - \pi/3 \approx 0.685$
49. (a) and (b) $9(2 - \sqrt{2}) \approx 5.272$
51. (a) and (b) $-(9/2)\ln(2\sqrt{7}/3 - 4\sqrt{3}/3 - \sqrt{21}/3 + 8/3) + 9\sqrt{3} - 2\sqrt{7} \approx 12.644$
53. $\sqrt{x^2-9} - 3\arctan(\sqrt{x^2-9}/3) + 1$
55. $\frac{1}{2}(x-15)\sqrt{x^2+10x+9} + 33\ln\left|\sqrt{x^2+10x+9} + (x+5)\right| + C$
57. $\frac{1}{2}\left(x\sqrt{x^2-1} + \ln\left|x + \sqrt{x^2-1}\right|\right) + C$
59. (a) Let $u = a\sin\theta$, $\sqrt{a^2-u^2} = a\cos\theta$, where $-\pi/2 \leq \theta \leq \pi/2$.
(b) Let $u = a\tan\theta$, $\sqrt{a^2+u^2} = a\sec\theta$, where $-\pi/2 < \theta < \pi/2$.
(c) Let $u = a\sec\theta$, $\sqrt{u^2-a^2} = \tan\theta$ if $u > a$ and $\sqrt{u^2-a^2} = -\tan\theta$ if $u < -a$, where $0 \leq \theta < \pi/2$ or $\pi/2 < \theta \leq \pi$.
61. $\ln\sqrt{x^2+9} + C$ **63.** True
65. False: $\int_0^{\sqrt{3}} \frac{dx}{(1+x^2)^{3/2}} = \int_0^{\pi/3} \cos\theta\,d\theta$
67. πab **69.** (a) $5\sqrt{2}$ (b) $25(1-\pi/4)$ (c) $r^2(1-\pi/4)$
71. $6\pi^2$ **73.** $\ln\left[\frac{5(\sqrt{2}+1)}{\sqrt{26}+1}\right] + \sqrt{26} - \sqrt{2} \approx 4.367$

75. Length of one arch of sine curve: $y = \sin x$, $y' = \cos x$

$$L_1 = \int_0^\pi \sqrt{1 + \cos^2 x}\, dx$$

Length of one arch of cosine curve: $y = \cos x$, $y' = -\sin x$

$$L_2 = \int_{-\pi/2}^{\pi/2} \sqrt{1 + \sin^2 x}\, dx$$

$$= \int_{-\pi/2}^{\pi/2} \sqrt{1 + \cos^2(x - \pi/2)}\, dx, \quad u = x - \pi/2,\, du = dx$$

$$= \int_{-\pi}^{0} \sqrt{1 + \cos^2 u}\, du = \int_0^\pi \sqrt{1 + \cos^2 u}\, du = L_1$$

77. (a) 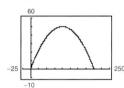   (b) 200

(c) $100\sqrt{2} + 50 \ln\left[\left(\sqrt{2} + 1\right)/\left(\sqrt{2} - 1\right)\right] \approx 229.559$

79. $(0, 0.422)$ **81.** $(\pi/32)\left[102\sqrt{2} - \ln(3 + 2\sqrt{2})\right] \approx 13.989$

83. (a) 187.2π lb   (b) $62.4\pi d$ lb

85. Proof   **87.** $12 + 9\pi/2 - 25 \arcsin(3/5) \approx 10.050$

Section 8.5   (page 559)

1. $\dfrac{A}{x} + \dfrac{B}{x - 10}$ **3.** $\dfrac{A}{x} + \dfrac{Bx + C}{x^2 + 10}$ **5.** $\dfrac{A}{x} + \dfrac{B}{x - 10}$

7. $\tfrac{1}{2}\ln|(x - 1)/(x + 1)| + C$ **9.** $\ln|(x - 1)/(x + 2)| + C$

11. $\tfrac{3}{2}\ln|2x - 1| - 2\ln|x + 1| + C$

13. $5\ln|x - 2| - \ln|x + 2| - 3\ln|x| + C$

15. $x^2 + \tfrac{3}{2}\ln|x - 4| - \tfrac{1}{2}\ln|x + 2| + C$

17. $1/x + \ln|x^4 + x^3| + C$

19. $2\ln|x - 2| - \ln|x| - 3/(x - 2) + C$

21. $\ln|(x^2 + 1)/x| + C$

23. $\tfrac{1}{6}\left[\ln|(x - 2)/(x + 2)| + \sqrt{2}\arctan(x/\sqrt{2})\right] + C$

25. $\tfrac{1}{16}\ln|(4x^2 - 1)/(4x^2 + 1)| + C$

27. $\ln|x + 1| + \sqrt{2}\arctan\left[(x - 1)/\sqrt{2}\right] + C$

29. $\ln 2$ **31.** $\tfrac{1}{2}\ln(8/5) - \pi/4 + \arctan 2 \approx 0.557$

33. $y = 3\ln|x - 3| - 9/(x - 3) + 9$

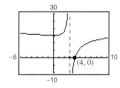

35. $y = \left(\sqrt{2}/2\right)\arctan(x/\sqrt{2}) - 1/[2(x^2 + 2)] + 5/4$

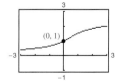

37. $y = \ln|x - 2| + \tfrac{1}{2}\ln|x^2 + x + 1|$
$\quad - \sqrt{3}\arctan\left[(2x + 1)/\sqrt{3}\right] - \tfrac{1}{2}\ln 13$
$\quad + \sqrt{3}\arctan(7/\sqrt{3}) + 10$

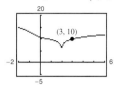

39. $y = \tfrac{1}{4}\ln|(x - 2)/(x + 2)| + \tfrac{1}{4}\ln 2 + 4$

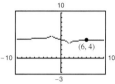

41. $\ln\left|\dfrac{\cos x}{\cos x - 1}\right| + C$ **43.** $\ln\left|\dfrac{-1 + \sin x}{2 + \sin x}\right| + C$

45. $\tfrac{1}{5}\ln\left|\dfrac{e^x - 1}{e^x + 4}\right| + C$ **47–49.** Proofs

51. $y = \tfrac{3}{2}\ln\left|\dfrac{2 + x}{2 - x}\right| + 3$ **53.** First divide x^3 by $(x - 5)$.

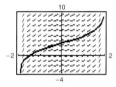

55. (a) Log Rule   (b) Partial fractions   (c) Inverse Tangent Rule

57. $12\ln\left(\tfrac{9}{8}\right) \approx 1.4134$ **59.** 4.90 or \$490,000

61. $V = 2\pi\left(\arctan 3 - \tfrac{3}{10}\right) \approx 5.963$; $(\bar{x}, \bar{y}) \approx (1.521, 0.412)$

63. $x = n[e^{(n+1)kt} - 1]/[n + e^{(n+1)kt}]$ **65.** $\pi/8$

Section 8.6   (page 565)

1. $-\tfrac{1}{2}x(2 - x) + \ln|1 + x| + C$

3. $\tfrac{1}{2}\left[e^x\sqrt{e^{2x} + 1} + \ln\left(e^x + \sqrt{e^{2x} + 1}\right)\right] + C$

5. $-\sqrt{1 - x^2}/x + C$

7. $\tfrac{1}{16}(6x - 3\sin 2x \cos 2x - 2\sin^3 2x \cos 2x) + C$

9. $-2\left(\cot\sqrt{x} + \csc\sqrt{x}\right) + C$ **11.** $x - \tfrac{1}{2}\ln(1 + e^{2x}) + C$

13. $\tfrac{1}{16}x^4(4\ln x - 1) + C$

15. (a) and (b) $e^x(x^2 - 2x + 2) + C$

17. (a) and (b) $\ln|(x + 1)/x| - 1/x + C$

19. $\tfrac{1}{2}\left[(x^2 + 1)\mathrm{arcsec}(x^2 + 1) - \ln\left(x^2 + 1 + \sqrt{x^4 + 2x^2}\right)\right] + C$

21. $\sqrt{x^2 - 4}/(4x) + C$ **23.** $\tfrac{2}{9}[\ln|1 - 3x| + 1/(1 - 3x)] + C$

25. $e^x \arccos(e^x) - \sqrt{1 - e^{2x}} + C$

27. $\tfrac{1}{2}(x^2 + \cot x^2 + \csc x^2) + C$

29. $\left(\sqrt{2}/2\right)\arctan\left[(1 + \sin\theta)/\sqrt{2}\right] + C$

31. $-\sqrt{2 + 9x^2}/(2x) + C$

33. $\tfrac{1}{4}(2\ln|x| - 3\ln|3 + 2\ln|x||) + C$

35. $(3x - 10)/[2(x^2 - 6x + 10)] + \tfrac{3}{2}\arctan(x - 3) + C$

37. $\tfrac{1}{2}\ln\left|x^2 - 3 + \sqrt{x^4 - 6x^2 + 5}\right| + C$

39. $-\tfrac{1}{3}\sqrt{4 - x^2}(x^2 + 8) + C$

41. $2/(1 + e^x) - 1/[2(1 + e^x)^2] + \ln(1 + e^x) + C$

43. $\frac{1}{2}(e - 1) \approx 0.8591$ **45.** $9 \ln(3) - \frac{26}{9} \approx 6.9986$
47. $\pi/2$ **49.** $\pi^3/8 - 3\pi + 6 \approx 0.4510$ **51–55.** Proofs
57. $y = -2\sqrt{1 - x}/\sqrt{x} + 7$

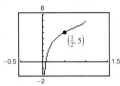

59. $y = \frac{1}{2}[(x - 3)/(x^2 - 6x + 10) + \arctan(x - 3)]$

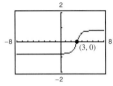

61. $y = -\csc \theta + \sqrt{2} + 2$

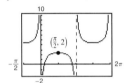

63. $\dfrac{1}{\sqrt{5}} \ln \left| \dfrac{2 \tan(\theta/2) - 3 - \sqrt{5}}{2 \tan(\theta/2) - 3 + \sqrt{5}} \right| + C$ **65.** $\ln 2$
67. $\frac{1}{2} \ln(3 - 2 \cos \theta) + C$ **69.** $2 \sin \sqrt{\theta} + C$ **71.** $\frac{40}{3}$
73. Use Formula 23 and let $a = 1$, $u = e^x$, and $du = e^x \, dx$, because it is in the form $\int \dfrac{1}{a^2 + u^2} \, du$.
75. Use Formula 81 and let $u = x^2$ and $du = 2x \, dx$, because it is in the form $\int e^u \, du$.
77. Impossible; there is no integration formula that fits.
79. (a) $\int x \ln x \, dx = \frac{1}{2} x^2 \ln x - \frac{1}{4} x^2 + C$

$\int x^2 \ln x \, dx = \frac{1}{3} x^3 \ln x - \frac{1}{9} x^3 + C$

$\int x^3 \ln x \, dx = \frac{1}{4} x^4 \ln x - \frac{1}{16} x^4 + C$

(b) $\int x^n \ln x \, dx = x^{n+1} \ln x/(n + 1) - x^{n+1}/(n + 1)^2 + C$

81. False. Substitutions may first have to be made to rewrite the integral in a form that appears in the table.
83. 1919.145 ft-lb
85. (a) $V = 80 \ln(\sqrt{10} + 3) \approx 145.5$ ft^3
   $W = 11{,}840 \ln(\sqrt{10} + 3) \approx 21{,}530.4$ lb
(b) $(0, 1.19)$
87. (a) $k = 30/\ln 7 \approx 15.42$
(b)

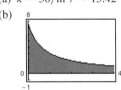

89. Putnam problem A3, 1980

Section 8.7   (page 574)

1.

x	-0.1	-0.01	-0.001	0.001	0.01	0.1
$f(x)$	2.4132	2.4991	2.500	2.500	2.4991	2.4132

2.5

3.

x	1	10	10^2	10^3	10^4	10^5
$f(x)$	0.9900	90,483.7	3.7×10^9	4.5×10^{10}	0	0

0

5. $\frac{1}{3}$ **7.** $\frac{1}{4}$ **9.** $\frac{5}{3}$ **11.** 3   **13.** 0   **15.** 2
17. ∞ **19.** $\frac{2}{3}$ **21.** 1   **23.** $\frac{3}{2}$ **25.** ∞
27. 0   **29.** 1   **31.** 0   **33.** 0   **35.** ∞
37. (a) Not indeterminate   **39.** (a) $0 \cdot \infty$
   (b) ∞ (b) 1
   (c) 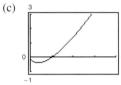   (c)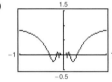

41. (a) Not indeterminate   **43.** (a) ∞^0
   (b) 0   (b) 1
   (c)   (c)
   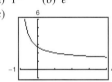   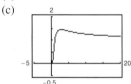

45. (a) 1^∞ (b) e **47.** (a) 0^0 (b) 3
   (c) 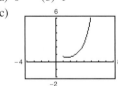   (c)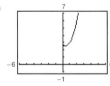

49. (a) 0^0 (b) 1   **51.** (a) $\infty - \infty$ (b) $-\frac{3}{2}$
   (c) 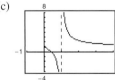   (c)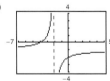

53. (a) $\infty - \infty$ (b) ∞ **55.** (a)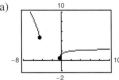
   (c)

   (b) $\frac{1}{2}$

57. (a)

59. $\dfrac{0}{0}, \dfrac{\infty}{\infty}, 0 \cdot \infty, 1^\infty, 0^0, \infty - \infty$

(b) $\frac{5}{2}$

61. Answers will vary. Examples:
  (a) $f(x) = x^2 - 25, g(x) = x - 5$
  (b) $f(x) = (x - 5)^2, g(x) = x^2 - 25$
  (c) $f(x) = x^2 - 25, g(x) = (x - 5)^3$

63.

x	10	10^2	10^4	10^6	10^8	10^{10}
$\dfrac{(\ln x)^4}{x}$	2.811	4.498	0.720	0.036	0.001	0.000

65. 0   **67.** 0   **69.** 0

71. Horizontal asymptote: **73.** Horizontal asymptote:
  $y = 1$ $y = 0$
  Relative maximum: $(e, e^{1/e})$ Relative maximum: $(1, 2/e)$

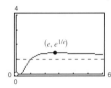 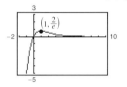

75. Limit is not of the form $0/0$ or ∞/∞.
77. Limit is not of the form $0/0$ or ∞/∞.
79. (a) $\lim\limits_{x\to\infty} \dfrac{x}{\sqrt{x^2+1}} = \lim\limits_{x\to\infty} \dfrac{\sqrt{x^2+1}}{x} = \lim\limits_{x\to\infty} \dfrac{x}{\sqrt{x^2+1}}$
  Applying L'Hôpital's Rule twice results in the original limit, so L'Hôpital's Rule fails.
  (b) 1
  (c)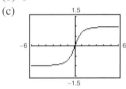

81. 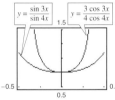 As $x \to 0$, the graphs get closer together.

83. $v = 32t + v_0$ **85.** Proof   **87.** $c = \dfrac{2}{3}$ **89.** $c = \pi/4$
91. False: L'Hôpital's Rule does not apply, **93.** True
  because $\lim\limits_{x\to 0}(x^2 + x + 1) \neq 0$.
95. $\dfrac{3}{4}$ **97.** $\dfrac{4}{3}$ **99.** $a = 1, b = \pm 2$ **101.** Proof
103. 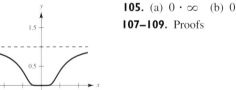   **105.** (a) $0 \cdot \infty$ (b) 0
  $g'(0) = 0$ **107–109.** Proofs

111. (a) 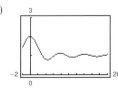   (b) $\lim\limits_{x\to\infty} h(x) = 1$
  (c) No

Section 8.8   (page 585)

1. Improper; $0 < \dfrac{2}{3} < 1$ **3.** Not improper; continuous on $[0, 1]$
5. Infinite discontinuity at $x = 0$; 4
7. Infinite discontinuity at $x = 1$; diverges
9. Infinite limit of integration; 1
11. Infinite discontinuity at $x = 0$; diverges
13. Infinite limit of integration; converges to 1   **15.** 1
17. Diverges   **19.** Diverges   **21.** 2   **23.** $\dfrac{1}{2}$
25. $1/[2(\ln 4)^2]$ **27.** π **29.** $\pi/4$ **31.** Diverges
33. Diverges   **35.** 6   **37.** $-\dfrac{1}{4}$ **39.** Diverges   **41.** $\pi/3$
43. $\ln(2 + \sqrt{3})$ **45.** 0   **47.** $2\pi\sqrt{6}/3$ **49.** $p > 1$
51. Proof   **53.** Diverges   **55.** Converges   **57.** Converges
59. Diverges   **61.** Converges
63. An integral with infinite integration limits, an integral with an infinite discontinuity at or between the integration limits
65. The improper integral diverges.   **67.** e **69.** π
71. (a) 1   (b) $\pi/2$ (c) 2π
73. 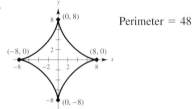 Perimeter = 48

75. $8\pi^2$ **77.** (a) $W = 20{,}000$ mile-tons   (b) 4000 mi
79. (a) Proof   (b) $P = 43.53\%$ (c) $E(x) = 7$
81. (a) \$757,992.41   (b) \$837,995.15   (c) \$1,066,666.67
83. $P = \left[2\pi NI\left(\sqrt{r^2+c^2}-c\right)\right]/(kr\sqrt{r^2+c^2})$
85. False. Let $f(x) = 1/(x + 1)$.   **87.** True
89. (a) $\displaystyle\int_1^\infty \dfrac{1}{x^n}\,dx$ will converge if $n > 1$ and diverge if $n \leq 1$.
  (b) 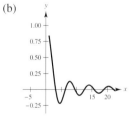   (c) Converges

91. (a) $\Gamma(1) = 1, \Gamma(2) = 1, \Gamma(3) = 2$ (b) Proof
  (c) $\Gamma(n) = (n - 1)!$
93. $1/s, s > 0$ **95.** $2/s^3, s > 0$ **97.** $s/(s^2 + a^2), s > 0$
99. $s/(s^2 - a^2), s > |a|$
101. (a) 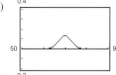   (b) ≈ 0.2525
  (c) 0.2525; same by symmetry

103. $c = 1; \ln(2)$
105. $8\pi[(\ln 2)^{2/3} - (\ln 4)/9 + 2/27] \approx 2.01545$ **107.** 0.6278

109. (a) 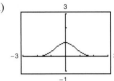 (b) Proof

17. $\dfrac{1/12}{x} + \dfrac{1/42}{x-3} + \dfrac{1/10}{x-1} + \dfrac{111/140}{x+4}$
19. Proof **21.** ≈ 0.0158

Chapter 9

Section 9.1 (page 602)

1. 2, 4, 8, 16, 32 **3.** $-\frac{1}{2}, \frac{1}{4}, -\frac{1}{8}, \frac{1}{16}, -\frac{1}{32}$ **5.** 1, 0, -1, 0, 1
7. $-1, -\frac{1}{4}, \frac{1}{9}, \frac{1}{16}, -\frac{1}{25}$ **9.** $5, \frac{19}{4}, \frac{43}{9}, \frac{77}{16}, \frac{121}{25}$
11. 3, 4, 6, 10, 18 **13.** 32, 16, 8, 4, 2
15. f **16.** a **17.** e **18.** b **19.** d **20.** c
21. **23.**

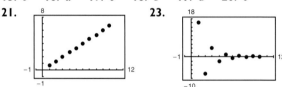

25. 14, 17; add 3 to preceding term
27. 80, 160; multiply preceding term by 2.
29. $\frac{3}{16}, -\frac{3}{32}$; multiply preceding term by $-\frac{1}{2}$ **31.** $10 \cdot 9 = 90$
33. $n+1$ **35.** $1/[(2n+1)(2n)]$ **37.** 5 **39.** 2 **41.** 0
43. **45.**

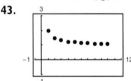 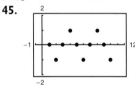

Converges to 1   Diverges
47. Diverges **49.** Converges to $\frac{3}{2}$ **51.** Converges to 0
53. Converges to 0 **55.** Converges to 0 **57.** Converges to 0
59. Diverges **61.** Converges to 0 **63.** Converges to 0
65. Converges to e^k **67.** Converges to 0
69. Answers will vary. Sample answer: $3n - 2$
71. Answers will vary. Sample answer: $n^2 - 2$
73. Answers will vary. Sample answer: $(n+1)/(n+2)$
75. Answers will vary. Sample answer: $(n+1)/n$
77. Answers will vary. Sample answer: $n/[(n+1)(n+2)]$
79. Answers will vary. Sample answer:
$$\dfrac{(-1)^{n-1}}{1 \cdot 3 \cdot 5 \cdots (2n-1)} = \dfrac{(-1)^{n-1}2^n n!}{(2n)!}$$
81. Answers will vary. Sample answer: $(2n)!$
83. Monotonic, bounded **85.** Monotonic, bounded
87. Not monotonic, bounded **89.** Monotonic, bounded
91. Not monotonic, bounded **93.** Not monotonic, bounded
95. (a) $\left|5 + \dfrac{1}{n}\right| \leq 6 \Rightarrow$ bounded (b)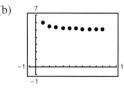

$a_n > a_{n+1} \Rightarrow$ monotonic
So, $\{a_n\}$ converges.

Limit = 5

Review Exercises for Chapter 8 (page 589)

1. $\frac{1}{3}(x^2 - 1)^{3/2} + C$ **3.** $\frac{1}{2}\ln|x^2 - 1| + C$
5. $\ln(2) + \frac{1}{2} \approx 1.1931$ **7.** $16 \arcsin(x/4) + C$
9. $\frac{1}{13}e^{2x}(2\sin 3x - 3\cos 3x) + C$
11. $\frac{2}{15}(x-5)^{3/2}(3x+10) + C$
13. $-\frac{1}{2}x^2 \cos 2x + \frac{1}{2}x \sin 2x + \frac{1}{4}\cos 2x + C$
15. $\frac{1}{16}\left[(8x^2 - 1)\arcsin 2x + 2x\sqrt{1-4x^2}\right] + C$
17. $\sin(\pi x - 1)[\cos^2(\pi x - 1) + 2]/(3\pi) + C$
19. $\frac{2}{3}[\tan^3(x/2) + 3\tan(x/2)] + C$ **21.** $\tan\theta + \sec\theta + C$
23. $3\pi/16 + \frac{1}{2} \approx 1.0890$ **25.** $3\sqrt{4-x^2}/x + C$
27. $\frac{1}{3}(x^2+4)^{1/2}(x^2-8) + C$ **29.** π
31. (a), (b), and (c) $\frac{1}{3}\sqrt{4+x^2}(x^2-8) + C$
33. $6\ln|x+2| - 5\ln|x-3| + C$
35. $\frac{1}{4}[6\ln|x-1| - \ln(x^2+1) + 6\arctan x] + C$
37. $x + \frac{9}{8}\ln|x-3| - \frac{25}{8}\ln|x+5| + C$
39. $\frac{1}{9}[2/(2+3x) + \ln|2+3x|] + C$ **41.** $1 - \sqrt{2}/2$
43. $\frac{1}{2}\ln|x^2 + 4x + 8| - \arctan[(x+2)/2] + C$
45. $\ln|\tan \pi x|/\pi + C$ **47.** Proof
49. $\frac{1}{8}(\sin 2\theta - 2\theta \cos 2\theta) + C$
51. $\frac{4}{3}[x^{3/4} - 3x^{1/4} + 3\arctan(x^{1/4})] + C$
53. $2\sqrt{1-\cos x} + C$ **55.** $\sin x \ln(\sin x) - \sin x + C$
57. $y = \frac{3}{2}\ln|(x-3)/(x+3)| + C$
59. $y = x\ln|x^2 + x| - 2x + \ln|x+1| + C$ **61.** $\frac{1}{5}$
63. $\frac{1}{2}(\ln 4)^2 \approx 0.961$ **65.** π **67.** $\frac{128}{15}$
69. $(\bar{x}, \bar{y}) = (0, 4/(3\pi))$ **71.** 3.82 **73.** 0 **75.** ∞ **77.** 1
79. $1000e^{0.09} \approx 1094.17$ **81.** Converges; $\frac{32}{3}$ **83.** Diverges
85. Converges; 1 **87.** (a) $6,321,205.59 (b) $10,000,000
89. (a) 0.4581 (b) 0.0135

P.S. Problem Solving (page 591)

1. (a) $\frac{4}{3}, \frac{16}{15}$ (b) Proof **3.** $\ln 3$
5. 2
7. (a) 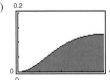 (b) $\ln 3 - \frac{4}{5}$
(c) $\ln 3 - \frac{4}{5}$

Area ≈ 0.2986
9. $\ln 3 - \frac{1}{2} \approx 0.5986$ **11.** Proof **13.** ≈ 0.8670
15. (a) ∞ (b) 0 (c) $-\frac{2}{3}$

97. (a) $\left|\frac{1}{3}\left(1 - \frac{1}{3^n}\right)\right| < \frac{1}{3} \Rightarrow$ bounded

$a_n < a_{n+1} \Rightarrow$ monotonic

So, $\{a_n\}$ converges.

(b) 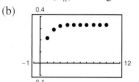 Limit $= \frac{1}{3}$

99. $\{a_n\}$ has a limit because it is bounded and monotonic; since $2 \le a_n \le 4$, $2 \le L \le 4$.

101. (a) No; $\lim_{n\to\infty} a_n$ does not exist.

(b)

n	1	2	3	4	5
A_n	$9041.25	$9082.69	$9124.32	$9166.14	$9208.15

n	6	7	8	9	10
A_n	$9250.35	$9292.75	$9335.34	$9378.13	$9421.11

103. (a) A sequence is a function whose domain is the set of positive integers.
(b) A sequence converges if it has a limit.
(c) A monotonic sequence is a sequence that has nonincreasing or nondecreasing terms.
(d) A bounded sequence is a sequence that has both an upper and a lower bound.

105. Answers will vary. Example: $a_n = 10n/(n+1)$

107. Answers will vary. Example: $a_n = (3n^2 - n)/(4n^2 + 1)$

109. (a) $2{,}500{,}000{,}000(0.8)^n$

(b)

Year	1	2
Budget	$2,000,000,000	$1,600,000,000

Year	3	4
Budget	$1,280,000,000	$1,024,000,000

(c) Converges to 0

111. (a) $a_n = -6.60n^2 + 151.7n + 387$ (b) 979 species

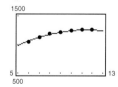

113. (a) $a_9 = a_{10} = 1{,}562{,}500/567$ (b) Decreasing
(c) Factorials increase more rapidly than exponentials.

115. 1, 1.4142, 1.4422, 1.4142, 1.3797, 1.3480; Converges to 1

117. True **119.** True

121. (a) 1, 1, 2, 3, 5, 8, 13, 21, 34, 55, 89, 144
(b) 1, 2, 1.5, 1.6667, 1.6, 1.6250, 1.6154, 1.6190, 1.6176, 1.6182
(c) Proof (d) $\rho = (1 + \sqrt{5})/2 \approx 1.6180$

123. (a) 1.4142, 1.8478, 1.9616, 1.9904, 1.9976
(b) $a_n = \sqrt{2 + a_{n-1}}$ (c) $\lim_{n\to\infty} a_n = 2$

125. (a) Proof (b) Proof (c) $\lim_{n\to\infty} a_n = \left(1 + \sqrt{1 + 4k}\right)/2$

127. (a) Proof (b) Proof

129. (a) Proof

(b)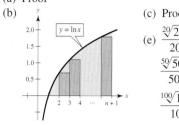

(c) Proof (d) Proof

(e) $\dfrac{\sqrt[20]{20!}}{20} \approx 0.4152$;

$\dfrac{\sqrt[50]{50!}}{50} \approx 0.3897$;

$\dfrac{\sqrt[100]{100!}}{100} \approx 0.3799$

131–133. Proofs **135.** Putnam Problem A1, 1990

Section 9.2 (page 612)

1. 1, 1.25, 1.361, 1.424, 1.464
3. 3, -1.5, 5.25, -4.875, 10.3125
5. 3, 4.5, 5.25, 5.625, 5.8125 **7.** Geometric series: $r = \frac{3}{2} > 1$
9. Geometric series: $r = 1.055 > 1$ **11.** $\lim_{n\to\infty} a_n = 1 \ne 0$
13. $\lim_{n\to\infty} a_n = 1 \ne 0$ **15.** $\lim_{n\to\infty} a_n = \frac{1}{2} \ne 0$ **17.** c; 3
18. b; 3 **19.** a; 3 **20.** d; 3 **21.** f; $\frac{34}{9}$ **22.** e; $\frac{5}{3}$
23. Telescoping series: $a_n = 1/n - 1/(n+1)$; Converges to 1.
25. Geometric series: $r = \frac{3}{4} < 1$
27. Geometric series: $r = 0.9 < 1$
29. (a) $\frac{11}{3}$

(b)

n	5	10	20	50	100
S_n	2.7976	3.1643	3.3936	3.5513	3.6078

(c) 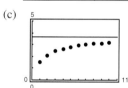 (d) The terms of the series decrease in magnitude relatively slowly, and the sequence of partial sums approaches the sum of the series relatively slowly.

31. (a) 20

(b)

n	5	10	20	50	100
S_n	8.1902	13.0264	17.5685	19.8969	19.9995

(c) 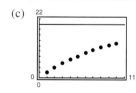 (d) The terms of the series decrease in magnitude relatively slowly, and the sequence of partial sums approaches the sum of the series relatively slowly.

33. (a) $\frac{40}{3}$

(b)

n	5	10	20	50	100
S_n	13.3203	13.3333	13.3333	13.3333	13.3333

(c) 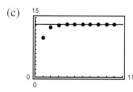 (d) The terms of the series decrease in magnitude relatively rapidly, and the sequence of partial sums approaches the sum of the series relatively rapidly.

35. $\frac{3}{4}$ **37.** 4 **39.** 2 **41.** $\frac{2}{3}$ **43.** $\frac{10}{9}$ **45.** $\frac{9}{4}$ **47.** $\frac{1}{2}$

49. $\dfrac{\sin(1)}{1 - \sin(1)}$ **51.** (a) $\sum_{n=0}^{\infty} \dfrac{4}{10}(0.1)^n$ (b) $\dfrac{4}{9}$

53. (a) $\sum_{n=0}^{\infty} \dfrac{81}{100}(0.01)^n$ (b) $\dfrac{9}{11}$

55. (a) $\sum_{n=0}^{\infty} \dfrac{3}{40}(0.01)^n$ (b) $\dfrac{5}{66}$ **57.** Diverges **59.** Converges

61. Diverges **63.** Converges **65.** Diverges **67.** Diverges
69. Diverges **71.** Diverges **73.** See definitions on page 606.

75. The series given by
$$\sum_{n=0}^{\infty} ar^n = a + ar + ar^2 + \cdots + ar^n + \cdots, \ a \neq 0$$
is a geometric series with ratio r. When $0 < |r| < 1$, the series converges to the sum $\sum_{n=0}^{\infty} ar^n = \dfrac{a}{1-r}$.

77. $\{a_n\}$ converges to 1 and $\sum_{n=1}^{\infty} a_n$ diverges. This fits Theorem 9.9, which states that if $\lim_{n \to \infty} a_n \neq 0$, then $\sum_{n=1}^{\infty} a_n$ diverges.

79. $-2 < x < 2$; $x/(2 - x)$ **81.** $0 < x < 2$; $(x - 1)/(2 - x)$
83. $-1 < x < 1$; $1/(1 + x)$
85. x: $(-\infty, -1) \cup (1, \infty)$; $x/(x - 1)$
87. (a) Yes. Answers will vary. (b) Yes. Answers will vary.
89. (a) x (b) $f(x) = 1/(1 - x)$, $|x| < 1$
  (c)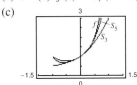

91. 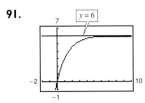 Horizontal asymptote: $y = 6$
The horizontal asymptote is the sum of the series.

93. The required terms for the two series are $n = 100$ and $n = 5$, respectively. The second series converges at a higher rate.
95. $80{,}000(1 - 0.9^n)$ units
97. $400(1 - 0.75^n)$ million dollars; Sum = \$400 million
99. 152.42 feet **101.** $\dfrac{1}{8}$; $\sum_{n=0}^{\infty} \dfrac{1}{2}\left(\dfrac{1}{2}\right)^n = \dfrac{1/2}{1 - 1/2} = 1$

103. (a) $-1 + \sum_{n=0}^{\infty}\left(\dfrac{1}{2}\right)^n = -1 + \dfrac{a}{1-r} = -1 + \dfrac{1}{1 - 1/2} = 1$
  (b) No (c) 2
105. (a) 126 in.^{2} (b) 128 in.^{2}
107. \$573,496.06; The \$1,000,000 sweepstakes has a present value of \$573,496.06. After accruing interest over the 20-year period, it attains its full value.
109. (a) \$5,368,709.11 (b) \$10,737,418.23 (c) \$21,474,836.47
111. (a) \$16,415.10 (b) \$16,421.83
113. (a) \$118,196.13 (b) \$118,393.43

115. (a) $a_n = 3484.1363e^{0.0490n}$ (b) \$50,809 million
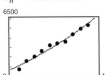 (c) \$50,815 million

117. False. $\lim_{n \to \infty} \dfrac{1}{n} = 0$, but $\sum_{n=1}^{\infty} \dfrac{1}{n}$ diverges.

119. False. $\sum_{n=1}^{\infty} ar^n = \left(\dfrac{a}{1 - r}\right) - a$
The formula requires that the geometric series begins with $n = 0$.
121. True **123.** Proof
125. Answers will vary. Example: $\sum_{n=0}^{\infty} 1$, $\sum_{n=0}^{\infty} (-1)$
127–131. Proofs
133. H = half-life of the drug
n = number of equal doses
P = number of units of the drug
t = equal time intervals
The total amount of the drug in the patient's system at the time the last dose is given is
$T_n = P + Pe^{kt} + Pe^{2kt} + \cdots + Pe^{(n-1)kt}$
where $k = -(\ln 2)/H$. One time interval after the last dose is given is
$T_{n+1} = Pe^{kt} + Pe^{2kt} + Pe^{3kt} + \cdots + Pe^{nkt}$
and so on. Because $k < 0$, $T_{n+s} \to 0$ as $s \to \infty$.
135. Putnam Problem A1, 1966

Section 9.3 (page 620)
1. Diverges **3.** Converges **5.** Converges **7.** Diverges
9. Diverges **11.** Diverges **13.** Converges **15.** Converges
17. Diverges **19.** Diverges
21. $f(x)$ is not positive for $x \geq 1$.
23. $f(x)$ is not always decreasing. **25.** Converges **27.** Diverges
29. Diverges **31.** Diverges **33.** Converges **35.** Converges
37. c; diverges **38.** f; diverges **39.** b; converges
40. a; diverges **41.** d; converges **42.** e; converges
43. (a)

n	5	10	20	50	100
S_n	3.7488	3.75	3.75	3.75	3.75

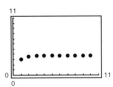 The partial sums approach the sum 3.75 very quickly.

(b)

n	5	10	20	50	100
S_n	1.4636	1.5498	1.5962	1.6251	1.635

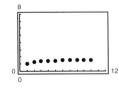 The partial sums approach the sum $\pi^2/6 \approx 1.6449$ more slowly than the series in part (a).

45. See Theorem 9.10 on page 617. Answers will vary. For example, convergence or divergence can be determined for the series $\sum_{n=1}^{\infty} \frac{1}{n^2 + 1}$.

47. No. Because $\sum_{n=1}^{\infty} \frac{1}{n}$ diverges, $\sum_{n=10{,}000}^{\infty} \frac{1}{n}$ also diverges. The convergence or divergence of a series is not determined by the first finite number of terms of the series.

49. 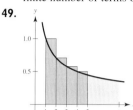 The series diverges because the area under the rectangles is greater than the infinite area under the curve $y = 1/\sqrt{x}$ for $x \geq 1$.

51. $p > 1$ **53.** $p > 1$ **55.** Diverges

57. Converges **59.** Proof

61. $S_6 \approx 1.0811$ **63.** $S_{10} \approx 0.9818$ **65.** $S_4 \approx 0.4049$
$R_6 \approx 0.0015$ $R_{10} \approx 0.0997$ $R_4 \approx 5.6 \times 10^{-8}$

67. $N \geq 7$ **69.** $N \geq 2$ **71.** $N \geq 1000$

73. (a) $\sum_{n=2}^{\infty} \frac{1}{n^{1.1}}$ converges by the p-Series Test since $1.1 > 1$.

$\sum_{n=2}^{\infty} \frac{1}{n \ln n}$ diverges by the Integral Test since $\int_{2}^{\infty} \frac{1}{x \ln x} dx$ diverges.

(b) $\sum_{n=2}^{\infty} \frac{1}{n^{1.1}} = 0.4665 + 0.2987 + 0.2176 + 0.1703$
$+ 0.1393 + \cdots$

$\sum_{n=2}^{\infty} \frac{1}{n \ln n} = 0.7213 + 0.3034 + 0.1803 + 0.1243$
$+ 0.0930 + \cdots$

(c) $n \geq 3.431 \times 10^{15}$

75. (a) Let $f(x) = 1/x$. f is positive, continuous, and decreasing on $[1, \infty)$.

$S_n - 1 \leq \int_1^n \frac{1}{x} dx = \ln n$

$S_n \geq \int_1^{n+1} \frac{1}{x} dx = \ln(n+1)$

So, $\ln(n+1) \leq S_n \leq 1 + \ln n$.

(b) $\ln(n+1) - \ln n \leq S_n - \ln n \leq 1$.
Also, $\ln(n+1) - \ln n > 0$ for $n \geq 1$. So, $0 \leq S_n - \ln n \leq 1$ and the sequence $\{a_n\}$ is bounded.

(c) $a_n - a_{n+1} = [S_n - \ln n] - [S_{n+1} - \ln(n+1)]$
$= \int_n^{n+1} \frac{1}{x} dx - \frac{1}{n+1} \geq 0$

So, $a_n \geq a_{n+1}$.

(d) Because the sequence is bounded and monotonic, it converges to a limit, γ.

(e) 0.5822

77. (a) Diverges (b) Diverges

(c) $\sum_{n=2}^{\infty} x^{\ln n}$ converges for $x < 1/e$.

79. Diverges **81.** Converges **83.** Converges **85.** Diverges
87. Diverges **89.** Converges

Section 9.4 (page 628)

1. (a)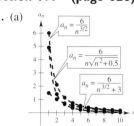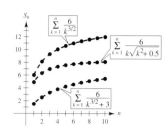

(b) $\sum_{n=1}^{\infty} \frac{6}{n^{3/2}}$; Converges

(c) The magnitudes of the terms are less than the magnitudes of the terms of the p-series. Therefore, the series converges.

(d) The smaller the magnitudes of the terms, the smaller the magnitudes of the terms of the sequence of partial sums.

3. Converges **5.** Diverges **7.** Converges **9.** Diverges
11. Converges **13.** Converges **15.** Diverges **17.** Diverges
19. Converges **21.** Diverges **23.** Converges **25.** Diverges
27. Diverges **29.** Diverges; p-Series Test

31. Converges; Direct Comparison Test with $\sum_{n=1}^{\infty} \left(\frac{1}{3}\right)^n$

33. Diverges; nth-Term Test **35.** Converges; Integral Test

37. $\lim_{n \to \infty} \frac{a_n}{1/n} = \lim_{n \to \infty} na_n$

$\lim_{n \to \infty} na_n \neq 0$, but is finite.

The series diverges by the Limit Comparison Test.

39. Diverges **41.** Converges

43. $\lim_{n \to \infty} n\left(\frac{n^3}{5n^4 + 3}\right) = \frac{1}{5} \neq 0$

So, $\sum_{n=1}^{\infty} \frac{n^3}{5n^4 + 3}$ diverges.

45. Diverges **47.** Converges

49. Convergence or divergence is dependent on the form of the general term for the series and not necessarily on the magnitude of the terms.

51. See Theorem 9.13 on page 626. Answers will vary. For example,

$\sum_{n=2}^{\infty} \frac{1}{\sqrt{n-1}}$ diverges because $\lim_{n \to \infty} \frac{1/\sqrt{n-1}}{1/\sqrt{n}} = 1$ and

$\sum_{n=2}^{\infty} \frac{1}{\sqrt{n}}$ diverges (p-series).

53.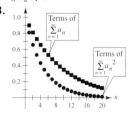

Because $0 < a_n < 1$, $0 < a_n^2 < a_n < 1$.

55. False. Let $a_n = 1/n^3$ and $b_n = 1/n^2$.

57. True **59.** True **61.** Proof **63.** $\sum_{n=1}^{\infty} \frac{1}{n^2}, \sum_{n=1}^{\infty} \frac{1}{n^3}$

65–69. Proofs **71.** Putnam Problem 1, afternoon session, 1953

Section 9.5 (page 636)

1. d   2. f   3. a   4. b   5. e   6. c

7. (a)

n	1	2	3	4	5
S_n	1.0000	0.6667	0.8667	0.7238	0.8349

n	6	7	8	9	10
S_n	0.7440	0.8209	0.7543	0.8131	0.7605

(b) [graph]

(c) The points alternate sides of the horizontal line $y = \pi/4$ that represents the sum of the series. The distances between the successive points and the line decrease.

(d) The distance in part (c) is always less than the magnitude of the next term of the series.

9. (a)

n	1	2	3	4	5
S_n	1.0000	0.7500	0.8611	0.7986	0.8386

n	6	7	8	9	10
S_n	0.8108	0.8312	0.8156	0.8280	0.8180

(b) [graph]

(c) The points alternate sides of the horizontal line $y = \pi^2/12$ that represents the sum of the series. The distances between the successive points and the line decrease.

(d) The distance in part (c) is always less than the magnitude of the next term of the series.

11. Converges   13. Converges   15. Diverges   17. Converges
19. Diverges   21. Diverges   23. Diverges   25. Converges
27. Converges   29. Converges   31. Converges
33. $2.3713 \leq S \leq 2.4937$ 35. $0.7305 \leq S \leq 0.7361$
37. (a) 7 terms (Note that the sum begins with $N = 0$.)
    (b) 0.368
39. (a) 3 terms (Note that the sum begins with $N = 0$.)
    (b) 0.842
41. (a) 1000 terms   (b) 0.693   43. 10   45. 7
47. Converges absolutely   49. Converges conditionally
51. Diverges   53. Converges conditionally
55. Converges absolutely   57. Converges absolutely
59. Converges conditionally   61. Converges absolutely
63. An alternating series is a series whose terms alternate in sign. See Theorem 9.14 on page 631 for the Alternating Series Test.
65. A series Σa_n is absolutely convergent if $\Sigma |a_n|$ converges. A series Σa_n is conditionally convergent if Σa_n converges and $\Sigma |a_n|$ diverges.
67. False. Let $a_n = \dfrac{(-1)^n}{n}$.   69. True   71. $p > 0$
73. (a) Proof
    (b) The converse is false. For example: Let $a_n = 1/n$.

75. $\sum_{n=1}^{\infty} \dfrac{1}{n^2}$ converges, hence so does $\sum_{n=1}^{\infty} \dfrac{1}{n^4}$.

77. (a) No; $a_{n+1} \leq a_n$ is not satisfied for all n. For example, $\tfrac{1}{9} < \tfrac{1}{8}$.   (b) Yes; 0.5
79. Converges; p-Series Test   81. Diverges; nth-Term Test
83. Converges; Geometric Series Test
85. Converges; Integral Test
87. Converges; Alternating Series Test
89. The first term of the series is 0, not 1. You cannot regroup series terms arbitrarily.
91. Putnam Problem 2, afternoon session, 1954

Section 9.6 (page 645)

1–3. Proofs   5. d   6. c   7. f   8. b   9. a   10. e
11. (a) Proof

(b)

n	5	10	15	20	25
S_n	9.2104	16.7598	18.8016	19.1878	19.2491

(c) 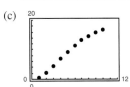   (d) 19.26

(e) The more rapidly the terms of the series approach 0, the more rapidly the sequence of partial sums approaches the sum of the series.

13. Diverges   15. Converges   17. Converges   19. Diverges
21. Converges   23. Diverges   25. Converges
27. Converges   29. Diverges   31. Converges
33–35. Proofs   37. Converges   39. Diverges   41. Converges
43. Diverges   45. Converges   47. Converges
49. Converges   51. Converges; Alternating Series Test
53. Converges; p-Series Test   55. Diverges; nth-Term Test
57. Diverges; Ratio Test
59. Converges; Limit Comparison Test with $b_n = 1/2^n$
61. Converges; Direct Comparison Test with $b_n = 1/2^n$
63. Converges; Ratio Test   65. Converges; Ratio Test
67. Converges; Ratio Test   69. a and c   71. a and b
73. $\sum_{n=0}^{\infty} \dfrac{n+1}{4^{n+1}}$ 75. (a) 9   (b) -0.7769
77. Diverges; $\lim_{n \to \infty} \left| \dfrac{a_{n+1}}{a_n} \right| > 1$
79. Converges; $\lim_{n \to \infty} \left| \dfrac{a_{n+1}}{a_n} \right| < 1$
81. Diverges; Lim $a_n \neq 0$ 83. Converges   85. Converges
87. $(-3, 3)$ 89. $(-2, 0]$ 91. $x = 0$
93. See Theorem 9.17 on page 639.
95. No; the series $\sum_{n=1}^{\infty} \dfrac{1}{n + 10{,}000}$ diverges.
97. Absolutely; by definition   99–103. Proofs
105. Putnam Problem 3, morning session, 1942

Section 9.7 (page 656)

1. d **2.** c **3.** a **4.** b
5. $P_1 = 6 - 2x$ **7.** $P_1 = \sqrt{2}x + \sqrt{2}(4 - \pi)/4$

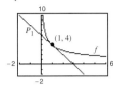
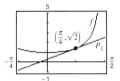

P_1 is the tangent line to the curve $f(x) = 4/\sqrt{x}$ at the point $(1, 4)$.

P_1 is the tangent line to the curve $f(x) = \sec x$ at the point $(\pi/4, \sqrt{2})$.

9.

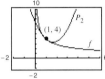

x	0	0.8	0.9	1	1.1
$f(x)$	Error	4.4721	4.2164	4.0000	3.8139
$P_2(x)$	7.5000	4.4600	4.2150	4.0000	3.8150

x	1.2	2
$f(x)$	3.6515	2.8284
$P_2(x)$	3.6600	3.5000

11. (a)

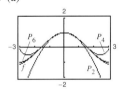

(b) $f^{(2)}(0) = -1 \quad P_2^{(2)}(0) = -1$
$f^{(4)}(0) = 1 \quad P_4^{(4)}(0) = 1$
$f^{(6)}(0) = -1 \quad P_6^{(6)}(0) = -1$

(c) $f^{(n)}(0) = P_n^{(n)}(0)$

13. $1 - x + \frac{1}{2}x^2 - \frac{1}{6}x^3$ **15.** $1 + 2x + 2x^2 + \frac{4}{3}x^3 + \frac{2}{3}x^4$
17. $x - \frac{1}{6}x^3 + \frac{1}{120}x^5$ **19.** $x + x^2 + \frac{1}{2}x^3 + \frac{1}{6}x^4$
21. $1 - x + x^2 - x^3 + x^4$ **23.** $1 + \frac{1}{2}x^2$
25. $1 - (x - 1) + (x - 1)^2 - (x - 1)^3 + (x - 1)^4$
27. $1 + \frac{1}{2}(x - 1) - \frac{1}{8}(x - 1)^2 + \frac{1}{16}(x - 1)^3 - \frac{5}{128}(x - 1)^4$
29. $(x - 1) - \frac{1}{2}(x - 1)^2 + \frac{1}{3}(x - 1)^3 - \frac{1}{4}(x - 1)^4$
31. (a) $P_3(x) = x + (1/3)x^3$
(b) $Q_3(x) = 1 + 2(x - \pi/4) + 2(x - \pi/4)^2 + \frac{8}{3}(x - \pi/4)^3$

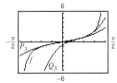

33. (a)

x	0	0.25	0.50	0.75	1.00
$\sin x$	0	0.2474	0.4794	0.6816	0.8415
$P_1(x)$	0	0.25	0.50	0.75	1.00
$P_3(x)$	0	0.2474	0.4792	0.6797	0.8333
$P_5(x)$	0	0.2474	0.4794	0.6817	0.8417

(b)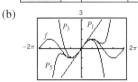

(c) As the distance increases, the polynomial approximation becomes less accurate.

35. (a) $P_3(x) = x + \frac{1}{6}x^3$

(b)
x	-0.75	-0.50	-0.25	0	0.25
$f(x)$	-0.848	-0.524	-0.253	0	0.253
$P_3(x)$	-0.820	-0.521	-0.253	0	0.253

x	0.50	0.75
$f(x)$	0.524	0.848
$P_3(x)$	0.521	0.820

(c)

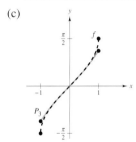

37. 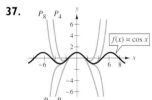 **39.**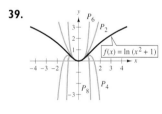

41. 0.6042 **43.** 0.1823 **45.** $R_4 \leq 2.03 \times 10^{-5}$; 0.000001
47. $R_3 \leq 7.82 \times 10^{-3}$ **49.** 3 **51.** 5
53. $n = 9$; $\ln(1.5) \approx 0.4055$ **55.** $n = 16$; $e^{-\pi(1.3)} \approx 0.16838$
57. $-0.3936 < x < 0$ **59.** $-0.9467 < x < 0.9467$
61. The graph of the approximating polynomial P and the elementary function f both pass through the point $(c, f(c))$, and the slope of P is the same as the slope of the graph of f at the point $(c, f(c))$. If P is of degree n, then the first n derivatives of f and P agree at c. This allows for the graph of P to resemble the graph of f near the point $(c, f(c))$.
63. See "Definitions of nth Taylor Polynomial and nth Maclaurin Polynomial" on page 650.
65. As the degree of the polynomial increases, the graph of the Taylor polynomial becomes a better and better approximation of the function within the interval of convergence. Therefore, the accuracy is increased.

67. (a) $f(x) \approx P_4(x) = 1 + x + (1/2)x^2 + (1/6)x^3 + (1/24)x^4$
$g(x) \approx Q_5(x) = x + x^2 + (1/2)x^3 + (1/6)x^4 + (1/24)x^5$
$Q_5(x) = xP_4(x)$
(b) $g(x) \approx P_6(x) = x^2 - x^4/3! + x^6/5!$
(c) $g(x) \approx P_4(x) = 1 - x^2/3! + x^4/5!$

69. (a) $Q_2(x) = -1 + (\pi^2/32)(x + 2)^2$
(b) $R_2(x) = -1 + (\pi^2/32)(x - 6)^2$
(c) No. Horizontal translations of the result in part (a) are possible only at $x = -2 + 8n$ (where n is an integer) because the period of f is 8.

71. Proof

73. As you move away from $x = c$, the Taylor polynomial becomes less and less accurate.

Section 9.8 (page 666)

1. 0 **3.** 2 **5.** $R = 1$ **7.** $R = \frac{1}{2}$ **9.** $R = \infty$
11. $(-2, 2)$ **13.** $(-1, 1]$ **15.** $(-\infty, \infty)$ **17.** $x = 0$
19. $(-4, 4)$ **21.** $(0, 10]$ **23.** $(0, 2]$ **25.** $(0, 6)$
27. $\left(-\frac{1}{2}, \frac{1}{2}\right)$ **29.** $(-\infty, \infty)$ **31.** $(-1, 1)$ **33.** $x = 3$
35. $R = c$ **37.** $(-k, k)$ **39.** $(-1, 1)$
41. $\sum_{n=1}^{\infty} \frac{x^{n-1}}{(n-1)!}$ **43.** $\sum_{n=1}^{\infty} \frac{x^{2n-1}}{(2n-1)!}$
45. (a) $(-2, 2)$ (b) $(-2, 2)$ (c) $(-2, 2)$ (d) $[-2, 2)$
47. (a) $(0, 2]$ (b) $(0, 2)$ (c) $(0, 2)$ (d) $[0, 2]$
49. c; $S_1 = 1$, $S_2 = 1.33$ **50.** a; $S_1 = 1$, $S_2 = 1.67$
51. b; diverges **52.** d; alternating
53. b **54.** c **55.** d **56.** a
57. A series of the form
$$\sum_{n=0}^{\infty} a_n(x - c)^n = a_0 + a_1(x - c) + a_2(x - c)^2 + \cdots + a_n(x - c)^n + \cdots$$
is called a power series centered at c, where c is a constant.

59. 1. A single point 2. An interval centered at c
3. The entire real line

61. Answers will vary.
63. (a) For $f(x)$: $(-\infty, \infty)$; For $g(x)$: $(-\infty, \infty)$
(b) Proof (c) Proof (d) $f(x) = \sin x$; $g(x) = \cos x$
65–69. Proofs
71. (a) Proof (b) Proof
(c) 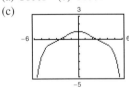 (d) 0.92

73. $f(x) = \cos x$ **75.** $f(x) = \dfrac{1}{1 + x}$
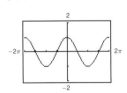 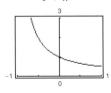

77. (a) $\frac{8}{5}$ (b) $\frac{8}{11}$
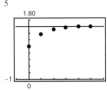 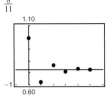

(c) The alternating series converges more rapidly. The partial sums of the series of positive terms approach the sum from below. The partial sums of the alternating series alternate sides of the horizontal line representing the sum.

(d)
M	10	100	1000	10,000
N	4	9	15	21

79. False. Let $a_n = (-1)^n/(n2^n)$ **81.** True **83.** Proof
85. (a) $(-1, 1)$ (b) $f(x) = (c_0 + c_1 x + c_2 x^2)/(1 - x^3)$
87. Proof

Section 9.9 (page 674)

1. $\sum_{n=0}^{\infty} \dfrac{x^n}{2^{n+1}}$ **3.** $\sum_{n=0}^{\infty} \dfrac{(-1)^n x^n}{2^{n+1}}$

5. $\sum_{n=0}^{\infty} \dfrac{(x - 5)^n}{(-3)^{n+1}}$ **7.** $-3 \sum_{n=0}^{\infty} (2x)^n$
(2, 8) $\left(-\frac{1}{2}, \frac{1}{2}\right)$

9. $-\dfrac{1}{11} \sum_{n=0}^{\infty} \left[\dfrac{2}{11}(x + 3)\right]^n$ **11.** $\dfrac{3}{2} \sum_{n=0}^{\infty} \left(-\dfrac{x}{2}\right)^n$
$\left(-\dfrac{17}{2}, \dfrac{5}{2}\right)$ $(-2, 2)$

13. $\sum_{n=0}^{\infty} \left[\left(-\dfrac{1}{2}\right)^n - 1\right] x^n$ **15.** $\sum_{n=0}^{\infty} x^n[1 + (-1)^n] = 2 \sum_{n=0}^{\infty} x^{2n}$
$(-1, 1)$ $(-1, 1)$

17. $2 \sum_{n=0}^{\infty} x^{2n}$ **19.** $\sum_{n=1}^{\infty} n(-1)^n x^{n-1}$ **21.** $\sum_{n=0}^{\infty} \dfrac{(-1)^n x^{n+1}}{n + 1}$
$(-1, 1)$ $(-1, 1)$ $(-1, 1]$

23. $\sum_{n=0}^{\infty} (-1)^n x^{2n}$ **25.** $\sum_{n=0}^{\infty} (-1)^n (2x)^{2n}$
$(-1, 1)$ $\left(-\dfrac{1}{2}, \dfrac{1}{2}\right)$

27.

x	0.0	0.2	0.4	0.6	0.8	1.0
S_2	0.000	0.180	0.320	0.420	0.480	0.500
$\ln(x + 1)$	0.000	0.182	0.336	0.470	0.588	0.693
S_3	0.000	0.183	0.341	0.492	0.651	0.833

29. (a) 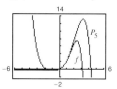 (b) $R = 1$
(c) -0.6931
(d) $\ln(0.5)$

31. c **32.** d **33.** a **34.** b **35.** 0.245 **37.** 0.125

39. $\sum_{n=1}^{\infty} nx^{n-1}, -1 < x < 1$ **41.** $\sum_{n=0}^{\infty} (2n+1)x^n, -1 < x < 1$

43. $E(n) = 2$. Because the probability of obtaining a head on a single toss is $\frac{1}{2}$, it is expected that, on average, a head will be obtained in two tosses.

45. Since $\dfrac{1}{1+x} = \dfrac{1}{1-(-x)}$, substitute $(-x)$ into the geometric series.

47. Since $\dfrac{5}{1+x} = 5\left(\dfrac{1}{1-(-x)}\right)$, substitute $(-x)$ into the geometric series and then multiply the series by 5.

49. Proof **51.** (a) Proof (b) 3.14

53. $\ln \frac{3}{2} \approx 0.4055$; See Exercise 21.

55. $\ln \frac{7}{5} \approx 0.3365$; See Exercise 53.

57. $\arctan \frac{1}{2} \approx 0.4636$; See Exercise 56.

59. $f(x) = \arctan x$ is an odd function (symmetric to the origin).

61. The series in Exercise 56 converges to its sum at a lower rate because its terms approach 0 at a much lower rate.

63. The series converges on the interval $(-5, 3)$ and perhaps also at one or both endpoints.

65. $\sqrt{3}\pi/6$

Section 9.10 (page 685)

1. $\sum_{n=0}^{\infty} \dfrac{(2x)^n}{n!}$ **3.** $\dfrac{\sqrt{2}}{2} \sum_{n=0}^{\infty} \dfrac{(-1)^{n(n+1)/2}}{n!}\left(x - \dfrac{\pi}{4}\right)^n$

5. $\sum_{n=0}^{\infty} \dfrac{(-1)^n(x-1)^{n+1}}{n+1}$ **7.** $\sum_{n=0}^{\infty} \dfrac{(-1)^n(2x)^{2n+1}}{(2n+1)!}$

9. $1 + x^2/2! + 5x^4/4! + \cdots$ **11–13.** Proofs

15. $\sum_{n=0}^{\infty} (-1)^n (n+1)x^n$

17. $\dfrac{1}{2}\left[1 + \sum_{n=1}^{\infty} \dfrac{(-1)^n 1 \cdot 3 \cdot 5 \cdots (2n-1)x^{2n}}{2^{3n}n!}\right]$

19. $1 + \dfrac{x^2}{2} + \sum_{n=2}^{\infty} \dfrac{(-1)^{n+1} 1 \cdot 3 \cdot 5 \cdots (2n-3)x^{2n}}{2^n n!}$

21. $\sum_{n=0}^{\infty} \dfrac{x^{2n}}{2^n n!}$ **23.** $\sum_{n=0}^{\infty} \dfrac{(-1)^n(3x)^{2n+1}}{(2n+1)!}$

25. $\sum_{n=0}^{\infty} \dfrac{(-1)^n x^{3n}}{(2n)!}$ **27.** $\sum_{n=0}^{\infty} \dfrac{x^{2n+1}}{(2n+1)!}$

29. $\dfrac{1}{2}\left[1 + \sum_{n=0}^{\infty} \dfrac{(-1)^n (2x)^{2n}}{(2n)!}\right]$ **31.** $\sum_{n=0}^{\infty} \dfrac{(-1)^n x^{2n+2}}{(2n+1)!}$

33. $\begin{cases} \sum_{n=0}^{\infty} \dfrac{(-1)^n x^{2n}}{(2n+1)!}, & x \neq 0 \\ 1, & x = 0 \end{cases}$ **35.** Proof

37. $P_5(x) = x + x^2 + \frac{1}{3}x^3 - \frac{1}{30}x^5 + \cdots$

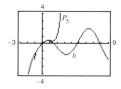

39. $P_5(x) = x - \frac{1}{2}x^2 - \frac{1}{6}x^3 + \frac{3}{40}x^5 + \cdots$

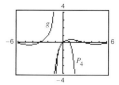

41. $P_4(x) = x - x^2 + \frac{5}{6}x^3 - \frac{5}{6}x^4 + \cdots$

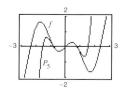

43. c; $f(x) = x \sin x$ **44.** d; $f(x) = x \cos x$ **45.** a; $f(x) = xe^x$

46. b; $f(x) = x^2\left(\dfrac{1}{1+x}\right)$ **47.** $\sum_{n=0}^{\infty} \dfrac{(-1)^{(n+1)}x^{2n+3}}{(2n+3)(n+1)!}$

49. 0.6931 **51.** 7.3891 **53.** 0 **55.** 0.9461
57. 0.2010 **59.** 0.7040 **61.** 0.3413

63. $P_5(x) = x - 2x^3 + \frac{2}{3}x^5$

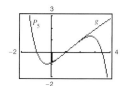

$\left[-\frac{3}{4}, \frac{3}{4}\right]$

65. $P_5(x) = (x-1) - \frac{1}{24}(x-1)^3 + \frac{1}{24}(x-1)^4 - \frac{71}{1920}(x-1)^5$

$\left[\frac{1}{4}, 2\right]$

67. See "Guidelines for Finding a Taylor Series" on page 680.

69. (a) Replace x with $-x$ in the series for e^x.
(b) Replace x with $3x$ in the series for e^x.
(c) Multiply the series for e^x by x.
(d) Replace x with $2x$ in the series for e^x. Then replace x with $-2x$ in the series for e^x. Then add the two together.

71. Proof

73. (a) 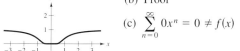 (b) Proof
(c) $\sum_{n=0}^{\infty} 0x^n = 0 \neq f(x)$

75. Proof **77.** 10 **79.** -0.0390625
81. $\sum_{n=0}^{\infty} \binom{k}{n} x^n$ **83.** Proof

Review Exercises for Chapter 9 (page 688)

1. $a_n = 1/n!$ **3.** a **4.** c **5.** d **6.** b
7.

Converges to 5

9. Converges to 0 **11.** Diverges
13. Converges to 0 **15.** Converges to 0
17. (a)

n	1	2	3	4
A_n	$5062.50	$5125.78	$5189.85	$5254.73

n	5	6	7	8
A_n	$5320.41	$5386.92	$5454.25	$5522.43

(b) $8218.10

19. (a)

k	5	10	15	20	25
S_k	13.2	113.3	873.8	6648.5	50,500.3

(b)
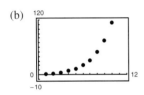

21. (a)

k	5	10	15	20	25
S_k	0.4597	0.4597	0.4597	0.4597	0.4597

(b)
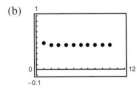

23. Converges **25.** Diverges **27.** 3 **29.** $\frac{1}{2}$
31. (a) $\sum_{n=0}^{\infty} (0.09)(0.01)^n$ (b) $\frac{1}{11}$ **33.** $45\frac{1}{3}$ m
35. $5087.14 **37.** Converges **39.** Diverges **41.** Converges
43. Diverges **45.** Converges **47.** Diverges **49.** Converges
51. Diverges
53. (a) Proof
(b)

n	5	10	15	20	25
S_n	2.8752	3.6366	3.7377	3.7488	3.7499

(c)
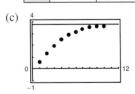

(d) 3.75

55. (a)

N	5	10	20	30	40
$\sum_{n=1}^{N} \frac{1}{n^p}$	1.4636	1.5498	1.5962	1.6122	1.6202
$\int_{N}^{\infty} \frac{1}{x^p}\,dx$	0.2000	0.1000	0.0500	0.0333	0.0250

(b)

N	5	10	20	30	40
$\sum_{n=1}^{N} \frac{1}{n^p}$	1.0367	1.0369	1.0369	1.0369	1.0369
$\int_{N}^{\infty} \frac{1}{x^p}\,dx$	0.0004	0.0000	0.0000	0.0000	0.0000

The series in part (b) converges more rapidly. This is evident from the integrals that give the remainders of the partial sums.

57. $P_3(x) = 1 - \dfrac{x}{2} + \dfrac{x^2}{8} - \dfrac{x^3}{48}$ **59.** 0.996 **61.** 0.560
63. (a) 4 (b) 6 (c) 5 (d) 10
65. $(-10, 10)$ **67.** $[1, 3]$ **69.** Converges only at $x = 2$
71. Proof **73.** $\sum_{n=0}^{\infty} \dfrac{2}{3}\left(\dfrac{x}{3}\right)^n$ **75.** $\sum_{n=0}^{\infty} \dfrac{2}{9}(n+1)\left(\dfrac{x}{3}\right)^n$
77. $f(x) = \dfrac{3}{3 - 2x}, \left(-\dfrac{3}{2}, \dfrac{3}{2}\right)$
79. $\dfrac{\sqrt{2}}{2} \sum_{n=0}^{\infty} \dfrac{(-1)^{n(n+1)/2}}{n!}\left(x - \dfrac{3\pi}{4}\right)^n$
81. $\sum_{n=0}^{\infty} \dfrac{(x \ln 3)^n}{n!}$ **83.** $-\sum_{n=0}^{\infty} (x+1)^n$
85. $1 + x/5 - 2x^2/25 + 6x^3/125 - 21x^4/625 + \cdots$
87. $\ln \frac{5}{4} \approx 0.2231$ **89.** $e^{1/2} \approx 1.6487$ **91.** $\cos \frac{2}{3} \approx 0.7859$
93. The series for Exercise 41 converges to its sum at a lower rate because its terms approach 0 at a lower rate.
95. $1 + 2x + 2x^2 + \dfrac{4}{3}x^3$ **97.** $\sum_{n=0}^{\infty} \dfrac{(-1)^n x^{2n+1}}{(2n+1)(2n+1)!}$
99. $\sum_{n=0}^{\infty} \dfrac{(-1)^n x^{n+1}}{(n+1)^2}$ **101.** 0

P.S. Problem Solving (page 691)

1. (a) 1 (b) Answers will vary. Example: $0, \frac{1}{3}, \frac{2}{3}$ (c) 0
3. $\pi/8$
5. (a) $R = 1$; Sum $= (3x^2 + 2x + 1)/(1 - x^3)$
(b) $R = 1$; Sum $= \dfrac{a_{p-1}x^{p-1} + a_{p-2}x^{p-2} + \cdots + a_1 x + a_0}{1 - x^p}$
7. (a) $\sum_{n=0}^{\infty} \dfrac{x^{n+1}}{n!}$; $\sum_{n=1}^{\infty} \dfrac{1}{n!(n+2)} = \dfrac{1}{2}$
(b) $\sum_{n=0}^{\infty} \dfrac{(n+1)x^n}{n!}$; $\sum_{n=0}^{\infty} \dfrac{n+1}{n!} = 2e \approx 5.4366$

9. Let $a_1 = \int_0^{\pi} \frac{\sin x}{x}\,dx = 1.8519$

$a_2 = \int_{\pi}^{2\pi} \frac{\sin x}{x}\,dx = -0.4338$

$a_3 = \int_{2\pi}^{3\pi} \frac{\sin x}{x}\,dx = 0.2566$

$a_4 = \int_{3\pi}^{4\pi} \frac{\sin x}{x}\,dx = -0.1826$.

It follows that the total area is

$\int_0^{\infty} \frac{\sin x}{x}\,dx = a_1 - a_2 + a_3 - a_4 + \cdots$.

Also, $\lim_{n\to\infty} a_n = 0$ and $0 < a_{n+1} \leq a_n$. Therefore, it follows by the Alternating Series Test that $\int_0^{\infty} f(x)\,dx$ converges.

11. (a) $a_1 = 3$, $a_2 \approx 1.7321$, $a_3 \approx 2.1753$, $a_4 \approx 2.2749$, $a_5 \approx 2.2967$, $a_6 \approx 2.3015$

Proof; $L = (1 + \sqrt{13})/2$

(b) Proof; $L = (1 + \sqrt{1+4a})/2$

13. (a) $1, \frac{9}{8}, \frac{11}{8}, \frac{45}{32}, \frac{47}{32}$ (b) Proof   (c) Proof

15. $S_6 = 240$; $S_7 = 440$; $S_8 = 810$; $S_9 = 1490$; $S_{10} = 2740$

17. (a) Diverges   (b) Converges

Chapter 10
Section 10.1   (page 704)

1. h   **2.** a   **3.** e   **4.** b   **5.** f   **6.** g   **7.** c   **8.** d

9. Vertex: $(0, 0)$
Focus: $\left(-\frac{3}{2}, 0\right)$
Directrix: $x = \frac{3}{2}$

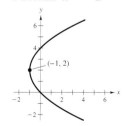

11. Vertex: $(-3, 2)$
Focus: $\left(-\frac{13}{4}, 2\right)$
Directrix: $x = -\frac{11}{4}$

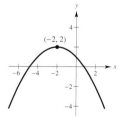

13. Vertex: $(-1, 2)$
Focus: $(0, 2)$
Directrix: $x = -2$

15. Vertex: $(-2, 2)$
Focus: $(-2, 1)$
Directrix: $y = 3$

17. Vertex: $\left(\frac{1}{4}, -\frac{1}{2}\right)$
Focus: $\left(0, -\frac{1}{2}\right)$
Directrix: $x = \frac{1}{2}$

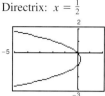

19. Vertex: $(-1, 0)$
Focus: $(0, 0)$
Directrix: $x = -2$

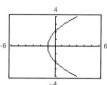

21. $y^2 - 4y + 8x - 20 = 0$ **23.** $x^2 - 24y + 96 = 0$
25. $x^2 + y - 4 = 0$ **27.** $5x^2 - 14x - 3y + 9 = 0$

29. Center: $(0, 0)$
Foci: $(\pm\sqrt{3}, 0)$
Vertices: $(\pm 2, 0)$
$e = \sqrt{3}/2$

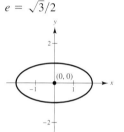

31. Center: $(1, 5)$
Foci: $(1, 9), (1, 1)$
Vertices: $(1, 10), (1, 0)$
$e = \frac{4}{5}$

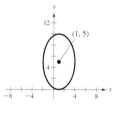

33. Center: $(-2, 3)$
Foci: $\left(-2, 3 \pm \sqrt{5}\right)$
Vertices: $(-2, 6), (-2, 0)$
$e = \sqrt{5}/3$

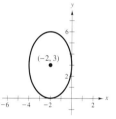

35. Center: $\left(\frac{1}{2}, -1\right)$
Foci: $\left(\frac{1}{2} \pm \sqrt{2}, -1\right)$
Vertices: $\left(\frac{1}{2} \pm \sqrt{5}, -1\right)$
To obtain the graph, solve for y and get
$y_1 = -1 + \sqrt{(57 + 12x - 12x^2)/20}$ and
$y_2 = -1 - \sqrt{(57 + 12x - 12x^2)/20}$.
Graph these equations in the same viewing window.

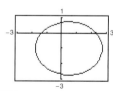

37. Center: $\left(\frac{3}{2}, -1\right)$
Foci: $\left(\frac{3}{2} - \sqrt{2}, -1\right), \left(\frac{3}{2} + \sqrt{2}, -1\right)$
Vertices: $\left(-\frac{1}{2}, -1\right), \left(\frac{7}{2}, -1\right)$
To obtain the graph, solve for y and get
$y_1 = -1 + \sqrt{(7 + 12x - 4x^2)/8}$ and
$y_2 = -1 - \sqrt{(7 + 12x - 4x^2)/8}$
Graph these equations in the same viewing window.

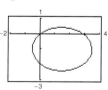

39. $x^2/9 + y^2/5 = 1$ **41.** $(x - 3)^2/9 + (y - 5)^2/16 = 1$
43. $x^2/16 + 7y^2/16 = 1$

45. Center: $(0, 0)$
Foci: $(0, \pm\sqrt{5})$
Vertices: $(0, \pm 1)$

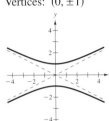

47. Center: $(1, -2)$
Foci: $(1 \pm \sqrt{5}, -2)$
Vertices: $(-1, -2), (3, -2)$

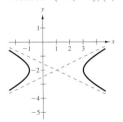

49. Center: $(2, -3)$
Foci: $(2 \pm \sqrt{10}, -3)$
Vertices: $(1, -3), (3, -3)$

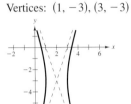

51. Degenerate hyperbola
Graph is two lines
$y = -3 \pm \frac{1}{3}(x + 1)$
intersecting at $(-1, -3)$.

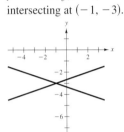

53. Center: $(1, -3)$
Foci: $(1, -3 \pm 2\sqrt{5})$
Vertices: $(1, -3 \pm \sqrt{2})$
Asymptotes:
$y = \frac{1}{3}x - \frac{1}{3} - 3$;
$y = -\frac{1}{3}x + \frac{1}{3} - 3$

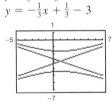

55. Center: $(1, -3)$
Foci: $(1 \pm \sqrt{10}, -3)$
Vertices: $(-1, -3), (3, -3)$
Asymptotes:
$y = \sqrt{6}x/2 - \sqrt{6}/2 - 3$;
$y = -\sqrt{6}x/2 + \sqrt{6}/2 - 3$

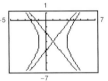

57. $x^2/1 - y^2/9 = 1$ **59.** $y^2/9 - (x-2)^2/(9/4) = 1$
61. $y^2/4 - x^2/12 = 1$ **63.** $(x-3)^2/9 - (y-2)^2/4 = 1$
65. (a) $(6, \sqrt{3})$: $2x - 3\sqrt{3}y - 3 = 0$
$(6, -\sqrt{3})$: $2x + 3\sqrt{3}y - 3 = 0$
(b) $(6, \sqrt{3})$: $9x + 2\sqrt{3}y - 60 = 0$
$(6, -\sqrt{3})$: $9x - 2\sqrt{3}y - 60 = 0$
67. Ellipse **69.** Parabola **71.** Circle
73. Circle **75.** Hyperbola
77. (a) A parabola is the set of all points (x, y) that are equidistant from a fixed line and a fixed point not on the line.
(b) For directrix $y = k - p$: $(x - h)^2 = 4p(y - k)$
For directrix $x = h - p$: $(y - k)^2 = 4p(x - h)$
(c) If P is a point on a parabola, then the tangent line to the parabola at P makes equal angles with the line passing through P and the focus, and with the line passing through P parallel to the axis of the parabola.
79. (a) A hyperbola is the set of all points (x, y) for which the absolute value of the difference between the distances from two distinct fixed points is constant.

(b) Transverse axis is horizontal: $\dfrac{(x - h)^2}{a^2} - \dfrac{(y - k)^2}{b^2} = 1$

Transverse axis is vertical: $\dfrac{(y - k)^2}{a^2} - \dfrac{(x - h)^2}{b^2} = 1$

(c) Transverse axis is horizontal:
$y = k + (b/a)(x - h)$ and $y = k - (b/a)(x - h)$
Transverse axis is vertical:
$y = k + (a/b)(x - h)$ and $y = k - (a/b)(x - h)$

81. $\frac{9}{4}$ m **83.** $y = 2ax_0 x - ax_0^2$ **85.** (a) Proof (b) Proof
87. $x_0 = 2\sqrt{3}/3$; Distance from hill: $2\sqrt{3}/3 - 1$
89. $[16(4 + 3\sqrt{3} - 2\pi)]/3 \approx 15.536$ ft^2
91. (a) $y = (1/180)x^2$
(b) $10\left[2\sqrt{13} + 9\ln\left(\dfrac{2 + \sqrt{13}}{3}\right)\right] \approx 128.4$ m

93.

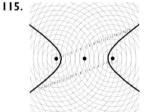

As p increases, the graph of $x^2 = 4py$ gets wider.

95. (a) $L = 2a$
(b) The thumbtacks are located at the foci and the length of string is the constant sum of distances from the foci.

97. **99.** Proof

101. $e \approx 0.9672$ **103.** $(0, \frac{25}{3})$
105. Minor-axis endpoints: $(-6, -2), (0, -2)$
Major-axis endpoints: $(-3, -6), (-3, 2)$
107. (a) Area $= 2\pi$
(b) Volume $= 8\pi/3$
Surface area $= [2\pi(9 + 4\sqrt{3}\pi)]/9 \approx 21.48$
(c) Volume $= 16\pi/3$
Surface area $= \dfrac{4\pi[6 + \sqrt{3}\ln(2 + \sqrt{3})]}{3} \approx 34.69$

109. 37.96 **111.** 40 **113.** $(x - 6)^2/9 - (y - 2)^2/7 = 1$
115.   **117.** Proof

119. $x = (-90 + 96\sqrt{2})/7 \approx 6.538$
$y = (160 - 96\sqrt{2})/7 \approx 3.462$

121. There are four points of intersection.

At $\left(\dfrac{\sqrt{2}\,ac}{\sqrt{2a^2-b^2}},\dfrac{b^2}{\sqrt{2}\sqrt{2a^2-b^2}}\right)$, the slopes of the tangent lines are $y'_e = -c/a$ and $y'_h = a/c$.

Since the slopes are negative reciprocals, the tangent lines are perpendicular. Similarly, the curves are perpendicular at the other three points of intersection.

123. False. See the definition of a parabola. **125.** True
127. True **129.** Putnam Problem B4, 1976

Section 10.2 (page 716)

1. (a)

t	0	1	2	3	4
x	0	1	$\sqrt{2}$	$\sqrt{3}$	2
y	1	0	-1	-2	-3

(b) and (c)   (d) $y = 1 - x^2$, $x \geq 0$

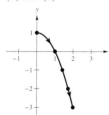

3. 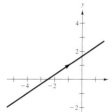   **5.**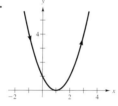

$2x - 3y + 5 = 0$ $y = (x-1)^2$

7.

$y = \tfrac{1}{2}x^{2/3}$

9. 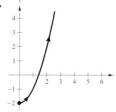   **11.**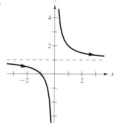

$y = x^2 - 2$, $x \geq 0$ $y = (x+1)/x$

13. 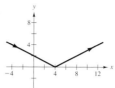   **15.**

$y = |x - 4|/2$ $y = x^3 + 1$, $x > 0$

17. 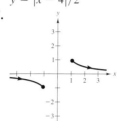   **19.**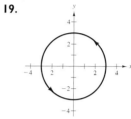

$y = 1/x$, $|x| \geq 1$ $x^2 + y^2 = 9$

21. 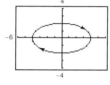   **23.**

$\dfrac{x^2}{16} + \dfrac{y^2}{4} = 1$ $\dfrac{(x-4)^2}{4} + \dfrac{(y+1)^2}{1} = 1$

25. 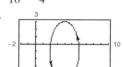   **27.**

$\dfrac{(x-4)^2}{4} + \dfrac{(y+1)^2}{16} = 1$ $\dfrac{x^2}{16} - \dfrac{y^2}{9} = 1$

29. 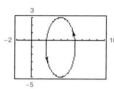   **31.**

$y = \ln x$ $y = 1/x^3$, $x > 0$

33. Each curve represents a portion of the line $y = 2x + 1$.

	Domain	Orientation	Smooth
(a)	$-\infty < x < \infty$	Up	Yes
(b)	$-1 \leq x \leq 1$	Oscillates	No, $\dfrac{dx}{d\theta} = \dfrac{dy}{d\theta} = 0$ when $\theta = 0, \pm\pi, \pm 2\pi, \ldots$
(c)	$0 < x < \infty$	Down	Yes
(d)	$0 < x < \infty$	Up	Yes

35. (a) and (b) represent the parabola $y = 2(1 - x^2)$ for $-1 \leq x \leq 1$. The curve is smooth. The orientation is from right to left in part (a) and in part (b).

37. (a)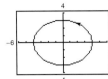

(b) The orientation is reversed.   (c) The orientation is reversed.

(d) Answers will vary. For example,
$$x = 2 \sec t \quad x = 2 \sec(-t)$$
$$y = 5 \sin t \quad y = 5 \sin(-t)$$
have the same graphs, but their orientations are reversed.

39. $y - y_1 = \dfrac{y_2 - y_1}{x_2 - x_1}(x - x_1)$ **41.** $\dfrac{(x-h)^2}{a^2} + \dfrac{(y-k)^2}{b^2} = 1$

43. $x = 5t$
$y = -2t$
(Solution is not unique.)

45. $x = 2 + 4\cos\theta$
$y = 1 + 4\sin\theta$
(Solution is not unique.)

47. $x = 5\cos\theta$
$y = 3\sin\theta$
(Solution is not unique.)

49. $x = 4\sec\theta$
$y = 3\tan\theta$
(Solution is not unique.)

51. $x = t$
$y = 3t - 2;$
$x = t - 3$
$y = 3t - 11$
(Solution is not unique.)

53. $x = t$
$y = t^3;$
$x = \tan t$
$y = \tan^3 t$
(Solution is not unique.)

55. 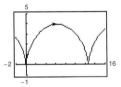   **57.**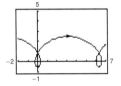

Not smooth when $\theta = 2n\pi$

59. 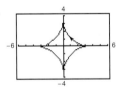   **61.**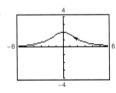

Not smooth when $\theta = \tfrac{1}{2}n\pi$

63. See page 709.   **65.** See page 714.

67. $x = a\theta - b\sin\theta;\ y = a - b\cos\theta$

69. False. The graph of the parametric equations is the portion of the line $y = x$ when $x \geq 0$.

71. (a) $x = \left(\dfrac{440}{3}\cos\theta\right)t;\ y = 3 + \left(\dfrac{440}{3}\sin\theta\right)t - 16t^2$

(b) 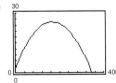   (c)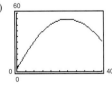

Not a home run   Home run

(d) $19.4°$

Section 10.3 (page 725)

1. $-2/t$ **3.** -1

5. $\dfrac{dy}{dx} = \dfrac{3}{2},\ \dfrac{d^2y}{dx^2} = 0;$ Neither concave upward nor concave downward

7. $dy/dx = 2t + 3,\ d^2y/dx^2 = 2$
At $t = -1,\ dy/dx = 1,\ d^2y/dx^2 = 2;$ Concave upward

9. $dy/dx = -\cot\theta,\ d^2y/dx^2 = -\csc^3\theta/2$
At $\theta = \pi/4,\ dy/dx = -1,\ d^2y/dx^2 = -\sqrt{2};$
Concave downward

11. $dy/dx = 2\csc\theta,\ d^2y/dx^2 = -2\cot^3\theta$
At $\theta = \pi/6,\ dy/dx = 4,\ d^2y/dx^2 = -6\sqrt{3};$
Concave downward

13. $dy/dx = -\tan\theta,\ d^2y/dx^2 = \sec^4\theta\csc\theta/3$
At $\theta = \pi/4,\ dy/dx = -1,\ d^2y/dx^2 = 4\sqrt{2}/3;$
Concave upward

15. $\left(-2/\sqrt{3}, 3/2\right):\ 3\sqrt{3}x - 8y + 18 = 0$
$(0, 2):\ y - 2 = 0$
$\left(2\sqrt{3}, 1/2\right):\ \sqrt{3}x + 8y - 10 = 0$

17. (a) and (d)   (b) At $t = 2,\ dx/dt = 2,$
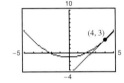 $dy/dt = 4,$ and $dy/dx = 2.$
(c) $y = 2x - 5$

19. (a) and (d)   (b) At $t = -1,\ dx/dt = -3,$
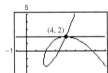 $dy/dt = 0,$ and $dy/dx = 0.$
(c) $y = 2$

21. $y = \pm\tfrac{3}{4}x$ **23.** $y = 3x - 5$ and $y = 1$

25. Horizontal: $(1, 0), (-1, \pi), (1, -2\pi)$
Vertical: $(\pi/2, 1), (-3\pi/2, -1), (5\pi/2, 1)$

27. Horizontal: $(1, 0)$ **29.** Horizontal: $(0, -2), (2, 2)$
Vertical: None   Vertical: None

31. Horizontal: $(0, 3), (0, -3)$ **33.** Horizontal: $(4, 0), (4, -2)$
Vertical: $(3, 0), (-3, 0)$ Vertical: $(2, -1), (6, -1)$

35. Horizontal: None
Vertical: $(1, 0), (-1, 0)$

37. Concave down: $-\infty < t < 0$
Concave up: $0 < t < \infty$

39. Concave up: $t > 0$

41. Concave down: $0 < t < \pi/2$
Concave up: $\pi/2 < t < \pi$

43. $\displaystyle\int_1^2 \sqrt{4t^2 + t + 4}\ dt$ **45.** $\displaystyle\int_{-2}^2 \sqrt{e^{2t} + 4}\ dt$

47. $2\sqrt{5} + \ln(2 + \sqrt{5}) \approx 5.916$ **49.** $\sqrt{2}(1 - e^{-\pi/2}) \approx 1.12$

51. $\tfrac{1}{12}\left[\ln(\sqrt{37} + 6) + 6\sqrt{37}\right] \approx 3.249$ **53.** $6a$ **55.** $8a$

57. (a) 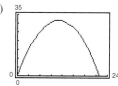 (b) 219.2 ft (c) 230.8 ft

59. (a) 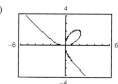 (b) $(0, 0)$, $(4\sqrt[3]{2}/3, 4\sqrt[3]{2}/3)$ (c) ≈ 6.557

61. (a)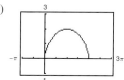

(b) The average speed of the particle on the second path is twice the average speed of the particle on the first path.
(c) 4π

63. $S = 2\pi \int_0^2 \sqrt{17}(t+1)\, dt = 8\pi\sqrt{17} \approx 103.625$

65. $S = 2\pi \int_0^{\pi/2} \left(\sin\theta \cos\theta \sqrt{4\cos^2\theta + 1}\right) d\theta = \dfrac{(5\sqrt{5}-1)\pi}{6}$
≈ 5.330

67. (a) $32\pi\sqrt{5}$ (b) $16\pi\sqrt{5}$ **69.** 32π **71.** $12\pi a^2/5$

73. See Theorem 10.7, Parametric Form of the Derivative, on page 719.

75. Answers will vary. Example:

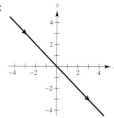

77. See Theorem 10.8, Arc Length in Parametric Form, on page 722.
79. Proof   **81.** $3\pi/2$ **83.** d   **84.** b   **85.** f   **86.** c
87. a   **88.** e   **89.** $\left(\tfrac{3}{4}, \tfrac{8}{5}\right)$ **91.** $V = 36\pi$
93. (a) $dy/dx = \sin\theta/(1-\cos\theta)$; $d^2y/dx^2 = -1/[a(\cos\theta-1)^2]$
(b) $y = (2+\sqrt{3})[x - a(\pi/6 - \tfrac{1}{2})] + a(1-\sqrt{3}/2)$
(c) $(a(2n+1)\pi, 2a)$
(d) Concave down on $(0, 2\pi)$, $(2\pi, 4\pi)$, etc.
(e) $s = 8a$
95. Proof
97. (a)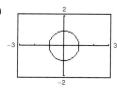

(b) Circle of radius 1 and center at $(0, 0)$ except the point $(-1, 0)$

(c) As t increases from -20 to 0, the speed increases, and as t increases from 0 to 20, the speed decreases.

99. False: $\dfrac{d^2y}{dx^2} = \dfrac{\dfrac{d}{dt}\left[\dfrac{g'(t)}{f'(t)}\right]}{f'(t)} = \dfrac{f'(t)g''(t) - g'(t)f''(t)}{[f'(t)]^3}$.

Section 10.4   (page 736)

1. 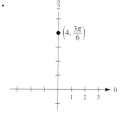 **3.**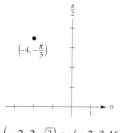

$(0, 4)$ $(-2, 2\sqrt{3}) \approx (-2, 3.464)$

5. 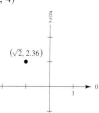 **7.**

$(-1.004, 0.996)$

9. 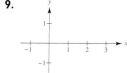 **11.**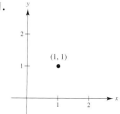

$\left(\sqrt{2}, \pi/4\right), \left(-\sqrt{2}, 5\pi/4\right)$

13. 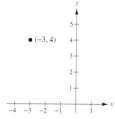 **15.**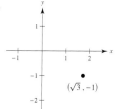

$(5, 2.214), (-5, 5.356)$ $(-2, 5\pi/6), (2, 11\pi/6)$
17. $(3.606, -0.588)$ **19.** $(2.833, 0.490)$
21. (a) 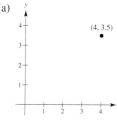 (b)

23. c   **24.** b   **25.** a   **26.** d

27. $r = a$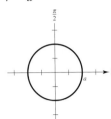

29. $r = 4 \csc \theta$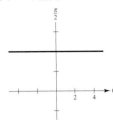

31. $r = \dfrac{-2}{3 \cos \theta - \sin \theta}$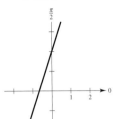

33. $r = 9 \csc^2 \theta \cos \theta$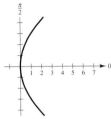

35. $x^2 + y^2 = 9$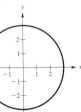

37. $x^2 + y^2 - y = 0$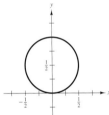

39. $\sqrt{x^2 + y^2} = \arctan(y/x)$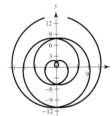

41. $x - 3 = 0$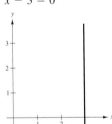

43.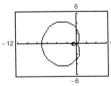
$0 \leq \theta < 2\pi$

45.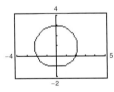
$0 \leq \theta < 2\pi$

47.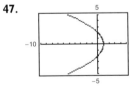
$-\pi < \theta < \pi$

49.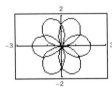
$0 \leq \theta < 4\pi$

51.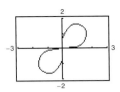
$0 \leq \theta < \pi/2$

53. $(x - h)^2 + (y - k)^2 = h^2 + k^2$
Radius: $\sqrt{h^2 + k^2}$
Center: (h, k)

55. $2\sqrt{5}$ **57.** ≈ 5.6

59. $\dfrac{dy}{dx} = \dfrac{2 \cos \theta (3 \sin \theta + 1)}{6 \cos^2 \theta - 2 \sin \theta - 3}$

$(5, \pi/2)$: $dy/dx = 0$

$(2, \pi)$: $dy/dx = -2/3$

$(-1, 3\pi/2)$: $dy/dx = 0$

61. (a) and (b)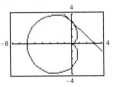

63. (a) and (b)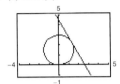

(c) $dy/dx = -1$ (c) $dy/dx = -\sqrt{3}$

65. Horizontal: $(2, 3\pi/2), (\tfrac{1}{2}, \pi/6), (\tfrac{1}{2}, 5\pi/6)$
Vertical: $(\tfrac{3}{2}, 7\pi/6), (\tfrac{3}{2}, 11\pi/6)$

67. $(5, \pi/2), (1, 3\pi/2)$

69.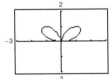

71.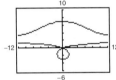

$(0, 0), (1.4142, 0.7854),$
$(1.4142, 2.3562)$

$(7, 1.5708), (3, 4.7124)$

73.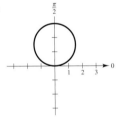
$\theta = 0$

75.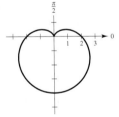
$\theta = \pi/2$

77.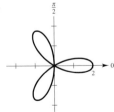
$\theta = \pi/6, \pi/2, 5\pi/6$

79.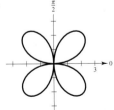
$\theta = 0, \pi/2$

81. 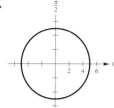 **83.**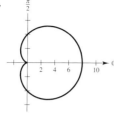

85. 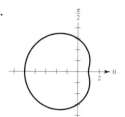 **87.**

89. 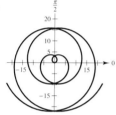 **91.**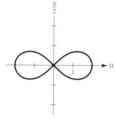

93. 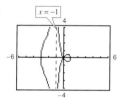 **95.**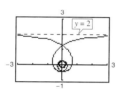

97. The rectangular coordinate system is a collection of points of the form (x, y), where x is the directed distance from the y-axis to the point and y is the directed distance from the x-axis to the point. Every point has a unique representation.

The polar coordinate system is a collection of points of the form (r, θ), where r is the directed distance from the origin O to a point P and θ is the directed angle, measured counterclockwise, from the polar axis to the segment $\overline{OP}$. Polar coordinates do not have unique representations.

99. $r = a$: Circle of radius a centered at the pole
$\theta = b$: Line passing through the pole

101. (a) 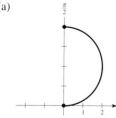 (b)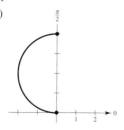

(c)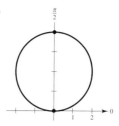

103. Proof

105. (a) $r = 2 - \sin(\theta - \pi/4)$ (b) $r = 2 + \cos\theta$
$ = 2 - \dfrac{\sqrt{2}(\sin\theta - \cos\theta)}{2}$

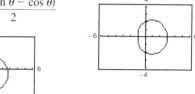

(c) $r = 2 + \sin\theta$ (d) $r = 2 - \cos\theta$

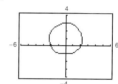 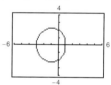

107. (a) 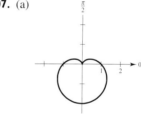 (b)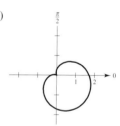

109. 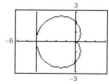   **111.**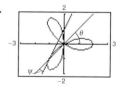

$\psi = \pi/2$ $\psi = \arctan\tfrac{1}{3} \approx 18.4°$

113. 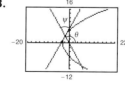   **115.** True   **117.** True

$\psi = \pi/3,\ 60°$

Section 10.5 (page 745)

1. $2\displaystyle\int_{\pi/2}^{\pi} \sin^2\theta\, d\theta$ **3.** $\dfrac{1}{2}\displaystyle\int_{\pi/2}^{3\pi/2} (1 - \sin\theta)^2\, d\theta$

5. (a) and (b) 16π **7.** $\pi/3$ **9.** $\pi/8$ **11.** $3\pi/2$

13.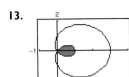
$(2\pi - 3\sqrt{3})/2$

15.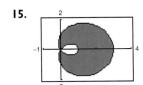
$\pi + 3\sqrt{3}$

17. $(1, \pi/2), (1, 3\pi/2), (0, 0)$

19. $\left(\dfrac{2 - \sqrt{2}}{2}, \dfrac{3\pi}{4}\right), \left(\dfrac{2 + \sqrt{2}}{2}, \dfrac{7\pi}{4}\right), (0, 0)$

21. $(\tfrac{3}{2}, \pi/6), (\tfrac{3}{2}, 5\pi/6), (0, 0)$ **23.** $(2, 4), (-2, -4)$

25. $(2, \pi/12), (2, 5\pi/12), (2, 7\pi/12), (2, 11\pi/12)$
$(2, 13\pi/12), (2, 17\pi/12), (2, 19\pi/12), (2, 23\pi/12)$

27. $(-0.581, \pm 2.607),$
$(2.581, \pm 1.376)$

29. $(0, 0), (0.935, 0.363),$
$(0.535, -1.006)$
The graphs reach the pole at different times (θ-values).

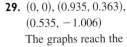

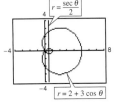

31.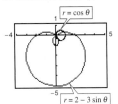

33.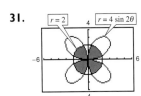

$\tfrac{4}{3}(4\pi - 3\sqrt{3})$

35. 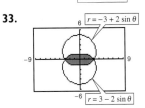 $11\pi - 24$

37. $5\pi a^2/4$ **39.** $(a^2/2)(\pi - 2)$

$\tfrac{2}{3}(4\pi - 3\sqrt{3})$

41. (a) $(x^2 + y^2)^{3/2} = ax^2$
(b) 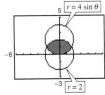   (c) $15\pi/2$

43. The area enclosed by the function is $\pi a^2/4$ if n is odd and is $\pi a^2/2$ if n is even.

45. $2\pi a$ **47.** 8

49.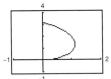
≈ 4.16

51.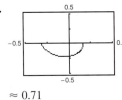
≈ 0.71

53.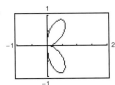
≈ 4.39

55. 36π

57. $\dfrac{2\pi\sqrt{1 + a^2}}{1 + 4a^2}(e^{\pi a} - 2a)$ **59.** 21.87

61. Area $= \dfrac{1}{2}\displaystyle\int_\alpha^\beta r^2\, d\theta$; Arc length $= \displaystyle\int_\alpha^\beta \sqrt{r^2 + \left(\dfrac{dr}{d\theta}\right)^2}\, d\theta$

63. a; Answers will vary.   **65.** $40\pi^2$

67. (a) 16π
(b)

θ	0.2	0.4	0.6	0.8	1.0	1.2	1.4
A	6.32	12.14	17.06	20.80	23.27	24.60	25.08

(c) and (d) $\tfrac{1}{4}$ of circle $(4\pi \approx 12.57) \approx 0.42$
$\tfrac{1}{2}$ of circle $(8\pi \approx 25.13) \approx 1.57 (\pi/2)$
$\tfrac{3}{4}$ of circle $(12\pi \approx 37.70) \approx 2.73$
(e) No. The results do not depend on the radius. Answers will vary.

69. Circle

71. (a)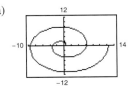
The graph becomes larger and more spread out. The graph is reflected over the y-axis.

(b) $(n\pi, an\pi)$ where $n = 1, 2, 3, \ldots$
(c) ≈ 21.26 (d) $4/3\pi^3$

73. $r = \sqrt{2}\cos\theta$

75. False. The graphs of $f(\theta) = 1$ and $g(\theta) = -1$ coincide.

77. Proof

Section 10.6   (page 753)

1.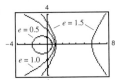
(a) Parabola (b) Ellipse
(c) Hyperbola

3.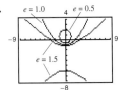
(a) Parabola (b) Ellipse
(c) Hyperbola

5. (a) 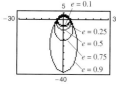 (b)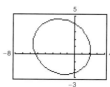

Ellipse   Parabola

As $e \to 1^-$, the ellipse becomes more elliptical, and as $e \to 0^+$, it becomes more circular.

(c) 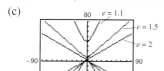 Hyperbola
As $e \to 1^+$, the hyperbola opens more slowly, and as $e \to \infty$, it opens more rapidly.

7. c **8.** f **9.** a **10.** e **11.** b **12.** d
13. $e = 1$ **15.** $e = \frac{1}{2}$
Distance $= 1$ Distance $= 6$

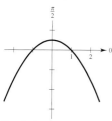 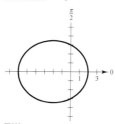

Parabola Ellipse

17. $e = \frac{1}{2}$ **19.** $e = 2$
Distance $= 4$ Distance $= \frac{5}{2}$

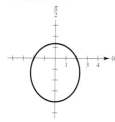 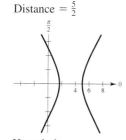

Ellipse Hyperbola

21. $e = 3$ **23.**
Distance $= \frac{1}{2}$

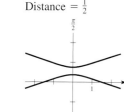 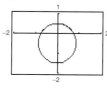

Hyperbola Ellipse

25. **27.**

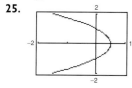 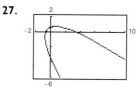

Parabola Rotated $\pi/4$ radian counterclockwise.

29.   **31.** $r = \dfrac{5}{5 + 3\cos\left(\theta + \dfrac{\pi}{4}\right)}$

Rotated $\pi/6$ radian counterclockwise.

33. $r = 1/(1 - \cos\theta)$ **35.** $r = 1/(2 + \sin\theta)$
37. $r = 2/(1 + 2\cos\theta)$ **39.** $r = 2/(1 - \sin\theta)$
41. $r = 16/(5 + 3\cos\theta)$ **43.** $r = 9/(4 - 5\sin\theta)$
45. If $0 < e < 1$, the conic is an ellipse.
If $e = 1$, the conic is a parabola.
If $e > 1$, the conic is a hyperbola.
47. (a) Hyperbola (b) Ellipse (c) Parabola (d) Hyperbola
49. Proof
51. $r^2 = \dfrac{9}{1 - (16/25)\cos^2\theta}$ **53.** $r^2 = \dfrac{-16}{1 - (25/9)\cos^2\theta}$

55. ≈ 10.88 **57.** $\dfrac{7979.21}{1 - 0.9372\cos\theta}$; 11,015 mi

59. $r = \dfrac{149{,}558{,}278.0560}{1 - 0.0167\cos\theta}$ **61.** $r = \dfrac{5{,}540{,}410{,}095.36}{1 - 0.2488\cos\theta}$

Perihelion: 147,101,680 km Perihelion: 4,436,587,200 km
Aphelion: 152,098,320 km Aphelion: 7,375,412,800 km

63. Answers will vary. Sample answers:
(a) 9.341×10^{18} km²; 21.867 yrs
(b) 0.8995 rad; Larger angle with the smaller ray to generate an equal area
(c) Part (a): 2.559×10^9 km; 1.17×10^8 km/yr
Part (b): 4.119×10^9 km; 1.88×10^8 km/yr

65. Proof
67. Let $r_1 = ed/(1 + \sin\theta)$ and $r_2 = ed/(1 - \sin\theta)$.

The points of intersection of r_1 and r_2 are $(ed, 0)$ and (ed, π). The slope of the tangent line to r_1 at $(ed, 0)$ is -1 and at (ed, π) is 1. The slope of the tangent line to r_2 at $(ed, 0)$ is 1 and at (ed, π) is -1. Therefore, at $(ed, 0)$, $m_1 m_2 = -1$ and at (ed, π), $m_1 m_2 = -1$ and the curves intersect at right angles.

Review Exercises for Chapter 10 (page 756)

1. e **2.** c **3.** b **4.** d **5.** a **6.** f
7. Circle **9.** Hyperbola
Center: $\left(\frac{1}{2}, -\frac{3}{4}\right)$ Center: $(-4, 3)$
Radius: 1 Vertices: $\left(-4 \pm \sqrt{2}, 3\right)$

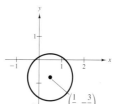 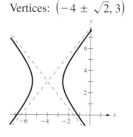

11. Ellipse
Center: $(2, -3)$
Vertices: $\left(2, -3 \pm \sqrt{2}/2\right)$

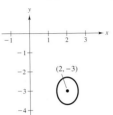

13. $y^2 - 4y - 12x + 4 = 0$ **15.** $(x-2)^2/25 + y^2/21 = 1$
17. $x^2/16 - y^2/20 = 1$ **19.** ≈ 15.87 **21.** $4x + 4y - 7 = 0$
23. (a) $(0, 50)$ (b) $\approx 38{,}294.49$

25.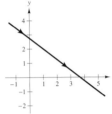
$4y + 3x - 11 = 0$

27.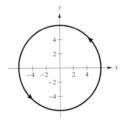
$x^2 + y^2 = 36$

29.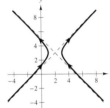
$(x-2)^2 - (y-3)^2 = 1$

31. $x = 5t - 2$
$y = 6 - 4t$

33. $x = 4\cos\theta - 3$
$y = 4 + 3\sin\theta$

35.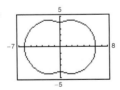

37. (a) $dy/dx = -\tfrac{3}{4}$;
Horizontal tangents: None
(b) $y = (-3x + 11)/4$
(c)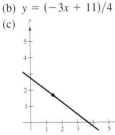

39. (a) $dy/dx = -2t^2$;
Horizontal tangents: None
(b) $y = 3 + 2/x$
(c)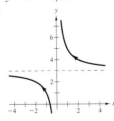

41. (a) $\dfrac{dy}{dx} = \dfrac{(t-1)(2t+1)^2}{t^2(t-2)^2}$;
Horizontal tangent: $\left(\tfrac{1}{3}, -1\right)$
(b) $y = \dfrac{4x^2}{(5x-1)(x-1)}$
(c)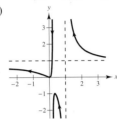

43. (a) $\dfrac{dy}{dx} = -\dfrac{5}{2}\cot\theta$;
Horizontal tangents: $(3, 7)$, $(3, -3)$
(b) $\dfrac{(x-3)^2}{4} + \dfrac{(y-2)^2}{25} = 1$
(c)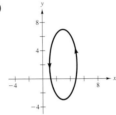

45. (a) $\dfrac{dy}{dx} = -4\tan\theta$;
Horizontal tangents: None
(b) $x^{2/3} + (y/4)^{2/3} = 1$
(c)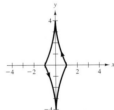

47. Horizontal tangent line: $t = 0$ $(4, 0)$
49. Horizontal tangent lines: $\theta = 0$ and $\theta = \pi$ $((2, 2)$ and $(2, 0))$
Vertical tangent lines: $\theta = \pi/2$ and $\theta = 3\pi/2$ $((4, 1)$ and $(0, 1))$
51. (a) and (c)
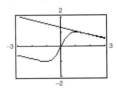
(b) $dx/d\theta = -4$, $dy/d\theta = 1$, $dy/dx = -\tfrac{1}{4}$

53. $\frac{1}{2}\pi^2 r$

55. (a) $s = 12\pi\sqrt{10} \approx 119.215$ **57.** $A = 3\pi$
(b) $s = 4\pi\sqrt{10} \approx 39.738$

59.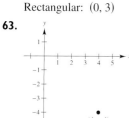
Rectangular: $(0, 3)$

61.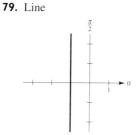
Rectangular: $(0.0187, 1.7320)$
$\left(4\sqrt{2}, \frac{7\pi}{4}\right), \left(-4\sqrt{2}, \frac{3\pi}{4}\right)$

63.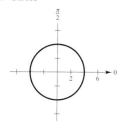

65. $x^2 + y^2 - 3x = 0$ **67.** $(x^2 + y^2 + 2x)^2 = 4(x^2 + y^2)$
69. $(x^2 + y^2)^2 = x^2 - y^2$ **71.** $y^2 = x^2[(4 - x)/(4 + x)]$
73. $r = a\cos^2\theta\sin\theta$ **75.** $r^2 = a^2\theta^2$
77. Circle   **79.** Line

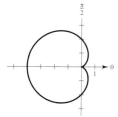

81. Cardioid   **83.** Limaçon

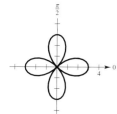

85. Rose curve   **87.** Rose curve

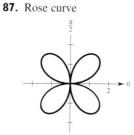

89. 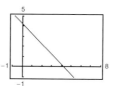   **91.**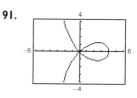

93. (a) $\pm\pi/3$
(b) Vertical: $(-1, 0), (3, \pi), \left(\frac{1}{2}, \pm 1.318\right)$
Horizontal: $(-0.686, \pm 0.568), (2.186, \pm 2.206)$
(c)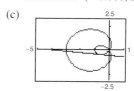

95. $\arctan(2\sqrt{3}/3) \approx 49.1°$ **97.** Proof

99. $A = 2\left(\frac{1}{2}\right)\int_0^\pi (2 + \cos\theta)^2\, d\theta \approx 14.14$

101. $A = 2\left(\frac{1}{2}\right)\int_0^{\pi/2} 4\sin 2\theta\, d\theta \approx 4.00$

103.

$A = 2\left(\frac{1}{2}\right)\int_0^{\pi/2} \sin^2\theta\cos^4\theta\, d\theta \approx 0.10$

105.

$A = 2\left[\frac{1}{2}\int_0^{\pi/12} 18\sin 2\theta + \frac{1}{2}\int_{\pi/12}^{5\pi/12} 9\, d\theta + \frac{1}{2}\int_{5\pi/12}^{\pi/2} 18\sin 2\theta\, d\theta\right]$
$\approx 1.2058 + 9.4248 + 1.2058 = 11.8364$

107. $4a$

109. $S = 2\pi\int_0^{\pi/2} (1 + 4\cos\theta)\sin\theta\sqrt{17 + 8\cos\theta}\, d\theta$
$= 34\pi\sqrt{17}/5 \approx 88.08$

111. Parabola   **113.** Ellipse

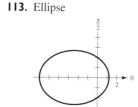

115. Hyperbola   **117.** $r = 10 \sin \theta$

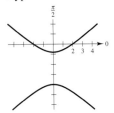

119. $r = 4/(1 - \cos \theta)$ **121.** $r = 5/(3 - 2 \cos \theta)$

P.S. Problem Solving (page 759)

1. (a)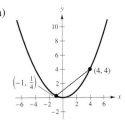

(b)–(c) Proofs

3. Proof

5. (a) $r = 2a \tan \theta \sin \theta$

(b) $x = 2at^2/(1 + t^2)$
$y = 2at^3/(1 + t^2)$

(c) $y^2 = x^3/(2a - x)$

7. (a) $y^2 = x^2[(1 - x)/(1 + x)]$

(b) $r = \cos 2\theta \cdot \sec \theta$

(c)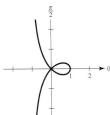

(d) $y = x, y = -x$

(e) $\left(\dfrac{\sqrt{5} - 1}{2}, \pm \dfrac{\sqrt{5} - 1}{2} \sqrt{-2 + \sqrt{5}} \right)$

9. (a)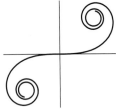

Generated by Mathematica

(b) Proof

(c) $a, 2\pi$

11. $A = \tfrac{1}{2}ab$ **13.** $r^2 = 2 \cos 2\theta$

15. (a) First plane: $x_1 = \cos 70°(150 - 375t)$
$y_1 = \sin 70°(150 - 375t)$
Second plane: $x_2 = \cos 45°(450t - 190)$
$y_2 = \sin 45°(190 - 450t)$

(b) $\{[\cos 45°(450t - 190) - \cos 70°(150 - 375t)]^2$
$+ [\sin 45°(190 - 450t) - \sin 70°(150 - 375t)]^2\}^{1/2}$

(c)

0.4145 hr; Yes

17.

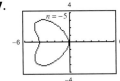 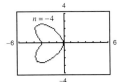

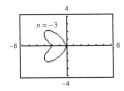 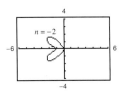

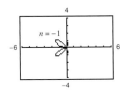 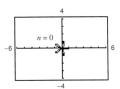

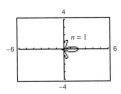 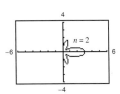

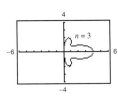 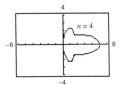

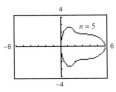

$n = 1, 2, 3, 4, 5$ produce "bells"; $n = -1, -2, -3, -4, -5$ produce "hearts."

Chapter 11
Section 11.1 (page 769)

1. (a) $\langle 4, 2 \rangle$ (b)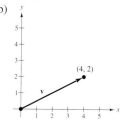

3. (a) $\langle -7, 0 \rangle$ (b)

5. $\mathbf{u} = \mathbf{v} = \langle 2, 4 \rangle$ **7.** $\mathbf{u} = \mathbf{v} = \langle 6, -5 \rangle$

9. (a) and (c) **11.** (a) and (c)

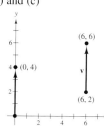   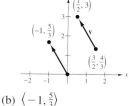

(b) $\langle 4, 3 \rangle$ (b) $\langle -4, -3 \rangle$

13. (a) and (c) **15.** (a) and (c)

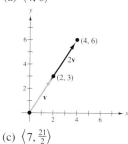   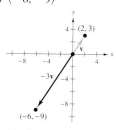

(b) $\langle 0, 4 \rangle$ (b) $\langle -1, \frac{5}{3} \rangle$

17. (a) $\langle 4, 6 \rangle$ (b) $\langle -6, -9 \rangle$

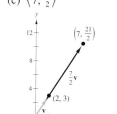   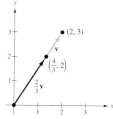

(c) $\langle 7, \frac{21}{2} \rangle$ (d) $\langle \frac{4}{3}, 2 \rangle$

19. 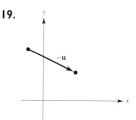   **21.**

23. (a) $\langle \frac{8}{3}, 6 \rangle$ (b) $\langle -2, -14 \rangle$ (c) $\langle 18, -7 \rangle$

25. $\langle 3, -\frac{3}{2} \rangle$ **27.** $\langle 4, 3 \rangle$

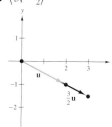   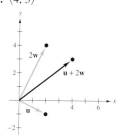

29. $(3, 5)$ **31.** 5 **33.** $\sqrt{61}$ **35.** 4

37. $\langle \sqrt{17}/17, 4\sqrt{17}/17 \rangle$ **39.** $\langle 3\sqrt{34}/34, 5\sqrt{34}/34 \rangle$

41. (a) $\sqrt{2}$ (b) $\sqrt{5}$ (c) 1 (d) 1 (e) 1 (f) 1

43. (a) $\sqrt{5}/2$ (b) $\sqrt{13}$ (c) $\sqrt{85}/2$ (d) 1 (e) 1 (f) 1

45.

$\|\mathbf{u}\| + \|\mathbf{v}\| = \sqrt{5} + \sqrt{41}$ and $\|\mathbf{u} + \mathbf{v}\| = \sqrt{74}$
$\sqrt{74} < \sqrt{5} + \sqrt{41}$

47. $\langle 2\sqrt{2}, 2\sqrt{2} \rangle$ **49.** $\langle 1, \sqrt{3} \rangle$ **51.** $\langle 3, 0 \rangle$ **53.** $\langle -\sqrt{3}, 1 \rangle$

55. $\left\langle \dfrac{2 + 3\sqrt{2}}{2}, \dfrac{3\sqrt{2}}{2} \right\rangle$ **57.** $\langle 2 \cos 4 + \cos 2, 2 \sin 4 + \sin 2 \rangle$

59. Answers will vary. Example: A scalar is a single real number such as 2. A vector is a line segment having both direction and magnitude. The vector $\langle \sqrt{3}, 1 \rangle$, given in component form, has a direction of $\pi/6$ and a magnitude of 2.

61. (a) Vector; has magnitude and direction
   (b) Scalar; has only magnitude

63. $a = 1, b = 1$ **65.** $a = 1, b = 2$ **67.** $a = \frac{2}{3}, b = \frac{1}{3}$

69. (a) $\pm(1/\sqrt{37})\langle 1, 6 \rangle$ **71.** (a) $\pm(1/\sqrt{10})\langle 1, 3 \rangle$
   (b) $\pm(1/\sqrt{37})\langle 6, -1 \rangle$ (b) $\pm(1/\sqrt{10})\langle 3, -1 \rangle$

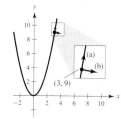   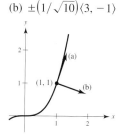

73. (a) $\pm\frac{1}{5}\langle -4, 3\rangle$
(b) $\pm\frac{1}{5}\langle 3, 4\rangle$

75. $\langle -\sqrt{2}/2, \sqrt{2}/2\rangle$

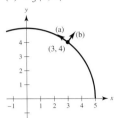

77. (a)–(c) Answers will vary.   **79.** 1.33, 132.5°
81. (a) Direction: $\alpha = 11.8°$
Magnitude: 440.2 N
(b) $M = \sqrt{(275 + 180\cos\theta)^2 + (180\sin\theta)^2}$
$\alpha = \arctan\left(\dfrac{180\sin\theta}{275 + 180\cos\theta}\right)$

(c)

θ	0°	30°	60°	90°	120°
M	455.0	440.2	396.9	328.7	241.9
α	0°	11.8°	23.1°	33.2°	40.1°

θ	150°	180°
M	149.3	95.0
α	37.1°	0°

(d)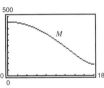

(e) M decreases because the forces change from acting in the same direction to acting in opposite directions as θ increases from 0° to 180°.

83. 71.3°, 228.5 lb
85. (a) $\theta = 0°$ (b) $\theta = 180°$
(c) No, the resultant can only be less than or equal to the sum.
87. $(-4, -1), (6, 5), (10, 3)$
89. Tension in cable $AC \approx 1758.8$ lb
Tension in cable $BC \approx 1305.4$ lb
91. Horizontal: 1193.43 ft/sec   **93.** 38.3° north of west
Vertical: 125.43 ft/sec   882.9 kph
95. True   **97.** True   **99.** False. $\|a\mathbf{i} + b\mathbf{j}\| = \sqrt{2}|a|$
101–103. Proofs   **105.** $x^2 + y^2 = 25$

Section 11.2 (page 778)

1.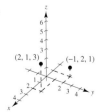

3.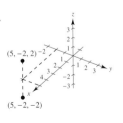

5. $A(2, 3, 4)$ **7.** $(-3, 4, 5)$ **9.** $(10, 0, 0)$ **11.** 0
$B(-1, -2, 2)$
13. Six units above the xy-plane
15. Four units in front of the yz-plane
17. To the left of the xz-plane and either above, below, or on the xy-plane and either in front of, behind, or on the yz-plane
19. Within three units of the xz-plane
21. Three units below the xy-plane, to the right of the xz-plane, and in front of the yz-plane, *or* three units below the xy-plane, to the left of the xz-plane, and behind the yz-plane
23. 1. Above the xy-plane and (a) to the right of the xz-plane and behind the yz-plane or (b) to the left of the xz-plane and in front of the yz-plane, *or*
2. Below the xy-plane and (a) to the right of the xz-plane and in front of the yz-plane or (b) to the left of the xz-plane and behind the yz-plane
25. $\sqrt{65}$ **27.** $\sqrt{61}$
29. $3, 3\sqrt{5}, 6$ **31.** $6, 6, 2\sqrt{10}$
Right triangle   Isosceles triangle
33. $(0, 0, 5), (2, 2, 6), (2, -4, 9)$
35. $\left(\frac{3}{2}, -3, 5\right)$ **37.** $(x - 0)^2 + (y - 2)^2 + (z - 5)^2 = 4$
39. $(x - 1)^2 + (y - 3)^2 + (z - 0)^2 = 10$
41. $(x - 1)^2 + (y + 3)^2 + (z + 4)^2 = 25$
Center: $(1, -3, -4)$
Radius: 5
43. $\left(x - \frac{1}{3}\right)^2 + (y + 1)^2 + z^2 = 1$
Center: $\left(\frac{1}{3}, -1, 0\right)$
Radius: 1
45. A solid sphere with center $(0, 0, 0)$ and radius 6
47. Interior of sphere of radius 4 centered at $(2, -3, 4)$
49. (a) $\langle -2, 2, 2\rangle$ **51.** (a) $\langle -3, 0, 3\rangle$
(b) 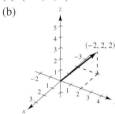   (b)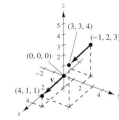

53. $\mathbf{u} = \langle 1, -1, 6\rangle$ **55.** $\mathbf{u} = \langle -1, 0, -1\rangle$
$\|\mathbf{u}\| = \sqrt{38}$ $\|\mathbf{u}\| = \sqrt{2}$
$\dfrac{\mathbf{u}}{\|\mathbf{u}\|} = \dfrac{1}{\sqrt{38}}\langle 1, -1, 6\rangle$ $\dfrac{\mathbf{u}}{\|\mathbf{u}\|} = \dfrac{1}{\sqrt{2}}\langle -1, 0, -1\rangle$

57. (a) and (c)   **59.** $(3, 1, 8)$

(b) $\langle 4, 1, 1\rangle$

61. (a) 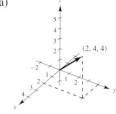 (b)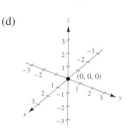

(c) (d)

63. $\langle -1, 0, 4 \rangle$ **65.** $\langle 6, 12, 6 \rangle$ **67.** $\langle \frac{7}{2}, 3, \frac{5}{2} \rangle$
69. a and b **71.** a **73.** Collinear **75.** Not collinear
77. $\overrightarrow{AB} = \langle 1, 2, 3 \rangle$
$\overrightarrow{CD} = \langle 1, 2, 3 \rangle$
$\overrightarrow{BD} = \langle -2, 1, 1 \rangle$
$\overrightarrow{AC} = \langle -2, 1, 1 \rangle$
Since $\overrightarrow{AB} = \overrightarrow{CD}$ and $\overrightarrow{BD} = \overrightarrow{AC}$, the given points form the vertices of a parallelogram.
79. 0 **81.** $\sqrt{14}$ **83.** $\sqrt{34}$
85. (a) $\frac{1}{3}\langle 2, -1, 2 \rangle$ (b) $-\frac{1}{3}\langle 2, -1, 2 \rangle$
87. (a) $(1/\sqrt{38})\langle 3, 2, -5 \rangle$ (b) $-(1/\sqrt{38})\langle 3, 2, -5 \rangle$
89. (a)–(d) Answers will vary. **91.** $\pm \frac{5}{3}$
93. $\langle 0, 10/\sqrt{2}, 10/\sqrt{2} \rangle$ **95.** $\langle 1, -1, \frac{1}{2} \rangle$
97. 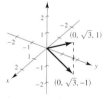 **99.** $(2, -1, 2)$

$\langle 0, \sqrt{3}, \pm 1 \rangle$

101. (a) 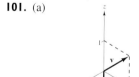 (b) $a = 0, a + b = 0, b = 0$
(c) $a = 1, a + b = 2, b = 1$
(d) Not possible

103. x_0 is directed distance to yz-plane.
y_0 is directed distance to xz-plane.
z_0 is directed distance to xy-plane.
105. $(x - x_0)^2 + (y - y_0)^2 + (z - z_0)^2 = r^2$ **107.** 0
109. (a) $T = 8L/\sqrt{L^2 - 18^2}, L > 18$
(b)

L	20	25	30	35	40	45	50
T	18.4	11.5	10	9.3	9.0	8.7	8.6

(c) 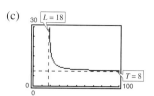 (d) Proof (e) 30 in.

111. $(\sqrt{3}/3)\langle 1, 1, 1 \rangle$
113. Tension in cable AB: 202.919 N
Tension in cable AC: 157.909 N
Tension in cable AD: 226.521 N
115. $(x - \frac{4}{3})^2 + (y - 3)^2 + (z + \frac{1}{3})^2 = \frac{44}{9}$

Section 11.3 (page 787)

1. (a) -6 (b) 25 (c) 25 (d) $\langle -12, 18 \rangle$ (e) -12
3. (a) -17 (b) 26 (c) 26 (d) $\langle 51, -34 \rangle$ (e) -34
5. (a) 2 (b) 29 (c) 29 (d) $\langle 0, 12, 10 \rangle$ (e) 4
7. (a) 1 (b) 6 (c) 6 (d) $\mathbf{i} - \mathbf{k}$ (e) 2
9. 20 **11.** $\pi/2$ **13.** $\arccos(-1/5\sqrt{2}) \approx 98.1°$
15. $\arccos(\sqrt{2}/3) \approx 61.9°$ **17.** $\arccos(-8\sqrt{13}/65) \approx 116.3°$
19. Neither **21.** Orthogonal **23.** Neither
25. Orthogonal **27.** Right triangle; answers will vary.
29. Acute triangle; answers will vary.
31. $\cos \alpha = \frac{1}{3}$ **33.** $\cos \alpha = 0$
$\cos \beta = \frac{2}{3}$ $\cos \beta = 3/\sqrt{13}$
$\cos \gamma = \frac{2}{3}$ $\cos \gamma = -2/\sqrt{13}$
35. $\alpha \approx 43.3°, \beta \approx 61.0°, \gamma \approx 119.0°$
37. $\alpha \approx 100.5°, \beta \approx 24.1°, \gamma \approx 68.6°$
39. Magnitude: 124.310 lb
$\alpha \approx 29.48°, \beta \approx 61.39°, \gamma \approx 96.53°$
41. $\alpha = 90°, \beta = 45°, \gamma = 45°$ **43.** $\langle 4, -1 \rangle$ **45.** $\langle 2, 1, 1 \rangle$
47. (a) $\langle \frac{5}{2}, \frac{1}{2} \rangle$ (b) $\langle -\frac{1}{2}, \frac{5}{2} \rangle$
49. (a) $\langle 0, \frac{33}{25}, \frac{44}{25} \rangle$ (b) $\langle 2, -\frac{8}{25}, \frac{6}{25} \rangle$
51. See "Definition of Dot Product," page 781.
53. (a) $\theta = \pi/2$ (b) $0 < \theta < \pi/2$ (c) $\pi/2 < \theta < \pi$
55. See the definitions of direction cosines and direction angles on page 784.
57. (a) The vectors are parallel. (b) The vectors are orthogonal.
59. \$12,351.25; Total revenue **61.** (a)–(c) Answers will vary.
63. Answers will vary. **65.** $\langle 0, 0 \rangle$
67. Answers will vary. Example: $\langle 4, 3 \rangle$ and $\langle -4, -3 \rangle$
69. Answers will vary. Example: $\langle 2, 0, 3 \rangle$ and $\langle -2, 0, -3 \rangle$
71. (a) 8335.1 lb (b) 47,270.8 lb **73.** 425 ft-lb
75. False. For example, $\langle 1, 1 \rangle \cdot \langle 2, 3 \rangle = 5$ and
$\langle 1, 1 \rangle \cdot \langle 1, 4 \rangle = 5$, but $\langle 2, 3 \rangle \neq \langle 1, 4 \rangle$.
77. $\arccos(1/\sqrt{3}) \approx 54.7°$
79. (a) To $y = x^2$ at $(1, 1)$: $\langle \pm\sqrt{5}/5, \pm 2\sqrt{5}/5 \rangle$
To $y = x^{1/3}$ at $(1, 1)$: $\langle \pm 3\sqrt{10}/10, \pm \sqrt{10}/10 \rangle$
To $y = x^2$ at $(0, 0)$: $\langle \pm 1, 0 \rangle$
To $y = x^{1/3}$ at $(0, 0)$: $\langle 0, \pm 1 \rangle$
(b) At $(1, 1), \theta = 45°$
At $(0, 0), \theta = 90°$

81. (a) To $y = 1 - x^2$ at $(1, 0)$: $\langle \pm\sqrt{5}/5, -2\sqrt{5}/5 \rangle$
To $y = x^2 - 1$ at $(1, 0)$: $\langle \pm\sqrt{5}/5, \pm 2\sqrt{5}/5 \rangle$
To $y = 1 - x^2$ at $(-1, 0)$: $\langle \pm\sqrt{5}/5, \pm 2\sqrt{5}/5 \rangle$
To $y = x^2 - 1$ at $(-1, 0)$: $\langle \pm\sqrt{5}/5, \mp -2\sqrt{5}/5 \rangle$
(b) At $(1, 0)$, $\theta = 53.13°$
At $(-1, 0)$, $\theta = 53.13°$

83. Proof

85. (a) 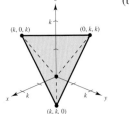 (b) $k\sqrt{2}$ (c) $60°$ (d) $109.5°$

87–89. Proofs

Section 11.4 (page 796)

1. $-\mathbf{k}$ 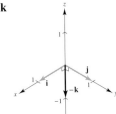   **3.** $\mathbf{i}$

5. $-\mathbf{j}$

7. (a) $-22\mathbf{i} + 16\mathbf{j} - 23\mathbf{k}$ **9.** (a) $17\mathbf{i} - 33\mathbf{j} - 10\mathbf{k}$
(b) $22\mathbf{i} - 16\mathbf{j} + 23\mathbf{k}$ (b) $-17\mathbf{i} + 33\mathbf{j} + 10\mathbf{k}$
(c) $\mathbf{0}$ (c) $\mathbf{0}$

11. $\langle -1, -1, -1 \rangle$ **13.** $\langle 0, 0, 54 \rangle$ **15.** $\langle -2, 3, -1 \rangle$

17. 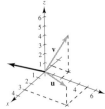   **19.**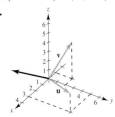

21. $\langle -70, -23, 57/2 \rangle$
$\langle -140/\sqrt{24{,}965}, -46/\sqrt{24{,}965}, 57/\sqrt{24{,}965} \rangle$

23. $\langle -\frac{71}{20}, -\frac{11}{5}, \frac{5}{4} \rangle$
$\langle -71/\sqrt{7602}, -44/\sqrt{7602}, 25/\sqrt{7602} \rangle$

25. Answers will vary.   **27.** 1   **29.** $6\sqrt{5}$ **31.** $2\sqrt{83}$

33. $3\sqrt{13}/2$ **35.** $\sqrt{16{,}742}/2$ **37.** $10 \cos 40 \approx 7.66$ ft-lb

39. (a) $90 \sin \theta$ (b) $45\sqrt{2} \approx 63.64$
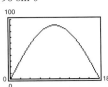   (c) $\theta = 90°$. This is what should be expected. When $\theta = 90°$, the pipe wrench is horizontal.

41. 1   **43.** 6   **45.** 2   **47.** 75

49. See "Definition of Cross Product of Two Vectors in Space," page 790.

51. The magnitude of the cross product will increase by a factor of 4.

53. False. The cross product of two vectors is not defined in a two-dimensional coordinate system.

55. True   **57–63.** Proofs

Section 11.5 (page 805)

1. (a)

(b) $P = (1, 2, 2)$, $Q = (10, -1, 17)$, $\overrightarrow{PQ} = \langle 9, -3, 15 \rangle$
(There are many correct answers.) The components of the vector and the coefficients of t are proportional because the line is parallel to $\overrightarrow{PQ}$.

(c) $\left(-\frac{1}{5}, \frac{12}{5}, 0\right)$, $(7, 0, 12)$, $\left(0, \frac{7}{3}, \frac{1}{3}\right)$

	Parametric Equations	Symmetric Equations	Direction Numbers
3.	$x = t$ $y = 2t$ $z = 3t$	$x = \dfrac{y}{2} = \dfrac{z}{3}$	1, 2, 3
5.	$x = -2 + 2t$ $y = 4t$ $z = 3 - 2t$	$\dfrac{x+2}{2} = \dfrac{y}{4} = \dfrac{z-3}{-2}$	2, 4, -2
7.	$x = 1 + 3t$ $y = -2t$ $z = 1 + t$	$\dfrac{x-1}{3} = \dfrac{y}{-2} = \dfrac{z-1}{1}$	3, -2, 1
9.	$x = 5 + 17t$ $y = -3 - 11t$ $z = -2 - 9t$	$\dfrac{x-5}{17} = \dfrac{y+3}{-11} = \dfrac{z+2}{-9}$	17, -11, -9
11.	$x = 2 + 8t$ $y = 3 + 5t$ $z = 12t$	$\dfrac{x-2}{8} = \dfrac{y-3}{5} = \dfrac{z}{12}$	8, 5, 12

13. $x = 2$
$y = 3$
$z = 4 + t$

15. $x = 2 + 3t$
$y = 3 + 2t$
$z = 4 - t$

17. $x = 5 + 2t$
$y = -3 - t$
$z = -4 + 3t$

19. $x = 2 - t$
$y = 1 + t$
$z = 2 + t$
21. $P(3, -1, -2)$; $\mathbf{v} = \langle -1, 2, 0 \rangle$
23. $P(7, -6, -2)$; $\mathbf{v} = \langle 4, 2, 1 \rangle$
25. $L_1 = L_2$ and is parallel to L_3.
27. $(2, 3, 1)$; $\cos \theta = 7\sqrt{17}/51$ **29.** Not intersecting
31.

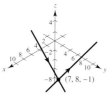

$(7, 8, -1)$

33. (a) $P = (0, 0, -1), Q = (0, -2, 0), R = (3, 4, -1)$
$\overrightarrow{PQ} = \langle 0, -2, 1 \rangle, \overrightarrow{PR} = \langle 3, 4, 0 \rangle$
(There are many correct answers.)
(b) $\overrightarrow{PQ} \times \overrightarrow{PR} = \langle -4, 3, 6 \rangle$
The components of the cross product are proportional to the coefficients of the variables in the equation. The cross product is parallel to the normal vector.

35. $x - 2 = 0$ **37.** $2x + 3y - z = 10$
39. $x - y + 2z = 12$ **41.** $3x + 9y - 7z = 0$
43. $4x - 3y + 4z = 10$ **45.** $z = 3$ **47.** $x + y + z = 5$
49. $7x + y - 11z = 5$ **51.** $y - z = -1$
53.   **55.** $x - z = 0$

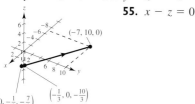

57. Orthogonal   **59.** Neither; 83.5°   **61.** Parallel
63.   **65.**

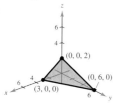 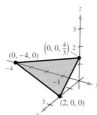

67.   **69.**

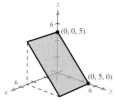 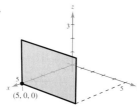

71.   **73.**

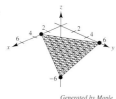 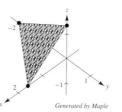

Generated by Maple   *Generated by Maple*

75. $P_1 = P_4$ and is parallel to P_2.
77. The planes have intercepts at $(c, 0, 0)$, $(0, c, 0)$, and $(0, 0, c)$ for each value of c.
79. If $c = 0$, $z = 0$ is xy-plane; If $c \neq 0$, plane is parallel to the x-axis and passes through $(0, 0, 0)$ and $(0, 1, -c)$.
81. $x = 2$ **83.** $(2, -3, 2)$ The line does not lie in the plane.
$y = 1 + t$
$z = 1 + 2t$
85. Not intersecting   **87.** $6\sqrt{14}/7$ **89.** $11\sqrt{6}/6$
91. $2\sqrt{26}/13$ **93.** $27\sqrt{94}/188$ **95.** $\sqrt{2533}/17$
97. $7\sqrt{3}/3$ **99.** $\sqrt{66}/3$
101. Parametric equations: $x = x_1 + at, y = y_1 + bt$, and $z = z_1 + ct$
Symmetric equations: $\dfrac{x - x_1}{a} = \dfrac{y - y_1}{b} = \dfrac{z - z_1}{c}$
You need a vector $\mathbf{v} = \langle a, b, c \rangle$ parallel to the line and a point $P(x_1, y_1, z_1)$ on the line.
103. Simultaneously solve the two linear equations representing the planes and substitute the values back into one of the original equations. Then choose a value for t and form the corresponding parametric equations for the line of intersection.
105. (a) Parallel if vector $\langle a_1, b_1, c_1 \rangle$ is a scalar multiple of $\langle a_2, b_2, c_2 \rangle$; $\theta = 0$.
(b) Perpendicular if $a_1 a_2 + b_1 b_2 + c_1 c_2 = 0$; $\theta = \pi/2$.
107. $cbx + acy + abz = abc$
109. (a)

Year	1994	1995	1996	1997	1998
z (approx.)	8.74	8.40	8.26	8.06	7.88

Year	1999	2000
z (approx.)	7.82	7.70

(b) Answers will vary.
111. (a) $\sqrt{70}$ in.
(b)

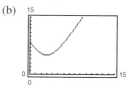

(c) The distance is never zero.
(d) 5 in.
113. $\left(\dfrac{77}{13}, \dfrac{48}{13}, -\dfrac{23}{13}\right)$ **115.** $\left(-\dfrac{1}{2}, -\dfrac{9}{4}, \dfrac{1}{4}\right)$ **117.** True   **119.** True

Section 11.6   (page 818)

1. c   **2.** e   **3.** f   **4.** b   **5.** d   **6.** a
7. Plane   **9.** Right circular cylinder

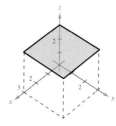 

11. Parabolic cylinder 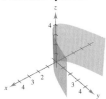 **13.** Elliptic cylinder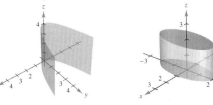

15. Cylinder 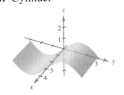 **17.** (a) $(20, 0, 0)$ (b) $(10, 10, 20)$ (c) $(0, 0, 20)$ (d) $(0, 20, 0)$

19. Ellipsoid 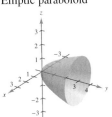 **21.** Hyperboloid of one sheet

23. Elliptic paraboloid 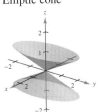 **25.** Hyperbolic paraboloid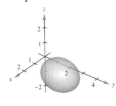

27. Elliptic cone  **29.** Ellipsoid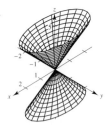

31. 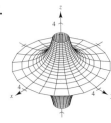 **33.**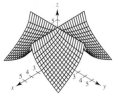

35. 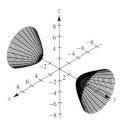 (continued from left column)

39. 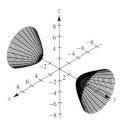 **41.**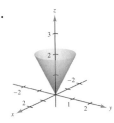

43. 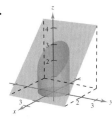 **45.** $x^2 + z^2 = 4y$

47. $4x^2 + 4y^2 = z^2$ **49.** $y^2 + z^2 = 4/x^2$
51. $y = \sqrt{2z}$ (or $x = \sqrt{2z}$)
53. Let C be a curve in a plane and let L be a line not in a parallel plane. The set of all lines parallel to L and intersecting C is called a cylinder. C is called the generating curve of the cylinder, and the parallel lines are called rulings.
55. See pages 812 and 813. **57.** $128\pi/3$
59. (a) Major axis: $4\sqrt{2}$ (b) Major axis: $8\sqrt{2}$
     Minor axis: 4   Minor axis: 8
     Foci: $(0, \pm 2, 2)$ Foci: $(0, \pm 4, 8)$
61. $x^2 + z^2 = 8y$; Elliptic paraboloid
63. $x^2/3963^2 + y^2/3963^2 + z^2/3950^2 = 1$
65. $x = at, y = -bt, z = 0$; **67.** True
     $x = at, y = bt + ab^2, z = 2abt + a^2b^2$
69. The Klein bottle does not have both an "inside" and an "outside." It is formed by inserting the small open end through the side of the bottle and making it contiguous with the top of the bottle.

Section 11.7 (page 825)

1. $(5, 0, 2)$ **3.** $(1, \sqrt{3}, 2)$ **5.** $(-2\sqrt{3}, -2, 3)$
7. $(5, \pi/2, 1)$ **9.** $(2, \pi/3, 4)$ **11.** $(2\sqrt{2}, -\pi/4, -4)$
13. $z = 5$ **15.** $r^2 + z^2 = 10$ **17.** $r = \sec\theta \tan\theta$
19. $r^2 \sin^2\theta = 10 - z^2$
21. $x^2 + y^2 = 4$ **23.** $x - \sqrt{3}y = 0$

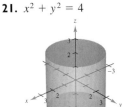 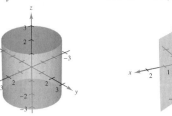

25. $x^2 + y^2 - 2y = 0$ **27.** $x^2 + y^2 + z^2 = 4$

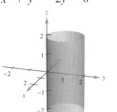 

29. $(4, 0, \pi/2)$ **31.** $(4\sqrt{2}, 2\pi/3, \pi/4)$ **33.** $(4, \pi/6, \pi/6)$
35. $(\sqrt{6}, \sqrt{2}, 2\sqrt{2})$ **37.** $(0, 0, 12)$ **39.** $(\frac{5}{2}, \frac{5}{2}, -5\sqrt{2}/2)$
41. $\rho = 3 \csc \phi \csc \theta$ **43.** $\rho = 6$
45. $\rho = 3 \csc \phi$ **47.** $\tan^2 \phi = 2$
49. $x^2 + y^2 + z^2 = 4$ **51.** $3x^2 + 3y^2 - z^2 = 0$

 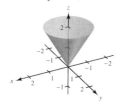

53. $x^2 + y^2 + (z - 2)^2 = 4$ **55.** $x^2 + y^2 = 1$

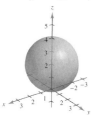 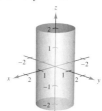

57. $(4, \pi/4, \pi/2)$ **59.** $(4\sqrt{2}, \pi/2, \pi/4)$
61. $(2\sqrt{13}, -\pi/6, \arccos[3/\sqrt{13}])$ **63.** $(13, \pi, \arccos[5/13])$
65. $(10, \pi/6, 0)$ **67.** $(36, \pi, 0)$
69. $(3\sqrt{3}, -\pi/6, 3)$ **71.** $(4, 7\pi/6, 4\sqrt{3})$

	Rectangular	Cylindrical	Spherical
73.	$(4, 6, 3)$	$(7.211, 0.983, 3)$	$(7.810, 0.983, 1.177)$
75.	$(4.698, 1.710, 8)$	$(5, \pi/9, 8)$	$(9.434, 0.349, 0.559)$
77.	$(-7.071, 12.247, 14.142)$	$(14.142, 2.094, 14.142)$	$(20, 2\pi/3, \pi/4)$
79.	$(3, -2, 2)$	$(3.606, -0.588, 2)$	$(4.123, -0.588, 1.064)$
81.	$(\frac{5}{2}, \frac{4}{3}, -\frac{3}{2})$	$(2.833, 0.490, -1.5)$	$(3.206, 0.490, 2.058)$
83.	$(-3.536, 3.536, -5)$	$(5, 3\pi/4, -5)$	$(7.071, 2.356, 2.356)$
85.	$(2.804, -2.095, 6)$	$(-3.5, 2.5, 6)$	$(6.946, 5.642, 0.528)$

87. d   **88.** e   **89.** c   **90.** a   **91.** f   **92.** b
93. Rectangular to cylindrical:
$r^2 = x^2 + y^2, \tan \theta = y/x, z = z$
Cylindrical to rectangular:
$x = r \cos \theta, y = r \sin \theta, z = z$
95. Rectangular to spherical:
$\rho^2 = x^2 + y^2 + z^2, \tan \theta = y/x, \phi = \arccos(z/\sqrt{x^2 + y^2 + z^2})$
Spherical to rectangular:
$x = \rho \sin \phi \cos \theta, y = \rho \sin \phi \sin \theta, z = \rho \cos \phi$
97. (a) $r^2 + z^2 = 16$ (b) $\rho = 4$
99. (a) $r^2 + (z - 1)^2 = 1$ (b) $\rho = 2 \cos \phi$
101. (a) $r = 4 \sin \theta$ (b) $\rho = 4 \sin \theta / \sin \phi = 4 \sin \theta \csc \phi$
103. (a) $r^2 = 9/(\cos^2 \theta - \sin^2 \theta)$
(b) $\rho^2 = 9 \csc^2 \phi / (\cos^2 \theta - \sin^2 \theta)$

105. 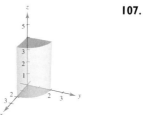   **107.**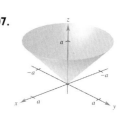

109.    **111.**

113. Rectangular: $0 \leq x \leq 10$ **115.** Spherical: $4 \leq \rho \leq 6$
$0 \leq y \leq 10$
$0 \leq z \leq 10$
117. Cylindrical: $r^2 + z^2 \leq 9, r \leq 3 \cos \theta, 0 \leq \theta \leq \pi$
119. False. $\theta = c$ represents a vertical half-plane.
121. False. See page 821.   **123.** Ellipse

Review Exercises for Chapter 11   (page 827)

1. (a) $\mathbf{u} = 3\mathbf{i} - \mathbf{j}$ (b) $2\sqrt{5}$ (c) $10\mathbf{i}$
   $\mathbf{v} = 4\mathbf{i} + 2\mathbf{j}$
3. $\mathbf{v} = \langle -4, 4\sqrt{3} \rangle$ **5.** $(-5, 4, 0)$
7. Above the xy-plane and to the right of the xz-plane *or* below the xy-plane and to the left of the xz-plane
9. $(x - 3)^2 + (y + 2)^2 + (z - 6)^2 = \frac{225}{4}$
11. $(x - 2)^2 + (y - 3)^2 + z^2 = 9$
Center: $(2, 3, 0)$
Radius: 3

13. 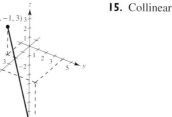   **15.** Collinear

$\mathbf{u} = \langle 2, 5, -10 \rangle$
17. $(1/\sqrt{38})\langle 2, 3, 5 \rangle$
19. (a) $\mathbf{u} = \langle -1, 4, 0 \rangle, \mathbf{v} = \langle -3, 0, 6 \rangle$ (b) 3 (c) 45
21. Orthogonal   **23.** $\theta = \arccos\left(\dfrac{\sqrt{2} + \sqrt{6}}{4}\right) = 15°$ **25.** π
27. Answers will vary. Example: $\langle -6, 5, 0 \rangle, \langle 6, -5, 0 \rangle$
29. $\mathbf{u} \cdot \mathbf{u} = 14 = \|\mathbf{u}\|^2$ **31.** $\langle -\frac{15}{14}, \frac{5}{7}, -\frac{5}{14} \rangle$
33. $(1/\sqrt{5})(-2\mathbf{i} - \mathbf{j})$ or $(1/\sqrt{5})(2\mathbf{i} + \mathbf{j})$
35. 4 **37.** $\sqrt{285}$ **39.** $100 \sec 20° \approx 106.4$ lb
41. (a) $x = 3 + 6t, y = 11t, z = 2 + 4t$
(b) $(x - 3)/6 = y/11 = (z - 2)/4$

Index of Applications

Engineering and Physical Sciences

Acceleration, 129, 157, 159, 177, 258, 922
Acceleration due to gravity, 125
Acid rain, 895
Adiabatic expansion, 156
Air pressure, 441
Air temperature, 770
Air traffic control, 155, 156, 866
Aircraft glide path, 196
Airplane separation, 258
Airplane speed, 152
Angle of elevation, 152, 156, 157
Angular rate of change, 379
Annual snowfall, 999
Antenna radiation, 746
Apparent temperature, 914
Archimedes Principle, 516
Architecture, 706
Area, 261
    of a lot, 270, 315
    of a pasture, 39
    of a polygon inscribed in a circle, 93
Asteroid Apollo, 752
Atmospheric pressure vs. altitude, 331, 358, 967
Automobile aerodynamics, 30
Average displacement, 533
Average speed, 40, 89
Average velocity, 113
Barn design, 1023
Beam deflection, 197, 705
Beam strength, 35, 225
Billiard balls and normal lines, 943
Boiling temperature, 36
Bond angle, 789
Bouncing ball problem, 611, 614, 689
Boyle's Law, 89, 127, 495
Braking load, 789
Breaking strength of a steel cable, 369
Bridge design, 706
Brine mixture, 208
Brinell hardness, 35
Buffon's Needle Experiment, 294
Building design, 455, 566, 1023, 1078
Building a pipeline, 967
Bulb design, 485
Buoyant force, 511
Cable tension, 772, 780
Cantor's disappearing table, 616
Capillary action, 1023
Car performance, 35, 36
Carbon dating, 419

Catenary, 396
Cavalieri's Theorem, 465
Center of mass
    of glass, 505
    of a section of a hull, 506
Center of pressure on a sail, 1016
Centripetal acceleration, 866
Centripetal force, 866, 880
Centroid
    of an industrial fan blade, 514
    of a parabolic spandrel, 505
    of a parallelogram, 505
    of a semicircle, 505
    of a semiellipse, 505
    of a trapezoid, 505
    of a triangle, 505
Charles's Law and absolute zero, 74
Chemical mixture problem, 436, 439, 440
Chemical reaction, 223, 397, 430, 560, 978
Circular motion
    of an automobile, 856
    of a stone, 856
Circumference of an ellipse, 315
Climb rate, 400
Comet Hale-Bopp, 755
Comparing two fluid forces, 548
Compressing a spring, 489
Construction
    of a building, 155
    of a semielliptical arch, 706
    of a wall, 780
Conveyor design, 16
Copper wire, 9
Coulomb's Law, 1055
Cycloidal motion of an automobile, 855
Deceleration, 258
Déjà vu, 82
Demolition crane, 494
Depth
    of a conical tank, 155
    of gasoline in a tank, 514
    of a swimming pool, 155
    of a trough, 155, 160
    of water in a vase, 29
Distance, 245, 965
Distance between two insects, 809
Distance between two ships, 244
Doppler effect, 139
Drag force, 978
Driving distance, 117
Earthquake intensity, 420
Electric circuits, 412, 438, 440
Electric force, 495
Electric force fields, 1055

Electric motor, 90
Electric potential, 894
Electrical charge, 1119
Electrical resistance, 188
Electricity, 156, 307
Electromagnetic theory, 587
Emptying a tank of oil, 491
Engine design, 1078
Engine efficiency, 207
Engine power, 234
Error
    in area of the end of a log, 240
    in area of a square, 240
    in area of a triangle, 240, 241
    in circumference of a circle, 240
    in surface area and volume of a sphere, 244
    in volume of a ball bearing, 237
    in volume and surface area of a cube, 240
    in volume and surface area of a sphere, 241
Escape velocity, 94, 257
Evaporation, 156
Explorer 18, 707, 755
Falling object, 34, 319, 437, 440
Ferris wheel, 882
Field strength, 551
Fire truck, 756
Firing a coast artillery gun, 772
Flight control, 157
Flow rate of a chemical, 359
Fluid flow, 160
Fluid force, 551
    on a circular plate, 512, 514
    of gasoline, 511, 512
    on a rectangular plate, 512
    on a stern of a boat, 512
    on a submerged sheet, 508, 511
    on a vertical surface, 509, 510, 514
    on a vertical wall, 514, 516
    of water, 511
Force, 292
    on a boat, 786
    on a concrete form, 511
    on legs of a tripod, 777
Free-falling object, 69, 82, 91
Frictional force, 874, 878
Fuel economy, 442
Gauss's Law, 1117
Gears, 130
Geology, seismic amplitudes, 895
Gravitational fields, 1055
Gravitational force, 128, 587

A67

Halley's comet, 707, 751
Hanging power cables, 391
Harmonic motion, 36, 38, 139, 242, 358, 400
Heat flow, 1137
Heat-seeking particle, 937
Heat-seeking path, 942, 950
Heat transfer, 340
Heaviside function, 39
Height
    of a basketball, 32
    of an oscillating object, 242
    of a tower, 976
Highway design, 171, 196, 880, 882
Honeycomb, 171
Hooke's Law, 34, 493
Horizontal motion, 159
Hours of daylight, 33
Hydraulic press, 495
Hydraulics, 1016
Hyperbolic detection system, 703
Hyperbolic mirror, 708
Ideal Gas Law, 895, 914, 930
Illumination, 226, 245
Inductance, 922
Inflating balloon, 151
Involute of a circle, 728
Irrigation canal gate, 512
Jerk function, 162
Kepler's Laws, 751, 752, 878
Kinetic and potential energy, 1088
Law of Conservation of Energy, 1085
Lawn sprinkler, 171
Length
    of a cable, 479
    of a catenary, 484, 514
    of Gateway Arch, 484
    of a hypotenuse, 30
    of pursuit, 484
    of a recording tape, 723
Linear and angular velocity, 160
Linear vs. angular speed, 157
Load supports, 780
Load-supporting cables, 788, 789
Lunar gravity, 257
Machine design, 156, 819, 1007
Machine part, 473
Magnetic field of Earth, 1138
Map of the ocean floor, 942
Mass
    of a spring, 1069
    on the surface of Earth, 496
Maximizing an angle, 375
Maximum angle, 930
Maximum area, 222, 223, 224, 225, 228, 244
    of a cross section of a trough, 965
    of an exercise room, 225

    of a Norman window, 224
    of two corrals, 224
Maximum cross-sectional area, of an irrigation canal, 227
Maximum volume, 224, 225, 227
    of a box, 218, 219, 223, 224, 960, 964, 975
    of a package, 225, 964, 965, 975
Mechanical design, 455, 550, 808
Medicine, 234
Meteorology, atmospheric pressure, 895, 942
Minimizing heat loss, 979
Minimum area, 225
    of a page, 224
    of a pasture, 223
Minimum distances among three factories, 227, 228
Minimum force, 227
Minimum length, 224, 244
    of a beam, 244
    of a pipe, 244
    of a power line, 226
    between two posts, 221
Minimum and maximum area of a triangle, 245
Minimum surface area, 225
    of a soft drink cylinder, 225
    of a tank, 979
Minimum time, 226
    Snell's Law of Refraction, 226
    between two points, 234
Moon, 125
Motion along a line, 189
Motion of a liquid, 1132, 1133
Motion of a particle, 726
Moving ladder, 155
Moving particle, 162
Moving shadow, 157, 160, 162
Moving a space module into orbit, 490
Muzzle velocity, 770
Navigation, 708, 760, 772
Newton's Law of Cooling, 118, 417, 420
Newton's Law and Einstein's Special Theory of Relativity, 207
Newton's Law of Gravitation, 1055
Noise level, 420
Number of cubic yards of earth, 992
Oblateness of Saturn, 475
Ohm's Law, 241
Optical illusions, 148
Orbit of Earth, 707
Orbit of the moon, 698, 699
Orbital speed, 866
Orbits of comets, 703
Parabolic reflector, 696
Particle motion, 129
Path of a ball, 854

Path of a baseball, 853, 854, 855
Path of a bicyclist, 47
Path of a bomb, 855, 881
Path of a dog, 760
Path of a football, 855
Path of a projectile, 185, 726, 854, 855, 980
Path of a shot, 855
Path of a stream, 485
Path of a swimmer, 82
Pendulum, 139, 241, 922
Planetary motion, 755
Planetary orbits, 699
Planimeter, 1136
Power, 171, 188, 922
Power lines, 542
Producing a machine part, 465
Product design, 1023
Projectile motion, 158, 159, 241, 551, 687, 718, 772, 852, 855, 863, 865, 866, 875, 880, 929
Projectile range, 226
Propulsion, 493, 587
Pumping diesel fuel, 494
Pumping gasoline, 494
Pumping water, 493, 494
Radio reception, 444
Radio and television reception, 706
Radioactive decay, 415, 419, 430, 441
Radioactive half-life, 360
Rainbows, 189
Rainfall at Seattle-Tacoma airport, 306
Ramp design, 12
Rate of change
    of a ladder moving down a house, 89
    of a light beam on a patrol car moving along a wall, 89
Rate a vehicle is traveling, 16
Rectilinear motion, 257, 294
Refraction of light, 975
Refrigeration, 160
Relativity, 89
Resistance, 922
Resultant force, 771
    on a hook, 771
    on a machine part, 771
    on an ocean liner, 768
Resultant speed and direction of an airplane, 769
Ripples, 29
Ripples in a pond, 150
Roadway design, 156
Rolling a ball bearing, 188
Roof area, 484
Rope tension, 830
Rotary engine, 757
Satellite antenna, 756
Satellite orbit, 707, 880, 882

Satellites, 128
Sending a space module into orbit, 581
Shadow length, 156
Shared load, 772
Shot-put throw, 856
Solar collector, 705
Sound intensity, 40, 331, 420
Sound location, 708
Specific gravity, 197
Speed, 177, 877, 967
Speed of sound, 287
Speeding truck, 175
St. Louis Arch, 398
Stacking blocks, 691
Stacking spheres, 692
Statics problems, 504
Stopping distance, 118, 129, 159
Strain distribution of a car door, 889
Stress test, 38
Submarine porthole, 512
Surface area
    of a dome, 1107
    of a golf green, 453
    of an oil spill, 453
    of a piece of tin, 1074
    of a pond, 516
    of a roof, 1050
    of a satellite-signal receiving dish, 706
    by the Second Theorem of Pappus, 506
Surveying, 241
Suspension bridge, 486
Tautochrone and brachistochrone problems, 715
Temperature, 177, 207, 348, 411, 443
    conversion, 18
    for Denver, Colorado, 139
    distribution, 894, 914, 936, 941, 942, 975, 979
    for Erie, Pennsylvania, 542
    for Honolulu and Chicago, 36
    of a house, 307
    at which water boils, 331
Tension in towlines, 830
Thermostat, 29
Throwing a dart, 270
Tidal energy, 495
Topographical map, 147
Topography, 887, 941, 942
Torque, 794, 796, 797, 827
Torricelli's Law, 443, 444
Tossing bales, 855
Tower guy wire, 780
Tractrix, 331, 394, 395, 398, 576, 728
Triangle measurements, 241
Velocity, 118, 177, 258, 292, 320
    of a diver, 114
    of a piston, 153
    of a rocket, 592

Velocity and acceleration, 316
    on the moon, 162
Velocity fields, 1055
Velocity in a resisting medium, 575
Vertical motion, 117, 158, 176, 177, 254, 257, 387, 398, 442
Vibrating spring, 158
Vibrating string, 533
Volume
    of a balloon, 154
    of a box, 30, 919
    of a conical sand pile, 154
    of a conical tank, 149
    of a fuel tank, 464
    of a goblet, 877
    of the Great Salt Lake, 1052
    of ice, 1008
    of a lab glass, 465
    of a pond, 474
    of a pontoon, 471
    of a propane tank, 894
    of a pyramid, 462
    of a shampoo bottle, 225
    of a shell, 241
    of a storage shed, 474
    of a storage tank, 550
    by the Theorem of Pappus, 503, 506
    of a trough, 922
    of a vase, 485
    of a water tank on a fire truck, 707
Water depth in a tank, 465
Water distribution, 1051
Water running into a vase, 196
Water supply, 307
Wave motion, 139
Weather map, 148
Weight of a car, 770
Wind chill, 922
Wind speed and direction, 1088
Work, 315, 514
    done by aircraft engines, 1135
    done in closing a door, 787
    done by a constant force, 493
    done by an expanding gas, 492
    done by a force, 1087
    done by a gravitational force field, 1087
    done in lifting a chain, 492, 494, 514
    done in lifting an object, 487
    done in moving an object, 493
    done in pulling an object, 789, 827
    done in pulling a wagon, 789
    done in pumping water, 514
    done in splitting a piece of wood, 495
    done in stretching a spring, 514
    done in walking up a staircase, 1077
    done in winding up cable, 514
Wrinkled and bumpy spheres, 1041

Business and Economics

Advertising costs, 234
Annuities, 615
Apartment rental, 18
Average cost, 197, 207
Average price, 340
Average production, 999
Average profit, 999
Average sales, 292
Break-even analysis, 37
Break-even point, 9
Capitalized cost, 587
Cash flow, 306
Cobb-Douglas production function, 889, 894, 971, 979
Compound interest, 365, 368, 400, 419, 576, 603, 688, 689
Construction cost, 894
Consumer price index, 9
Consumer and producer surpluses, 516
Cost, 140, 294, 348
    of removing a chemical from waste water, 560
Customers entering a store, 293
Declining sales, 416
Delivery charges, 92
Demand, 966
Demand function, 244
Depreciation, 306, 357, 368, 400, 614, 688
Diminishing returns, 227
Eliminating budget deficits, 454
End-of-year assets for the Medicare Hospital Insurance Trust Fund, 188
Government expenditures, 604
Home mortgage, 331, 402
Inflation, 367, 604
Inventory cost, 171, 197, 243
Inventory management, 82, 118
Inventory replenishment, 127
Investment, 894, 914
Investment growth, 439
Lorenz curve, 454
Manufacturing, 308, 461, 466
Marginal costs, 914
Marginal productivity, 914
Marketing, 614
Maximum profit, 227, 961, 965, 978
Maximum revenue, 965
Minimum cost, 965, 975, 978
    of a delivery trip, 244
    of an industrial tank, 225
    of laying pipe, 227
    of manufacturing a product, 227
Multiplier effect, 614
Present value, 533, 590, 614
Production level, 975, 978
Profit, 38, 188, 241, 455

Rate of change
　of price of a new machine, 37
　of revenue, 16
Receipts and expenditures for the Old-Age and Survivors Insurance Trust Fund, 454
Reimbursed expenses, 18
Reorder costs, 176
Revenue, 454, 788
Salary, 615, 689
Sales, 177, 307, 340, 441, 443
　Avon Products, Inc., 604, 615
　Wal-Mart, 895
Sales growth, 197, 243
Service revenues for cellular telephone industry, 513
Shares traded on the New York Stock Exchange, 353
Stock price, 770
Straight-line depreciation, 18
Telephone charges, 56, 81
Value of a mid-sized sedan, 358

Social and Behavioral Sciences

Air conditioner use, 895
Amount given to philanthropy, 369
Amount of time women spend watching television, 706
Automobile costs, 35
Energy consumption and gross national product, 34
Fuel cost, 118, 318
Health maintenance organizations, 35
Illegal drugs, 89
Learning curve, 419, 439
Learning theory, 368
Marginal utility, 914
Memory model, 533
Net receipts and amounts required to service the national debt, 420
Number of bankruptcies, 189
Number of motor homes in the United States, 128
Number of single and married women in the civilian work force, 157

Outlays for national defense, 243
Per capita consumption of milk, 808, 914, 921
Population, 1008
　of countries, 419
　of Kentucky, 12
　of United States, 16, 420
Population growth, 439, 441
Public medical expenditures, 915
Speed limit, 246
Traffic control, 223
Women in the work force, 966
World population, 967
World record times for running one mile, 207

Life Sciences

Age and systolic blood pressure, 966
Bacteria in a culture, 217
Bacterial culture growth, 365, 419, 431
Biomass, 444
Blood flow, 292
Carbon dioxide concentration, 7
Carcinogens, 34
Circulatory system, 139
Concentration of a tracer drug in a fluid, 444
Connecticut River, 228
DNA molecule, 833
Endangered species, 431
Environment, 92
Epidemic model, 560
Farm size, 9, 30
Forest defoliation, 368
Forestry, 420, 894
Hardy-Weinberg Law, 965, 975
Height vs. arm span, 31
Intravenous feeding, 439
Normal probability, of American men's height, 588
Number of endangered and threatened species in the United States, 604
Oxygen level in a pond, 203
Population, 566

Population growth, 692
　of bacteria, 127, 256, 340
　of brook trout, 442
　of coyotes, 425
　of elk, 428
　of fish, 369
　of fruit flies, 416
Probability of a warbler's length, 590
Respiratory cycle, 292, 318
Timber yield, 368
Trachea contraction, 188
Tree growth, 256
Weight gain of a calf, 430
Weight loss, 440
Wheat yield, 966
Wildflowers, 951

General

Anamorphic art, 738
Applicants to a university, 914
Average quiz and test scores, 18
Average typing speed, 197, 207
Boating, 39, 155
Building blocks, 270
Career choice, 18
Dental inlays, 830
Folding paper, 246
Geography, 819, 830
Jewelry, 57
Job offers, 454, 513
Monte Carlo Method, 270
Near point, 967
Playground slide, 838
Probability, 307, 359, 587, 614, 675, 1000, 1008, 1049
Queuing model, 894
Quiz scores, 34
Sailing, 431
School commute, 27, 28
Security camera, 157
Solera method, 630
Sphereflake, 615
Spiral staircase, 879
Sports, 57, 156, 922
Swimming pool, 81

Index

A

Abel, Niels Henrik (1802–1829), 232
Absolute convergence, 634
Absolute maximum of a function, 164
   of two variables, 952
Absolute minimum of a function, 164
   of two variables, 952
Absolute value
   derivative involving, 328
   function, 22
Absolute Value Theorem for sequences, 598
Absolute zero, 74
Absolutely convergent, 634
Acceleration, 125, 849, 873
   centripetal component of, 861
   tangential and normal components of, 861, 875
   vector, 860, 875
Accumulation function, 288
Addition
   of ordinates, 389
   of vectors
      in the plane, 762
      in space, 775
Additive Identity Property of Vectors, 765
Additive Interval Property, 276
Additive Inverse Property of Vectors, 765
Agnesi, Maria Gaetana (1718–1799), 201
d'Alembert, Jean Le Rond (1717–1783), 906
Algebraic function(s), 24, 25, 376
   derivatives of, 136
Algebraic properties of the cross product, 791
Alternating series, 631
   geometric series, 631
   harmonic series, 632, 634
   remainder, 633
Alternating Series Test, 631
Alternative form
   of the derivative, A6
   of the directional derivative, 934
   of Green's Theorem, 1094, 1095
Angle
   between two nonzero vectors, 782
   between two planes, 800
   of incidence, 696
   of inclination of a plane, 947
   of reflection, 696
Angular speed, 1014
Antiderivative, 248
   of f with respect to x, 249
   general, 249
   representation of, 248
   of a vector-valued function, 844

Antidifferentiation, 249
   of a composite function, 295
Aphelion, 707, 755
Apogee, 707
Approximating zeros
   bisection method, 78
   Intermediate Value Theorem, 77
Approximation
   linear, 235, 918
   tangent line, 235
Arc length, 476, 477
   function, 868
   parameter, 868, 869
   in parametric form, 722
   of a polar curve, 743
   of a space curve, 867
   in the xy-plane, 1018
Arccosecant function, 371
Arccosine function, 371
Arccotangent function, 371
Archimedes (287–212 B.C.), 261
Archimedes Principle, 516
Arcsecant function, 371
Arcsine function, 371
   series for, 682
Arctangent function, 371
   series for, 682
Area
   line integral for, 1092
   of a parametric surface, 1102
   in polar coordinates, 739
   problem, 45
   of a rectangle, 261
   of a region between two curves, 447
   of a region in the plane, 265, 984
   of a surface of revolution, 481
      in parametric form, 724
      in polar coordinates, 744
   of the surface S, 1018
   in the xy-plane, 1018
Associative Property of Vector Addition, 765
Astroid, 146
Asymptote(s)
   horizontal, 199
   of a hyperbola, 701
   slant, 211
   vertical, 84, 85, A6
Average rate of a change, 12
Average value
   of a continuous function over a solid region Q, 1034
   of a function on an interval, 286
   of a function over a region R, 999
Average velocity, 113
Axis
   conjugate, of a hyperbola, 701

   major, of an ellipse, 697
   minor, of an ellipse, 697
   of a parabola, 695
   of revolution, 456
   transverse, of a hyperbola, 701

B

Barrow, Isaac (1630–1677), 145
Base(s), 325
   of a natural logarithm, 325
   other than e, derivatives for, 362
Basic differentiation rules for elementary functions, 376
Basic equation, 554
   guidelines for solving, 558
Basic integration rules, 250, 383, 520
   procedures for fitting integrands to, 521
Basic limits, 59
Basic types of transformations, 23
Bearing, 769
Bernoulli equation, 434
   general solution of, 434
Bernoulli, James (1654–1705), 715
Bernoulli, John (1667–1748), 552
Bifolium, 146
Binomial series, 681
Bisection method, 78
Boundary point of a region R, 896
Bounded
   above, 601
   below, 601
   monotonic sequence, 601
   region R, 952
   sequence, 601
Brachistochrone problem, 715
Breteuil, Emilie de (1706–1749), 488
Bullet-nose curve, 138

C

Cantor, Georg (1845–1918), 691
Capillary action, 1023
Cardioid, 734, 735
Carrying capacity, 427
Catenary, 391
Cauchy, Augustin-Louis (1789–1857), 75
Cauchy-Riemann differential equations, 930
Cauchy-Schwarz Inequality, 789
Center
   of curvature, 872
   of an ellipse, 697
   of gravity, 499
      of a one-dimensional system, 498
      of a two-dimensional system, 498

of a hyperbola, 701
of mass, 497, 498
    of a one-dimensional system, 498
    of a planar lamina, 500
    of a planar lamina of variable density, 1011, 1029
    of a two-dimensional system, 499
of a power series, 659
Centered at c, 648
Central force field, 1055
Centripetal component of acceleration, 861
Centripetal force, 866
Centroid, 501
    of a simple region, 1011
Chain Rule, 130, 131, 136, A7
    implicit differentiation, 928
    one independent variable, 923, A19
    two independent variables, 925
Change in x, 97
Change in y, 97
Change of variables, 298
    for definite integrals, 301
    for double integrals, 1044
    guidelines for making, 299
    for homogeneous equations, 424
    to polar form, 1003
    using a Jacobian, 1042
Charles's Law, 74
Circle, 146, 694, 735
Circle of curvature, 161, 872
Circulation of $\mathbf{F}$ around C, 1131
Circumscribed rectangle, 263
Cissoid, 146
    of Diocles, 759
Classification of conics by eccentricity, 748, A18
Closed
    curve, 1084
    disk, 896
    region R, 896
    surface, 1120
Cobb-Douglas production function, 889
Coefficient
    correlation, 31
    leading, 24
Collinear, 17
Common logarithmic function, 361
Common types of behavior associated with nonexistence of a limit, 51
Commutative Property
    of the dot product, 781
    of vector addition, 765
Comparison Test
    Direct, 624
    Limit, 626
Complete, 601
Completeness, 77
Completing the square, 381
Component of acceleration

centripetal, 861
normal, 861, 875
tangential, 861, 875
Component form of a vector in the plane, 763
Component functions, 832
Components of a vector, 785
    along $\mathbf{v}$, 785
    in the direction of $\mathbf{v}$, 786
    orthogonal to $\mathbf{v}$, 785
    in the plane, 763
Composite function, 25
    antidifferentiation of, 295
    continuity of, 75
    limit of, 61, A4
    of two variables, 885
        continuity of, 901
Composition of functions, 25
Compound interest formulas, 364
Concave downward, 190
Concave upward, 190
Concavity, 190
    interpretation, A8
    test for, 191
Conditional convergence, 634
Conditionally convergent, 634
Conic(s), 694
    circle, 694
    classification by eccentricity, 748, A18
    degenerate, 694
    directrix of, 748
    eccentricity of, 748
    ellipse, 694, 697
    focus of, 748
    hyperbola, 694, 701
    parabola, 694, 695
    polar equations of, 749
Conic section, 694
Conjugate axis of a hyperbola, 701
Connected region, 1082
Conservative vector field, 1057, 1079
    independence of path, 1082
    test for, 1058, 1061
Constant
    force, 487
    function, 24
    of integration, 249
    Multiple Rule, 110, 136
        differential form, 238
    of proportionality, 414
    Rule, 107, 136
    term of a polynomial function, 24
Constraint, 968
Continuity
    of a composite function, 75
    of a composite function of two variables, 901
    differentiability implies, 103
    and differentiability of inverse

functions, 345, A12
implies integrability, 273
properties of, 75
of a vector-valued function, 836
Continuous, 70
    at c, 59, 70
    on the closed interval $[a, b]$, 73
    compounding, 364
    everywhere, 70
    function of two variables, 900
    on an interval, 836
    from the left and from the right, 73
    on an open interval (a, b), 70
    in the open region R, 900, 902
    at a point, 836
    (x_0, y_0), 900
    (x_0, y_0, z_0), 902
    vector field, 1054
Continuously differentiable, 476
Contour lines, 887
Converge, 231, 595, 606
Convergence
    absolute, 634
    conditional, 634
    of a geometric series, 608
    of improper integral with infinite discontinuities, 581
    of improper integral with infinite integration limits, 578
    interval of, 660, 664
    of p-series, 619
    of a power series, 660, A17
    radius of, 660, 664
    of a sequence, 595
    of a series, 606
    of Taylor series, 678
    tests for series
        Alternating Series Test, 631
        Direct Comparison Test, 624
        geometric series, 608
        guidelines, 643
        Integral Test, 617
        Limit Comparison Test, 626
        p-series, 619
        Ratio Test, 639
        Root Test, 642
        summary of, 644
Convergent series, nth term of, 610
Convex limaçon, 735
Coordinate conversion, 730
    cylindrical to rectangular, 820
    cylindrical to spherical, 823
    rectangular to cylindrical, 820
    rectangular to spherical, 823
    spherical to cylindrical, 823
    spherical to rectangular, 823
Coordinate planes, 773
    xy-plane, 773
    xz-plane, 773

yz-plane, 773
Coordinate system
　cylindrical, 820
　polar, 729
　spherical, 823
　three-dimensional, 773
Coordinates, polar, 729
Copernicus, Nicolaus (1473–1543), 697
Cornu spiral, 759, 881
Correlation coefficient, 31
Cosecant function
　derivative of, 123, 136
　integral of, 337
　inverse of, 371
Cosine function, 22
　derivative of, 112, 136
　integral of, 337
　inverse of, 371
　series for, 682
Cotangent function
　derivative of, 123, 136
　integral of, 337
　inverse of, 371
Coulomb's Law, 489, 1055
Critical number(s)
　of a function, 166
　relative extrema occur only at, 166
Critical point
　of a function of two variables, 953
　relative extrema occur at, 953
Cross product of two vectors in space, 790
　algebraic properties of, 791
　determinant form, 790
　geometric properties of, 792
　torque, 794
Cruciform, 146
Cubic function, 24
Cubing function, 22
Curl of a vector field, 1060
　and divergence, 1062
Curtate cycloid, 717
Curvature, 870
　center of, 872
　circle of, 872
　formulas for, 871, 875
　radius of, 872
　in rectangular coordinates, 872, 875
　related to acceleration and speed, 873
Curve
　closed, 1084
　lateral surface area over, 1077
　level, 887
　natural equation for, 881
　orientation of, 1065
　piecewise smooth, 1065
　plane, 709, 832
　pursuit, 393, 395
　rectifiable, 476
　rose, 732, 735

　simple, 1089
　smooth, 476, 714, 842, 857
　　piecewise, 714
　space, 832
　tangent line to, 857
Cusps, 842
Cycloid, 714, 718
　curtate, 717
　prolate, 721
Cylinder, 810
　directrix of, 810
　equations of, 810
　generating curve of, 810
　right, 810
　rulings of, 810
Cylindrical coordinate system, 820
　pole of, 820
Cylindrical coordinates
　converting to rectangular coordinates, 820
　converting to spherical coordinates, 820
Cylindrical surface, 810

D

Decomposition of $N(x)/D(x)$ into partial fractions, 553
Decreasing function, 179
　test for, 179
Definite integral(s), 273
　as the area of a region, 274
　change of variables, 301
　evaluation of a line integral as a, 1067
　properties of, 277
　two special, 276
　of a vector-valued function, 844
Degenerate conic, 694
　line, 694
　point, 694
　two intersecting lines, 694
Degree of a polynomial function, 24
Delta, δ, 896
　δ-neighborhood, 896
Demand, 18
Density, 500
Density function ρ, 1009, 1029
Dependent variable, 19
　of a function of two variables, 884
Derivative(s)
　of algebraic functions, 136
　alternative form, A6
　for bases other than e, 362
　Chain Rule, 130, 131, 136
　　implicit differentiation, 928
　　one independent variable, 923, A19
　　two independent variables, 925
　Constant Multiple Rule, 110, 136
　Constant Rule, 107, 136
　of cosecant function, 123, 136

　of cosine function, 112, 136
　of cotangent function, 123, 136
　Difference Rule, 111, 136
　directional, 931, 932, 939
　of an exponential function, base a, 362
　of a function, 99
　General Power Rule, 132, 136
　higher-order, 125
　of hyperbolic functions, 390
　implicit, 142
　of an inverse function, 345, A13
　of inverse trigonometric functions, 374
　involving absolute value, 328
　from the left and from the right, 101
　of a logarithmic function, base a, 362
　of the natural exponential function, 352
　of the natural logarithmic function, 326
　notation, 99
　parametric form, 719
　partial, 906
　　first, 906
　Power Rule, 108, 136
　of power series, 664
　Product Rule, 119, 136
　Quotient Rule, 121, 136
　of secant function, 123, 136
　second, 125
　Simple Power Rule, 108, 136
　of sine function, 112, 136
　Sum Rule, 111, 136
　of tangent function, 123, 136
　third, 125
　of trigonometric functions, 123, 136
　of a vector-valued function, 840
　　properties of, 842
Descartes, René (1596–1650), 2
Determinant form of cross product, 790
Difference quotient, 20, 97
Difference Rule, 111, 136
　differential form, 238
Difference of two vectors, 764
Differentiability implies continuity, 103, 919
　of a function of two variables, 919
Differentiable at x, 99
Differentiable, continuously, 476
Differentiable function
　on the closed interval $[a, b]$, 101
　on an open interval (a, b), 99
　in a region R, 917
　of three variables, 918
　of two variables, 917
　vector-valued, 840
Differential, 236
　function of three variables, 918
　function of two variables, 916
　of x, 236
　of y, 236
Differential equation, 249
　Bernoulli equation, 434

Cauchy-Riemann, 930
direction field, 406
doomsday, 443
Euler's Method, 408
first-order linear, 432
general solution of, 249
homogeneous, 423
   change of variables, 424
initial condition, 253, 405
logistic, 245, 427
order of, 404
particular solution of, 253
slope field, 406
solution curves, 405
solutions of, 404
   Bernoulli, 434
   first-order linear, 433
   general, 404
   particular, 405
   singular, 404
summary of first-order, 435
Differential form, 238
   of a line integral, 1073
Differential formulas, 238
   constant multiple, 238
   product, 238
   quotient, 238
   sum or difference, 238
Differential operator, 1060, 1062
   Laplacian, 1064
Differentiation, 99
   implicit, 141
      Chain Rule, 928
      guidelines for, 142
   involving inverse hyperbolic functions, 394
   logarithmic, 327
   numerical, 103
   partial, 906
   of a vector-valued function, 841
Differentiation rules
   Chain, 130, 131, 136
   Constant, 107, 136
   Constant Multiple, 110, 136
   cosecant function, 123, 136
   cosine function, 112, 136
   cotangent function, 123, 136
   Difference, 111, 136
   for elementary functions, 376
   general, 136
   General Power, 132, 136
   Power, 108, 136
      for Real Exponents, 363
   Product, 119, 136
   Quotient, 121, 136
   secant function, 123, 136
   Simple Power, 108, 136
   sine function, 112, 136
   Sum, 111, 136
   summary of, 136
   tangent function, 123, 136
Dimpled limaçon, 735
Direct Comparison Test, 624
Direct substitution, 59, 60
Directed distance, 695
Directed line segment, 762
   equivalent, 762
   initial point of, 762
   length of, 762
   magnitude of, 762
   terminal point of, 762
Direction angles of a vector, 784
Direction cosines of a vector, 784
Direction field, 256, 323, 406
Direction of motion, 848
Direction numbers, 798
Direction vector, 798
Directional derivative, 931, 932
   alternative form, 934
   of f in the direction of $\mathbf{u}$, 932, 939
   of a function of three variables, 939
Directrix
   of a conic, 748
   of a cylinder, 810
   of a parabola, 695
Dirichlet, Peter Gustav (1805–1859), 51
Dirichlet function, 51
Discontinuity, 71
   infinite, 578
   nonremovable, 71
   removable, 71
Disk, 456, 896
   closed, 896
   method, 457
   open, 896
Distance
   between a point and a line in space, 804
   between a point and a plane, 803
   directed, 695
Distance Formula in space, 774
Distributive Property
   for the dot product, 781
   for vectors, 765
Diverge, 595, 606
Divergence
   of improper integral with infinite discontinuities, 581
   of improper integral with infinite integration limits, 578
   of a sequence, 595
   of a series, 606
   tests for series
      Direct Comparison Test, 624
      geometric series, 608
      guidelines, 643
      Integral Test, 617
      Limit Comparison Test, 626
      nth-Term Test, 610
      p-series, 619
      Ratio Test, 639
      Root Test, 642
      summary of, 644
   of a vector field, 1062
   and curl, 1062
Divergence-free vector field, 1062
Divergence Theorem, 1095, 1120
Divide out like factors, 63
Domain
   of a function, 19
      of two variables, 884
   of a vector-valued function, 833
Doomsday equation, 443
Dot product
   Commutative Property of, 781
   Distributive Property for, 781
   form of work, 787
   projection using the, 786
   properties of, 781
   of vectors, 781
Double integral, 990, 991, 992
   change of variables for, 1044
   of f over R, 992
   properties of, 992
Doyle Log Rule, 894
Dummy variable, 275
Dyne, 487

E

e, the number, 325
Eccentricity, 699, 748
   classification of conics by, 748, A18
   of an ellipse, 699
   of a hyperbola, 702
Eight curve, 161
Electric force field, 1055
Elementary function(s), 24, 376
   basic differentiation rules for, 376
   power series for, 682
Eliminating the parameter, 711
Ellipse, 694, 697
   center of, 697
   eccentricity of, 699
   foci of, 697
   major axis of, 697
   minor axis of, 697
   reflective property of, 699
   rotated, 146
   standard equation of, 697
   vertices of, 697
Ellipsoid, 811, 812
Elliptic cone, 811, 813
Elliptic paraboloid, 811, 813
Endpoint extrema, 164
Energy
   kinetic, 1085
   potential, 1085

Epicycloid, 718, 722
Epsilon-delta, ε-δ, 52
　definition of limit, 52
Equal vectors
　in space, 775
　in the plane, 763
Equality of mixed partial derivatives, 911
Equation of a plane in space
　general form, 799
　standard form, 799
Equation(s)
　basic, 554
　　guidelines for solving, 558
　Bernoulli, 434
　of cylinders, 810
　doomsday, 443
　Gompertz, 443
　graph of, 2
　harmonic, 1064
　Laplace's, 1064
　of a line
　　general form, 14
　　horizontal, 14
　　point-slope form, 11, 14
　　slope-intercept form, 13, 14
　　in space, parametric, 798
　　in space, symmetric, 798
　　summary, 14
　　vertical, 14
　parametric, 709, 1098
　primary, 218, 219
　related-rate, 149
　secondary, 219
　separable, 421
　solution point of, 2
　of tangent plane, 944
Equilibrium, 497
Equipotential
　curves, 426
　lines, 887
Equivalent
　conditions, 1084
　directed line segments, 762
Error
　in approximating a Taylor polynomial, 654
　in measurement, 237
　　percent error, 237
　　propagated error, 237
　　relative error, 237
　in Simpson's Rule, 313
　in Trapezoidal Rule, 313
Escape velocity, 94
Euler, Leonhard (1707–1783), 24
Euler's Method, 408
Evaluate a function, 19
Evaluating a flux integral, 1114
Evaluating a surface integral, 1108
Evaluation by iterated integrals, 1025

Evaluation of a line integral as a definite integral, 1067
Even function, 26
　integration of, 303
　test for, 26
Everywhere continuous, 70
Evolute, 877
Existence of an inverse function, 343
Existence of a limit, 73
Existence theorem, 77, 164
Expanded about c, 648
Explicit form of a function, 19, 141
Exponential decay, 414
Exponential function, 24
　to base a, 360
　　derivative of, 362
　　integration rules, 354
　natural, 350
　　derivative of, 352
　　properties of, 351
　　operations with, 350
　　series for, 682
Exponential growth, 414
Exponential growth and decay model, 414
　initial value, 414
　proportionality constant, 414
Extended Mean Value Theorem, 245, 568, A14
Extrema
　endpoint, 164
　of a function, 164
　guidelines for finding, 167
　relative, 165
Extreme Value Theorem, 164, 952
Extreme values of a function, 164

F

Factorial, 597
Family of functions, 273
Famous curves
　astroid, 146
　bifolium, 146
　bullet-nose curve, 138
　circle, 146, 694, 735
　cissoid, 146
　cruciform, 146
　eight curve, 161
　folium of Descartes, 146, 747
　kappa curve, 145, 147
　lemniscate, 40, 144, 147, 735
　parabola, 2, 146, 694, 695
　quartic, 161
　rotated ellipse, 146
　rotated hyperbola, 146
　serpentine, 127
　top half of circle, 138
　witch of Agnesi, 127, 146, 839
Faraday, Michael (1791–1867), 1085

Fermat, Pierre de (1601–1665), 166
Field
　central force, 1055
　electric force, 1055
　force, 1054
　gravitational, 1055
　inverse square, 1055
　vector, 1054
　　over Q, 1054
　　over R, 1054
　velocity, 1054, 1055
Finite Fourier series, 542
First Derivative Test, 181
First moments, 1013, 1029
First-order differential equations, summary of, 435
First-order linear differential equation, 432
　integrating factor, 432
　solution of, 433
　standard form, 432
First partial derivatives, 906
　notation for, 907
Fixed plane, 878
Fluid force, 508
Fluid pressure, 507
Flux integral
　evaluating, 1114
　of **F** across S, 1114
Focal chord of a parabola, 695
Focus
　of a conic, 748
　of an ellipse, 697
　of a hyperbola, 701
　of a parabola, 695
Folium of Descartes, 146, 747
Force, 487
　constant, 487
　exerted by a fluid, 508
　of friction, 874
　resultant, 768
　variable, 488
Force field, 1054
　central, 1055
　electric, 1055
　work, 1070
Form of a convergent power series, 676
Formulas
　for curvature, 871, 875
　summation, 260, A9
Fourier, Joseph (1768–1830), 669
Fourier Sine Series, 533
Frenet-Serret formulas, 882
Fresnel function, 319
Friction, 874
Fubini's Theorem, 994
　for a triple integral, 1025
Function, 6, 19
　absolute maximum of, 164
　absolute minimum of, 164

absolute value, 22
accumulation, 288
addition of, 25
algebraic, 376
antiderivative of, 248
arc length, 476, 477, 868
average value of, 286
Cobb-Douglas production, 889
common logarithmic, 361
component, 832
composite, 25
    of two variables, 885
composition of, 25
concave downward, 190
concave upward, 190
constant, 24
continuous, 70
continuously differentiable, 476
cosine, 22
critical number of, 166
cubic, 24
cubing, 22
decreasing, 179
    test for, 179
defined by power series, properties of, 664
density, 1009, 1029
derivative of, 99
difference of, 25
Dirichlet, 51
domain of, 19
elementary, 24, 376
    algebraic, 24, 25
    exponential, 24
    logarithmic, 24
    trigonometric, 24
evaluate, 19
even, 26
explicit form, 19, 141
exponential to base a, 360
extrema of, 164
extreme values of, 164
family of, 273
Fresnel, 319
graph of, guidelines for analyzing, 209
greatest integer, 72
Heaviside, 39
homogeneous, 423
hyperbolic, 388
    cosecant, 388
    cosine, 388
    cotangent, 388
    secant, 388
    sine, 388
    tangent, 388
identity, 22
implicit form, 19
implicitly defined, 141
increasing, 179
    test for, 179
inner product of, 542
integrable, 273
inverse, 341
inverse hyperbolic, 392
    cosecant, 392
    cosine, 392
    cotangent, 392
    secant, 392
    sine, 392
    tangent, 392
inverse trigonometric, 371
    cosecant, 371
    cosine, 371
    cotangent, 371
    secant, 371
    sine, 371
    tangent, 371
limit of, 48
linear, 24
logarithmic to base a, 361
logistic growth, 365
natural exponential, 350
natural logarithmic, 322
notation, 19
odd, 26
one-to-one, 21
onto, 21
orthogonal, 542
point of inflection, 191, 193
polynomial, 24, 60
    of two variables, 885
position, 113, 853, 1057
potential, 1057
product of, 25
pulse, 94
quadratic, 24
quotient of, 25
radius, 816
range of, 19
rational, 22, 25
    of two variables, 885
real-valued, 19
relative extrema of, 166
relative maximum of, 165
relative minimum of, 165
Riemann zeta, 623
signum, 82
sine, 22
sine integral, 320
square root, 22
squaring, 22
standard normal probability density, 353
step, 72
strictly monotonic, 180, 343
of three variables
    continuity of, 902
    directional derivative, 939
    gradient of, 939
transcendental, 25, 376
transformation of a graph of, 23
    horizontal shift, 23
    reflection about origin, 23
    reflection about x-axis, 23
    reflection about y-axis, 23
    reflection in the line $y = x$, 342
    vertical shift, 23
of two variables, 884
    absolute maximum of, 952
    absolute minimum of, 952
    continuity of, 900
    critical point of, 953
    dependent variable, 884
    differentiability implies continuity, 919
    differentiable, 917
    differential of, 916
    domain of, 884
    gradient of, 934
    graph of, 886
    independent variables, 884
    limit of, 897
    maximum of, 952
    minimum of, 952
    nonremovable discontinuity of, 900
    partial derivative of, 906
    range of, 884
    relative extrema of, 952
    relative maximum of, 952, 955
    relative minimum of, 952, 955
    removable discontinuity of, 900
    total differential of, 916
unit pulse, 94
vector-valued, 832
Vertical Line Test, 22
of x and y, 884
zero of, 26
Functions that agree at all but one point, 62, A5
Fundamental Theorem
    of Algebra, 1120
    of Calculus, 282
        guidelines for using, 283
        Second, 289
    of Line Integrals, 1079, 1080

G

Gabriel's Horn, 584, 1100
Galilei, Galileo (1564–1642), 376
Galois, Evariste (1811–1832), 232
Gauss, Carl Friedrich (1777–1855), 1120
Gauss's Law, 1117
Gauss's Theorem, 1120
General antiderivative, 249
General differentiation rules, 136
General form
    of the equation of a line, 14
    of the equation of a plane in space, 799

of a second-degree equation, 694
General harmonic series, 619
General Power Rule
    for differentiation, 132, 136
    for integration, 300
General second-degree equation, 694
General solution
    of a Bernoulli equation, 434
    of a differential equation, 249, 404
Generating curve of a cylinder, 810
Geometric properties of the cross product, 792
Geometric property of triple scalar product, 795
Geometric series, 608
    alternating, 631
    convergence of, 608
Gibbs, Josiah Willard (1839–1903), 1065
Golden ratio, 604
Gompertz equation, 443
Grad, 934
Gradient, 1054, 1057
    of a function of three variables, 939
    of a function of two variables, 934
    normal to level curves, 938
    normal to level surfaces, 948
    properties of, 935
    recovering a function from, 1061
Graph(s)
    of absolute value function, 22
    of common functions, 22
    of cosine function, 22
    of cubing function, 22
    of an equation, 2
    of a function
        guidelines for analyzing, 209
        transformation of, 23
        of two variables, 886
    of identity function, 22
    intercept of, 4
    of parametric equations, 709
    of rational function, 22
    of sine function, 22
    of square root function, 22
    of squaring function, 22
    symmetry of, 5
Gravitational field, 1055
Greatest integer function, 72
Green, George (1793–1841), 1090
Green's Theorem, 1089
    alternative forms, 1094, 1095
Gregory, James (1638–1675), 664
Guidelines
    for analyzing the graph of a function, 209
    for evaluating integrals involving secant and tangent, 537
    for evaluating integrals involving sine and cosine, 534
    for finding extrema on a closed interval, 167
    for finding intervals on which a function is increasing or decreasing, 180
    for finding an inverse function, 344
    for finding limits at infinity of rational functions, 201
    for finding a Taylor series, 680
    for implicit differentiation, 142
    for integration, 335
    for integration by parts, 525
    for making a change of variables, 299
    for solving applied minimum and maximum problems, 219
    for solving the basic equation, 558
    for solving related-rate problems, 150
    for testing a series for convergence or divergence, 643
    for using the Fundamental Theorem of Calculus, 283
Gyration, radius of, 1014

H

Half-life, 415
Hamilton, Isaac William Rowan (1805–1865), 764
Harmonic equation, 1064
Harmonic series, 619
    alternating, 632, 634
    general, 619
Heaviside, Oliver (1850–1925), 39
Heaviside function, 39
Helix, 833
Herschel, Caroline (1750–1848), 703
Higher-order derivative, 125
Homogeneous of degree n, 423
Homogeneous differential equation, 423
Homogeneous equation, change of variables, 424
Homogeneous function, 423
Hooke's Law, 489
Horizontal asymptote, 199
Horizontal component of a vector, 767
Horizontal line, 14
Horizontal line test, 343
Horizontal shift of a graph of a function, 23
    to the left, 23
    to the right, 23
Horizontally simple region of integration, 984
Huygens, Christian (1629–1695), 476
Hypatia (370–415 A.D.), 694
Hyperbola, 694, 701
    asymptotes of, 701
    center of, 701
    conjugate axis of, 701
    eccentricity of, 702
    foci of, 701
    rotated, 146
    standard equation of, 701
    transverse axis of, 701
    vertices of, 701
Hyperbolic functions, 388
    derivatives of, 390
    graph of, addition of ordinates, 389
    identities, 389, 390
    integrals of, 390
    inverse, 392
        differentiation involving, 394
        integration involving, 394
Hyperbolic identities, 389, 390
Hyperbolic paraboloid, 811, 813
Hyperboloid
    of one sheet, 811, 812
    of two sheets, 811, 812
Hypocycloid, 718

I

Identities, hyperbolic, 389, 390
Identity function, 22
Image of x under f, 19
Implicit derivative, 142
Implicit differentiation, 141, 928
    Chain Rule, 928
    guidelines for, 142
Implicit form of a function, 19
Implicitly defined function, 141
Improper integral, 578
    with infinite discontinuities, 581
        convergence of, 581
        divergence of, 581
    with infinite integration limits, 578
        convergence of, 578
        divergence of, 578
    special type, 584
Incidence, angle of, 696
Inclination, angle of, 947
Incompressible, 1062, 1125
Increasing function, 179
    test for, 179
Increment of z, 916
Increments of x and y, 916
Indefinite integral, 249
    of a vector-valued function, 844
Indefinite integration, 249
Independence of path and conservative vector fields, 1082
Independent of path, 1082
Independent variable, 19
    of a function of two variables, 884
Indeterminate form, 63, 85, 200, 567
Index of summation, 259
Inductive reasoning, 599
Inequality
    Cauchy-Schwarz, 789
    preservation of, 278, A10

triangle, 767
Inertia
  moment of, 1013, 1029
    polar, 1013
Infinite discontinuity, 578
Infinite interval, 198
Infinite limit(s), 83
  at infinity, 204
  from the left and from the right, 83
  properties of, 87
Infinite series (or series), 606
  alternating, 631
  convergence of, 606
  divergence of, 606
  geometric, 606
  harmonic, alternating, 632, 634
  nth partial sum, 606
  properties of, 610
  p-series, 619
  sum of, 606
  telescoping, 607
  terms of, 606
Infinity, limit at, 198, 199, A8
Inflection point, 192, 193
Initial condition, 253, 405
Initial point of a directed line segment, 762
Initial value, 414
Inner partition, 990, 1024
  polar, 1002
Inner product
  of two functions, 542
  of two vectors, 781
Inner radius of a solid of revolution, 459
Inscribed rectangle, 263
Inside limits of integration, 983
Instantaneous rate of change, 12
Integrability and continuity, 273
Integrable function, 273, 992
Integral(s)
  definite, 273
    properties of, 277
    two special, 276
  double, 990, 991, 992
  flux, 1114
  of hyperbolic functions, 390
  improper, 578
  indefinite, 249
  involving inverse trigonometric functions, 380
  involving secant and tangent, guidelines for evaluating, 537
  involving sine and cosine, guidelines for evaluating, 534
  iterated, 983
  line, 1066
  Mean Value Theorem, 285
  of $p(x) = Ax^2 + Bx + C$, 311
  single, 992

  of the six basic trigonometric functions, 337
  surface, 1108
  triple, 1024
Integral Test, 617
Integrating factor, 432
Integration
  additive interval property, 276
  basic rules of, 250, 520
  change of variables, 298
  constant of, 249
  of even and odd functions, 303
  guidelines for, 333
  indefinite, 249
  involving inverse hyperbolic functions, 394
  Log Rule, 332
  lower limit of, 273
  of power series, 664
  preservation of inequality, 278, A10
  region R of, 983
  rules for exponential functions, 354
  upper limit of, 273
  of a vector-valued function, 844
Integration by parts, 525
  guidelines for, 525
  summary of common integrals using, 530
  tabular method, 530
Integration by tables, 561
Integration formulas
  reduction formulas, 563
  special, 547
  summary of, 1132
Integration rules
  basic, 383
  General Power Rule, 300
Integration techniques
  basic integration rules, 520
  integration by parts, 525
  method of partial fractions, 552
  substitution for rational functions of sine and cosine, 564
  tables, 561
  trigonometric substitution, 543
Intercept(s), 4
  x-intercept, 4
  y-intercept, 4
Interior point of a region R, 896, 902
Intermediate Value Theorem, 77
Interpretation of concavity, A8
Interval
  of convergence, 660, 664
  infinite, 198
Inverse function, 341
  continuity and differentiability of, 345, A12
  derivative of, 345, A13
  existence of, 343

  guidelines for finding, 344
  Horizontal Line Test, 343
  properties of, 361
  reflective property of, 342
Inverse hyperbolic functions, 392
  differentiation involving, 394
  graphs of, 393
  integration involving, 394
Inverse square field, 1055
Inverse trigonometric functions, 371
  derivatives of, 374
  graphs of, 372
  integrals involving, 380
  properties of, 373
Irrotational vector field, 1060
Isobars, 148, 887
Isothermal curves, 426
Isothermal surface, 889
Isotherms, 887
Iterated integral, 983
  evaluation by, 1025
  inside limits of integration, 983
  outside limits of integration, 983
Iteration, 229
ith term of a sum, 259

J

Jacobi, Carl Gustav (1804–1851), 1042
Jacobian, 1042

K

Kappa curve, 145, 147
Kepler, Johannes (1571–1630), 751
Kepler's Laws, 751
Kinetic energy, 1085
Kovalevsky, Sonya (1850–1891), 896

L

Lagrange, Joseph-Louis (1736–1813), 174, 969
Lagrange form of the remainder, 654
Lagrange multiplier, 968, 969
Lagrange's Theorem, 969
Lambert, Johann Heinrich (1728–1777), 388
Lamina, planar, 500
Laplace, Pierre Simon de (1749–1827), 1035
Laplace's equation, 1064
Laplacian, 1064
Lateral surface area over a curve, 1077
Latus rectum, of a parabola, 695
Law of Conservation of Energy, 1085
Leading coefficient
  of a polynomial function, 24
  test, 24

Least squares
  method of, 962
  regression, 7
    line, 962, 963
Least upper bound, 601
Left-handed orientation, 773
Legendre, Adrien-Marie (1752–1833), 963
Leibniz, Gottfried Wilhelm (1646–1716), 238
Leibniz notation, 238
Lemniscate, 40, 144, 147, 735
Length
  of an arc, 476, 477
  of a directed line segment, 762
  of the moment arm, 497
  of a scalar multiple, 766
  of a vector in the plane, 763
  of a vector in space, 775
  on x-axis, 1018
Level curve, 887
  gradient is normal to, 938
Level surface, 889
  gradient is normal to, 948
L'Hôpital, Guillaume (1661–1704), 568
L'Hôpital's Rule, 568, A15
Limaçon, 735
  convex, 735
  dimpled, 735
  with inner loop, 735
Limit(s), 45, 48
  basic, 59
  of a composite function, 61, A4
  definition of, 52
  ε-δ definition of, 52
  evaluating
    direct substitution, 59, 60
    divide out like factors, 63
    rationalize the numerator, 63
  existence of, 73
  of a function involving a radical, 60, A4
  of a function of two variables, 897
  indeterminate form, 63, 85
  infinite, 83
    from the left and from the right, 83
    properties of, 87
  at infinity, 198, 199, A8
    infinite, 204
    of a rational function, guidelines for finding, 201
  of integration
    inside, 983
    lower, 273
    outside, 983
    upper, 273
  involving e, 364, A13
    from the left and from the right, 72
  of the lower and upper sums, 265
  nonexistence of, common types of behavior, 51
  of nth term of a convergent series, 610
  one-sided, 72
  of polynomial and rational functions, 60
  properties of, 59, A2
  of a sequence, 595
    properties of, 596
  strategy for finding, 62
  of trigonometric functions, 61
  two special trigonometric, 65
  of a vector-valued function, 835
Limit Comparison Test, 626
Line(s)
  contour, 887
  equation of
    general form, 14
    horizontal, 14
    point-slope form, 11, 14
    slope-intercept form, 13, 14
    summary, 14
    vertical, 14
  equipotential, 887
  least squares regression, 962, 963
  moment about, 497
  normal, 943, 944
    at a point, 147
  parallel, 14
  perpendicular, 14
  secant, 45, 97
  slope of, 10
  in space
    direction number of, 798
    direction vector of, 798
    parametric equations of, 798
    symmetric equations of, 798
  tangent, 45, 97
    with slope m, 97
    vertical, 99
Line of impact, 943
Line integral, 1066
  for area, 1092
  differential form of, 1073
  evaluation of as a definite integral, 1067
  of f along C, 1066
  independent of path, 1082
  summary of, 1117
  of a vector field, 1070
Line segment, directed, 762
Linear approximation, 235, 918
Linear combination, 767
Linear function, 24
Locus, 694
Log Rule for Integration, 332
Logarithmic differentiation, 327
Logarithmic function, 24
  to base a, 361
  common, 361
  natural, 322
    derivative of, 326
    properties of, 323, A11
Logarithmic properties, 323, A11
Logarithmic spiral, 747
Logistic curve, 427, 560
Logistic differential equation, 245, 427
  carrying capacity, 427
Logistic growth function, 365
Lorenz curves, 454
Lower bound of a sequence, 601
Lower bound of summation, 259
Lower limit of integration, 273
Lower sum, 263
  limit of, 265
Lune, 551

M

Macintyre, Sheila Scott (1910–1960), 534
Maclaurin, Colin (1698–1746), 676
Maclaurin polynomial, 650
Maclaurin series, 677
Magnitude
  of a directed line segment, 762
  of a vector in the plane, 763
Major axis of an ellipse, 697
Marginal productivity of money, 971
Mass, 496, 1114
  center of, 497, 498
    of a planar lamina of variable density, 1011, 1029
    two-dimensional system, 499
  moments of, 1011
  of a planar lamina of variable density, 1009
Mathematical model, 7, 962
Maximum
  absolute, 164
  of f on I, 164
  of a function of two variables, 952
  relative, 165
Mean Value Theorem, 174
  Extended, 245, 568, A14
  for Integrals, 285
Method of Lagrange Multipliers, 968, 969
Method of least squares, 962
Method of partial fractions, 552
  basic equation, 554
  guidelines for solving, 558
Midpoint Rule, 269
  in space, 774
Minimum
  absolute, 164
  of f on I, 164
  of a function of two variables, 952
  relative, 165
Minor axis of an ellipse, 697
Mixed partial derivatives, 910
  equality of, 911
Möbius Strip, 1107

Model, mathematical, 7
Moment(s)
  about a line, 497
  about the origin, 497, 498
  about a point, 497
  about the x-axis, 499
  about the x- and y-axes, 500
  about the y-axis, 499
  arm, length of, 497
  first, 1029
  of a force about a point, 794
  of inertia, 1013, 1029, 1137
    polar, 1013
    for a space curve, 1078
  of mass, 1011
    of a one-dimensional system, 498
    of a planar lamina, 500
  second, 1013, 1029
  of a two-dimensional system, 499
Monotonic sequence, 600
  bounded, 601
Mutually orthogonal, 426

N

n factorial, 597
Napier, John (1550–1617), 322
Natural equation for a curve, 881
Natural exponential function, 350
  derivative of, 352
  integration rules, 354
  operations with, 351
  properties of, 351
  series for, 682
Natural logarithmic base, 325
Natural logarithmic function, 322
  base of, 325
  derivative of, 326
  properties of, 323, A11
  series for, 682
Negative of a vector, 764
Newton, Isaac (1642–1727), 96
Newton's Law of Cooling, 417
Newton's Law of Gravitation, 1055
Newton's Law of Universal Gravitation, 489
Newton's Method, 229
  for approximating the zeros of a function, 229
  convergence of, 231
  iteration, 229
Newton's Second Law of Motion, 852
Nodes, 842
Noether, Emmy (1882–1935), 766
Nonexistence of a limit, common types of behavior, 51
Nonremovable discontinuity, 71
  of a function of two variables, 900
Norm
  of a partition, 272, 990, 1002, 1024
    polar, 1002

of a vector in the plane, 763
Normal component
  of acceleration, 860, 875
  of a vector field, 1114
Normal line, 943, 944
  at a point, 147
  to S at P, 944
Normal vectors, 783
  principal unit, 857, 875
  to a smooth parametric surface, 1101
Normalization of $\mathbf{v}$, 766
Notation
  derivative, 99
  for first partial derivatives, 907
  function, 19
  Leibniz, 238
  sigma, 259
nth Maclaurin polynomial for f at c, 650
nth partial sum, 606
nth Taylor polynomial for f at c, 650
nth term
  of a convergent series, 610
  of a sequence, 594
nth-Term Test for Divergence, 610
Number, critical, 166
Number e, 325
  limit involving, 364, A13
Numerical differentiation, 103

O

Octants, 773
Odd function, 26
  integration of, 303
  test for, 26
Ohm's Law, 241
One-dimensional system
  center of mass of, 498
  moment of, 498
One-sided limit, 72
One-to-one function, 21
Onto function, 21
Open disk, 896
Open interval
  continuous on, 70
  differentiable on, 99
Open region R, 896, 902
  continuous in, 900, 902
Open sphere, 902
Operations with exponential functions, 351
Operations with power series, 671
Order of a differential equation, 404
Orientable surface, 1113
Orientation
  of a curve, 1065
  of a plane curve, 710
  of a space curve, 832
Oriented surface, 1113
Origin
  moment about, 497, 498

of a polar coordinate system, 729
  reflection about, 23
  symmetry, 5
Orthogonal, 542
  graphs, 147
  trajectory, 426
  vectors, 783
Outer radius of a solid of revolution, 459
Outside limits of integration, 983

P

Pappus
  Second Theorem of, 506
  Theorem of, 503
Parabola, 2, 146, 694, 695
  axis of, 695
  directrix of, 695
  focal chord of, 695
  focus of, 695
  latus rectum of, 695
  reflective property of, 696
  standard equation of, 695
  vertex of, 695
Parabolic spandrel, 505
Parallel
  lines, 14
  planes, 800
  vectors, 776
Parameter, 709
  arc length, 868, 869
  eliminating, 711
Parametric equations, 709
  graph of, 709
  of a line in space, 798
  for a surface, 1098
Parametric form
  of arc length, 722
  of area of a surface of revolution, 724
  of the derivative, 719
Parametric surface, 1098
  area of, 1102
  equations for, 1098
  partial derivatives of, 1101
  smooth, 1101
    normal vector to, 1101
  surface area of, 1102
Partial derivative(s), 906
  equality of mixed, 911
  first, 906
    notation for, 907
  of a function of two variables, 906
  mixed, 910
  notation for, 907
  of a parametric surface, 1101
  of $\mathbf{r}$, 1101
Partial differentiation, 906
Partial fractions, 552
  decomposition of $N(x)/D(x)$ into, 553
  method of, 552

Partial sums, sequence of, 606
Particular solution of a differential equation, 253, 405
Partition
   inner, 990, 1024
      polar, 1002
   norm of, 272, 990, 1024
      polar, 1002
   regular, 272
Pascal, Blaise (1623–1662), 507
Pascal's Principle, 507
Path, 897, 1065
Pear-shaped quartic, 161
Percent error, 237
Perigee, 707
Perihelion, 707, 755
Perpendicular
   lines, 14
   planes, 800
   vectors, 783
Piecewise smooth curve, 714, 1065
Planar lamina, 500
   center of mass of, 500
   moment of, 500
Plane
   angle of inclination of, 947
   distance between a point and, 803
   region, simply connected, 1089
   tangent, 944
      equation of, 944
   vector in, 762
Plane curve, 709, 832
   orientation of, 710
   smooth, 1065
Plane in space
   angle between two, 800
   equation of
      general form, 799
      standard form, 799
   parallel, 800
      to the axis, 802
      to the coordinate plane, 802
   perpendicular, 800
   trace of, 802
Planimeter, 1136
Point
   of diminishing returns, 227
   of inflection, 192, 193
   of intersection, 6
   moment about, 497
   in a vector field
      incompressible, 1125
      sink, 1125
      source, 1125
Point-slope equation of a line, 11, 14
Polar axis, 729
Polar coordinate system, 729
   origin of, 729
   polar axis of, 729

pole, 729
Polar coordinates, 729
   area in, 739
   area of a surface of revolution in, 744
   converting to rectangular coordinates, 730
Polar curve, arc length of, 743
Polar equations of conics, 749
Polar form of slope, 733
Polar moment of inertia, 1013
Polar sectors, 1001
Pole, 729
   of cylindrical coordinate system, 820
   tangent lines at, 734
Polynomial
   Maclaurin, 650
   Taylor, 161, 650
Polynomial approximation, 648
   centered at c, 648
   expanded about c, 648
Polynomial function, 24, 60
   constant term of, 24
   degree, 24
   leading coefficient of, 24
   limit of, 60
   of two variables, 885
Position function, 113
   for a projectile, 853
Potential energy, 1085
Potential function for a vector field, 1057
Pound mass, 496
Power Rule
   for differentiation, 108, 136
   for integration, 300
   for Real Exponents, 363
Power series, 659
   centered at c, 659
   convergence of, 660, A17
   convergent form, 676
   derivative of, 664
   for elementary functions, 682
   integration of, 664
   interval of convergence of, 660
   operations with, 671
   properties of functions defined by, 664
      interval of convergence of, 664
      radius of convergence of, 664
   radius of convergence of, 660
Preservation of inequality, 278, A10
Pressure, 507
   fluid, 507
Primary equation, 218, 219
Principal unit normal vector, 857, 858, 875
Probability density function, 587
Procedures for fitting integrands to basic rules, 521
Product Rule, 119, 136
   differential form, 238
Projectile, position function for, 853

Projection form of work, 787
Projection of **u** onto **v**, 785
   using the dot product, 786
Prolate cycloid, 721
Propagated error, 237
Properties
   of continuity, 75
   of the cross product
      algebraic, 791
      geometric, 792
   of definite integrals, 277
   of the derivative of a vector-valued function, 842
   of the dot product, 781
   of double integrals, 992
   of functions defined by power series, 664
   of the gradient, 935
   of infinite limits, 87
   of infinite series, 610
   of inverse functions, 361
   of inverse trigonometric functions, 373
   of limits, 59, A2
   of limits of sequences, 596
   logarithmic, 323, A11
   of the natural exponential function, 323, 351
   of the natural logarithmic function, 323, A11
   of vector operations, 765
Proportionality constant, 414
p-series, 619
   convergence of, 619
   harmonic, 619
Pulse function, 94
   unit, 94
Pursuit curve, 393, 395

Q

Quadratic function, 24
Quadric surface, 811
   ellipsoid, 811, 812
   elliptic cone, 811, 813
   elliptic paraboloid, 811, 813
   hyperbolic paraboloid, 811, 813
   hyperboloid of one sheet, 811, 812
   hyperboloid of two sheets, 811, 812
   standard form of the equations of, 811, 812, 813
Quaternions, 764
Quotient Rule, 121, 136
   differential form, 238
Quotient, difference, 20, 97

R

Radial lines, 729
Radical, limit of a function involving a, 60, A4

Radius
    of convergence, 660, 664
    of curvature, 872
    function, 816
    of gyration, 1014
Ramanujan, Srinivasa (1887–1920), 673
Range of a function, 19
    of two variables, 884
Raphson, Joseph (1648–1715), 229
Rate of change, 12, 909
    average, 12
    instantaneous, 12
Ratio, 12
Ratio Test, 639
Rational function, 22, 25
    guidelines for finding limits at infinity of, 201
    limit of, 60
    of two variables, 885
Rationalize the numerator, 63
Real exponents, Power Rule, 363
Real-valued function f of a real variable x, 19
Recovering a function from its gradient, 1061
Rectangle
    area of, 261
    circumscribed, 263
    inscribed, 263
    representative, 446
Rectangular coordinates
    converting to cylindrical coordinates, 820
    converting to polar coordinates, 730
    converting to spherical coordinates, 823
    curvature in, 872, 875
Rectifiable curve, 476
Recursively defined sequence, 594
Reduction formulas, 563
Reflection
    about the origin, 23
    about the x-axis, 23
    about the y-axis, 23
    angle of, 696
    in the line $y = x$, 342
Reflective property
    of an ellipse, 699
    of inverse functions, 342
    of a parabola, 696
Reflective surface, 696
Refraction, 226, 975
Region of integration R, 983
    horizontally simple, 984
    r-simple, 1003
    θ-simple, 1003
    vertically simple, 984
Region in the plane
    area of, 265, 984
        between two curves, 447

centroid of, 501
connected, 1082
Region R
    boundary point of, 896
    bounded, 952
    closed, 896
    differentiable in, 917
    interior point of, 896, 902
    open, 896, 902
        continuous in, 900, 902
    simply connected, 1089
Regular partition, 272
Related-rate equation, 149
Related-rate problems, guidelines for solving, 150
Relation, 19
Relationship between divergence and curl, 1062
Relative error, 237
Relative extrema
    First Derivative Test for, 181
    of a function, 165
        of two variables, 952
    occur only at critical numbers, 166
    occur only at critical points, 953
    Second Derivative Test for, 194
    Second Partials Test for, 955
Relative maximum
    at $(c, f(c))$, 165
    First Derivative Test for, 181
    of a function, 165
        of two variables, 952, 955
    Second Derivative Test for, 194
    Second Partials Test for, 955
Relative minimum
    at $(c, f(c))$, 165
    First Derivative Test for, 181
    of a function, 165
        of two variables, 952, 955
    Second Derivative Test for, 194
    Second Partials Test for, 955
Remainder
    alternating series, 633
    of a Taylor polynomial, 654
        Lagrange form, 654
Removable discontinuity, 71
    of a function of two variables, 900
Representation of antiderivatives, 248
Representative element, 451
    disk, 456
    rectangle, 446
    washer, 459
Resultant force, 768
Resultant vector, 764
Return wave method, 542
Review of basic integration rules, 520
Revolution
    axis of, 456
    solid of, 456

surface of, 480
    area of, 481
Riemann, Georg Friedrich Bernhard (1826–1866), 272
Riemann sum, 272
Riemann zeta function, 623
Right cylinder, 810
Right-handed orientation, 773
Rolle, Michel (1652–1719), 172
Rolle's Theorem, 172
Root Test, 642
Rose curve, 732, 735
Rotated ellipse, 146
Rotated hyperbola, 146
Rotation of $\mathbf{F}$ about $\mathbf{N}$, 1131
r-simple region of integration, 1003
Rule(s)
    basic integration, 250, 520
        procedures for fitting integrands to, 520
    Midpoint, 269
    Simpson's, 312
    Trapezoidal, 310
Rulings of a cylinder, 810

S

Saddle point, 955
Scalar, 762
    field, 887
    multiple, 764
    multiplication, 764, 775
    product of two vectors, 781
    quantity, 762
Secant function
    derivative of, 123, 136
    integral of, 337
    inverse of, 371
Secant line, 45, 97
Second derivative, 125
Second Derivative Test, 194
Second Fundamental Theorem of Calculus, 289
Second moment, 1013, 1029
Second Partials Test, 955
Second Theorem of Pappus, 506
Secondary equation, 219
Separable equations, 421
Sequence, 594
    Absolute Value Theorem, 598
    bounded, 601
    bounded above, 601
    bounded below, 601
    convergence of, 595
    divergence of, 595
    least upper bound of, 601
    limit of, 595
        properties of, 596
    lower bound of, 601

monotonic, 600
nth term of, 594
of partial sums, 606
recursively defined, 594
Squeeze Theorem, 597
terms of, 594
upper bound of, 601
Series, 606
absolutely convergent, 634
alternating, 631
binomial, 681
conditionally convergent, 634
convergence of, 606
divergence of, 606
nth-term test for, 610
geometric, 608
alternating, 631
convergence of, 608
guidelines for testing for convergence or divergence, 643
harmonic, alternating, 632, 634
infinite, 606
properties of, 610
Maclaurin, 677
nth partial sum, 606
nth term of convergent, 610
power, 659
p-series, 619
sum of, 606
summary of tests for, 644
Taylor, 676, 677
telescoping, 607
terms of, 606
Serpentine, 127
Shell method, 467, 468
Shift of a graph
horizontal, 23
to the left, 23
to the right, 23
vertical, 23
downward, 23
upward, 23
Sigma notation, 259
index of summation, 259
ith term, 259
lower bound of summation, 259
upper bound of summation, 259
Signum function, 82
Simple curve, 1089
Simple Power Rule, 108, 136
Simple solid region, 1121
Simply connected plane region, 1089
Simpson's Rule, 312
error in, 313
Sine function, 22
derivative of, 112, 136
integral of, 337
inverse of, 371
series for, 682

Sine integral function, 320
Single integral, 992
Singular solution of a differential equation, 404
Sink, 1125
Slant asymptote, 211
Slope(s)
field, 256, 304, 323, 406
of the graph of f at $x = c$, 97
of a line, 10
in polar form, 733
of the surface in the x- and y-directions, 907
of a tangent line, 97
Slope-intercept equation of a line, 13, 14
Smooth
curve, 476, 714, 842, 857
on an open interval, 842
piecewise, 714
parametric surface, 1101
plane curve, 1065
space curve, 1065
Snell's Law of Refraction, 226, 975
Solenoidal, 1062
Solid of revolution, 456
inner radius of, 459
outer radius, 459
Solid region, simple, 1121
Solution
curves, 405
of a differential equation, 404
Bernoulli, 434
Euler's Method, 408
first-order linear, 433
general, 404
particular, 405
singular, 404
of an equation, by radicals, 232
point of an equation, 2
by radicals, 232
Some basic limits, 59
Somerville, Mary Fairfax (1780–1872), 884
Source, 1125
Space curve, 832
arc length of, 867
moment of inertia for, 1078
smooth, 1065
Special integration formulas, 547
Special polar graphs, 735
Special type of improper integral, 584
Speed, 114, 848, 849, 873, 875
angular, 1014
Sphere, 774
open, 902
standard equation of, 774
Spherical coordinate system, 823
converting to rectangular coordinates, 823
converting to cylindrical coordinates, 823

Spiral
of Archimedes, 723, 731, 747
cornu, 759, 881
logarithmic, 747
Square root function, 22
Squared errors, sum of, 962
Squaring function, 22
Squeeze Theorem, 65, A5
for Sequences, 597
Standard equation
of an ellipse, 697
of a hyperbola, 701
of a parabola, 695
of a sphere, 774
Standard form
of the equation of a plane in space, 799
of the equations of quadric surfaces, 811, 812, 813
of a first-order linear differential equation, 432
Standard normal probability density function, 353
Standard position of a vector, 763
Standard unit vector, 767
notation, 775
Step function, 72
Stokes, George Gabriel (1819–1903), 1128
Stokes's Theorem, 1094, 1128
Strategy for finding limits, 62
Strictly monotonic function, 180, 343
Strophoid, 759
Substitution for rational functions of sine and cosine, 564
Sufficient condition for differentiability, 917, A18
Sum
ith term of, 259
lower, 263
limit of, 265
Riemann, 272
Rule, 111, 136
differential form, 238
of a series, 606
of the squared errors, 962
of two vectors, 764
upper, 263
limit of, 265
Summary
of common integrals using integration by parts, 530
of compound interest formulas, 364
of differentiation rules, 136
of equations of lines, 14
of first-order differential equations, 435
of integration formulas, 1132
of line and surface integrals, 1117
of tests for series, 644
of velocity, acceleration, and curvature, 875

Summation
  formulas, 260, A9
  index of, 259
  lower bound of, 259
  upper bound of, 259
Surface
  closed, 1120
  cylindrical, 810
  isothermal, 889
  level, 889
  orientable, 1113
  oriented, 1113
  parametric, 1098
  parametric equations for, 1098
  quadric, 811
  reflective, 696
  trace of, 811
Surface area
  of a parametric surface, 1102
  of a solid, 1017, 1018
Surface integral, 1108
  evaluating, 1108
  of f over S, 1108
  summary of, 1117
Surface of revolution, 480, 816
  area of, 481
    parametric form, 724
    polar form, 774
Symmetric equations of a line in space, 798
Symmetry
  tests for, 5
  with respect to the origin, 5
  with respect to the point (a, b), 401
  with respect to the x-axis, 5
  with respect to the y-axis, 5

T

Table of values, 2
Tables, integration by, 561
Tabular method for integration by parts, 530
Tangent function
  derivative of, 123, 136
  integral of, 337
  inverse of, 371
Tangent line(s), 45, 97
  approximation, 235
  to a curve, 858
  at the pole, 734
  problem, 45
  slope of, 97
  with slope m, 97
  vertical, 99
Tangent plane, 944
  equation of, 944
  to S at P, 944
Tangent vector, 848

Tangential component of acceleration, 860, 861, 875
Tautochrone problem, 715
Taylor, Brook (1685-1731), 650
Taylor polynomial, 161, 650
  error in approximating, 654
  remainder, Lagrange form of, 654
Taylor series, 676, 677
  convergence of, 678
  guidelines for finding, 680
Taylor's Theorem, 654, A16
Telescoping series, 607
Terminal point of a directed line segment, 762
Terms
  of a sequence, 594
  of a series, 606
Test(s)
  for concavity, 191
  conservative vector field in the plane, 1058
  conservative vector field in space, 1061
  for convergence
    Alternating Series Test, 631
    Direct Comparison Test, 624
    geometric series, 608
    guidelines, 643
    Integral Test, 617
    Limit Comparison Test, 626
    p-series, 619
    Ratio Test, 639
    Root Test, 642
    summary of, 744
  for even and odd functions, 26
  for increasing and decreasing functions, 179
  for symmetry, 5
Theorem, existence, 164
Theorem of Pappus, 503
  second, 506
Theta, θ
  simple region of integration, 1003
Third derivative, 125
Three-dimensional coordinate system, 773
  left-handed orientation, 773
  right-handed orientation, 773
Top half of circle, 138
Topographic map, 887
Torque, 498, 794
Torricelli's Law, 443
Torsion, 882
Total differential, 916
Total mass
  of a one-dimensional system, 498
  of a two-dimensional system, 498
Trace
  of a plane in space, 802
  of a surface, 811
Tractrix, 331, 393, 394

Transcendental function, 25, 376
Transformation, 23, 1043
Transformation of a graph of a function, 23
  basic types, 23
  horizontal shift, 23
  reflection about origin, 23
  reflection about x-axis, 23
  reflection about y-axis, 23
  reflection in the line $y = x$, 342
  vertical shift, 23
Transverse axis, of a hyperbola, 701
Trapezoidal Rule, 310
  error in, 313
Triangle inequality, 767
Trigonometric function(s), 24
  cosine, 22
  derivative of, 123, 136
  integrals of the six basic, 337
  inverse, 371
    derivatives of, 374
    integrals involving, 380
    properties of, 373
  limit of, 61
  sine, 22
Trigonometric substitution, 543
Triple integral, 1024
  of f over Q, 1024
Triple scalar product, 794
  geometric property of, 795
Two-dimensional system
  center of mass of, 499
  moment of, 499
Two-point Gaussian quadrature approximation, 319
Two special definite integrals, 276
Two special trigonometric limits, 65

U

Unit pulse function, 94
Unit tangent vector, 857, 875
Unit vector, 763
  in the direction of $\mathbf{v}$
    in the plane, 766
    in space, 775
  standard, 767
Upper bound
  least, 601
  of a sequence, 601
  of summation, 259
Upper limit of integration, 273
Upper sum, 263
  limit of, 265
u-substitution, 295

V

Value of f at x, 19
Variable
  dependent, 19

dummy, 275
force, 488
independent, 19
Vector(s)
   acceleration, 860, 875
   addition
      associative property of, 765
      commutative property of, 765
      in the plane, 762
      in space, 775
   Additive Identity Property, 765
   Additive Inverse Property of, 765
   angle between two, 782
   component
      of **u** along **v**, 785
      of **u** orthogonal to **v**, 785
   component form of, 763
   components, 763, 785
   cross product of, 790
   difference of two, 764
   direction, 798
   direction angles of, 784
   direction cosines of, 784
   Distributive Property, 765
   dot product of, 781
   equal, 763, 775
   horizontal component of, 767
   initial point, 762
   inner product of, 781
   length of, 763, 775
   linear combination of, 767
   magnitude of, 763
   negative of, 764
   norm of, 763
   normal, 783
   normalization of, 766
   operations, properties of, 765
   orthogonal, 783
   parallel, 776
   perpendicular, 783
   in the plane, 762
   principal unit normal, 857, 875
   product of two vectors in space, 790
   projection of, 785
   resultant, 764
   scalar multiplication, 764
   scalar product of, 781
   in space, 775
   space, 766
      axioms, 766
   standard position, 763
   standard unit notation, 775
   sum, 764
   tangent, 848
   terminal point, 762
   triple scalar product, 794
   unit, 763
      in the direction of **v**, 766, 775
      standard, 767
   unit tangent, 857, 875
   velocity, 848, 875
   vertical component of, 767
   zero, 763, 775
Vector field, 1054
   circulation of, 1131
   conservative, 1057, 1079
      test for, 1058, 1061
   continuous, 1054
   curl of, 1060
   divergence of, 1062
   divergence-free, 1062
   incompressible, 1125
   irrotational, 1060
   line integral of, 1070
   normal component of, 1114
   over Q, 1054
   over R, 1054
   potential function for, 1057
   rotation of, 1131
   sink, 1125
   solenoidal, 1062
   source, 1125
Vector-valued function(s), 832
   antiderivative of, 844
   continuity of, 836
   continuous on an interval, 836
   continuous at a point, 836
   definite integral of, 844
   derivative of, 840
      properties of, 842
   differentiation, 841
   domain of, 833
   indefinite integral of, 844
   integration of, 844
   limit of, 835
Velocity, 114, 849
   average, 113
   field, 1054, 1055
      incompressible, 1062
   potential curves, 426
   vector, 848, 875
Vertéré, 201
Vertex
   of an ellipse, 697
   of a hyperbola, 701
   of a parabola, 695
Vertical asymptote, 84, 85, A6
Vertical component of a vector, 767
Vertical line, 14
Vertical Line Test, 22
Vertical shift of a graph of a function, 23
   downward, 23
   upward, 23
Vertical tangent line, 99
Vertically simple region of integration, 984
Volume of a solid
   disk method, 457
   with known cross sections, 461
   shell method, 467, 468
   washer method, 459
Volume of a solid region, 992, 1024

W

Wallis, John (1616–1703), 536
Wallis's Formulas, 536
Washer, 459
Washer method, 459
Weierstrass, Karl (1815–1897), 953
Witch of Agnesi, 127, 146, 839
Work, 787
   done by a constant force, 487
   done by a variable force, 488
   dot product form, 787
   force field, 1070
   projection form, 787

X

x-axis
   moment about, 499
   reflection about, 23
   symmetry, 5
x-intercept, 4
xy-plane, 773
xz-plane, 773

Y

y-axis
   moment about, 499
   reflection about, 23
   symmetry, 5
y-intercept, 4
Young, Grace Chisholm (1868–1944), 42
yz-plane, 773

Z

Zero factorial, 597
Zero of a function, 26
   approximating
      bisection method, 78
      Intermediate Value Theorem, 77
      with Newton's Method, 229
Zero vector
   in the plane, 763
   in space, 775

ALGEBRA

Factors and Zeros of Polynomials

Let $p(x) = a_n x^n + a_{n-1} x^{n-1} + \cdots + a_1 x + a_0$ be a polynomial. If $p(a) = 0$, then a is a *zero* of the polynomial and a solution of the equation $p(x) = 0$. Furthermore, $(x - a)$ is a *factor* of the polynomial.

Fundamental Theorem of Algebra

An nth degree polynomial has n (not necessarily distinct) zeros. Although all of these zeros may be imaginary, a real polynomial of odd degree must have at least one real zero.

Quadratic Formula

If $p(x) = ax^2 + bx + c$, and $0 \le b^2 - 4ac$, then the real zeros of p are $x = \left(-b \pm \sqrt{b^2 - 4ac}\right)/2a$.

Special Factors

$x^2 - a^2 = (x - a)(x + a)$ $x^3 - a^3 = (x - a)(x^2 + ax + a^2)$

$x^3 + a^3 = (x + a)(x^2 - ax + a^2)$ $x^4 - a^4 = (x^2 - a^2)(x^2 + a^2)$

Binomial Theorem

$(x + y)^2 = x^2 + 2xy + y^2$ $(x - y)^2 = x^2 - 2xy + y^2$

$(x + y)^3 = x^3 + 3x^2y + 3xy^2 + y^3$ $(x - y)^3 = x^3 - 3x^2y + 3xy^2 - y^3$

$(x + y)^4 = x^4 + 4x^3y + 6x^2y^2 + 4xy^3 + y^4$ $(x - y)^4 = x^4 - 4x^3y + 6x^2y^2 - 4xy^3 + y^4$

$(x + y)^n = x^n + nx^{n-1}y + \dfrac{n(n-1)}{2!}x^{n-2}y^2 + \cdots + nxy^{n-1} + y^n$

$(x - y)^n = x^n - nx^{n-1}y + \dfrac{n(n-1)}{2!}x^{n-2}y^2 - \cdots \pm nxy^{n-1} \mp y^n$

Rational Zero Theorem

If $p(x) = a_n x^n + a_{n-1} x^{n-1} + \cdots + a_1 x + a_0$ has integer coefficients, then every *rational zero* of p is of the form $x = r/s$, where r is a factor of a_0 and s is a factor of a_n.

Factoring by Grouping

$acx^3 + adx^2 + bcx + bd = ax^2(cx + d) + b(cx + d) = (ax^2 + b)(cx + d)$

Arithmetic Operations

$ab + ac = a(b + c)$ $\dfrac{a}{b} + \dfrac{c}{d} = \dfrac{ad + bc}{bd}$ $\dfrac{a + b}{c} = \dfrac{a}{c} + \dfrac{b}{c}$

$\dfrac{\left(\dfrac{a}{b}\right)}{\left(\dfrac{c}{d}\right)} = \left(\dfrac{a}{b}\right)\left(\dfrac{d}{c}\right) = \dfrac{ad}{bc}$ $\dfrac{\left(\dfrac{a}{b}\right)}{c} = \dfrac{a}{bc}$ $\dfrac{a}{\left(\dfrac{b}{c}\right)} = \dfrac{ac}{b}$

$a\left(\dfrac{b}{c}\right) = \dfrac{ab}{c}$ $\dfrac{a - b}{c - d} = \dfrac{b - a}{d - c}$ $\dfrac{ab + ac}{a} = b + c$

Exponents and Radicals

$a^0 = 1, \quad a \ne 0$ $(ab)^x = a^x b^x$ $a^x a^y = a^{x+y}$ $\sqrt{a} = a^{1/2}$ $\dfrac{a^x}{a^y} = a^{x-y}$ $\sqrt[n]{a} = a^{1/n}$

$\left(\dfrac{a}{b}\right)^x = \dfrac{a^x}{b^x}$ $\sqrt[n]{a^m} = a^{m/n}$ $a^{-x} = \dfrac{1}{a^x}$ $\sqrt[n]{ab} = \sqrt[n]{a}\sqrt[n]{b}$ $(a^x)^y = a^{xy}$ $\sqrt[n]{\dfrac{a}{b}} = \dfrac{\sqrt[n]{a}}{\sqrt[n]{b}}$

© Houghton Mifflin Company, Inc.

FORMULAS FROM GEOMETRY

Triangle
$h = a \sin \theta$
Area $= \dfrac{1}{2}bh$
(Law of Cosines)
$c^2 = a^2 + b^2 - 2ab \cos \theta$

Sector of Circular Ring
(p = average radius,
w = width of ring,
θ in radians)
Area $= \theta p w$

Right Triangle
(Pythagorean Theorem)
$c^2 = a^2 + b^2$

Ellipse
Area $= \pi ab$
Circumference $\approx 2\pi \sqrt{\dfrac{a^2 + b^2}{2}}$

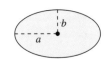

Equilateral Triangle
$h = \dfrac{\sqrt{3}\,s}{2}$
Area $= \dfrac{\sqrt{3}\,s^2}{4}$

Cone
(A = area of base)
Volume $= \dfrac{Ah}{3}$

Parallelogram
Area $= bh$

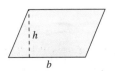

Right Circular Cone
Volume $= \dfrac{\pi r^2 h}{3}$
Lateral Surface Area $= \pi r \sqrt{r^2 + h^2}$

Trapezoid
Area $= \dfrac{h}{2}(a + b)$

Frustum of Right Circular Cone
Volume $= \dfrac{\pi(r^2 + rR + R^2)h}{3}$
Lateral Surface Area $= \pi s(R + r)$

Circle
Area $= \pi r^2$
Circumference $= 2\pi r$

Right Circular Cylinder
Volume $= \pi r^2 h$
Lateral Surface Area $= 2\pi rh$

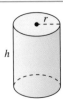

Sector of Circle
(θ in radians)
Area $= \dfrac{\theta r^2}{2}$
$s = r\theta$

Sphere
Volume $= \dfrac{4}{3}\pi r^3$
Surface Area $= 4\pi r^2$

Circular Ring
(p = average radius,
w = width of ring)
Area $= \pi(R^2 - r^2)$
$= 2\pi p w$

Wedge
(A = area of upper face,
B = area of base)
$A = B \sec \theta$

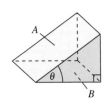

© Houghton Mifflin Company, Inc.